Classics in Mathematics

Iosif I. Gikhman • Anatoli V. Skorokhod

The Theory of Stochastic Processes I

Springer
Berlin
Heidelberg
New York
Hong Kong
London
Milan
Paris
Tokyo

Iosif Ilyich Gikhman was born on the 26th of May 1918 in the city of Uman, Ukraine. He studied in Kiev, graduating in 1939, then remained there to teach and do research under the supervision of N. Bogolyubov, defending a "candidate" thesis on the influence of random processes on dynamical systems in 1942 and a doctoral dissertation on Markov processes and mathematical statistics in 1955.

I.I. Gikhman is one of the founders of the theory of stochastic differential equations and also contributed significantly to mathematical statistics, limit theorems, multidimensional martingales, and stochastic control. He died in 1985, in Donetsk.

Anatoli Vladimirovich Skorokhod was born on September 10th, 1930 in the city Nikopol, Ukraine. He graduated from Kiev University in 1953, after which his graduate studies at Moscow University were directed by E.B. Dynkin.

From 1956 to 1964 Anatoli Skorokhod was a professor of Kiev University. Thereafter he worked at the Institute of Mathematics of the Ukrainian Academy of Sciences, but he has also, since 1993, been professor of Statistics and Probability at Michigan State University. Skorokhod was elected to the Ukrainian Academy of Sciences in 1985 and became a Fellow of American Academy of Arts and Sciences in 2000.

His mathematical research interests are the theory of stochastic processes, stochastic differential equations, Markov processes, randomly perturbed dynamical systems.

Iosif I. Gikhman • Anatoli V. Skorokhod

The Theory of Stochastic Processes I

Reprint of the 1974 Edition

Authors
Iosif Ilyich Gikhman

Anatoli Vladimirovich Skorokhod
Department of Statistics
and Probability
Michigan State University
East Lansing, MI 48824-1027, USA

Translator
Samuel Kotz

Originally published as Vol. 210 of the
Grundlehren der mathematischen Wissenschaften

Mathematics Subject Classification (2000): 60-02, 60Gxx

Cataloging-in-Publication Data applied for

A catalog record for this book is available from the Library of Congress.

Bibliographic information published by Die Deutsche Bibliothek
Die Deutsche Bibliothek lists this publication in the Deutsche Nationalbibliografie; detailed bibliographic data is available in the Internet at <http://dnb.ddb.de>.

ISSN 1431-0821
ISBN 3-540-20284-6 Springer-Verlag Berlin Heidelberg New York

Springer-Verlag is a part of Springer Science+Business Media
springeronline.com

Printed in Germany

Printed on acid-free paper 41/3142LK-5 4 3 2 1 0

I. I. Gihman A. V. Skorohod

The Theory of Stochastic Processes I

Translated from the Russian
by S. Kotz

Corrected Printing of the First Edition

Springer-Verlag
Berlin Heidelberg New York 1980

Iosif Il'ich Gihman
Academy of Sciences of the Ukrainian SSR
Institute of Applied Mathematics and Mechanics
Donetsk/USSR

Anatoliĭ Vladimirovich Skorohod
Academy of Sciences of the Ukrainian SSR
Institute of Mathematics
Kiev/USSR

Translator:

Samuel Kotz
Department of Management and Statistics
University of Maryland
College Park, MD 20742/USA

Title of the Russian Original Edition: Teoriya sluchainyh protsessov, Vol. I. Publisher: Nauka, Moscow 1971

AMS Subject Classification (1980): 60-02, 60 G xx

ISBN 3-540-06573-3 Springer-Verlag Berlin Heidelberg New York
ISBN 0-387-06573-3 Springer-Verlag New York Heidelberg Berlin

Library of Congress Catalog Card Number 74-2552
Printed in Germany

Typesetting: D. Reidel, Book Manufacturers, Dordrecht · Holland.
Printing and bookbinding: Konrad Triltsch, Würzburg.
2141/3140-543210

Preface

We have endeavoured in this planned three volume work to present an exposition of the basic results, methods and applications of the theory of random processes. The various branches of the theory are, however, not treated in equal detail.

This volume should be of value principally to mathematicians who are interested in studying the theory of random processes. We hope that researchers who apply the methods of the theory of random processes will also find the book interesting and useful. Prerequisites to the study of this book are basic courses in probability theory, measure theory and integration, and functional analysis.

The first volume of *"The Theory of Random Processes"* is devoted to general problems of the theory of random functions and measure theory in function spaces. Some of the material presented in the authors' book *"Introduction to the Theory of Random Processes"* (Ergebnisse der Mathematik Band 72) is utilized here. Chapters III, IV, V and IX of the *Introduction* have been revised and now constitute the contents of Chapters I, III, IV and VI respectively.

In volume II, the following topics are treated: the general theory of Markov processes, the theory of processes with independent increments, jump Markov processes, semi-Markov processes and branching processes.

The third volume deals with the theory of martingales, stochastic integrals, stochastic differential equations, diffusion processes and limit theorems associated with stochastic differential equations.

I. I. Gihman and A. V. Skorohod

Table of Contents

Chapter I

Basic Notions of Probability Theory

§1. Axioms and Definitions

Events. The basic notions of probability theory are experiment, event and probability of events.

A formal description of these notions is usually based on the set-theoretical model of probability theory developed by A. N. Kolmogorov in 1929.

The experiments studied in probability theory (referred to as stochastic experiments) are carried out when a certain set of conditions Y is satisfied. This set of conditions does not uniquely determine the results of the experiment (also called the outcome or realization). This means that if the experiment is repeated (provided that the set of conditions Y is accurately satisfied) the results of the experiment will generally be different.

When formalizing the notions of probability theory the first fundamental assumption is that the results of a collection of experiments under investigation in a given situation can be described by means of a certain set Ω. Every meaningful event (occurring or not during the given experiment) corresponds to a certain subset A of Ω in such a manner that the probabilistic operations on events correspond to set-theoretical operations on the corresponding subsets of Ω.

Moreover, the points $\omega \in \Omega$ correspond to atoms – namely, every event is a sum of points while each point ω cannot be represented as a sum of other events. For this reason the points belonging to Ω are called elementary events.

In relation to Ω, an experiment is completely characterized by the class of those events (subsets of Ω) such that one can assert in each case whether it did or did not occur during the given experiment. These events are called observable (in the given experiment).

Henceforth we shall adhere to this model of probability theory and identify events with the corresponding subsets of Ω. The resulting dual terminology is presented below in a glossary translating set-theoretic notions into probabilistic notions.

Set theory	Probability theory
Space Ω	Sure event
ω a point of Ω	Elementary event
$\emptyset$ the empty set	Impossible event
A a subset of Ω, $A \subset \Omega$	Event
The set A is contained in B ($A \subset B$)	Event A implies B
C the sum (union) of sets A and B ($C = A \cup B$)	C the sum (union) of events A and B
C the intersection of sets A and B ($C = A \cap B$)	C the intersection (or product) of events A and B
$\bar{A}$ the complement of set A	$\bar{A}$ the contrary event of A
C the difference of two sets A and B ($C = A \setminus B$)	C the difference of events A and B
Sets A and B are without common points ($A \cap B = \emptyset$)	Events A and B are disjoint

We note that any arbitrary subset of Ω is called an event. However, from both a practical as well as a purely mathematical point of view it does not make sense to regard any arbitrary subsets of Ω as events worthy of interest. Therefore one must select out of Ω a suitable class of events. This class should be sufficiently wide and contain all the events which may arise during the solution of various practical problems. On the other hand, the size of this class is limited by the feasibility of effective utilization of mathematical techniques. Obviously, the problem of selecting the corresponding class of events should be solved individually in each case, however, we shall always assume subsequently that this class forms a σ-algebra of events.

Definition 1. A class of events $\mathfrak{A}$ is called an *algebra of events* if it contains the sure event Ω, the impossible event $\emptyset$ and together with each pair of events A and B belonging to the class, their sum as well as the contrary event $\bar{A}$.

Two events Ω and $\emptyset$ constitute the *trivial algebra*.

The minimal algebra containing event A consists of four events: Ω, $\emptyset$, A and $\bar{A}$.

Definition 2. An algebra of events which contains a sequence of events along with their sum is called a *σ-algebra*.

It is clear that in the definitions and properties above we could have referred to algebras and σ-algebras of sets of a certain abstract space Ω.

Definition 3. The space Ω along with the σ-algebra of sets $\mathfrak{A}$ defined on it is called the *measurable space* $\{\Omega, \mathfrak{A}\}$ and the subsets of Ω belonging to $\mathfrak{A}$ are called *$\mathfrak{A}$-measurable sets* ($\mathfrak{A}$-measurable events) or simply *measurable sets* (events) if no ambiguity arises concerning the σ-algebra under consideration.

The σ-algebra of all the events under consideration in a given situation is usually denoted by the letter $\mathfrak{S}$. With respect to the measurable space $(\Omega, \mathfrak{S})$ any given stochastic experiment is completely characterized by the class of events $\mathfrak{F}$ observed during this experiment. Clearly, (this class is contained in Ω and it is also evident that the class $\mathfrak{F}$ is closed with respect to the operations of addition, intersection and complementation. It is therefore natural to consider $\mathfrak{F}$ a σ-algebra of events. Therefore, formally a stochastic experiment is determined by a certain σ-algebra $\mathfrak{F}$ of $\mathfrak{S}$-measurable events. We call it the *σ-algebra corresponding to the given experiment*.

Probability. Definition 4. A triple $(\Omega, \mathfrak{S}, \mathsf{P})$ consisting of a space of elementary events Ω, a selected σ-algebra of events $\mathfrak{S}$ in Ω, and a measure P defined on $\mathfrak{S}$ such that $\mathsf{P}(\Omega)=1$ is called a probability space and the measure P is called the probability.

Probability spaces are the initial objects of probability theory. This, however, does not contradict the fact that when solving many specific problems the probability space is not given explicitly.

We present below several of the simplest well known properties of probability which easily follow from its definition (S and S_n, $n=1, 2, \ldots,$ as given below all belong to $\mathfrak{S}$):

a) $\mathsf{P}(\emptyset)=0$;

b) if $S_k \cap S_r = \emptyset$, $k \neq r$, then $\mathsf{P}\left(\bigcup_1^\infty S_k\right) = \sum_1^\infty \mathsf{P}(S_k)$;

c) if $S_1 \subset S_2$, then $\mathsf{P}(S_2 \setminus S_1) = \mathsf{P}(S_2) - \mathsf{P}(S_1)$;

d) $\mathsf{P}(\bar{S}) = 1 - \mathsf{P}(S)$;

e) if $S_n \subset S_{n+1}$, $n=1, 2, \ldots,$ then $\mathsf{P}\left(\bigcup_1^\infty S_n\right) = \lim \mathsf{P}(S_n)$;

f) if $S_n \supset S_{n+1}$, $n=1, 2, \ldots,$ then $\mathsf{P}\left(\bigcap_1^\infty S_n\right) = \lim \mathsf{P}(S_n)$.

Random variables. The concept of a random variable corresponds to the description of a stochastic experiment which measures a certain numerical quantity ξ. It is assumed that for any pair of numbers a and b $(a<b)$ the event $A(a, b)$ expressing that $\xi \in (a, b)$ is an observable event.

The minimal σ-algebra $\mathfrak{F}_\xi$ containing all the events $A(a, b)$, $-\infty < a < b < \infty$ is the σ-algebra corresponding to this stochastic experiment.

Let $A_x(-\infty < x < \infty)$ denote the event $\xi = x$. This event is measurable. Indeed $A_x = \bigcap_{n=1}^{\infty} A\left(x-\frac{1}{n}, x+\frac{1}{n}\right)$. Moreover, if $x_1 \neq x_2$, events A_{x_1} and A_{x_2} are disjoint (this follows from the single-valuedness of the measurement results) and the union of all A_x, $-\infty < x < \infty$, is the set Ω, since the measurement result is always represented by some real number. We now define a single-valued real function $f(\omega)$, $\omega \in \Omega$ by setting $f(\omega) = x$ if $\omega \in A_x$. It follows from the definition, that $\xi = f(\omega)$ in each experiment and, moreover, that the set $\{\omega : a < f(\omega) < b\} = A(a, b)$ is measurable. Recall that a real-valued function $f(\omega)$ defined on a measurable space $\{\Omega, \mathfrak{S}\}$ is called *measurable* ($\mathfrak{S}$-measurable) if for any two real numbers a and b the set $\{\omega : a < f(\omega) < b\} \in \mathfrak{S}$. Therefore, a random variable ξ can be identified with a certain measurable function on the probability space $(\Omega, \mathfrak{S}, \mathsf{P})$.

Definition 5. A $\mathfrak{S}$-measurable real-valued function of elementary events ω is called a *random variable* ξ (on a given probability space $\{\Omega, \mathfrak{S}, \mathsf{P}\}$).

Henceforth, we shall occasionally consider measurable functions on $\{\Omega, \mathfrak{S}, \mathsf{P}\}$ which may possibly take on the values $\pm\infty$ also, or functions which are defined only on a measurable subset of $\{\Omega, \mathfrak{S}, \mathsf{P}\}$. These functions are called generalized random variables.

We note the following point connected with the definition of a random variable. It is commonly assumed that from the empirical point of view one cannot distinguish between events which differ on an event of probability zero. It would therefore be natural to identify two random variables ξ and η which are equal to each other with probability 1 and hence interpret a random variable as a class of measurable functions, in which each pair of functions may differ only on a set of probability 0. Such functions are called equivalent (or P-equivalent). This point of view is also justified by the fact that the majority of notions introduced here as well as the relationships obtained refer essentially to classes of equivalent functions. However, a consistent adherence to this point of view presents certain technical as well as basic problems. For this reason it would seem more convenient to regard random variables as individual functions and use special notation for their equivalent classes.

Definition 6. Random variables ξ and η are called *equivalent* (P-equivalent) if $\mathsf{P}\{\xi \neq \eta\} = 0$. The P-equivalence of 2 random variables ξ and η is denoted by $\xi = \eta \pmod{\mathsf{P}}$.

Equivalent random variables are also referred to as satisfying $\xi = \eta$ almost surely (a.s.) or $\xi = \eta$ with probability 1.

Analogous terminology and notation is also used in more general cases. We thus say that a certain function (or certain other objects) possess property H almost surely (for almost all ω or for all ω (mod P)) if the set of ω for which this property is not satisfied is of probability 0. For example, if a sequence of random variables $\xi_n = f_n(\omega)$ converges to $\xi = f(\omega)$ for each ω except for a certain set N and $\mathsf{P}(N)=0$, we say that ξ_n converges to ξ almost surely or that

$$\xi = \lim \xi_n \pmod{\mathsf{P}}.$$

We now present a number of basic properties of random variables which follow directly from the corresponding properties of arbitrary measurable functions. It is assumed that the random variables are defined on a fixed probability space $\{\Omega, \mathfrak{S}, \mathsf{P}\}$.

a) If $h(t_1, t_2, \ldots, t_n)$ is an arbitrary Borel function of n real variables $t_1, \ldots, t_n$, and $\xi_1, \xi_2, \ldots, \xi_n$ are random variables, then $h(\xi_1, \xi_2, \ldots, \xi_n)$ is also a random variable.

b) If $\{\xi_n; n=1, 2, \ldots\}$ is a sequence of random variables, then $\sup \xi_n$, $\inf \xi_n$, $\overline{\lim}\, \xi_n$, $\underline{\lim}\, \xi_n$ are also random variables.

Hence a very wide class of analytic operations commonly performed on functions transforms a random variable into a random variable independently of the specific form of the σ-algebra $\mathfrak{S}$. It is easy to see that these operations do not interfere with the equivalence relations between the random variables. More precisely:

c) If ξ_n and η_n are equivalent ($n=1, 2, \ldots$), and $h(t_1, t_2, \ldots, t_n)$ is a Borel function of n real variables, then $h(\xi_1, \xi_2, \ldots, \xi_n)$ and $h(\eta_1, \eta_2, \ldots, \eta_n)$ are also equivalent. Moreover, the following pairs of random variables are equivalent as well: $\sup \xi_n$ and $\sup \eta_n$, $\inf \xi_n$ and $\inf \eta_n$, $\overline{\lim}\, \xi_n$ and $\overline{\lim}\, \eta_n$, $\underline{\lim}\, \xi_n$ and $\underline{\lim}\, \eta_n$.

d) Let ξ_n, $n=1, 2, \ldots$ be a sequence of random variables. The event $S = \{\lim \xi_n \text{ exists}\}$ is $\mathfrak{S}$-measurable. It is easy to verify that this event can be represented as:

$$S = \bigcap_{k=1}^{\infty} \bigcup_{n=1}^{\infty} \bigcap_{m_1, m_2 > n} \left\{\omega : |\xi_{m_1} - \xi_{m_2}| < \frac{1}{k}\right\}.$$

Indicators of events serve as an important example of random variables. The *indicator of an event* A is a random variable $\chi_A = \chi_A(\omega)$ defined as follows:

$$\chi_A(\omega) = 1 \quad \text{if} \quad \omega \in A$$
$$\chi_A(\omega) = 0 \quad \text{if} \quad \omega \notin A.$$

If $A \in \mathfrak{S}$, then $\chi_A(\omega)$ is $\mathfrak{S}$-measurable.

Note the correspondence between set-theoretical operations on events and the analogous algebraic operations on indicators:

$$\chi_{\bigcup\limits_{k=1}^{\infty} A_k}(\omega)=\sum_{k=1}^{\infty}\chi_{A_k}(\omega), \quad \text{if} \quad A_k\cap A_r=\emptyset \quad \text{for} \quad k\neq r,$$

$$\chi_{A\cap B}(\omega)=\chi_A(\omega)\,\chi_B(\omega),$$

$$\chi_{A\setminus B}(\omega)=\chi_A(\omega)-\chi_B(\omega), \quad \text{if} \quad B\subset A,$$

$$\chi_{\overline{\lim} A_n}(\omega)=\overline{\lim}\,\chi_{A_n}(\omega), \qquad \chi_{\underline{\lim} A_n}(\omega)=\underline{\lim}\,\chi_{A_n}(\omega).$$

A random variable ξ is called *discrete* if it admits only a finite or countable number of distinct values. Such a variable can be expressed as $\xi=\sum_k c_k\chi_{A_k}(\omega)$, where A_k are $\mathfrak{S}$-measurable sets pairwise disjoint and $\bigcup_k A_k=\Omega$. For each ω only one summand is nonzero in the r.h.s. of the last equality and $\xi=c_k$ if $\omega\in A_k$. For an arbitrary random variable ξ one can always construct a sequence ξ_n of discrete random variables taking on only a finite number of possible values and converging to ξ for each ω. To prove this assertion it is sufficient to set

$$\xi_n=\sum_{j=-n}^{n-1}\sum_{k=1}^{n}\left(j+\frac{k-1}{n}\right)\chi_{A_{jk}},$$

where

$$A_{jk}=\left\{\omega: j+\frac{k-1}{n}\leqslant\xi<j+\frac{k}{n}\right\}.$$

It then follows that $|\xi-\xi_n|<\frac{1}{n}$, if $|\xi|<n$.

It is easy to verify that for a non-negative ξ one can construct a monotonically increasing sequence of non-negative discrete random variables (taking on a countable number of values) uniformly converging to ξ. Indeed, in this case we set

$$\xi_n=\sum_{k=0}^{\infty}\frac{k}{2^n}\chi_{A_{kn}}, \quad \text{where} \quad A_{kn}=\left\{\omega: \frac{k}{2^n}\leqslant\xi<\frac{k+1}{2^n}\right\}.$$

Then $0\leqslant\xi-\xi_n<2^{-n}$ for all ω.

Random elements. The notion of a random variable can be generalized to the notion of a random element with the values in an arbitrary measurable space $\{\mathscr{X},\mathfrak{B}\}$. Let $\{\Omega,\mathfrak{S}\}$ and $\{\mathscr{X},\mathfrak{B}\}$ be two measurable spaces. The mapping $g:\omega\to x$ $(x\in\mathscr{X})$ is called a measurable mapping of $\{\Omega,\mathfrak{S}\}$ into $\{\mathscr{X},\mathfrak{B}\}$ if $g^{-1}(B)=\{\omega: g(\omega)\in B\}\in\mathfrak{S}$ for an arbitrary $B\in\mathfrak{B}$.

Definition 7. *A random element ξ with values in a measurable space* $\{\mathscr{X},\mathfrak{B}\}$ is a measurable mapping of $\{\Omega,\mathfrak{S},\mathsf{P}\}$ into $\{\mathscr{X},\mathfrak{B}\}$.

If $\mathscr{X}$ is a metric space then $\mathfrak{B}$ is always assumed to be a σ-algebra of Borel sets (unless stipulated otherwise). If $\mathscr{X}$ is a vector space, then ξ is called a *random vector*.

Let a sequence of random elements $\{\xi_k; k=1, 2, \ldots, n\}$ be given, defined on a fixed probability space $\{\Omega, \mathfrak{S}, \mathsf{P}\}$ with values in the spaces $\{\mathscr{X}_k, \mathfrak{B}_k\}$ correspondingly. This sequence can be considered as a single random element ζ, which will be called the direct product of random elements $\xi_1, \ldots, \xi_n$, with values in a measurable space $\{\mathscr{Y}, \mathfrak{B}\}$ where $\mathscr{Y} = \prod_{k=1}^{n} \mathscr{X}_k$ is the product of the spaces $\mathscr{X}_1, \mathscr{X}_2, \ldots, \mathscr{X}_n$ and $\mathfrak{B} = \prod_1^n \mathfrak{B}_k$ is the product of the σ-algebras $\mathfrak{B}_1, \mathfrak{B}_2, \ldots, \mathfrak{B}_n$.

The last remark is also valid in the more general case of an arbitrary set of random elements ξ_α, $\alpha \in A$, with the values in $\{\mathscr{X}_\alpha, \mathfrak{B}_\alpha\}$ where A is a set of indices. Here the product $\mathscr{Y} = \prod_{\alpha \in A} \mathscr{X}_\alpha$ represents the space of all the mappings $y = y(\alpha)$: $\alpha \to x_\alpha$; $x_\alpha \in \mathscr{X}_\alpha$, $\alpha \in A$, i.e. the space of all functions defined on A admitting a value in $\mathscr{X}_\alpha$ for each $\alpha \in A$.

A cylindrical set in $\mathscr{Y}$ is called a set C of all $y \in \mathscr{Y}$ satisfying the relations of the type

$$y(\alpha_k) \in B_{\alpha_k}, \quad k=1, \ldots, n, \quad B_{\alpha_k} \in \mathfrak{B}_{\alpha_k}.$$

Here n is an arbitrary integer and α_k are arbitrary elements of A. More precisely, we call $C = C_{\alpha_1 \ldots \alpha_n}(B_{\alpha_1} \times \ldots \times B_{\alpha_n})$ a cylindrical set with the bases $B_{\alpha_1} \times B_{\alpha_2} \times \ldots \times B_{\alpha_n}$ over the coordinates $\alpha_1, \alpha_2, \ldots, \alpha_n$. The minimal σ-algebra containing all the cylindrical sets is denoted by $\mathfrak{B}$ and is called the product of σ-algebras $\mathfrak{B}_\alpha$, $\mathfrak{B} = \prod_{\alpha \in A} \mathfrak{B}_\alpha$. It is easy to observe that the mapping $g: \omega \to y(\alpha)$ defined by the relations $g(\omega) = g(\omega, \alpha) = f_\alpha(\omega)$ where $f_\alpha(\omega) = \xi_\alpha$ is a measurable mapping of $\{\Omega, \mathfrak{S}\}$ into $\{\mathscr{Y}, \mathfrak{B}\}$. If all $\mathscr{X}_\alpha$ are the same, $\mathscr{X}_\alpha = \mathscr{X}$, then $\mathscr{Y} = \mathscr{X}^A$ represents the space of all functions with values in $\mathscr{X}$ defined on A and the mapping $g(\omega)$ associates a function from $\mathscr{X}^A$ with each elementary event ω; in other words the mapping $g(\omega)$ is a random function. Thus, the family of random variables $\{\xi_\alpha, \alpha \in A\}$ may be regarded as a random function.

Let $\xi = f(\omega)$ be a random element with the values in $\{\mathscr{Y}, \mathfrak{B}\}$.

Definition 8. *A σ-algebra generated by a random element* ξ is a σ-algebra σ_ξ or $\sigma(\xi)$ consisting of all sets of the form $\{f^{-1}(B); B \in \mathfrak{B}\}$.

Clearly the class of sets $\{f^{-1}(B); B \in \mathfrak{B}\}$ is a σ-algebra.

The following statement is an equivalent formulation of the above: the σ-algebra σ_ξ is the minimal σ-algebra in Ω with respect to which the random element ξ is measurable.

It is intuitively clear that measurability of a certain random variable η with respect to σ_ξ means that η is a function of ξ.

Lemma 1. *Let $\xi = f(\omega)$ be a random element on $(\Omega, \mathfrak{S}, \mathsf{P})$ with values in $\{\mathcal{X}, \mathfrak{B}\}$ and η be a σ_ξ-measurable random variable. Then there exists a $\mathfrak{B}$-measurable real valued function $g(x)$ such that $\eta = g(\xi)$.*

Proof. Assume that η is a discrete random variable admitting values a_n, $n = 1, 2, \ldots$. Let $A_n = \{\omega : \eta = a_n\}$. Then there exists $B_n \in \mathfrak{B}$ such that $f^{-1}(B_n) = A_n$. Put $B'_n = B_n \setminus \bigcup_{k=1}^{n-1} B_k$. The sets $B'_n \in \mathfrak{B}$ are disjoint, $f^{-1}(B'_n) =$

$$= A_n \setminus \bigcup_{k=1}^{n-1} A_k = A_n, \text{ and } f^{-1}\left(\bigcup_1^\infty B'_n\right) = \bigcup_1^\infty A_n = \Omega, \text{ i.e. } f(\Omega) \subset \bigcup_1^\infty B'_n. \text{ Now}$$

put $g(x) = a_n$ if $x \in B'_n$. Then $\eta = g(\xi)$.

We now consider the general case. There exists a sequence of discrete σ_ξ-measurable random variables η_n, convergent to η for each ω. Therefore $\eta_n = g_n(\xi)$, where $g_n(x)$ is $\mathfrak{B}$-measurable. The set of points S on which the functions $g_n(x)$ converge to a certain point is $\mathfrak{B}$-measurable, it contains $f(\Omega)$ and $\lim g_n(x) = \lim \eta_n = \eta$ for $x \in f(\Omega)$. Putting $g(x) = \lim g_n(x)$ for $x \in S$ and $g(x) = 0$ for $x \notin S$ we obtain $\eta = g(\xi)$. □

Mathematical expectation. The mathematical expectation of a random variable is its most important numerical characteristic. This notion corresponds to the intuitive notion of the value of the arithmetic mean of observations on a random variable in a long sequence of identical stochastic experiments.

By definition the *mathematical expectation* of a random variable $\xi = f(\omega)$ is equal to the integral of $f(\omega)$ with respect to the measure P. We denote it as

$$\mathsf{E}\xi = \int_\Omega f(\omega)\,\mathsf{P}(d\omega) = \int_\Omega \xi\, d\mathsf{P}.$$

Often the designation Ω of the region of integration is omitted. Mathematical expectation possesses a number of properties which are well known from the theory of abstract integration.

Convergence in probability. Various types of convergence of sequences of random variables play an important role in probability theory. The definition of convergence with probability 1 (almost surely) was presented earlier.

Definition 9. If there exists a random variable ξ such that for any $\varepsilon > 0$

$$\mathsf{P}\{|\xi_n - \xi| > \varepsilon\} \to 0 \quad \text{as} \quad n \to \infty,$$

we say that the *sequence* $\{\xi_n; n=1, 2, \ldots\}$ *converges in probability to the random variable* ξ and denote

$$\xi = \mathsf{P}\text{-}\lim \xi_n .$$

In measure theory convergence in probability corresponds to convergence in measure. The following corollaries follow from the general results of measure theory:

a) *If a sequence* $\{\xi_n; n=1, 2, \ldots\}$ *converges almost surely it converges in probability. The converse is generally not true. However, a subsequence which converges almost surely can be selected from a sequence of random variables convergent in probability.*

b) *A necessary and sufficient condition for convergence in probability of a sequence of random variables is as follows*: for arbitrary $\varepsilon>0$ and $\delta>0$ an $n_0=n(\varepsilon, \delta)$ can be found such that for n and $n'>n_0$

$$\mathsf{P}\{|\xi_{n'}-\xi_n|>\varepsilon\}<\delta .$$

This condition is called the *condition of fundamentality in probability of the sequence* $\{\xi_n, n=1, 2, \ldots\}$.

c) If $\xi=\mathsf{P}\text{-}\lim \xi_n$ and $\eta=\mathsf{P}\text{-}\lim \xi_n$ then $\xi=\eta \pmod{\mathsf{P}}$.

d) *Let* $\eta_k=\mathsf{P}\text{-}\lim \xi_{kn} (k=1, 2, \ldots, m)$ *and let the function* $\varphi(t_1, t_2, \ldots, t_m)$ *be everywhere continuous in the m-dimensional Euclidean space* $\mathscr{R}^m$, *except possibly on a Borel set* $D (D\subset\mathscr{R}^m)$ *such that*

$$\mathsf{P}\{(\eta_1, \eta_2, \ldots, \eta_m)\in D\}=0 .$$

Then the sequence $\xi_n=\varphi(\xi_{1n}, \xi_{2n}, \ldots, \xi_{mn})$ *converges in probability to* $\eta=\varphi(\eta_1, \eta_2, \ldots, \eta_m)$. *In particular, if the sequences* ξ_{kn} *are convergent in probability, so are the sequences* $\xi_{1n}+\xi_{2n}$, $\xi_{1n}\xi_{2n}$ *and* ξ_{1n}/ξ_{2n}, *the latter under the assumption that* $\mathsf{P}\{\mathsf{P}\text{-}\lim \xi_{2n}=0\}=0$ *and, moreover*

$$\begin{aligned}\mathsf{P}\text{-}\lim(\xi_{1n}+\xi_{2n})&=\mathsf{P}\text{-}\lim \xi_{1n}+\mathsf{P}\text{-}\lim \xi_{2n}, \\ \mathsf{P}\text{-}\lim(\xi_{1n}\cdot\xi_{2n})&=\mathsf{P}\text{-}\lim \xi_{1n}\mathsf{P}\text{-}\lim \xi_{2n},\end{aligned} \qquad \mathsf{P}\text{-}\lim \frac{\xi_{1n}}{\xi_{2n}}=\frac{\mathsf{P}\text{-}\lim \xi_{1n}}{\mathsf{P}\text{-}\lim \xi_{2n}} .$$

A sufficient condition for convergence with probability 1 as stated below is useful in various specific problems:

Lemma 2. If there exists a sequence $\varepsilon_n>0$, such that

$$\sum_{n=1}^{\infty} \mathsf{P}\{|\xi_{n+1}-\xi_n|>\varepsilon_n\}<\infty, \quad \sum_{n=1}^{\infty} \varepsilon_n<\infty ,$$

then ξ_n *converges with probability* 1 *to a certain random variable* ξ. *If for any* $\varepsilon>0$,

$$\sum_{n=1}^{\infty} \mathsf{P}\{|\xi-\xi_n|>\varepsilon\}<\infty ,$$

then ξ_n converges to ξ with probability 1.

Proof. Let A_n denote the event $|\xi_{n+1}-\xi_n|>\varepsilon_n$. Then

$$\mathsf{P}(\overline{\lim}\, A_n)=\mathsf{P}\left(\bigcap_{m=1}^{\infty}\bigcup_{n=m}^{\infty} A_n\right)\leqslant \lim_{m\to\infty}\sum_{m}^{\infty}\mathsf{P}(A_n)=0.$$

Therefore, the terms of the series $\xi_1+\sum_{1}^{\infty}(\xi_{n+1}-\xi_n)$ starting with some index $m=m(\omega)$ are dominated with probability 1 by the terms of the convergent series $\sum_{n=1}^{\infty}\varepsilon_n$. This proves the first assertion. Next, let

$$B_{Nn}=\left\{|\xi-\xi_n|>\frac{1}{N}\right\}.$$

Then

$$\mathsf{P}\{\overline{\lim}\,|\xi-\xi_n|>0\}=\mathsf{P}\left\{\bigcup_{N=1}^{\infty}\bigcap_{m=1}^{\infty}\bigcup_{n=m}^{\infty} B_{Nn}\right\}\leqslant \lim_{N\to\infty}\lim_{m\to\infty}\sum_{n=m}^{\infty}\mathsf{P}(B_{Nn})=0,$$

which proves the second assertion. □

$\mathscr{L}_p$-spaces. By $\mathscr{L}_p=\mathscr{L}_p(\Omega,\mathfrak{S},\mathsf{P})$ $(p\geqslant 1)$ we denote a linear normed space of random variables ξ on $(\Omega,\mathfrak{S},\mathsf{P})$ satisfying $\mathsf{E}\,|\xi|^p<\infty$. The norm in $\mathscr{L}_p$ is defined by

$$\|\xi\|=\{\mathsf{E}\,|\xi|^p\}^{1/p}.$$

The convergence of the sequence ξ_n to its limit ξ in $\mathscr{L}_p$ (the $\mathscr{L}_p$-convergence) signifies that

$$\mathsf{E}\,|\xi-\xi_n|^p\to 0 \quad \text{as} \quad n\to\infty.$$

The $\mathscr{L}_p$-convergence implies convergence in probability. This fact follows directly from Chebyshev's inequality

$$\mathsf{P}\{|\xi_n-\xi|>\varepsilon\}\leqslant\frac{\mathsf{E}\,|\xi-\xi_n|^p}{\varepsilon^p}.$$

The space $\mathscr{L}_p$ is complete. The most important $\mathscr{L}_p$-spaces are $\mathscr{L}_1=\mathscr{L}$ and $\mathscr{L}_2$. We shall now discuss $\mathscr{L}_2$ in some detail. Note that all the definitions above and the theorems in this section are valid with no modifications for the complex-valued random variables.

The space $\mathscr{L}_2=\mathscr{L}_2(\Omega,\mathfrak{S},\mathsf{P})$ of complex-valued random variables becomes a Hilbert space if we define in $\mathscr{L}_2$, for each pair of random variables ξ and η, their scalar product putting it equal to $\mathsf{E}\xi\bar{\eta}$.

Two random variables ζ and η are called orthogonal if $\mathsf{E}\zeta\bar{\eta}=0$. In the case when ζ and η are real and $\mathsf{E}\zeta=\mathsf{E}\eta=0$, orthogonality is equivalent to the property that variables are uncorrelated. Convergence of the sequence $\{\zeta_n; n=1, 2, \ldots\}$ in $\mathscr{L}_2$ to a random variable ζ means that

$$\|\zeta-\zeta_n\|^2=\mathsf{E}\,|\zeta-\zeta_n|^2\to 0 \quad \text{as} \quad n\to\infty.$$

This type of convergence is called the mean-square convergence and is designated as $\zeta=\text{l.i.m.}\,\zeta_n$.

Note that the scalar product is a continuous function of its arguments. In certain cases it is convenient to express the condition of convergence in $\mathscr{L}_2$ in terms of the covariance of the family of random variables.

Definition 10. The function $B(t_1, t_2)=\mathsf{E}\zeta_{t_1}\bar{\zeta}_{t_2}$ is called the *covariance of a set of random variables* $\{\zeta_t;\, t\in T\}$, $\zeta_t\in\mathscr{L}_2$, $t_i\in T$.

In this definition T designates an arbitrary set.

Let a non-negative function $\psi(t)$ be defined on T which takes on arbitrarily small values.

The random variable $\eta\,(\eta\in\mathscr{L}_2)$ is called the *limit of the family* $\{\zeta_t;\, t\in T\}$ *in* $\mathscr{L}_2$ *(m.s. limit)* as $\psi(t)\to 0$, if for any $\varepsilon>0$ a $\delta>0$ can be found such that

$$\mathsf{E}\,|\eta-\zeta_t|^2\leqslant\varepsilon$$

for all t such that $0<\psi(t)<\delta$.

Lemma 3. *For the existence of the limit of a collection of random variables* $\{\xi_t;\, t\in T\}$ *as* $\psi(t)\to 0$, *it is necessary and sufficient that the limit of the covariance* $B(t, t')=\mathsf{E}\zeta_t\bar{\zeta}_{t'}$ *exist as* $\psi(t)+\psi(t')\to 0$. *If this condition is satisfied and* $\eta=\underset{\psi(t)\to 0}{\text{l.i.m.}}\,\zeta_t$, *then*

$$\mathsf{E}\,|\eta|^2=\lim_{\psi(t)\to 0} B(t, t).$$

Proof. Necessity. The necessity follows from the continuity of the scalar product.

Sufficiency. Let there exist $\lim B(t_1, t_2)=B_0$ as $\psi(t_1)+\psi(t_2)\to 0$. Note that B_0 is non-negative ($B_0=\lim B(t, t)$ as $\psi(t)\to 0$). Therefore

$$\mathsf{E}\,|\zeta_{t_1}-\zeta_{t_2}|^2=B(t_1, t_1)-2\,\mathrm{Re}\,B(t_1, t_2)+B(t_2, t_2)\to 0$$

for $\psi(t_1)+\psi(t_2)\to 0$. It follows from the completeness of $\mathscr{L}_2$ that there exists $\text{l.i.m.}\,\zeta_t=\eta$ for $\psi(t)\to 0$. Moreover,

$$\big|\|\eta\|^2-\|\zeta_t\|^2\big|\leqslant\|\eta-\zeta_t\|\;\|\eta\|+\|\eta-\zeta_t\|\;\|\zeta_t\|\to 0 \quad \text{for} \quad \psi(t)\to 0,$$

i.e. $\|\eta\|^2=\mathsf{E}\,|\eta|^2=\lim B(t, t)$ for $\psi(t)\to 0$; this completes the proof of the lemma. □

The Hilbert space $\mathscr{L}_2^m=\mathscr{L}_2^m(\Omega, \mathfrak{S}, \mathsf{P})$ of random vectors with values in the m-dimensional complex space $\mathscr{Z}^m$ is defined in an analogous manner. This space consists of random vectors ζ with the values in $\mathscr{Z}^m$

for which $\mathsf{E}|\zeta|^2<\infty$. Here the scalar product of two random vectors ζ and η is defined as $\mathsf{E}(\zeta, \eta)$ where (x, y) is the scalar product in $\mathscr{X}^m$, $|x|^2 = (x, x)$. Lemma 3 is also valid in the space $\mathscr{L}_2^m$ if we define $B(t, t')$ as $\mathsf{E}(\zeta_t, \zeta_{t'})$.

Distributions of random vectors. Let ξ be a random element with values in the measurable space $\{\mathscr{X}, \mathfrak{B}\}$. *The distribution of the random element* ξ is the measure μ induced in $\{\mathscr{X}, \mathfrak{B}\}$ by ξ, i.e.

$$\mu(B)=\mathsf{P}\{\xi\in B\}, \qquad B\in\mathfrak{B}.$$

Values of any statistical characteristic of a random element ξ may be determined by means of its distributions. Indeed

$$\mathsf{E}f(\xi)=\int_{\mathscr{X}} f(x)\,\mu(dx) \tag{1}$$

for any $\mathfrak{B}$-measurable function $f(x)$ such that one of the sides of equality (1) is meaningful. Formula (1) is the rule of change of variables in abstract integrals.

Distributions in metric spaces are studied in Chapter V. In the present section we consider distributions in $\mathscr{R}^m$. Here $\mathfrak{B}^m$ denotes a σ-algebra of Borel sets in $\mathscr{R}^m$. Distributions in $\mathscr{R}^m$ are defined by means of distribution functions. Henceforth we shall write $a<b$ $(a\leqslant b)$, $a=(a^1, a^2, \dots, a^m)\in\mathscr{R}^m$, $b=(b^1, b^2, \dots, b^m)\in\mathscr{R}^m$, if $a^i<b^i$ $(a^i\leqslant b^i)$ $(i=1, \dots, m)$. The set $\{x: x<a\}$ will be denoted by I_a. The function

$$F(x)=\mu(I_x)=\mathsf{P}\{\xi<x\}$$

is called the distribution function of a random vector ξ (or the distribution function of the measure μ).

Sets of the type $I[a, b)=\{x: a\leqslant x<b\}$ are called intervals in $\mathscr{R}^m$. We now express the probability that a vector ξ falls into an interval in terms of its distribution function. We introduce the notation

$$\Delta^{(k)}_{(a,b)}G(x)=G(x^1, \dots, x^{k-1}, b, x^{k+1}, \dots, x^m)- \\ -G(x^1, \dots, x^{k-1}, a, x^{k+1}, \dots, x^m)$$

for any function $G(x)$, $x\in\mathscr{R}^m$.

The quantity $\Delta^k_{[a,b)}F(x)$ is the probability of the event

$$\xi^1<x^1, \dots, \xi^{k-1}<x^{k-1}, a^k\leqslant\xi^k<b^k, \xi^{k+1}<x^{k+1}, \dots, \xi^m<x^m.$$

It is easy to verify that

$$\mu(I_{[a,b)})=\Delta^{(1)}_{[a^1,b^1)}\,\Delta^{(2)}_{[a^2,b^2)}\cdots\Delta^{(m)}_{[a^m,b^m)}\,F(x). \tag{2}$$

In addition to intervals $I[a, b)$ we shall also consider closed intervals $I[a, b]=\{x: a^j\leqslant x^j\leqslant b^j, j=1, 2, \dots m\}$ as well as open intervals $I(a, b)=$

$=\{x: a^j < x^j < b^j, j=1, 2, \dots m\}$. We note several properties of distribution functions:

1) $0 \leqslant F(x) \leqslant 1$;
2) if $x \leqslant y$, then $F(x) \leqslant F(y)$;
3) $\mu[a, b) \geqslant 0$, where $\mu[a, b) = \mu(I_{[a, b)})$ is determined by formula (2);
4) $F(x-0) = F(x)$;
5) $F(x) \to 0$, provided that at least one of the coordinates of point x approaches $-\infty$;
6) $F(+\infty, +\infty, \dots, +\infty) = 1$.

Lemma 4. *For an arbitrary function $F(x)$ in $\mathscr{R}^m$ satisfying conditions* 1)–6) *there exists a unique probability measure on $\mathfrak{B}^m$ whose distribution function coincides with $F(x)$.*

Consider the class $\mathfrak{M}$ of intervals $I[a, b)$ in $\mathscr{R}^m$. This class forms a semi-ring. Define on $\mathfrak{M}$ a set function $F(I[a, b))$ by the expression appearing in the r.h.s. of formula (2). The function $F(I[a, b))$ is an additive function on $\mathfrak{M}$.

In order that F be extended to a measure on $\mathfrak{B}^m$ it is necessary and sufficient that it satisfy the semi-additivity property, i.e. the inequality

$$F(I[a_0, b_0)) \leqslant \sum_{k=1}^{M} F(I[a_k, b_k)) \tag{3}$$

should be satisfied for any system of semi-intervals $I[a_k, b_k)$ $(k=1, 2, \dots)$ such that $\bigcup_{1}^{\infty} I[a_k, b_k) \supset I[a_0, b_0)$. This extension on $\mathfrak{B}^m$ is unique. We now verify condition (3) in the case under consideration.

Since $F(x)$ is continuous from the left, for any $\eta > 0$, $\varepsilon^k > 0$ can be found such that $0 \leqslant F(I[a_k - \bar{\varepsilon}_k, b_k)) - F(I[a_k, b_k)) < \frac{\eta}{2^k}$, where $\bar{\varepsilon}_k = (\varepsilon^k, \dots, \varepsilon^k)$ $(k=1, 2, \dots)$. The open intervals $(a_k - \bar{\varepsilon}, b_k)$ cover the closed interval $[a_0, b_0 - \bar{\varepsilon}]$, $\bar{\varepsilon} > 0$. In view of the Heine-Borel theorem, a finite subcover can be extracted, for example $\{(a_k - \bar{\varepsilon}_k, b_k)\}$, $k=1, 2, \dots, n$. Thus the sequence of intervals $\{(a_k - \bar{\varepsilon}_k, b_k)\}$, $k=1, 2, \dots, n$, covers the interval $[a_0, b_0 - \bar{\varepsilon})$. The disjoint sets

$$[a_0, b_0 - \bar{\varepsilon}) \cap \left\{[a_k - \bar{\varepsilon}_k, b_k) \setminus \bigcup_{i=1}^{k-1} [a_i - \bar{\varepsilon}_i, b_i, b_i)\right\}, \quad k=1, \dots, n,$$

are sums of disjoint intervals $\Delta_j^{(k)}$ $(j=1, 2, \dots, m_k)$. Thus

$$[a_0, b_0 - \bar{\varepsilon}) = \bigcup_{k=1}^{n} \bigcup_{j=1}^{m_k} \Delta_j^{(k)},$$

$$F(I[a_0, b_0-\bar{\varepsilon})) = \sum_{k=1}^{n} \sum_{j=1}^{m_k} F(\Delta_j^{(k)}) \leqslant$$

$$\leqslant \sum_{k=1}^{n} F(I[a_k - \bar{\varepsilon}_k, b_k)) \leqslant \sum_{k=1}^{\infty} F(I[a_k - \bar{\varepsilon}_k, b_k)) \leqslant \sum_{k=1}^{\infty} F(I_k) + \eta .$$

Approaching the limit as $\varepsilon \to 0$, we obtain

$$F(I[a_0, b_0)) \leqslant \sum_{k=1}^{\infty} F(I_k) + \eta .$$

Since $\eta > 0$ is arbitrary, inequality (3) and the lemma are proved. □

Definition 11. A sequence of finite measures μ_n on $\mathfrak{B}^m$ is called *weakly convergent* to the measure μ (on $\mathfrak{B}^m$) if for an arbitrary bounded and continuous function $f(x)$

$$\int_{\mathscr{R}^m} f(x)\, \mu_n(dx) \to \int_{\mathscr{R}^m} f(x)\, \mu(dx). \tag{4}$$

A family of measures is called *weakly compact* if a weakly convergent subsequence can be extracted from an arbitrary sequence.

The following theorem is valid.

Theorem 1. *In order that a sequence of measures μ_n on $\{\mathscr{R}^m, \mathfrak{B}^m\}$ be weakly compact it is necessary and sufficient that* a) $\mu_n(\mathscr{R}^m) \leqslant c$ *and* b) *for any $\varepsilon > 0$ any interval $I[a, b)$ can be found such that*

$$\varliminf_{n\to\infty} \mu_n(I[a, b)) > \mu_n(\mathscr{R}^m) - \varepsilon . \tag{5}$$

The proof of this theorem is given in Section 1 of Chapter VI.

Characteristic functions. *The function $J(u)$, $u = (u_1, u_2, \ldots, u_m)$,* defined by

$$J(u) = \mathsf{E}\, e^{i(u, \xi)} = \int_{\mathscr{R}^m} e^{i(u, x)}\, \mu(dx)$$

is called *the characteristic function of random vector ξ* (or of the corresponding distribution μ) in $\mathscr{R}^m$.

The following properties of characteristic functions are obvious:

1) $J(0) = 1$, $|J(u)| \leqslant 1$
2) $J(u)$ is uniformly continuous, $u \in \mathscr{R}^m$
3) for any n, any complex numbers z_j and any $u_j \in \mathscr{R}^m$ ($j = 1, \ldots, n$)

$$\sum_{j,k=1}^{n} J(u_j - u_k)\, z_j \bar{z}_k \geqslant 0 .$$

Conversely if a function possesses properties 1)–3), it is then the characteristic function of a certain distribution. The proof of this assertion is given in Section 2 of Chapter IV.

One can define distributions on $\mathscr{R}^m$ by means of characteristic functions because the latter uniquely determines a distribution. For example, given a distribution with density $f(x)$, the characteristic function

$$J(u)=\int_{\mathscr{R}^m} e^{i(u,x)} f(x)\,dx$$

is a Fourier transform of the function $f(x)$. Moreover, if $f(x)$ satisfies certain additional conditions (which are discussed in detail in the theory of Fourier integrals), then $f(x)$ can be recovered from $J(u)$ using the formula

$$f(x)=\frac{1}{(2\pi)^m}\int_{\mathscr{R}^m} e^{-i(u,x)} J(u)\,du.$$

An analogous inversion formula can also be given for the distribution function $F(x)$ in the general case. However, we now present a theorem concerning the uniqueness of determination of a distribution function from its characteristic function without using the inversion formula.

Theorem 2. *If*

$$\int_{\mathscr{R}^m} e^{i(u,x)}\mu_1(dx)=\int_{\mathscr{R}^m} e^{i(u,x)}\mu_2(dx),\ u\in\mathscr{R}^m,$$

where μ_i are measures on $\{\mathscr{R}^m, \mathfrak{B}^m\}$ then $\mu_1=\mu_2$.

Proof. Denote by K the class of complex-valued bounded Borel functions for which

$$\int_{\mathscr{R}^m} f(x)\,\mu_1(dx)=\int_{\mathscr{R}^m} f(x)\,\mu_2(dx). \tag{6}$$

We show that K contains all the bounded Borel functions. Clearly, K is a linear class. Since it contains functions $e^{i(u,x)}$ it also contains all the possible linear combinations of these functions $\mathsf{P}(x)=\sum_k c_k\, e^{i(a_k,x)}$. Since K is closed with respect to the operation of taking limits of sequences of uniformly bounded functions converging everywhere to a certain limit, and since in view of Weierstrass' theorem an arbitrary bounded continuous function $f(x)$ can be approximated by a uniformly bounded sequence $P_n(x)$ convergent to $f(x)$ for any $x \in \mathcal{R}^m$, it follows that K contains all the continuous functions. In view of K's closure with respect to taking limits, it now follows that K contains all the bounded Borel functions. Putting in (6) $f(x)=\chi_B(x)$, $B \in \mathfrak{B}^m$, we obtain that $\mu_1(B)=\mu_2(B)$. □

Next we establish the connection between the weak convergence of the distributions μ_n (to μ) and the convergence of their characteristic functions. Let $J(u)$ and $J_n(u)$ denote the characteristic functions of distributions μ and μ_n. It follows by definition from the weak convergence of μ_n to μ that $J_n(u) \to J(u)$.

A more profound result is contained in the following theorem:

Theorem 3. *If $J_n(u)$ are convergent for each u to a certain function $\varphi(u)$ and $\varphi(u)$ is continuous at $u=0$, then the distribution μ_n converges weakly to a certain distribution μ and $\varphi(u)$ is the characteristic function of the distribution μ.*

Proof. We first show that μ_n is a weakly compact sequence of measures. Let $A=(a, \dots, a)$. We have

$$\frac{1}{(2a)^m} \int_{[-A,A]} (1-J_n(u))\,du = \frac{1}{(2a)^m} \int_{[-A,A]} \int_{\mathcal{R}^m} (1-e^{-i(u,x)})\,\mu_n(dx)\,du =$$

$$= \int_{\mathcal{R}^m} \left(1-\prod_{k=1}^{m} \frac{\sin a x_k}{a x_k}\right) \mu_n(dx) \geqslant$$

$$\geqslant \frac{1}{2} \int_{\overline{[-A_1, A_1]}} \mu_n(dx) = \frac{1}{2}\mu_n(\overline{[-A_1, A_1]}),$$

where A_1 denotes the vector $\left(\frac{2}{a}, \frac{2}{a}, \dots, \frac{2}{a}\right)$. Approaching the limit in the inequality obtained as $n \to \infty$ and using Lebesgue's dominated convergence theorem we obtain

$$\overline{\lim}\, \mu_n(\overline{[-A_1, A_1]}) \leqslant \frac{2}{(2a)^m} \int_{[-A,A]} (1-\varphi(u))\,du.$$

In view of the continuity of $\varphi(u)$ at $u=0$, the r.h.s. of the last equality approaches zero as $a \to 0$. It thus follows from Theorem 1 that the sequence μ_n is weakly compact. Next we show that μ_n weakly converges to a certain limit. Indeed, there exists a subsequence μ_{n_j} weakly convergent to a distribution μ_0. If μ_n does not converge weakly to μ_0, then one can find another subsequence μ_{k_j} which is weakly convergent to a limit μ_0' different from μ_0. However, it follows from the remark above that $\varphi(u)$ is the characteristic function of distribution μ_0 as well as of distribution μ_0'. On the other hand the characteristic function uniquely determines the distribution, thus $\mu_0 \equiv \mu_0'$. The contradiction obtained shows that μ_n is weakly convergent to μ_0. The theorem is thus proved. □

We note some additional oft used properties of characteristic functions.

If ξ_1 and ξ_2 are independent random vectors in $\mathscr{R}^m$, $\xi_3 = \xi_1 + \xi_2$ and $J_i(u)$ are the characteristic functions of $\xi_i (i = 1, 2, 3)$, then

$$J_3(u) = J_1(u)\, J_2(u). \tag{7}$$

Next, let $\xi_i (i = 1, 2)$ be random vectors with the values in $\mathscr{R}^{m_i}$ and $\xi_3 = (\xi_1, \xi_2)$ the composite vector with values in $\mathscr{R}^{m_1} \times \mathscr{R}^{m_2}$. In order that ξ_1 and ξ_2 be independent it is necessary and sufficient that

$$J_3(u, v) = J_1(u)\, J_2(v), \tag{8}$$

where

$$J_3(u, v) = \mathsf{E} e^{i[(u, \xi_1) + (v, \xi_2)]}, \; J_1(u) = J_3(u, 0), \; J_2(v) = J_3(0, v).$$

The necessity of this condition is obvious.

The sufficiency follows from the uniqueness of a distribution function with a given characteristic function.

Definition 12. *The moment* $m_{j_1 j_2 \ldots j_s}$ of an s-dimensional vector $\xi = (\xi^1, \xi^2, \ldots, \xi^s)$ is the quantity

$$m_{j_1 j_2 \ldots j_s} = \mathsf{E} (\xi^1)^{j_1} (\xi^2)^{j_2} \ldots (\xi^s)^{j_s},$$

provided the expectation in the r.h.s. of the equality is finite. The value $q = j_1 + j_2 + \cdots + j_s$ is called the *order of the moment*.

It is easy to verify that if $\mathsf{E}\, |\xi^k|^p < \infty$, $k = 1, 2, \ldots, s$, then all the moments of orders $q \leqslant p$ are finite. Indeed, if follows from the relationship between the arithmetic and geometric means $(q = j_1 + j_2 + \cdots + j_s)$ that

$$\prod_{k=1}^{s} |\xi^k|^{j_k} = \prod_{k=1}^{s} |\xi^k|^{q \cdot \frac{j_k}{q}} \leqslant \sum_{k=1}^{s} \frac{j_k}{q} |\xi^k|^q,$$

hence

$$\mathsf{E} \prod_{k=1}^{s} |\xi^k|^{j_k} \leqslant \sum_{k=1}^{s} (j_k/q)\, \mathsf{E}\, |\xi^k|^q \leqslant \sum_{k=1}^{\infty} (j_k/q) (\mathsf{E} |\xi^k|^p)^{q/p}.$$

Moments $m_{j_1 \ldots j_s}$ with integer-valued indices may be computed from the characteristic functions by means of differentiation. Indeed,

$$m_{j_1 \dots j_s} = (-1)^q \left. \frac{\partial^q J(u)}{\partial u_2^{j_1} \partial u_2^{j_2} \dots \partial u_s^{j_s}} \right|_{u=0} \tag{9}$$

for $q \leqslant p$, if $\mathsf{E}|\xi^k|^p < \infty$. The proof of the formula follows from the fact that one may differentiate under the sign of mathematical expectation in the formula

$$J(u) = \mathsf{E}\, e^{i(u,\,\xi)}.$$

In certain cases it is necessary to use the converse assertion. The latter, however, holds only for moments with even indices. Let Δ_k be the operation of taking the symmetric finite difference in the variable u_k and let Δ_k^j be its j-th power. Then

$$\Delta_k J(u_1, \dots, u_s) = J(u_1, \dots, u_k + h_k, \dots, u_s) - J(u_1, \dots, u_k - h_k, \dots, u_s),$$

$$\Delta_k^j J(u_1, \dots, u_s) = \sum_{r=0} (-1)^r C_j^r J(u_1, \dots, u_k + (j-2r)\, h_k, u_{k+1}, \dots, u_s).$$

Hence,

$$\Delta_1^{2j_1} \Delta_2^{2j_2} \dots \Delta_s^{2j_s} J(u_1, \dots, u_s)|_{u=0} = \mathsf{E} \prod_{k=1}^{s} \sum_{r=0}^{2j_k} (-1)^r C_{2j_k}^r e^{i2(j_k - r) h_k \xi^k} =$$

$$= \mathsf{E} \prod_{k=1}^{s} (e^{ih_k\xi^k} - e^{-ih_k\xi^k})^{2j_k} = \prod_{k=1}^{s} h_k^{2j_k} (2i)^{2q}\, \mathsf{E} \prod_{k=1}^{s} \left(\frac{\sin h_k \xi^k}{h_k \xi^k} \right)^{2j_k} [\xi^k]^{2j_k}$$

or

$$\prod_{k=1}^{s} \left. \frac{\Delta_k^{2j_k} J}{(2h_k)^{2j_k}} \right|_{u=0} = (-1)^q\, \mathsf{E} \prod_{k=1}^{s} \left(\frac{\sin h_k \xi^k}{h_k \xi^k} \right)^{2j_k} [\xi^k]^{2j_k}.$$

Using Fatou's lemma we obtain

$$\lim_{\substack{h_k \to 0 \\ k = 1, 2, \dots, s}} \left. \frac{(-1)^q \prod_{k=1}^{s} \Delta_k^{2j_k} J}{\prod_{k=1}^{s} (2h_k)^{2j_k}} \right|_{u=0} \geqslant \mathsf{E} \prod_{k=1}^{s} [\xi^k]^{2j_k}.$$

The expression in the left-hand-side of the obtained inequality is equal (up to the sign) to the derivative $\partial^{2q} J / \partial u_1^{2j_1} \dots \partial u_s^{2j_s}$ evaluated at $u=0$, provided J is differentiable $2q$ times.

We have thus obtained the following theorem.

Theorem 4. *If the characteristic function $J(u_1, \ldots u_s)$ is p differentiable (where p is an even integer) at point $u=0$, then there exist moments of order $q \leqslant p$ which may be evaluated using formula* (9).

Random time. Suppose that stochastic experiments are carried out continually. Let a certain event A be considered. The realization of event A can be determined by observing the results of the experiments up to a certain random instant of time. Such an instant of time is called random time. Occasionally it is referred to as a random variable independent of the future, as a Markov moment or as stopping time.

The formal definition is as follows:

Let T denote the set of real numbers corresponding to the times at which the stochastic experiments were carried out.

Definition 13. A monotonically decreasing family of σ-algebras $\{\mathfrak{F}_t; t \in T\}$, $\mathfrak{F}_t \subset \mathfrak{S}$, $\mathfrak{F}_{t_1} \subset \mathfrak{F}_{t_2}$ if $t_1 < t_2$ on a given probability space $\{\Omega, \mathfrak{S}, \mathsf{P}\}$ is called a *current of σ-algebras* (a *current of experiments*).

Here $\mathfrak{F}_t$ are interpreted as the class of all the observed events in the experiments carried out up to the moment t inclusively.

Definition 14. *A random time on a current of σ-algebras* $\{\mathfrak{F}_t, t \in T\}$ is a function $\tau = f(\omega)$ with the values in T defined on a subset Ω_τ of the space Ω and such that $\{\tau \leqslant t\} \in \mathfrak{F}_t$ for any $t \in T$.

The condition $\{\tau \leqslant t\} \in \mathfrak{F}_t$ means the following: one may infer about the occurence of the random moment τ before the time t by observing the results of the experiment in the instant s of times s, $s \in T$, $s \leqslant t$. The set Ω_τ corresponds to the event that τ occurred during the time period T. Obviously, Ω_τ is $\mathfrak{S}$ measurable. If T possesses the maximal value $t_{\max}$, then $\Omega_\tau = \{\tau \leqslant t_{\max}\} \in \mathfrak{F}_{t\,\max}$. If no such maximal value exists and $t_k \uparrow \sup\{t, t \in T\}$, then $\Omega_\tau = \bigcup_k \{\tau \leqslant t_k\}$.

Note that if $\Omega_\tau = \Omega$ the condition $\{\tau \leqslant t\} \in \mathfrak{F}_t$ is equivalent to the requirement that $\{\tau > t\} \in \mathfrak{F}_t$, or in the case of a countable T to the requirement that $\{\tau = t\} \in \mathfrak{F}_t$ for any $t \in T$.

One can associate with the random time τ a minimal σ-algebra of events such that by observing the results of experiments up to times τ inclusive one may infer about the realization of these events. Denote by $\mathfrak{F}_\tau$ the class of events B ($B \in \mathfrak{S}$) such that $B \cap \{\tau \leqslant t\} \in \mathfrak{F}_t$ for any $t \in T$. It is easy to verify that $\mathfrak{F}_\tau$ is a σ-algebra. We call $\mathfrak{F}_\tau$ a σ-algebra generated by the random time τ.

Clearly the random variable τ is $\mathfrak{F}_\tau$-measurable.

As an example consider the case $\tau = t_0$, $t_0 \in T$. Then $\{\tau \leqslant t\}$ is either $\emptyset$ or Ω so that $\tau = t_0$ is a particular case of a random time. Moreover,

$B\cap\{\tau\leqslant t\}$ is either $\emptyset$ or B so that $\mathfrak{F}_\tau=\mathfrak{F}_{t_0}$, i.e. the notation introduced for the σ-algebra generated by a random time, agrees in the particular case $\tau=t_0$ with the previous notations.

We present a number of properties of the random time τ on a fixed current of σ-algebras $\{\mathfrak{F}_t;\, t\in T\}$.

a) if K is a Borel set on the real line and $\{\sup x : x\in K\}\leqslant t$ then the event $\{\tau\in K\}$ is $\mathfrak{F}_t$-measurable.

b) If $\theta(t)$ is a real Borel function, which maps T into T and $\theta(t)\geqslant t$ $(t\in T)$, then $\theta(\tau)$ is a random time.

This property follows from the previous one.

c) If τ_i are random times $(i=1, 2)$, then $\min(\tau_1, \tau_2)$ and $\max(\tau_1, \tau_2)$ are also random times. In particular, the quantity $\min(\tau, t_0)$ $t\in T$ is a random time provided that τ is such.

The proof of these assertions follows from the fact that

$$\{\max(\tau_1, \tau_2)\leqslant t\}=\{\tau_1\leqslant t\}\cap\{\tau_2\leqslant t\}$$

and

$$\{\min(\tau_1, \tau_2)\leqslant t\}=\{\tau_1\leqslant t\}\cup\{\tau_2\leqslant t\};$$

d) If $\tau_i (i=1, 2)$ are random times and $\tau_1\leqslant\tau_2$, $\Omega_{\tau_1}\subset\Omega_{\tau_2}$, then $\mathfrak{F}_{\tau_1}\subset\mathfrak{F}_{\tau_2}$. Indeed, let $A\in\mathfrak{F}_{\tau_1}$. Since $\{\tau_1\leqslant t\}\supset\{\tau_2\leqslant t\}$, then $A\cap\{\tau_2\leqslant t\}=A\cap\{\tau_1\leqslant t\}$ $\cap\{\tau_2\leqslant t\}=B\cap\{\tau_2\leqslant t\}\in\mathfrak{F}_t$ in view of the fact that $B=A\cap\{\tau_1\leqslant t\}\in\mathfrak{F}_t$ and $\{\tau_2\leqslant t\}\in\mathfrak{F}_t$.

Let T be a finite or a countable set, $\{\mathfrak{F}_t;\, t\in T\}$ be a current of σ-algebras, τ be a random time on $\{\mathfrak{F}_t;\, t\in T\}$, $\Omega_\tau=\Omega$. Consider the set of random variables $\{\xi_t;\, t\in T\}$ such that ξ_t is $\mathfrak{F}_t$-measurable for each $t\in T$. Set $\xi_\tau=\xi_t$ if $\tau=t$. The quantity ξ_τ is thus defined for all $\omega\in\Omega$.

Lemma 5. *The variable ξ_τ is $\mathfrak{F}_\tau$-measurable.*

Indeed, let c_k, $k=1, 2,\dots$ be the set of possible values of τ. Then

$$\begin{aligned}\{\omega:\xi_\tau<x\}\cap\{\omega:\tau\leqslant t\}&=\bigcup_{c_k\leqslant t}(\{\omega:\xi_\tau<x\}\cap\{\omega:\tau=c_k\})=\\&=\bigcup_{c_k\leqslant t}(\{\omega:\xi(c_k)<x\}\cap\{\omega:\tau=c_k\})\in\mathfrak{F}_t,\end{aligned}$$

since each event in the last sum belongs to $\mathfrak{F}_{c_k}\subset\mathfrak{F}_t$ □

§ 2. Independence

Definitions. Let $(\Omega, \mathfrak{S}, \mathsf{P})$ be a fixed probability space. Unless otherwise specified, events in this section are understood to be $\mathfrak{S}$-measurable subsets of Ω.

Two events A and B are called *independent* if $\mathsf{P}(A \cap B) = \mathsf{P}(A)\,\mathsf{P}(B)$. The statements below follow directly from the definition:

a) Ω and A, where A is an arbitrary event, are independent.

b) If $\mathsf{P}(N) = 0$ and A is arbitrary, then N and A are independent.

c) If A and B_i $(i = 1, 2)$ are independent, $B_1 \supset B_2$, then A and $B_1 \backslash B_2$ are independent. In particular, A and $\bar{B}_1$ are independent.

d) If A and B_i are independent $(i = 1, 2, \ldots, n)$, and $B_1, B_2, \ldots, B_n$ are pairwise disjoint, then A and $\bigcup_{i=1}^{n} B_i$ are also independent.

Note that without the stipulation that B_i's are pairwise disjoint, the last assertion does not generally hold.

e) A is independent of A if and only if $\mathsf{P}(A) = 0$ or $\mathsf{P}(A) = 1$.

Let I be a set, let $\{\mathfrak{M}_i, i \in I\}$ be the set of classes of events enumerated by means of the index i, taking on values from I.

Definition 1. The classes of the events $\{\mathfrak{M}_i, i \in I\}$ are called *independent (or jointly independent)* if for arbitrary pairwise different indices $i_1, i_2, \ldots, i_n$ $(i_k \in I)$ and arbitrary A_{i_k}, $A_{i_k} \in \mathfrak{M}_{i_k}$, $k = 1, 2, \ldots, n$,

$$\mathsf{P}(A_{i_1} \cap A_{i_2} \cap \cdots \cap A_{i_n}) = \mathsf{P}(A_{i_1})\,\mathsf{P}(A_{i_2}) \ldots \mathsf{P}(A_{i_n}).$$

Note that for an infinite collection of classes of events the definition of independence is equivalent to the requirement that an arbitrary finite subcollection of classes of events will consist of independent classes of events.

Henceforth, the notation $\sigma\{\mathfrak{M}\}$ will denote the minimal σ-algebra containing $\mathfrak{M}$.

A class of events $\mathfrak{A}$ is called the *π-class* if it is closed with respect to the operation of intersection of events ($A_k \in \mathfrak{A}$, $k = 1, 2$ implies $A_1 \cap A_2 \in \mathfrak{A}$) and the class of events is called the *λ-class* if

a) $A_k \in \mathfrak{A} : k = 1, 2, \ldots,$ and $A_k \cap A_r = \emptyset$ for $k \neq r$ implies that

$$\bigcup_{k=1}^{\infty} A_k \in \mathfrak{A};$$

b) $\Omega \in \mathfrak{A}$ and $B_2 \supset B_1$, $B_k \in \mathfrak{A}$ $k = 1, 2$ implies $B_2 \backslash B_1 \in \mathfrak{A}$. Evidently if $\mathfrak{A}$ is simultaneously a π-class and λ-class it is also a σ-algebra.

Lemma 1. *If the λ-class $\mathfrak{A}$ contains the π-class $\mathfrak{M}$, then $\sigma(\mathfrak{M})$ is contained in $\mathfrak{A}$.*

Proof. Denote by $\mathfrak{A}_1$ the minimal λ-class containing $\mathfrak{M}$ ($\mathfrak{A}_1$ is the intersection of all the λ-classes containing $\mathfrak{M}$). We show that $\mathfrak{A}_1 = \sigma\{\mathfrak{M}\}$. Let $\mathfrak{A}(B)$ denote the class of all events A in $\mathfrak{A}_1$ for which $A \cap B = \mathfrak{A}_1$.

It is easy to verify that $\mathfrak{A}(B)$ is a λ-class. If $B \in \mathfrak{M}$, then $\mathfrak{A}(B) \supset \mathfrak{M}$ (since $\mathfrak{M}$ is a π-class). Therefore $\mathfrak{A}(B) = \mathfrak{A}_1$ $(B \in \mathfrak{M})$. But this means that

$\mathfrak{A}(A) \supset \mathfrak{M}$ for any $A \in \mathfrak{A}_1$ i.e. $\mathfrak{A}(A) = \mathfrak{A}_1$. Thus, $\mathfrak{A}_1$ is a π-class. In view of the previous remark $\mathfrak{A}_1 = \sigma\{\mathfrak{M}\}$. □

Theorem 1. *Let $\{\mathfrak{M}_i, i \in I\}$ be a collection of independent π-classes. Then the minimal σ-algebras $\sigma\{\mathfrak{M}_i\}$, $i \in I$, are independent.*

Proof. Without loss of generality, a finite number of classes $\mathfrak{M}_1, \ldots, \mathfrak{M}_n$ will be considered. It is sufficient to show that if one of the classes i.e. $\mathfrak{M}_1$ is replaced by $\sigma(\mathfrak{M}_1)$, then the new sequence of classes will also be independent.

Denote by $\mathfrak{A}$ the class of all those events which do not depend on $\mathfrak{M}_2, \ldots, \mathfrak{M}_n$. By definition, $\mathfrak{M}_1 \subset \mathfrak{A}$ and $\mathfrak{A}$ possess the following properties: it is closed with respect to summation of non-intersecting events and taking the difference $B_2 \backslash B_1$ under the condition that $B_2 \supset B_1$. Thus $\mathfrak{A}$ is a λ-class and in view of the preceding lemma $\mathfrak{A} \supset \sigma\{\mathfrak{M}_1\}$. The theorem is thus proved. □

Theorem 2. *Let $\{\mathfrak{M}_i, i \in I\}$ be a collection of independent classes of events each of which is closed with respect to the intersection operation and let $I = I_1 \cup I_2$, $(I_1 \cap I_2) = \emptyset$. Denote by $\mathfrak{B}_j$ $(j = 1, 2)$ the minimal σ-algebra containing all the $\mathfrak{M}_i$, $i \in I_j$. Then $\mathfrak{B}_1$ and $\mathfrak{B}_2$ are independent.*

In view of the previous theorem, it is sufficient to consider the case when $\mathfrak{M}_i$ are σ-algebras. Consider classes $\mathfrak{A}_j (j = 1, 2)$ consisting of all possible events of the form $A_{i_1} \cap A_{i_2} \cap \cdots \cap A_{i_n}$, $A_{i_k} \in \mathfrak{M}_{i_k}$ where n is arbitrary and $i_k \in I_j$. These are closed with respect to intersections, $\mathfrak{A}_j$ contains all $\mathfrak{M}_i$, $i \in I_j$ and $\mathfrak{A}_1$ and $\mathfrak{A}_2$ are independent. In view of the previous theorem $\sigma\{\mathfrak{A}_1\} = \sigma\{\mathfrak{M}_i, i \in I_1\}$ and $\sigma\{\mathfrak{A}_2\} = \sigma\{\mathfrak{M}_i, i \in I_2\}$ are independent.

Corollary. *If I is subdivided into arbitrary collection of subsets $I = \bigcup_{j \in M} |I_j|$ pairwise disjoint then the σ-algebras $\{\mathfrak{B}_j = \sigma(\mathfrak{M}_i, i \in I_j), j \in M\}$ are jointly independent.*

Independent random variables. Definition 2. Random variables $\{\zeta_i, i \in I\}$ are called *independent* (jointly independent) if the classes of events $\mathfrak{M}_i$, $i \in I$, where $\mathfrak{M}_i$ consists of the events of the type $\{\omega: \zeta_i < a\}$, $-\infty < a < \infty$, are independent.

The definition of independence of a collection of random variables is equivalent to the following: random variables ζ_i $(i \in I)$ are independent if for any n and any $i_k \in I$, $k = 1, \ldots, n$, the joint distribution function of the variables $\zeta_{i_1}, \zeta_{i_2}, \ldots, \zeta_{i_n}$ is equal to the product of the distribution functions of the variables ζ_{i_k}:

$$\mathsf{P}\{\zeta_{i_1} < a_1, \zeta_{i_2} < a_2, \ldots, \zeta_{i_n} < a_n\} = \prod_{k=1}^{n} \mathsf{P}\{\zeta_{i_k} < a_k\}.$$

The independence of a collection of classes of random variables is stated analogously.

Consider a collection of classes of random variables $\{\zeta_i^\mu, i\in I_\mu\}$ where μ is a fixed index and i runs over the set I_μ depending on the index μ. For convenience, we refer to this set as a class and consider a collection of such classes enumerated by means of the index μ running over the set M.

Definition 3. Classes of random variables $\{\zeta_i^\mu, i\in I_\mu\}$ $(\mu\in M)$ are called *(mutually) independent* if the sets of events $\mathfrak{M}_\mu(\mu\in M)$ are mutually independent. Here $\mathfrak{M}_\mu$ consists of all the events of the form

$$\{\omega:\zeta_{i_1}^\mu<a_1,\ldots,\zeta_{i_n}^\mu<a_n\}, \qquad (1)$$
$$n=1,2,\ldots,\ i_k\in I_\mu,\ -\infty<a_k<\infty.$$

Definition 4. A σ-algebra of events $\sigma\{\zeta_i, i\in I\}$, generated by the events of the form

$$\{\omega:\zeta_{i_1}<a_1,\ldots,\zeta_{i_n}<a_n\},$$
$$n=1,2,\ldots,\ i_k\in I,\ -\infty<a_k<\infty,$$

is called a *σ-algebra generated by the class of random variables* $\{\zeta_i, i\in I\}$. The closure of the σ-algebra $\sigma\{\zeta_i, i\in I\}$ is denoted by $\tilde{\sigma}\{\zeta_i, i\in I\}$.

In other words, $\sigma\{\zeta_i, i\in I\}$ is the minimal σ-algebra of events with respect to which all the ζ_i are random variables (i.e. it is the minimal σ-algebra of sets, with respect to which all the functions $\zeta_i=f_i(\omega)$ are measurable).

We note in particular that the σ-algebra of events generated by a single random variable ζ is the minimal σ-algebra containing events of the form $\{\omega:\zeta<a\}$ $(-\infty<a<\infty)$.

Theorem 3. *If classes of random variables $\{\zeta_i^\mu, i\in I_\mu\}$, $\mu\in M$, are independent, then the collections of σ-algebras $\sigma\{\zeta_i^\mu, i\in I_\mu\}$, $\mu\in M$ and their closures $\tilde{\sigma}\{\zeta_i^\mu, i\in I_\mu\}$ are also independent.*

The proof follows from the fact that the classes introduced in Definition 3 are closed with respect to intersections of events inside the class and from Theorem 1. □

Corollary. *Let $g_\mu(t_1, t_2,\ldots, t_s)$ $(\mu\in M)$ be a set of finite Borel functions of s real variables. If the sequences of random variables $\{(\zeta_1^\mu, \zeta_2^\mu,\ldots,\zeta_s^\mu), \mu\in M\}$ are jointly independent, then the random variables $\xi_\mu=g_\mu(\zeta_1^\mu, \zeta_2^\mu,\ldots,\zeta_s^\mu)$, $\mu\in M$ are also independent.*

The notion of independence of random variables and the theorem proved above are easily generalized to the case of random elements in an arbitrary measurable space $\{\mathscr{X}, \mathfrak{B}\}$.

Let $\zeta_i = f_i(\omega)$, $i \in I$ be a set of random elements in $\{\mathscr{X}_i, \mathfrak{B}_i\}$. The elements $\{\zeta_i\ i \in I\}$ are called independent (or jointly independent) if for an arbitrary n, $n=1, 2, \ldots$ and arbitrary $B_{i_k} \in \mathfrak{B}_{i_k}$, $i_k \in I$,

$$\mathsf{P}\left\{\bigcap_{k=1}^{n} \{\zeta_{i_k} \in B_{i_k}\}\right\} = \prod_{k=1}^{n} \mathsf{P}\{\zeta_{i_k} \in B_{i_k}\}. \tag{2}$$

A collection of independent classes of random elements is defined analogously.

An arbitrary collection of random elements generates the minimal σ-algebra of sets in Ω, with respect to which every random element is measurable. From the independence of a certain collection of classes of random elements follows the independence of minimal σ-algebras (and their closures) generated by the corresponding classes of random elements. The proof of this assertion is analogous to the one for random variables.

Let a sequence of random elements $\xi_k = f_k(\omega)$ in $\{\mathscr{X}_k, \mathfrak{B}_k\}$, $k=1, 2, \ldots, n$ be given. This sequence may be considered as a random element with values in $\prod_{k=1}^{n} \mathscr{X}_k$. Indeed, denote by $\mathfrak{B}^{(n)}$ the product of σ-algebras $\mathfrak{B}_1, \ldots, \mathfrak{B}_n$. If $C = A_1 \times A_2 \times \ldots \times A_n$, $A_i \in \mathfrak{B}_i$, $i=1, 2, \ldots, n$, then

$$\{\omega : (f_1(\omega), \ldots, f_n(\omega)) \in C\} = \bigcap_{i=1}^{n} \{\omega : f_i(\omega) \in A_i\},$$

i.e. the preimage of C is $\mathfrak{S}$-measurable. Hence, the preimage of any set belonging to the minimal σ-algebra containing all C, i.e. the preimage of any set in $\mathfrak{B}^{(n)}$ is $\mathfrak{S}$-measurable. Denote by $m_{1, 2, \ldots, n}$ the measure in $\left\{\prod_{k=1}^{n} \mathscr{X}_k, \mathfrak{B}^{(n)}\right\}$ induced by the sequence $(\xi_1, \ldots, \xi_n)$, i.e.

$$m_{1, 2, \ldots, n}(C) = \mathsf{P}\{(\xi_1, \ldots, \xi_n) \in C\}.$$

Assume that the elements ξ_k, $k=1, \ldots, n$ are independent. Then formula (2) shows that

$$m_{1, 2, \ldots, n}(A_1 \cap A_2 \cap \ldots \cap A_n) = m_1(A_1)\, m_2(A_2) \ldots m_n(A_n),$$

where $m_k(A_k) = \mathsf{P}\{\xi_k \in A_k\}$. In view of the uniqueness of continuation of a measure from a semiring of sets onto a minimal σ-algebra the measure $m_{1, 2, \ldots, n}$ is the product of measures $m_1, m_2, \ldots, m_n$. The converse is trivial: if the measure $m_{1, 2, \ldots, n}$ coincides with the product of measures $m_1, m_2, \ldots, m_n$, then the random elements $\xi_1, \xi_2, \ldots, \xi_n$ are independent. We have thus obtained

Theorem 4. *The random elements $\xi_1, \xi_2, \ldots, \xi_n$ are independent if and only if the measure $m_{1,2,\ldots,n}$ induced by sequence $(\xi_1, \xi_2, \ldots, \xi_n)$ on the σ-algebra $\mathfrak{B}^{(n)}$ is the product of measures m_k $(k=1, \ldots, n)$ induced by the elements ξ_k on $\mathfrak{B}_k$.*

Theorem 5. *Let $g(x_1, x_2)$ be a $\mathfrak{B}^{(2)}$-measurable finite function and ξ_1 and ξ_2 be independent random elements and let, moreover,*

$$\mathsf{E}g(\xi_1, \xi_2) < \infty.$$

Then $\varphi(x_1) = \mathsf{E}g(x_1, \xi_2)$ is a *$\mathfrak{B}_1$-measurable function of x_1* and

$$\mathsf{E}g(\xi_1, \xi_2) = \mathsf{E}\varphi(\xi_1)$$

or

$$\mathsf{E}g(\xi_1, \xi_2) = \int_{\mathscr{X}_1} m_1(dx_1) \int_{\mathscr{X}_2} g(x_1, x_2)\, m_2(dx_2).$$

The same formula remains valid for an arbitrary $\tilde{\mathfrak{B}}^{(2)}$-measurable function where the sign ~ indicates the completion of a σ-algebra (or measure) if the measures m_1 and m_2 are assumed to be complete. The theorem is a direct consequence of the theorem on change of variables for abstract integrals, which yields that

$$\mathsf{E}g(\xi_1, \xi_2) = \int_{\mathscr{X}_1 \times \mathscr{X}_2} g(x_1, x_2)\, m_{1,2}(d(x_1, x_2)),$$

and of Fubini's theorem. □

Corollary. *If ξ_1 and ξ_2 are independent random variables with finite expectations, then*

$$\mathsf{E}\xi_1\xi_2 = \mathsf{E}\xi_1 \cdot \mathsf{E}\xi_2$$

Zero-one law. Let A_n, $n=1, 2, \ldots$, be a sequence of events.

Theorem 6. *If $\sum_{n=1}^{\infty} \mathsf{P}(A_n) < \infty$, then the event $\overline{\lim}\, A_n$ is of probability* 0.

The proof follows from formula $\overline{\lim}\, A_n = \bigcap_{n=1}^{\infty} \bigcup_{k=n}^{\infty} A_k$ which yields that

$$\mathsf{P}(\overline{\lim}\, A_n) = \lim_{n\to\infty} \mathsf{P}\left(\bigcup_{k=n}^{\infty} A_k\right) \leq \lim_{n\to\infty} \sum_{k=n}^{\infty} \mathsf{P}(A_k) = 0. \quad \square$$

The following stronger result is valid in case of a sequence of *independent* events.

Theorem 7. (Borel-Cantelli's lemma). *If events $\{A_n, n=1, 2, \ldots\}$ are in-*

dependent then the probability of the event $\overline{\lim} A_n$ *is either* 0 *or* 1 *depending on whether the series* $\sum_{n=1}^{\infty} \mathsf{P}(A_n)$ *is convergent or divergent.*

One need prove only that if $\sum_{n=1}^{\infty} \mathsf{P}(A_n)=\infty$, then $\mathsf{P}(\overline{\lim} A_n)=1$. If $A^*= \overline{\lim} A_n$, then

$$\Omega \backslash A^* = \bigcup_{n=1}^{\infty} \bigcap_{k=n}^{\infty} (\Omega \backslash A_k)$$

and

$$\mathsf{P}(\Omega \backslash A^*) = \lim_{n\to\infty} \mathsf{P}\left(\bigcap_{k=n}^{\infty} (\Omega \backslash A_k)\right) = \lim_{n\to\infty} \prod_{k=n}^{\infty} \mathsf{P}(\Omega \backslash A_k) = \lim_{n\to\infty} \prod_{k=n}^{\infty} (1-\mathsf{P}(A_k))=0$$

in view of the divergence of the series $\sum_{k=1}^{\infty} \mathsf{P}(A_k)$. □

Consider now an arbitrary sequence of independent σ-algebras $\mathfrak{S}_n$, $n=1, 2, \ldots$. In view of the Borel-Cantelli lemma the event $A=\overline{\lim} A_n$ where A_n is an arbitrary sequence such that $A_n \in \mathfrak{S}_n$ is of probability 0 or 1. This result can be generalized to arbitrary events generated by the collection of all σ-algebras $\mathfrak{S}_n$, $n=1, 2, \ldots$ and which are independent of the arbitrary finite sequence of σ-algebras $\mathfrak{S}_1, \mathfrak{S}_2, \ldots, \mathfrak{S}_n$. More precisely, let $\sigma\{\mathfrak{S}_k, \mathfrak{S}_{k+1}, \ldots, \mathfrak{S}_n, \ldots\} = \mathfrak{B}_k$ be a σ-algebra generated by the sequence $\mathfrak{S}_n$, $n=k, k+1, \ldots$; $\mathfrak{B}_k$ form a monotonically decreasing sequence of σ-algebras. Their intersection $\mathfrak{B} = \bigcap_{k=1}^{\infty} \mathfrak{B}_k$ is also a σ-algebra. Define

$$\overline{\lim} \mathfrak{S}_n = \mathfrak{B} = \bigcap_{k=1}^{\infty} \sigma\{\mathfrak{S}_k, \mathfrak{S}_{k+1}, \ldots\}.$$

Clearly the σ-algebra $\overline{\lim} \mathfrak{S}_n$ remains unchanged if we replace an arbitrary finite number of σ-algebras $\mathfrak{S}_1, \ldots, \mathfrak{S}_n$ by some other σ-algebras.

Theorem 8. (Kolmogorov's general zero-one law) *If* $\mathfrak{S}_n$, $n=1, 2, \ldots$, *are mutually independent σ-algebras, then any event in* $\overline{\lim} \mathfrak{S}_n$ *has probability either* 0 *or* 1.

Indeed, let $A \in \overline{\lim} \mathfrak{S}_n$. Then $A \in \mathfrak{B}_k$ for any k and thus A and $\sigma\{\mathfrak{S}_1, \ldots, \mathfrak{S}_{k-1}\}$ are independent. Since $A \in \sigma\{\mathfrak{S}_1, \ldots, \mathfrak{S}_n, \ldots\}$ A does not depend on A. But this is possible only if either $\mathsf{P}(A)=0$ or $\mathsf{P}(A)=1$. □

Corollary. *Let* $\{\xi_n, n=1, 2, \ldots\}$ *be a sequence of independent random elements in a fixed metric space* $\{\mathscr{X}, \mathfrak{B}\}$, $\mathfrak{S}_n = \sigma\{\xi_n\}$ *be a σ-algebra generated by* ξ_n, $\mathfrak{B}_n = \sigma\{\mathfrak{S}_n, \mathfrak{S}_{n+1}, \ldots\}$. *Then*

a) *the limit of the sequence* $\{\xi_n, n=1,2,\ldots\}$ *exists with either probability* 0 *or with probability* 1;

b) *if* $\mathscr{X}$ *is separable and complete, then the limit of the sequence* $\{\xi_n, n=1,2,\ldots\}$ *is constant* (mod P) *with probability* 1 *provided it exists.*

c) *If* $z=f(x_1, x_2, \ldots, x_n, \ldots)$ *is a function of infinitely many arguments* $x_n \in \mathscr{X}$, $n=1,2,\ldots$ *and for every* n $f(\xi_1, \xi_2, \ldots, \xi_n, \ldots)$ *is* $\mathfrak{B}_n$*-measurable, then the function is constant with probability* 1.

Proof. a) If $\varrho(x, y)$ is the distance in $\mathscr{X}$ then the set of points on which ξ_n is convergent can be written in the form

$$D=\bigcap_{k=1}^{\infty} \bigcup_{n=1}^{\infty} \bigcap_{n', n'' \geqslant n} \left\{\omega : \varrho(\xi_{n'}, \xi_{n''})<\frac{1}{k}\right\}.$$

Since the events $A_n=\bigcap_{n', n'' \geqslant n} \left\{\varrho(\xi_{n'}, \xi_{n''})<\frac{1}{k}\right\}$ are monotonically increasing, $\bigcup_{n=1}^{\infty} A_n=\mathfrak{B}_r$ for every r so that $D \in \mathfrak{B}_r$ for every r and the general zero-one law is applicable. □

b) Let F be a closed set, $F \subset \mathscr{X}$, denoted by F_k the open set $F_k=\left\{x : \varrho(x, F)<\frac{1}{k}\right\}$. Then the event $A=D \cap \{\lim \xi_n \in F\}$ can be represented as $A=D \cap \left[\bigcap_{k=1}^{\infty} \bigcup_{n=1}^{\infty} \bigcap_{n' \geqslant n} \{\xi_{n'} \in F_k\}\right]$. Hence, in view of the same considerations as those used in the proof of a), $A \in \mathfrak{B}_r$, $r=1,2,\ldots$. Therefore $P\{\lim \xi_n \in F\}=0$ or 1 for any closed F. But the class of events for which an analogous assertion is valid is a σ-algebra, therefore $\mathrm{P}\{\lim \xi_n \in B\}=0$ or 1 for any $B \in \mathfrak{B}$. From here it easily follows that in the case of a separable and complete space $\mathscr{X}$ the measure m induced on $\mathfrak{B}$ by the random element $\lim \xi_n$ is concentrated on a single atom. Indeed, since $m(\mathscr{X})=1$ a sphere S_1 of radius 1 can be found such that $m(S_1)=1$. If there were no such sphere then all the spheres in $\mathscr{X}$ of radius 1 would have been of measure 0 which is impossible since $\mathscr{X}$ is covered by a countable number of such spheres. Analogously, a sphere S_2 of radius $\frac{1}{2}$ can be found such that $S_2 \subset S_1$ and $m(S_2)=1$. Continuing this argument we obtain a sequence of nested spheres S_n of measure 1 with radii approaching zero. These spheres have only one point x in common and $m\{x\}=\lim m(S_n)=1$. □

c) By assumption, the events $A=\{\omega : f(\xi_1, \ldots, \xi_n, \ldots)<a\} \in \mathfrak{B}_n$; therefore $A \in \overline{\lim}\, \mathfrak{S}_n$ so that A is of probability either 0 or 1. Therefore, the distribution function of the random variable $\zeta=f(\xi_1, \ldots, \xi_n, \ldots)$ takes on the values 0 or 1 only and the variable ζ is a constant with probability one. □

§ 3. Conditional Probabilities and Conditional Expectations

Definitions. We first recall the definition of the conditional probability and conditional expectation in the elementary case. The conditional probability $\mathsf{P}(A \mid B)$ of an event A given B such that $\mathsf{P}(B) \neq 0$ is defined by the relationship

$$\mathsf{P}(A \mid B) = \frac{\mathsf{P}(A \cap B)}{\mathsf{P}(B)}.$$

For a fixed B the conditional probability $\mathsf{P}(A \mid B)$ is a normed measure defined on the same σ-algebra of events as the "unconditional" probability $\mathsf{P}(A)$. Correspondingly the conditional mathematical expectation of a certain random variable $\xi = f(\omega)$, given B, is defined by the formula

$$\mathsf{E}\{\xi \mid B\} = \int_{\Omega} f(\omega)\,\mathsf{P}(d\omega \mid B).$$

Taking into account the definition of conditional probability this relationship can be rewritten as follows:

$$\mathsf{P}(B)\,\mathsf{E}\{\xi \mid B\} = \int_{B} \xi\, d\mathsf{P}. \tag{1}$$

To be able to define conditional mathematical expectations and conditional probabilities given events of probability zero it is necessary to reconsider these notions. Firstly, we note that if ξ is the indicator of event A then $\mathsf{E}\{\xi \mid B\} = \mathsf{P}(A \mid B)$. Therefore conditional probabilities are particular cases of conditional expectations and we shall consider meanwhile only the latter. Let $\mathfrak{M}$ be a countable class of disjoint events $\{B_i;\, i = 1, 2, \ldots,\, B_i \in \mathfrak{S}\}$ and $\bigcup_{i=1}^{\infty} B_i = \Omega$. Define a random variable $\mathsf{E}\{\xi \mid \mathfrak{M}\}$ to be equal to $\mathsf{E}\{\xi \mid B_i\}$ if $\omega \in B_i$ and call it the conditional mathematical expectation of ξ given the class of sets $\mathfrak{M}$. This variable is defined only for the values of ω which belong to B_i such that $\mathsf{P}(B_i) \neq 0$, i.e. the random variable $\mathsf{E}\{\xi \mid \mathfrak{M}_i\}$ is defined with probability 1. This variable is constant on the set $B_i \in \mathfrak{M}$ such that $\mathsf{P}(B_i) \neq 0$ and is equal to the conditional mathematical expectation of ξ given B_i. Observe that knowing $\mathsf{E}\{\xi \mid \mathfrak{M}\}$ one can define not only $\mathsf{E}\{\xi \mid B_i\}$ for $B_i \in \mathfrak{M}$, $\mathsf{P}(B_i) \neq 0$, but also the conditional mathematical expectation of the random variable given an arbitrary B, with $\mathsf{P}(B) \neq 0$, belonging to $\sigma\{\mathfrak{M}\}$. Indeed if $B = \bigcup_{k=1}^{\infty} B_{j_k}$, then

$$\mathsf{P}(B)\,\mathsf{E}\{\xi \mid B\} = \sum_{k=1}^{\infty} \mathsf{P}(B_{j_k})\,\mathsf{E}\{\xi \mid B_{j_k}\}. \tag{2}$$

This formula shows how one can compute conditional mathematical expectations for given countable sums of sets by knowing the conditional mathematical expectations for the given B_i's $(i=1,2,\ldots)$ and hence any conditional probability given an arbitrary set from the smallest σ-algebra containing all B_i. Note that relation (2) can be written as follows:

$$\int_B \xi\, d\mathsf{P} = \int_B \mathsf{E}\{\xi \mid \mathfrak{M}\}\, \mathsf{P}(d\omega).$$

This relation holds for an arbitrary B in the σ-algebra generated by $\mathfrak{M}$, and the random variable $\eta = \mathsf{E}\{\xi \mid \mathfrak{M}\}$ is measurable with respect to this σ-algebra.

It is easy to verify that these properties define the conditional mathematical expectation uniquely $(\operatorname{mod} \mathsf{P})$. Indeed, if there exist two $\mathfrak{F}$-measurable random variables η_i $(i=1,2)$ for which

$$\int_B \eta_1\, d\mathsf{P} = \int_B \eta_2\, d\mathsf{P}$$

for any $B \in \mathfrak{F}$ ($\mathfrak{F}$ is a σ-algebra) then η_1 and η_2 coincide P-almost everywhere.

The last observation can be used for the definition of conditional, mathematical expectation in the general case. Let a certain experiment described by a σ-algebra of events $\mathfrak{B}$ be carried out. It is required to determine the conditional mathematical expectation of a random variable ξ under the assumption that the result of the experiment is known. This conditional mathematical expectation is considered to be a function of the result of the experiment, i.e. as a $\mathfrak{B}$-measurable random variable satisfying the relationship just obtained.

Definition 1. Let $\mathfrak{B}$ be an arbitrary σ-algebra of events contained in $\mathfrak{S}$ and ξ an arbitrary random variable for which the mathematical expectation exists. *The conditional mathematical expectation of a random variable ξ given σ-algebra* $\mathfrak{B}$ is called the random variable $\mathsf{E}\{\xi \mid \mathfrak{B}\}$ measurable with respect to $\mathfrak{B}$ and satisfying the equality

$$\int_B \mathsf{E}\{\xi \mid \mathfrak{B}\}\, d\mathsf{P} = \int_B \xi\, d\mathsf{P} \tag{3}$$

for any $B \in \mathfrak{B}$.

The existence and uniqueness (up to an equivalence) of the random variable $\mathsf{E}\{\xi \mid \mathfrak{B}\}$ follows directly from the Radon-Nikodym theorem. Indeed, the r.h.s. of (3) is a σ-finite countably-additive set function on $\mathfrak{B}$ which is absolutely continuous with respect to measure P. Therefore there exists a $\mathfrak{B}$-measurable function $g(\omega)$ such that

$$\int_B \xi \, d\mathsf{P} = \int_B g(\omega) \mathsf{P}(d\omega).$$

The function $g(\omega)$ is unique (up to an equivalence). This function is by definition the conditional mathematical expectation of ξ given the σ-algebra $\mathfrak{B}$.

Remark. Let $\tilde{\mathfrak{B}}$ be the completion of $\mathfrak{B}$ with respect to probability P. It is easy to verify that

$$\mathsf{E}\{\xi \mid \mathfrak{B}\} = \mathsf{E}\{\xi \mid \tilde{\mathfrak{B}}\} \quad (\text{mod}\, \mathsf{P}).$$

Since the class of $\tilde{\mathfrak{B}}$-measurable functions is wider than the class of $\mathfrak{B}$-measurable functions, it is sometimes expedient to consider the conditional mathematical expectation given the completed σ-algebras.

The conditional probabilities $\mathsf{P}\{A \mid \mathfrak{B}\}$ given the σ-algebra $\mathfrak{S}$ are defined as a particular case of the conditional mathematical expectation by putting $\xi = \chi_A(\omega)$.

Definition 2. For a fixed A the conditional probability $\mathsf{P}\{A \mid \mathfrak{B}\}$ is a $\mathfrak{B}$-measurable random variable satisfying for every $B \in \mathfrak{B}$ the equation

$$\int_B \mathsf{P}\{A \mid \mathfrak{B}\} \, d\mathsf{P} = \mathsf{P}(A \cap B). \tag{4}$$

Properties of conditional expectations and conditional probabilities. In this section we shall always assume that the random variables under consideration possess finite or infinite expectations and that the assertions stated or proved are valid with probability 1.

a) *If* $\xi \geqslant 0$, *then* $\mathsf{E}\{\xi \mid \mathfrak{B}\} \geqslant 0$.

b) *If* ξ *is a* $\mathfrak{B}$*-measurable random variable then*

$$\mathsf{E}\{\xi \mid \mathfrak{B}\} = \xi$$

In particular if event B *is* $\mathfrak{B}$*-measurable, then*

$$\mathsf{P}\{B \mid \mathfrak{B}\} = \chi_B$$

c) $\mathsf{E}\mathsf{E}\{\xi \mid \mathfrak{B}\} = \mathsf{E}\xi$.

d) *If* $\mathsf{E}\xi_i \neq \infty$, $i = 1, 2$, *then*

$$\mathsf{E}\{a\xi_1 + b\xi_2 \mid \mathfrak{B}\} = a\mathsf{E}\{\xi_1 \mid \mathfrak{B}\} + b\mathsf{E}\{\xi_2 \mid \mathfrak{B}\}.$$

To prove the last assertion it is sufficient to verify that its r.h.s. satisfies the definition of the conditional mathematical expectation of the random variable $a\xi_1 + b\xi_2$.

Setting $\xi_i = \chi_{B_i}$, $B_1 \cap B_2 = \emptyset$ we obtain as a particular case, the additivity of the conditional probability

$$\mathsf{P}\{B_1 \cup B_2 \mid \mathfrak{B}\} = \mathsf{P}\{B_1 \mid \mathfrak{B}\} + \mathsf{P}\{B_2 \mid \mathfrak{B}\}.$$

e) *If the sequence $\{\xi_n, n=1, 2, \ldots\}$ is a monotonically decreasing sequence of non-negative random variables, then*

$$\lim \mathsf{E}\{\xi_n \mid \mathfrak{B}\} = \mathsf{E}\{\lim \xi_n \mid \mathfrak{B}\}.$$

The proof of this assertion is an immediate consequence of the application of the Lebesque monotone convergence theorem to the equality

$$\int_B \mathsf{E}\{\xi_n \mid \mathfrak{B}\}\, d\mathsf{P} = \int_B \xi_n \, d\mathsf{P}. \quad \square$$

For conditional probabilities the above result yields:

if $\{B_n, n=1, 2, \ldots\}$ is a monotonically increasing sequence of events, then

$$\lim \mathsf{P}\{B_n \mid \mathfrak{B}\} = \mathsf{P}\left\{\bigcup_{n=1}^{\infty} B_n \mid \mathfrak{B}\right\};$$

if A_n, $n=1, 2, \ldots$, are pairwise disjoint, then

$$\sum_{n=1}^{\infty} \mathsf{P}\{A_n \mid \mathfrak{B}\} = \mathsf{P}\left\{\bigcup_{n=1}^{\infty} A_n \mid \mathfrak{B}\right\}. \tag{5}$$

Remark. The last property of conditional probabilities does not mean that they can be considered for a fixed ω as countably-additive set functions. For a given sequence A_n the probability that equality (5) is not satisfied is zero, but the corresponding exceptional event depends on the choice of this sequence. Therefore it is possible that there is not a single ω for which (5) is valid for all the sequences A_n in $\mathfrak{S}$.

f) *If the random variable ξ and the σ-algebra $\mathfrak{B}$ are independent, then*

$$\mathsf{E}\{\xi \mid \mathfrak{B}\} = \mathsf{E}\xi. \tag{6}$$

The independence of the random variable ξ and the σ-algebra $\mathfrak{B}$ by definition means that σ-algebras $\sigma\{\xi\}$ and $\mathfrak{B}$ are independent. Therefore for an arbitrary $B \in \mathfrak{B}$

$$\int_B \xi \, d\mathsf{P} = \mathsf{E}\xi\chi_B = \mathsf{E}\xi\mathsf{P}(B),$$

and equality (3) will be satisfied if we set $\mathsf{E}\{\xi \mid \mathfrak{B}\} = \mathsf{E}\xi$.

It follows from the property just proven that if event A does not depend on σ-algebra $\mathfrak{B}$, then

$$\mathsf{P}\{A \mid \mathfrak{B}\} = \mathsf{P}(A). \tag{7}$$

g) *If η is a $\mathfrak{B}$-measurable random variable, then*

$$\mathsf{E}\{\xi\eta \mid \mathfrak{B}\} = \eta \mathsf{E}\{\xi \mid \mathfrak{B}\}. \tag{8}$$

To prove this property it is sufficient to consider the case when η is non-negative.

If $\eta = \chi_{B_1}$, $B_1 \in \mathfrak{B}$, then

$$\int_B \eta \mathsf{E}\{\xi \mid \mathfrak{B}\}\, d\mathsf{P} = \int_{B \cap B_1} \mathsf{E}\{\xi \mid \mathfrak{B}\}\, d\mathsf{P} = \int_{B \cap B_1} \xi\, d\mathsf{P} = \int_B \eta\xi\, d\mathsf{P},$$

so that equality (8) is satisfied. Since the class of random variables satisfying (8) is linear and closed with respect to passage to the limits of the monotone non-negative sequences, it contains arbitrary $\mathfrak{B}$-measurable non-negative random variables. □

Since the conditional expectation $\mathsf{E}\{\xi \mid \mathfrak{B}\}$ is a random variable, one can examine the conditional expectation of this variable given another σ-algebra $\mathfrak{B}_1$. We thus arrive at the iterated conditional mathematical expectation $\mathsf{E}\{\mathsf{E}\{\xi \mid \mathfrak{B}\} \mid \mathfrak{B}_1\}$. We now establish an important property of this operation. Note that if $\mathfrak{B}$ and $\mathfrak{B}_1$ are two σ-algebras and $\mathfrak{B}_1 \subset \mathfrak{B}$ it follows from the definition of the conditional mathematical expectation that $\mathsf{E}\{\mathsf{E}\{\xi \mid \mathfrak{B}_1\} \mid \mathfrak{B}\} = \mathsf{E}\{\xi \mid \mathfrak{B}_1\}$.

The following property is more profound:

h) *Let $\mathfrak{B} \subset \mathfrak{B}_1$; then $\mathsf{E}\{\mathsf{E}\{\xi \mid \mathfrak{B}_1\} \mid \mathfrak{B}\} = \mathsf{E}\{\xi \mid \mathfrak{B}\}$.*

Indeed if $B \in \mathfrak{B}$ then $B \in \mathfrak{B}_1$, hence

$$\int_B \mathsf{E}\{\mathsf{E}\{\xi \mid \mathfrak{B}_1\} \mid \mathfrak{B}\}\, d\mathsf{P} = \int_B \mathsf{E}\{\xi \mid \mathfrak{B}_1\}\, d\mathsf{P} = \int_B \xi\, d\mathsf{P} = \int_B \mathsf{E}\{\xi \mid \mathfrak{B}\}\, d\mathsf{P}.$$

Comparing the extreme terms in this chain of equalities, we obtain

$$\mathsf{E}\{\mathsf{E}\{\xi \mid \mathfrak{B}_1\} \mid \mathfrak{B}\} = \mathsf{E}\{\xi \mid \mathfrak{B}\}. \quad \square$$

It follows from the result just proved that if $\mathfrak{B} \subset \mathfrak{B}_1$ and η is a $\mathfrak{B}_1$-measurable random variable, then

$$\mathsf{E}\{\xi\eta \mid \mathfrak{B}\} = \mathsf{E}\{\eta \mathsf{E}\{\xi \mid \mathfrak{B}_1\} \mid \mathfrak{B}\}. \tag{9}$$

The following assertion which generalizes g) is often used. Let $\zeta = h(\omega)$, $\eta = s(\omega)$ be two measurable mappings of $\{\Omega, \mathfrak{S}\}$ into $\{\mathscr{X}, \mathfrak{A}\}$ and $\{\mathscr{Z}, \mathfrak{C}\}$

correspondingly. Let $g(x, z)$ be a numerical function defined on $\mathscr{X} \times \mathscr{Z}$ measurable with respect to $\sigma\{\mathfrak{A} \times \mathfrak{C}\}$ and let the mathematical expectation of $g(\zeta, \eta)$ be finite.

i) *If* η *is* $\mathfrak{B}$*-measurable* ($\mathfrak{B} \subset \mathfrak{S}$), *then we can define* $\mathsf{E}\{g(\zeta, z) \mid \mathfrak{B}\}$ *such that*

$$\mathsf{E}\{g(\zeta, \eta) \mid \mathfrak{B}\} = \mathsf{E}\{g(\zeta, z) \mid \mathfrak{B}\}|_{z=\eta}.$$

To prove this assertion we note that in view of g) the last formula is valid for the functions of the form $g(x, z) = \sum_{k=1}^{n} g_k(x)\, v_k(z)$. For arbitrary functions $g(x, z)$ such that $\mathsf{E}|g(\zeta, \eta)| < \infty$ the assertion follows from the existence of a sequence of functions $h_n(x, z)$ of the previous type and such that $h_n(\zeta, \eta)$ converges to $g(\zeta, \eta)$ with probability 1 also in $\mathscr{L}$. □

Conditional mathematical expectation given a random variable. Let ζ be a random variable taking on values $z_1, z_2, \ldots, z_n, \ldots$, $\mathsf{P}(\zeta = z_n) > 0$, B_n denotes the event $\{\zeta = z_n\}$ and

$$\mathsf{P}_n(A) = \frac{\mathsf{P}(A \cap B_n)}{\mathsf{P}(B_n)}$$

is the conditional probability of A given $\zeta = z_n$. The conditional mathematical expectation of variable ξ given $\zeta = z_n$ is defined by the formula

$$\mathsf{E}\{\xi | \zeta = z_n\} = \int_{\Omega} \xi \, d\mathsf{P}_n = \frac{1}{\mathsf{P}(B_n)} \int_{B_n} \xi \, d\mathsf{P}.$$

Regarding this sequence of numbers as a function of the outcome of the experiment to determine the value of ζ, we arrive at the notion of the conditional mathematical expectation of ξ given the random variable ζ. This is the random variable $\mathsf{E}\{\xi \mid \zeta\}$ taking on values $\mathsf{E}\{\xi \mid \zeta = z_n\}$ if $\zeta = z_n$. This definition coincides with the previously stated definition of the conditional mathematical expectation given a countable partition $\mathfrak{M}$ of the space Ω. The place of $\mathfrak{M}$ is taken by events $\{B_n;\, n = 1, 2, \ldots\}$. The last remark leads us to a general definition.

Consider a measurable mapping $\zeta = g(\omega)$ of the space $\{\Omega, \mathfrak{S}\}$ into the measurable space $\{\mathscr{X}, \mathfrak{B}\}$. Thus ζ is a random element with the values in $\mathscr{X}$. Let $\mathfrak{F}_\zeta$ be a σ-algebra generated by the mapping $\zeta: \mathfrak{F}_\zeta = \{S: S = g^{-1}(B),\, B \in \mathfrak{B}\}$.

Definition 3. The random variable $\mathsf{E}\{\xi \mid \mathfrak{F}_\zeta\}$ is called *the conditional mathematical expectation* $\mathsf{E}\{\xi \mid \zeta\}$ *of a random variable* ξ *given a random element* ζ.

This definition is equivalent to the following: for any $B \in \mathfrak{B}$

$$\int_{g^{-1}(B)} \mathsf{E}\{\xi \mid \zeta\}\, d\mathsf{P} = \int_{g^{-1}(B)} \xi\, d\mathsf{P}. \tag{10}$$

The conditional probability given ζ is defined analogously:

$$\mathsf{P}\{A \mid \zeta\} = \mathsf{P}\{A \mid \mathfrak{F}_\zeta\}.$$

Theorem 1. *The conditional mathematical expectation given the random element ζ is a $\mathfrak{B}$-measurable function of ζ:*

$$\mathsf{E}\{\xi \mid \zeta\} = s(\zeta),$$

where $s(x)$ is a $\mathfrak{B}$-measurable function.

Proof. Let ξ be non-negative. We have

$$\int_{g^{-1}(B)} \mathsf{E}\{\xi \mid \zeta\}\, d\mathsf{P} = \int_{g^{-1}(B)} \xi\, d\mathsf{P} = q(B). \tag{11}$$

Clearly, $q(B)$ is a σ-finite measure on $\mathfrak{B}$. Moreover, $q(B)=0$ if $\mathsf{P}\{g^{-1}(B)\} = 0$, i.e. q is absolutely continuous with respect to measure P_g, where $\mathsf{P}_g(A) = \mathsf{P}\{g^{-1}(A)\}$. In view of the Radon-Nikodym theorem there exists a $\mathfrak{B}$-measurable non-negative function $s(x)$ such that

$$q(B) = \int_B s(x)\, \mathsf{P}_g(dx).$$

Applying the rule of change of variables, we obtain

$$q(B) = \int_{g^{-1}(B)} s(g(\omega))\, \mathsf{P}(d\omega).$$

Comparing the last equation with (11) we obtain the equality

$$\mathsf{E}\{\xi \mid \zeta\} = s(g(\omega)) = s(\zeta). \quad \square$$

The theorem just proven shows that the conditional mathematical expectation given the random element ζ can be considered as a function of a variable x in the measurable space $\{\mathscr{X}, \mathfrak{B}, \mathsf{P}_g\}$, while in the initial definition the conditional mathematical expectation is a function of an elementary event ω. The function $s(x)$ is uniquely determined by relation

$$\int_{g^{-1}(B)} \xi\, d\mathsf{P} = \int_B s(x)\, \mathsf{P}_g(dx), \tag{12}$$

which is valid for $B \in \mathfrak{B}$.

We now present several propositions which follow directly from the previously obtained properties of conditional mathematical expectations.

a) *Let* ξ *and* ζ *be independent. Then* $\mathsf{E}\{\xi \mid \zeta\} = \mathsf{E}\xi$.

b) *If* ξ *is a* $\mathfrak{F}_\zeta$*-measurable random variable, then*

$$\mathsf{E}\{\xi \mid \zeta\} = \xi.$$

c) *If* $\eta_i = g_i(\omega)$ *are measurable mappings of* $\{\Omega, \mathfrak{S}\}$ *into* $\{\mathscr{X}_i, \mathfrak{B}_i\}$ $(i=1, 2)$ *then*

$$\mathsf{E}\{\mathsf{E}\{\xi \mid (\eta_1, \eta_2)\} \mid \eta_1\} = \mathsf{E}\{\xi \mid \eta_1\},$$

where (η_1, η_2) *denotes the mapping* $\omega \to (g_1(\omega), g_2(\omega))$ *of the space* $\{\Omega, \mathfrak{S}\}$ *into the product space* $\{\mathscr{X}_1 \times \mathscr{X}_2, \sigma\{\mathfrak{B}_1 \times \mathfrak{B}_2\}\}$.

Regular probabilities. As we pointed out previously, one cannot in general consider conditional probabilities as measures depending on an elementary event. However, in a number of cases this interpretation is valid. We now state the problem more precisely.

The conditional probability $\mathsf{P}\{A \mid \mathfrak{B}\} = h(\omega, A)$ is a function of $\omega \in \Omega$ and $A \in \mathfrak{S}$ defined for each fixed A only up to an event with probability 0. The question arises whether it is possible to find a function $p(\omega, A)$ $(\omega \in \Omega, A \in \mathfrak{S})$ such that:

a) for a fixed ω the function $p(\omega, A)$ is the probability on the σ-algebra $\mathfrak{S}$;

b) $h(\omega, A) = p(\omega, A)$ almost surely for an arbitrary fixed A.

Definition 4. If there exists a function $p(\omega, A)$ satisfying conditions a) and b), then the family of conditional probabilities $\mathsf{P}\{A \mid \mathfrak{B}\}$ is called *regular*. In this case $\mathsf{P}\{A \mid \mathfrak{B}\}$ is identified with $p(\omega, A)$.

In the regular case the conditional mathematical expectations, as would be expected, are expressed by means of integrals with conditional probabilities serving as measures.

Theorem 2. *If* $\mathsf{P}\{A \mid \mathfrak{B}\}$ *is a regular conditional probability,* $\xi = f(\omega)$, *then*

$$\mathsf{E}\{\xi \mid \mathfrak{B}\} = \int f(\omega) \mathsf{P}(d\omega \mid \mathfrak{B}) \quad (\mathrm{mod}\, \mathsf{P}). \tag{13}$$

It is not difficult to prove this assertion. Firstly, the assertion is valid in the case when ξ is the indicator of an event $A \in \mathfrak{S}$. In view of the linearity of both sides of equality (13) regarded as functionals of f, it follows that the assertion is valid also for simple functions. By taking the limits of monotonically increasing sequences of simple functions we prove (13) for an non-negative random variable ξ. Finally, repeated utilization of the linearity of both sides of (13) concludes the proof. □

In certain cases it would be necessary to emphasize that the conditional probability is a function of an elementary event. In such a case we write $\mathsf{P}\{A \mid \mathfrak{B}\} = \mathsf{P}_{\mathfrak{B}}(\omega, A)$ or simply, $\mathsf{P}(\omega, A)$ if the σ-algebra $\mathfrak{B}$ is fixed. Analogously, $\mathsf{P}_{\xi}(\omega, A)$ will serve as an alternative notation for $\mathsf{P}\{A \mid \xi\}$.

Since the property of regularity of conditional probabilities does not always hold, it seems desirable to somewhat extend this notion.

Let $\{\mathscr{X}, \mathfrak{B}\}$ be a measurable space, ζ be a random element in $\{\mathscr{X}, \mathfrak{B}\}$ and $\mathfrak{F}$ be a σ-algebra, $\mathfrak{F} \subset \mathfrak{S}$.

Definition 5. The function $Q(\omega, B)$ defined on $\Omega \times \mathfrak{B}$ is called the *regular conditional distribution of a random element ζ given a σ-algebra $\mathfrak{F}$*, provided the following is valid:

a) for a fixed $B \in \mathfrak{B}$, $Q(\omega, \mathfrak{B})$ is $\mathfrak{F}$-measurable,

b) for a fixed ω, $Q(\omega, B)$ is a probability measure on $\mathfrak{B}$ with probability 1,

c) for each $B \in \mathfrak{B}$, $Q(\omega, B) = \mathsf{P}\{(\zeta \in B) \mid \mathfrak{F}\}\,(\operatorname{mod} \mathsf{P})$.

The last requirement is equivalent to the following: for an arbitrary $F \in \mathfrak{F}$

$$\int_F Q(\omega, B)\,\mathsf{P}(d\omega) = \mathsf{P}\{(\zeta \in B) \cap F\}. \tag{14}$$

Theorem 3. *Let $\mathscr{X}$ be a complete separable metric space, $\mathfrak{B}$ be a σ-algebra of Borel sets in $\mathscr{X}$, ζ be a random element in $\{\mathscr{X}, \mathfrak{B}\}$ and $\mathfrak{F}$ an arbitrary σ-algebra, $\mathfrak{F} \subset \mathfrak{S}$. Then there exists a regular conditional distribution of the random element ζ given the σ-algebra $\mathfrak{F}$.*

Proof. Set $q(B) = \mathsf{P}(\{\zeta \in B\})$, $B \in \mathfrak{B}$. Using a theorem in measure theory one can find a compact set K_n in $\mathscr{X}$ for any n such that $q(\mathscr{X} \setminus K_n) < 1/n$ (c.f. Chapter V, Section 2, Remark 1 to Theorem 1).

Denote by $\mathscr{C}(\mathscr{X})$ – where $\mathscr{X}$ is a metric space – the space of real-valued functions, continuous and bounded on $\mathscr{X}$ with the metric $\varrho(f, g) = \|f(x) - g(x)\| = \{\sup |f(x) - g(x)|, x \in \mathscr{X}\}$.

The space $\mathscr{C}(K_n)$ is separable. Let $\{f_{nk}(x), k = 1, 2, \ldots\}$ be a countable everywhere dense net in $\mathscr{C}(K_n)$. We extend $f_{nk}(x)$ to the whole $\mathscr{X}$ in such a manner that $\{\sup |f_{nk}(x)|\ x \in \mathscr{X}\} = \{\max |f_{nk}(x)|, x \in K_n\}$. Set $\chi_n = \chi_n(\zeta)$ where $\chi_n(x)$ is the indicator of the compact set K_n. It follows from the properties of mathematical expectations that there exists $D_0 \in \mathfrak{S}$ such that $\mathsf{P}(D_0) = 0$ and for $\omega \notin D_0$ the following relationships are valid:

if $f_{nk}(x) \geqslant 0$ then $\mathsf{E}\{f_{nk}(\zeta) \mid \mathfrak{F}\} \geqslant 0$;

if $|f_{nk}(x) - f_{nj}(x)| < r$ then $\mathsf{E}\{|f_{nk}(\zeta) - f_{nj}(\zeta)|\ \mathfrak{F}\} \leqslant r$;

$$\mathsf{E}\{r f_{nk}(\zeta) \mid \mathfrak{F}\} = r\mathsf{E}\{f_{nk}(\zeta)) \mid \mathfrak{F}\},$$

$$\mathsf{E}\{f_{nk}(\zeta)\pm f_{nj}(\zeta)\mid\mathfrak{F}\}=\mathsf{E}\{f_{nk}(\zeta)\mid\mathfrak{F}\}\pm\mathsf{E}\{f_{nj}(\zeta)\mid\mathfrak{F}\},$$
$$\lim_{m\to\infty}\mathsf{E}\{(1-\chi_m)\,f_{nk}(\zeta)\mid\mathfrak{F}\}=0$$

for all n, k and j and rational r. On the other hand for an arbitrary $F\in\mathfrak{F}$

$$\left|\int_F(\mathsf{E}\{f(\zeta)\mid\mathfrak{F}\}-\mathsf{E}\{f_{nk}(\zeta)\mid\mathfrak{F}\})\,d\mathsf{P}\right|\leqslant\int_F|f(x)-f_{nk}(x)|\,q(dx). \tag{15}$$

Therefore if one chooses an arbitrary sequence $f_{nk_n}(x)$ such that $\|\chi_n(f(x)-f_{nk_n}(x))\|\to 0$ (clearly, for any $f\in\mathscr{C}(\mathscr{X})$ such a sequence can be selected), then the functions $f_{nk_n}(x)$ will be uniformly bounded and

$$\int_{\mathscr{X}}|f(x)-f_{nk_n}(x)|\,q(dx)\to 0;$$

also in view of (15)

$$\mathsf{E}\{f(\zeta)\mid\mathfrak{F}\}=\lim\mathsf{E}\{f_{nk_n}(\zeta)\mid\mathfrak{F}\}\ (\mathrm{mod}\,\mathsf{P}),$$

where the limit on the right exists and is independent of the choice of the approximating sequence $(\omega\notin D_0)$. Since conditional mathematical expectations are defined only (up to mod P), we stipulate that for an arbitrary $f\in\mathscr{C}(\mathscr{X})$ $\mathsf{E}\{f(\zeta)\mid\mathfrak{F}\}$ is defined by the last relationship.

Using this definition, the conditional mathematical expectation possesses the following properties (for all $\omega\notin D_0$):

$$\mathsf{E}\{f(\zeta)\mid\mathfrak{F}\}\geqslant 0\quad\text{if}\quad f\geqslant 0,$$
$$\mathsf{E}\{\alpha_1 f_1(\zeta)+\alpha_2 f_2(\zeta)\mid\mathfrak{F}\}=\alpha_1\mathsf{E}\{f_1(\zeta)\mid\mathfrak{F}\}+\alpha_2\mathsf{E}\{f_2(\zeta)\mid\mathfrak{F}\},$$
$$\mathsf{E}\{f(\zeta)(1-\chi_n(\zeta)\mid\mathfrak{F}\}\to 0\quad\text{as}\quad n\to\infty.$$

Thus, $L_\omega(f)=\mathsf{E}\{f(\zeta)\mid\mathfrak{F}\}$ represents a positive linear functional on $\mathscr{C}(\mathscr{X})$ and in view of a theorem on the representation of linear functionals (cf. Chapter V, §2, Theorem 1), there exists on $\{\mathscr{X},\mathfrak{B}\}$ a measure $q_\omega(B)$ such that

$$\mathsf{E}\{f(\zeta)\mid\mathfrak{F}\}=\int_{\mathscr{X}}f(x)\,q_\omega(dx).$$

It is easy to verify that this formula can be considered as a definition of the conditional mathematical expectation for an arbitrary $\mathfrak{B}$-measurable non-negative (or q-integrable) function. Putting $f(x)=\chi_B(x)$, we obtain $\mathsf{P}\{B\mid\mathfrak{F}\}=q_\omega(B)$ for $\omega\notin D_0$ and $\mathsf{P}\{B\mid\mathfrak{F}\}=q(B)$ for $\omega\in D_0$. Hence, the required assertion follows. □

Consider random elements ζ_1 and ζ_2 in $\{\mathscr{Y}_1,\mathfrak{B}_1\}$ and $\{\mathscr{Y}_2,\mathfrak{B}_2\}$

correspondingly, where $\mathscr{Y}_i$ is a complete separable metric space, $\mathfrak{B}_i$ is the σ-algebra of Borel sets $\mathscr{Y}_i (i=1, 2)$. We set

$$\mathscr{Y}^{(1,2)} = \mathscr{Y}_1 \times \mathscr{Y}_2, \quad \mathfrak{B}^{(1,2)} = \sigma\{\mathfrak{B}_k, k=1, 2\}. \tag{16}$$

The sequence $\zeta^{(1,2)} = (\zeta_1, \zeta_2)$ can be considered as a random element in $\{\mathscr{Y}^{(1,2)}, \mathfrak{B}^{(1,2)}\}$ and $\mathscr{Y}^{(1,2)}$ is a complete metric separable space.

Let q_i denote the distribution of ζ_i $(i=1, 2)$, $q^{(1,2)}$ – the distribution of $\zeta^{(1,2)}$ and $q^{(2|1)}$ – the regular conditional distribution of ζ_2 given the σ-algebra $\mathfrak{F}_{\zeta_1}$, generated by the random element ζ_1. Since $q^{(2|1)}$ is a $\mathfrak{F}_{\zeta_1}$-measurable function, it follows that

$$g^{(2|1)}(B_2, \omega) = q(B_2, \zeta_1),$$

where $B_2 \in \mathfrak{B}_2$ and the function $q(B_2, y)$ is $\mathfrak{B}_1$-measurable. It follows from the definition of conditional probabilities that

$$\int_{g_1^{-1}(B_1)} q(B_2, \zeta_1)\, d\mathsf{P} = q^{(1,2)}(B_1 \times B_2),$$

where B_1 is an arbitrary set in $\mathfrak{B}_1$ and $\zeta_1 = g_1(\omega)$.

Using the rule of change of variables we can rewrite this equality in the form

$$q^{(1,2)}(B_1 \times B_2) = \int_{B_1} q(B_2, y_1)\, dq_1,$$

or

$$q^{(1,2)}(B_1 \times B_2) = \int_{\mathscr{Y}_1} \chi_{B_1}^{(1)}(y_1) \left(\int_{\mathscr{Y}_2} \chi_{B_2}^{(2)}(y_2)\, q(dy_2, y_1) \right) dq_1,$$

where $\chi^{(i)}$ are the indicators of the sets in the space $\mathscr{Y}_i$. It follows from the last formula that

$$\int_{\mathscr{Y}^{(1,2)}} f(y_1, y_2)\, dq^{(1,2)} = \int_{\mathscr{Y}_1} \left(\int_{\mathscr{Y}_2} f(y_1, y_2)\, q(dy_2, y_1) \right) dq_1 \tag{17}$$

for any $\mathfrak{B}^{(1,2)}$-measurable non-negative functions. Indeed, the class of functions f, for which formula (17) is valid, is linear and closed with respect to passage to the limit of monotone sequences. Since it contains functions of the form $\chi^{(1)}\chi^{(2)}$ it also contains their linear combinations. On the other hand, an arbitrary $\mathfrak{B}^{(1,2)}$-measurable function can be approximated by monotonically increasing sequences of linear combinations of functions of the form $\chi^{(1)}\chi^{(2)}$.

Note that formula (17) is valid also for functions f with alternating signs provided only one of the sides of equality (17) is meaningful. It follows from formula (17) that

$$\mathsf{E}\{f(\zeta_1,\zeta_2)\mid\mathfrak{F}_{\zeta_1}\}=\int_{\mathscr{Y}_2} f(\zeta_1,y_2)\,q(dy_2,\zeta_1). \tag{18}$$

The result obtained can be presented in the following more general form. Let ζ_k be random elements in $\{\mathscr{Y}_k,\mathfrak{B}_k\}$ where $\mathscr{Y}_k$ is a complete separable metric space. Set $\mathscr{Y}^{(1,s)}=\prod_{k=1}^{s}\mathscr{Y}_k$, $\mathfrak{B}^{(1,s)}=\sigma\{\mathfrak{B}_k,\,k=1,\dots,s\}$, $\eta_s=(\zeta_1,\zeta_2,\dots,\zeta_s)$ and let q_k be the distribution of the element ζ_k in $\{\mathscr{Y}_k,\mathfrak{B}_k\}$, $q^{(s)}=q^{(s)}(B_s,\zeta_1,\dots,\zeta_{s-1})$ be the regular conditional distribution of the element ζ_s given σ-algebra $\mathfrak{F}_{\eta_{s-1}}=\mathfrak{F}_{(\zeta_1,\dots\zeta_{s-1})}$.

Applying repeatedly relation (18) we obtain from the formula

$$\mathsf{E}\{f\mid\mathfrak{F}_{\zeta_1}\}=\mathsf{E}\{\dots\{\mathsf{E}\{f\mid\mathfrak{F}_{\eta_{n-1}}\}\mid\mathfrak{F}_{\eta_{n-2}}\}\dots\mid\mathfrak{F}_{\eta_1}\}$$

the following relations:

$$\begin{aligned}\mathsf{E}\{f(\zeta_1,\dots,\zeta_n)\mid\mathfrak{F}_{\zeta_1}\}&=\int_{\mathscr{Y}_2}\dots\int_{\mathscr{Y}_{n-1}}\Big(\int_{\mathscr{Y}_n} f(\zeta_1,y_2,\dots,y_n)\times\\&\times q^{(n)}(dy_n,\zeta_1,y_2,\dots,y_{n-1})\Big)\,q^{(n-1)}(dy_{n-1},\zeta_1,y_2,\dots,y_{n-2})\times\\&\times\dots\times q^{(2)}(dy_2,\zeta_1),\end{aligned} \tag{19}$$

$$\begin{aligned}\mathsf{E}\{f(\zeta_1,\dots,\zeta_n)\}&=\int_{\mathscr{Y}_1}\dots\int_{\mathscr{Y}_n} f(y_1,\dots,y_n)\,q^{(n)}(dy_n,y_1,\dots,y_{n-1})\times\\&\times q^{(n-1)}(dy_{n-1},y_1,\dots,y_{n-2})\times\dots\times q^{(2)}(dy_2,y_1)\,q_1(dy_1).\end{aligned} \tag{20}$$

Conditional densities. Let $\{\mathscr{X},\mathfrak{A},m\}$ be a measure space and let $\zeta=g(\omega)$ be a measurable mapping of $\{\Omega,\mathfrak{S}\}$ into $\{\mathscr{X},\mathfrak{A}\}$.

We say that a random element possesses *density function* $\varrho(x)$ (with respect to measure m), if for an arbitrary $A\in\mathfrak{A}$

$$\mathsf{P}(\zeta\in A)=\int_A \varrho(x)\,m(dx).$$

In view of the Radon-Nikodym theorem a random element ζ possesses a density function iff the measure P_g is absolutely continuous with respect to m.

Let $\{\mathscr{Y},\mathfrak{B},q\}$ be some other measurable space and let $\eta=h(\omega)$ be a random element in $\{\mathscr{Y},\mathfrak{B}\}$. The pair (ζ,η) can be considered as a measurable mapping of $\{\Omega, \mathfrak{S}\}$ into the product space $\{\mathscr{X} \times \mathscr{Y}, \sigma\{\mathfrak{A} \times \mathfrak{B}\}\}$. Indeed, if the set C in $\sigma\{\mathfrak{A} \times \mathfrak{B}\}$ is of the form $C = A \times B$, $A \in \mathfrak{A}$, $B \in \mathfrak{B}$ then the events $\{(\zeta, \eta) \in C\} = \{\zeta \in A\} \cap \{\eta \in B\} \in \mathfrak{S}$, therefore any event in the minimal σ-algebra containing events of the form $\{(\zeta, \eta) \in A \times B\}$, i.e. any event of the form $\{(\zeta, \eta) \in C\}$, $C \in \sigma\{\mathfrak{A} \times \mathfrak{B}\}$ will be $\mathfrak{S}$-measurable. Assume that the pair (ζ, η) possesses density function $\varrho(x, y)$ with respect to measure $m \times q$. Then for any $A \in \mathfrak{A}$ and $B \in \mathfrak{B}$,

$$\mathsf{P}\{\zeta \in A, \eta \in B\} = \int_A \int_B \varrho(x, y)\, m(dx)\, q(dy).$$

The function $\varrho(x, y)$ is called the *joint density function* of the random elements ζ and η. The existence of the joint density function implies the fact that each one of the random elements ζ and η possesses density with respect to the corresponding measure. Indeed,

$$\mathsf{P}(\zeta \in A) = \int_A \int_{\mathscr{Y}} \varrho(x, y)\, m(dx)\, q(dy) = \int_A \varrho_\zeta(x)\, m(dx),$$

where

$$\varrho_\zeta(x) = \int_{\mathscr{Y}} \varrho(x, y)\, q(dy).$$

Analogously,

$$\mathsf{P}(\eta \in B) = \int_B \varrho_\eta(y)\, q(dy),$$

where

$$\varrho_\eta(y) = \int_{\mathscr{X}} \varrho(x, y)\, m(dx).$$

We now show how to compute the conditional mathematical expectation $\mathsf{E}\{f(\eta) \mid \zeta\}$ if (ζ, η) possesses the probability density $\varrho(x, y)$. From the definition of the conditional mathematical expectation we obtain

$$\int_{g^{-1}(A)} \mathsf{E}\{f(\eta) \mid \zeta\}\, d\mathsf{P} = \int_{g^{-1}(A)} f(\eta)\, d\mathsf{P} = \mathsf{E} f(\eta)\, \chi_A(\zeta) =$$

$$= \int_{\mathscr{X}} \int_{\mathscr{Y}} f(y)\, \chi_A(x)\, \varrho(x, y)\, m(dx)\, q(dy) =$$

$$=\int_A \left(\int_{\mathscr{Y}} f(y)\frac{\varrho(x,y)}{\varrho_\zeta(x)}q(dy)\right)\varrho_\zeta(x)\,m(dx)=\int_{g^{-1}(A)} \bar{f}(\zeta)\,d\mathsf{P},$$

where

$$\bar{f}(x)=\int_{\mathscr{Y}} f(y)\frac{\varrho(x,y)}{\varrho_\zeta(x)}q(dy).$$

Thus

$$\mathsf{E}\{f(\eta)\mid\zeta\}=\int_{\mathscr{Y}} f(y)\frac{\varrho(\zeta,y)}{\varrho_\zeta(\zeta)}q(dy).$$

The quantity $\frac{\varrho(x,y)}{\varrho_\zeta(x)}=\varrho(y\mid x)$ is called *the conditional density function of the random element* η *given* $\xi=x$. Using this density the conditional mathematical expectation given ζ is computed by the formula

$$\mathsf{E}\{f(\eta)\mid\zeta\}=\int_{\mathscr{Y}} f(y)\,\varrho(y\mid\zeta)\,q(dy).$$

§4. Random Functions and Random Mappings

Definitions. Let $\{\Omega, \mathfrak{S}, \mathsf{P}\}$ be a given probability space. If the realization of an experiment is described by means of a function $f(x)$ of a definite argument x, $x\in X$, we say that a random function is defined on $\{\Omega, \mathfrak{S}, \mathsf{P}\}$. Thus a random function is the mapping: $\omega\to f(x)=f(x,\omega)$, $\omega\in\Omega$. Additionally it is required that the function $f(x,\omega)$ for a fixed x will be a random variable (or a random element).

The general definition is as follows. Let X be a set and $\{\mathscr{Y}, \mathfrak{B}\}$ be a measurable space.

Definition 1. *A random mapping* $\zeta(x)$ of a set X into a measurable space $\{\mathscr{Y}, \mathfrak{B}\}$ is called the mapping of $X\times\Omega$ into $\mathscr{Y}$ which is, for arbitrary fixed x, a measurable mapping of $\{\Omega, \mathfrak{S}\}$ into $\{\mathscr{Y}, \mathfrak{B}\}$ i.e. such that for any $B\in\mathfrak{B}$

$$\{\omega:\zeta(x)\in B\}\in\mathfrak{S}.$$

In place of the term "random mapping" we shall henceforth use the term "random function with the values in $\mathscr{Y}$" here as well; x is referred to as the argument of a random function. In the case when X is a real line or a segment of a line and the argument of the random function is interpreted as time, we shall use letters T and t in place of X and x, respectively, and we shall call the random function a random process. If the argument of the random function takes on integral non-negative values ($X=T_+=\{0, 1, 2, \ldots, n, \ldots\}$) or arbitrary integral values ($X=T=\{\ldots, -n, n+1, \ldots, -1, 0, 1, \ldots, n, \ldots\}$) the random function is then called a discrete parameter random process. If X is a finite-dimensional Euclidean space $\mathscr{R}^m$ or a region in $\mathscr{R}^m$, then $\zeta(x)$ is sometimes called a random field.

The following particular case of the general definition is of interest. Assume that Ω is a functional space, $\omega=\omega(x)$, $x\in X$, and the σ-algebra $\mathfrak{S}$ contains all the sets of the space Ω of the form

$$\{\omega:\omega(x_0)\in B\},$$

for any $x_0\in X$ and $B\in\mathfrak{B}$ and P is an arbitrary probability measure on $\mathfrak{S}$. It is natural to associate the random function $g(x, \omega)=\omega(x)$ with such a probability space. In certain problems it is convenient to identify the random function $g(x, \omega)=\omega(x)$ with the probability space $\{\Omega, \mathfrak{S}, \mathsf{P}\}$ of the type described here.

It is easy to see that the general definition of a random function can be reduced to the above described particular case. Indeed, if a random function $\zeta(x)$ is defined as a function of two variables $\zeta(x)=g(x, \omega)$ then by putting $u=g(x, \omega)$, where ω is fixed, $\omega\in\Omega$ and denoting by U the set of all functions $\{u:u=u_\omega(x)=g(x, \omega), \omega\in\Omega\}$ we obtain a mapping S of the set Ω onto U. Here the σ-algebra $\mathfrak{S}$ of the sets in Ω is mapped into a σ-algebra $\mathfrak{S}'$ of the sets in U, and the probability measure P on $\mathfrak{S}$ is mapped into the probability measure P' on $\mathfrak{S}'$. For any fixed x the set $\{u:u(x)\in B\}$, $B\in\mathfrak{B}$, belongs to $\mathfrak{S}'$, since

$$S^{-1}\{u:u(x)\in B\}=\{\omega:g(x, \omega)\in B\}\in\mathfrak{S}.$$

Thus a probability space $\{U, \mathfrak{S}', \mathsf{P}'\}$ is obtained where U is a set of functions $u=u(x)$ and for any $n, x_1, x_2, \ldots, x_n$ ($x_k\in X$, $k=1, \ldots, n$) the distribution of the sequence of random elements on $\{\Omega, \mathfrak{S}, \mathsf{P}\}$

$$g(x_1, \omega), g(x_2, \omega), \ldots, g(x_n, \omega)$$

coincides with the distribution of the sequence

$$u(x_1), u(x_2), \ldots, u(x_n).$$

Hence, the random function is equivalent in a definite sense, to a certain functional space with a probability measure, i.e. to a probability space in which the space of elementary events is given by a certain set of functions.

Let $\zeta(x)$, $x\in X$ be a random function with values in $\{\mathscr{Y}, \mathfrak{B}\}$, n is an arbitrary integer and x_k ($k=1, 2, \ldots, n$) are arbitrary points in X. Consider a random element in $\{\mathscr{Y}^n, \mathfrak{B}^n\}$ determined by the sequence

$$\{\zeta(x_1), \zeta(x_2), \ldots, \zeta(x_n)\}. \tag{1}$$

The corresponding measure $\mathsf{P}_{x_1x_2\dots x_n}(B)$ on $\mathfrak{B}^n$ is

$$\mathsf{P}_{x_1x_2\dots x_n}(B)=\mathsf{P}\{\omega:(\zeta(x_1), \zeta(x_2), \dots, \zeta(x_n))\in B\}, \ B\in\mathfrak{B}^n. \qquad (2)$$

The measures (2) are termed *marginal distributions* of the random function $\zeta(x)$. The family of marginal distributions of a random function possesses two obvious properties:

1. $$\mathsf{P}_{x_1x_2\dots x_{n+m}}(B\times\mathscr{Y}^m)=\mathsf{P}_{x_1x_2\dots x_n}(B), \qquad (3)$$

where $B\in\mathfrak{B}^n$.

2: *Let s be a pointwise mapping into X^n acting according to the following rule: $s(x_1, x_2, \dots, x_n)=(x_{i_1}, x_{i_2}, \dots, x_{i_n})$ where $(i_1, i_2, \dots, i_n)$ is a certain permutation of the indices $(1, 2, \dots, n)$ and S the corresponding mapping of the sets into $\mathscr{Y}^n$.*

Then

$$\mathsf{P}_{s(x_1, x_2, \dots, x_n)}(SB)=\mathsf{P}_{x_1x_2\dots x_n}(B), \qquad (4)$$

for any B and n.

Properties (3) and (4) are called *the conditions of compatibility* for the family of marginal distributions.

We now return to the general definition of a random function. The arguments concerning the actual indistinguishability of equivalent random variables discussed above are also important in the case of random functions. It is customary to assume that from a practical point of view an experiment allows us to distinguish only between hypotheses which refer to marginal distributions of a random function.

Thus, one cannot, using experimental data, distinguish between two random functions which have marginal distributions coinciding for any n and $x_k\in X$. In this connection we stipulate the following

Definition 2. The random functions $\zeta(x)$ and $\zeta'(x)$, $x\in X$ with the values in $\mathscr{Y}$ defined possibly on two different probability spaces $\{\Omega, \mathfrak{S}, \mathsf{P}\}$ and $\{\Omega', \mathfrak{S}', \mathsf{P}'\}$ are called *stochastically equivalent in the wide sense* if for any integer $n\geq 1$ and any $x_k\in X$, $k=1, 2, \dots, n$ their marginal distributions coincide:

$$\mathsf{P}\{\omega:(\zeta(x_1), \zeta(x_2), \dots, \zeta(x_n))\in B\}=\mathsf{P}'\{\omega':(\zeta'(x_1), \dots, \zeta'(x_n))\in B\}.$$

In what follows we shall often utilize the notion of stochastic equivalence of random functions in a narrower sense.

Definition 3. Two random functions $g_1(x, \omega)$ and $g_2(x, \omega)$ $(x\in X, \omega\in\Omega)$ defined on the same probability space $\{\Omega, \mathfrak{S}, \mathsf{P}\}$ are called stochastically equivalent if for any $x\in X$

$$\mathsf{P}\{g_1(x, \omega)\neq g_2(x, \omega)\}=0.$$

Clearly, if $g_1(x, \omega)$ and $g_2(x, \omega)$ are stochastically equivalent, then they are stochastically equivalent in the wide sense.

How then can one define random functions in specific problems? Firstly, a random function can be defined using the general definition by explicitly stating the probability space $\{\Omega, \mathfrak{S}, \mathsf{P}\}$ and the function $\zeta(x) = g(x, \omega)$ both of which should be as simple as possible. Another method is to define a measure on a certain functional space U whose elements are functions on X and to consider the functions $\zeta(u) = u(x)$ on U. This method of definition and investigation is studied in Chapter V. The difficulty in this method is due to the complexity of the specific description of a measure in a functional space. This difficulty can sometimes be alleviated by considering a given random function $\zeta(u) = u(x)$ as a result of some transformation s defined on a more-or-less simple functional space $\mathscr{V}$ with a relatively simple measure $\mu: u(x) = S[v]$, $v \in \mathscr{V}$ where $\{\mathscr{V}, \mathfrak{S}, \mu\}$ is a measure space.

Such an approach is discussed in Chapters 4 and 8 dealing with linear and nonlinear transformations of random functions correspondingly.

The third method, possibly the most prevalent, of defining a random function is based on the description of the family of its marginal distributions. This is due to the fact that, firstly, in many practical problems, random functions are characterized by their marginal distributions and the corresponding probability space is not usually given at all. Secondly, in many cases, it is simpler to define marginal distributions than the corresponding probability spaces and functions $g(x, \omega)$. Next as it turns out, it is sufficient to know only the marginal distributions of random functions to solve many important problems. On the other hand, as it will be proved shortly, under very broad assumptions, given an arbitrary family of distributions $\mathsf{P}_{x_1 x_2 \ldots x_n}(B)$ defined for any integer valued n and $x_k \in X$ defined on $\{\mathscr{Y}^n, \mathfrak{B}^n\}$ one can construct a probability space $\{\Omega, \mathfrak{S}, \mathsf{P}\}$ and a random function $\zeta(x) = g(x, \omega)$ whose marginal distributions coincide with the given family.

Definition 4. A compatible choice of distributions $\{\mathsf{P}_{x_1 x_2 \ldots x_n}(B^n)$, $n = 1, 2, \ldots, x_k \in X, B^n \in \mathfrak{B}^n\}$ where $\mathfrak{B}$ is a σ-algebra of sets in the space $\mathscr{Y}$ and $\mathfrak{B}^n$ is its n-th power is called a *random function in the wide sense* with values in $\mathscr{Y}$.

The standard notation $\xi(x), \eta(x), \ldots$ is utilized for random functions in the wide sense, x is called its argument and the distribution of the sequence $\{\xi(x_1), \ldots, \xi(x_n)\}$ is identified with $\mathsf{P}_{x_1 x_2 \ldots x_n}(B^n)$. It follows from the above that each random function in the wide sense (here $\mathscr{Y}$ is a complete separable metric space and X is arbitrary) can be considered also as a random function in the sense of the basic definition 1 (cf. Theorem 2 of the present paragraph). On the other hand a random function in the wide sense can be identified with a class of all stochastically equivalent (in the wide sense) random functions possessing the given marginal distributions.

Construction of a random function according to its marginal distributions. In view of the definition of stochastically equivalent (in the wide sense) random functions the most typical feature for random functions is not the probability space or the form of the function $g(x, \omega)=\zeta(x)$ but the family of its marginal distributions. This means that we can change (arbitrarily) the probability space and the form of the function $g(x, \omega)$ as long as the family of marginal distributions remains unchanged. This very important fact is widely used for obtaining a possibly simpler and more convenient representation of a random function. In this connection the following problem immediately arises: let a family of distributions

$$\{\mathsf{P}_{x_1 x_2 \ldots x_n}(B^{(n)});\ n=1, 2, \ldots, x_k \in X, B^{(n)} \in \mathfrak{B}^n\}, \tag{5}$$

be given where X is an arbitrary set and $\{\mathscr{Y}, \mathfrak{B}\}$ a measurable space. Does a random function exist for which the given family of distributions is the family of its marginal distributions?

Clearly, the family of distributions (5) cannot be completely arbitrary. It should at least satisfy the compatibility conditions (3) and (4).

Definition 5. If there exists a probability space $\{\Omega, \mathfrak{S}, \mathsf{P}\}$ and a function of two variables $g(x, \omega)$ with the values in $\mathscr{Y}$, defined on $X \times \Omega$, which is $\mathfrak{S}$-measurable for each fixed $x \in X$ and such that the marginal distributions of the random function $g(x, \omega)$ coincide with the given family (5), i.e. for each $B^{(n)} \in \mathfrak{B}^n$

$$\mathsf{P}\{\omega:(g(x_1, \omega), g(x_2, \omega), \ldots, g(x_n, \omega)) \in B^{(n)}\} = \mathsf{P}_{x_1 \ldots x_n}(B^{(n)}), \tag{6}$$

then the probability space $\{\Omega, \mathfrak{S}, \mathsf{P}\}$ and the function $g(x, \omega)$ are called *the representation of the family of distributions* (5).

It will be shown, that under sufficiently broad assumptions a compatible family of distributions (5) admits a certain representation. The space Ω is replaced here by the space of all functions defined on X with the values in $\mathscr{Y}$ and the elementary events are functions in x, $\omega=\omega(x)$ and $g(x, \omega)=\omega(x)$.

Definition 6. Let Ω be the space of all functions $\omega=\omega(x)$ defined on the set X with values in some measurable space $\{\mathscr{Y}, \mathfrak{B}\}$ and $B^{(n)} \in \mathfrak{B}^n$. The set of functions $\omega(x) \in \Omega$ for which the point $\{\omega(x_1), \ldots, \omega(x_n)\}$ from $\mathscr{Y}^n$ belongs to $B^{(n)}$, i.e. the set

$$C_{x_1 \ldots x_n}(B^{(n)}) = \{\omega:(\omega(x_1), \omega(x_2), \ldots, \omega(x_n)) \in B^{(n)}\}$$

is called *the cylindrical set in Ω with the basis $B^{(n)}$ over the coordinates* $x_1, x_2, \ldots, x_n$ or simply *the cylindrical set* (or *cylinder*).

A few remarks concerning cylindrical sets and the operations involving these sets are in order. If n and the points $x_1, x_2, \ldots, x_n$ are fixed, then there exists an isomorphism between the cylindrical sets over the coordinates $x_1, x_2, \ldots, x_n$ and the sets in $\mathfrak{B}^n$: each set $B^{(n)} \in \mathfrak{B}^n$ determines a cylindrical set $C_{x_1 \ldots x_n}(B^{(n)})$ for which it serves as a basis; different cylindrical sets correspond to different bases. The sum, difference and intersection of cylindrical sets correspond to the sum, difference and intersection of the bases. This follows directly from the definition of cylindrical sets.

Considering now the operations on cylindrical sets in the general case one should keep in mind that the same set can be defined for different choices of coordinates. It is clear that

$$C_{x_1 \ldots x_2}(B^{(n)}) = C_{x_1 \ldots x_n x_{n+1} \ldots x_{n+m}}(B^{(n)} \times \mathscr{Y}^m).$$

It is also easy to see that any two cylindrical sets $C = C_{x_1 \ldots x_n}(B^{(n)})$ and $C' = C_{x'_1 \ldots x'_m}(B^{(m)})$ can always be regarded as cylindrical sets over the same sequence of coordinates $x''_1, \ldots, x''_p$ containing $x_1, x_2, \ldots, x_n$ as well as $x'_1, x'_2, \ldots, x'_m$. It follows from here that when discussing algebraic operations on a finite number of cylindrical sets it can be assumed that they are defined on a fixed sequence of coordinates. Therefore the following theorem is valid:

Theorem 1. *The class $\mathfrak{C}$ of all cylindrical sets forms an algebra of sets.*

Moreover, if X contains infinitely many points and $\mathscr{Y}$ at least two points, then $\mathfrak{C}$ is not a σ-algebra. Indeed, the set

$$\bigcup_{k=1}^{\infty} C_{x_k}(\{y_k\}),$$

where $\{y_k\}$ $k = 1, 2, \ldots$, is a sequence of points in $\mathscr{Y}$, is not a cylindrical set.

We now prove the following theorem.

Theorem 2. (*Kolmogorov's*). *Let $\mathscr{Y}$ be a complete separable space. The family of distributions* (5) *satisfying the compatibility conditions* (3) *and* (4) *admits a certain representation.*

We first define a set function $\mathsf{P}'(C)$, $C \in \mathfrak{C}$ on the algebra of cylindrical sets $\mathfrak{C}$ of the space Ω by putting

$$\mathsf{P}'(C) = \mathsf{P}_{x_1 \ldots x_n}(B^{(n)}),$$

where C is a cylindrical set with the base $B^{(n)}$ over the coordinates $x_1, x_2, \ldots, x_n$. The compatibility conditions guarantee the uniqueness of the definition of function $\mathsf{P}'(C)$. Let C_k, $k = 1, 2, \ldots, n$, be a sequence of cylindrical sets. Without loss of generality it can be assumed that the sets C_k are defined by the bases $B_k^{(p)}$ over the same sequence of coordinates $x_1, x_2, \ldots, x_p$. The same algebraic operations over the bases $B_k^{(p)}$ corresponds exactly to the algebraic operations over the sets C_k. Since the measure $\mathsf{P}_{x_1 \ldots x_p}(B^{(p)})$ is countably additive on $\mathscr{Y}^p$ it follows that the set function $\mathsf{P}'(C)$ is finitely-additive on $\mathfrak{C}$. It remains only to extend the function $\mathsf{P}'(C)$ defined on algebra $\mathfrak{C}$ to a measure $\tilde{\mathsf{P}}$ on a certain σ-algebra $\tilde{\mathfrak{C}}$. For this purpose it is sufficient, based on the well known theorem on the extension of measures, to check that for an arbitrary $C \in \mathfrak{C}$ and any cover $\{C_k\}$, $k=1, 2, \ldots, n, \ldots$, $C_k \in \mathfrak{C}$ of the set C, $C \subset \bigcup_{k=1}^{\infty} C_k$, the inequality

$$\mathsf{P}'(C) \leqslant \sum_{k=1}^{\infty} \mathsf{P}(C_k) \tag{7}$$

is satisfied.

We now show that if

$$\bigcup_{k=1}^{\infty} C_k = C \quad (C \in \mathfrak{C},\ C_k \in \mathfrak{C}, \quad k=1, 2, \ldots)$$

and $C_k \cap C_r = \emptyset$, $(k \neq r)$, then

$$\mathsf{P}'(C) = \sum_{k=1}^{\infty} \mathsf{P}'(C_k). \tag{8}$$

The validity of (7) follows from here for an arbitrary cover of the cylindrical set C by means of the sets belonging to $\mathfrak{C}$. Now set

$$C \setminus \bigcup_{k=1}^{n} C_k = D_n.$$

The sets D_n form a monotonically decreasing sequence of cylindrical sets with a void intersection

$$\bigcap_{n=1}^{\infty} D_n = C \setminus \bigcup_{k=1}^{\infty} C_k = \emptyset. \tag{9}$$

The equality $\mathsf{P}'(C) = \sum_{k=1}^{n} \mathsf{P}'(C_k) + \mathsf{P}'(D_n)$ follows from the additivity of P'. To prove (8) it is therefore sufficient to show that

$$\lim_{n \to \infty} \mathsf{P}'(D_n) = 0.$$

Assume the contrary that

$$\lim_{n \to \infty} \mathsf{P}'(D_n) = L > 0.$$

Denote by B_n the basis of the cylinder D_n and let D_n be situated over the coordinates $x_1, x_2, \ldots, x_{m_n}$. It is assumed here that as n increases the collection of the corresponding points $x_1, x_2, \ldots, x_{m_n}$ does not decrease. As it was shown above, this assumption does not restrict the generality.

For each B_n a compact set K_n $(K_n \subset B_n)$ can be found such that

$$\mathsf{P}_{x_1 \ldots x_{m_n}}(B_n \backslash K_n) < \frac{L}{2^{n+1}}, \quad n=1, 2, \ldots$$

Let Q_n be a cylinder over the coordinates $x_1, x_2, \ldots, x_n$ with the basis K_n, $G_n = \bigcap_{r=1}^{n} Q_r$ and let M_n be the basis of the set G_n. Clearly M_n is a compact set being an intersection of closed sets such that at least one of them, namely K_n, is compact.

Since the sets G_n are monotonically decreasing it follows from the relation $\omega(x) \in G_{n+p}$ $(p>0)$ that $\omega(x) \in G_n$. Therefore if

$$\{y_1, y_2, \ldots, y_n, \ldots, y_{m_{n+p}}\} \in M_{n+p}, \quad (p>0),$$

then

$$\{y_1, y_2, \ldots, y_{m_n}\} \in M_n.$$

The sets G_n are clearly non-empty. Moreover, since

$$D_n \backslash G_n = \bigcup_{r=1}^{n} (D_n \backslash Q_r) \subset \bigcup_{r=1}^{n} (D_r \backslash Q_r),$$

then

$$\mathsf{P}'(D_n \backslash G_n) \leqslant \sum_{r=1}^{n} \mathsf{P}'(D_r \backslash Q_r) = \sum_{r=1}^{n} \mathsf{P}_{x_1 \ldots x_{m_r}}(B_r \backslash K_r) \leqslant \frac{L}{2},$$

thus it follows from here that

$$\lim_{n\to\infty} \mathsf{P}'(G_n) = \lim_{n\to\infty} \mathsf{P}'(D_n) - \lim_{n\to\infty} \mathsf{P}'(D_n \backslash G_n) \geqslant \frac{L}{2}.$$

Select a point

$$\{y_1^{(n)}, \ldots, y_{m_n}^{(n)}\}.$$

from each one of the sets M_n.

In view of the above for any k the sequence of the points $\{y_k^{(n)}\}$, $n=1, 2, \ldots$, belongs to a compact set in $\mathscr{Y}$ and the sequence

$$\{y_1^{(n+r)}, \ldots, y_{m_n}^{(n+r)}\}, \quad r=0, 1, 2, \ldots$$

is in M_n. By means of the diagonal process a sequence of indices n_j can be found such that for each k the sequence $y_k^{(n_j)}$ converges to a certain limit $y_k^{(0)}$. Since the sets M_n are closed it follows that

$$\{y_1^{(0)}, \dots, y_{m_n}^{(0)}\} \in M_n$$

for each n.

Define a function $\omega(x)$ by putting $\omega(x_k) = y_k^{(0)}$, $k = 1, 2, \dots, n, \dots$, and supplementing the definition arbitrarily at other points. Then for each n we have $\omega(x) \in G_n \subset D_n$. Hence $\bigcap_{n=1}^{\infty} D_n$ is not empty which contradicts (9). Therefore inequality (10) cannot be valid and

$$\lim_{n \to \infty} \mathsf{P}'(D_n) = 0 .$$

Consequently function P' satisfies inequality (7) and can be extended to a full measure $(\tilde{\mathfrak{C}}, \tilde{\mathsf{P}})$, $\tilde{\mathfrak{C}} \supset \mathfrak{C}$. Define the function $g(x, \omega)$, $\omega \in \Omega$, $x \in X$ by the equality $g(x, \omega) = \omega(x)$. We have for an arbitrary Borel set $B^{(n)}$ in $\mathscr{Y}^n$ and any $n, x_1, \dots, x_n$

$$\tilde{\mathsf{P}}\{(g(x_1, \omega), g(x_2, \omega), \dots, g(x_n, \omega)) \in B^{(n)}\} =$$
$$= \tilde{\mathsf{P}}\{(\omega(x_1), \omega(x_2), \dots, \omega(x_n)) \in B^{(n)}\} = \mathsf{P}_{x_1 \dots x_n}(B^{(n)}) .$$

Therefore the required representation stated in the theorem has been constructed for the family of distributions (5). The theorem is thus proved. □

Chapter II

Random Sequences

§ 1. Preliminary Remarks

Sequences of random variables $\xi_1, \xi_2, \ldots, \xi_n, \ldots$ may be regarded as a discrete-time random process. Random sequences play an important role in the general theory.

Firstly, many probabilistic problems involve the discrete parameter (time).

Secondly, investigation of discrete parameter processes utilizes in a certain sense simpler methods, while these processes can be used to approximate or study arbitrary continuous-parameter processes.

The basic problems studied in this chapter pertain to the asymptotic behavior of a random sequence as the value of the parameter increases to infinity. These are the problems dealing with the existence of limits of sequences, the behavior of the arithmetic mean, the asymptotic character of the distribution of the terms of a divergent sequence and so on. This class of problems interconnects the classical topics of probability theory (laws of large numbers, limiting theorems for the sum of random terms, etc.) with the general theory of random processes.

Evidently, to obtain significant results which differ from the general criteria of convergence it is required to impose definite restrictions on the random sequences under investigation. Correspondingly certain important classes of random sequences are introduced and investigated for which there are available nontrivial results related to the problems mentioned above.

Let $\{\Omega, \mathfrak{S}, \mathsf{P}\}$ be a fixed probability space and $\{\mathscr{X}, \mathfrak{B}\}$ be a measurable space. In this chapter T denotes either the sequence of non-negative integers $T_+ = \{0, 1, 2, \ldots, n, \ldots\}$ or an ordered set of all integers $\tilde{T} = \{\ldots, -n, \ldots, -1, 0, 1, \ldots, n, \ldots\}$ (unless otherwise stipulated).

The function $\{\xi(\cdot) = \xi(t) = \xi(t, \omega), t \in T, \omega \in \Omega\}$ with the values in X is called a random sequence or a *random discrete parameter process* if, for any $B \in \mathfrak{B}$ and $t \in T$, $\{\omega : \xi(t, \omega) \in B\} \in \mathfrak{S}$.

The values $\xi(t)$ are sometimes called the *states* of a certain stochastic system Σ and the space $\mathscr{X}$ the *phase space* of the system Σ.

In the case when $\mathscr{X}$ is a metric space, $\mathfrak{B}$ will always denote a σ-algebra of Borel sets of $\mathscr{X}$.

Let $\{\mathscr{X}^s, \mathfrak{B}^s\}$ be the s-th power of the measurable space $\{\mathscr{X}, \mathfrak{B}\}$. For an arbitrary choice of integers $n_1, n_2, \dots, n_s$, $0 \leqslant n_1 < n_2 < \dots < n_s < \infty$ the random sequence $\{\xi(t), t \in T\}$ defines a probability measure $\mathsf{P}_{n_1 n_2 \dots n_s}(\cdot)$ on $\{\mathscr{X}^s, \mathfrak{B}^s\}$:

$$\mathsf{P}_{n_1 n_2 \dots n_s}(B^{(s)}) = \mathsf{P}\{(\xi(n_1), \xi(n_2), \dots, \xi(n_s)) \in B^{(s)}\},$$

where $B^{(s)}$ is a set in $\mathfrak{B}^s$. These measures are called *marginal distributions* of the random sequence.

Marginal distributions completely determine in a certain sense, the corresponding random sequence. The precise meaning of this proposition is as follows.

Consider the space $\mathscr{X}^T$ of all possible sequences $\bar{x} = \{x_t, t \in T\}$. Denote by $\mathfrak{C}_0$ the algebra of cylindrical sets C in the space $\mathscr{X}^T$:

$$\mathfrak{C}_0 = \{C = C_{t_1 t_2 \dots t_s}(B^{(s)});\ t_k \in T,\ B^{(s)} \in \mathfrak{B}^s\}, \qquad s = 1, 2, \dots$$
$$C_{t_1 t_2 \dots t_s}(B^{(s)}) = \{\bar{x} : (x_{t_1}, x_{t_2}, \dots, x_{t_s}) \in B^{(s)}\}.$$

The mapping $\omega \to \bar{x}$ determined by the random sequence

$$\{\xi(t), t \in T\} : \bar{x} = \{x_t, t \in T\} = \{\xi(t, \omega), t \in T\},$$

induces a transformation of the probability measure P into the probability measure P' defined on a certain σ-algebra $\mathfrak{C}'$ of the space $\mathscr{X}^T$ containing all the cylinders. The measure P' coincides on the cylinders with the marginal distributions of the random sequence, namely

$$\mathsf{P}'(C_{t_1 t_2 \dots t_s}(B^{(s)})) = \mathsf{P}_{t_1 t_2 \dots t_s}(B^{(s)}),$$

and hence the marginal distributions uniquely determine the measure P' on a minimal σ-algebra $\mathfrak{C}$ ($\mathfrak{C} \subset \mathfrak{C}'$) containing the algebra of cylindrical sets.

To solve the problems arising in the theory of random sequences it is often sufficient to know the probabilities of events in $\mathfrak{C}$. There is therefore no reason to distinguish between random sequences $\{\xi_i(t), t \in T\}$ $i = 1, 2$, with the same phase space $\{\mathscr{X}, \mathfrak{B}\}$ defined on different probability spaces $\{\Omega_i, \mathfrak{C}_i, \mathsf{P}_i\}$ if the probabilities P_i' induced by these sequences coincide on the cylindrical sets of the space $\mathscr{X}^T$. In this connection we shall agree to refer to a random sequence $\{\tilde{\xi}(t); t \in T\}$ stochastically equivalent to $\{\xi(t); t \in T\}$ and defined on $\{\mathscr{X}^T, \mathfrak{C}, \mathsf{P}'\}$ and such that $\tilde{\xi}(t) = \xi(t, x) = x_t$ as the *natural representation* of the *random sequence* $\{\xi(t); t \in T\}$.

The rationale behind the introduction of the notion of a natural representation of a random sequence is the fact that in many problems marginal distributions are defined in a certain manner. On the other hand, given an arbitrary family of marginal distributions and if $\mathscr{X}$ is a complete and separable metric space, one can always construct a random sequence in its natural representation whose marginal distributions coincide with the given ones. This is a direct corollary of Kolmogorov's theorem (Chapter I, Section 4, Theorem 2).

§ 2. Semi-Martingales and Martingales

Definitions and basic properties. Martingales and semi-martingales form an important class of random sequences with numerous applications.

To avoid repetitions in future, we present definitions relating not only to sequences but to more general random processes.

Let T be an arbitrary ordered set and $\{\mathfrak{F}_t, t\in T\}$ be a current of σ-algebras, $\mathfrak{F}_t \subset \mathfrak{S}$.

Introduce the following notation

$$a^+ = \max(a, 0), \quad a^- = \max(-a, 0).$$

Definition 1. The family $\{\xi(t), \mathfrak{F}_t; t\in T\}$ in which the random variables $\xi(t)$ are $\mathfrak{F}_t$-measurable for each $t\in T$ is called a *martingale* if

$$\mathsf{E}|\xi(t)| < \infty, \tag{1}$$

$$\mathsf{E}\{\xi(t) \mid \mathfrak{F}_s\} = \xi(s), \quad s<t, \quad s, t\in T. \tag{2}$$

The family is called a *sub-martingale*, if

$$\mathsf{E}\xi^+(t) < \infty, \quad \mathsf{E}\{\xi(t) \mid \mathfrak{F}_s\} \geq \xi(s) \quad s<t, \quad s, t\in T, \tag{3}$$

and *supermartingale* if

$$\mathsf{E}\xi^-(t) < \infty, \quad \mathsf{E}\{\xi(t) \mid \mathfrak{F}_s\} \leq \xi(s), \quad s<t, s, t\in T. \tag{4}$$

Super and sub-martingales are also called *semi-martingales*.

In certain cases, when the family of σ-algebras $\{\mathfrak{F}_t, t\in T\}$ is fixed and no confusion can possibly arise, we shall use the corresponding term to denote the family of random variables $\{\xi(t), t\in T\}$ themselves.

It follows from the definition that $\mathfrak{F}_t$ always contains the σ-algebra generated by the random variables $\{\xi(s), s\leq t\}$. Sometimes this σ-algebra is taken as $\mathfrak{F}_t$ in the definition of martingales (semi-martingales).

We now present a number of properties of martingales and sub-martingales. Since the replacement of $\xi(t)$ by $-\xi(t)$ transforms a sub-martingale into a supermartingale the properties of submartingales are easily stated for supermartingales.

a) *Relations* (2) *and* (3) *are equivalent to the following* ($s<t$, $s, t\in T$):

$$\int_{B_s} \xi(s)\,\mathsf{P}(d\omega)=\int_{B_s} \xi(t)\,\mathsf{P}(d\omega), \tag{5}$$

$$\int_{B_s} \xi(s)\,\mathsf{P}(d\omega)\leqslant\int_{B_s} \xi(t)\,\mathsf{P}(d\omega), \tag{6}$$

where B_s is an arbitrary $\mathfrak{F}_s$-measurable set.

Indeed, (5) and (6) are obtained by integrating (2) and (3) respectively. It is easy to verify that conversely, (5) and (6) yield (2) and (3) correspondingly.

b) *If $\{\xi(t), t\in T\}$ is a submartingale, then $\mathsf{E}\xi(t)$ is a monotonically non-decreasing function of t; if $\{\xi(t), t\in T\}$ is a martingale, then $\mathsf{E}\xi(t)$ is constant.*

c) *If $\{\xi(t), \mathfrak{F}_t, t\in T\}$ is a submartingale, $f(x)$ is a continuous and monotonically non-decreasing convex function on the real line and $\mathsf{E}f(\xi(t))<\infty$ for $t\in T$, then $\{f(\xi(t)), \mathfrak{F}_t, t\in T\}$ is also a submartingale.*

This assertion follows from the definition of a submartingale and the Jensen inequality. Indeed,

$$\mathsf{E}\{f(\xi(t))\mid\mathfrak{F}_s\}\geqslant f(\mathsf{E}\{\xi(t)\mid\mathfrak{F}_s\})\geqslant f(\xi(s)). \tag{7}$$

In particular,

d) *If $\{\xi(t), \mathfrak{F}_t, t\in T\}$ is a submartingale, then $\{(\xi(t)-a)^+, \mathfrak{F}_t, t\in T\}$ is also a submartingale.*

e) *If $\{\xi(t), \mathfrak{F}_t, t\in T\}$ is a martingale and $f(x)$ is a continuous convex function and $\mathsf{E}|f(\xi(t))|<\infty$ then $\{f(\xi(t)), \mathfrak{F}_t, t\in T\}$ is a submartingale.*

For the proof it is sufficient to observe that the string of inequalities in (7) is retained in the case under consideration, the only difference being that the second $\geqslant$ is replaced by an equality and that monotonicity of $f(x)$ is not used in this case.

Properties b) and e) yield

f) *If $\xi(t)$ is a martingale, then $\mathsf{E}|\xi(t)|$ is monotonically nondecreasing on T.*

Lemma 1. *If T possesses a maximal element $t_{\max}$ and $\{\xi(t), \mathfrak{F}_t, t\in T\}$ is a submartingale, then the family of the random variables $\{\xi^+(t), t\in T\}$ is uniformly integrable.*

Proof. From the inequalities

$$NP(\xi(t)>N) \leqslant \int_{B_t} \xi(t)\, P(d\omega) \leqslant$$

$$\leqslant \int_{B_t} \xi(t_{\max})\, P(d\omega) \leqslant \int_{\Omega} \xi^+(t_{\max})\, P(d\omega), \qquad N>0,$$

where $B_t=\{\omega : \xi(t)>N\}$ it follows that $P(B_t)\to 0$ for $N\to\infty$ uniformly in t. Therefore for any $\varepsilon>0$ an N_0 can be found not depending on t, such that

$$0 \leqslant \int_{B_t} \xi^+(t_{\max})\, P(d\omega) < \varepsilon$$

for $N>N_0$. In view of the inequalities in the previous string

$$E\chi(\xi(t)>N)\cdot\xi(t)<\varepsilon$$

for all $N>N_0$ which proves the uniform integrability of the family $\{\xi^+(t), t\in T\}$. □

Some inequalities. In the present section it is assumed that $T=\{0, 1, 2, \dots, n\}$, $\{\mathfrak{F}_k; k\in T\}$ is a monotonically non-decreasing sequence of σ-algebras, $\mathfrak{F}_k \in \mathfrak{S}$, $\xi_k (k\in T)$ is an $\mathfrak{F}_k$-measurable random variable and $E\xi_k^+<\infty$. Let τ_i $(i=1, 2)$ be random times on $\{\mathfrak{F}_k, k\in T\}$ (τ_i take on values from T) and let $\tau_1 \leqslant \tau_2$ with probability 1. Set

$$\eta_i = \sum_{k=1}^{\tau_i} \xi_k$$

where $\tau_i=0$ implies $\eta_i=0$.

Let $\mathfrak{F}_i^*$ denote the σ-algebra of events induced by the random time τ_i. Recall that (Chapter I, Section 1) $\mathfrak{F}_i^*$ consists of those sets $E\in\mathfrak{S}$ for which

$$E\cap\{\tau_i\leqslant k\}\in\mathfrak{F}_k, \qquad k=0, 1, \dots, n. \tag{8}$$

Lemma 2. *Let*

$$E\{\xi_k \mid \mathfrak{F}_{k-1}\} \geqslant 0, \qquad k=1, 2, \dots, n, \tag{9}$$

and $A\in\mathfrak{F}_1^*$. *Then*

$$\int_A \eta_1 P(d\omega) \leqslant \int_A \eta_2 P(d\omega). \tag{10}$$

If however

$$E\{\xi_k \mid \mathfrak{F}_{k-1}\} = 0, \tag{11}$$

then

$$\int_A \eta_1 \mathsf{P}(d\omega) = \int_A \eta_2 \mathsf{P}(d\omega). \tag{12}$$

Proof. We first note that $\mathfrak{F}_1^* \subset \mathfrak{F}_2^*$ (Chapter I, Section 1). To prove inequality (10) it is sufficient to consider the case when τ_1 is constant on A, since in the general case $A = \bigcup_{j=0}^{n} A_j$, where $A_j = A \cap \{\tau_1 = j\} \in \mathfrak{F}_1^*$. Let $\tau_1 = j$ on A. Then $A \in \mathfrak{F}_j$ and $\tau_2 \geqslant j$ on A. We have

$$\int_A \eta_2 \mathsf{P}(d\omega) = \int_A \eta_1 \mathsf{P}(d\omega) + \int_{A \cap \{\tau_2 > j\}} \sum_{k=j+1}^{\tau_2} \xi_k \mathsf{P}(d\omega) =$$

$$= \int_A \eta_1 \mathsf{P}(d\omega) + \int_{A \setminus \{\tau_2 \leqslant j\}} \xi_{j+1} \mathsf{P}(d\omega) +$$

$$+ \int_{A \setminus \{\tau_2 \leqslant j+1\}} \xi_{j+2} \mathsf{P}(d\omega) + \dots + \int_{A \setminus \{\tau_2 \leqslant n-1\}} \xi_n \mathsf{P}(d\omega).$$

Since $A \setminus \{\tau_2 \leqslant k\} \in \mathfrak{F}_k$ $(k \geqslant j)$, it follows that

$$\int_{A \setminus \{\tau_2 \leqslant k\}} \xi_{k+1} \mathsf{P}(d\omega) = \int_{A \setminus \{\tau_2 \leqslant k\}} \mathsf{E}\{\xi_{k+1} \mid \mathfrak{F}_k\} \mathsf{P}(d\omega) \geqslant 0,$$

which yields (10).

The argument presented above also shows that if inequality (9) is replaced by equality (11) then relation (12) will hold. □

Lemma 2 can also be formulated in the following manner: If $\{\zeta_k, \mathfrak{F}_k, k \in T\}$ is a submartingale, τ_j are random times on $\{\mathfrak{F}_k, k \in T\}$ and $\tau_1 \leqslant \tau_2$, then for any $A \in \mathfrak{F}_1^*$

$$\int_A \zeta_{\tau_1} \mathsf{P}(d\omega) \leqslant \int_A \zeta_{\tau_2} \mathsf{P}(d\omega), \quad A \in \mathfrak{F}_1^*.$$

Corollary 1. *If $\tau_1, \tau_2, \dots, \tau_s$ is a sequence of random times on $\{\mathfrak{F}_k, k = 0, \dots, 1\}$ and $\tau_1 \leqslant \tau_2 \leqslant \dots \tau_s$, then $\{\eta_k, \mathfrak{F}_k^*, k = 1, \dots, s\}$, $\eta_k = \zeta_{\tau_k}$ forms a submartingale. If however $\{\zeta_k, \mathfrak{F}_k, k \in T\}$ is a martingale, then $\{\eta_k, \mathfrak{F}_k^*, k = 1, \dots, s\}$ is also a martingale.*

Therefore martingales (semi-martingales) observable at random instants of time are martingales (semi-martingales) as well.

Corollary 2. *If under the assumptions of Lemma 2 condition* (9) *is omitted and is replaced by assumption* $\mathsf{E}\psi_k^- < \infty$ *where* $\psi_k = \mathsf{E}\{\xi_k \mid \mathfrak{F}_{k-1}\}$, *then*

$$\int_A \left(\eta_1 - \sum_{k=1}^{n} \psi_k^-\right) \mathsf{P}(d\omega) \leqslant \int_A \eta_2 \mathsf{P}(d\omega). \tag{13}$$

Indeed, (13) is a consequence of (10) if ξ_k is replaced by $\xi_k - \psi_k$. □

Lemma 3. *Assume that random variables* ξ_k *are* $\mathfrak{F}_k$*-measurable,* $\mathsf{E}\xi_k^+ < \infty$, $\{\mathfrak{F}_k, k \in T\}$ *is a current of* σ*-algebras, and* $C > 0$. *Set*

$$\zeta_0 = 0, \quad \zeta_k = \sum_{j=1}^{k} \xi_j, \quad k = 1, 2, \ldots, n.$$

Then

$$\mathsf{P}\{\max_{0 \leqslant k \leqslant n} \zeta_k \geqslant C\} \leqslant \frac{1}{C} \mathsf{E}(\zeta_n^+ + \varrho_n), \tag{14}$$

where $\varrho_n = \sum_{k=1}^{n} \psi_k^-$. *If, moreover, for some* $p > 1$

$$\mathsf{E}(\zeta_n^+ + \varrho_n)^p < \infty$$

then

$$\mathsf{E}(\max_{0 \leqslant k \leqslant n} \zeta_k)^p \leqslant \left(\frac{p}{p-1}\right)^p \mathsf{E}(\zeta_n^+ + \varrho_n)^p. \tag{15}$$

Proof. Let τ_1 be the smallest index k such that $\zeta_k \geqslant C$ $(k = 1, 2, \ldots, n)$ and set $\tau_1 = n$ if such an index does not exist. Let $\tau_2 = n$ and A be the event $\{\eta \geqslant C\}$ where $\eta = \max_{0 \leqslant k \leqslant n} \zeta_k$.

Here τ_1 and τ_2 are random times on $\mathfrak{F}_k$, A is $\mathfrak{F}_1^*$-measurable and $\tau_1 \leqslant \tau_2$. Applying (13) we obtain

$$C\mathsf{P}(A) \leqslant \int_A \zeta_{\tau_1} \mathsf{P}(d\omega) \leqslant \int_A (\zeta_n + \varrho_n) \mathsf{P}(d\omega) \leqslant \mathsf{E}(\zeta_n^+ + \varrho_n),$$

which proves inequality (14). Moreover, if $\chi(C)$ denotes the indicator of event A, then

$$\eta^p = p \int_0^\infty \chi(C)\, C^{p-1}\, dC.$$

As we have just seen

$$\mathsf{E}C\chi(C) \leqslant \mathsf{E}\chi(C)(\zeta_n^+ + \varrho_n),$$

therefore

$$\mathsf{E}\eta^p \leqslant p\mathsf{E}\int_0^\infty (\zeta_n^+ + \varrho_n)\,\chi(C)\,C^{p-2}dC = \frac{p}{p-1}\mathsf{E}(\zeta_n^+ + \varrho_n)\,\eta^{p-1}.$$

Utilizing Hölder's inequality we obtain

$$\mathsf{E}\eta^p \leqslant \frac{p}{p-1}\{\mathsf{E}\eta^p\}^{\frac{p-1}{p}}\{\mathsf{E}(\zeta_n^+ + \varrho_n)^p\}^{\frac{1}{p}},$$

which yields (15). □

Corollary. *If* $\{\zeta_k, k=1, \ldots, n\}$ *is a submartingale, then*

$$\mathsf{P}\{\max_{1\leqslant k\leqslant n} \zeta_k^+ \geqslant C\} \leqslant \frac{1}{C}\mathsf{E}\zeta_n^+, \tag{16}$$

moreover if $\mathsf{E}(\zeta_n^+)^p < \infty$ *for some* $p > 1$, *then*

$$\mathsf{E}\left(\max_{1\leqslant k\leqslant n} \zeta_k^+\right)^p \leqslant \left(\frac{p}{p-1}\right)^p \mathsf{E}(\zeta_n^+)^p. \tag{17}$$

Next, if $\{\eta_k, k\in T\}$ *is also a martingale, then*

$$\mathsf{P}\left\{\max_{1\leqslant k\leqslant n} |\zeta_k| \geqslant C\right\} \leqslant \frac{1}{C^p}\mathsf{E}|\zeta_n|^p. \tag{18}$$

The numbers of crossings $\nu[a, b)$ of the half interval $[a, b)$ by a family of random variables $\{\zeta(t), t\in T\}$ where T is an ordered set is defined as the least upper bound of numbers s such that there exists a sequence $\{t_i, i=1, 2, \ldots, 2s\}$ $t_i < t_{i+1}$, $t_i \in T$, for which

$$\zeta(t_1) \geqslant b, \zeta(t_2) < a, \zeta(t_3) \geqslant b, \ldots, \zeta(t_{2s}) < a.$$

We estimate the mathematical expectation of the number of crossings of the half-interval $[a, b)$ by the sequence $\{\zeta_k, k=1, 2, \ldots, n\}$ defined in Lemma 3.

Introduce the sequence of random integer-valued variables $j_1, j_2, \ldots, j_k, \ldots$, which may not always exist. Let A_k denote the event: the number j_k exists. Moreover, let j_1 be the smallest integer such that $\zeta_{j_1} \geqslant b$, $j_1 \leqslant n$; j_2 be the smallest integer larger than j_1 such that $j_2 \leqslant n$, $\zeta_{j_2} < a$ and so on, j_{2m-1} be the smallest integer larger than j_{2m-2} such that $j_{2m-1} \leqslant n$

and $\zeta_{j_{2m-1}} \geqslant b$; j_{2m} be the smallest integer larger than j_{2m-1} such that $\zeta_{j_{2m}} < a$, $j_{2m} \leqslant n$.

The numbers $j_1, j_2, \ldots, j_k, \ldots$ form a monotonically non-decreasing sequence of sampling variables, $j_k \leqslant n$. We extend the definition of j_k on the whole set Ω by putting $j_k = n$ if $\omega \notin A_k$. It follows from the definition of j_k and the inequality (13) that

$$0 \leqslant \int_{A_{2m-1}} (\zeta_{j_{2m-1}} - b) \mathsf{P}(d\omega) \leqslant \int_{A_{2m-1}} (\zeta_{j_{2m}} - b) \mathsf{P}(d\omega) +$$

$$+ \sum_{k \geqslant 1} \int_{B_{m,k}} \psi^-_{j_{2m-1}+k} \mathsf{P}(d\omega) \leqslant (a-b) \mathsf{P}(A_{2m}) +$$

$$+ \int_{A_{2m-1} \setminus A_{2m}} (\zeta_{j_{2m}} - b) \mathsf{P}(d\omega) + \sum_{k \geqslant 1} \int_{B_{m,k}} \psi^-_{j_{2m-1}+k} \mathsf{P}(d\omega),$$

where $B_{m,k} = A_{2m-1} \cap \{j_{2m} - j_{2m-1} \geqslant k\}$. Hence

$$(b-a) \mathsf{P}(A_{2m}) \leqslant \int_{A_{2m-1} \setminus A_{2m}} (\zeta_n - b) \mathsf{P}(d\omega) + \sum_{k \geqslant 1} \int_{B_{m,k}} \psi^-_{j_{2m-1}+k} \mathsf{P}(d\omega) \leqslant$$

$$\leqslant \int_{A_{2m-1} \setminus A_{2m}} (\zeta_n - b)^+ \mathsf{P}(d\omega) + \sum_{k \geqslant 1} \int_{B_{m,k}} \psi^-_{j_{2m-1}+k} \mathsf{P}(d\omega).$$

Summing up these inequalities over all $m \geqslant 1$, we obtain

$$(b-a) \sum_{m \geqslant 1} \mathsf{P}(A_{2m}) \leqslant \mathsf{E}[(\zeta_n - b)^+ + \varrho_n], \quad \varrho_n = \sum_{k=1}^{n} \psi^-_k .$$

Note that $\nu[a, b) = \sum_{m \geqslant 1} \chi(A_{2m})$ where $\chi(A)$ is the indicator of event A. Therefore

$$\mathsf{E}\nu[a, b) = \sum_{m \geqslant 1} \mathsf{P}(A_{2m}).$$

We have thus proved the following

Lemma 4. *If a sequence $\{\xi_k, \mathfrak{F}_k; k = 1, 2, \ldots, n\}$ satisfies the conditions of Lemma 3, then*

$$\mathsf{E}\nu[a, b) \leqslant \frac{\mathsf{E}[(\zeta_n - b)^+ + \varrho_n]}{b-a}. \tag{19}$$

In the case of submartingales the last inequality becomes

$$\mathsf{E}\nu[a, b) \leqslant \frac{\mathsf{E}(\zeta_n - b)^+}{b-a}. \tag{20}$$

These inequalities are easily generalized for the case when T is a countable sequence. Thus if $T=\{1, 2, \ldots, n, \ldots\}$ and $T'=\{\ldots, -n, -n+1, \ldots, -1\}$ then inequality (14) yields

$$\mathsf{P}\{\sup_{n\in T}\zeta_n \geqslant C\} \leqslant \frac{1}{C}\sup_{n\in T}\mathsf{E}(\zeta_n^+ + \varrho_n) \tag{21}$$

and

$$\mathsf{P}\{\sup_{n\in T'}\zeta_n \geqslant C\} \leqslant \frac{1}{C}\mathsf{E}(\zeta_{-1}^+ + \varrho'), \tag{22}$$

where $\varrho' = \sum_{k=1}^{\infty} \psi_{-k}^-$

The *proof* of these relations follows from the fact that $\sup_{n\in T}\zeta_n = \lim_{n\to\infty}\max_{1\leqslant k\leqslant n}\zeta_k$, and thus

$$\mathsf{P}(\sup_{n\in T}\zeta_n \geqslant C) = \lim_{n\to\infty}\mathsf{P}\{\max_{1\leqslant k\leqslant n}\zeta_k \geqslant C\}.$$

Analogously if $\nu_\infty[a, b)$ and $\nu'_\infty[a, b)$ denote the number of times the sequences $\{\zeta_n, n\in T\}$ and $\{\zeta_n, n\in T'\}$ correspondingly cross a half interval $[a, b)$ from left to right and $\nu_n[a, b)$ and $\nu'_n[a, b)$ denote the number of times the truncated sequences $\{\zeta_k, k=1, \ldots, n\}$ and $\{\zeta_{-k}, k=1, \ldots, n\}$ correspondingly cross the half-interval $[a, b)$, then in view of the fact that $\nu_n[a, b)$ and $\nu'_n[a, b)$ form monotonically non-decreasing sequences, $\nu_\infty[a, b) = \lim \nu_n[a, b)$, $\nu'_\infty[a, b) = \lim \nu'_n[a, b)$, and property f), yields that inequalities

$$(b-a)\,\mathsf{E}\nu_\infty[a, b) \leqslant \sup_n \mathsf{E}[(\zeta_n - b)^+ + \varrho_n], \tag{23}$$

$$(b-a)\,\mathsf{E}\nu'_\infty[a, b) \leqslant \mathsf{E}[(\zeta_{-1} - b)^+ + \varrho'] \tag{24}$$

are valid.

In the case of submartingales the same argument is applicable also when T is an arbitrary countable set of reals. Here one should introduce a monotonically increasing sequence of sets T_n consisting of a finite number of points, apply to the sequences $\{\zeta(t), \mathfrak{F}_t, t\in T\}$ the inequalities obtained and pass on to the limit as $n\to\infty$.
We then obtain that

$$\mathsf{P}\{\sup_{t\in T}\zeta^+(t) > C\} \leqslant \frac{\sup_{t\in T}\mathsf{E}\zeta^+(t)}{C} \tag{25}$$

$$\mathsf{E}[\sup_{t\in T}\zeta^+(t)]^p \leqslant \left(\frac{p}{p-1}\right)^p \sup_{t\in T}\mathsf{E}[\zeta^+(t)]^p, \tag{26}$$

$$\mathsf{E}\nu[a,b) \leqslant \frac{\sup_{t\in T}\mathsf{E}(\zeta(t)-b)^+}{b-a}. \tag{27}$$

Moreover if the set T possesses the maximal element $t_{\max}$ then the least upper bounds in the r.h.s. of the above inequalities are actually attained at $t = t_{\max}$.

Existence of the limit. Consider the sequence $\{\zeta_n, \mathfrak{F}_n; n\in T\}$ where $T = \{\dots, -n, -n+1, \dots, -1, 0, 1, \dots, n \dots\}$, $\mathfrak{F}_n$ is a monotonically non-decreasing family of σ-algebras, ζ_n are $\mathfrak{F}_n$-measurable and let moreover $\mathsf{E}\zeta_n^+ < \infty$, $\xi_n = \zeta_n - \zeta_{n-1}$ and $\mathsf{E}\{\xi_n \mid \mathfrak{F}_{n-1}\} = \psi_n$.

Theorem 1. a) *If*

$$\sup_{n\geqslant 1}\mathsf{E}(\zeta_n^+ + \varrho_n) < \infty, \qquad \varrho_n = \sum_{k=1}^{n} \psi_k^-,$$

then there exists with probability 1 *the finite limit* $\zeta_\infty = \lim_{n\to\infty} \zeta_n$,

b) *if* $\sup_{n\geqslant 1}\mathsf{E}\varrho_n' < \infty$ *where* $\varrho_n' = \sum_{k=1}^{n} \psi_{-k}^-$, *then there exists with probability* 1 *the finite limit* $\zeta_{-\infty} = \lim_{n\to\infty} \zeta_{-n}$.

Proof. From Fatou's inequality

$$\mathsf{E} \varlimsup_{n\to+\infty} \zeta_n^+ \leqslant \varlimsup_{n\to+\infty} \mathsf{E}\zeta_n^+ < \infty$$

it follows that the relation $\varlimsup_{n\to+\infty} \zeta_n^+ = +\infty$ holds only with probability 0. On the other hand it follows from inequality (23) that

$$\varliminf_{a\to-\infty} \mathsf{E}\nu[a,b) = \mathsf{E} \lim_{a\to-\infty} \nu[a,b) = 0,$$

where $\nu[a,b)$ is the number of crossings of the half-interval $[a,b)$ by the sequence $\{\zeta_n; n\geqslant 1\}$. Therefore if $\zeta_1 > -\infty$ then with probability 1 there exists $a = a(\omega)$ such that $\zeta_n > a$ for all $n \geqslant 1$. Hence $\varliminf_{n\to+\infty} \zeta_n > -\infty$ a.s.

Assume now that the finite $\lim_{n\to+\infty} \zeta_n$ does not exist with a positive probability. Then, with a positive probability $\varliminf_{n\to+\infty} \zeta_n < \varlimsup_{n\to+\infty} \zeta_n$. Therefore a pair of numbers a and b can be found such that with a positive probability

$$\varliminf_{n\to+\infty} \zeta_n < a < b < \varlimsup_{n\to+\infty} \zeta_n.$$

Then, however, the number of crossings $\nu[a, b)$ of the half interval $[a, b)$ by the sequence $\{\zeta_n, n \geqslant 1\}$ is equal to ∞ with no lesser probability. This contradicts inequality (23). Thus with probability 1 $\varliminf_{n\to+\infty} \zeta_n = \varlimsup_{n\to+\infty} \zeta_n$.

The proof of assertion b) is analogous. We merely note that if we let $b \to +\infty$ in (24) then we obtain that $\lim_{b\to+\infty} \nu'[a, b) = 0$ with probability 1, where $\nu'[a, b)$ is the number of crossings of the half-interval $[a, b)$ by the sequence $\{\zeta_n, n \leqslant -1\}$. From here it follows that $\varliminf \zeta_n < +\infty$ with probability 1. All the other arguments given in the proof of assertion 1 remain valid with obvious modifications and ζ_{-1} takes the place of ζ_n. □

When applied to semi-martingales the theorem just proved yields

Corollary. *If* $\{\zeta_k, \mathfrak{F}_k, k = \ldots -n, -n+1, \ldots, -1\}$ *is a submartingale then the limit* $\lim_{n\to-\infty} \zeta_n = \zeta_{-\infty}$ *exists with probability* 1. *If* $\{\zeta_k, \mathfrak{F}_k, k = 1, 2, \ldots, n, \ldots\}$ *is a submartingale and* $\sup \mathsf{E}\zeta_n^+ < \infty$, *then the limit* $\lim_{n\to+\infty} \zeta_n = \zeta_{+\infty}$ *also exists with probability* 1.

Definition 2. A random variable $\bar{\xi}$ ($\underline{\xi}$) is called a *closure from the right (left) of the submartingale* $\{\xi(t), \mathfrak{F}_t, t \in T\}$, if $\mathsf{E}\bar{\xi}^+ < \infty$ ($\mathsf{E}\underline{\xi}^+ < \infty$), $\bar{\xi}$ is measurable with respect to $\bar{\mathfrak{F}} = \sigma\{\mathfrak{F}_t, t \in T\}$ ($\underline{\xi}$ is measurable with respect to $\underline{\mathfrak{F}} = \bigcap_{t \in T} \mathfrak{F}_t$) and for all $t \in T$

$$\xi(t) \leqslant \mathsf{E}\{\bar{\xi} \mid \mathfrak{F}_t\}, \quad (\underline{\xi} \leqslant \mathsf{E}\{\xi(t) \mid \underline{\mathfrak{F}}\}).$$

Theorem 2. *A submartingale* $\{\zeta_n, \mathfrak{F}_n, n = 1, 2, \ldots\}$ *has a closure from the right if and only if the sequence* $\{\zeta_n^+, n = 1, 2, \ldots\}$ *is uniformly integrable.*

Proof. If the submartingale $\{\zeta_n, \mathfrak{F}_n, n = 1, 2, \ldots\}$ possesses a closure from the right then putting $T = \{1, 2, \ldots, n, \ldots\} \cup \{\infty\}$ $\mathfrak{F}_\infty = \sigma\{\mathfrak{F}_n, n = 1, 2, \ldots\}$, $\zeta_\infty = \bar{\zeta}$, we obtain that $\{\zeta_t, \mathfrak{F}_t, t \in T\}$ forms a submartingale and the set T possesses the maximal element ∞. Therefore (in view of Lemma 1) the family $\{\zeta_n^+, n = 1, 2, \ldots\}$ is uniformly integrable. Assume now that the family $\{\zeta_n^+, n = 1, 2, \ldots\}$ is uniformly integrable. Since $\sup \mathsf{E}\zeta_n^+ < \infty$, there exists with probability 1 the limit $\bar{\zeta} = \lim \zeta_n$. Let $\xi_n^N = \max\{\zeta_n, -N\}$, $N > 0$. Since for any N the sequence $\{\xi_n^N, n = 1, 2, \ldots\}$ is a submartingale, it follows from the definition of submartingales that $\mathsf{E}\xi_n^N \chi(A) < \mathsf{E}\xi_{n+m}^N \chi(A)$ for any $A \in \mathfrak{F}_n$, $m > 0$. Approaching to the limit as $m \to \infty$ in the r.h.s. of the inequality and taking into account that the sequence $\{\xi_n^N, n = 1, 2, \ldots\}$ is uniformly integrable (being a sum of two sequences one of which is, by assumption, uniformly integrable while the second is bounded in its absolute value by a constant N and hence also uniformly integrable) we obtain

$$\mathsf{E}\xi_n^N \chi(A) \leqslant \mathsf{E}\bar{\zeta}_n \chi(A)$$

Approaching to the limit as $N \to \infty$ we obtain the inequality

$$\mathsf{E}\zeta_n \chi(\Omega) \leqslant \mathsf{E}\bar{\zeta}_n \chi(A)$$

moreover, $\mathsf{E}\bar{\zeta}^+ \leqslant \underline{\lim}\, \mathsf{E}\zeta_n^+ \leqslant \infty$ so that $\bar{\zeta}$ is a closure of the martingale. □

Theorem 3. *Let* $\{\zeta_n, \mathfrak{F}_n; n=1, 2, \ldots\}$ *be a martingale. The following conditions are equivalent:*

a) *the family* $\{\zeta_n, n=1, 2, \ldots\}$ *is uniformly integrable.*

b) *the martingale* $\{\zeta_n, \mathfrak{F}_n, n=1, 2, \ldots\}$ *possesses a closure from the right*

c) $\mathsf{E}|\zeta_n - \zeta_{n'}| \to 0$ *for* $n', n \to \infty$.

If one of these conditions is satisfied then $\lim \zeta_n = \zeta$ *exists with probability* 1 *and this limit is the closure from the right of the martingale in the sense of convergence in* $\mathscr{L}_1$.

If for some $p > 1$, $\mathsf{E}|\zeta_n|^p \leqslant C$, *then* a), b) *and* c) *are valid and* $\zeta = \lim \zeta_n$ *in the sense of convergence in* $\mathscr{L}_p$.

Proof. The equivalence of a) and b) follows from the preceding theorem. The equivalence of uniform integrability and convergence in $\mathscr{L}_1$ is a general result in measure theory. If the assertion stated in the second part of the theorem is fulfilled then the sequence ζ_n is uniformly integrable. In view of (17) sequence $|\zeta_n|^p$ possesses an integrable majorant $\sup|\zeta_n|^p$, and hence sequences $|\zeta_n|^p$ and $|\zeta - \zeta_n|^p$ where $\zeta = \lim \zeta_n$ are uniformly integrable. It follows from here that $\mathsf{E}|\zeta_n - \zeta|^p \to 0$ for $n \to \infty$. □

Some applications. Let $\mathfrak{F}_1 \subset \mathfrak{F}_2 \subset \ldots \subset \mathfrak{F}_n \subset \ldots$ and $\mathfrak{F} = \sigma\{\mathfrak{F}_n, n=1, 2, \ldots\}$, ξ be a random variable with $\mathsf{E}|\xi| < \infty$. Put

$$\xi_n = \mathsf{E}\{\xi \mid \mathfrak{F}_n\}.$$

Theorem 4. *The sequence* $\{\xi_n, \mathfrak{F}_n, n=1, 2, \ldots\}$ *is a martingale and*

$$\lim \mathsf{E}\{\xi \mid \mathfrak{F}_n\} = \mathsf{E}\{\xi \mid \mathfrak{F}\}$$

with probability 1.

Proof. We have (Chapter I, Section 3)

$$\mathsf{E}\{\xi_{n+1} \mid \mathfrak{F}_n\} = \mathsf{E}\{\mathsf{E}\{\xi \mid \mathfrak{F}_{n+1}\} \mid \mathfrak{F}_n\} = \mathsf{E}\{\xi \mid \mathfrak{F}_n\} = \xi_n,$$

and moreover ξ_n is $\mathfrak{F}_n$-measurable so that the sequence $\{\xi_n, \mathfrak{F}_n, n=1, 2, \ldots\}$ is a martingale. Also

$$\mathsf{E}\{\mathsf{E}\{\xi \mid \mathfrak{F}\} \mid \mathfrak{F}_n\} = \mathsf{E}\{\xi \mid \mathfrak{F}_n\} = \xi_n.$$

The last equality means that $\mathsf{E}\{\xi \mid \mathfrak{F}\}$ is the closure of this martingale. The assertion of the theorem now follows from Theorem 3. □

Corollary 1. *If* $A \in \mathfrak{F}$, *then with probability* 1 $\lim_{n\to\infty} \mathsf{P}\{A \mid \mathfrak{F}_n\} = \chi(A)$.

Let $\{\mathscr{X}, \mathfrak{B}\}$ be a measurable space. The sequence of subdivisions $\{A_{nk}, k=1, 2, \dots\}$ $n=1, 2, \dots$, of the space $\mathscr{X}$ is called *exhaustive* if

a) $A_{nk} \in \mathfrak{B}$, $A_{nk} \cap A_{nr} = \emptyset$ for $k \neq r$, $\bigcup_{k=1}^{\infty} A_{nk} = \mathscr{X}$, $n=1, 2, \dots$;

b) the $n+1$-st subdivision is a refinement of the n-th one, i.e. for any j $A_{n+1\,j} \subset A_{nk}$ for some $k = k(j)$;

c) the minimal σ-algebra which contains all A_{nk} $k=1, 2, \dots$, $n=1, 2, \dots$, coincides with $\mathfrak{B}$.

Corollary 2. Let $\{A_{nk}, k=1, 2, \dots, n\dots\}$ $n=1, 2, \dots$, be an exhaustive sequence of subdivisions of $\{\mathscr{X}, \mathfrak{B}\}$ and m be a measure on $\mathfrak{B}$, $m(\mathscr{X})=1$. Denote by $A_n(x)$ the set A_{nk} which contains the point x. Then for any $\mathfrak{B}$-measurable and μ-integrable function $f(x)$ we have

$$\lim_{n\to\infty} \frac{\int_{A_n(x)} f(u)\, m(du)}{m(A_n(x))} = f(x)$$

m-almost for all x.

The last assertion can be regarded as an analogue of the basic theorem of integral calculus for abstract integrals. The validity of this assertion follows from the fact that if $\mathfrak{F}_n$ is viewed as a σ-algebra generated by a finite or countable number of sets $\{A_{nk}, k=1, 2, \dots\}$, then $\sigma\{\mathfrak{F}_n, n=1, 2, \dots\} = \mathfrak{B}$ and $\mathsf{E}\{f \mid \mathfrak{F}_n\} = \dfrac{\int_{A_{nk}} f(u)\, m(du)}{m(A_{nk})}$ if $x \in A_{nk}$ (Chapter I, Section 3). Moreover, the right-hand-side of the last relation is not defined if $m(A_{nk})=0$. But m – the measure of the set of x's such that the latter holds, if only for one n, – is zero.

Using similar reasoning one can obtain a "direct" proof (in a certain sense) of Radon's theorem on absolute continuity of measures.

Lemma 5. *Let* $\{\mathscr{X}, \mathfrak{B}\}$ *be a measurable space and let the* σ-*algebra* $\mathfrak{B}$ *be generated by a countable sequence of sets,* $\mathfrak{B} = \sigma\{B_1, B_2, \dots, B_n, \dots\}$. *Then there exists in* $\{\mathscr{X}, \mathfrak{B}\}$ *an exhaustive sequence of subdivisions.*

Proof. Let the sequence $\{A_{1k}, k=1, 2, \dots\}$ consist of two sets B_1 and $\bar{B}_1$. If $\{A_{nk}, k=1, 2, \dots\}$ is already constructed then the sequence $\{A_{n+1\,k},$

$k=1, 2, \ldots\}$ is defined as a collection of all the sets of the type $A_{nk} \cap B_{n+1}$ and $A_{nk} \cap \bar{B}_{n+1}$ $(k=1, 2, \ldots)$. □

Theorem 5. *Let $\{\mathscr{X}, \mathfrak{B}, m\}$ be a probability space, $q(\cdot)$ be a measure on $\{\mathscr{X}, \mathfrak{B}\}$, $q(\mathscr{X})<\infty$; let the measure be absolutely continuous with respect to m and let $\{A_{nk}, k=1, 2, \ldots\}$, $n=1, 2, \ldots$, be an arbitrary exhaustive sequence of subdivisions of $\mathscr{X}$. Put*

$$g_n(x)=\frac{q(A_{nk}(x))}{m(A_{nk}(x))}$$

for $m(A_{nk}(x))>0$, where $A_{nk}(x)$ is that set of the sequence $\{A_{nk}, k=1, 2, \ldots\}$ which contains point x. If $m(A_{nk}(x))=0$, we put $g_n(x)=0$. Then:

a) *the sequence $\{g_n, \mathfrak{F}_n, n=1, 2, \ldots\}$ where $\mathfrak{F}_n=\sigma\{A_{n1}, A_{n2}, \ldots\}$ forms a martingale;*

b) *there exists the limit $g(x)=\lim_{n\to\infty} g_n(x)$ (mod m) independent (mod m) from the choice of the exhaustive sequence $\{A_{nk}, k=1, 2, \ldots\}$, $n=1, 2, \ldots$;*

c) *for an arbitrary $B \in \mathfrak{B}$*

$$q(B)=\int_B g(x)\, m(dx). \tag{28}$$

Proof. The function $g_n(x)$ is $\mathfrak{F}_n$-measurable and takes on at most a countable number of values. Therefore

$$\mathsf{E}\{g_{n+1} \mid \mathfrak{F}_n\}=\sum_j \frac{q(A_{n+1\,j})}{m(A_{n+1\,j})} \frac{m(A_{n+1\,j} \cap A_{nk}(x))}{m(A_{nk}(x))}=$$

$$=\sum_{\substack{j' \\ A_{n+1\,j'} \subset A_{nk}(x)}} \frac{q(A_{n+1\,j'})}{m(A_{n+1\,j'})} \frac{m(A_{n+1\,j'})}{m(A_{nk}(x))}=\frac{q(A_{nk}(x))}{m(A_{nk}(x))}=g_n(x),$$

which proves a). Furthermore

$$\int_{\mathscr{X}} |g_n(x)|\, m(dx)=\int_{\mathscr{X}} g_n(x)\, m(dx)=q(\mathscr{X})<\infty$$

and

$$\int_A g_n(x)\, m(dx)=\sum_k \int_{A \cap A_{nk}} g_n(x)\, m(dx)=\sum_k q(A_{nk} \cap A)=q(A).$$

Since q is absolutely continuous with respect to measure m, for any $\varepsilon>0$ a $\delta>0$ can be found such that $m(A)<\delta$ implies $q(A)<\varepsilon$. From here it follows that the sequence $g_n(x)$ is uniformly integrable (with respect to measure m). Therefore there exists m-almost for all x the lim $g_n(x)=g(x)$ and $g(x)$ is the closure of the martingale. Hence for each $A_n \in \mathfrak{F}_n$

$$\int_{A_n} g(x)\, m(dx) = \lim_{k\to\infty} \int_{A_n} g_k(x)\, m(dx) = q(A_n).$$

Since the class of sets A for which formula (28) is valid is monotonic and contains algebra $\bigcup_n \mathfrak{F}_n$, formula (28) is valid also for $\sigma\{\mathfrak{F}_n, n=1,2,\dots\}$ i.e. it is valid for any $B \in \mathfrak{F}$. Finally assertion c) implies the independence of function $g(x)$ from the choice of the exhaustive sequence. If there exists two functions g' and g'' for which c) holds then $\int_B [g'(x)-g''(x)]\, md(x) = 0$ for any $B \in \mathfrak{B}$, which is possible if and only if $g'(x)=g''(x)$ (mod m). □

§3. Series

Some general criteria for convergence of series. In the present section conditions are investigated for convergence with probability 1 of series with random terms.

Let the series

$$\xi_1 + \xi_2 + \dots + \xi_n + \dots. \tag{1}$$

be given.

Theorem 1. *If there exists a sequence of numbers* $\varepsilon_n > 0$, $n=1,2,\dots$ *such that*

$$\sum_{n=1}^{\infty} \varepsilon_n < \infty, \quad \sum_{n=1}^{\infty} \mathsf{P}\{|\xi_n| > \varepsilon_n\} < \infty, \tag{2}$$

then series (1) *is absolutely convergent with probability* 1.

Proof. Let $A_n = \{|\xi_n| > \varepsilon_n\}$. From the convergence of the second series in (2) and Theorem 6 of Section 2 in Chapter I it follows that $\mathsf{P}(\overline{\lim}\, A_n)=0$ i.e. with probability 1 only a finite number of events A_n can occur. Therefore there exists $N=N(\omega)$ such that for $n > N(\omega)$, $|\xi_n| < \varepsilon_n$ and series (1) converges. □

For random variables ξ_n with finite moments one can formulate the following sufficient condition for the convergence of series (1):

Theorem 2. *If*

$$\sum_{n=1}^{\infty} \mathsf{E}|\xi_n| < \infty, \tag{3}$$

then series (1) *is absolutely convergent with probability* 1.

The proof follows from Lebesgue's theorem which assures that

$$\mathsf{E}\sum_{1}^{\infty} \xi_n^+ = \sum_{1}^{\infty} \mathsf{E}\xi_n^+, \quad \mathsf{E}\sum_{1}^{\infty} \xi_n^- = \sum_{1}^{\infty} \mathsf{E}\xi_n^-,$$

and hence with probability 1 the series

$$\sum_{1}^{\infty} (\xi_n^+ + \xi_n^-) = \sum_{1}^{\infty} |\xi_n|.$$

is convergent. □

Corollary. *If there exists a sequence $c_n > 0$, $n = 1, 2, \dots$ and $p > 1$ such that the series*

$$\sum_{n=1}^{\infty} c_n^{-q}, \quad \sum_{n=1}^{\infty} c_n^p \mathsf{E}|\xi_n|^p, \quad \frac{1}{p} + \frac{1}{q} = 1,$$

is convergent, then series (1) *converges with probability* 1 *in* $\mathscr{L}_p$ *as well.*

To prove this assertion we note that in view of Hölder's and Jensen's inequalities we have

$$\begin{aligned}\sum_{m+1}^{m+n} \mathsf{E}|\xi_k| &= \sum_{m+1}^{m+n} c_k^{-1}\mathsf{E}c_k|\xi_k| \leqslant \\ &\leqslant \left(\sum_{m+1}^{m+n} c_k^{-q}\right)^{1/q} \left(\sum_{m+1}^{m+n} c_k^p (\mathsf{E}|\xi_k|)^p\right)^{1/p} \leqslant \\ &\leqslant \left(\sum_{m+1}^{m+n} c_k^{-q}\right)^{1/q} \left(\sum_{m+1}^{m+n} c_k^p \mathsf{E}|\xi_k|^p\right)^{1/p},\end{aligned}$$

and the convergence of series (3) follows from this inequality by taking into account the premise of the corollary. □

Stronger results are valid for semi-martingales. Put

$$\zeta_n = \xi_1 + \xi_2 + \dots + \xi_n, \quad \zeta_0 = 0.$$

Theorem 3. *Let ξ_n be $\mathfrak{F}_n$-measurable, $\{\mathfrak{F}_n, n = 0, 1, \dots\}$ be a current of σ-algebras. Then:*

a) *if* $\mathsf{E}\{\xi_n \mid \mathfrak{F}_{n-1}\} \geqslant 0$ *and* $\sup \mathsf{E}\zeta_n^+ < \infty$, *series* (1) *converges with probability* 1;

b) *if* $\mathsf{E}\{\xi_n \mid \mathfrak{F}_{n-1}\}=0$ *and for some* $p>1$

$$\sup_n \mathsf{E}\,|\zeta_n|^p<\infty,$$

then series (1) *is convergent with probability* 1 *in* $\mathscr{L}_p$ *as well.*

Condition a) is equivalent to the assumption that $\{\zeta_n, \mathfrak{F}_n, n=1,2,\dots\}$ is a submartingale. The corresponding assertion is therefore a corollary of the theorem on convergence of submartingales. Condition b) means that $\{\zeta_n, \mathfrak{F}_n, n=1,2,\dots\}$ is a martingale and the assertion of this part of the theorem follows from Theorem 3 of Section 2. □

Corollary 1. *If* $\mathsf{E}\{\xi_n \mid \mathfrak{F}_{n-1}\}=0$ *and*

$$\sum_{n=1}^{\infty} \mathsf{E}\xi_n^2<\infty$$

then series (1) *converges with probability* 1 *in* $\mathscr{L}_2$ *as well.*

The *proof* follows from the facts that for $k<n$

$$\mathsf{E}\xi_k\xi_n=\mathsf{E}\{\xi_k\mathsf{E}\{\xi_n \mid \mathfrak{F}_{n-1}\}\}=0,$$

$$\mathsf{E}\zeta_n^2=\mathsf{E}\left(\sum_{k=1}^{n}\xi_k\right)^2=\sum_{k=1}^{n}\mathsf{E}\xi_k^2+2\sum_{j=2}^{n}\sum_{k<j}\mathsf{E}\xi_k\xi_j=\sum_{k=1}^{n}\mathsf{E}\xi_k^2,$$

and from assertion b) of the theorem. □

For series with independent terms the last result is known as Kolmogorov's theorem.

Corollary 2. (Kolmogorov's theorem) *If* $\{\xi_n, n=1,2,\dots\}$ *are independent random variables,* $\mathsf{E}\xi_k=0$ *and the series* $\sum_{k=1}^{\infty}\mathsf{V}\xi_k<\infty$, *then series* (1) *converges with probability* 1.

This assertion follows from Corollary 1 if the σ-algebra generated by the random variables $\xi_1, \xi_2, \dots, \xi_n$ is taken for $\mathfrak{F}_n$, and realizing that in view of the independence of the random variables ξ_n

$$\mathsf{E}\{\xi_n \mid \mathfrak{F}_{n-1}\}=\mathsf{E}\xi_n=0. \quad \square$$

Series of independent random variables. We shall now discuss in some detail the convergence of series with independent terms. As we have seen previously such a series is convergent either with probability 0 or with probability 1 (Theorem 8, Section 2, Chapter I).

The *following* bound on the distribution of the maximum of a sum of independent terms will be needed in the sequel:

Theorem 4. *If* $\{\xi_k, k=1,2,\dots,n\}$ *are independent* $\mathsf{E}\xi_k=0$ *and* $|\xi_k|<c$ *with probability* 1, *where c is a constant, then*

$$\mathsf{P}\{\max_{1\leqslant k\leqslant n}|\zeta_k|\leqslant t\}\leqslant(c+t)^2/\sum_{k=1}^{n}\sigma_k^2, \tag{4}$$

where $\sigma_k^2=\mathsf{E}\xi_k^2=\mathsf{V}\xi_k$.

Denote by E_n the events $\{\max_{0\leqslant k\leqslant n}|\zeta_k|\leqslant t\}$, $n=1,2,\dots$. These events form a monotonically decreasing sequence. We have

$$\mathsf{E}\chi(E_n)\zeta_n^2=\sum_{k=1}^{n}\mathsf{E}\{\chi(E_k)\zeta_k^2-\chi(E_{k-1})\zeta_{k-1}^2\}=$$

$$=\sum_{k=1}^{n}\mathsf{E}\chi(E_{k-1})(\zeta_k^2-\zeta_{k-1}^2)-\sum_{k=1}^{n}\mathsf{E}\{\chi(E_{k-1}\backslash E_k)\zeta_k^2. \tag{5}$$

Next,

$$\mathsf{E}\chi(E_{k-1}\backslash E_k)\zeta_k^2=\mathsf{E}\chi(E_{k-1}\backslash E_k)(\zeta_{k-1}+\xi_k)^2\leqslant$$
$$\leqslant(t+c)^2\,\mathsf{E}\chi(E_{k-1}\backslash E_k),$$

$$\sum_{k=1}^{n}\mathsf{E}\chi(E_{k-1}\backslash E_k)\zeta_k^2\leqslant(t+c)^2\sum_{k=1}^{n}\mathsf{E}\chi(E_{k-1}\backslash E_k)=(t+c)^2\,[1-\mathsf{P}(E_n)]. \tag{6}$$

Moreover,

$$\mathsf{E}\chi(E_{k-1})(\zeta_k^2-\zeta_{k-1}^2)=\mathsf{E}\chi(E_{k-1})(2\zeta_{k-1}\xi_k+\xi_k^2)=$$
$$=2\mathsf{E}\chi(E_{k-1})\zeta_{k-1}\mathsf{E}\xi_k+\mathsf{E}\chi(E_{k-1})\mathsf{E}\xi_k^2=\sigma_k^2\mathsf{E}\chi(E_{k-1}). \tag{7}$$

Relations (5), (6) and (7) yield

$$t^2\mathsf{P}(E_n)\geqslant\mathsf{E}\chi(E_n)\zeta_n^2\geqslant\sum_{k=1}^{n}\sigma_k^2\mathsf{E}\chi(E_{k-1})-(t+c)^2\,(1-\mathsf{P}(E_n))\geqslant$$

$$\geqslant\mathsf{P}(E_n)\left\{\sum_{k=1}^{n}\sigma_k^2+(t+c)^2\right\}-(t+c)^2$$

or

$$(t+c)^2\geqslant\mathsf{P}(E_n)\left\{\sum_{k=1}^{n}\sigma_k^2+c^2+2ct\right\},$$

and relation (4) follows from here. □

In the general case of series with independent terms the problem of convergence of series (1) is fully solved using the following theorem.

Theorem 5. (Kolmogorov's three-series criterion). *For the convergence* *of series* (1) *of independent random variables with probability* 1 *it is necessary for each* $c>0$ *and sufficient that for some* $c>0$ *the following series*

$$\sum_{n=1}^{\infty} \mathsf{P}\{|\xi_n|>c\}, \tag{8}$$

$$\sum_{n=1}^{\infty} \mathsf{E}\xi'_n, \tag{9}$$

$$\sum_{n=1}^{\infty} \mathsf{V}\xi'_n, \tag{10}$$

will be convergent, where $\xi'_n=\xi_n$ *for* $|\xi_n|<c$ *and* $\xi'_n=0$ *for* $|\xi_n|>c$.

Proof. Sufficiency. In view of Theorem 3, Corollary 2 the series

$$\sum_{n=1}^{\infty} (\xi'_n-\mathsf{E}\xi'_n)$$

is convergent with probability 1. Taking into account the convergence of series (9) it follows from here that series $\sum_{n=1}^{\infty} \xi'_n$ is convergent. It follows from condition (8) and Borel-Cantelli's lemma that only a finite number of terms in series $\sum_{n=1}^{\infty} (\xi_n-\xi'_n)$ is non-zero. Therefore series (1) converges with probability 1.

Necessity. Let series (1) be convergent with probability 1. Then its general term tends to zero with probability 1 since only a finite number of terms of this series exceeds c $(c>0)$ in absolute value. Therefore the series $\sum_{n=1}^{\infty} \xi'_n$ is convergent with probability 1. Denote by $\{\eta_n\}$ $n=1, 2,\ldots$ the sequence of independent random variables which do not depend on the sequence $\{\xi'_n\}$ $n=1, 2,\ldots$ having the same distribution as ξ'_n. Set $\tilde{\xi}_n=\xi'_n-\eta_n$. Then the series $\sum_{n=1}^{\infty} \tilde{\xi}_n$ converges with probability 1, $\mathsf{E}\tilde{\xi}_n=0$, $|\tilde{\xi}_n|\leqslant 2c$ and $\mathsf{V}\tilde{\xi}_n=2\mathsf{V}\xi'_n$. It follows from the convergence of the series $\sum_{n=1}^{\infty} \xi_n$, that

$$\mathsf{P}\left\{\sup_{1\leqslant n\leqslant\infty}\left|\sum_{k=1}^{n} \tilde{\xi}_k\right|<\infty\right\}=1.$$

Therefore

$$\mathsf{P}\left\{\sup_{1\leqslant n\leqslant\infty}\left|\sum_{k=1}^{n} \tilde{\xi}_k\right|\leqslant t\right\}=a>0$$

for some t. It follows from inequality (4) that for any n

$$2\sum_{k=1}^{n} \mathsf{V}\xi'_k = \sum_{k=1}^{n} \mathsf{V}\tilde{\xi}_k \leqslant \frac{(2c+t)^2}{a},$$

which proves the convergence of series (10). It now follows from Theorem 3, Corollary 2 that series $\sum_{n=1}^{\infty} (\xi'_n - \mathsf{E}\xi'_n)$ converges with probability 1. In turn it follows from this fact that series (9) is convergent. Convergence of series (8) follows from Borel-Cantelli's lemma since if series (1) is convergent then only a finite number of terms of series (1) can be found with probability 1 such that $|\xi_n| > c$. The theorem is thus proved. □

Corollary. *For convergence of series* (1) *of independent nonnegative random variables it is necessary that for any* $c > 0$ *and sufficient that for some* $c > 0$ *the following series*

$$\sum_{n=1}^{\infty} \mathsf{P}\{\xi_n > c\}, \quad \sum_{n=1}^{\infty} \mathsf{E}\xi'_n$$

will be convergent.

Indeed, for non-negative variables ξ_n we have $\mathsf{E}\xi'^2_n \leqslant c\mathsf{E}\xi'_n$ so that the convergence of series (10) follows from the convergence of series (9). □

Lévy obtained an interesting result which states that convergence in probability of a series of independent random variables implies convergence with probability 1.

To prove this assertion an inequality is required similar to the inequalities previously obtained for submartingales but without assuming the existence of any mathematical expectations.

Theorem 6. *Let* $\{\xi_n, n = 1, 2, \ldots\}$ *be independent random variables,* $\zeta_n = \xi_1 + \xi_2 + \ldots + \xi_n$, $\zeta_0 = 0$. *If* $\mathsf{P}\{|\zeta_n - \zeta_k| \leqslant t\} \geqslant \alpha$, $k = 0, 1, \ldots, n$, *then*

$$\mathsf{P}\{\max_{1 \leqslant k \leqslant n} |\zeta_k| > 2t\} \leqslant \frac{1-\alpha}{\alpha}. \tag{11}$$

Proof. We introduce the events $A_k = \{|\zeta_1| \leqslant 2t, \ldots, |\zeta_{k-1}| \leqslant 2t, |\zeta_k| > 2t\}$, $B_k = \{|\zeta_n - \zeta_k| \leqslant t\}$, $k = 1, \ldots, n$. Then

$$\{|\zeta_n| > t\} \supset \bigcup_{k=1}^{n} (A_k \cap B_k),$$

where the events A_k, $k = 1, 2, \ldots, n$ are pairwise disjoint and the events A_k, B_k (k fixed, $k = 1, \ldots, n$) are independent. Therefore

$$1 - a > \mathsf{P}\{|\zeta_n| > t\} \geqslant \mathsf{P}\left\{\bigcup_{k=1}^{n} (A_k \cap B_k)\right\} =$$

$$= \sum_{k=1}^{n} \mathsf{P}(A_k)\,\mathsf{P}(B_k) \geqslant \alpha \sum_{k=1}^{n} \mathsf{P}(A_k) = \alpha\ \mathsf{P}\{\max_{1\leqslant k\leqslant n} |\zeta_k| > 2t\},$$

which implies (11). □

Theorem 7. *If series* (1), *where* $\{\xi_n,\ n=1,2,\ldots\}$ *are mutually independent, converges in probability, then it also converges with probability* 1.

Let $A_{n,N}$ denote the event $\left\{\sup_{n',n''\geqslant n} |\zeta_{n'}-\zeta_{n''}| > \frac{1}{N}\right\}$. Series (1) diverges on the set $D = \bigcup_{N=1}^{\infty} \bigcap_{n=1}^{\infty} A_{n,N}$. We bound the probability of D. Let ε and η be arbitrary positive numbers. The convergence of series (1) in probability implies the existence of $n_0 = n_0(\varepsilon, \eta)$ such that $\mathsf{P}\{|\zeta_{n'} - \zeta_{n''}| > \varepsilon\} < \eta$ for $n', n'' > n_0$. Applying Theorem 6 to the variables $\zeta'_k = \zeta_k - \zeta_{n_0}$, $k > n_0$, we obtain

$$\mathsf{P}\{\max_{n_0\leqslant k\leqslant n'} |\zeta'_k| > 2\varepsilon\} = \mathsf{P}\{\max_{n_0\leqslant k\leqslant n'} |\zeta_k - \zeta_{n_0}| > 2\varepsilon\} \leqslant \frac{\eta}{1-\eta},$$

where n' is an arbitrary integer greater than n_0. Therefore

$$\mathsf{P}\{\sup_{n_0\leqslant k} |\zeta_k - \zeta_{n_0}| > 2\varepsilon\} \leqslant \frac{\eta}{1-\eta},$$

where η is arbitrarily small. It thus follows that

$$\mathsf{P}(A_{n,N}) \leqslant \frac{\eta}{1-\eta} \quad \text{and} \quad \mathsf{P}\left(\bigcap_{n=1}^{\infty} A_{n,N}\right) = 0$$

for any N. Therefore $P(D) = 0$ and the theorem is proved. □

Applications to the strong law of large numbers. Using a simple transformation, one can construct from the theorem on convergence of series with probability 1 theorems on the strong law of large numbers type (i.e. theorems on convergence with probability 1 of means of random variables).

Lemma 1. *If the series* $\sum_{n=1}^{\infty} z_n$ *converges and* $a_n > 0$, $a_n \to \infty$ *is a monotonically increasing sequence, then*

$$\frac{1}{a_n} \sum_{k=1}^{n} a_k z_k \to 0.$$

Proof. Let

$$S_n = \sum_{k=1}^{n} z_k \quad (S_0 = 0), \quad |S_n| \leqslant c,$$

where c is a constant. Set $a_k - a_{k-1} = \Delta_k$, $k = 1, 2, \ldots, a_0 = 0$. Then

$$\sum_{k=1}^{n} a_k z_k = \sum_{k=1}^{n} (\Delta_1 + \Delta_2 + \ldots + \Delta_k) z_k = \sum_{k=1}^{n} \Delta_k (S_n - S_{k-1}).$$

Therefore

$$\left|\frac{1}{a_n}\sum_{k=1}^{n} a_k z_k\right| \leqslant \left|\frac{1}{a_n}\sum_{k=1}^{n_0} \Delta_k (S_n - S_{k-1})\right| + \sup_{n_0 \leqslant k \leqslant n} |S_n - S_{k-1}| \leqslant$$

$$\leqslant 2C \frac{a_{n_0}}{a_n} + \sup_{n_0 \leqslant k \leqslant n} |S_n - S_{k-1}| < \varepsilon$$

for any $\varepsilon > 0$ provided n and n_0 are chosen sufficiently large.

From the lemma just proved and the theorems presented in the previous subsection the following assertions follow.

Theorem 8. a) *If* $\{\xi_n, n = 1, 2, \ldots\}$ *is an arbitrary sequence of random variables with finite moments of the first order and*

$$\sum_{n=1}^{\infty} \frac{1}{n} \mathsf{E}|\xi_n - a_n| < \infty, \quad a_n = \mathsf{E}\xi_n,$$

then

$$\lim_{n\to\infty} \frac{1}{n} \sum_{k=1}^{n} (\xi_k - a_k) = 0$$

with probability 1.

b) *If* $\{\xi_n, n = 1, 2, \ldots\}$ *is a sequence such that the partial sums* $\{\zeta_n = \xi_1 + \xi_2 + \ldots + \xi_n\}$ *form a martingale and for some* $p \geqslant 1$

$$\sup_n \mathsf{E}\left|\sum_{1}^{n} \frac{1}{k} \xi_k\right|^p < \infty,$$

then with probability 1

$$\lim_{n\to\infty} \frac{1}{n} \sum_{k=1}^{n} \xi_k = 0.$$

c) *If* $\{\xi_n, n = 1, 2, \ldots\}$ *are independent and*

$$\sum_{n=1}^{\infty} \frac{1}{n^2} \mathsf{V}\xi_n < \infty,$$

then

$$\lim_{n\to\infty} \frac{1}{n} \sum_{k=1}^{n} (\xi_k - \mathsf{E}\xi_k) = 0$$

with probability 1.

For identically distributed random variables ξ_n more powerful results will be obtained later as corollaries of general ergodic theorems.

§ 4. Markov Chains

Generalizing the notion of a random walk one can arrive at the much broader notion of a Markov chain (Markov process) which plays an important role in the theory of random processes. Before presenting the formal definition we shall discuss a simple but quite general model resulting in a Markov chain.

Systems with random effects. Assume that a stochastic system Σ is considered such that the states of this system are represented by points of a certain measurable space $\{\mathscr{X}, \mathfrak{B}\}$. Assume that the transition of the system from the state $\xi(t)$ attained at time t to a new state at time $t+1$ is completely determined by the value of t, the state $\xi(t)$ and some random factor α_t which is independent of the state of the system Σ up to time t inclusive and which forms a process with independent values in time. Thus

$$\xi(t+1) = f(t, \xi(t), \alpha_t), \tag{1}$$

where $f(t, x, \alpha)$ is a function of arguments $t \in T$, $x \in \mathscr{X}$, $\alpha \in A$, where A is a measurable space. Using formula (1) one can express the state of the system Σ at any instant of time s starting from the state $\xi(t)$ $(t < s)$:

$$\xi(s) = g_{t,s}(\xi(t), \alpha_t, \alpha_{t+1}, \dots, \alpha_{s-1}). \tag{2}$$

If at the initial time $t=0$, $\xi(0)$ is independent of the sequence $\{\alpha_t, t \in T\}$ then $\xi(t)$ is independent of the sequence $\{\alpha_t, \alpha_{t+1}, \dots, \alpha_n, \dots\}$.

Let $\{\Omega, \mathfrak{S}, \mathsf{P}\}$ be a probability space on which the random elements α_t are defined. Assume that for any fixed t and $s (s > t)$ the function $g_{t,s}(x, \alpha_t, \alpha_{t+1}, \dots, \alpha_{s-1})$ is $\mathfrak{B} \times \mathfrak{S}$-measurable. Then if the motion of the system Σ starts at the time t and the state of the system $\xi(t) = x$ is known then formula (2) allows us to determine the probability that the system Σ will find itself in an arbitrary set $A \in \mathfrak{B}$ at the instant of time $s > t$. This probability is called the *transition probability* and is denoted by $\mathsf{P}\{t, x, s, A\}$. If $\chi_A(x)$ denotes the indicator of the set A, then

$$\mathsf{P}\{t, x, s, A\} = \mathsf{E}\chi_A\{g_{t,s}(x, \alpha_t, \dots, \alpha_{s-1})\}.$$

Let $t<u<v$. The equalities

$$\mathsf{P}\{t, x, v, A\}=\mathsf{E}\chi_A[g_{u,v}(\xi(u), \alpha_u, \ldots, \alpha_{v-1})]=$$
$$=\mathsf{E}\,[\{\mathsf{E}\chi_A[g_{u,v}(y, \alpha_u, \ldots, \alpha_{v-1})]\}_{y=\xi(u)}]=\mathsf{E}\mathsf{P}\{u, \xi(u), v, A\},$$

follow from formula (2) and the independence of the variables α_t, $\alpha_{t+1} \ldots \alpha_{v-1}$.

This can be denoted in the following manner:

$$\mathsf{P}(t, x, v, A)=\int \mathsf{P}(u, y, v, A)\, \mathsf{P}(t, x, u, dy),\ t<u<v. \tag{3}$$

Relation (3) is called the *Chapman-Kolmogorov equation.* It expresses an important property of the system under consideration – the absence of an after-effect: if the state of the system in a given instant of time u is known, then the transition probabilities from this state are independent of the behavior of the system in the instants of time preceeding u. Systems with such properties are called Markovian (or Markov). They often occur in various problems of the natural sciences and engineering. It follows from the definition of transition probabilities that for any non-negative $\mathfrak{B}$-measurable function $f(x)$ we have

$$\mathsf{E}f(g_{t,s}(x, \alpha_t, \alpha_{t+1}, \ldots, \alpha_{s-1}))=\int f(y)\, \mathsf{P}(t, x, s, dy).$$

Taking into account the independence of $\xi(t)$ from $\alpha_t, \alpha_{t+1}, \ldots, \alpha_{s-1}$ (Chapter I, §3) we thus obtain

$$\mathsf{E}f(g_{t,s}(\xi(t), \alpha_t, \ldots, \alpha_{s-1}))=\mathsf{E}\int f(y)\, \mathsf{P}(t, \xi(t), s, dy).$$

The last formula is easily generalized for arbitrary $\mathfrak{B}^m$-measurable non-negative functions $f(x_1, x_2, \ldots, x_n)$, $x_k \in \mathscr{X}$. Let $t_1<t_2<\ldots<t_m$. Then

$$\mathsf{E}f(\xi(t_1), \xi(t_2), \ldots, \xi(t_m))=$$
$$=\mathsf{E}f(\xi(t_1), \ldots, \xi(t_{m-1}), g_{t_{m-1}t_m}(\xi(t_{m-1}), \alpha_{t_{m-1}} \ldots \alpha_{t_m-1}))=$$
$$=\mathsf{E}\int f(\xi(t_1), \ldots, \xi(t_{m-1}), y_m)\, \mathsf{P}(t_{m-1}, \xi(t_{m-1}), t_m, dy_m)=$$
$$=\mathsf{E}\int\int f(\xi(t_1), \ldots, \xi(t_{m-2}), y_{m-1}, y_m)\times$$
$$\times \mathsf{P}(t_{m-1}, y_{m-1}, t_m, dy_m)\, \mathsf{P}(t_{m-2}, \xi(t_{m-2}), t_{m-1}, dy_{m-1})=\ldots.$$

Thus

$$\mathsf{E}f(\xi(t_1), \xi(t_2), \ldots, \xi(t_m)) = \mathsf{E} \int \mathsf{P}(t_1, \xi(t_1), t_2, dy_2) \times$$
$$\times \int \mathsf{P}(t_2, y_2, t_3, dy_3) \ldots \int \mathsf{P}(t_{m-1}, y_{m-1}, t_m, dy_m)\, f(\xi(t_1), y_2, \ldots, y_m). \tag{4}$$

If we assume that the initial state of the system is non-random, $\xi(0)=x$, applying formula (4) to sequence $\xi(1), \xi(2), \ldots, \xi(n)$ and to function $f(x_1, x_2, \ldots, x_n) = \chi_{B^{(n)}}$ where $B^{(n)}$ is an arbitrary set in $\mathfrak{B}^n$ we obtain a family of marginal distributions $\{\mathsf{P}^{(x)}_{1,2,\ldots,n}(\cdot), n=1, 2, \ldots\}$ defined by formula

$$\mathsf{P}^{(x)}_{1,2,\ldots,n}(B^{(n)}) = \int \cdots \int_{B(n)} \mathsf{P}_1(x, dy_1)\, \mathsf{P}_2(y_1, dy_2) \ldots \mathsf{P}_n(y_{n-1}, dy_n), \tag{5}$$

where $\mathsf{P}_k(x, A) = \mathsf{P}(k-1, x, k, A)$ is the one-step transition probability. In the case when the initial state of the system $\xi(0)$ has an arbitrary distribution m (where m is a probability measure on $\mathfrak{B}$) we obtain from (4), in place of (5), the following system of marginal distributions:

$$\mathsf{P}^{(m)}_{0,1,\ldots,n}(B^{(n+1)}) =$$
$$= \int \cdots \int_{B(n+1)} m(dx)\, \mathsf{P}_1(x, dy_1)\, \mathsf{P}_2(y_1, dy_2) \ldots \mathsf{P}_n(y_{n-1}, dy_n). \tag{6}$$

Moreover

$$\mathsf{P}^{(m)}_{0,1,\ldots,n}(B^{(n+1)}) = \mathsf{P}\{(\xi(0), \xi(1), \ldots, \xi(n)) \in B^{(n+1)}\}.$$

Formula (6) may be used as the basis of a general definition of a Markov chain. However, one should first analyze the meaning of the integrals of type (6) when the family of measures $\mathsf{P}_k(x, A)$ is defined independently with no connection to the auxiliary variables α_t and functions $f(t, x, \alpha)$.

Stochastic kernels. Let two measurable spaces $\{\mathscr{X}, \mathfrak{A}\}$ and $\{\mathscr{Y}, \mathfrak{B}\}$ be given.

Definition 1. *A stochastic kernel* on $\{\mathscr{X}, \mathfrak{B}\}$ is a function $\mathsf{P}(x, B)$ ($x \in \mathscr{X}$, $B \in \mathfrak{B}$) satisfying the following conditions:

a) for a fixed x the function $\mathsf{P}(x, \cdot)$ is a probability measure of $\mathfrak{B}$.

b) for a fixed B the function $\mathsf{P}(\cdot, B)$ is $\mathfrak{A}$-measurable. If $\mathsf{P}(x, \cdot)$ is a measure and $\mathsf{P}(x, \mathscr{Y}) \leqslant 1$, then $\mathsf{P}(x, B)$ is called a *semistochastic kernel.*

Lemma 1. *Let $f(x, y)$ be a non-negative $\sigma\{\mathfrak{A} \times \mathfrak{B}\}$ measurable function and $\mathsf{P}(\cdot, \cdot)$ be a stochastic (semistochastic) kernel on $\{\mathscr{X}, \mathfrak{B}\}$. Then the* *function*

$$g(x)=\int_{\mathscr{X}} f(x, y)\, \mathsf{P}(x, dy)$$

is $\mathfrak{A}$-measurable.

Proof. For a fixed x the function $f(x, \cdot)$ is $\mathfrak{B}$-measurable so that the integral appearing in the r.h.s. of the equality is meaningful. Classes of the non-negative functions $f(x, \cdot)$ for which the lemma is valid are cones, and, by virtue of Lebesgue's theorem, are monotone classes. The K class of functions is called monotone if for $0 \leqslant f_1 \leqslant f_2 \leqslant \ldots$, $f_n \in K$ $\lim f_n \in K$. Since it contains indicators of the sets of the type $A \times B$, where $A \in \mathfrak{A}$ and $B \in \mathfrak{B}$, it therefore contains all the non-negative $\sigma\{\mathfrak{A} \times \mathfrak{B}\}$-measurable functions as well. □

The following assertion may be considered as a generalization of the well-known Fubini's theorem.

Theorem 1. Let $\{\mathscr{X}, \mathfrak{A}\}$, $\{\mathscr{Y}, \mathfrak{B}\}$, $\{\mathscr{Z}, \mathfrak{C}\}$ be measurable spaces, $Q_1(x, B)$, $Q_2(y, C)$ be stochastic (or semistochastic) kernels on $\{\mathscr{X}, \mathfrak{B}\}$, $\{\mathscr{Y}, \mathfrak{C}\}$ respectively. There exists a unique stochastic (semistochastic) kernel $Q_3(x, D)$ on $\{\mathscr{X}, \sigma\{\mathfrak{B} \times \mathfrak{C}\}\}$ such that

$$Q_3(x, B \times C)=\int_B Q_1(x, dy)\, Q_2(y, C). \tag{7}$$

Moreover for an arbitrary non-negative $\sigma\{\mathfrak{B} \times \mathfrak{C}\}$-measurable function $f(y, z)$ we have

$$\int_{\mathscr{Y} \times \mathscr{Z}} f(y, z)\, Q_3(x, dy \times dz)=\int_{\mathscr{Y}}\left(\int_{\mathscr{Z}} f(y, z)\, Q_2(y, dz)\right) Q_1(x, dy). \tag{8}$$

To prove the first part of the theorem it is sufficient to show that formula (7) determines an elementary measure on the semi-ring of rectangles in the space $\mathscr{Y} \times \mathscr{Z}$. Let $D_1=B_1 \times C_1$, $D_2=B_2 \times C_2$ and $D_2 \subset D_1$. Then $B_2 \subset B_1$, $C_2 \subset C_1$ and $D_1=D_2 \cup D' \cup D''$, where $D'=B_2 \times (C_1 \backslash C_2)$ and $D''=(B_1 \backslash B_2) \times C_1$.

The sets D_2, D' and D'' are pairwise disjoint. If we apply formula (7) repeatedly to the sets D_2, D' and D'', we obtain

$$Q_3(x, D_2)+Q_3(x, D')+Q_3(x, D'')=$$
$$=\int_{B_1} Q_1(x, dy)\, Q_2(y, C_2)+\int_{B_2} Q_1(x, dy)\, Q_2(y_1, C_1 \backslash C_2)+$$

$$+\int_{B_1\setminus B_2} Q_1(x, dy)\, Q_2(y, C) = \int_{B_1} Q_1(x, dy)\, Q_2(y, C) = Q_3(x, D_1).$$

Thus the function $Q_3(x, D)$ is additive on these special subdivisions of the set D_3. In particular if $D_3 = D_1 \cup D_2$, where D_i are rectangles and $D_1 \cap D_2 = \emptyset$ then $Q_3(x, D_1) + Q_3(x, D_2) = Q_3(x, D_1)$ and $Q_3(x, \mathscr{Y} \times \mathscr{Z}) = 1$. (If Q_1 and Q_2 are semistochastic kernels, then $Q_3(x, \mathscr{Y} \times \mathscr{Z}) \leqslant 1$). The additivity of the function $Q_3(x, \cdot)$ on the semi-ring of all rectangles in the general case can easily be obtained using the induction argument.

Let $D = \bigcup_{k=1}^{n} D_k$, where D_k are pairwise disjoint rectangles. Then $D \setminus D_n = D' \cup D'' = \bigcup_{k=1}^{n-1} D_k$, where D' and D'' are determined by the preceeding formulas. As it has already been shown

$$Q_3(x, D) = Q_3(x, D_n) + Q_3(x, D') + Q_3(x, D'').$$

Using the induction assumption we obtain

$$Q_3(x, D') = Q_3\left(x, D' \cap \left(\bigcup_{k=1}^{n-1} D_k\right)\right) = \sum_{k=1}^{n-1} Q_3(x, D' \cap D_k)$$

and an analogous expression for $Q_3(x, D'')$. Therefore

$$Q_3(x, D) = Q_3(x, D_n) + \sum_{k=1}^{n-1} \left[Q_3(x, D' \cap D_k) + Q_3(x, D'' \cap D_k)\right].$$

Since D' and D'' are disjoint rectangles jointly covering D_k, it follows that $D' \cap D_k$ and $D'' \cap D_k$ are also rectangles and $(D' \cap D_k) \cup (D'' \cap D_k) = D_k$. Therefore: $Q_3(x, D' \cap D_k) + Q_3(x, D'' \cap D_k) = Q_3(x, D_k)$, and hence

$$Q_3(x, D) = \sum_{k=1}^{n} Q_3(x, D_k).$$

We have thus proved the additivity of $Q_3(x, \cdot)$. We now verify the property of countable semi-additivity for $Q_3(x, \cdot)$. Let $D_0 \subseteq \bigcup_{1}^{\infty} D_k$, $D_k = B_k \times C_k$, $k = 0, 1, \ldots$. Then

$$\chi_{D_0}(y, z) \leqslant \sum_{k=1}^{\infty} \chi_{D_k}(y, z).$$

Since

$$\chi_{D_k}(y, z) = \chi_{B_k}(y)\, \chi_{C_k}(z),$$

it follows that

$$\chi_{B_0}(y)\,\chi_{C_0}(z) \leqslant \sum_{k=1}^{\infty} \chi_{B_k}(y)\,\chi_{C_k}(z)$$

Integrating both sides of this inequality with respect to measure $Q_2(y,\cdot)$ over the space $\mathscr{Z}$, we obtain

$$\chi_{B_0}(y)\,Q_2(y, C_0) \leqslant \sum_{k=1}^{\infty} \chi_{B_k}(y)\,Q_2(y, C_k).$$

Once more integrating the relation obtained with respect to the measure $Q_1(x,\cdot)$ over the space $\mathscr{Y}$, we arrive at inequality

$$Q_3(x, D_0) \leqslant \sum_{k=1}^{\infty} Q_3(x, D_k),$$

which shows that $Q_3(x, D_k)$ is countably semiadditive. From here it follows that $Q_3(x, B\times C)$ admits a unique extension on $\sigma\{\mathfrak{B}\times\mathfrak{C}\}$. In order to prove formula (8) we first note that in view of the preceeding lemma the inner integral on the right-hand-side of (8) is a $\mathfrak{B}$-measurable function, so that the double integral in the r.h.s. of (8) is meaningful. Next, the class of functions f $(f\geqslant 0)$ for which formula (8) is valid is a cone and a monotone class. Moreover, in view of formula (7) it contains the indicators of the rectangles. Therefore it contains all the $\sigma\{\mathfrak{B}\times\mathfrak{C}\}$-measurable non-negative functions. The theorem is thus proved. □

In the same manner one can prove the following theorem.

Theorem 2. *Let* $\{\mathscr{X},\mathfrak{A}\}, \{\mathscr{Y}_1,\mathfrak{B}_1\},\dots \{\mathscr{Y}_s,\mathfrak{B}_s\}$ *be measurable spaces and* $Q_1(x, B^{(1)}), Q_2(y_1, B^{(2)}),\dots, Q_s(y_{s-1}, B^{(s)})$ *be stochastic (semistochastic) kernels,* $y_k\in\mathscr{Y}_k$, $B^{(k)}=\mathfrak{B}_k$ $(k=1,\dots,s)$. *There exists a unique stochastic (semistochastic) kernel* $Q^{(1,s)}(x, D)$ *on* $\{\mathscr{X},\mathfrak{D}\}$ where $\mathfrak{D}=\sigma\{\mathfrak{B}_1\times\times\mathfrak{B}_2\times\dots\times\mathfrak{B}_s\}$ such that

$$Q^{(1,s)}(x, B^{(1)}\times\dots\times B^{(s)}) = \int_{B^{(1)}} Q_1(x, dy_1)\int_{B^{(2)}} Q_2(y_1, dy_2)\dots \int_{B^{(s-1)}} Q_s(y_{s-1}, B^{(s)})\,Q_{s-1}(y_{s-2}, dy_{s-1}). \quad (9)$$

Moreover for an arbitrary non-negative $\mathfrak{D}$*-measurable function* $f(y_1,\dots,y_n)$

$$\int_{\mathscr{Y}_1\times\dots\times\mathscr{Y}_s} f(y_1,\dots,y_s)\,Q^{(1,s)}(x, dy_1\times\dots\times dy_s) =$$

$$\int_{\mathscr{Y}_1} Q_1(x, dy_1) \ldots \int_{\mathscr{Y}_s} f(y_1, \ldots, y_s)\, Q_s(y_{s-1}, dy_s). \tag{10}$$

Remark. Formulas (8) and (10) were proved for non-negative functions. They are obviously valid for arbitrary f also provided only one of the functions f^+ of f^- is integrable. An analogous situation holds also in other theorems where for the sake of brevity only non-negative functions are mentioned.

The kernel $Q^{(1,s)}$ is called the *direct product* of kernels $Q_1, Q_2, \ldots, Q_s$ and is denoted as $Q^{(1,s)} = Q_1 \times Q_2 \times \ldots \times Q_s$.

If in (9) we put $B^{(1)} = \mathscr{Y}_1, B^{(2)} = \mathscr{Y}_2, \ldots, B^{(s-1)} = \mathscr{Y}_{s-1}$ then a new probability kernel in $\{\mathscr{X}, \mathfrak{B}_s\}$ is obtained:

$$Q^{*(1,s)}(x, B^{(s)}) = Q^{(1,s)}(x, \mathscr{Y}_1 \times \mathscr{Y}_2 \times \ldots \times \mathscr{Y}_{s-1} \times B^{(s)}). \tag{11}$$

This kernel is called the *convolution* of *kernels* and is denoted as

$$Q^{*(1,s)} = Q_1 * Q_2 * \ldots * Q_s.$$

We now apply formula (10) to function $f(y_1, y_2, \ldots, y_s) = f(y_s) = \chi_{B^{(s)}}(y^{(s)})$ and compare it with (11). We thus obtain:

$$\int_{\mathscr{Y}_1 \times \mathscr{Y}_2 \times \ldots \times \mathscr{Y}_s} f(y_s)\, Q^{(1,s)}(x, dy_1 \times dy_2 \times \ldots \times dy_s) = \int_{\mathscr{Y}_s} f(y_s)\, Q^{*(1,s)}(x, dy_s). \tag{12}$$

Since the class of non-negative functions for which formula (12) is valid is a cone and a monotone class, (12) is valid for an arbitrary non-negative $\mathfrak{B}_s$-measurable function. In turn, it follows from here that for an arbitrary non-negative $\sigma\{\mathfrak{B}_{m_1} \times \mathfrak{B}_{m_2} \times \ldots \times \mathfrak{B}_{m_r} \times \mathfrak{B}_s\}$-measurable function of $r+1$ variables of the form

$$f(y_{m_1}, y_{m_2}, \ldots, y_{m_r}, y_s), \quad (y_m \in \mathscr{Y}_m, 0 \leqslant m_1 < m_2 < \ldots < m_r < s)$$

we have

$$\int_{\mathscr{Y}_{m_1} \times \mathscr{Y}_{m_2} \times \ldots \times \mathscr{Y}_{m_r} \times \mathscr{Y}_s} f(y_{m_1}, y_{m_2}, \ldots, y_{m_r}, y_s) \times$$

$$\times\; Q^{(1,s)}(x, dy_{m_1} \times dy_{m_2} \times \ldots \times dy_s) = \int_{\mathscr{Y}_{m_1}} Q^{*(1,m_1)}(x, dy_{m_1}) \times$$

$$\times \int_{\mathscr{Y}_{m_2}} Q^{*(m_1+1,m_2)}(y_{m_1}, dy_{m_2}) \dots \int_{\mathscr{Y}_s} f(y_{m_1}, \dots, y_{m_r}, y_s) \times$$
$$\times Q^{*(m_r+1,s)}(y_{m_r}, dy_s). \tag{13}$$

A particular case of formula (13) is the relation

$$Q^{*(1,s)} = Q^{*(1,m_1)} * Q^{*(m_1+1,m_2)} * \dots * Q^{*(m_r+1,s)},$$

which shows that the convolution operation is associative.

Consider infinite products of stochastic kernels. Let $\{\mathscr{X}_n, \mathfrak{B}_n\}$ $n=0, 1, 2, \dots$, be an infinite sequence of measurable spaces and $\mathsf{P}_n(\cdot, \cdot)$, $n=1, 2, \dots$ be a sequence of stochastic kernels defined on $\{\mathscr{X}_{n-1}, \mathfrak{B}_n\}$. In accordance with Theorem 2 we construct direct products of kernels

$$\mathsf{P}^{(1,n)} = \mathsf{P}_1 \times \mathsf{P}_2 \times \dots \times \mathsf{P}_n,$$

$$\mathsf{P}^{(1,n)} = \mathsf{P}^{(1,n)}(x_0, D), \quad x_0 \in \mathscr{X}_0, \quad D \in \mathfrak{C}_n,$$

where $\mathfrak{C}_n$ is the minimal σ-algebra containing rectangles $B_1 \times B_2 \times \dots \times B_n$ $(B_k \in \mathfrak{B}_k)$, $\mathfrak{C}_n = \sigma\{\mathfrak{B}_1 \times \mathfrak{B}_2 \times \dots \times \mathfrak{B}_n\}$.

We introduce the space $\mathscr{X}^\infty = \prod_{n=1}^{\infty} \mathscr{X}_n$, with the elements being the infinite sequences $\omega = (x_1, x_2, \dots, x_n, \dots)$ $x_n \in \mathscr{X}_n$. Denote by $\mathfrak{C}^0$ the algebra of cylindrical sets in $\mathscr{X}^\infty$ and define on $\mathfrak{C}^0$ a family of set functions $\mathsf{P}^{(x_0)}$ depending on the parameter x_0 $(x_0 \in \mathscr{X}_0)$ as follows: if C is a cylindrical set

$$C = \{\omega : (x_0, x_1, \dots, x_n) \in D\}, \quad D \in \mathfrak{C}_n$$

we put

$$\mathsf{P}^{(x_0)}(C) = \mathsf{P}^{(1,n)}(x_0, D).$$

These set functions are uniquely defined. Indeed, if

$$C = \{\omega : (x_0, x_1, \dots, x_n) \in D'\}, \quad D' \in \mathfrak{C}_{n'}$$

and if, for example, $n' > n$, then $D' = D \times \mathscr{X}_{n+1} \times \dots \times \mathscr{X}_{n'}$ and

$$\mathsf{P}^{(1,n')}(x_0, D') = \int \dots \int_{\mathscr{X}_1 \times \dots \times \mathscr{X}_{n'}} \mathsf{P}_1(x_0, dx_1)\, \mathsf{P}_2(x_1, dx_2) \dots$$
$$\dots \mathsf{P}_{n'}(x_{n-1}, dx_{n'})\, \chi_{D'}(x_1, \dots, x_{n'}),$$

where $\chi_{D'}(x_1, x_2, \dots, x_{n'})$ is the indicator of D'. Noting that $\chi_{D'}(x_1, \dots, x_{n'}) = \chi_D(x_1, \dots, x_n)$ and that $\mathsf{P}_k(x, \mathscr{X}_k) = 1$, it follows from the last expression that

$$\mathsf{P}^{(1,n')}(x_0, D') = \mathsf{P}^{(1,n)}(x_0, D).$$

The additivity of function $\mathsf{P}^{(x_0)}$ on $\mathfrak{C}_0$ is obvious.

Theorem 3. *There exists on $\{\mathscr{X}^\infty, \mathfrak{C}\}$, where $\mathfrak{C}$ is a σ-algebra generated by the cylinders of space $\mathscr{X}^\infty$ a unique family of measures $\mathsf{P}^{(x_0)}$ such that*

$$\mathsf{P}^{(x_0)}\{\omega: x_k \in B_k, k=1,\dots,n\}=$$
$$=\int_{B_1} \mathsf{P}_1(x_0, dx_1) \int_{B_2} \mathsf{P}_2(x_1, dx_2)\dots \int_{B_{n-1}} \mathsf{P}_{n-1}(x_{n-2}, dx_{n-1})\, \mathsf{P}_n(x_{n-1}, B_n).$$

Proof. It is sufficient to show that the measure $\mathsf{P}^{(x_0)}$ introduced on $\mathfrak{C}_0$ satisfies the continuity condition: for any monotonically decreasing sequence of cylindrical sets C_n such that $\bigcap_{n=1}^{\infty} C_n=\emptyset$ we have $\mathsf{P}^{(x_0)}(C_n)\to 0$. Assume the contrary: that $\mathsf{P}^{(x_0)}(C_n)\geqslant\varepsilon$ for some x_0; denote the bases of the cylindrical sets C_n by D_n, the indicator of D_n by $\chi(D_n, x_1, x_2, \dots, x_{m_n})=\chi(D_n)$ and let D_n be situated over the coordinates $(1, 2, \dots, m_n)$. Define the sequence of sets in $\mathfrak{B}_1$ by

$$B_n^{(1)}=\left\{x_1: \int_{\mathscr{X}^{(2,m_n)}} \chi(D_n; x_1, x_2, \dots, x_{m_n})\, \mathsf{P}^{(2,m_n)}(x_1, dx_2\times\dots\times dx_{m_n})>\frac{\varepsilon}{2}\right\},$$

where $\mathscr{X}^{(s,m)}$ denotes the product of the spaces $\mathscr{X}_s\times\mathscr{X}_{s+1}\times\dots\times\mathscr{X}_m$.

Since C_n are decreasing it follows that $B_n^{(1)}$ are also monotonically decreasing. Moreover if $\chi(B_n^{(1)})$ is the indicator of $B_n^{(1)}$ and $\bar{\chi}(B_n^{(1)})=1-\chi(B_n^{(1)})$, then

$$\varepsilon\leqslant \mathsf{P}^{(x_0)}(C_n)=\int_{\mathscr{X}_1}\int_{\mathscr{X}^{(2,m_n)}} (\chi(B_n^{(1)})+\bar{\chi}(B_n^{(1)}))\times$$
$$\times\chi(D_n)\, \mathsf{P}_1(x_0, dx_1)\, \mathsf{P}^{(2,m_n)}(x_1, dx_2\times\dots dx_{m_n})\leqslant$$
$$\leqslant \mathsf{P}_1(x_0, B_n^{(1)})+\frac{\varepsilon}{2}\int_{\mathscr{X}_1} \bar{\chi}(B_n^{(1)})\, \mathsf{P}_1(x_0, dx_1)\leqslant \mathsf{P}_1(x_0, B_n^{(1)})+\frac{\varepsilon}{2}.$$

Therefore $\mathsf{P}_1(x_0, B_n^{(1)})>\varepsilon/2$. Since $\mathsf{P}_1(x_0, \cdot)$ is a measure it follows that $\bigcap_{n=1}^{\infty} B_n^{(1)}=\emptyset$. Let $\bar{x}_1\in B_n^{(1)}$, $n=1, 2, \dots$. Then

$$\int_{\mathscr{X}^{(2,m_n)}} \chi(D_n; \bar{x}_1, x_2, \dots, x_{m_n})\, \mathsf{P}^{(2,m_n)}(\bar{x}_1, dx_2\times\dots\times dx_{m_n})>\frac{\varepsilon}{2}.$$

The above arguments can be applied to the kernel $\mathsf{P}^{(3,m_n)}(x_2, dx_3\times\dots\times dx_{m_n})$ and the measure $\mathsf{P}_2(\bar{x}_1, dx_2)$.

This will prove the existence of a point $\bar{x}_2$ such that for any D_n

$$\int_{\mathscr{X}^{(3,m_n)}} \chi(D_n; \bar{x}_1, \bar{x}_2, x_3, \dots, x_{m_n})\, \mathsf{P}^{(3,m_n)}(\bar{x}_2, dx_3\times\dots\times dx_{m_n})>\frac{\varepsilon}{4}.$$

We therefore construct a sequence $(\bar{x}_1, \bar{x}_2, \ldots, \bar{x}_n, \ldots)$ where $\bar{x}_n \in \mathscr{X}_n$ and such that for arbitrary s and D_n,

$$\int_{\mathscr{X}^{(s+1, m_n)}} \chi(D_n, \bar{x}_1, \bar{x}_2, \ldots, \bar{x}_s, x_{s+1}, \ldots, x_{m_n}) \times$$

$$\times \mathsf{P}^{(s+1, m_n)}(\bar{x}_s, dx_{s+1} \times \ldots \times dx_{m_n}) > \frac{\varepsilon}{2^s}.$$

Consider an arbitrary set C_k. Assume that its basis D_k is situated over the coordinates $(1, 2, \ldots, s)$. The last inequality shows that $(\bar{x}_1, \bar{x}_2, \ldots, \bar{x}_s) \in D_k$ (otherwise we would have $\chi(D_k, \bar{x}_1, \bar{x}_2, \ldots, \bar{x}_s, x_{s+1}, \ldots, x_{m_n}) \equiv 0$ for all $(x_{s+1}, \ldots, x_m)$). Therefore $(\bar{x}_1, \bar{x}_2, \ldots, \bar{x}_s, \ldots) \in C_k$ for any C_k and hence $\bigcap_{k=1}^{\infty} C_k \neq \emptyset$ which contradicts the initial assumption. The theorem is thus proved. □

Corollary. *Let a countable sequence of probability spaces $\{\mathscr{X}_n, \mathfrak{B}_n, q_n\}$, $n=1, 2, \ldots$ be given. Let $\mathscr{X}^\infty$ be the space of all sequences $\omega = (x_1, x_2, \ldots, x_n, \ldots)$, $x_n \in \mathscr{X}_n$ and let $\mathfrak{C}$ be the σ-algebra generated by the cylindrical sets in $\mathscr{X}^\infty$. There exists on $\{\mathscr{X}^\infty, \mathfrak{C}\}$ a unique probability measure Q such that*

$$\mathsf{Q}\{\omega : x_k \in B_k, k=1, 2, \ldots, n\} = \prod_{k=1}^{n} q_k(B_k), \quad B_k \in \mathfrak{B}_k.$$

In other words, if a sequence of probability spaces $\{\mathscr{X}_n, \mathfrak{B}_n, q_n\}$ $n=1, 2, \ldots$ is given, then there always exists a probability space $\{\Omega, \mathfrak{S}, \mathsf{Q}\}$ and a sequence of mappings f_n of the space Ω into $\mathscr{X}_n$ such that the random elements $\xi_n = f_n(\omega)$ have the given distributions q_n on $\mathfrak{B}_n$ and $\{\xi_n, n=1, 2, \ldots\}$ are jointly independent.

Remark. The theorem just proved, unlike Kolmogorov's theorem (Chapter I, Section 4, Theorem 2) does not require any topological assumptions on the nature of the spaces $\mathscr{X}_n$. On the other hand it is less general than Kolmogorov's theorem since it applies only to a special construction of measures in the product space.

Definition of a Markov chain. Definition 2. A *Markov chain* with phase space $\{\mathscr{X}, \mathfrak{B}\}$ is a family of random processes with discrete time $t \in T_+$, depending on an arbitrary measure m on $\{\mathscr{X}, \mathfrak{B}\}$ which serves as a parameter, with marginal distributions defined by the formula

$$\mathsf{P}^{(m)}\{\xi(k) \in B_k, \quad k=0, 1, \ldots, n\} = \int_{B_0} m(dx) \int_{B_1} \mathsf{P}_1(x, dy_1) \ldots \int_{B_{n-1}} \mathsf{P}_n(y_{n-1}, B_n), \tag{14}$$

where $\{\mathsf{P}_t(x, B), t=1, 2, \ldots\}$ is a system of stochastic kernels on $\{\mathscr{X}, \mathfrak{B}\}$.

The stochastic kernels $\mathsf{P}_t(x, B)$ are called *one-step transition probabilities* and the measure m *the initial distribution* of the chain. Fixing the measure m, we obtain a random sequence with values in $\mathscr{X}$ which is called a *Markov process* corresponding to the initial distribution m.

The marginal distributions of this process are denoted by $\mathsf{P}^{(m)}_{t_1, t_2, \ldots t_n}$ and the operation of taking the mathematical expectation of a certain function of the process with respect to probability measure $\mathsf{P}^{(m)}$ will be denoted by the symbol E_m.

If measure m is concentrated at a fixed point x of the phase space we call x the initial state of the process and the marginal distributions, the measure in $\{\mathscr{X}^T, \mathfrak{C}\}$ and the mathematical expectation of a function of the process with respect to a corresponding measure and denoted by $\mathsf{P}^{(m)}_{t_1, t_2, \ldots, t_n}$, $\mathsf{P}^{(x)}$, and E_x respectively.

Put

$$\mathsf{P}(k, x, r, B) = \int_{\mathscr{X}} \mathsf{P}_{k+1}(x, dy_{k+1}) \int_{\mathscr{X}} \mathsf{P}_{k+2}(y_{k+1}, dy_{k+2}) \times \ldots$$

$$\ldots \times \int_{\mathscr{X}} \mathsf{P}_{r-1}(y_{r-2}, dy_{r-1}) \, \mathsf{P}_r(y_{r-1}, B).$$

From an analytical point of view $\mathsf{P}(k, \cdot, r, \cdot)$ is a stochastic kernel which is a convolution of the transition probabilities $\mathsf{P}_{k+1} * \mathsf{P}_{k+2} * \ldots * \mathsf{P}_r$. It is also called a transition probability. More precisely $\mathsf{P}(k, x, r, B)$ is the transition probability from the state x during the time interval (k, r) into set B. From the associativity of a convolution of kernels the equality

$$\mathsf{P}(k, x, s, B) = \int_{\mathscr{X}} \mathsf{P}(k, x, r, dy) \, \mathsf{P}(r, y, s, B), \; k < r < s, \tag{15}$$

follows, which is the Chapman-Kolmogorov equation and formula (13) yields

$$\mathsf{E}_m f(\xi(t_1), \xi(t_2), \ldots, \xi(t_s)) =$$

$$= \int m(dx) \int \mathsf{P}(0, x, t_1, dy_1) \int \mathsf{P}(t_1, y_1, t_2, dy_2) \times \ldots$$

$$\ldots \times \int f(y_1, y_2, \ldots, y_s) \, \mathsf{P}(t_{s-1}, y_{s-1}, t_s, dy_s). \tag{16}$$

We have thus obtained the same formulas as before (cf. (3) and (4)), but now they follow from the general definition of Markov chains. On the other hand the arguments presented earlier show that the random sequence $\{\xi(t), t\in T_+\}$ obtained by means of the recursive relationship

$$\xi(t+1)=f(t, \xi(t), \alpha_t), \quad t=0, 1, 2, \ldots,$$

where

$$\xi(0), \alpha_1, \alpha_2, \ldots, \alpha_n, \ldots$$

are jointly independent variables and $\xi(0)$ has an arbitrary distribution m on $\mathfrak{B}$ under minimal assumptions on the measurability of the function $f(t, \cdot, \cdot)$, forms a Markov chain.

Formula (16) allows us to render a more precise probabilistic meaning to the notion of transition probabilities. For this purpose we compute the conditional mathematical expectation of a non-negative function $f(\xi(s), \xi(s+1), \ldots, \xi(s+n))$ (here $f(y_0, y_1, \ldots, y_n)$ is a Borel function of $n+1$ variables) given σ-algebra $\mathfrak{F}_{[0,t]}$ $(t\leqslant s)$ generated by the variables $\xi(0), \xi(1), \ldots, \xi(t)$. The corresponding conditional mathematical expectation is denoted by Ψ. By definition, Ψ is a unique $\mathfrak{F}_{[0,t]}$-measurable random variable such that for any non-negative function $g(x_0, x_1, \ldots, x_t)$ the following equality

$$\mathsf{E}_m g(\xi(0), \xi(1), \ldots, \xi(t))\, f(\xi(s), \xi(s+1), \ldots, \xi(s+n))= \\ =\mathsf{E}_m g(\xi(0), \xi(1), \ldots, \xi(t))\, \Psi.$$

is fulfilled.

On the other hand it follows from (16) that

$$\mathsf{E}_m g(\xi(0), \xi(1), \ldots, \xi(t))\, f(\xi(s), \xi(s+1), \ldots, \xi(s+n))= \\ =\mathsf{E}_m g(\xi(0), \xi(1), \ldots, \xi(t))\, \hat{f},$$

where

$$\hat{f}=\hat{f}(\xi(t))=\int \mathsf{P}(t, \xi(t), s, dy_0)\int \mathsf{P}_{s+1}(y_0, dy_1)\times\ldots \\ \ldots\times\int f(y_0, y_1, \ldots, y_n)\,\mathsf{P}_{s+n}(y_{n-1}, dy_n).$$

Thus $\Psi=\hat{f}$.

The formula obtained leads to the following conclusions.

Theorem 4. *The conditional mathematical expectation of an arbitrary non-negative function $f(\xi(s), \xi(s+1), \ldots, \xi(s+n))$ given $\mathfrak{F}_{[0,t]}(t<s)$ does not depend on the initial distribution m, the transition probability preceding the instant of time t and the values $\xi(0), \xi(1), \ldots, \xi(t-1)$. It is given by the*

expression

$$\mathsf{E}_m\{f(\xi(s),\xi(s+1),\ldots,\xi(s+n)\mid\mathfrak{F}_{[0,t]}\}=$$
$$=\int \mathsf{P}(t,\xi(t),s,dy_0)\int \mathsf{P}_{s+1}(y_0,dy_1)\times\ldots$$
$$\ldots\times\int f(y_0,y_1,\ldots,y_n)\,\mathsf{P}_{s+n}(y_{n-1},dy_n). \tag{17}$$

The conditional distribution of the variables $\xi(s),\xi(s+1),\ldots,\xi(s+n)$ in $\{\mathcal{X}^{n+1},\mathfrak{B}^{n+1}\}$ given $\mathfrak{F}_{[0,t]}$ coincides with the direct product of the kernels

$$\mathsf{P}(t,\xi(t),s,\cdot),\mathsf{P}_{s+1}(\cdot,\cdot),\ldots,\mathsf{P}_{s+n}(\cdot,\cdot).$$

In particular, the transition probability $\mathsf{P}(t,\xi(t),s,B)$ coincides with the conditional probability of the system falling at time s into the set B if the states $\xi(0),\xi(1),\ldots,\xi(t)$ are known. This probability depends only on the state $\xi(t)$ at the last known instant of time and does not depend on the values of m $\xi(0),\xi(1),\ldots,\xi(t-1)$ or on the transition probabilities $\mathsf{P}_1(\cdot,\cdot),\mathsf{P}_2(\cdot,\cdot),\ldots,\mathsf{P}_t(\cdot,\cdot)$. This last property of a Markov chain is called, as mentioned above, the absence of an after-effect and it is the basic qualitative characteristic of a Markov chain.

Remark. Let a measurable space $\{\mathcal{X},\mathfrak{B}\}$ and a system of stochastic kernels $\mathsf{P}_n(x,B)$ $n=1,2,\ldots$ defined on it be given. Then there exists a Markov chain for which $\mathsf{P}_n(x,B)$ are the one-step transition probabilities. The proof of this assertion and the construction of the corresponding probability space is given in Theorem 3.

A Markov chain is called *homogeneous* if the one-step transition probabilities are independent of the time:

$$\mathsf{P}_t(x,B)=\mathsf{P}(x,B).$$

In this case the transition probabilities for the time interval (t,s) depend only on the length of this interval

$$\mathsf{P}(t,x,s,B)=\int_{\mathcal{X}}\mathsf{P}(x,dy_1)\int_{\mathcal{X}}\mathsf{P}(y_1,dy_2)\times\ldots$$
$$\ldots\times\int_{\mathcal{X}}\mathsf{P}(y_{s-1},B)\,\mathsf{P}(y_{s-2},dy_{s-1})=\mathsf{P}^{(s-t)}(x,B).$$

For a homogeneous chain the Chapman-Kolmogorov equation becomes

$$\mathsf{P}^{(s+m)}(x,B)=\int_{\mathcal{X}}\mathsf{P}^{(s)}(x,dy)\,\mathsf{P}^{(m)}(y,B).$$

Let a Markov chain be homogeneous. Formula (16) shows that

$$\mathsf{E}_m(f(\xi(s+1), \xi(s+2), \dots, \xi(s+n)) = \mathsf{E}_{m_s} f(\xi(1), \xi(2), \dots, \xi(n)), \quad (18)$$

where

$$m_s(B) = \int \mathsf{P}(0, x, s, B)\, m(dx) = \int \mathsf{P}^{(s)}(x, B)\, m(dx).$$

If quantity (18) does not depend on s for any function $f(\cdot)$, then a homogeneous Markov process corresponding to a given initial distribution m is called *stationary*. For stationarity of a process it is necessary and sufficient that the measure m satisfy condition

$$m(B) = \int \mathsf{P}^{(s)}(x, B)\, m(dx). \quad (19)$$

This condition is equivalent to a (seemingly) simpler condition

$$m(B) = \int \mathsf{P}(x, B)\, m(dx). \quad (20)$$

Indeed (20) is a particular case of (19). If, however (20) is satisfied, then

$$m(B) = \int \mathsf{P}(x, B) \int \mathsf{P}(y, dx)\, m(dy) =$$

$$= \int \mathsf{P}^{(2)}(y, B)\, m(dy) = \dots = \int \mathsf{P}^{(s)}(y_s, B)\, m(dy_s).$$

Probability measures m satisfying (19) are called *invariant* or, more explicitly, *invariant measures* corresponding to a given stochastic kernel.

Therefore if for a given stochastic kernel there exists an invariant probability measure, then there exists an initial distribution for a homogeneous Markov chain to which a stationary Markov process corresponds. The given kernel serves as the one-step transition probability for this process. If the invariant measure is unique then there exists a unique stationary process for the given chain.

Let $\mathfrak{F}_t$ denote the minimal σ-algebra with respect to which the variables $\xi(0), \xi(1), \dots, \xi(t)$ $(t=0, 1, 2, \dots)$ are measurable, let τ be a random time on $\{\mathfrak{F}_t, t=0, 1, \dots\}$ and Ω_τ the domain of definition of τ. Consider the following problem. Let $\xi(t)$ be a homogeneous Markov chain. How does the process $\xi_\tau(t) = \xi(t+\tau)$ behave on Ω_τ? It is natural to expect that under the hypothesis $\xi(\tau) = x$ the random process $\xi_\tau(t)$ behaves exactly in the same manner as the Markov process $\xi(t)$ under the hypothesis $\xi(0) = x$. We now state this assertion rigorously and prove it. (This property is called the strong Markov property).

Clearly, $\xi(t+\tau)$ is defined on Ω_τ and it follows from Lemma 5 of Section 1, Chapter I that $\xi(t+\tau)$ $(t\geqslant 0)$ is $\mathfrak{S}$-measurable. Put $\mathsf{P}^{(\tau)}(x, A)=$ $=\mathsf{P}^{(x)}\{\Omega_\tau\cap(\xi(t)\in A)\}$. Then $\mathsf{P}^{(\tau)}(x, A)$ is a semistochastic kernel on $\{\mathscr{X}, \mathfrak{B}\}$. Indeed,

$$\mathsf{P}^{(\tau)}(x, A)=\sum_{s=1}^{\infty} \mathsf{P}^{(x)}\{[\tau=s]\cap[\xi(s)\in A]\}.$$

From here it follows immediately that $\mathsf{P}^{(\tau)}(x, A)$ is a measure on $\mathfrak{B}$ and

$$\mathsf{P}^{(\tau)}(x, \mathscr{X})=\mathsf{P}^{(x)}\{\Omega_\tau\}\leqslant 1.$$

On the other hand there exists a set $B^{(s)}\in\mathfrak{B}^{(s)}$ such that the event $\{\tau=s\}$ is equivalent to the event $\{\xi(0), \xi(1), \ldots, \xi(s)\}\in B^{(s)}$. Therefore

$$\mathsf{P}^{(x)}\{[\tau=s]\cap[\xi(s)\in A]\}=\mathsf{P}^{(x)}\{(\xi(0), \xi(1), \ldots, \xi(s))\in B^{(s)}\cap A^{(s)}\},$$

where $A^{(s)}=\mathscr{X}\times\mathscr{X}\times\ldots\times\mathscr{X}\times A$ (the $(s-1)$-th factor equals $\mathscr{X}$), and it follows from the properties of semistochastic kernels that this probability, as well as $\mathsf{P}^{(\tau)}(x, A)$ are $\mathfrak{B}$-measurable functions.

Denote by $\mathfrak{F}_\tau$ the σ-algebra induced by the random time τ.

Theorem 5. If $D\in\mathfrak{F}_\tau$ and $D\subset\Omega_\tau$, then

$$\mathsf{P}^{(x)}\left\{D\cap\left(\bigcap_{k=1}^{r}[\xi(t_k+\tau)\in A_k]\right)\right\}=$$
$$=\int_{\mathscr{X}} \mathsf{P}^{(y)}\left(\bigcap_{k=1}^{r}[\xi(t_k)\in A_k]\right)\mathsf{P}^{(\tau)}(x, D, dy), \tag{21}$$

where

$$\mathsf{P}^{(\tau)}(x, D, A)=\mathsf{P}^{(x)}(D\cap[\xi(\tau)\in A]).$$

Proof. Since $D\subset\Omega_\tau$, it follows that

$$\mathsf{P}^{(x)}\left\{D\cap\left(\bigcap_{k=1}^{r}[\xi(t_k+\tau)\in A_k]\right)\right\}=\sum_{s=1}^{\infty}\mathsf{P}^{(x)}\left\{D_s\cap\left(\bigcap_{k=1}^{r}[\xi(t_n+\tau)\in A_k]\right)\right\},$$

where $D_s=D\cap[\tau=s]$. Let $\chi(D_s)$ be the indicator of event D_s. Taking into account the properties of conditional probabilities of a Markov chain (Theorem 4), we have

$$\mathsf{P}^{(x)}\left\{D_s\cap\left(\bigcap_{k=1}^{r}[\xi(t_k+\tau)\in A_k]\right)\right\}=$$
$$=\mathsf{E}_x\left\{\chi(D_s)\,\mathsf{P}^{(x)}\left(\bigcap_{k=1}^{r}[\xi(t_k+s)\in A_k]\mid\mathfrak{F}_s\right)\right\}=$$

$$=\mathsf{E}_x\left\{\chi(D_s)\,\mathsf{P}^{(\xi(s))}\left(\bigcap_{k=1}^{r}\,[\xi(t_k+s)\in A_k]\right)\right\}.$$

In view of the fact that the chain is homogeneous, the r.h.s. of the last equality is equal to

$$\int_{D_s}\mathsf{P}^{(\xi(s))}\left\{\bigcap_{k=1}^{r}\,[\xi(t_k)\in A_k)\right\}d\mathsf{P}^{(x)}=$$

$$=\int_{\mathscr{X}}\mathsf{P}^{(y)}\left\{\bigcap_{k=1}^{r}\,[\xi(t_k)\in A_k]\right\}\mathsf{P}(s,x,D,dy),\qquad(22)$$

where $\mathsf{P}(s, x, D, \cdot)$ is a measure defined on $\{\mathscr{X}, \mathfrak{B}\}$ by the relation

$$\mathsf{P}(s,x,D,A)=\mathsf{P}^{(x)}\{D\cap[\tau=s]\cap[\xi(s)\in A]\}.$$

If we now introduce measure

$$\mathsf{P}^{(\tau)}(x,D,A)=\mathsf{P}^{(x)}\{D\cap[\xi(\tau)\in A]\}=\sum_{s=1}^{\infty}\mathsf{P}(s,x,D,A)$$

and sum up equation (22) with respect to s, we obtain the required assertion. □

§5. Markov Chains with a Countable Number of States

Reducibility and irreducibility. Let X be a finite or a countable set. In this case the set of all subsets of X will always be considered as the σ-algebra of the measurable sets of X. Here arbitrary functions on X are found to be measurable.

Points of the space X will be denoted by the letters $i, j, \ldots$. Consider a homogeneous Markov chain with values in X. It is defined by the one-step transition probabilities $p(i, j)$, $i, j\in X$ into singletons $\{j\}$. The one-step transition probability into an arbitrary set B is expressed in terms of $p(i, j)$ by the formula

$$P(i,B)=\sum_{j\in B}p(i,j)$$

and integration with respect to the measure corresponding to the stochastic kernel $\mathsf{P}(i, B)$ becomes summation

$$\int_X f(j)\,P(i,dj)=\sum_{j\in X}p(i,j)\,f(j).$$

The expression for the n-step transition probabilities into singletons j becomes

$$p^{(n)}(i,j)=\sum_{j_1,j_2,\ldots,j_{n-1}\in X} p(i,j_1)\,p(j_1,j_2)\ldots p(j_{n-1},j). \tag{1}$$

Introducing matrix $P^{(n)}$ (with a finite or infinite number of rows) whose elements are the n-step transition probabilities $P^{(n)}=\{p^{(n)}(i,j)\}_{i,j\in X}$, we obtain from formula (1) that

$$P^{(n)}=P^n,$$

where P^n is the n-th power of the matrix $P=P^{(1)}$ which is the matrix of the one-step transition probabilities. The matrix $P=\{p(i,j)\}$ has the properties

$$\text{a) } p(i,j)\geqslant 0, \qquad \text{b) } \sum_{j\in X} p(i,j)=1. \tag{2}$$

A matrix P possessing properties a) and b) is called a *stochastic* matrix. It follows from the equality $P^{n+m}=P^nP^m$ that

$$p^{(n+m)}(i,j)=\sum_{k\in X} p^{(n)}(i,k)\,p^{(m)}(k,j). \tag{3}$$

On the other hand, formula (3) is the Chapman-Kolmogorov equation ((15), Section 4) for this particular case.

Definition 1. The state $j\in X$ is *accessible* from the state i if the transition probability from i to j in a number of steps is positive. If j is accessible from i and i is accessible from j then the states i and j are called *communicative*. By definition the state i always communicates with i.

The fact that i and j are communicative states is denoted symbolically by $i\leftrightarrow j$. If j is accessible from i and k from j, then k is accessible from i. This follows from the inequality $p^{(n+m)}(i,k)\geqslant p^{(m)}(i,j)\,p^{(m)}(j,k)$. The relation $\leftrightarrow$ is an equivalence relation:

a) $i\leftrightarrow i$;

b) if $i\leftrightarrow j$, then $j\leftrightarrow i$;

c) if $i\leftrightarrow j$ and $j\leftrightarrow k$, then $i\leftrightarrow k$.

Indeed, a) follows from the fact that $p^{(0)}(i,i)=1$, b) follows from the symmetry of i and j in the definition of communicative states and finally c) follows from the fact that

$$p^{(n+m)}(i,k)\geqslant p^{(n)}(i,j)\,p^{(m)}(j,k)>0,$$
$$p^{(n_1+m_1)}(k,i)\geqslant p^{(n_1)}(k,j)\,p^{(m_1)}(j,i)>0,$$

if

$$p^{(n)}(i,j)>0,\; p^{(m_1)}(j,i)>0;\; p^{(m)}(j,k)>0,\; p^{(n_1)}(k,j)>0.$$

An arbitrary Markov chain can be decomposed into disjoint classes X_α of communicative states. This decomposition may be carried out as follows. Choose an arbitrary state i_1 and denote by X_{i_1} the totality of all states which communicate with i_1. It follows from property c) of the relation "$\leftrightarrow$" that any pair of states in X_{i_1} are communicative. If X_{i_1} does not exhaust X we choose a state $i_2 \notin X_{i_1}$ and construct the class X_{i_2} analogously. Since i_1 and i_2 do not communicate the classes X_{i_1} and X_{i_2} are disjoint. We continue the construction of the sets X_{i_k} until the whole space X is exhausted. The classes X_α so constructed possess the following properties:

1) the number of classes X_α is at most countable
2) each element of X occurs in one and only one class X_α.
3) each pair of states in X_α are communicative
4) any pair of states belonging to the different classes do not communicate.

The last two properties can also be stated as follows: when given an arbitrary state i of a given class X_α it is possible to reach with positive probability in a certain number of steps any other state of the same class. It may also occur that a system in a given class may leave it eventually but the probability that having left the class it will ever return is zero.

Definition 2. A Markov chain is called *irreducible* if it consists only of one class of communicative states. If any state j accessible from i communicates with i, then the state i is called *essential*. Otherwise it is called *unessential**.

It is easy to observe that only essential states are accessible from an essential state. Indeed, let i be essential and j be accessible from i. If k is accessible from j, then k is accessible from i and since i is essential i is accessible from k. But then j also is accessible from k, i.e. j is essential.

The following corollary is thus valid: *In a class of communicative states all the states* are either *essential* or *unessential*.

Recurrency. Let $\xi(n)$ be a state of a Markov system at the instant of time n. Denote by $\tau_j=\tau_j(n)$ the number of steps required by the system starting from time n to reach state j for the first time. Thus, $\tau_j(n)$ is determined by the string of relations

$$\xi(n+1)\neq j, \ldots, \xi(n+\tau_j-1)\neq j, \xi(n+\tau_j)=j.$$

We introduce a family of σ-algebras $\{\mathfrak{F}_{[n,t]}, t=0, 1, \ldots\}$ where $F_{[n,t]}$ is the minimal σ-algebra with respect to which the functions $\xi(n)$, $\xi(n+l)$, ..., $\xi(n+t)$ are measurable.

* *transient* using Feller's terminology. *Translators Remark.*

The variable $\tau_j(n)$ is a random time for this family. Put

$$f^{(s)}(i,j)=\mathsf{P}(\tau_j(n)=s \mid \xi(n)=i), \quad s=1,2,\ldots$$
$$f^{(0)}(i,j)=0.$$

Moreover

$$f^{(1)}(i,j)=p^{(1)}(i,j)=p(i,j).$$

Since the chain is homogeneous it follows that the probabilities $f^{(s)}(i,j)$ do not depend on n. For $i\neq j$ these probabilities are called the *probabilities of first passage* through state j and for $i=j$ the *first recurrence probabilities* into state i.

The sum

$$F(i,j)=\sum_{s-1}^{\infty} f^{(s)}(i,j), \quad i\neq j,$$

is the probability that the system leaving the state i will eventually reach the state j. Analogously, $F(i,i)$ is the probability that the system leaving the state i will return to this state in a finite number of steps. For $F(i,j)<1$ the r.v. τ_j is improper.

Definition 3. The state i is called *recurrent** if $F(i,i)=1$, and is called *non-recurrent* if $F(i,i)<1$.

It is easy to establish the connection between the transition probabilities and the probabilities of the first passage. These are given by relationship

$$p^{(n)}(i,j)=\sum_{s=1}^{n} f^{(s)}(i,j)\, p^{(n-s)}(j,j), \quad n\geqslant 1, \tag{4}$$

where $p^{(0)}(i,j)=\delta_{ij}$. Indeed, let τ_j be the (waiting) time up to the first passage through j starting from a certain initial moment. Then

$$P^{(n)}(i,j)=\mathsf{P}^{(i)}\left\{\bigcup_{s=1}^{n}[\tau_j=s]\cap[\xi(n)=j]\right\}=$$
$$=\sum_{s=1}^{n}\mathsf{P}^{(i)}\{[\tau_j=s]\cap[\xi(n)=j]\}=$$
$$=\sum \mathsf{P}^{(i)}\{\tau_j=s\}\,\mathsf{P}^{(i)}\{\xi(n)=j\mid\tau_j=s\}=\sum_{s=1}^{n} f^{(s)}(i,j)\,P^{(n-s)}(j,j).$$

Formula (4) is thus proved. We note its particular case:

$$p^{(n)}(i,i)=\sum_{s=1}^{n} f^{(s)}(i,i)\,p^{(n-s)}(i,i), \tag{5}$$

---

* *persistent* in Feller's terminology. *Translator's Remark.*

which can also be rewritten in the form

$$f^{(n)}(i,i)=p^{(n)}(i,i)-\sum_{s=1}^{n-1} f^{(s)}(i,i)\,p^{(n-s)}(i,i).$$

This last relationship allows us to calculate recursively the recurrence probabilities given the transition probabilities. Note that in order to evaluate the recurrence probability into state i it is sufficient to know the transition probabilities into that state only.

We introduce generating functions $P_{ij}(z)$, $F_{ij}(z)$ of the sequences $\{p^{(n)}(i,j), n=0,1,2,\ldots\}$, $\{f^{(n)}(i,j), n=0,1,2,\ldots\}$:

$$P_{ij}(z)=\sum_{n=0}^{\infty} p^{(n)}(i,j)\,z^n, \qquad F_{ij}(z)=\sum_{n=0}^{\infty} f^{(n)}(i,j)\,z^n.$$

It follows from formula (5) that

$$\begin{aligned}P_{ii}(z)&=p^{(0)}(i,i)+\sum_{n=1}^{\infty}\sum_{k=1}^{n} f^{(k)}(i,i)\,z^k p^{(n-k)}(i,i)\,z^{n-k}=\\ &=1+\sum_{k=1}^{\infty}\sum_{n=k}^{\infty} f^{(k)}(i,i)\,z^k p^{(n-k)}(i,i)\,z^{n-k}=\\ &=1+\sum_{k=1}^{\infty} f^{(k)}(i,i)\,z^k P_{ii}(z)\end{aligned}$$

or that

$$P_{ii}(z)=1+P_{ii}(z)\,F_{ii}(z).$$

One can interchange the order of summation and integration in the above calculations since the series are absolutely convergent for $|z|<1$. The last formula can also be written in the form

$$P_{ii}(z)=\frac{1}{1-F_{ii}(z)}. \tag{6}$$

The following equality can be derived from (4) analogously:

$$P_{ij}(z)=P_{jj}(z)\,F_{ij}(z),\ i\neq j. \tag{7}$$

Let z be a real number and $z\uparrow 1$. The functions $P_{ii}(z)$ and $F_{ii}(z)$ are monotonically increasing functions and in view of Abel's theorem the limit $\lim_{z\uparrow 1} F_{ii}(z)$ exists and moreover $\lim_{z\uparrow 1} F_{ii}(z)=F_{ii}(1)=F(i,i)$. Set $\lim_{z\uparrow 1} P_{ii}(z)=G(i,i)=P_{ii}(1)$.

Using (6) we obtain the following:

Theorem 1. *The state i is recurrent if* $G(i, i)=\sum_{n=0}^{\infty} p^{(n)}(i, i)=\infty$ *and is non-recurrent if* $G(i, i)=\sum p^{(n)}(i, i)<\infty$. *In the non-recurrent case*

$$G(i, i)=\frac{1}{1-F(i, i)}.$$

Theorem 2. *If the states i and j are communicative then they are either both recurrent or both non-recurrent.*

Proof. Since $i \leftrightarrow j$, one can find m_1 and m_2 such that $p^{(m_1)}(i, j)>0$, $p^{(m_2)}(j, i)>0$. Since

$$p^{(m_1+m_2+n)}(j, j) \geqslant p^{(m_2)}(j, i)\, p^{(n)}(i, i)\, p^{(m_1)}(i, j),$$

then

$$\sum_{n=m_1+m_2}^{\infty} p^{(n)}(j, j) \geqslant p^{(m_2)}(j, i)\, p^{(m_1)}(i, j) \sum_{n=0}^{\infty} p^{(n)}(i, j).$$

Moreover the series $G(j, j)$ is divergent if the series $G(i, i)$ is divergent. Exchanging the roles of i and j we obtain that either both $G(i, i)$ and $G(j, j)$ are finite or they are both infinite. □

Thus the recurrence property for a Markov chain is hardly the property of the state but rather the property of the class of communicative states.

Intuitive considerations indicate that a recurrence during an infinite time interval into a recurrent state should take place infinitely often, while only finitely often into a non-recurrent state. These assertions can easily be proved.

Let $Q_j(m)$ be an event that the system reaches the j-th state at least m times and let τ_j be the number of steps until the first passage through the state j. Then

$$Q_j(m)=\bigcup_{n=1}^{\infty} Q_j(m) \cap \{\tau_j=n\}.$$

Let $q_{ij}(m)$ be the probability of the event $Q_j(m)$ given $\xi(0)=i$. We have

$$q_{ij}(m)=\sum_{n=1}^{\infty} \mathsf{P}(Q_j(m) \cap [\tau_j=n] \mid \xi(0)=i)=$$

$$\sum_{n=1}^{\infty} \mathsf{P}^{(i)}(Q_j(m) \mid \tau_j=n)\, \mathsf{P}^{(i)}(\tau_j=n \mid \xi(0)=i)=$$

$$=\sum_{n=1}^{\infty} f^{(n)}(i, j)\, \mathsf{P}^{(i)}(Q_j(m) \mid \tau_j=n).$$

It is easy to verify that

$$\mathsf{P}^{(i)}(Q_j(m)\,|\,\tau_j=n)=\mathsf{P}^{(j)}(Q_j(m-1))=q_{jj}(m-1).$$

Thus

$$q_{ij}(m)=F(i,j)\,q_{jj}(m-1). \tag{8}$$

Let $q_{ij}=q_{ij}(\infty)$ be the probability that the system after leaving the i-th state reaches the j-th state infinitely often. Since $q_{ij}=\lim_{m\to\infty} q_{ij}(m)$ it follows from (8) that

$$q_{ij}=F(i,j)\,q_{jj}. \tag{9}$$

Theorem 3. *If j is a recurrent state, then $q_{ij}=F(i,j)$ and in particular $q_{jj}=1$ if, however, j is a non-recurrent state then $q_{ij}=0$ for every i.*

Proof. If $F(j,j)<1$, we obtain that $q_{jj}=0$ by putting $i=j$ in (9). Also from (9) we have $q_{ij}=0$. If $F(j,j)=1$ it follows from (8) that $q_{jj}(m)==[F(j,j)]^{m-1}=1$ and hence $q_{jj}=1$. It then follows from (9) that $q_{ij}==F(i,j)$. □

Let $F(i,j)=1$. In view of the strong Markov property (cf. Section 4, Theorem 5) we obtain

$$\mathsf{P}^{(i)}\left(B\cap\left(\bigcap_{k=1}^{r}\{\xi(\tau_j+t_k)=j_k\}\right)\right)=\mathsf{P}^{(i)}(B)\,\mathsf{P}^{(j)}\left(\bigcap_{k=1}^{r}\{\xi(t_k)=j_k\}\right)$$

for any $B\in\mathfrak{F}_{\tau_j}$. This relationship yields

Theorem 4. *If $F(i,j)=1$, then the random process $\xi'(t)=\xi(\tau_j+t)\,(\xi(0)=i)$ is stochastically equivalent to the process $\xi(t)$ with the initial state $\xi(0)=j$ and does not depend on the σ-algebra $\mathfrak{F}_{\tau_j}$.*

Corollary. *Let $\xi(0)=i$, where i is a recurrence state, ξ_1 is the number of steps up to the first recurrence to i, ξ_2 the number of steps between the first and the second recurrence to i, and so on.*

The random variables $\xi_1, \xi_2, \ldots, \xi_n$ are identically distributed and independent.

Theorem 5. *If the state i is recurrent and $F(i,j)>0$, then the system after leaving i will reach the state j infinitely many times ($q_{ij}=1$) and $F(j,i)>0$. In particular in this case $F(i,j)=1$.*

It follows from Theorem 3 that there are infinitely many returns to the state i. Let C_k denote the event that the system will reach state j between its $(k-1)$-th and k-th recurrence to state i. In view of the strong Markov property of the process the events C_k are independent and have the same probability. Since $\mathsf{P}\left(\bigcup_{k=1}^{\infty} C_k\right)$ is the probability that the system eventually reaches state j, it follows that $\mathsf{P}(C_k)>0$ and $\sum_{k=1}^{\infty} \mathsf{P}(C_k)=\infty$.

From Borel-Cantelli's lemma we obtain that with probability 1 infinitely many events C_k will occur. Moreover if the system reaches state j it will then reach state i infinitely many times.

Corollary 1. *Only recurrent states can be reached from a recurrent state. Recurrent states are essential.*

This corollary sharpens Theorem 2 which was obtained previously using the method of generating functions.

Corollary 2. *In a class of communicative states containing a recurrent state, all other states are also recurrent and a point belonging to this class will, eventually with probability one, find itself in all the other states of this class and, moreover, this will happen infinitely often.*

A class of recurrent communicative states is called a *recurrent class*.

Now set

$$G(i,j)=\sum_{n=0}^{\infty} p^{(n)}(i,j).$$

The meaning of this series was explained for the case $i=j$. We now establish the following relation:

$$\lim_{N\to\infty} \sum_{n=1}^{N} p^{(n)}(i,j)/\sum_{n=0}^{N} p^{(n)}(j,j)=F(i,j). \tag{10}$$

The proof is based on formula (4). Setting in (4) $n=1, 2, \ldots, N$ and summing up the equalities obtained we have

$$\sum_{n=1}^{N} p^{(n)}(i,j)=\sum_{n=1}^{N}\sum_{s=0}^{n-1} f^{(n-s)}(i,j)\,p^{(s)}(j,j)=\sum_{s=0}^{N-1}\sum_{n=s+1}^{N} f^{(n-s)}(i,j)\times$$
$$\times p^{(s)}(j,j)=\sum_{s=0}^{N-1} p^{(s)}(j,j)\,F_{N-s},$$

where $F_{N-s}=\sum_{n=1}^{N-s} f^{(n)}(i,j)$ and $F_N\to F(i,j)$ for $N\to\infty$. Therefore

$$\frac{\sum_{n=1}^{N} p^{(n)}(i,j)}{\sum_{n=0}^{N} p^{(n)}(j,j)}=\sum_{s=0}^{N} F_{N-s}\frac{p^{(s)}(j,j)}{\sum_{n=0}^{N} p^{(n)}(j,j)}.$$

The validity of formula (10) now follows from the following lemma

Lemma 1. *If* $\{b_n, n=0, 1, \ldots, N\}$ *is a sequence of non-negative numbers and* $\frac{b_N}{\sum_{s=0}^{N} b_s} \to 0$, *then we have for an arbitrary convergent sequence* $\{c_n, n=1, 2, \ldots\}$:

$$\lim_{N\to\infty} \frac{\sum_{k=0}^{N} b_k c_{N-k}}{\sum_{k=0}^{N} b_k} = \lim_{n\to\infty} c_n .$$

Proof. If $c=\lim c_n$, then

$$\frac{\sum_{k=0}^{N} b_k c_{N-k}}{\sum_{k=0}^{N} b_k} - c = \frac{\sum_{k=0}^{N-n} b_k (c_{N-k}-c)}{\sum_{k=0}^{N} b_k} - c\,\frac{\sum_{k=N-n+1}^{N} b_k}{\sum_{k=0}^{N} b_k} + \frac{\sum_{k=N-n+1}^{N} b_k c_{N-k}}{\sum_{k=0}^{N} b_k}. \quad (11)$$

If the index is chosen so that for $n' \geqslant n$, $|c-c_{n'}|<\varepsilon$, where $\varepsilon>0$ is arbitrary, then the first term in the r.h.s. of equality (11) will be less than ε. Since c_n are bounded, the second and third terms for fixed n also approach 0 as $N\to\infty$. □

This proves equality (10) since the conditions of the lemma are always applicable to the case under consideration in view of the fact that $p^{(n)}(i, j)$ are bounded. □

From formula (10) we obtain

Theorem 6. *In a recurrent class* $G(i, j) = +\infty$; *if, however,* j *is not recurrent, then* $G(i, j) < \infty$ *for all* i.

Indeed if j is a nonrecurrent state, then the denominator in the left-hand-side of (10) tends to a finite limit, and therefore the limit of the numerator is also finite. If, however, j is recurrent then the limit of the denominator is ∞, and if $F(i, j)>0$ then the limit of the numerator is also ∞. □

Periodicity. Note that if $p^{(n)}(i, i)>0$, then $p^{(kn)}(i, i)>0$ also. Indeed $p^{(kn)}(i, i) \geqslant p^{(n)}(i, i)\, p^{(n)}(i, i) \ldots p^{(n)}(i, i)$. Denote by $d(i)$ the greatest common divisor of all n such that $p^{(n)}(i, i)>0$. If $p^{(n)}(i, i)=0$ for all $n\geqslant 1$ we shall assume that $d(i)=\infty$.

Theorem 7. *If* $i \leftrightarrow j$, *then* $d(i)=d(j)$.

Proof. Firstly if $i \leftrightarrow j$, then $d(i)$ and $d(j)$ are finite. Let $p^{(s)}(i, i) > 0$. There are $n > 0$ and $m > 0$ such that $p^{(n)}(i, j) > 0$ and $p^{(m)}(j, i) > 0$. Hence, $p^{(n+m+s)}(j, j) \geqslant p^{(m)}(j, i)\, p^{(s)}(i, i) \cdot p^{(n)}(i, j) > 0$. Analogously $p^{(n+m+ks)}(j, j) > 0$. Therefore $d(j)$ divides $(n+m+2s)-(n+m+s)=s$. It follows from here that $d(j) \leqslant d(i)$. Interchanging the role of i and j, we have $d(i) \leqslant d(j)$ in view of symmetry, i.e. $d(i)=d(j)$. □

Corollary. *In any class of communicative states the quantity $d(i)$ is constant.*

In particular, for an irreducible Markov chain the quantity $d=d(i)$ is independent of the state.

Definition 4. If in an irreducible chain $d=1$, then the Markov chain is called *aperiodic*; if $d>1$, the chain is called *periodic* and the number d is its *period.*

The next lemma is a number-theoretic result.

Lemma 2. *Let d be the least common divisor of a sequence of positive integers $n_1, n_2, \ldots, n_s$. There exists a number $m_0 > 0$ such that for all integer-valued $m \geqslant m_0$ the indeterminate equation*

$$md = \sum_{j=1}^{s} c_j n_j$$

has a solution in non-negative integers c_j.

Proof. Let A be the set of all numbers admitting representation $x=\sum_1^s a_j n_j$, where a_j are integers (positive, negative or zero). Every x is divisible by d. Let d_0 be the smallest positive integer belonging to A. Since $x - kd_0 \in A$ for any integral k, for any x a k can be found, such that $x=kd_0$. (Otherwise a k_1 could be found such that $x_1 = x - k_1 d_0$ would satisfy $0 < x_1 < d_0$ which contradicts the definition of d_0) Thus d_0 is the greatest common divisor of numbers in A. Next let $B=\left\{x : x=\sum_{j=1}^{s} b_j n_j\right\}$, where b_j are non-negative integers and let $d_1=\sum_1^s n_j$. The number d_0 can be represented in the form $d_0 = N_1 - N_2$, where $N_i \in B$. Let c be the largest *integer-valued* coefficient of n_j contained in N_2. For any integer $m>0$ we set $m=kd_1+m_1$ where $0 \leqslant m_1 \leqslant d_1$. Then $md_0 = kd_0 d_1 + m_1 d_0 \in B$ provided $kd_0 > m_1 c$ which is evidently satisfied if either $k > d_1 c/d_0$, or $m > d_1^2 c/d_0 + d_1$. The lemma is thus proved. □

Theorem 8. *If $d(i) < \infty$, an n_0 can be found such that for $n > n_0$*

$$p^{(nd(i))}(i, i) > 0 .$$

Proof. Let $n_k (k=1, 2, \ldots, s)$ be a sequence of numbers such that $p^{(n_k)}(i, i)>0$ and let the greatest common divisor of the numbers $n_1, n_2, \ldots, n_s$ be equal to $d(i)$. In view of the preceding lemma, n_0 can be found such that for $n \geqslant n_0$ we have $nd(i) = \sum_{k=1}^{s} c_k n_k$. Therefore

$$p^{(nd(i))}(i, i) \geqslant [p^{(n_1)}(i, i)]^{c_1} [p^{(n_2)}(i, i)]^{c_2} \ldots [p^{(n_s)}(i, i)]^{c_s} > 0. \quad \square$$

Corollary. If $p^{(m)}(j, i)>0$, then for all n sufficiently large

$$p^{(m+nd(i))}(j, i)>0.$$

Indeed

$$p^{(m+nd(i))}(j, i) \geqslant p^{(m)}(j, i)\, p^{(nd(i))}(i, i).$$

When studying Markov chains it is often more convenient to investigate aperiodic chains first and then generalize the results obtained for periodic ones.

We now show that the period of a state can be computed from the probability of the first recurrence.

Lemma 3. *The period of the i-th state coincides with the greatest common divisor of the collection of n such that* $f^{(n)}(i, i)>0$.

Proof. Let Z_N and Z'_N be the set of n such that $n \leqslant N$ and such that $p^{(n)}(i, i)>0$ and $f^{(n)}(i, i)>0$ respectively and let d_N and d'_N be their common divisors. Clearly, $Z'_N \subset Z_N$ and hence $d'_N \geqslant d_N$. Moreover $d'_1 = d_1$. Let there exist N such that $d'_n = d_n$ for $n \leqslant N$ and $d'_{N+1} > d_{N+1}$. Then $f^{(N+1)}(i, i)=0$ and $p^{(N+1)}(i, i)>0$. In view of the equality $p^{(N+1)}(i, i) = f^{(N+1)}(i, i) + \sum_{k=1}^{N} f^{(k)}(i, i)\, p^{(N+1-k)}(i, i)$ we have for some s, $0<s \leqslant N$, the inequality $f^{(s)}(i, i)\, p^{(N+1-s)}(i, i)>0$, i.e. s and $N+1-s$ are divisible by d_N and therefore $N+1$ is divisible by d_N which contradicts the inequality $d_{N+1} < d'_{N+1} = d_N$. The lemma is thus proved. $\square$

Theorem 9. *Any class K of communicative states of period d* $(d<\infty)$ *can be subdivided into d pairwise disjoint subclasses* $K_0, K_1, \ldots, K_{d-1}$ *such that in one step from* K_s $(s<d-1)$ *one can move only to* K_{s+1} *and from* K_{d-1} *only to* K_0. *Moreover, if* $i \in K_r$, $j \in K_s$, *then an* $N = N(i, j)$ *can be found such that* $p^{nd+s-r}(i, j)>0$ *for* $n>N$.

Proof. Let K_0 be the set of all the states j such that we have $p^{(kd)}(i, j)>0$ at least for one positive integer k where i is an arbitrarily chosen state in K. Then $i \in K_0$. Since i and j communicate, an m can be found such that $p^{(m)}(j, i)>0$. The number m is a multiple of d. Indeed, $p^{(kd+m)}(i, i) \geqslant p^{(kd)}(i, j)\, p^{(m)}(j, i)>0$ and hence $kd+m$ is divisible by d. Since m is divisible by d, K_0 would remain unchanged if one were to take an arbitrary j for which $p^{(kd)}(i,j)>0$ for some k in place of i in the definition of K_0. We now define K_1 as the set of j, $j\in K$, such that $\sum_{i\in K_0} p(i,j)>0$, K_2 as the set of all those j such that $\sum_{i\in K_1} p(i,j)>0$, $j\in K$, and so on. It follows from the definition of the sets K_s that $K_{rd+s}\subset K_s$ for any r and s. On the other hand, if $j\in K_s$, then $j_0, j_1, \dots, j_s=j$ can be found such that $j_r\in K_r$, $r\leqslant s$ and $p(j_{r-1}, j_r)>0$, i.e. $p^{(s-r)}(j_r, j)>0$. The converse is also true: If $p^{(s-r)}(j_r, j)>0$, $j_r\in K_r$, $j\in K$, then $j\in K_s$ (since from $j_r\to j_{r+1}, j_{r+1}\to j$ and $j_r\leftrightarrow j$ it follows that $j_{r+1}\leftrightarrow j_r$). We now show that the classes K_r and K_s, $0\leqslant r<s<d$, are disjoint. Indeed, let $j\in K_r$ and $j\in K_s$. Then i_1 and $i_2\in K_0$ can be found such that $p^{(r)}(i_1, j)>0$ and $p^{(s)}(i_2, j)>0$. Since i_2 and j communicate, it follows that for some m $p^{(m)}(j, i_2)>0$. Therefore $p^{m+s}(i_2, i_2) > p^{(s)}(i_2, j)\, p^{(m)}(j, i_2)>0$, and hence $m+s$ is divisible by d, i.e. $m=kd-s$, where s is an integer. But then

$$0<p^{(r)}(i_1, j)\, p^{(m)}(j, i_2)\leqslant p^{(kd-s+r)}(i_1, i_2),$$

which is impossible as it was shown above since the transitions from i_1 into i_2, $i_1, i_2\in K$, are possible only for a number of steps which is a multiple of d. Next, let $i\in K_r$ and $j\in K_s$. One can find m such that $p^{(m)}(i,j)>0$. This m is of the form $m=k_0d+(s-r)$. On the other hand, in view of Theorem 8, $p^{(nd)}(i,i)>0$ for all $n\geqslant n_0(i)$. Therefore, $p^{((n+k_0)d+s-r)}(i,j)\geqslant$ $\geqslant p^{(nd)}(i,i)\, p^{(k_0d+s-r)}(i,j)>0$ for all $n>n_0(i)$.

The theorem is thus proved. □

We shall refer to the sets $K_0, K_1, \dots, K_{d-1}$ as *subclasses* of a periodic class of communicative states.

The basic theorem of renewal theory. In order to study the asymptotic behavior of the transition probabilities $p^{(n)}(i,j)$ as $n\to\infty$ we shall use a theorem which is often called the basic theorem of renewal theory. We shall confine ourselves to the version required in the sequel which, however, is not the most general one. To explain the terminology assume that we are considering the performance of a piece of equipment which may fail from time to time. When a piece fails it is immediately replaced by a new one. The duration of the survival period τ_n of the n-th piece is a random variable taking on values 1, 2, ..., and the random variables τ_n ($n=0, 1, \dots$) are mutually independent and identically distributed. Set

$$p_k=\mathsf{P}\{\tau_n=k\}, \quad k=1,2,\dots, \sum_{k=1}^{\infty} p_k=1.$$

The sum $\tau_0+\tau_1+\dots+\tau_{n-1}$ is called the instant of the n-th renewal, while the variable τ_n is called the duration of the n-th renewal. Denote by $G(n)$ the probability that n is an instant of renewal. The events

$$\{\tau_0=n\},\ \{\tau_0+\tau_1=n\},\dots,\ \{\tau_0+\tau_1+\dots+\tau_{k-1}=n\},\dots$$

are pairwise disjoint, therefore

$$G(n)=\mathsf{P}\{\tau_0=n\}+\mathsf{P}\{\tau_0+\tau_1=n\}+\dots+\mathsf{P}\{\tau_0+\tau_1+\dots+\tau_{k-1}=n\}+\dots$$

and $G(n)\leqslant 1$ for $n\geqslant 1$. Set $G(0)=1$. The function $G(n)$ is called the *renewal function.*

The theorem which characterizes the asymptotic behavior of $G(n)$ as $n\to\infty$ is called the basic theorem of renewal theory.

Denote by d the greatest common divisor of those n for which $p_n>0$. If $d=1$ the renewal process is called aperiodic; if $d>1$ the renewal process is periodic and d is called the renewal period. It can be easily seen that in the case of an aperiodic renewal $G(n)>0$ for all n starting with some n_0, $n\geqslant n_0$. If, however, $d>1$ then for all k sufficiently large, $k\geqslant k_0$, $G(kd)>0$. These assertions follow from the arithmetical Lemma 2. It turns out that if the renewal is aperiodic, then $G_\infty=\lim\limits_{n\to\infty} G(n)=\dfrac{1}{m}$, where $m=\mathsf{E}\tau_k$ (for $\mathsf{E}\tau_k=\infty$, $G_\infty=0$).

We first prove the existence of the limit G_∞ and then find its value.

Lemma 4. *Let τ be a random variable taking on values n ($n=0, \pm 1, \pm 2,\dots$) with probabilities p_n, and let $J(u)$ be the characteristic function of the variable τ. If $d=1$ then $J(u)\neq 1$ for $|u|<2\pi$, $u\neq 0$.*

Proof. We have

$$J(u)=\mathsf{E}e^{iu\tau}=\sum_{-\infty}^{\infty} p_n e^{iun}.$$

Let $J(u_0)=1$, $|u_0|<2\pi$, $u_0\neq 0$. We have

$$0=1-\operatorname{Re} J(u_0)=\sum_{-\infty}^{\infty} (1-\cos nu_0)\, p_n.$$

Therefore $\cos nu_0=1$ for all those n such that $p_n>0$ or $nu_0=2\pi k$. Select a sequence of integers $n_1, n_2,\dots, n_s$ such that $p_{n_r}>0$ and such that their greatest common divisor is 1. Then $n_r u_0=2\pi k_r$ ($r=1, 2,\dots, s$). On the other hand equation $\sum\limits_{r=1}^{s} a_r n_r=1$ has a solution in integral a_r. Therefore

$$u_0=\sum_{r=1}^{s} a_r n_r u_0=2\pi\sum_{r=1}^{s} a_r k_r=2\pi k_0,$$

where k_0 is an integer which contradicts the condition that $|u_0|<2\pi$. □

Lemma 5. *If the renewal is aperiodic, then the limit $G_\infty = \lim_{n\to\infty} G(n)$ exists.*

Proof. Set

$$G(z, n) = \sum_{s=0}^{\infty} z^s p_n(s), \quad n \geqslant 0, \quad 0 \leqslant z \leqslant 1,$$

where $p_n(s) = \mathsf{P}(\eta_s = n)$ and $\eta_s = \tau_0 + \tau_1 + \ldots + \tau_{s-1}$ for $s \geqslant 1$ and $p_n(0) = 0$. It follows from Abel's theorem in the theory of power series that

$$G(n) = \lim_{z \uparrow 1} G(z, n).$$

Since the characteristic function of random variable η_s is equal to $[J(u)]^s$,

$$[J(u)]^s = \sum_{n=1}^{\infty} p_n(s)\, e^{inu},$$

where $J(u)$ is the characteristic function of the random variable τ_0, it follows that

$$p_n(s) = \frac{1}{2\pi} \int_{-\pi}^{\pi} e^{-inu} [J(u)]^s \, du.$$

Therefore

$$G(z, n) = \frac{1}{2\pi} \int_{-\pi}^{\pi} \frac{e^{-inu}\, du}{1 - zJ(u)}, \quad n \geqslant 0.$$

The integral in the r.h.s. of the last formula vanishes for $n < 0$. Therefore

$$G(z, n) = \frac{1}{\pi} \int_{-\pi}^{\pi} \frac{\cos nu \, du}{1 - zJ(u)}.$$

Set $h(z, u) = \frac{1}{\pi} \operatorname{Re}(1 - zJ(u))^{-1}$. Since $G(z, n)$ is a real-valued function, we have

$$G(z, n) = \int_{-\pi}^{\pi} h(z, u) \cos nu \, du.$$

In view of the aperiodicity of the renewal and Lemma 4, the kernel $h(z, u)$ $(z \in [0, 1], 0 < |u| < 2\pi)$ is positive and continuous. Therefore for any $\varepsilon>0$

$$G(n)=\lim_{z\uparrow 1}\int_{-\varepsilon}^{\varepsilon} h(z,u)\cos nu\,du+\int_{\varepsilon\leqslant|u|\leqslant\pi} h(1,u)\cos nu\,du. \tag{12}$$

Putting $n=0$ here we observe that the limit

$$h_\varepsilon=\lim_{z\uparrow 1}\int_{-\varepsilon}^{\varepsilon} h(z,u)\,du$$

exists and that $h_\varepsilon\leqslant G(0)$. Since h_ε decreases as $\varepsilon\downarrow 0$ the $\lim_{\varepsilon\to 0} h_\varepsilon=h_0$ also exists. Therefore the double limit

$$\lim_{\varepsilon\to 0}\lim_{z\uparrow 1}\int_{-\varepsilon}^{\varepsilon} h(z,u)\cos nu\,du=h.$$

also exists. Returning now to formula (12) we see that $h(1,u)$ is an integrable function (in the Cauchy sense) on the interval $(-\pi,\pi)$ and that

$$G(n)=h+\int_{-\pi}^{\pi} h(1,u)\cos nu\,du.$$

Since $h(1,u)$ is integrable, it follows from the Riemann-Lebesgue theorem that

$$\lim_{n\to\infty}\int_{-\pi}^{\pi} h(1,u)\cos nu\,du=0.$$

Therefore the existence of $\lim_{n\to\infty} G(n)=h$ is proved.

Theorem 10. *If the renewal is aperiodic, then*

$$\lim_{n\to\infty} G(n)=\frac{1}{m},\quad m=\mathsf{E}\tau_k,$$

moreover, if $\mathsf{E}\tau_k=\infty$, *then* $\lim_{n\to\infty} G(n)=0$.

Proof. Since, in view of the previous lemma, the limit $\lim_{n\to\infty} G(n)=h$ exists, we will obtain, using Abel's theorem on power series that

$$h=\lim_{z\uparrow 1}\left(1+\sum_{n=1}^{\infty} z^n[G(n)-G(n-1)]\right)=$$

$$=\lim_{z\uparrow 1}\sum_{n=0}^{\infty} z^n(1-z)\,G(n)=\lim_{z\uparrow 1}(1-z)\,\Phi(z),$$

where $\Phi(z)=\sum_{n=0}^{\infty} z^n G(n)$ is the generating function of the sequence $\{G(n), n=0, 1, \ldots\}$. From the fact that τ_r are independent and identically distributed it follows that $G(n)$ satisfies the equation

$$G(n)=\delta(n)+\sum_{k=1}^{n} G(n-k)\,p_k, \qquad n\geqslant 0 \tag{13}$$

($\delta(n)=0$ for $n>0$, $\delta(0)=1$). Multiplying this relation by z^n and summing up for all $n\geqslant 0$ we obtain

$$\Phi(z)=1+F(z)\,\Phi(z), \quad |z|<1,$$

where $F(z)=\sum_{n=1}^{\infty} p_n z^n$. Thus

$$\Phi(z)=[1-F(z)]^{-1}$$

and

$$h=\lim_{z\uparrow 1}\left[\frac{1-F(z)}{1-z}\right]^{-1}.$$

If $m=\infty$, then for any $N>0$

$$\lim_{z\to 1}\frac{1-F(z)}{1-z}\geqslant \lim_{z\uparrow 1}\sum_{n=1}^{N} p_n\frac{1-z^n}{1-z}=\sum_{n=1}^{N} p_n n,$$

From here it follows that $h=0$. If, however $m<\infty$ then taking into account the inequality $\left|\frac{1-z^n}{1-z}\right|<n$ for $|z|<1$, we obtain

$$\lim_{z\uparrow 1}\frac{1-F(z)}{1-z}=\lim_{z\uparrow 1}\sum_{1}^{\infty} p_n\frac{1-z^n}{1-z}=\sum_{n=1}^{\infty} p_n n=m.$$

The theorem is thus proved. □

Corollary. *If the renewal is of the period d, then*

$$\lim_{n\to\infty} G(nd)=d/m, \quad m=\mathsf{E}\tau_k. \tag{14}$$

Indeed, if the given renewal is periodic and d is its period, then the new renewal with the duration of $\tau_n' = \tau_n/d$ is aperiodic. If $G'(n)$ is its renewal function, then $G'(n) = G(nd)$. On the other hand $\mathsf{E}\tau_n' = \frac{\mathsf{E}\tau_n}{d} = \frac{m}{d}$. Formula (14) now follows from the theorem just proved. □

Limit theorems for transition probabilities.

Theorem 11. *Let $p^{(n)}(i, j)$ be the transition probabilities of an irreducible recurrent aperiodic Markov chain. Denote by m_i the average number of steps until the first recurrence of the i-th state,*

$$m_i = \sum_{n=1}^{\infty} n f^{(n)}(i, i).$$

Then for each j

$$\lim_{n \to \infty} p^{(n)}(j, i) = \frac{1}{m_i}. \tag{15}$$

Proof. Let τ_0 be the number of steps until the first recurrence of the i-th state, τ_1-the number of steps between the second and first recurrence of this state and so on. In view of the corollary to Theorem 4, the variables $\tau_0, \tau_1, \ldots, \tau_n, \ldots$ are mutually independent, identically distributed and take on integer-values greater or equal to 1, and, moreover,

$$\mathsf{P}\{\tau_k = n\} = f^{(n)}(i, i), \quad \sum_{n=1}^{\infty} f^{(n)}(i, i) = 1.$$

The mathematical expectation of the variables τ_n is equal to m_i. Consider a renewal process in which τ_n is the duration of the n-th renewal. The quantities p_n and $G(n)$ are replaced here by $f^{(n)}(i, i)$ and $p^{(n)}(i, i)$ respectively. Since the chain is aperiodic, in view of Lemma 3, the renewal is also aperiodic.

From Theorem 10 the following equality is obtained

$$\lim_{n \to \infty} p^{(n)}(i, i) = \frac{1}{m_i}.$$

This is a particular case of (15) for $j = i$. The general case can now be dealt with easily. Using formula 4 we obtain

$$\frac{p^{(n)}(j, i)}{\sum_{k=1}^{n} f^{(k)}(j, i)} = \sum_{k=1}^{n} \bar{f}^{(k)}(j, i)\, p^{(n-k)}(i, i), \quad \bar{f}^{(n)}(j, i) = \frac{f^{n}(j, i)}{\sum_{k=1}^{n} f^{(k)}(j, i)}.$$

Noting that $f^{(n)}(j, i) \to 0$, $\sum_{k=1}^{n} f^{(k)}(j, i) \to 1$ as $n \to \infty$ (in view of the irreducibility and recurrency of the chain) and applying Lemma 1 we obtain equality (15) in the general case. □

We have just proved the ergodic theorem for Markov chains. More on ergodic theorems is presented in Section 8.

Theorem 12. *If an irreducible recurrent Markov chain is periodic with period d, then*

$$\lim_{n\to\infty} p^{(nd)}(i, i) = \frac{d}{m_i}. \tag{16}$$

If K_s are subclasses introduced in Theorem 9 and $i\in K_r$, $j\in K_s$, then

$$\lim_{n\to\infty} p^{(nd+l)}(i, j) = \begin{cases} \dfrac{d}{m_j}, & l = s-r \pmod d, \\ 0, & l \neq s-r \pmod d. \end{cases} \tag{17}$$

Proof. It follows from Lemma 3 that the period of an irreducible Markov chain coincides with the period of the renewal process introduced in the proof of the previous theorem. Therefore equality (16) follows directly from the corollary of Theorem 10. From Theorem 9 we have:

$$p^{(nd+l)}(i, j) = 0 \quad \text{for} \quad i\in K_r, \quad j\in K_s$$

and $l \neq s-r \pmod d$. Therefore if, for example, $r<s$, then

$$p^{(nd+s-r)}(i, j) = \sum_{k=0}^{n} f^{(kd+s-r)}(i, j)\, p^{(n-k)d}(j, j).$$

Referring to Lemma 1 the proof of formula (17) is completed as in the proof of Theorem 11. □

Definition 5. A recurrent state j is called a *null* state if $\lim_{n\to\infty} p^{(nd_j)}(j, j) = 0$ and is called a *positive* state if $\lim_{n\to\infty} p^{(nd_j)}(j, j) > 0$.

In a recurrent class of states the states are either all positive or all null. Indeed, if $i\leftrightarrow j$, then it follows from inequality $p^{(m+nd_j+s)}(i, i) \geqslant p^{(m)}(i, j)\, p^{(nd_j)}(j, j)\, p^{(s)}(j, i)$ (where m and s are such that $p^{(m)}(i, j) > 0$, $p^{(s)}(j, i) > 0$) that

$$\lim p^{(nd)}(i, i) \geqslant \lim p^{(nd)}(j, j), \quad d = d_i = d_j.$$

Interchanging the roles of i and j we obtain the proof of the assertion.

The results obtained can be summarized as follows:

Theorem 13. a) *In order that the state j be non-recurrent it is necessary* *and sufficient that* $G_{jj}=\sum_{n=1}^{\infty} p^{(n)}(j,j)<\infty$. *Moreover for all i*

$$G_{ij}=\sum_{n=1}^{\infty} p^{(n)}(i,j)\leqslant G_{jj}<\infty, \qquad \lim_{n\to\infty} p^{(n)}(i,j)=0.$$

b) *Let j be a recurrent state with period d and the mean return time* m_j. *If i is accessible from j, then i is also a recurrent state with the same period d, and is either null or positive depending on whether j is null or positive and there exists k,* $0\leqslant k<d$, *depending on i and j only, such that*

$$\lim p^{(md+r)}(i,j)=\begin{cases} \dfrac{d}{m_j} & \text{if } \ r=k, \\ 0 & \text{if } \ r\not\equiv k \pmod d. \end{cases} \tag{18}$$

c) *If the states i and j belong to the same recurrent class then*

$$\lim_{N\to\infty}\frac{1}{N}\sum_{n=1}^{N} p^{(n)}(i,j)=\frac{1}{m_j}. \tag{19}$$

The last assertion is a direct consequence of b). On the other hand unlike assertion b) formula (19) does not reflect the distinction between periodic and aperiodic classes of states. An irreducible recurrent Markov chain is called *positive* (*null*) if its states are positive (null).

Criteria for recurrency. Stationary distributions. The property of a Markov chain's being recurrent (positive or null) is closely connected with non-trivial solutions of the system of linear equations.

$$\sum_{j\in I} p(j,i)\,x_j=x_i \qquad i\in I, \tag{20}$$

and its transpose

$$\sum_{j\in I} p(i,j)\,x_j=x_i \qquad i\in I \tag{21}$$

If the system (20) admits a non-negative and summable solution, i.e. $x_i\geqslant 0$ and $\sum x_i<\infty$, we may then assume that $\sum x_i=1$ and such a solution may be interpreted as an invariant initial distribution $x_i=\mathsf{P}\{\xi(0)=i\}=$ $=\mathsf{P}\{\xi(1)=i\}=\ldots$, which generates a stationary Markov process. On the other hand, the existence of a stationary Markov process with given transition probabilities is equivalent to the existence of a non-negative summable solution of the system (20).

As far as the transposed system (21) is concerned the existence of a non-trivial solution $x_i=c$ for this system is evident. A characteristic feature of a recurrent Markov chain is that (21) does not admit other non-trivial non-negative solutions. Moreover the following theorem holds:

Theorem 14. *An irreducible Markov chain is recurrent if and only if the system of inequalities*

$$\sum_{j\in I} p(i,j)\,x_j \leqslant x_i, \quad i\in I \tag{22}$$

admits no non-negative solutions other than solutions of the form $x_i=c$, $i\in I$.

Proof. Assume that a chain is recurrent and $x_i \geqslant 0$ and $x_i\,(i\in I)$ constitute a solution of the system (22). We choose an arbitrary $x_l>0$ (if there is no such x_l, then all $x_i\equiv 0$). It follows from (22) that

$$x_i \geqslant \sum_{j\in I} p(i,j) \sum_{k\in I} p(j,k)\,x_k = \sum_{k\in I} p^{(2)}(i,k)\,x_k,$$

and by induction

$$x_i \geqslant \sum_{k\in I} p^{(n)}(i,k)\,x_k.$$

For each i, an n can be found such that $p^{(n)}(i,l)>0$; therefore $x_i \geqslant p^{(n)}(i,l)\,x_l>0$. Thus $x_i>0$ for all $i\in I$. Set $y_i=\dfrac{x_i}{x_l}$, where l is an arbitrarily chosen state. We have $y_i \geqslant \sum_{j\in I} p(i,j)\,y_j \geqslant p(i,l)+\sum_{j\neq l} p(i,j)\,y_j$. Applying this inequality to the quantities y_j appearing in the r.h.s., we obtain

$$\begin{aligned} y_i &\geqslant p(i,l)+\sum_{j\neq l} p(i,j)\,p(j,l)+\sum_{j\neq l}\sum_{k\neq l} p(i,j)\,p(j,k)\,y_k= \\ &= f^{(1)}(i,l)+f^{(2)}(i,l)+\sum_{k\neq l} {}_l p^{(2)}(i,k)\,y_k, \end{aligned}$$

where ${}_l p^{(2)}(i,k)=\sum_{j\neq l} p(i,j)\,p(j,k)$ is the probability of hitting the k-th state on the second step, after leaving the i-th state, and not entering the l-th state. Iterating this method we arrive at the inequality

$$y_i \geqslant \sum_{n=1}^{N} f^{(n)}(i,l)+\sum_{k\neq l} {}_l p^{(N)}(i,k)\,y_k,$$

where ${}_l p^{(N)}(i,k)$ is the N-step transition probability from the i-th state into the k-th state not entering the l-th state. Approaching $N\to\infty$ in the last inequality we obtain

$$y_i \geqslant \sum_{n=1}^{\infty} f^{(n)}(i,l)=1,$$

i.e. $x_i \geqslant x_l$.

Since i and l are arbitrary integers, it follows that $x_i = x_l = \text{const}$, i.e. the system of inequalities (22) admits no non-negative solutions other than $x_i = c$, $i \in I$ for which the sign of equality holds in all the equations of system (22).

Now let the chain possess at least one non-recurrent state (the irreducibility of the chain is not utilized here). Set $x_l = 1$, $x_i = F(i, l)$ for $i \neq l$, where l is an arbitrary non-recurrent state. Note that $F(i, l) = 1$ holds not for all i, $i \neq l$. Indeed, otherwise we would have

$$F(l, l) = \sum_{k \neq l} p(l, k)\, F(k, l) + p(l, l) = \sum_{k \in I} p(l, k) = 1,$$

which contradicts the fact that the state l is non-recurrent. Thus the non-negative numbers x_i defined above are not all equal. We have for $i \neq l$

$$x_i = F(i, l) = \sum_{k \neq l} p(i, k)\, F(k, l) + p(i, l) = \sum_{k \in I} p(l, k)\, x_k$$

and

$$x_l = 1 > F(l, l) = \sum_{k \in I} p(i, k)\, x_k,$$

i.e. $\{x_i, i \in I\}$ forms a non-negative solution of the system (22) which is not a constant. The theorem is thus proved. □

We now investigate the connection between the existence of invariant initial distributions and the recurrence properties of a Markov chain, i.e. we shall study the problem of solvability of system (20) for a recurrent chain.

Theorem 15. *Let a Markov chain be irreducible and recurrent. The system of equations* (20) *can have no more than one solution satisfying the conditions*

$$\sum_{i \in I} |x_i| < \infty, \quad \sum_{i \in I} x_i = 1. \tag{23}$$

If the chain is positive-recurrent the solution of a system (20) *satisfying relations* (23) *is of the form*

$$x_i = v_i = \lim_{N \to \infty} \frac{1}{N} \sum_{n=1}^{N} p^{(n)}(j, i). \tag{24}$$

If, however, the chain is null-recurrent, then the only absolutely summable solution of system (20) *is the trivial one* ($x_i = 0$).

Proof. We first prove the uniqueness of the solution of system (20) under conditions (23). Let such a solution exist. Multiplying (20) by $p(i, k)$ and summing up over all i, we obtain

$$x_k = \sum_{i \in I} x_i p(i, k) = \sum_{i \in I} \sum_{j \in I} x_j p(j, i)\, p(i, k) =$$

$$=\sum_{j\in I} x_j \sum_{i\in I} p(j, i)\, p(i, k) = \sum_{j\in I} x_j p^{(2)}(j, k).$$

The interchange of order of summations is permissible since the corresponding double series is absolutely convergent. Analogously we obtain

$$x_k = \sum_{j\in I} x_j p^{(n)}(j, k). \tag{25}$$

Set

$$s_N(j, k) = \frac{1}{N} \sum_{n=1}^{N} p^{(n)}(j, k);$$

then

$$x_k = \sum_{j\in I} x_j s_N(j, k),$$

Taking into account the fact that $s_N(j, k) \to m_k^{-1}$ and the absolute convergence of the series $\sum_{j\in I} x_j$ and approaching the limit in the last equality we obtain

$$x_k = \sum_{j\in I} x_j m_k^{-1} = m_k^{-1}, \tag{26}$$

which proves the uniqueness of the solution of the system (20) (23). It also follows from here that if the chain is null-recurrent, then $x_k = 0$ for all $k \in I$.

We now prove that for a positive-recurrent chain the quantities (24) constitute the required solution of system (20). Let I' be an arbitrary finite subset of I. It follows from the inequality

$$p^{(n+1)}(k, i) \geq \sum_{j\in I'} p^{(n)}(k, j)\, p(j, i).$$

that

$$s_{N+1}(k, i) - \frac{1}{N+1} p(k, i) \geq \frac{N}{N+1} \sum_{j\in I'} s_N(k, j)\, p(j, i).$$

Approaching to the limit with $N \to \infty$ we obtain that

$$v_i \geq \sum_{j\in I'} v_j p(j, i).$$

Assuming now the $I' \to I$ we obtain $v_i \geq \sum_{j\in I} v_j p(j, i)$. Multiplying the last inequality by $p(i, k)$ and summing up with respect to k we arrive at the inequalities

$$v_k \geq \sum_{i\in I} v_i p(i, k) \geq \sum_{i\in I} v_i p^{(2)}(i, k)$$

and continuing this process we arrive at the inequalities

$$v_k \geqslant \sum_{i \in I} v_i p^{(n)}(i, k)$$

for any $n \geqslant 1$. If there were a sign of strict inequality for at least one k in the last relation, we would have

$$\sum_{k \in I} v_k > \sum_{i \in I} v_i \sum_{k \in I} p^{(n)}(i, k) = \sum_{i \in I} v_i,$$

which is impossible. Therefore

$$v_k = \sum_{i \in I} v_i p^{(n)}(i, k), \quad k \in I, \quad n = 1, 2, \dots \tag{27}$$

In particular the quantities v_i form a solution of the system (20). From (27) we obtain

$$v_k = \sum_{i \in I} v_i s_N(i, k). \tag{28}$$

Note that the inequality $\sum_{k \in I'} p^{(n)}(i, k) \leqslant 1$ yields the relations $\sum_{k \in I'} s_N(i, k) \leqslant 1$ and $\sum_{k \in I'} v_k \leqslant 1$ for any finite $I' \subset I$. Hence, $\sum_{k \in I} v_k \leqslant 1$. Therefore we may approach the limit in (28) as $N \to \infty$ which will result in the equality $v_k = \sum_{i = I} v_i v_k$, so that $\sum_{i \in I} v_i = 1$. Therefore the solution v_i of system (20) satisfies conditions (23). The theorem is thus proved. □

Remark. If a Markov chain is arbitrary, $\{x_i, i \in I\}$ is an absolutely summable solution of system (20), and k is a non-recurrent state, then $x_k = 0$.

This assertion follows from the fact that one may approach to the limit as $n \to \infty$ in equation (25) and from the relation $\lim_{n \to \infty} p^{(n)}(j, k) = 0$ which holds for an arbitrary non-recurrent k.

Corollaries.

1. *In order that an irreducible Markov chain be positive-recurrent it is necessary and sufficient that system* (20) *admits a non-trivial absolutely summable solution* $\{x_i, i \in I\}$. *Moreover,* $x_i = c v_i$, *where* c *is a constant and* $v_i > 0$.

2. *An irreducible Markov chain possesses an invariant initial distribution if and only if it is positive-recurrent.*

3. *If the chain is positive-recurrent and aperiodic then the unique solution of system* (20) *satisfying* (23) *is of the form*

$$x_i = v_i = \lim_{n \to \infty} p^{(n)}(j, i). \tag{29}$$

The last assertion follows from the fact that for a positive aperiodic chain the limits $\lim_{n \to \infty} p^{(n)}(j, i)$ exist so that (24) yields (29). □

It follows from the previous theorem that system (20) cannot admit, for a null-recurrent chain, a non-trivial absolutely summable solution. However, it possesses an important non-negative non-summable solution. To obtain this solution we introduce taboo- probabilities. This notion is a generalization of the notion of the probability of the first hit. It has already been encountered in the course of the proof of Theorem 14. The taboo-probability ${}_lp^{(n)}(i,j)$ is the probability to hit the j-th state at the n-th step starting from the i-th state without entering the l-th state during the times $1, 2, \ldots$, and $n-1$. Thus

$$ {}_lp^{(n)}(i,j) = \sum_{\substack{j_1, j_2, \ldots, j_{n-1}, \\ j_r \neq l,\, r=1,\ldots,n-1}} p(i,j_1)\, p(j_1,j_2) \ldots p(j_{n-1},j),\ n \geqslant 1\,. $$

Clearly,

$$ {}_lp^{(1)}(i,j) = p(i,j), \qquad {}_jp^{(n)}(i,j) = f^{(n)}(i,j)\,. $$

We also set

$$ {}_lp^{(0)}(i,j) = \delta(i,j)\,. $$

The taboo-probabilities ${}_Hp^{(i,j)}$ are introduced analogously. Here "the prohibited part" is a certain set of states H. If l and j are prohibited it would be logical to denote the taboo-probability ${}_{\{l,j\}}p^{(n)}(i,j)$ by ${}_lf^{(n)}(i,j)$. This is the probability that starting from the initial state i we hit the state j for the first time at the n-th step without entering the state l.

We note the two following equations:

$$ {}_lp^{(n)}(i,j) = \sum_{k=1}^{n} {}_lf^{(k)}(i,j)\, {}_lp^{(n-k)}(j,j)\,, \tag{30} $$

$$ {}_lp^{(n)}(i,j) = \sum_{k=1}^{n} {}_lp^{(k)}(i,i)\, {}_{\{i,l\}}p^{(n-k)}(i,j)\,. \tag{31} $$

Each summand in the r.h.s. of (30) represents the probability to hit the j-th state at the n-th step starting from the initial state i and to hit the j-th state *for the first time* at the k-th step ($k \leqslant n$) without entering the state l during the first n steps.

The l.h.s. of (30) represents the summation of these probabilities with respect to k. The summands in the r.h.s. of (31) have the following meaning. They equal the probability of hitting the j-th state starting from the initial state i, not entering the state l during this time, but entering the state i for the last time – before the n-th step on the k-th step ($k \leqslant n$). In particular it follows from formula (31) that (for $l=j$)

$$ f^{(n)}(i,j) = \sum_{k=1}^{n} {}_jp^{(k)}(i,i)\, {}_if^{(n-k)}(i,j)\,. \tag{32} $$

We introduce the following generating functions

$$ {}_lP_{ij}(z)=\sum_{n=0}^{\infty} {}_lp^{(n)}(i,j)\,z^n, $$

$$ {}_lF_{ij}(z)=\sum_{n=0}^{\infty} {}_lf^{(n)}(i,j)\,z^n, \qquad {}_lf^{(0)}(i,j)=0. $$

The r.h.s. of equations (30) and (32) are convolutions of two sequences, therefore

$$ {}_lP_{ij}(z)={}_lF_{ij}(z)\,{}_lP_{jj}(z), \qquad F_{ij}(z)={}_jP_{ii}(z)\,{}_iF_{ij}(z), \tag{33} $$

Note that the series ${}_lF_{ij}(z)$ are convergent for $z=1$ and, moreover, if the states i and j communicate then ${}_iF_{ij}(1)>0$. Under this assumption the second of the equations (33) shows that there exists a finite limit ${}_jP_{ii}(z)$ as $z\to 1$ and therefore ${}_jP_{ii}(1)<\infty$.

Define

$$ {}_lG(i,j)=\sum_{n=0}^{\infty} {}_lp^{(n)}(i,j). \tag{34} $$

Thus if the states i and j communicate, then

$$ {}_jG(i,i)=\frac{F_{ij}(1)}{{}_iF_{ij}(1)}<\infty. \tag{35} $$

On the other hand, the first of the relations (33) yields

$$ {}_iG(i,j)={}_iF_{ij}(1)\,{}_iG(j,j). $$

Hence

$$ {}_iG(i,j)\leqslant {}_iG(j,j)<\infty. \tag{36} $$

Returning to the solution of the system (20) we prove the following theorem.

Theorem 16. *Let l be an arbitrary state of an irreducible recurrent Markov chain. The system* (20) *admits a non-negative solution*

$$ x_l=1, \qquad x_i={}_lG(l,i) \quad (i\neq l), \qquad i\in I. $$

Proof. Set

$$ u_l=1, \qquad u_i={}_lG(l,i) \quad (i\neq l). \tag{37} $$

We have for $i\neq l$

$$ \sum_{j\in I} u_j p(j,i)=p(l,i)+\sum_{j\neq l} {}_lG(l,j)\,p(j,i)= $$

$$=p(l, i)+\sum_{j\neq l}\sum_{n=1}^{\infty} {}_l p^{(n)}(l, j)\, p(j, i)=$$

$$=p(l, i)+\sum_{n=1}^{\infty} {}_l p^{(n+1)}(l, i)={}_l G(l, i)=u_i,$$

if however $i=l$, then

$$\sum_{j\in I} u_j p(j, l)=p(l, l)+\sum_{n=1}^{\infty} f^{(n+1)}(l, l)=\sum_{n=1}^{\infty} f^{(n)}(l, l)=u_l.$$

and the theorem is proved. □

We now consider the problem of uniqueness of the solution of system (20) satisfying conditions $u_l=1$ and $u_i\geqslant 0$. For this purpose we utilize a method connected with the introduction of an inverted Markov chain.

First we shall assume that the chain is positive recurrent and let $\{v_j, j\in I\}$ be the invariant initial distribution.

Consider a stationary Markov process corresponding to the initial distribution $\{v_j, j\in I\}$. Denote by $\mathsf{P}^{(v)}$ the probability measure corresponding to this process. We introduce the following conditional probabilities

$$q_i(j_1, j_2, \ldots, j_n)=\mathsf{P}^{(v)}\{\xi(t-1)=j_1, \xi(t-2)=j_2, \ldots, \xi(t-n)=j_n \mid \xi(t)=i\},$$

where $t>n$; we have

$$q_i(j_1, j_2, \ldots, j_n)=\frac{v_{j_n} p(j_n, j_{n-1})\, p(j_{n-1}, j_{n-2})\ldots p(j_1, i)}{v_i}=$$

$$=q(i, j_1)\, q(j_1, j_2)\ldots q(j_{n-1}, j_n),$$

where

$$q(i, j)=p(j, i)\frac{v_j}{v_i}.$$

Thus in a stationary positive-recurrent Markov chain the conditional transition probabilities obtained from the change in time direction (from present to past) corresponds to a certain Markov chain. Moreover it should be noted that all $v_i>0$ and therefore

$$q(i, j)\geqslant 0, \qquad \sum_{j\in I} q(i, j)=\frac{1}{v_i}\sum_{j\in I} v_j p(j, i)=\frac{v_i}{v_i}=1.$$

The above construction is applicable not only for a positive but also for an arbitrary recurrent chain (i.e. for a null-recurrent chain as well). For this purpose we consider an arbitrary positive solution $\{x_j, j\in I\}$ of system (20) (it will be shown below that such a solution exists) and set

$$q(i,j)=p(j,i)\cdot\frac{x_j}{x_i}. \tag{38}$$

As in the case above,

$$q(i,j)\geqslant 0, \quad \sum_{j\in I} q(i,j)=1.$$

A Markov chain with transition probabilities (38) is called the *inversion of the initial chain (the inverse chain)*.

We note the following formulas for n-step transition probabilities in an inverted chain. We have

$$q^{(n)}(i,j)=\sum_{j_1,j_2,\ldots,j_{n-1}} q(i,j_1)\,q(j_1,j_2)\ldots q(j_{n-1},j)=$$
$$=\sum_{j_1,j_2,\ldots,j_{n-1}} p(j_1,i)\,p(j_2,j_1)\ldots p(j,j_{n-1})\,\frac{x_j}{x_i},$$

i.e.

$$q^{(n)}(i,j)=\frac{x_j}{x_i}\,p^{(n)}(j,i). \tag{39}$$

The following corollary follows from the above:

If the initial chain is irreducible, recurrent, positive or null, then the inverted chain possesses the same properties.

From the limit theorem for ratios we obtain in the case of a recurrent chain

$$\lim_{N\to\infty}\frac{\sum_{n=1}^{N} q^{(n)}(i,j)}{\sum_{n=0}^{N} q^{(n)}(j,j)}=1.$$

Using formulas (39) we have

$$\lim_{N\to\infty}\frac{\sum_{n=1}^{N} p^{(n)}(j,i)}{\sum_{n=0}^{N} p^{(n)}(j,j)}=\frac{x_i}{x_j}. \tag{40}$$

The following theorem is a corollary of the relationship obtained:

Theorem 17. *For an irreducible recurrent chain a non-negative solution* *of system* (20) *such that* $x_l = 1$ *is unique. Moreover* $x_i = {}_lG(l, i)$ *and*

$$\lim_{N\to\infty} \frac{\sum_{n=1}^{N} p^{(n)}(j, i)}{\sum_{n=0}^{N} p^{(n)}(j, j)} = {}_jG(j, i). \tag{41}$$

Formula (41) follows from the uniqueness of the solution of system (20) and Theorem 16, while the uniqueness follows from formula (40) and the assumption that $x_j > 0$ for all j. Therefore in view of Theorem 16, it is sufficient to show that if $\{x_j, j \in I\}$ is a non-negative non-trivial solution of system (20), then $x_j > 0$. This can be obtained as follows. We have for a non-negative solution of system (20)

$$x_i = \sum_j x_j p(j, i) = \sum_j \sum_k x_k p(k, j)\, p(j, i) =$$
$$= \sum_k x_k \sum_j p(k, j)\, p(j, i) = \sum_k x_k p^{(2)}(k, i).$$

Using induction one can easily obtain that $x_i = \sum_{k\in I} x_k p^{(n)}(k, i)$. Let $x_l > 0$; one can find n such that for any i, $p^{(n)}(l, i) > 0$ and therefore $x_i \geq x_l p^{(n)}(l, i) > 0$. Constructing an inverted chain for the given solution of system (20) and putting $x_l = 1$ we obtain from (40) the uniqueness of x_i, $i \in I$. In view of Theorem 16, $x_i = {}_lG(l, i)$.

Remark. Formula (40) is a generalization of the relationship $\lim_{N\to\infty} N^{-1} \times \sum_{n=1}^{N} p^{(n)}(j, i) = v_i$, (valid for a positive recurrent chain) where $\{v_i\}$ is the invariant initial distribution.

The following theorem is a refinement of Theorem 17.

Theorem 18. *For an irreducible non-recurrent Markov chain the system of inequalities*

$$x_i \geq \sum_{j\in I} x_j p(j, i), \quad x_i \geq 0, \quad x_l = 1, \tag{42}$$

admits a unique solution and moreover $x_i = \sum x_j p(j, i)$, $i \in I$.

In view of Theorem 16 it is sufficient to prove the uniqueness of the solution for system (42). We introduce the inverted Markov chain with transition probabilities $q(i, j) = p(j, i)\frac{u_j}{u_i}$ where u_i is the positive solution of system (20). This chain is irreducible and recurrent. We have

$$\sum_j q(i, j)\frac{x_j}{u_j} = \sum_j p(j, i)\frac{x_j}{u_i} \leq \frac{x_i}{u_i}, \qquad \frac{x_l}{u_l} = 1.$$

But in view of Theorem 14 the system of inequalities

$$\sum_j q(i,j)\, y_j \leqslant y_i, \qquad y_l = 1$$

admits the unique non-negative solution $y_i = 1$. Hence, $x_i = u_i$ for all $i \in I$. The theorem is thus proved. □

§6. Random Walks on a Lattice

Irreducibility. Definition 1. A set Z of vectors $z = \sum_{i=1}^{s} a_i e_i$, where e_i $(i = 1, \ldots, s)$ are linearly independent vectors in $\mathscr{R}^m$ and a_i – integers $(a_i = 0, \pm 1, \pm 2, \ldots)$ is called a *lattice*.

Clearly, Z is the minimal additive group containing vectors $e_1, e_2, \ldots, e_s$. The number s is the dimension of the lattice and the vectors $e_1, e_2, \ldots, e_s$ are its basis. If $s < m$ the lattice is called degenerate; for $s = m$ the lattice is non-degenerate.

A random walk $\{\zeta(n), n = 0, 1, 2, \ldots\}$ on the lattice Z is defined by the formula $\zeta(n) = x + \xi_1 + \ldots + \xi_n$, $n \geqslant 1$, $\zeta(0) = x$, where x is a non-random vector being the initial position of the random walk, $x \in Z$ and $\xi_1, \xi_2, \ldots, \ldots, \xi_n \ldots$ are identically distributed independent random vectors with the values in Z. Put $p(x) = \mathsf{P}\{\xi_k = x\}$, $x \in Z$. If $x_k \in Z$ $(k = 0, 1, \ldots, n)$, we obtain from the definition of random walks

$$\mathsf{P}\{\zeta(0) = x_0, \zeta(1) = x_1, \ldots, \zeta(n) = x_n\} = \\ = \delta(x_0 - x) \prod_{k=1}^{n} p(x_k - x_{k-1}).$$

Therefore a random walk on a lattice is a particular case of a homogeneous Markov chain with a countable number of states and with one-step transition probabilities satisfying $p(x, y) = p(y - x)$. The basic characteristic feature of random walks which distinguish them from the general Markov chains with a countable number of states is the spatial homogeneity of transition probabilities:

$$p(x+z, y+z) = p(x, y) = p(y - x).$$

This property is just another expression of independence of the displacement vector of the walk $\xi_{n+1} = \zeta(n+1) - \zeta(n)$ from its position at the given moment. Evidently the spatial homogeneity holds also for the n-step transition probabilities:

$$p^{(n)}(x+z, y+z) = \mathsf{P}\{\zeta(n) = y+z \mid \zeta(0) = x+z\} = \\ = \mathsf{P}\{\zeta(n) - \zeta(0) = y - x \mid \zeta(0) = x\} = p^{(n)}(y - x),$$

where $p^{(n)}(x) = \mathsf{P}\{\xi_1 + \xi_2 + \ldots + \xi_n = x\}$ is the probability that the sum of n independent identically distributed random vectors takes on value x.

It follows from the spatial homogeneity of the walk that the set K_x of all points Z accessible from a given point x can be represented in the form $K_0 + x$, where K_0 is the set of points accessible from 0 $(0 \in K_0)$. To describe set K_0 we introduce set D consisting of all those $x \in Z$ for which $p(x) > 0$. The set D is called the support of the distribution of random vectors ξ_k. Only points belonging to D are accessible in one step from 0. Only those points of Z which admit representation $x = x_1 + x_2$ where $x_i \in D$ are accessible in two steps from 0. Let H_+ denote the collection of all points in $\mathscr{R}^m$ of the form $x = n_1 x_1 + \ldots + n_s x_s$, where $s \geqslant 0$ and n_k $(k = 1, \ldots, s)$ are arbitrary positive integers and $x_k \in D$. Clearly,

$$K_0 = H_+$$

i.e. H_+ is the set of all points accessible from 0.

Two points x and y belonging to Z are termed *communicating* if $x - y \in H_+$ and $y - x \in H_+$. Set

$$H_* = H_+ \cap \{-H_+\}.$$

The decomposition into classes of communicating states described in Section 5 in the present case, consists of the following: H_* is the class of states containing point zero, all the other classes of communicating states are of the form $H_k = x_k + H_*$, where x_k is an arbitrary sequence belonging to Z such that $x_k - x_j \notin H_*$ $(k \neq j)$.

It follows from the spatial homogeneity of random walks that different classes of communicative states are either all essential or all non-essential, so that the property of being essential or non-essential is related to random walks as a whole.

The condition of being essential is equivalent to the requirement that $H_+ = \{-H_+\}$ which means that H_+ is a group.

Therefore in order that the states of a random walk be essential it is necessary and sufficient that subset H_+ of points Z be a group.

We introduce the set H of points z which can be represented as $z = x - y$, where $x, y \in H_+$. This set is the minimal group containing points $z \in Z$ accessible from zero. It will be shown that H is a lattice in $\mathscr{R}^m$ (possibly of a lower dimension). It thus follows that when studying random walks one may assume that H coincides with the lattice Z of all vectors with integer-valued coordinates in the space $\mathscr{R}^m$. Denote by Z^m the lattice of all integer-valued vectors in space $\mathscr{R}^m$.

The following theorem is purely algebraic in nature and can be formulated as

Theorem 1. *The r-dimensional additive group H $(H \subset Z^m)$ of vectors in a linear space $\mathscr{R}^m$ is an r-dimensional lattice.*

Proof. Let r be the maximal number of linearly independent vectors in H. We shall prove that there exists r linearly independent vectors $x_k \in H$ such that H coincides with the set of vectors of the form $a_1 x_1 + a_2 x_2 + \ldots + a_r x_r$ where a_k are arbitrary integers ($a_k = 0, \pm 1, \pm 2, \ldots, k = 1, 2, \ldots r$). Let $x_1^*, x_2^*, \ldots, x_r^*$ be an arbitrary maximal system of linearly independent vectors in H. Then each vector $x \in H$ can be represented in the form

$$x = \sum_{k=1}^{r} b_k x_k^*, \tag{1}$$

where b_k are real numbers. On the other hand, $x_k^* = \sum_{j=1}^{m} c_{kj} e_j$, where c_{kj} are integers and the rank of the matrix $\{c_{kj}\}$ equals r. Representation (1) is equivalent to the system of linear equations $\sum_{k=1}^{r} b_k c_{kj} = a_j, j = 1, 2, \ldots, m$, where a_i are the integervalued coordinates of x in the basis $\{e_k, k = 1, \ldots, m\}$. It thus follows that for $0 \leqslant b_k < 1$ there may exist only a finite number of vectors of form (1) with b_k being rational numbers. Therefore if B is the least common denominator of all b_k, then representation (1) can be written in the form

$$x = \sum_{k=1}^{r} c_k y_k, \quad y_k = \frac{x_k^*}{B},$$

where c_k are integers. Consider now an arbitrary linear transformation $z_k = \sum_{j=1}^{r} n_{kj} y_j$ ($k = 1, \ldots, r$) with integer-valued coordinates and the determinant

$$V(z_1, z_2, \ldots, z_r) = \begin{vmatrix} n_{11} \ldots n_{1r} \\ n_{21} \ldots n_{2r} \\ \ldots\ldots\ldots \\ n_{r1} \ldots n_{rr} \end{vmatrix}, \tag{2}$$

consisting of coordinates of the vectors $z_1, \ldots, z_r$ in the basis $\{y_j, j = 1, \ldots, r\}$. The determinant will be integer-valued (different from zero) if and only if the system of vectors $z_1, \ldots, z_r$ is linearly independent. We choose a system of vectors $z_1, \ldots, z_r$ so that $z_k \in H$ and the determinant (2) will be of the smallest positive value. Such a system exists. We denote the corresponding vectors by $l_1, \ldots, l_r$. Suppose that for some $x \in H$ in the expansion $x = \sum_{k=1}^{r} d_k l_k$ not all d_k are integers, then there exists a vector

$l' \in H$ such that $l' = \sum d'_k l_k$, $0 \leqslant d'_k < 1$, $d'_j > 0$ for some j. Moreover

$$\begin{aligned} V(l_1, \dots, l_{j-1}, l', l_{j+1}, \dots, l_r) = \\ &= V(l_1, \dots, l_{j-1}, d'_j l_j, l_{j+1}, \dots, l_r) = \\ &= d'_j V(l_1, \dots, l_{j-1}, l_j, l_{j+1}, \dots, l_r), \end{aligned}$$

which contradicts the fact that the determinant $V(l_1, \dots, l_r)$ attains the minimal value.

Therefore the lattice for which the system of vectors $\{l_1, \dots, l_r\}$ is the basis coincides with H. The theorem is thus proved. □

Definition 2. A random walk on an integer-valued lattice Z^m is called *irreducible* if $H = Z^m$ and is called *reducible* if $H \neq Z^m$.

Note that the notion of irreducibility of a random walk just introduced bears no relation to the definition of irreducibility of a Markov chain.

The preceding theorem shows that by means of an affine transformation of the space one can always assure that the m-dimensional random walk will be m-dimensional irreducible. The following criterion of irreducibility of a random walk can be given in terms of characteristic functions. Let

$$J(u) = \mathsf{E} e^{i(u, \xi_1)} = \sum_{x \in Z^m} p(x)\, e^{i(u, x)} \tag{3}$$

be the characteristic function of the vector $\xi_1 = \zeta(1) - \zeta(0)$ representing one step in the random walk.

Theorem 2. *In order that a random walk be irreducible it is necessary and sufficient that $J(u) \neq 1$ for $u \neq 2\pi x$, $x \in Z^m$.*

Proof. Sufficiency. Let the walk be reducible. If the dimension of H is less than m, there exists a vector e orthogonal to H, such that $(ce, \xi_1) = 0$ with probability 1 for any c and the condition of the theorem is not satisfied. Assume now that the dimension of H is m. Choose in H the basis $l_1, \dots, l_m$ and let T be a linear transformation changing the basis $\{e_k, k = 1, \dots, m\}$ to $\{l_k, k = 1, \dots, m\}$, $l_k = Te_k$. The matrix of the transformation T in the basis $\{e_k, k = 1, \dots, m\}$ consists of the coordinates of the vectors l_k and has therefore integer-valued entries. Its determinant however does not equal ± 1. Indeed, if it were ± 1, the inverse matrix T^{-1} also would have had integer-valued entries and each point in Z would have been a point in H which contradicts the fact that the walk is reducible. Consider the set Z' of all vectors $v \in \mathscr{R}^m$ such that $T^* v \in Z^m$, where T^* is the conjugate of T. Clearly Z' is an additive group and $Z^m \subset Z'$. On the other hand, $Z' \neq Z^m$ otherwise the integer-valued transformation T^* would have had an integer-valued inverse which contradicts the relations

$$1=\mathrm{Det}(T^*T^{*-1})=\mathrm{Det}(T)\,\mathrm{Det}(T^{*-1}),$$

since $\mathrm{Det}(T)\neq\pm 1$. Therefore there exists a vector v such that $v\in Z'$, $v\notin Z^m$ and $T^*v\in Z^m$. Hence the number $(v, l_k')=(v, Te_k)=(T^*v, e_k)$ is an integer for any k and therefore (v, ξ_1) is an integer with probability 1, so that $J(2\pi v)=\mathsf{E}\exp\{2\pi i(v, \xi_1)\}=1$ for $v\notin Z^m$.

We have thus proved the sufficiency part of the theorem.

Necessity. Let the walk be irreducible and let $J(2\pi v)=1$. Then $\mathsf{E}[1-\exp\{2\pi i(v, \xi_1)\}]=0$, which is possible only if (v, ξ_1) is an integer with probability 1. From the irreducibility of the walk it follows that (v, l_k) is an integer $(k=1,\dots,m)$ i.e. $v\in Z^m$. This completes the proof. □

Recurrent walks. Let $f^{(s)}(x, y)$ be the probability that a random walk with the initial state x passes the state y for the first time at the instant s, and let $F(x, y)=\sum_{s=1}^{\infty} f^{(s)}(x, y)$.

From relation (4) of Section 5 we have

$$f^{(n)}(x, y)=p^{(n)}(x, y)-\sum_{s=1}^{n-1} f^{(s)}(x, y)\,p^{(n-s)}(y, y)$$

and it follows from the spatial homogeneity of the random walk that

$$f^{(n)}(x, y)=f^{(n)}(0, y-x)=f^{(n)}(y-x);$$

also the previous equation can be rewritten as

$$f^{(n)}(x)=p^{(n)}(x)-\sum_{s=1}^{n-1} f^{(s)}(x)\,p^{(n-s)}(0).$$

Moreover the function $F(x, y)$ depends only on the difference $y-x$, and we may put $F(y, y+x)=F(x)$. In particular, $F(x, x)=F(0, 0)$, so that the states of the random walk are either all recurrent or all non-recurrent. Therefore we shall henceforth refer to recurrent or non-recurrent random walks.

Let

$$F_x(z)=\sum_{n=1}^{\infty} f^{(n)}(x)\,z^n, \qquad P_x(z)=\sum_{n=0}^{\infty} p^{(n)}(x)\,z^n.$$

The functions $F_x(z)$ and $\mathsf{P}_x(z)$ are related as follows (cf. Section 5, (6)).

$$P_0(z)=(1-F_0(z))^{-1}, \qquad P_x(z)=P_0(z)\,F_x(z), \qquad (x\neq 0),$$

We thus obtain the following assertion:

for a recurrent random walk it is necessary and sufficient that

$$G(0)=\sum_{n=0}^{\infty} p^{(n)}(0)=\infty.$$

Recall that in view of the results presented in Section 5 if a walk is recurrent then the return to the initial state occurs with probability 1 infinitely often during an infinite interval of time. Relation (10) of Section 5 becomes

$$F(x)=\lim_{N\to\infty}\frac{\sum_{n=1}^{N} p^{(n)}(x)}{\sum_{n=0}^{N} p^{(n)}(0)}. \tag{4}$$

Therefore if state x is accessible from zero and the walk is recurrent, then $G(x)=\infty$, where

$$G(x)=\sum_{n=0}^{\infty} p^{(n)}(x).$$

If, however, the walk is not recurrent, then

$$G(x)\leqslant G(0)<\infty.$$

The function $G(x)$ has the following probabilistic meaning. It equals the mean value (mathematical expectation) of the number of "visits" by a random walk which started at point 0 to state x during the interval $(0, \infty)$. In the case of a recurrent walk $G(x)$ is either 0 or ∞. In the non-recurrent case $G(x)$ is called the Green function of a random walk.

The following criterion for non-recurrency of a random walk is a simple corollary of the strong law of large numbers.

Let the step of a random walk possess a finite mathematical expectation different from zero. Then the walk is non-recurrent.

Indeed, with probability 1

$$\lim_{n\to\infty}\frac{\zeta(n)}{n}=\mathsf{E}\xi_1=m\neq 0.$$

Therefore, for almost all ω, $n_0=n_0(\omega)$ can be found such that $|\zeta(n)|>\frac{|m|}{2}n$ for $n\geqslant n_0$, so that the return to the point 0 is impossible starting from the moment n_0.

One can obtain a number of other criteria of recurrency and non-recurrency by using the characteristic function $J(u)$ of the step in a random walk. It is easy to see that

$$G(0)=\lim_{t\uparrow 1}\frac{1}{(2\pi)^m}\int_C \operatorname{Re}(1-tJ(u))^{-1}\,du, \tag{5}$$

where m is the dimension of a lattice, C is a cube in $\mathscr{R}^m$, $C=\{u:|u^i|<\pi, i=1,\dots,m\}$, $0<t<1$. Indeed, first of all

$$G(0)=\lim_{t\uparrow 1}\sum_{n=0}^{\infty} p^n(0)\,t^n.$$

On the other hand, expression (3) for the characteristic function of a random walk shows that $p(x)$ are the Fourier coefficients of the Fourier-series expansion of $J(u)$. Therefore

$$p(x)=\frac{1}{(2\pi)^m}\int_C J(u)\,e^{-i(u,x)}\,du,$$

and

$$p^{(n)}(x)=\frac{1}{(2\pi)^m}\int_C J^n(u)\,e^{-i(u,x)}\,du; \tag{6}$$

so that for $0<t<1$

$$P_0(t)=\frac{1}{(2\pi)^m}\int_C (1-tJ(u))^{-1}\,du.$$

Since $P_0(t)$ is real, one can replace the integrand in the last integral by its real part. Approaching the limit as $t\to 1$ we obtain formula (5). Putting $J_c(u)=\operatorname{Re}J(u)$ we can rewrite formula (5) as

$$G(0)=\lim_{t\uparrow 1}\frac{1}{(2\pi)^m}\int_C \frac{1-tJ_c(u)}{|1-tJ(u)|^2}\,du.$$

Utilizing this formula one may obtain a number of special criteria of recurrency. For example we shall now prove that a one-dimensional random walk for which $m=\mathsf{E}\xi_1=0$ is recurrent.

Indeed

$$\frac{1-J(u)}{u}\to m=0 \quad\text{as}\quad u\to 0.$$

Therefore for any $\varepsilon>0$ a $\delta>0$ can be found such that $|1-J(u)|<\varepsilon|u|$

for $|u|<\delta$. Therefore

$$G(0)\geqslant\overline{\lim_{t\uparrow 1}}\frac{1}{2\pi}\int_{-\delta}^{\delta}\frac{1-t}{2[(1-t)^2+|1-J(u)|^2]}du\geqslant$$

$$\geqslant\overline{\lim_{t\uparrow 1}}\frac{1}{2\pi}\int_{0}^{\delta}\frac{1-t}{(1-t)^2+\varepsilon^2u^2}du=\overline{\lim_{t\uparrow 1}}\frac{1}{2\pi}\frac{1}{\varepsilon}\operatorname{arctg}\frac{\varepsilon\delta}{1-t}=\frac{1}{4\varepsilon},$$

hence $G(0)=\infty$ as claimed.

To obtain analogous results for multidimensional walks certain bounds on characteristic functions are required.

Lemma 1. *For a non-recurrent random walk of dimension $m\geqslant 2$ there exists a constant k such that $1-J_c(u)\geqslant k|u|^2$ for $u\in C$, where C is the cube, $C=\{u\colon \max\limits_{1\leqslant i\leqslant m}|u^i|\leqslant\pi\}$.*

Proof. Since

$$1-J_c(u)=\sum_{x\in Z^m}[1-\cos(u,x)]\,p(x)$$

and

$$1-\cos(u,x)=2\sin^2\frac{(u,x)}{2}\geqslant 2\left(\frac{2}{\pi}\frac{(u,x)}{2}\right)^2=\frac{2}{\pi^2}(u,x)^2$$

for $|(u,x)|<\pi$, we have

$$1-J_c(u)\geqslant\frac{2}{\pi^2}\sum{}'(u,x)^2\,p(x),$$

where $\sum'$ denotes the summation over all $x\in Z^m$ satisfying the condition $|(u,x)|\leqslant\pi$. Since a walk is irreducible one can choose in the set $\{x\colon p(x)>0\}$ a basis of $\mathscr{R}^m$. Let $\{e_1,\ldots,e_m\}$ be the vectors of this basis and let $N=\max\{|e_k|,\,k=1,2,\ldots,m\}$. Moreover let $|u|\leqslant\pi N^{-1}$ then

$$1-J_c(u)\geqslant\frac{2}{\pi^2}\sum_{k=1}^{m}(u,e_k)^2\,p(e_k).$$

The quadratic form appearing in the r.h.s. of the inequality is positive definite. Therefore there exists a constant k_1 such that $\sum\limits_{k=1}^{m}(u,e_k)^2\,p(e_k)\geqslant$ $\geqslant k_1|u|^2$. Thus

$$1-J_c(u)\geqslant\frac{2}{\pi^2}k_1|u|^2\quad\text{for}\quad|u|\leqslant\pi N^{-1}.$$

In view of Theorem 2 in Section 1, $J(u)$ is different from the one in the region $C_1 = C\setminus\{u: |u| < \pi N^{-1}\}$, and therefore $\min\limits_{u\in C_1}[1 - J_c(u)] = k_2 > 0$. But then $1 - J_c(u) \geqslant k_2 N^2 \pi^{-2} |u|^2$ for $|u| \geqslant \pi N^{-1}$, $u \in C$, which proves the assertion.

We now return to the problem of recurrent random walks. Assume that the two-dimensional random walk is irreducible, $\mathsf{E}\xi_1 = 0$ and $\mathsf{E}|\xi_1|^2 \times \infty$. In view of Fatou's lemma

$$\lim_{t\uparrow 1} \int_C \frac{1 - tJ_c(u)}{|1 - tJ(u)|^2}\, du \geqslant \int_C \frac{1 - J_c(u)}{|1 - J(u)|^2}\, du. \tag{7}$$

It follows from Lemma 1 that $1 - J_c(u) \geqslant k|u|^2$, $u \in C$. On the other hand, since ξ_1 possesses finite second moments,

$$J(u) = 1 - \tfrac{1}{2}\mathsf{E}(\xi_1, u)^2 + o(|u|^2).$$

Hence $|1 - J(u)| \leqslant k_1 |u|^2$ in some neighbourhood of the point $u = 0$. Therefore the integrand in the r.h.s. of inequality (7) is not less than $\dfrac{k|u|^2}{k_1 |u|^4} \sim \dfrac{1}{|u|^2}$ in some neighbourhood of point $u = 0$ and the corresponding integral diverges. Therefore the random walk under consideration is recurrent. As far as the random walk of dimension $\geqslant 3$ is concerned it is always non-recurrent.
Indeed, since

$$I = \lim_{t\uparrow 1} \int_C \frac{1 - tJ_c(u)}{|1 - tJ(u)|^2}\, du \leqslant \lim_{t\uparrow 1} \int_C \frac{du}{1 - tJ_c(u)} \leqslant \int_C \frac{du}{1 - J_c(u)},$$

using Lemma 1, we have

$$I \leqslant \int_C \frac{du}{k|u|^2},$$

and the last integral converges if the dimension of the space is $m \geqslant 3$. The results obtained can be summarized as follows:

Theorem 3. *A random walk of dimension $m \geqslant 3$ is always non-recurrent. It is also non-recurrent if there exists $\mathsf{E}\xi_1$ and $\mathsf{E}\xi_1 \neq 0$. If $\mathsf{E}\xi_1 = 0$ then the walk is recurrent in the one dimensional case. If in addition to $\mathsf{E}\xi_1 = 0$ the condition $\mathsf{E}|\xi_1|^2 < \infty$ is also satisfied then the random walk is recurrent in the two-dimensional case.*

§7. Local Limit Theorems for Lattice Walks

The asymptotic behavior as $n \to \infty$ of the probability $p^{(n)}(x)$ of hitting the lattice point x during n steps of a random walk is studied in the present section. Analytically the problem consists of investigating the asymptotic behavior of the integral (cf. Section 6 (6))

$$p^{(n)}(x) = \frac{1}{(2\pi)^m} \int_C [J(u)]^n e^{-i(u,x)} du \tag{1}$$

as $n \to \infty$. We shall consider only irreducible walks. Moreover it is assumed that the walk possesses the property of complete irreducibility as defined below.

Definition 1. An irreducible walk is called *completely irreducible* if for any point $x_0 \in D$ the random walk with the step (of the walk) $\eta_1 = \xi_1 - x_0$ is also irreducible.

Utilizing Theorem 2 of Section 6 it is easy to formulate a criterion of complete irreducibility of a random walk.

Theorem 1. *In order that a walk be completely irreducible it is necessary and sufficient that*

$$|J(u)| \neq 1, \quad \text{if} \quad u \neq 2\pi x, \quad x \in Z^m.$$

Proof. Let the walk be completely irreducible and let $J(u) = e^{it}$, where t is a real number. It follows from the equalities

$$\sum_{x \in Z^m} e^{i(x,u)} p(x) = e^{it}, \qquad \sum_{x \in Z^m} p(x) = 1$$

that $(x, u) = t + 2\pi n$, $n = n(x)$ for each x such that $p(x) > 0 (x \in D)$. Let $x_0 \in D$, then

$$1 = \sum_{x \in Z^m} p(x)\, e^{i(x - x_0, u)} =$$
$$\sum_{x \in Z^m} p(x + x_0)\, e^{i(x,u)} = \sum_{x \in Z^m} q(x)\, e^{i(x,u)},$$

where $q(x)$ is the distribution of a random vector $\eta_1 = \xi_1 - x_0$. It follows from the irreducibility of the walk with the step η_1 and Theorem 2 of Section 6 that the last relationship can hold only if $u = 2\pi x$, $x \in Z^m$. On the other hand let $J(u_0) = e^{it}$ for $u_0 \neq 2\pi x$ $(x \in Z^m)$. It follows from the above that the walk with the distribution of the first step $q(x)$ will be reducible. The theorem is thus proved. □

We now proceed to bound integral (1). Firstly, transforming it using the transformation

$$x = na + \sqrt{n}\, x_n, \quad a = \mathsf{E}\xi_1, \quad x \in Z^m$$

we get

$$p^{(n)}(x) = \frac{1}{(2\pi)^m n^{m/2}} \int\limits_{\sqrt{n}C} \left[J\left(\frac{u}{\sqrt{n}}\right) e^{-i\left(a, \frac{u}{\sqrt{n}}\right)} \right]^n e^{-i(u, x_n)}\, du, \tag{2}$$

where $\sqrt{n}\, C$ denotes the cube $\{u: |u^j| < \sqrt{n}\,\pi, j = 1, 2, \ldots, m\}$. We shall assume that ξ_1 possesses finite moments of orders $r+2$, $r>0$. Expanding $e^{i(u,x)}$ by means of Taylor's formula we obtain

$$J(u) = 1 + iA_1(u) + i^2 A_2(u) + \ldots + i^{r+2} A_{r+2}(u) + o(|u|^{r+2}), \tag{3}$$

where $A_k(u)$ is a homogeneous form in u of order k and

$$A_1(u) = \mathsf{E}(u, \xi_1), \quad A_2(u) = \tfrac{1}{2}\mathsf{E}(u, \xi_1)^2.$$

Hence it follows that in a neighbourhood of the point $u=0$ one can define a single-valued continuous function $\ln J(u)$ which satisfies

$$\ln J(u) = iS_1(u) + i^2 S_2(u) + \ldots + i^{r+2} S_{r+2}(u) + o(|u|^{r+2}).$$

Here $S_k(u)$ are also homogeneous forms in u of order k,

$$S_1(u) = A_1(u) = \mathsf{E}(u, \xi_1) = (u, a),$$
$$S_2(u) = \tfrac{1}{2}\mathsf{V}(u, \xi_1) = \tfrac{1}{2}(Bu, u),$$

where B is the variance-covariance matrix of vector ξ_1. Thus putting $J_1(u) = J(u)\, e^{-i(u,a)}$ we obtain

$$J_1^n\left(\frac{u}{\sqrt{n}}\right) = e^{-S_2(u) + I_n},$$

where

$$I_n = \sum_{k=1}^{r} \frac{i^{k+2}}{\sqrt{n^k}} S_{k+2}(u) + o\left(\frac{|u|^{r+2}}{\sqrt{n^r}}\right).$$

Since

$$e^{I_n} = \sum_{j=0}^{r} \frac{1}{j!} I_n^j + O(I_n^{r+1})$$

and $I_n^{r+1} = O\left(\dfrac{\ln^{3r+3} n}{\sqrt{n^{r+1}}}\right)$ in the region $|u| \leqslant k \ln n$, it follows that

$$\left[J_1\left(\frac{u}{\sqrt{n}}\right)\right]^n = e^{-\frac{1}{2}(Bu,u)}\left(1+\frac{1}{\sqrt{n}}l_1(u)+\frac{1}{n}l_2(u)+\dots\right.$$

$$\left.\dots+\frac{1}{\sqrt{n^r}}l_r(u)+\frac{Q^*_{nr}(u)}{\sqrt{n^{r+1}}}\right)+O\left(\frac{\ln^{3r+3}n}{\sqrt{n^{r+1}}}\right), \tag{4}$$

where $Q^*_{nr}(u)$ is a polynomial in u of a fixed degree whose coefficients remain bounded as $n\to\infty$ and $l_j(u)$ are polynomials in u of degree $3j$.

Denote by D^k_j the operation of the k-time partial differentiation with respect to u^j. Note that relation (3) may be differentiated at least $r+2$ times in the sense that

$$D^{k_1}_{j_1}D^{k_2}_{j_2}\dots D^{k_s}_{j_s}J(u)=D^{k_1}_{j_1}D^{k_2}_{j_2}\dots D^{k_s}_{j_s}\left(\sum_{k=0}^{r+2} i^k A_k(u)\right)+o(|u|^{r+2-p}),$$

where $p=k_1+\dots+k_s$. From here it follows that a similar assertion holds for the function $\ln J(u)$ as well. Using this fact it can be shown that relation (4) can be differentiated term by term with respect to u so that

$$D^{(p)}J^n_1\left(\frac{u}{\sqrt{n}}\right)=D^{(p)}\,e^{-\frac{1}{2}(Bu,u)}\left(1+\frac{1}{\sqrt{n}}l_1(u)+\dots\right.$$

$$\left.\dots+\frac{1}{\sqrt{n^r}}l_r(u)+\frac{O^*_{nr}(u)}{\sqrt{n^{r+1}}}\right)+O\left(\frac{\ln^{3r+3}n}{\sqrt{n^{r+1}}}\right), \tag{5}$$

where

$$D^{(p)}=D^{k_1}_{j_1}D^{k_2}_{j_2}\dots D^{k_s}_{j_s},\quad k_1+k_2+\dots+k_s\le r+2.$$

We now proceed to bound the integral

$$I=\int\limits_{\sqrt{n}C}\left|D^{(p)}J^n_1\left(\frac{u}{\sqrt{n}}\right)-\right.$$

$$\left.-D^{(p)}\,e^{-\frac{1}{2}(Bu,u)}\left(1+\frac{1}{\sqrt{n}}l_1(u)+\dots+\frac{1}{\sqrt{n^r}}l_r(u)\right)\right|du.$$

For this purpose certain inequalities for the function $J(u)$ will be required. From the facts that the walk is irreducible and that moments of the second order are finite (cf. Section 6) the existence of a $\delta>0$ follows such that for $|u|<\delta$

$$|J(u)|<1-\tfrac{1}{4}(Bu,u),$$

and since the distribution of ξ_1 is non-degenerate, $|J(u)|<1-c|u|^2<$

$<e^{-c|u|^2}$ for $|u|<\delta$, where c is a constant. On the other hand it follows from the complete irreducibility of the walk that $|J(u)|<1-\varrho$ for $|u|\geqslant\delta$, $u\in C$, where $0<\varrho<1$. These two bounds for $J(u)$ taken together yield

$$|J(u)|\leqslant \varrho e^{-c|u|^2}+1-\varrho, \quad u\in C. \tag{6}$$

The same bound is also applicable to $J_1(u)$.

We now return to integral I. We have

$$I\leqslant I_1+I_2+I_3,$$

where

$$I_1=\int_{B_n}\left|D^{(p)}J_1^n\left(\frac{u}{\sqrt{n}}\right)-D^{(p)}\,e^{-\frac{1}{2}(Bu,u)}\sum_{k=0}^{r} n^{-k/2}l_k(u)\right|du,$$

$$I_2=\int_{B_n^*}\left|D^{(p)}J_1^n\left(\frac{u}{\sqrt{n}}\right)\right|du,$$

$$I_3=\int_{B_n^*}\left|D^{(p)}\,e^{-\frac{1}{2}(Bu,u)}\sum_{k=0}^{r} n^{-k/2}l_k(u)\right|du$$

and

$$B_n=\{u:|u|\leqslant\sqrt{k\ln n}\}\cap(\sqrt{n}\,C),$$
$$B_n^*=\{u:|u|>\sqrt{k\ln n}\}\cap(\sqrt{n}\,C).$$

It follows from (5) that

$$I_1\leqslant\int_{B_n}\left[D^p\left(e^{-\frac{1}{2}(Bu,u)}\frac{Q_{nr}^*(u)}{\sqrt{n^{r+1}}}\right)+O\left(\frac{\ln^{3r+3}n}{\sqrt{n^{r+1}}}\right)\right]du.$$

Taking into account the uniform boundedness of the coefficients of the polynomial $Q_{nr}^*(x)$ regarded as a function of n and the convergence of the integrals $\int_{\mathscr{R}^m} P(u)\,e^{-\frac{1}{2}(Bu,u)}\,du$ for any polynomial $P(u)$, we obtain

$$I_1=O\left(\frac{1}{\sqrt{n^{r+1}}}\right)+O\left(\frac{\ln^{3n+3}}{\sqrt{n^{r+1}}}\right)O(\sqrt{k\ln n})^m=o(n^{-r/2}).$$

Furthermore,

$$I_2=\int_{B_n^*}\left|D^{(p)}J_1^n\left(\frac{u}{\sqrt{n}}\right)\right|du=n^{m/2}\int_{C\cap\left\{|u|>\sqrt{\frac{k\ln n}{n}}\right\}}|D^{(p)}J_1^n(u)|\,du.$$

The expression $D^{(p)}J_1^n(u)$ is a polynomial in $J_1(u)$ and its partial derivatives; moreover $J_1(u)$ enters into this polynomial in the $n-p$-th power or higher, while partial derivatives of $J_1(u)$ are of order $p\leqslant r+2$ or lower and later in powers p or lower, and finally the coefficients of this polynomial are of order n^p. Therefore

$$|D^{(p)}J_1^n(u)|\leqslant An^p\,|J_1(u)|^{n-p},$$

where A is independent of n. Hence it follows from (6) that

$$I_2\leqslant An^{p+\frac{m}{2}}\left(1+\varrho\left(e^{-\frac{ck\ln n}{n}}-1\right)\right)^{n-p}\int\limits_C du\leqslant$$

$$\leqslant(2\pi)^{m/2}\,An^{p+\frac{m}{2}}\,e^{\varrho(n-p)}\left(e^{-\frac{ck\ln n}{n}}-1\right)=O\left(\frac{1}{n^{\varrho ck-p-\frac{m}{2}}}\right).$$

Therefore a constant k can be chosen independently of n such that

$$I_2\leqslant o\left(\frac{1}{n^{r/2}}\right).$$

It remains to bound the integral I_3. We have

$$I_3=\int\limits_{B_n^*} e^{-\frac{1}{2}(Bu,u)}\,|\bar{Q}_{nrp}(u)|\,du\leqslant e^{-\frac{1}{4}ck\ln n}\int\limits_{\mathscr{R}^m} e^{-\frac{1}{4}c|u|^2}\,|\bar{Q}_{nrp}(u)|\,du,$$

where $\bar{Q}_{nrp}$ is a polynomial whose degree depends only on r and p while the coefficients depend also on n, but remain bounded as $n\to\infty$. Again, it follows from the convergence of the integral in the r.h.s. of the last inequality that a k can be chosen such that

$$I_3=o\left(\frac{1}{n^{r/2}}\right).$$

Thus we have proved that for a suitable choice of k

$$I=o\left(\frac{1}{n^{r/2}}\right). \tag{7}$$

Consider now integral

$$L=\int\limits_{\mathscr{R}^m} e^{-i(u,x)}D^{(p)}\,e^{-\frac{1}{2}(Bu,u)}\sum_{k=0}^{r} n^{-k/2}l_k(u)\,du.$$

For $r=0$ and $p=0$ this integral is the Fourier transform of the characteristic function of an m-dimensional normal distribution. Therefore

$$\int_{\mathscr{R}^m} e^{-i(u,x)-\frac{1}{2}(Bu,u)} du = (2\pi)^{\frac{m}{2}} \sqrt{|B|^{-1}}\, e^{-\frac{1}{2}(B^{-1}x,x)},$$

where $|B|$ is the determinant of matrix B and B^{-1} is its inverse. In this formula it is permissible to differentiate with respect to x an arbitrary number of times and moreover in the l.h.s. of this equality this differentiation may be carried out under the sign of the integral. Therefore we have for an arbitrary polynomial $P(u)$

$$\int_{\mathscr{R}^m} P(u)\, e^{-i(u,x)-\frac{1}{2}(Bu,u)}\, du = (2\pi)^{m/2} \sqrt{|B|^{-1}}\, e^{-\frac{1}{2}(B^{-1}x,x)}\, Q(x),$$

where $Q(x)$ is a polynomial in x of the same degree as polynomial P. Next, using integration by parts we obtain

$$\int_{\mathscr{R}^m} e^{-i(u,x)}\, D^{(p)}(e^{-\frac{1}{2}(Bu,u)}\, P(u))\, du =$$

$$= (ix^1)^{k_1} (ix^2)^{k_2} \dots (ix^m)^{k_m} \int_{\mathscr{R}^m} e^{-i(u,x)-\frac{1}{2}(Bu,u)}\, P(u)\, du,$$

provided $D^{(p)} = \dfrac{\partial^p}{(\partial u^1)^{k_1} (\partial u^2)^{k_2} \dots (\partial u^m)^{k_m}}$. Thus

$$L = (ix^1)^{k_1} (ix^2)^{k_2} \dots (ix^m)^{k_m} (2\pi)^{m/2} \sqrt{|B|^{-1}}\, e^{-\frac{1}{2}(B^{-1}x,x)} \sum_{k=0}^{r} \frac{Q_k(x)}{n^{k/2}},$$

where the polynomials $Q(x)$ are of the same degree as polynomials $l_k(u)$, i.e. of degree $3k$ and $Q_0(x)=1$.

The proof of the following theorem is now almost completed:

Theorem 2. *Let $\zeta(n)$ be a completely irreducible walk, $\zeta(n)=\xi_1+\xi_2+\dots+\xi_n$, where ξ_k are mutually independent and identically distributed vectors with values in Z^m with finite moments of orders $r+2$, $r \geq 0$.*
Then

$$n^{m/2}\, \mathsf{P}\{\zeta(n) = na + \sqrt{n}\, x_n\} =$$

$$= \frac{1}{(2\pi)^{m/2} \sqrt{|B|}}\, e^{-\frac{1}{2}(B^{-1}x_n, x_n)} \left(1 + \sum_{1}^{r} \frac{Q_k(x_n)}{n^{k/2}}\right) + \varepsilon_n,$$

where

$$\varepsilon_n = \frac{1}{(1+|x_n|^{r+2})}\, o\left(\frac{1}{n^{r/2}}\right),$$

uniformly in x, and a is the vector of mean values of the step of the random walk, and B is the correlation matrix.

Indeed, we have

$$(2\pi\sqrt{n})^m (ix^1)^{k_1}(ix^2)^{k_2}\dots(ix^m)^{k_m} p^{(n)}(x) =$$
$$= \int\limits_{\sqrt{n}C} e^{-i(u,x_n)} D^{(p)} J_1^n\left(\frac{u}{\sqrt{n}}\right) du =$$
$$= \int\limits_{\mathscr{R}^m} e^{-i(u,x_n)} D^{(p)}\left[e^{-\frac{1}{2}(Bu,u)} \sum_{k=0}^{r} \frac{l_k(u)}{n^{k/2}}\right] du + I_4 + I_5,$$

where

$$I_4 = \int\limits_{\sqrt{n}C} e^{-i(u,x_n)} D^{(p)}\left[J_1^n\left(\frac{u}{\sqrt{n}}\right) - e^{-\frac{1}{2}(Bu,u)} \sum_{k=0}^{r} \frac{l_k(u)}{n^{k/2}}\right] du,$$

$$I_5 = -\int\limits_{\mathscr{R}^m \setminus \sqrt{n}C} e^{-i(u,x_n)} D^{(p)}\left[e^{-\frac{1}{2}(Bu,u)} \sum_{k=0}^{r} \frac{l_k(u)}{n^{k/2}}\right] du.$$

However it follows from inequality (7) that

$$|I_4| \leqslant I = o\left(\frac{1}{n^{r/2}}\right).$$

Moreover,

$$|I_5| \leqslant \int\limits_{\mathscr{R}^m \setminus \sqrt{n}C} \left| D^{(p)}\left[e^{-\frac{1}{2}(Bu,u)} \sum_{k=0}^{r} \frac{l_k(u)}{n^{k/2}}\right]\right| du \leqslant e^{-\frac{1}{4}n} \int\limits_{\mathscr{R}^m} e^{-\frac{1}{4}(Bu,u)} R(u)\, du,$$

where $R(u)$ is a polynomial; therefore $|I_5| = O(\varrho^n) = o(n^{-k})$ $(\varrho = e^{-\frac{1}{4}})$ for any k. Thus we have for an arbitrary polynomial of degree at most $r+2$:

$$(2\pi\sqrt{n})^m P(x_n)\, p^{(n)}(x_n) =$$
$$= P(x_n)\,(2\pi)^{m/2}\sqrt{|B|^{-1}}\, e^{-\frac{1}{2}(B^{-1}x_n, x_n)} \sum_{k=0}^{r} \frac{Q_k(x_n)}{n^{k/2}} + o\left(\frac{1}{n^{r/2}}\right).$$

It is clear that the second summand in the r.h.s. of the last formula depends on the choice of $P(x)$. Set $P(x) \geqslant 1 + m^{r/2} \sum_{j=1}^{m} |x^j|^{r+2}$. Taking into account that $\mathsf{P}(x)$ may be replaced by a function which takes on smaller values and that $m^{r/2}\sum_{j=1}^{m}|x^j|^{r+2}\geqslant|x|^{r+2}$, we obtain the required result. □

§8. Ergodic Theorems

Measure-preserving transformations. Definition 1. A random process $\{\xi(t), t\in T\}$ with values in a measurable space $\{\mathscr{X}, \mathfrak{B}\}$ is called *stationary* if for any $n, t_1, t_2, \ldots, t_n$ and t such that $t_k+t\in T\,(k=1,\ldots,n)$, the joint distribution in $\{\mathscr{X}^n, \mathfrak{B}^n\}$ of the sequence

$$\xi(t_1+t), \xi(t_2+t), \ldots, \xi(t_n+t)$$

does not depend on t.

This definition of a stationary process is equivalent to the following: for an arbitrary bounded $\mathfrak{B}^n$-measurable function $f(x_1,\ldots,x_n)$, $x_k\in\mathscr{X}$ the mathematical expectation

$$\mathsf{E}f(\xi(t_1+t), \xi(t_2+t), \ldots, \xi(t_n+t))$$

is independent of t for any choice of $n, t_1, \ldots, t_n\,(t_k+t\in T)$. It follows from here that if $h(x_1,\ldots,x_n)$ is a measurable mapping $\{\mathscr{X}^n, \mathfrak{B}^n\}\to\{\mathscr{Y}, \mathfrak{C}\}$, then $\eta(t)=h(\xi(t_1+t),\ldots,\xi(t_n+t))$ is a stationary process on the set of values t for which $\eta(t)$ is defined.

In the present section we shall consider stationary sequences, i.e. stationary processes defined on the set $T=\{t: t=0, \pm 1, \pm 2, \ldots\}$ with values in $\{\mathscr{X}, \mathfrak{B}\}$.

Let $\mathscr{X}^T$ be the space of all sequences $u=\{\ldots, x_{-n}, x_{-n+1}, \ldots, x_0, x_1, \ldots, x_n, \ldots\}$, $\mathfrak{C}$ be the minimal σ-algebra containing all the cylinders in $\mathscr{X}^T$, P_ξ be the measure induced on $\mathfrak{C}$ by the sequence $\{\xi(t), t\in T\}$. Thus the probability space $\{\mathscr{X}^T, \mathfrak{C}, \mathsf{P}_\xi\}$ is a natural representation of the process $\{\xi(t), t\in T\}$. Denote by $\{\mathscr{X}^T, \tilde{\mathfrak{C}}, \tilde{\mathsf{P}}_\xi\}$ the space with the completed measure. Introduce in $\mathscr{X}^T$ the shift operation $S: u'=Su$, if $x'_n=x_{n+1}$, $n\in T$ where $u=\{x_n, n\in T\}$, $u'=\{x'_n, n\in T\}$. The operation S possesses an inverse S^{-1} and moreover if $u''=S^{-1}u$, then $u''=\{x''_n, n\in T\}$, $x''_n=x_{n-1}$. The condition of stationarity of the sequence $\xi(t)$ means that for an arbitrary cylinder C

$$\mathsf{P}_\xi(C)=\mathsf{P}_\xi(SC). \tag{1}$$

Since a measure on cylinders uniquely determines a measure on $\mathfrak{C}$ and on its completion $\tilde{\mathfrak{C}}$, the equality remains valid for an arbitrary $A\in\tilde{\mathfrak{C}}$,

$$\mathsf{P}_\xi(A)=\mathsf{P}_\xi(SA), \quad A\in\tilde{\mathfrak{C}}. \tag{2}$$

Definition 2. Let $\{\mathcal{U}, \mathfrak{F}, \mu\}$ be a measure space, S be a measurable transformation of $\{\mathcal{U}, \mathfrak{F}\}$ into $\{\mathcal{U}, \mathfrak{F}\}$. The transformation S is called *measure-preserving* if for any $A \in \mathfrak{F}$

$$\mu(S^{-1}A) = \mu(A),$$

where $S^{-1}A$ is the complete pre-image of the set A.

A transformation S is called *invertable* if there exists a measurable transformation S^{-1} such that $SS^{-1} = S^{-1}S = I$, where I is the identity transformation. In this case the transformation S^{-1} is called the *inverse* transformation of S. The definition of a stationary sequence is equivalent to the following: a sequence $\{\xi(t), t \in T\}$ is stationary if the shift operator S in $\mathscr{X}^T$ preserves the measure P_ξ.

Therefore the problem of studying stationary sequences is a particular case of the problem of studying measure-preserving invertable transformations (automorphisms) of a certain measure space. Consider the problem of the asymptotic behavior of the mean

$$\frac{1}{n}\sum_{k=0}^{n-1} f(S^k u), \quad n \to \infty, \tag{3}$$

where S^k is the k-th power of transformation S, $f(u)$ is an arbitrary $\mathfrak{F}$-measurable function, $\{\mathcal{U}, \mathfrak{F}, \mu\}$ is a space with measure μ and $\mu(\mathcal{U}) \leqslant \infty$. To clarify the meaning of the problem consider the case when $\{\mathcal{U}, \mathfrak{F}, \mu\}$ coincides with $\{\mathscr{X}^T, \mathfrak{C}, \mathsf{P}_\xi\}$ and S is the shift operator. Let $\xi_k = \xi(k, u) = x_k$, $f(u) = \chi_B(x_0)$, where $\chi_B(x)$ is the indicator of the set $B \in \mathfrak{B}$. Then

$$f(S^k u) = \chi_B(S^k u) = \chi_B(\xi(k))$$

and

$$\frac{1}{n}\sum_{k=0}^{n-1} f(S^k u) = \frac{v_n(B, u)}{n}, \tag{4}$$

where $v_n(B, u)$ is the number of terms in the sequence $\xi(0), \xi(1), \ldots, \xi(n-1)$ whose values fall into the set B, i.e. $v_n(B, u)$ is the frequency of hitting the set B by the first n terms of the sequence $\xi(t)$ $(t = 0, 1, \ldots, n-1)$. Therefore the problem under consideration is a particular case of a problem concerning the behavior of the frequency with which values of a random variable $\xi(t)$ fall into an arbitrary set B. Firstly, we shall prove that there exists with probability 1 the limit as $n \to \infty$ of the mean value given in (3). This proposition is the well-known Birkhoff-Khinchin theorem.

Lemma 1. *If S preserves measure μ, $D \in \mathfrak{F}$ and $f(u)$ is an $\mathfrak{F}$-measurable non-negative (μ-integrable function), then*

$$\int_{S^{-1}D} f(Su)\,\mu(du) = \int_D f(u)\,\mu(du). \tag{5}$$

Proof. If we put $f(u)=\chi_A(u)$ formula (5) becomes the equality $\mu(S^{-1}(A\cap D))=\mu(A\cap D)$, which is valid for any A and $D\in\mathfrak{F}$. It follows from here that formula (5) is valid for arbitrary $\mathfrak{F}$-measurable non-negative and μ-integrable functions. □

The following lemma is of an elementary arithmetic nature. Let $a_1, a_2, \ldots, a_n$ be a sequence of real numbers, and p be an integer. We refer to the term a_k of the sequence as p-marked* if in the sequence of sums

$$a_k, a_k+a_{k+1}, \ldots, a_k+a_{k+1}+\ldots+a_{k+p-1}$$

at least one sum is non-negative (a_k is 1-marked iff it is non-negative).

Lemma 2. *The sum of all p-marked elements is non-negative.*

Proof. Let a_{k_1} be a p-marked element of the sequence with the lowest index and let $a_{k_1}+a_{k_1+1}+\ldots+a_{k_1+r}(r\leqslant p-1)$ be the non-negative sum with the smallest number of summands. For $h<r$, $a_{k_1}+a_{k_1+1}+\ldots+$ $+a_{k_1+h}<0$, hence $a_{k_1+h+1}+\ldots+a_{k_1+r}\geqslant 0$, i.e. all the terms of the sequence $a_{k_1}, a_{k_1+1}, \ldots, a_{k_1+r}$ are p-marked and their sum is non-negative. These considerations may be applied to the sequence starting with its a_{k_1+r+1}-th term. Thus the whole sequence is subdivided into parts where each part ends with a group of p-marked terms and the sum of the p-marked elements in each part is non-negative. The set of p-marked elements in the whole sequence coincides with the union of sets of the p-marked elements contained in its parts, which completes the proof of the lemma. □

The following lemma is the basic step in the proof of Birkhoff-Khinchin's theorem.

Lemma 3. *Let $f(u)$ be a μ-integrable function, S be a measurable transformation of $\{\mathcal{U}, \mathfrak{F}\}$ into $\{\mathcal{U}, \mathfrak{F}\}$ preserving the measure μ and let*

$$\mathsf{E}=\bigcup_{n=1}^{\infty}\left\{u: \sum_{k=1}^{n} f(S^{k-1}u)\geqslant 0\right\}.$$

Then

$$\int_E f(u)\,\mu(du)\geqslant 0. \tag{6}$$

Proof. Consider the sequence $f(u), f(Su), \ldots, f(S^{N+p-1}u)$ and denote by $s(u)$ the sum of all p-marked elements of this sequence. In view of Lemma 2 $s(u)\geqslant 0$. Let $D_k=\{u: f(S^k u)$ be a p-marked element$\}$, $\chi_k(u)$ be the indicator

* or *p-non-negative* using Loeve's terminology. *Translator's Remark.*

of the set D_k. Note that

$$D_0=\left\{u: \sup_{n\leqslant p} \sum_{k=1}^{n} f(S^{k-1}(u))\geqslant 0\right\} \quad \text{and} \quad D_k=S^{-1}D_{k-1} \quad \text{for} \quad k\leqslant N.$$

Hence $D_k=S^{-k}D_0\,(k\leqslant N)$. We thus have:

$$0\leqslant \int_{\mathscr{U}} s(u)\,\mu(du)=\int_{\mathscr{U}} \sum_{k=0}^{N+p-1} f(S^k u)\,\chi_k(u)\,\mu(du)= = \sum_{k=0}^{N+p-1} \int_{D_k} f(S^k u)\,\mu(du).$$

In view of Lemma 1

$$\int_{D_k} f(S^k u)\,\mu(du)=\int_{S^{-k}D_0} f(S^k u)\,\mu(du)=\int_{D_0} f(u)\,\mu(du), \quad k\leqslant N.$$

Consequently,

$$N\int_{D_0} f(u)\,\mu(du)+\sum_{k=N+1}^{N+p-1} \int_{D_k} f(S^k u)\,\mu(du)\geqslant 0. \tag{7}$$

Since

$$\left|\int_{D_k} f(S^k u)\,\mu(du)\right|\leqslant \int_{\mathscr{U}} |f(S^k u)|\,\mu(du)=\int_{\mathscr{U}} |f(u)|\,\mu(du)<\infty,$$

dividing inequality (7) by N and approaching with N to ∞, we obtain

$$\int_{D_0} f(u)\,\mu(du)\geqslant 0. \tag{8}$$

The sets $D_0=D_0(p)$ $(p=1, 2, \dots)$ form a monotonically increasing sequence and

$$\lim_{p\to\infty} D_0(p)=\bigcup_{p=1}^{\infty} D_0(p)=E.$$

Approaching to the limit in (8) as $p\to\infty$, we obtain (6). □

Lemma 4. (Maximal ergodic theorem). *If $f(u)$ is μ-integrable, λ is a real number and*

$$E_\lambda=\bigcup_{n=1}^{\infty} \left\{u: \frac{1}{n} \sum_{k=1}^{n} f(S^{k-1}u)\geqslant \lambda\right\},$$

then

$$\int_{E_\lambda} f(u)\,\mu(du) \geqslant \lambda \mu(E_\lambda). \tag{9}$$

Applying Lemma (3) to the function $f(u)-\lambda$ we obtain the proof of the theorem.

Theorem 1. (The Birkhoff-Khinchin theorem). *Let $\{\mathscr{U}, \mathfrak{F}, \mu\}$ be a measure space, S be a measurable mapping of $\{\mathscr{U}, \mathfrak{F}\}$ into $\{\mathscr{U}, \mathfrak{F}\}$ preserving measure μ and $f(u)$ be an arbitrary μ-integrable function. Then there exists the limit*

$$\lim_{n\to\infty} \frac{1}{n} \sum_{k=0}^{n-1} f(S^k u) = f^*(u), \tag{10}$$

μ-almost everywhere in $\mathscr{U}$, the function $f^*(u)$ is S-invariant, i.e.

$$f^*(Su) = f^*(u) \pmod{\mu}, \tag{11}$$

the function $f^*(u)$ is integrable and, if $\mu(\mathscr{U})<\infty$, then

$$\int_{\mathscr{U}} f^*(u)\,\mu(du) = \int_{\mathscr{U}} f(u)\,\mu(du). \tag{12}$$

Proof. We may assume without loss of generality that $f(u)$ is finite and non-negative.

Set

$$g^*(u) = \overline{\lim} \frac{1}{n} \sum_{k=0}^{n-1} f(S^k u), \qquad g_*(u) = \underline{\lim} \frac{1}{n} \sum_{k=0}^{n-1} f(S^k u).$$

It is required to show that $g^*(u) = g_*(u) \pmod{\mu}$. Let

$$K_{\alpha\beta} = \{u : g^*(u) > \beta,\ g_*(u) < \alpha\}, \qquad 0 \leqslant \alpha < \beta.$$

It is sufficient to show that $\mu(K_{\alpha\beta}) = 0$, for all α, β, (α, $\beta \in R$). (Indeed $\{u : g^*(u) > g_*(u)\} = \bigcup_{\substack{\alpha<\beta \\ \alpha,\beta\in R}} K_{\alpha\beta}$, where R is the set of non-negative rational numbers.) Note that

$$g^*(Su) = \overline{\lim} \left\{ \frac{n+1}{1} \frac{1}{n+1} \sum_{k=0}^{n} f(S^k u) - \frac{f(u)}{n} \right\} = g^*(u)$$

and analogously $g_*(Su) = g_*(u)$. This means in particular that $S^{-1}K_{\alpha\beta} = K_{\alpha\beta}$. Therefore Lemma 4 is applicable to the space with measure

$\{K_{\alpha\beta}, \mathfrak{F}\cap K_{\alpha\beta}, \mu\}$. It thus follows that

$$\int_{K_{\alpha\beta}} f(u)\,\mu(du)\geqslant\beta\mu(K_{\alpha\beta}). \tag{13}$$

Applying Lemma 4 to the function $-f(u)$ we obtain

$$\int_{K_{\alpha\beta}} f(u)\,\mu(du)\leqslant\alpha\mu(K_{\alpha\beta}). \tag{14}$$

Since $\beta>0$, it follows from (13) that $\mu(K_{\alpha\beta})<\infty$ but then inequality (14) may hold only if $\mu(K_{\alpha\beta})=0$. Thus the existence (mod μ) of limit (10) is verified. Set $f^*(u)=g^*(u)$. Then equation (10) is satisfied, and function $f^*(u)$ is S-invariant everywhere on $\mathscr{U}$.

In order to prove formula (12) we set $A_{kn}=\left\{u:\frac{k}{2^n}\leqslant f^*(u)<\frac{k+1}{2^n}\right\}$.

We have $\mathscr{U}=\bigcup_{k=-\infty}^{\infty} A_{kn}$, $S^{-1}A_{kn}=\left\{u:\frac{k}{2^n}\leqslant f^*(Su)<\frac{k+1}{2^n}\right\}=A_{kn}$. We now apply Lemma 4 to set A_{kn}. For any $\varepsilon>0$ we obtain $\int_{A_{kn}} f(u)\,\mu(du)>$ $>\left(\frac{k}{2^n}-\varepsilon\right)\mu(A_{kn})$. Now as $\varepsilon\to 0$ we have the inequality $\int_{A_{kn}} f(u)\,\mu(du)\geqslant$ $\geqslant\frac{k}{2^n}\mu(A_{kn})$. Analogously, $\int_{A_{kn}} f(u)\,\mu(du)\leqslant\frac{k+1}{2^n}\mu(A_{kn})$, and thus

$$\left|\int_{A_{kn}} f(u)\,\mu(du)-\int_{A_{kn}} f^*(u)\,\mu(du)\right|\leqslant\frac{1}{2^n}\mu(A_{kn}).$$

Summing up these inequalities over all k, we have

$$\left|\int_{\mathscr{U}} f(u)\,\mu(du)-\int_{\mathscr{U}} f^*(u)\,\mu(du)\right|<\frac{1}{2^n}\mu(\mathscr{U}).$$

Taking into account that n can be chosen arbitrarily in the case when $\mu(\mathscr{U})<\infty$, we obtain formula (12). The theorem is thus proved. □

Some corollaries of Birkhoff-Khinchin's theorem.

Corollary 1. *Let* $\mu(\mathscr{U})<\infty$, $f(u)\in\mathscr{L}_p\{\mathscr{U}, \mathfrak{F}, \mu\}$.

Then

$$\int_{\mathscr{U}} \left| \frac{1}{n} \sum_{k=0}^{n-1} f(S^k u) - f^*(u) \right|^p \mu(du) \to 0 \quad \text{as} \quad n \to \infty. \tag{15}$$

To prove this statement we consider an arbitrary bounded function $f_0(u)$ and let $\|f(u)-f_0(u)\|_p=\delta$, where $\|f\|_p$ is the norm of the element f in $\mathscr{L}_p\{\mathscr{U}, \mathfrak{F}, \mu\}$. Then

$$\left\| \frac{1}{n} \sum_{k=0}^{n-1} f(S^k u) - f^*(u) \right\|_p \leqslant \left\| \frac{1}{n} \sum_{k=0}^{n-1} [f(S^k u) - f_0(S^k u)] \right\|_p +$$

$$+ \left\| \frac{1}{n} \sum_{k=0}^{n-1} f_0(S^k u) - f_0^*(u) \right\|_p + \|f_0^*(u) - f^*(u)\|_p.$$

In view of Jensen's inequality and Lemma 1

$$\left\| \frac{1}{n} \sum_{k=0}^{n-1} [f(S^k u) - f_0(S^k u)] \right\|_p =$$

$$= \left\{ \int_{\mathscr{U}} \left[\frac{1}{n} \sum_{k=0}^{n-1} (f(S^k u) - f_0(S^k u)) \right]^p \mu(du) \right\}^{1/p} \leqslant$$

$$\leqslant \left\{ \int_{\mathscr{U}} \frac{1}{n} \sum_{k=0}^{n-1} |f(S^k u) - f_0(S^k u)|^p \mu(du) \right\}^{1/p} =$$

$$= \left\{ \frac{1}{n} \sum_{k=0}^{n-1} \int_{\mathscr{U}} |f(u) - f_0(u)|^p \mu(du) \right\}^{1/p} = \delta.$$

Utilizing Fatou's lemma we obtain

$$\|f_0^*(u) - f^*(u)\|_p = \left\{ \int_{\mathscr{U}} \lim \left| \frac{1}{n} \sum_{k=0}^{n-1} [f(S^k u) - f_0(S^k u)] \right|^p \mu(du) \right\}^{1/p} \leqslant$$

$$\leqslant \underline{\lim} \left\| \frac{1}{n} \sum_{k=0}^{n-1} [f(S^k u) - f_0(S^k u)] \right\|_p \leqslant \delta.$$

Next, since $f_0(u)$ is bounded all its means are bounded by the same constant. Therefore in the expression

$$\left\| \frac{1}{n} \sum_{k=0}^{n-1} f_0(S^k u) - f_0^*(u) \right\|_p = \left\{ \int_{\mathscr{U}} \left| \frac{1}{n} \sum_{k=0}^{n-1} f_0(S^k u) - f_0^*(u) \right|^p \mu(du) \right\}^{1/p},$$

in view of Lebesgue's theorem, it is permissible to pass to the limit as

$n \to \infty$ under the sign of the integral. Hence this expression tends to zero, and for n sufficiently large it becomes less than δ. Therefore,

$$\left\| \frac{1}{n} \sum_{k=0}^{n-1} f(S^k u) - f^*(u) \right\|_p < 3\delta, \qquad n \geqslant n_0 = n_0(\delta),$$

where the number δ can be chosen arbitrarily small ($\delta > 0$). The proof of (15) is thus completed. □

Definition 3. The set $A \in \mathfrak{F}$ is called *S-invariant* if $\mu((S^{-1}A)\, \Delta A) = 0$, where Δ denotes the symmetric difference of sets.

It is easy to verify that the class of all S-invariant sets forms a σ-algebra of $\mathfrak{F}$-measurable sets. Next, if $g(u)$ is an S invariant function, then the sets $\{u: g(u) \geqslant c\}$ $\{u: g(u) = c\}$ are S-invariant. On the other hand if A is S-invariant, then $\chi_A(u)$ is an S-invariant function. Denote the σ-algebra of S-invariant sets by I. Let $\mu(\mathscr{U}) = 1$. We consider $\{\mathscr{U}, \mathfrak{F}, \mu\}$ to be a probability space and let the symbol E denote the integration with respect to measure μ.

Corollary 2. $f^*(u) = \mathsf{E}\{f(u) \mid I\}$ (mod μ). Clearly $\mathsf{E}\{f(u) \mid I\}$ is an S-invariant function. Therefore to prove Corollary 2 it is sufficient to verify that for an arbitrary bounded S-invariant function $g(u)$

$$\mathsf{E} g(u)\,(f^*(u) - \mathsf{E}\{f(u) \mid I\}) = 0$$

or that $\mathsf{E}(g(u)\, f^*(u) - g(u)\, f(u)) = 0$. The latter however follows from (12) since

$$(g(u)\, f(u))^* = \lim \frac{1}{n} \sum_{k=0}^{n-1} g(S^k u)\, f(S^k u) = g(u)\, f^*(u) \quad (\text{mod}\, \mu). \quad \square$$

Ergodic stationary sequences. We now return to stationary sequences.

Let $\{\xi(t), t \in T\}$ be a stationary sequence and $\{\mathscr{X}^T, \mathfrak{C}, \mathsf{P}\}$ be its natural representation.

Corollary 3. *If f is a measurable function in $\{\mathscr{X}^m, \mathfrak{B}^m\}$ and $\mathsf{E} f(\xi(0), \xi(1), \ldots, \xi(m-1)) \neq \infty$, then with probability 1 as $n \to \infty$,*

$$\frac{1}{n} \sum_{k=0}^{n-1} f(\xi(k), \xi(k+1), \ldots, \xi(k+m-1)) \to$$
$$\to \mathsf{E}\{f(\xi(0), \xi(1), \ldots, \xi(m-1)) \mid I\},$$

where I is a σ-algebra of events in $\mathfrak{C}$, invariant under shift-transformations.

Consider an arbitrary event $A \in \mathfrak{C}$ and a sequence of events obtained from A by means of the "shifts": $A, S^{\pm 1}A, S^{\pm 2}A, \ldots$. If χ_n is the indicator of the event $S^n A$, then $\chi_n (n=0, \pm 1, \ldots)$ forms a stationary sequence of random variables and $\frac{1}{n}\sum_{k=0}^{n-1} \chi_k$ is the frequency of occurrences of event A evaluated from a single realization of the sequence $\{\xi(t), t=0, 1, 2, \ldots\}$ under $n-1$ consecutive shifts from the "origin",

$$\frac{1}{n}\sum_{k=1}^{n-1} \chi_k = \frac{\nu_n(A)}{n}.$$

In view of the Birkhoff-Khinchin theorem there exists with probability 1 the limit

$$\lim_{n\to\infty} \frac{\nu_n(A)}{n} = \pi(A) = \mathsf{E}\{\chi_A \mid I\} \quad \text{and} \quad \mathsf{E}\pi(A) = \mathsf{P}(A).$$

Quantity $\pi(A)$ may be called the empirical probability of event A. This quantity is a random variable and is determined from a single realization of the infinite sequence $\{\xi(t), t=0, 1, 2, \ldots\}$. The question arises: under what circumstances is the empirical probability $\pi(A)$ independent of chance and coincides with the probability $\mathsf{P}(A)$?

Stationary sequences which possess this property are called *ergodic*. The following definition is more general:

Definition 4. Let $\{\mathscr{U}, \mathfrak{F}, \mu\}$ be a probability space, let S be a measure-preserving transformation of $\mathscr{U}$ into itself, $\nu_n(A) = \nu_n(A, u)$ be the number of terms of the sequence $\{u, Su, \ldots, S^{n-1}u\}$ falling into the set A. The transformation S is called *ergodic* if for any $A \in \mathfrak{F}$

$$\lim_{n\to\infty} \frac{\nu_n(A, u)}{n} = \mu(A) \pmod{\mu}.$$

The transformation S is called *metrically transitive* if any S-invariant set has measure 0 or 1.

Theorem 2. *In order that a transformation S in the probability space $\{U, \mathfrak{F}, \mu\}$ be ergodic it is necessary and sufficient that one of the following two conditions be satisfied:*

a) *S is metrically transitive*

b) *For any $\mathfrak{F}$-measurable μ-integrable function $f(u)$, the function*

$$f^*(u) = \lim \frac{1}{n}\sum_{k=0}^{n-1} f(S^k u)$$

is constant with probability 1.

Proof. Let A be an S-invariant set, $0 < \mu(A) < 1$. The sets $A, SA, S^2A, \ldots,$

differ on a set of measure 0 and $\nu_n(A)=n\chi_A(u)\ (\mathrm{mod}\,\mu)$. Hence $\lim_{n\to\infty}\frac{\nu_n(A)}{n}$ cannot be a constant $(\mathrm{mod}\,\mu)$. Therefore ergodicity yields metrical transitivity. Now let S be metrically transitive. Since the function $f^*(u)$ is S-invariant the symmetric difference of the sets

$$S^{-1}\{u: f^*(u)<x\}=\{u: f^*(Su)<x\} \quad \text{and} \quad \{u: f^*(u)<x\}$$

is of μ-measure 0. From here it follows that $\mu\{u: f^*(u)<x\}=0$ or 1 for any real x, i.e. $f^*(u)=\mathrm{const}\,(\mathrm{mod}\,\mu)$. Therefore a) implies b). Finally, the condition of ergodicity is a particular case of condition b) namely the case when $f(u)$ is the indicator of a certain event. □

We now present a few corollaries of ergodicity.

Let $\{\mathscr{X}^T, \mathfrak{C}, \mathsf{P}\}$ be a natural representation of a stationary sequence $\xi(n)$, S be the shift transformation in $\mathscr{X}^T$, $\mathscr{L}_2=\mathscr{L}_2\{\mathscr{X}^T, \bar{\mathfrak{C}}, \mathsf{P}\}$.

It follows from Corollary 1 of Theorem 1 that for arbitrary functions $f(u)$ and $g(u)$ in $\mathscr{L}_2$

$$\lim_{n\to\infty}\int_{\mathscr{X}^T}\frac{1}{n}\sum_{k=0}^{n-1} f(S^k u)\,g(u)\,\mathsf{P}(du)=\int_{\mathscr{X}^T} f^*(u)\,g(u)\,\mathsf{P}(du). \tag{16}$$

We say that the sequence $\{\xi(n), n=0, \pm 1, \ldots\}$ is *ergodic* if the transformation S is ergodic. Set $g(u)=\eta$, $f(S^k u)=\zeta_k$ and assume that the initial stationary sequence $\{\xi(n), n=0, \pm 1, \ldots\}$ is ergodic. Relation (16) now becomes

$$\lim_{n\to\infty}\mathsf{E}\frac{1}{n}\sum_{k=0}^{n-1}\zeta_k\eta=\mathsf{E}\zeta_0\,\mathsf{E}\eta. \tag{17}$$

Let

$$g(u)=\chi_B(u), \quad f(u)=\chi_A(u), \quad A \text{ and } B\in\bar{\mathfrak{C}}.$$

It follows from (17) that

$$\lim_{n\to\infty}\frac{1}{n}\sum_{k=0}^{n-1}\mathsf{P}(S^{-k}A\cap B)=\mathsf{P}(A)\,\mathsf{P}(B) \tag{18}$$

or (if $\mathsf{P}(B)\neq 0$)

$$\lim_{n\to\infty}\frac{1}{n}\sum_{k=0}^{n-1}\mathsf{P}(S^{-k}A\mid B)=\mathsf{P}(A), \tag{19}$$

where $\mathsf{P}(S^{-k}A\mid B)$ is the conditional probability of the event $S^{-k}A$ given B.

Lemma 5. *The validity of equality* (18) (*or* (19)) *for any sets* $A, X\in\bar{\mathfrak{C}}$ *is equivalent to ergodicity.*

It is sufficient to show that (18) implies ergodicity. Let C be an arbitrary S-invariant event. Set in (18) $A=B=C$. This equality then becomes $\mathsf{P}(C)=\mathsf{P}^2(C)$, i.e. $\mathsf{P}(C)=0$ or 1 and the lemma follows from Theorem 2. □

Equation (19) has the following probabilistic meaning. Let A and B be two events in $\mathfrak{C}$. If event A is shifted indefinitely in time, then on the average, events $S^{-n}A$ and B become independent for any event B. Condition (19) may be replaced by a more stringent requirement

$$\lim_{n\to\infty} \mathsf{P}(S^{-n}A \mid B)=\mathsf{P}(A), \tag{20}$$

which is called the *mixing condition*. Condition (20) is a particular case of equality

$$\lim_{n\to\infty} \mathsf{E}\zeta_n\eta=\mathsf{E}\zeta_0\mathsf{E}\eta, \tag{21}$$

where $\zeta_n=f(S^n u)$, $\eta=g(u)$, $f(u)$ and $g(u)$ are arbitrary functions in $\mathscr{L}_2$. On the other hand (20) implies (21) for simple functions f and g. Approximating arbitrary functions $f(u)$ and $g(u)$ in $\mathscr{L}_2$ by means of sequences of simple functions $f_n(u)$ and $g_n(u)$ converging in $\mathscr{L}_2$ to $f(u)$ and $g(u)$ respectively, it is easy to see that the mixing condition is equivalent to condition (21) ($f(u)$ and $g(u)$ are arbitrary functions in $\mathscr{L}_2$). On the other hand it is sufficient to check condition (21) for a certain set of functions whose linear span is everywhere dense in $\mathscr{L}_2$. Indicators of cylindrical sets can be chosen as such a set of functions.

Consider a sequence $\{\xi_n, n=0, \pm 1, \ldots\}$ of independent identically distributed random variables such that $\mathsf{E}|\xi_n|<\infty$. Such a sequence is stationary. In view of Birkhoff-Khinchin's theorem

$$\lim_{n\to\infty} \frac{1}{n}\sum_{k=0}^{n-1} \xi_k=\xi^* \quad (\operatorname{mod} \mathsf{P}), \qquad \mathsf{E}\xi^*=\mathsf{E}\xi.$$

Evidently, the random variable ξ^* does not depend on any finite set of variables $\xi_0, \xi_1, \ldots, \xi_p$. Therefore it is measurable with respect to $\overline{\lim}\, \tilde{\sigma}\{\xi_k\}$ and in view of the zero-one law is a constant, i.e. $\xi^*=c(\operatorname{mod} \mathsf{P})$, and moreover $c=\mathsf{E}\xi$. We thus obtain the following theorem:

Theorem 3. *(The strong law of large numbers). If $\{\xi_n, n=0, \pm 1, \ldots\}$ is a sequence of independent identically distributed random variables and $\mathsf{E}|\xi_n|<\infty$, then with probability* 1

$$\lim_{n\to\infty} \frac{1}{n}\sum_{k=0}^{n-1} \xi_k=\mathsf{E}\xi_0. \tag{22}$$

The theorem just proved is a corollary of the ergodicity of a sequence of independent identically distributed random variables. One can, however, prove a stronger result – namely, that the shift operator in $\mathscr{X}^T$ is a mixing with respect to the measure induced in $\mathscr{X}^T$ by a sequence of independent random variables. This in turn follows from a more general assertion. Let $\{\zeta_n, n=0, \pm 1, \pm 2, \ldots\}$ be a stationary sequence of random elements in $\{\mathscr{X}, \mathfrak{B}\}$, $\mathfrak{F}_n$ be the σ-algebra generated by the random elements ξ_n, $\xi_{n+1}, \ldots, \mathfrak{F}_\infty = \bigcap_n \mathfrak{F}_n = \overline{\lim}\, \mathfrak{F}_n$. We say that the zero-one law is applicable to the sequence $\{\xi_n, n=0, \pm 1, \ldots\}$ if the σ-algebra $\mathfrak{F}_\infty$ contains only those events whose probability is 0 or 1.

Theorem 4. *If the sequence $\{\xi_n, n=0, \pm 1, \ldots\}$ satisfies the zero-one law then the shift transformation is a mixing.*

Set $\zeta_{-n} = \mathsf{P}\{B \mid \mathfrak{F}_n\}$. The sequence $\{\zeta_n, \mathfrak{F}_n, n = \ldots -k, -k+1 \ldots, 0\}$, $\mathfrak{F}_{-n} = \mathfrak{F}_n$, is a martingale (cf. Theorem 4, Chapter II, Section 2) and $\mathsf{P}\{B \mid \mathfrak{F}_{-\infty}\}$ is its closure on the left. Since σ-algebra $\mathfrak{F}_{-\infty}$ is trivial, it follows that $\mathsf{P}\{B \mid \mathfrak{F}_{-\infty}\} = \text{const} = \mathsf{P}(B) \pmod{\mathsf{P}}$. In view of the theorem on convergence of martingales (Theorem 1, Corollary, Section 2 in Chapter II) $\lim \mathsf{P}\{B \mid \mathfrak{F}_n\} = \mathsf{P}(B)$ with probability 1. Let A be a cylinder over the coordinates $n=0, 1, 2, \ldots$ Then $S^{-n}A \in \mathfrak{F}_n$. Therefore for $n \to \infty$

$$\mathsf{P}(B \cap S^{-n}A) = \int_{S^{-n}A} \mathsf{P}\{B \mid \mathfrak{F}_n\}\, \mathsf{P}(du) \sim \mathsf{P}(B)\, \mathsf{P}(S^{-n}A) = \mathsf{P}(B)\, \mathsf{P}(A).$$

Clearly this relationship holds for any $A \in \mathfrak{C}$. This yields relation (21) as was noted earlier. The theorem is thus proved. □

Consider as another example of a process satisfying the mixing condition the stationary Gaussian sequence with correlation coefficient approaching zero. Let $\{\xi_n, n=0, \pm 1, \pm 2, \ldots\}$ be a stationary Gaussian sequence, $\mathsf{E}\xi_n = m$,

$\mathsf{E}(\xi_n - m)(\xi_0 - m) = R_n$; $f(u) = f(x_0, x_1, \ldots, x_p)$ and $g(u) = g(x_0, x_1, \ldots, x_p)$

be bounded sufficiently smooth functions of $p+1$ variables possessing an absolutely integrable Fourier transforms $f^*(\lambda_0, \ldots, \lambda_p)$, $g^*(\lambda_0, \ldots, \lambda_p)$. Then

$$\begin{aligned}
&\mathsf{E}f(\xi_n, \xi_{n+1}, \ldots, \xi_{n+p})\, g(\xi_0, \xi_1, \ldots, \xi_p) = \\
&\quad = \mathsf{E} \int_{-\infty}^{\infty} \cdots \int_{-\infty}^{\infty} \exp\left\{ i \left(\sum_{k=0}^{p} \lambda_k \xi_{n+k} + \sum_{k=0}^{p} \mu_k \xi_k \right) \right\} \times \\
&\quad\quad \times f^*(\lambda_0, \ldots, \lambda_p)\, g^*(\mu_0, \ldots, \mu_p)\, d\lambda_0 \ldots d\lambda_p d\mu_0 \ldots d\mu_p =
\end{aligned}$$

$$= \int_{-\infty}^{\infty} \dots \int_{-\infty}^{\infty} \exp\left[-\tfrac{1}{2}\left\{\sum_{k,r=0}^{p} R_{k-r}(\lambda_k\lambda_r + \mu_k\mu_r) + \sum_{k,r=0}^{p} R_{n+k-r}\lambda_k\mu_r\right\}\right] \times$$

$$\times f^*(\lambda_0, \dots, \lambda_p)\, g^*(\mu_0, \dots, \mu_p)\, d\lambda_0 \dots d\lambda_p\, d\mu_0 \dots d\mu_p.$$

If $\lim_{n\to\infty} R_n = 0$, then approaching in this relation to the limit as $n\to\infty$ we obtain

$$\lim_{n\to\infty} \mathsf{E} f(\xi_n, \xi_{n+1}, \dots, \xi_{n+p})\, g(\xi_0, \xi_1, \dots, \xi_p) =$$
$$= \mathsf{E} f(\xi_0, \xi_1, \dots, \xi_p)\, \mathsf{E} g(\xi_0, \xi_1, \dots, \xi_p). \tag{23}$$

Since the class of functions f and g for which the last relation has been proved is everywhere dense in $\mathscr{L}_2$, it follows that relation (23) is valid for arbitrary f and g belonging to $\mathscr{L}_2$. We have thus proved the following result.

Theorem 5. *A stationary Gaussian sequence with correlation coefficient $R_n \to 0$ as $n\to\infty$ satisfies the mixing condition.*

We now present a number of corollaries and remarks related to ergodic properties of Markov chains.

Consider an irreducible Markov chain with a countable number of states. It possesses an invariant initial distribution if and only if it is positive-recurrent (Corollary 2 of Theorem 15, Section 5). In turn for the last property to be valid, it is necessary and sufficient that the system of equations

$$\sum_k x_k p(k, j) = x_j$$

possess a non-trivial absolutely summable solution. The solution of this system satisfying condition $\sum_k x_k = 1$, provided it exists, is of the form

$$x_k = v_k = \lim_{N\to\infty} \frac{1}{N} \sum_{n=1}^{N} p^{(n)}(j, k). \tag{24}$$

Moreover, if the chain is also aperiodic, then

$$x_k = v_k = \lim_{n\to\infty} p^{(n)}(j, k)$$

(cf. Section 5, Theorem 15). Assume that the chain is irreducible and positive-recurrent. The invariant initial distribution for this chain is unique and the corresponding stationary Markov process is $\{\xi(t), t=0, \pm 1, \dots\}$. The condition of ergodicity of this process may be represented in the form

$$\lim_{N\to\infty}\frac{1}{N}\sum_{n=1}^{N}\mathsf{P}\{\xi(1)=i_1,\xi(2)=i_2,\dots,\xi(s)=i_s,$$
$$\xi(n+1)=j_1,\dots,\xi(n+r)=j_r\}=\mathsf{P}\{\xi(1)=i_1,\dots\xi(s)=i_s\}\times$$
$$\times\mathsf{P}\{\xi(1)=j_1,\dots,\xi(r)=j_r\}. \qquad (25)$$

Indeed, on one hand this condition is a particular case of condition (18); on the other hand, however, it is easy to verify that (25) implies (18) for arbitrary cylinders A and B in $\mathscr{X}^T$. Furthermore condition (25) is equivalent to equalities

$$\lim_{N\to\infty} v_i\frac{1}{N}\sum_{n=1}^{N}p^{(n)}(i,j)=v_iv_j,$$

i.e. to equalities (24).

Thus the following theorem holds:

Theorem 6. *A stationary process which corresponds to an invariant initial distribution of an irreducible, positive-recurrent Markov chain is ergodic.*

Remark. One can verify analogously that the mixing condition for a stationary Markov process reduces to the requirement $\lim_{n\to\infty}p^{(n)}(j,k)=v_k$, therefore a stationary process which corresponds to a positive-recurrent and aperiodic Markov chain possesses the mixing property.

Chapter III

Random Functions

§ 1. Some Classes of Random Functions

Gaussian random functions. Definition 1. A real random function $\xi(x)$, $x \in X$ is called *Gaussian* if for any integer $n \geqslant 1$ and any x_k, $k=1, 2, \ldots, n$, $x_k \in X$, the sequence $\{\xi(x_1), \xi(x_2), \ldots, \xi(x_n)\}$ has a joint normal distribution.

It follows from the definition, that the characteristic function of this distribution is of the form

$$J(x_1, x_2, \ldots, x_n, u^1, \ldots, u^n) = \\ = \exp\left\{i \sum_{k=1}^{n} u^k a_k - \tfrac{1}{2} \sum_{k,r=1}^{n} b_{kr} u^k u^r\right\}, \tag{1}$$

where the constants a_k and b_{kr} satisfy

$$a_k = \mathsf{E}\xi(x_k), \quad b_{kr} = \mathsf{E}(\xi(x_k) - a_k)(\xi(x_r) - a_r). \tag{2}$$

Thus all the marginal distributions of a Gaussian random function are determined by two real functions – the mean value $a(x)$ and the correlation function $b(x_1, x_2)$

$$a(x) = \mathsf{E}\xi(x), \quad b(x_1, x_2) = \mathsf{E}(\xi(x_1) - a(x_1))(\xi(x_2) - a(x_2)).$$

The correlation function $b(x, y)$ possesses the following properties:

1) $b(x, y) = b(y, x)$,
2) for any n, any real numbers u_k and points $x_k \in X$

$$\sum_{k,r=1}^{n} b(x_k, x_r)\, u_k u_r \geqslant 0.$$

Real functions possessing these properties are called positive-definite kernels on X^2.

This definition is equivalent to the requirement that for any x_r and $x_k \in X$ the matrix $\|b(x_k, x_r)\|$ $(k, r = 1, 2, \ldots, n)$ is real, symmetric and non-negative-definite.

Note that for an arbitrary set X, real-valued function $a(x)$, $x \in X$ and a non-negative-definite real kernel $b(x_1, x_2)$ on X^2, there exists a Gaussian random function for which $a(x)$ is its mean value and $b(x_1, x_2)$ is its correlation function. To prove this assertion consider a family of distributions $\{p_{x_1 \dots x_n}(\cdot), n=1, 2, \dots, x_k \in X\}$ with the characteristic functions given by relations (1). It is easy to verify that this family satisfies compatability conditions. It now remains to apply Kolmogorov's theorem (Theorem 2, Section 4, Chapter 1).

A vector Gaussian process with real components is defined analogously. Let $\xi(x)$, $x \in X$ be a random function with values in the m-dimensional space $\mathscr{R}^m$. This function is called Gaussian if the joint distribution of all components of the sequence $\{\xi(x_1), \xi(x_2), \dots, \xi(x_n)\}$ for any $n \geqslant 1$ and any $x_k \in X$ is normal. The corresponding characteristic function is of the form

$$J(x_1, \dots, x_n, u_1^{(1)}, \dots, u_m^{(1)}, \dots, u_1^{(n)}, \dots, u_m^{(n)}) = \\ = \exp\left\{ i \sum_{r=1}^{m} \sum_{k=1}^{n} u_r^{(k)} a^{(r)}(x_k) - \tfrac{1}{2} \sum_{k,l=1}^{n} \sum_{r,s=1}^{m} b^{rs}(x_k, x_l)\, u_r^{(k)} u_s^{(l)} \right\}.$$

To simplify this expression we introduce vectors $u^{(k)} = (u_1^{(k)}, u_2^{(k)}, \dots, u_m^{(k)})$, $a(x) = (a^1(x), \dots, a^m(x))$ with values in $\mathscr{R}^m$ and the matrix $b(x_1, x_2)$ with elements $b^{rs}(x_1, x_2)$ $r, s = 1, 2, \dots, m$. Then the previous expression becomes

$$J(x_1, \dots, x_n, u^{(1)}, \dots, u^{(n)}) = \\ = \exp\left\{ i \sum_{k=1}^{n} (u^{(k)}, a(x_k)) - \tfrac{1}{2} \sum_{k,l=1}^{n} (b(x_k, x_l)\, u^{(l)}, u^{(k)}) \right\}.$$

Here the vector function $a(x)$ may be arbitrary and the real matrix function $b(x_1, x_2)$ should be symmetric and should satisfy the condition: for any integer $n \geqslant 1$, any $x_k \in X$ and $u^{(k)} \in \mathscr{R}^m$,

$$\sum_{k,l=1}^{n} (b(x_k, x_l)\, u^{(l)}, u^{(k)}) \geqslant 0. \tag{3}$$

The converse is also obvious: Given an arbitrary function $a(x)$ with values in $\mathscr{R}^m$ and a matrix function $b(x_1, x_2)$ satisfying condition (3) there exists a Gaussian random function $\xi(x) = (\xi^1(x), \dots, \xi^m(x))$ for which

$$a^{(k)}(x) = \mathsf{E}\xi^k(x),$$
$$b^{rs}(x_1, x_2) = \mathsf{E}(\xi^r(x_1) - a^r(x_1))\,(\xi^s(x_2) - a^s(x_2)).$$

In certain problems, moments of a Gaussian random function may be useful. These can be obtained from the series expansion of characteristic functions. We shall confine ourselves to central moments of a scalar random function. Set

$$a(x)=0, \quad u=(u^{(1)}, u^{(2)}, \dots, u^{(n)}),$$
$$B=|b(x_k, x_r)|, \quad k, r=1, \dots, n.$$

Then

$$J(x_1, x_2, \dots, x_n, tu)=e^{-t^2/2\,(Bu,\,u)}=$$
$$=1-\frac{t^2}{2}(Bu, u)+\frac{t^1}{2!\,2^2}(Bu, u)^2+\dots$$
$$\dots+(-1)^n\frac{t^{2n}}{2^n n!}(Bu, u)^n+\dots,$$

from here it follows that

$$\mathsf{E}\left(\sum_{k=1}^{n} u^{(k)}\,\xi(x_k)\right)^{2n}=(2n-1)!!\,(Bu, u)^n \tag{4}$$

and

$$\mathsf{E}\left(\sum_{k=1}^{n} u^{(k)}\xi(x_k)\right)^{2n-1}=0.$$

We introduce the n-point moment functions

$$m_{j_1 j_2 \dots j_n}(x_1, x_2, \dots, x_n)=\mathsf{E}\,[\xi(x_1)]^{j_1}\,[\xi(x_2)]^{j_2}\dots[\xi(x_n)]^{j_n}.$$

The quantity $j_1+j_2+\dots+j_n$ is called the order of the moment function. Moment functions of odd order are equal to zero:

$$m_{j_1 j_2 \dots j_n}(x_1, x_2, \dots, x_n)=0 \quad \text{for} \quad \sum_1^n j_k=2s-1.$$

Formula (4) can be written in the form

$$m_{j_1 j_2 \dots j_n}(x_1 x_2 \dots x_n)=\frac{\partial^{2n}}{\partial u_1^{j_1}\partial u_2^{j_2}\dots\partial u_n^{j_n}}\frac{1}{2^n n!}(Bu, u)^n. \tag{5}$$

Moment functions of the second order coincide with the correlation function

$$m_{11}(x_1, x_2)=b(x_1, x_2), \quad m_2(x)=m_{11}(x, x)=b(x, x).$$

For moment functions of the fourth order the following formulas are available:

$$m_4(x)=3b^2(x, x), \quad m_{22}(x_1, x_2)=2b^2(x_1, x_2),$$
$$m_{31}(x_1, x_2)=3b(x_1, x_1)\,b(x_1, x_2),$$

$$m_{211}(x_1, x_2, x_3) = b(x_1, x_1)\, b(x_2, x_3) + 2b(x_1, x_2)\, b(x_1, x_3),$$
$$m_{1111}(x_1, x_2, x_3, x_4) =$$
$$= b(x_1, x_2)\, b(x_3, x_4) + b(x_1, x_3)\, b(x_2, x_4) + b(x_1, x_4)\, b(x_2, x_4).$$

Generally the following relation

$$m_{j_2 \dots j_n}(x_1, x_2, \dots, x_n) = \sum \prod b(x_p, x_q) \tag{6}$$

holds. The structure of this formula can be described as follows: we write down the points $x_1, x_2, \dots, x_n$ into a sequence where x_k is written j_k times. This sequence is then subdivided into arbitrary pairs. The product in the r.h.s. of formula (6) is taken over all the pairs of this subdivision and the sum is taken over all the subdivisions (those pairs which are permutations of one another are counted once). This assertion follows directly from formula (5).

Complex-valued Gaussian random functions are considered in a number of problems. Their definition involves a feature which distinguishes them from general vector Gaussian functions with real components. We shall discuss only functions $\{\zeta(x), x \in X\}$ with values in $\mathscr{Z}^1$. Set $\zeta(x) = \xi(x) + i\eta(x)$, where $\xi(x)$ and $\eta(x)$ are real.

Definition 2. A random function $\{\zeta(x), x \in X\}$ is called a *complex Gaussian random function* if the real vector function $\{(\xi(x), \eta(x)), x \in X\}$ is Gaussian and $\mathsf{E}(\zeta(x) - a(x))\,(\zeta(y) - a(y)) = 0$, $a(x) = \mathsf{E}\zeta(x)$, for any x, $y \in X$.

It may be assumed without loss of generality that $a(x) = 0$. It is easy to verify that the condition $\mathsf{E}\zeta(x)\,\zeta(y) = 0$ is equivalent to conditions

$$\mathsf{E}\xi(x)\,\xi(y) = \mathsf{E}\eta(x)\,\eta(y), \qquad \mathsf{E}\xi(x)\,\eta(y) = -\,\mathsf{E}\xi(y)\,\eta(x). \tag{7}$$

On the other hand if equalities (7) are satisfied, then

$$b(x, y) = \mathsf{E}\zeta(x)\,\overline{\zeta(y)} = 2(b_{11}(x, y) - ib_{12}(x, y)), \tag{8}$$

where $b_{11}(x, y) = \mathsf{E}\xi(x)\,\xi(y)$, $b_{12}(x, y) = \mathsf{E}\xi(x)\,\eta(y)$. From conditions (7) it follows in particular that $b_{12}(x, x) = \mathsf{E}\xi(x)\,\eta(x) = 0$, and since $(\xi(x), \eta(x))$ have joint Gaussian distributions, variables $\xi(x)$ and $\eta(x)$ are independent. If we now put $\zeta(x) = \varrho(x)\, e^{i\varphi(x)}$ then, as it is easy to verify, the variables $\varrho(x)$ and $\varphi(x)$ will be independent, $\varphi(x)$ having the uniform distribution on $(-\pi, \pi)$ and $\varrho(x)$ has the density given by

$$\frac{u}{\sigma^2(x)}\, e^{-\frac{u^2}{2\sigma^2(x)}}, \; u > 0, \qquad \sigma^2(x) = \mathsf{V}\xi(x) = b_{11}(x, x).$$

In relation (8) the function $b_{11}(x, y)$ is a non-negative-definite kernel, and $b_{12}(x, y)$ possess the property that $b_{12}(x, y) = -b_{12}(y, x)$. Utilizing

Consider the "truncated" variables $\alpha_{nk}(x)$ and their moments

$$\alpha_{nk}^{\varepsilon}(x)=\chi_{\varepsilon}(\alpha_{nk}(x))\,\alpha_{nk}(x), \qquad a_{nk}^{\varepsilon}(x)=\mathsf{E}\alpha_{nk}^{\varepsilon}(x),$$
$$b_{nk}^{\varepsilon}(x_1, x_2)=\mathsf{E}\,[\alpha_{nk}^{\varepsilon}(x_1)-a_{nk}^{\varepsilon}(x_1)]\,[\alpha_{nk}^{\varepsilon}(x_2)-a_{nk}^{\varepsilon}(x_2)],$$

where $\varepsilon>0$ and $\chi_{\varepsilon}(x)$ is the indicator of the interval $(-\varepsilon, \varepsilon)$.

Theorem 2. *Let the functions $\alpha_{n1}(x), \alpha_{n2}(x), \dots, \alpha_{nm_n}(x)$ be mutually independent for each n and satisfy conditions:*

1) *for any $\varepsilon>0$*

$$\sum_{k=1}^{m_n} \mathsf{P}\{|\alpha_{nk}(x)|>\varepsilon\}\to 0 \quad \text{as} \quad n\to\infty;$$

2) *for some $\varepsilon=\varepsilon_0=\varepsilon_0(x)>0$,*

$$\sum_{k=1}^{m_n} a_{nk}^{\varepsilon_0}(x)\to a(x), \qquad \sum_{k=1}^{m_n} b_{nk}^{\varepsilon_0}(x_1, x_2)\to b(x_1, x_2) \tag{9}$$

as $n\to\infty$. Then the marginal distributions of the random function $\eta_n(x)$ as $n\to\infty$ converge weakly to the corresponding marginal distributions of a Gaussian random function with mathematical expectation $a(x)$ and correlation function $b(x_1, x_2)$.

Processes with independent increments. Let T be a finite or infinite interval closed on the left, $a=\min T>-\infty$.

Definition 3. A random process $\{\xi(t), t\in T\}$ with values in $\mathscr{R}^m$ is a *process with independent increments* if for any n, $t_k\in T$, $t_1<t_2<\dots t_n$, the random vectors $\xi(a), \xi(t_1)-\xi(a), \dots, \xi(t_n)-\xi(t_{n-1})$ are mutually independent. The vector $\xi(a)$ is called the *initial state (value) of the processes* and its distribution is called *the initial distribution of the process.*

To define a process with independent increments in the wide sense it is sufficient to define the initial distribution $\mathsf{P}_0(B)$ and a family of probabilities $\mathsf{P}(t, h, B)$ $(t\geqslant 0, h>0, B\in\mathfrak{B}^m)$ where $\mathfrak{B}^m$ is the σ-algebra of Borel sets in $\mathscr{R}^m$ and $\mathsf{P}(t, h, B)$ is the distribution of the vector $\xi(t+h)-\xi(t)$. Indeed, if these distributions are given, then arbitrary joint distributions of vectors $\xi(t_1), \xi(t_2), \dots, \xi(t_n)$ are uniquely determined by the formula

$$\mathsf{P}\left(\bigcap_{k=0}^{n}\{\xi(t_k)\in B_k\}\right)=$$
$$=\int_{B_0}\mathsf{P}_0(dy_0)\int_{B_1-y_0}\mathsf{P}(0, t_1, dy_1)\int_{B_2-(y_0+y_1)}\mathsf{P}(t_1, t_2-t_1, dy_2)\dots$$
$$\dots\int_{B_0-(y_0+\dots+y_n)}\mathsf{P}(t_{n-1}, t_n-t_{n-1}, dy_n). \tag{10}$$

this fact it is easy to verify that function $b(x, y)$, determined by formula (8) in which $b_{11}(x, y)$ and $b_{12}(x, y)$ are arbitrary functions possessing these properties, satisfies the relation

$$\sum_{k, r=1}^{n} b(x_k, x_r)\, z_k \bar{z}_r \geqslant 0$$

for arbitrary n, $x_k \in X$ and arbitrary complex numbers z_k. Functions possessing these properties are called non-negative – definite kernels on X^2.

Theorem 1. *For any non-negative-definite kernel $b(x, y)$ $(x, y \in X)$ there exists a complex Gaussian random function $\zeta(x)$ for which $\mathsf{E}\zeta(x)=0$ and $\mathsf{E}\zeta(x)\overline{\zeta(y)}=b(x, y)$.*

To prove this assertion we introduce a real matrix function of second order $B(x, y)=\|b_{ik}(x, y)\|$ $(i, k=1, 2)$ putting

$$b_{11}(x, y)=b_{22}(x, y)=\frac{b'(x, y)}{2},$$
$$b_{12}(x, y)=-b_{21}(x, y)=-\tfrac{1}{2}b''(x, y),$$

where $b'(x, y)=\mathrm{Re}\, b(x, y)$, $b''(x, y)=\mathrm{Im}\, b(x, y)$. Since $b(x, y)$ is a non-negative-definite kernel, $b(x, y)=\overline{b(y, x)}$ and it follows from here that $b_{11}(x, y)=b_{11}(y, x)$, $b_{12}(x, y)=-b_{12}(y, x)$. Construct a two-dimensional Gaussian random function $(\xi(x), \eta(x))$ with correlation matrix $B(x, y)$. In view of the previous remarks $\zeta(x)=\xi(x)+i\eta(x)$ is a complex Gaussian random function and

$$\mathsf{E}\zeta(x)\overline{\zeta(y)}=2(b_{11}(x, y)-ib_{12}(x, y))=b'(x, y)+ib''(x, y)=b(x, y). \quad \square$$

The fact that Gaussian random functions play an important role in practical problems may often be explained as follows. Under very general conditions the sum of a large number of independent small (in magnitude) random functions is approximately a Gaussian random function independently of the probabilistic nature of the components (summands). This assertion is the so-called theorem on normal correlation which is a multivariate generalization of the central limit theorem. We now present one version of this theorem.

Let a double sequence of random functions $\{\alpha_{nk}(x), x \in X\}$, $k=1, 2, \ldots, m_n$, $n=1, 2, \ldots$ be given. Set

$$\eta_n(x)=\sum_{k=1}^{m_n} \alpha_{nk}(x).$$

Here $B-z$ denotes the set $\{x : x = y - z, y \in B\}$. As far as the initial distribution is concerned it may be chosen arbitrarily. On the other hand, one cannot guarantee that a process exists with independent increments which corresponds to an arbitrarily defined family of distributions $\mathsf{P}(t, h, B)$.

In order that this be the case, it is necessary and sufficient that $\mathsf{P}(t, h, B)$ possess the following property: for an arbitrary n and any $a = t_0 < t_1 < \ldots < t_n = t + h$, $\mathsf{P}(t, h, B)$ is a distribution of sums of independent random vectors $\xi_1, \xi_2, \ldots, \xi_n$, where ξ_k is distributed according to $\mathsf{P}(t_{k-1}, t_k - t_{k-1}, B)$.

Indeed if this condition is satisfied, then the family of distributions (10) satisfies the compatability conditions. Therefore Kolmogorov's theorem is applicable and there exists a random process with marginal distributions (10). The form of these distributions indicates that the process has independent increments.

It is convenient to study processes with independent increments using characteristic functions.

Set

$$J(t, h, u) = \int_{\mathcal{R}^m} e^{i(u, x)} \mathsf{P}(t, h, dx).$$

Function $J(t, h, u)$ is called the characteristic function of a process with independent increments. This function completely determines the joint distribution of the differences

$$\xi(t_1) - \xi(a), \xi(t_2) - \xi(t_1), \ldots, \xi(t_n) - \xi(t_{n-1}). \tag{11}$$

Indeed the joint distribution of the sequence of vectors (11) has its characteristic function $J(t_1, t_2, \ldots, t_n, u^1, u^2, \ldots, u^n)$ equal to

$$J(t_1, t_2, \ldots, t_n, u^1, u^2, \ldots, u^n) = \prod_{k=1}^{n} J(t_{k-1}, \Delta t_k, u_k),$$
$$\Delta t_k = t_k - t_{k-1}, \; t_0 = a.$$

Therefore to define a process with independent increments in the wide sense it is sufficient to define $J(t, h, u)$ (in addition to $\mathsf{P}_0(B)$). The necessary and sufficient condition on $\mathsf{P}(t, h, B)$ stated above means that the characteristic function $J(t, h, u)$ considered as a function of the interval $[t, t+h)$ must be multiplicative:

$$J(t, h_1 + h_2, u) = J(t, h_1, u)\, J(t + h_1, h_2, u).$$

In turn this condition is necessary and sufficient in order that $J(t, h, u)$ be the characteristic function of a process with independent increments.

Definition 4. A process with independent increments is called *homogeneous* if the differences $\xi(t+h)-\xi(t)$ are distributed independently of t, i.e. $\mathsf{P}(t, h, B)=\mathsf{P}(h, B)$. A homogeneous process is called *stochastically continuous* if

$$\lim_{h\to 0} \mathsf{P}(h, \bar{S}_\varepsilon)=0$$

for any sphere $S_\varepsilon=\{x:|x|<\varepsilon\}$, $\varepsilon>0$.

(See Section 2 for additional details concerning the condition of stochastic continuity and its significance.) If a homogeneous process is stochastically continuous then for any t the difference $\xi(t+h)-\xi(t)$ converges in probability to zero and hence the distribution of $\xi(t+h)-$ $-\xi(t)$ is weakly convergent to zero (as $h\downarrow 0$). In view of the continuity of the correspondence between distributions and their characteristic functions it follows that stochastic continuity is equivalent to the following property: for $h\downarrow 0$ $J(h, u)\to 1$ uniformly in any bounded region $|u|\leqslant N$.

We note a few properties of characteristic functions of homogeneous stochastically continuous processes with independent increments.

a) *The characteristic function of a homogeneous process with independent increments satisfies equation*

$$J(h_1+h_2, u)=J(h_1, u)\, J(h_2, u). \tag{12}$$

In particular for any integral n

$$J(nh, u)=[J(h, u)]^n.$$

b) *The characteristic function $J(h, u)$ of a homogeneous stochastically continuous process nowhere vanishes.*

Indeed, for an arbitrary u one can find t_0 such that $|J(h, u)|\geqslant\frac{1}{2}$ for $0<h\leqslant t_0$. If t is arbitrary and $t=t_0(n+\theta)$, where $0\leqslant\theta<1$, then $J(t, u)=$ $=J(t_0 n, u)\times J(t_0\theta, u)=[J(t_0, u)]^n\times J(t_0\theta, u)$; thus $|J(t, u)|\geqslant(\frac{1}{2})^{n+1}$. Since $J(h, u)\to 1$ as $h\downarrow 0$ uniformly in an arbitrary sphere $|u|\leqslant N$, it is possible to define a single-valued function $g_1(t, u)=\ln J(t, u)$ in the region $t\in[0, h]$, $|u|\leqslant N$, $h=h(N)$, and this function is also jointly continuous in its variables t and u in the region under consideration. It follows from (12) that $g_1(t, u)$ satisfies equation

$$g_1(t_1+t_2, u)=g_1(t_1, u)+g_1(t_2, u), \qquad |u|\leqslant N,$$
$$t_i>0, \qquad t_1+t_2\leqslant h.$$

Therefore $g_1(t, u)=tg(u)$ and $J(t, u)=e^{tg(u)}$. It is easy to verify that the last equality should be satisfied for all t and u. Indeed, if this equality holds for given u and for all $t \leqslant h_0$, $t>0$, then for an arbitrary t

$$J(t, u) = \left[J\left(\frac{t}{n}, u\right)\right]^n = \left[e^{\frac{t}{n} g(u)}\right]^n = e^{tg(u)} \quad \text{for} \quad n > \frac{t}{h_0}.$$

Hence

$$J(t, u) = e^{tg(u)}, \tag{13}$$

where $g(u)$ is a single-valued continuous function.

This simple result completely characterizes the dependence of the characteristic function $J(t, u)$ on t. Clearly, that characteristic function of form (13) satisfies condition (12). The structure of function $g(u)$ remains to be determined. It follows from the above that $g(u)$ can be arbitrary provided that $e^{tg(u)}$ is the characteristic function of a certain distribution for each t. It follows from (13) that

$$g(u) = \lim_{t \downarrow 0} \frac{J(t, u) - 1}{t}, \tag{14}$$

and the convergence is uniform in every bounded sphere $|u| \leqslant N$, $0 < N < \infty$.

Theorem 3. *Let $J(t, u)$, $t>0$, $u \in \mathscr{R}^m$ be a family of characteristic functions such that the limit* (14) *exists uniformly in an arbitrary sphere $|u| \leqslant N$, $N>0$. Then there exists in $\{\mathscr{R}^m, \mathfrak{B}^m\}$ a finite measure $\Pi(B)$, a non-negative-definite operator b defined in $\mathscr{R}^m$ and a vector a such that*

$$g(u) = i(a, u) - \tfrac{1}{2}(bu, u) + \int_{\mathscr{R}^m} \left[e^{i(u,z)} - 1 - \frac{i(u, z)}{1+|z|^2}\right] \frac{1+|z|^2}{|z|^2} \Pi(dz). \tag{15}$$

Proof. Let $\{Q_t(\cdot), \mathfrak{B}^m\}$ be the distribution corresponding to characteristic function $J(t, u)$. Set

$$\Pi_t(B) = \frac{1}{t} \int_B \frac{|z|^2}{1+|z|^2} Q_t(dz), \qquad B \in \mathfrak{B}^m.$$

It will be shown below that the family of measures $\{\Pi_t(\cdot), t>0\}$ is weakly compact. Choose a sequence $t_n \downarrow 0$ such that Π_{t_n} converge weakly to a certain measure Π' on $\mathfrak{B}^m$. Next

$$\frac{J(t, u) - 1}{t} = \int_{\mathscr{R}^m} (e^{i(u,z)} - 1) \frac{1+|z|^2}{|z|^2} \Pi_t(dz) =$$

$$= iA_t(u) - \tfrac{1}{2} B_t(u) + \int_{\mathscr{R}^m} f(u, z)\, \Pi_t(dz), \tag{16}$$

where

$$A_t(u)=\int_{\mathscr{R}^m}\frac{(u,z)}{|z|^2}\,\Pi_t(dz),\qquad B_t(u)=\int_{\mathscr{R}^m}\frac{(u,z)^2}{|z|^2}\,\Pi_t(dz),$$

$$f(u,z)=\left(e^{i(u,z)}-1-\frac{i(u,z)}{1+|z|^2}+\frac{1}{2}\frac{(u,z)^2}{1+|z|^2}\right)\frac{1+|z|^2}{|z|^2}.$$

If we define $f(u,0)=0$, then $f(u,z)$ becomes a continuous and bounded function. Therefore

$$\lim\int_{\mathscr{R}^m}f(u,z)\,\Pi_{t_n}(dz)=\int_{\mathscr{R}^m}f(u,z)\,\Pi'(dz).$$

Since the limit in the l.h.s. of equation (16) exists at $t=t_n$ as $n\to\infty$, the limits

$$\lim A_{t_n}(u)=a(u),\qquad \lim B_{t_n}(u)=B(u)$$

also exist, where, moreover, $a(u)$ is a linear function, $B(u)$ is a positive definite quadratic form i.e. $a(u)=(a,u)$ and $B(u)=(b'u,u)$ where b' is a positive-definite symmetric operator. Approaching through sequence t_n to the limit in (16) we obtain

$$g(u)=i(a,u)-\tfrac{1}{2}(b'u,u)+\int_{\mathscr{R}^m}f(u,z)\,\Pi'(dz). \tag{17}$$

Let $\Pi(A)=\Pi'(A-\{0\})$ ($\{0\}$ is a singleton containing point 0). In the r.h.s. of equality (17) the measure $\Pi'(\cdot)$ appearing in the integral may be replaced by measure $\Pi(\cdot)$.

On the other hand, the integral

$$\frac{1}{2}\int_{\mathscr{R}^m}\frac{(u,z)^2}{|z|^2}\,\Pi(dz)$$

exists and represents a certain positive-definite quadratic form $(b''u,u)$. It is easy to verify that $(b'u,u)\geqslant(b''u,u)$. Therefore the operator $b=b'-b''$ is a positive-definite symmetric operator. We thus obtain

$$g(u)=i(a,u)-\tfrac{1}{2}(bu,u)+\int_{\mathscr{R}^m}\left(f(u,z)-\frac{1}{2}\frac{(u,z)^2}{|z|^2}\right)\Pi(dz),$$

which proves (15).

We proceed to the verification of the weak compactness of the family

$\{\Pi_t, t>0\}$. It is required to show that

a) $\Pi_t(\mathscr{R}^m)\leqslant C$, b) $\lim_{N\to\infty} \overline{\lim_{t\downarrow 0}} \Pi_t\{\bar{S}_N\}=0$,

where $\bar{S}_N=\{z:|z|>N\}$.

Let $|u|\leqslant N_1$, N_1 be arbitrary. It follows from the conditions of the theorem and (16) that for any $\delta>0$ a $t_0=t_0(N_1, \delta)$ can be found such that

$$-\operatorname{Re} g(u)+\delta\geqslant\int_{S_1}\frac{1-\cos(u, z)}{|z|^2}\Pi_t(dz), \quad t<t_0 \tag{18}$$

and for $c\geqslant 1$

$$-\operatorname{Re} g(u)+\delta\geqslant\int_{\bar{S}_c}[1-\cos(u, z)]\,\Pi_t(dz), \quad t<t_0. \tag{19}$$

Since $1\text{-}\cos x\geqslant\frac{x^2}{2!}-\frac{x^4}{4!}$ for all x, it follows from (18) that

$$-\operatorname{Re} g(u)+\delta\geqslant\int_{S_1}\left[\frac{(u, z)^2}{2!}-\frac{(u, z)^4}{4!}\right]\frac{1}{|z|^2}\Pi_t(dz). \tag{20}$$

To obtain the required bounds the values of the following integral are needed:

$$J(\varrho)=\int_{S_\varrho} e^{i(u,z)}du, \quad J_k(\varrho)=\int_{S_\varrho}(u, z)^k\,du, \quad k=2, 4.$$

These are:

$$J(\varrho)=\left(\frac{2\pi\varrho}{|z|}\right)^{m/2} I_{m/2}(\varrho|z|),$$

$$I_2(\varrho)=\frac{\pi^{m/2}\varrho^{m+2}|z|^2}{2\Gamma\left(\frac{m}{2}+2\right)}, \quad I_4(\varrho)=\frac{3\pi^{m/2}\varrho^{m+4}|z|^4}{4\Gamma\left(\frac{m}{2}+3\right)}. \tag{21}$$

Integrating inequalities (19) and (20) with respect to $u\in S_\varrho$ and dividing them by the volume S_ϱ, where $S_\varrho=\Omega_m\varrho^m$, $\Omega_m=\dfrac{\pi^{m/2}}{\Gamma\left(\frac{m}{2}+1\right)}$, we obtain

$$-\frac{1}{\Omega_m\varrho^m}\int_{S_\varrho}\operatorname{Re} g(u)\,du+\delta\geqslant\int_{S_1}\frac{\varrho^2}{2(m+2)}\left(1-\frac{\varrho^2|z|^2}{4(m+4)}\right)\Pi_t(dz) \tag{22}$$

and

$$-\frac{1}{\Omega_m \varrho^m}\int\limits_{S_\varrho} \operatorname{Re} g(u)\,du+\delta \geqslant$$

$$\geqslant \int\limits_{\bar{S}_c}\left[1-\Gamma\left(\frac{m}{2}+1\right)\left(\frac{2}{\varrho|z|}\right)^{m/2} I_{m/2}(\varrho|z|)\right]\Pi_t(dz). \qquad (23)$$

Choosing in (22) the value of ϱ from the condition $\varrho^2=2(m+4)$ and taking $N_1>\varrho$ we obtain

$$\Pi_t(S_1)\leqslant 2\left[\delta-\frac{1}{\Omega_m \varrho^m}\int\limits_{S_\varrho} \operatorname{Re} g(u)\,du\right].$$

Since the function $I_{m/2}(x)$ is bounded, one can choose for any $c>0$ the value $\varrho=\varrho_1$ from condition

$$(\varrho_1 c)^{m/2}\geqslant 2^{m+2/2}\Gamma\left(\frac{m}{2}+1\right)\sup_{x>0}|I_{m/2}(x)|. \qquad (24)$$

We thus obtain

$$\Pi_t(\bar{S}_c)\leqslant 2\left[\delta-\frac{1}{\Omega_m \varrho_1^m}\int\limits_{S_{\varrho 1}} \operatorname{Re} g(u)\,du\right],$$

which shows that $\Pi_t(\mathscr{R}^m)<K$. Finally noting that for $\varrho\to 0$ we have

$$-\frac{1}{\Omega_m \varrho^m}\int\limits_{S_\varrho} \operatorname{Re} g(u)\,du\to g(0)=0,$$

we first choose $\varrho=\varrho_2$ sufficiently small so that the l.h.s. of inequality (23) is less or equal to 2δ and then choose $c=N=N_\delta$ in such a manner that equality (24) is satisfied. We then obtain

$$\Pi_t(\bar{S}_{N_\delta})<4\delta,$$

and these inequalities are valid independently of $t\in[0, t_0)$, $t_0=t_0(N_1, \delta)$. The theorem is thus proved. □

From the results obtained follows

Theorem 4. *If $\xi(t)$, $t\geqslant 0$ is a homogeneous stochastically continuous process with values in $\mathscr{R}^m$, then the characteristic function $J(t, u)$ of the difference $\xi(s+t)-\xi(s)$ is of the form*

$$J(t, u)=e^{tg(u)}, \qquad (25)$$

where $g(u)$ is given by formula (15).

Consider now a few particular cases of formula (25).

a) $b=0$, $\Pi(B)\equiv 0$.

In this case $J(t, u)=e^{it(a,u)}$, which corresponds to the characteristic function of a degenerate distribution concentrated at point $ta\in\mathcal{R}^m$. Thus $\xi(t)=\xi(0)+at$ with probability 1 and point $\xi(t)$ moves uniformly with a constant velocity a.

b) $\Pi(B)\equiv 0$.

In this case the increments $\xi(t+s)-\xi(s)$ are normally distributed with the mean a and correlation matrix bt so that if for example $\xi(0)=0$, the process $\xi(t)$ is Gaussian. In Section 5 of the present Chapter it will be shown that in this and only this case the process with independent increments is stochastically equivalent to the process with continuous (with probability 1) sampling functions. The process under consideration is called Brownian motion.

As it is known, if one observes a small particle of colloidal dimensions immersed in a liquid through a very powerful microscope then he will notice that such a particle is in constant motion and its path represents a very complicated broken line with randomly oriented segments. This phenomenon is due to the collision between the molecules of the liquid and the colloidal particle. The measurements of the particle are large as compared with molecules of the liquid and a huge number of molecules collide with the particle during a time period of one second. The result of each single collision is impossible to detect. The motion of the particle is called Brownian motion. As a rough approximation it is assumed that the changes in position of the particle as a result of the collision with the molecules of the medium are independent and Brownian motion is considered as a continuous process with independent increments. In view of the above, such a process is Gaussian. If $\xi(t)$ is one-dimensional, $b=1$, $a=0$, then the Brownian motion is called a Wiener process.

c) $a=0$, $b=0$, the measure Π represents a mass of magnitude q concentrated at point z_0.

In this case the characteristic function (25) is of the form

$$J(t, u)=\exp\left\{\frac{qt(1+|z_0|^2)}{|z_0|^2}\left(e^{i(u, z_0)}-1-\frac{i(u, z_0)}{1+|z_0|^2}\right)\right\}. \tag{26}$$

It is easy to verify that the increment $\xi(t)-\xi(0)$ can be represented in the form

$$\xi(t)-\xi(0)=z_0\left(v(t)-\frac{qt}{|z_0|^2}\right),$$

where $v(t)$ is a Poisson process with mean value $\mathsf{E}v(t)=\dfrac{q(1+|z_0|^2)}{|z_0|^2}t$.

d) Let $b=0$ and the measure Π satisfy

$$\int_{\mathscr{R}^m} \frac{\Pi(dz)}{|z|^2} < \infty. \tag{27}$$

In this case $g(u)$ can be represented in the form

$$g(u) = i(\tilde{a}, u) + q \int_{\mathscr{R}^m} (e^{i(u,z)} - 1)\, \Pi_0(dz), \tag{28}$$

where $q>0$ and Π_0 is a probability measure on $\{\mathscr{R}^m, \mathfrak{B}^m\}$. The interpretation of this measure is as follows: We have

$$J(t,u) = e^{i(\tilde{a}t, u)} \sum_{n=0}^{\infty} e^{-qt} \frac{(qt)^n}{n!} \left[\int_{\mathscr{R}^m} e^{i(u,z)}\, \Pi_0(dz) \right]^n,$$

which represents the characteristic function of the sum

$$\tilde{a}t + \xi_1 + \xi_2 + \dots + \xi_{\nu(t)},$$

where $\xi_1, \xi_2, \dots, \xi_n$ are independent and identically distributed random vectors with values in $\mathscr{R}^m$ distributed according to Π_0, $\tilde{a}$ is a constant vector, $\nu(t)$ is an integer-valued random variable, independent of the family $\{\xi_k, k=1,2,\dots\}$, obeying the Poisson distribution with parameter qt:

$$\mathsf{P}\{\nu(t)=n\} = e^{-qt} \frac{(qt)^n}{n!}.$$

This process is called the *generalized Poisson process in* $\mathscr{R}^m$.

Note that for any function $g(u)$ defined by formula (15) a sequence convergent to it of function $g(u)$ of form (28) can be constructed. Since the members of the sequence determine characteristic functions of certain distributions, the function $e^{tg(u)}$, where $g(u)$ is an arbitrary function of form (15) is the characteristic function of a certain distribution. We thus have the following

Theorem 5. *In order that the process $\xi(t)$ be a homogeneous stochastically continuous process with independent increments it is necessary and sufficient that its characteristic function be represented by formulas* (25) *and* (15) *where a is an arbitrary vector, b an arbitrary positive-definite operator and $\Pi(B)$ an arbitrary finite measure on $\{\mathscr{R}^m, \mathfrak{B}^m\}$ and $\Pi\{z=0\}=0$.*

Markov processes. Markov processes play a most important role in modern probability theory and its applications. They are studied in detail in Volume II. Here we give only the simplest definition of this class of processes. The notion of a discrete parameter Markov process was introduced and discussed in Section 4 of Chapter II.

The notion of a Markov process (a Markov system) is based on the representation of a system whose behavior in the future depends only on the present state of the system (i.e. does not depend on the past behavior of the system). Let $\xi(t)$, $t\in T$, where T is a finite or infinite interval of time, be a random process with values in a complete metric space $\mathcal{Y}$ and let $\mathfrak{B}$ be a σ-algebra of Borel sets in $\mathcal{Y}$.

The space $\mathcal{Y}$ is called the phase space of the system, $\xi(t)$ is its state at time t. The hypothesis of "independence of the future from the past" or equivalently "the absence of after-effect" can be most simply described using conditional probabilities as follows:

$$\mathsf{P}\{\xi(t)\in B \mid \xi(t_1), \xi(t_2), \dots, \xi(t_n)\} = \mathsf{P}\{\xi(t)\in B \mid \xi(t_n)\} \pmod{\mathsf{P}} \tag{29}$$

for any $B\in\mathfrak{B}$ and $t_1<t_2<\dots t_n<t$. Since the conditional probability given a random variable can be regarded as a function of this variable, we set

$$\mathsf{P}\{\xi(t)\in A \mid \xi(s)\} = \mathsf{P}(s, \xi(s), t, A) \quad (s<t).$$

It follows from formula (19) Section 3 Chapter I that for $t_1<t_2<\dots<t_n$ the following equality holds for an arbitrary bounded Borel function $g(x_1, x_2, \dots, x_n)$ $(x_k\in\mathcal{Y}, k=1, 2, \dots, n)$:

$$\begin{aligned}\mathsf{E}\{g(\xi(t_1), \xi(t_2), \dots, \xi(t_n)) \mid \xi(t_1)\} &= \\ &= \int \mathsf{P}(t_1, \xi(t_1), t_2, dy_2) \int \mathsf{P}(t_2, y_2, t_3, dy_3) \dots \\ &\dots \int \mathsf{P}(t_{n-1}, y_{n-1}, t_n, dy_n)\, g(\xi(t_1), y_1, \dots, y_n) \pmod{\mathsf{P}}.\end{aligned} \tag{30}$$

In particular if we put $g=\chi_B(x_3)$, where $\chi_B(\cdot)$ is the indicator of the set $B\in\mathfrak{B}$, then it follows from (30) that with probability 1

$$\mathsf{P}(t_1, \xi(t_1), t_3, B) = \int \mathsf{P}(t_2, y_2, t_3, B)\, \mathsf{P}(t_1, \xi(t_1), t_2, dy_2). \tag{31}$$

The equality obtained has already been encountered in Section 4 of Chapter 11 as the Chapman-Kolmogorov equations.

Definition 5. A random process $\xi(t)$ $(t\in T)$ with values in $\mathcal{Y}$ is called *Markov* (or *Markovian*) if

a) for any $t_1<t_2<\dots<t_n<t$, $t_k\in T$ $(k=1, \dots, n)$, $t\in T$ equality (29) is satisfied.

b) There exists a function $\mathsf{P}(s, y, t, B)$, $\mathfrak{B}$-measurable with respect to

y for fixed s, t, B, which is, for fixed s, y and t, a probability measure on $\mathfrak{B}$ satisfying the Chapman-Kolmogorov equation

$$\mathsf{P}\{t_1, y, t_3, B\} = \int \mathsf{P}(t_2, y_2, t_3, B)\, \mathsf{P}(t_1, y, t_2, dy_2) \tag{32}$$

and which coincides with probability 1 with the conditional probabilities

$$\mathsf{P}(s, \xi(s), t, A) = \mathsf{P}\{\xi(t) \in A \mid \xi(s)\}.$$

Functions $\mathsf{P}(t, y, s, B)$ are called *transition probabilities* of a Markov process.

Therefore, it follows from the definition that the family of conditional probabilities (29) is regular and process $\xi(t)$ does not depend on the "past". The property of a process expressed by equality (29) is called the Markov property or the absence of after-effect.

We now show that certain stronger assertions can be deduced from the Markov property. Applying again formula (19) of Section 3, Chapter 1 and equality (30), we obtain that for $t_1 < t_2 < \ldots < t_m < \ldots < t_{m+n}$, $t_k \in T$ $(k = 1, \ldots, n+m)$

$$\begin{aligned} &\mathsf{E}\{g(\xi(t_{m+1}), \xi(t_{m+2}), \ldots, \xi(t_{n+m})) \mid \xi(t_1), \xi(t_2), \ldots, \xi(t_m)\} = \\ &\quad = \int \mathsf{P}(t_m, \xi(t_m), t_{m+1}, dy_1) \ldots \times \\ &\quad \times \ldots \int \mathsf{P}(t_{n+m-1}, y_{n-1}, t_{n+m}, dy_n)\, g(y_1, \ldots, y_n) = \\ &\quad = \mathsf{E}\{g(\xi(t_{m+1}), \ldots, \xi(t_{n+m})) \mid \xi(t_m)\} \pmod{\mathsf{P}}. \end{aligned}$$

If we set $g(y_1, \ldots, y_n) = \chi_{B^{(n)}}(y_1, \ldots, y_n)$ where $B^{(n)}$ is a Borel set in $\mathscr{Y}^n$, then the equality which generalizes the Markov property of a process:

$$\begin{aligned} &\mathsf{P}\{[\xi(t_{m+1}), \ldots, \xi(t_{m+n})] \in B^{(n)} \mid \xi(t_1), \ldots, \xi(t_m)\} = \\ &\qquad = \mathsf{P}\{[\xi(t_{m+1}), \ldots, \xi(t_{m+n})] \in B^{(n)} \mid \xi(t_m)\} \pmod{\mathsf{P}} \end{aligned}$$

will follow for any $t_1 < t_2 < \ldots < t_{n+m} (\in T)$ and any n and m. Denote by $\mathfrak{F}_t$ the σ-algebra of events generated by the random variables $\xi(s)$, $s \in T$, $s \leqslant t$, and by $\mathfrak{F}_t^*$ the σ-algebra generated by the variables $\xi(s)$, $s \in T$, $s > t$. We then have for any cylindrical set $C \in \mathfrak{F}_t^*$ with $t_1 < t_2 < \ldots t_n \leqslant t$

$$\mathsf{P}\{C \mid \xi(t_1), \ldots, \xi(t_n)\} = \mathsf{P}\{C \mid \xi(t_n)\} \pmod{\mathsf{P}}. \tag{33}$$

Let Λ be the class of events for which (33) holds. In view of the properties of conditional probabilities (Section 3, Chapter I), Λ is a λ-class and contains the Π-class of events. Therefore $\Lambda \supset \mathfrak{F}_t^*$. On the other hand, let $\mathfrak{N}$ be the class of events N for which for any $S \in \mathfrak{F}_t^*$

$$\int_N \mathsf{P}(S \mid \mathfrak{F}_t)\, d\mathsf{P} = \int_N \mathsf{P}(S \mid \xi(t))\, d\mathsf{P}. \tag{34}$$

In view of (33) all the cylinders from $\mathfrak{N}$ are included in $\mathfrak{F}_t$. Since the r.h.s. and the l.h.s. of equality (34) are countably-additive functions on $\mathfrak{F}_t$ the fact that they coincide on the cylindrical sets of $\mathfrak{F}_t$ yields that they are identical on $\mathfrak{F}_t$. We thus have the following

Theorem 6. *For an arbitrary* $S \in \mathfrak{F}_t^*$

$$\mathsf{P}(S \mid \mathfrak{F}_t) = \mathsf{P}(S \mid \xi(t)) \pmod{\mathsf{P}}. \tag{35}$$

Relation (35) shows that the conditional probability of an arbitrary event S, which is determined by the behavior of a Markov process in the "future" if the "past" is completely specified, depends on the "present".

A family consisting of a probability measure μ_0 on $\{\mathscr{Y}, \mathfrak{B}\}$ and of transition probabilities $\mathsf{P}(t, y, s, B)$ $(t<s, t, s\in T, B\in\mathfrak{B})$, satisfying conditions b) of definition 5 is called *a wide-sense Markov process defined on* $T=[0, b]$ or $T=[0, \infty]$.

The measure μ_0 is called the initial distribution of the system.

For an arbitrary bounded Borel function $f(y_1, \ldots, y_n)$ of n variables $y_k \in \mathscr{Y}$ and for arbitrary $t_k \in T$ $(k=1, \ldots, n,\ 0<t_1<\ldots t_n)$ we set

$$F_{t_1 t_2 \ldots t_n}[f] = \int \mu_0(dy_0) \int \mathsf{P}(0, y_0, t_1, dy_1) \times \ldots$$
$$\ldots \times \int f(y_1, y_2, \ldots, y_n)\, \mathsf{P}(t_{n-1}, y_{n-1}, t_n, dy_n) \tag{36}$$

and

$$\mathsf{P}_{t_1 t_2 \ldots t_n}(A^{(n)}) = F_{t_1 t_2 \ldots t_n}[\chi_{A^{(n)}}], \tag{37}$$

where $\chi_{A^{(n)}}$ is the indicator of the set $A^{(n)} \in \mathfrak{B}^n$, and $\mathfrak{B}^n$ is the σ-algebra of Borel sets in $\mathscr{Y}^n$. Note that for an arbitrary Borel function $f(y_1, \ldots, y_n)$ the function

$$f_1(y_1, y_2, \ldots, y_{n-1}) = \int f(y_1, y_2, \ldots, y_n)\, \mathsf{P}(t, y_{n-1}, s, dy_n) \quad (t<s)$$

is also Borel, since the integral is represented as the limit of integrals of simple functions, and the latter are Borel functions of the variables $y_1, y_2, \ldots, y_{n-1}$. In view of the properties of the integral, $\mathsf{P}_{t_1 \ldots t_n}(B^{(n)})$ is a measure on $\mathfrak{B}^n$. Clearly the family of measures $\mathsf{P}_{t_1 \ldots t_n}(B^{(n)})$ satisfies the compatability conditions and in view of Kolmogorov's theorem (Theorem 2 of Section 4, Chapter 1) – in the case when $\mathscr{Y}$ is a complete separable metric space – admits a certain representation $\{\Omega, \mathfrak{S}, \mathsf{P}\}$, where Ω is the space of all functions $\omega(t)$, $t \in T$ with values in $\mathscr{Y}$. Let $\xi(t)$ be an arbitrary process stochastically equivalent to $\{\Omega, \mathfrak{S}, \mathsf{P}\}$. We check that

$$\mathsf{P}\{\xi(t)\in B \mid \xi(t_1), \xi(t_2), \ldots, \xi(t_n)\} = P(t_n, \xi(t_n), t, B) \quad (\mathrm{mod}\, \mathsf{P}),$$

i.e. $\xi(t)$ is a Markov process with given transition probabilities. For this purpose it is sufficient to verify equality

$$\int_{B^{(n)}} \mathsf{P}(t_n, y_n, t, B)\, \mathsf{P}_{t_1 t_2 \ldots t_n}(dy_1, dy_2, \ldots, dy_n) = \mathsf{P}_{t_1, t_2, \ldots t_n, t}(B^{(n)} \times B)$$

for arbitrary $B^{(n)} \in \mathfrak{B}, B \in \mathfrak{B}$ and $t_1 < t_2 < t_3 < \ldots t_n < t_n$. But this equality follows directly from formulas (36), (37) and Theorem 2 of Section 4 Chapter II.

Therefore if $\mathscr{Y}$ is a complete metric separable space then a certain representation exists for an arbitrary wide-sense Markov process.

§2. Separable Random Functions

The basic theorem. Let a random function $\zeta(x) = g(x, \omega)$ be defined on the probability space $\{\Omega, \mathfrak{S}, \mathsf{P}\}$, where $x \in \mathscr{X}$ and the values of the function are in a certain measurable space $\{\mathscr{Y}, \mathfrak{B}\}$. We shall assume that $\{\Omega, \mathfrak{S}, \mathsf{P}\}$ is a complete probability space.

In many problems events of the form

$$\{\omega : \zeta(x) \in F \quad \text{for all} \quad x \in G\}. \tag{1}$$

play a significant role.

Unfortunately, if G is uncountable one cannot in general assert that event (1) is $\mathfrak{S}$-measurable. Nevertheless, it is often required to consider random functions for which this event is measurable for a wide class of sets F and G.

The feasibility of overcoming the difficulties connected with the uncountability of event G is based on the following remark. Assume that there exists in $\mathscr{X}$ a countable set of points I and a ω-set N such that $\mathsf{P}\{N\} = 0$ and the symmetric difference of set (1) and set

$$\{\omega : \zeta(x) \in F \quad \text{for all} \quad x \in G \cap I\} = \bigcap_{x \in G \cap I} \{\omega : \zeta(x) \in F\} \tag{2}$$

is contained in N for all $G \in \mathfrak{G}$ and $F \in \mathfrak{F}$. Then set (1) is measurable. Random functions which satisfy the formulated assumption are called separable (relative to classes $\mathfrak{G}$ and $\mathfrak{F}$). Intuitively it is clear that in order that a random function be separable the sets in $\mathfrak{G}$ should contain a sufficiently large number of points of I so that it would be plausible to regard sets (1) and (2) as insignificantly different from one another.

For example if $\mathscr{X}$ and $\mathscr{Y}$ are metric spaces, $\mathscr{X}$ is a separable space, $\mathfrak{G}$ is the class of open sets, $\mathfrak{F}$ is the class of closed sets in $\mathscr{Y}$ and the function $\zeta(x)=g(x, \omega)$ is continuous for almost all ω, then an arbitrary countable everywhere dense set in $\mathscr{X}$ is chosen for I. For this choice sets (1) and (2) coincide for each ω for which $\zeta(x)$ is continuous.

In the present section it is assumed that $\mathscr{X}$ and $\mathscr{Y}$ are metric spaces with distances $r(x_1, x_2)$ and $\varrho(y_1, y_2)$ correspondingly, $\mathscr{X}$ is a separable space and the separability property of a random function is considered relative to the classes $\mathfrak{G}$ and $\mathfrak{F}$ of open sets in $\mathscr{X}$ and closed sets in $\mathscr{Y}$.

Definition 1. A random function $\zeta(x)=g(x, \omega)$ is called *separable* if there exists in $\mathscr{X}$ an everywhere dense countable set I of points $\{x_j\}, j=1, 2, \ldots$ and in Ω a set N of probability 0 such that for an arbitrary open set $G\subset\mathscr{X}$ and an arbitrary closed set $F\subset\mathscr{Y}$ the two sets

$$\{\omega: g(x, \omega)\in F \quad \text{for all} \quad x\in G\},$$
$$\{\omega: g(x, \omega)\in F \quad \text{for all} \quad x\in G\cap I\}$$

differ from each other only on the subset of N.

The countable set I of point x_j which appears in this definition is called the separability set of a random function. It turns out that the separability property is not a stringent restriction imposed on a random function. Under sufficiently broad assumptions pertaining only to the nature of the domain of the definition of $\mathscr{X}$ and the region of values $\mathscr{Y}$ of a random function there exists a separable random function which is stochastically equivalent to that given. It should, however be noted, that when constructing the equivalent separable random function it may sometimes be necessary to extend the range of values of the function so that it will become a compact set.

We first present a criterion of separability of a random function. Let $\mathscr{Y}$ be compact, $\tilde{g}(x, \omega)$ be a separable random function with values in $\mathscr{Y}$, I be the separability set, N be the corresponding exceptional set of points ω.

Denote by V the class of all open spheres of the space $\mathscr{X}$ with rational radii and center at the points of a fixed countable everywhere dense set in $\mathscr{X}$. The class V is countable. On the other hand an arbitrary open set G in $\mathscr{X}$ can be represented as a sum (of a countable number) of spheres in V.

Let $A(G, \omega)$ be the closure of the set of values of the function $\tilde{g}(x, \omega)$

where x runs through the set $I \cap G$ and

$$A(x, \omega) = \cap A(S, \omega)$$

is the intersection of all $A(S, \omega)$ when S runs through the collection of spheres in V each containing point x. The family of closed sets $A(S, \omega)$ $(x \in S)$ is "centered", i.e. an arbitrary finite number of sets of this family has common points and in view of the compactness of $\mathscr{Y}$ their intersection is non-void. It follows from the separability of function $\tilde{g}(x, \omega)$ that

$$\tilde{g}(x, \omega) \in A(x, \omega), \quad \omega \notin N. \tag{3}$$

Conversely if (3) is satisfied for each $\omega \notin N$ with $\mathsf{P}\{N\} = 0$, then $\tilde{g}(x, \omega)$ is a separable random function. Indeed, if $\tilde{g}(x, \omega) \in F$ for all $x \in I \cap S$, where F is a closed set in $\mathscr{Y}$ and $S \subset V$, then $A(x, \omega) \in A(S, \omega)$ for each $x \in S$ and consequently $\tilde{g}(x, \omega) \in F$ for all $x \in S$.

Let G be an arbitrary open set in $\mathscr{X}$. We represent it as the sum $G = \bigcup_k S_k$ of sets in V. In view of the remark just made, it follows from relation

$$\tilde{g}(x, \omega) \in F \quad \text{for all} \quad x \in I \cap G, \quad \omega \notin N,$$

that

$$\tilde{g}(x, \omega) \in F \quad \text{for any} \quad x \in G.$$

We state the result obtained as follows:

Lemma 1. *In order that a random function $\tilde{g}(x, \omega)$ with values in a compact space $\mathscr{Y}$ be separable it is necessary and sufficient that there exist a set N with $\mathsf{P}\{N\} = 0$ such that for $\omega \notin N$ inclusion* (3) *be satisfied.*

Thus to construct a separable stochastically equivalent function for $g(x, \omega)$ it is sufficient to find a function $\tilde{g}(x, \omega)$ satisfying (3) which coincides with probability 1 with the function $g(x, \omega)$:

$$\mathsf{P}\{\tilde{g}(x, \omega) \neq g(x, \omega)\} = 0.$$

Lemma 2. *Let B be an arbitrary Borel set in $\mathscr{Y}$, where $\mathscr{Y}$ is compact. There exists a finite or countable sequence of points $x_1, x_2, \dots$ such that the set*

$$N(x, B) = \{\omega : g(x_k, \omega) \in B, k = 1, 2, \dots, g(x, \omega) \notin B\}$$

has probability 0 *for any $x \in \mathscr{X}$.*

Proof. Let x_1 be arbitrary. If $x_1, x_2, \dots, x_k$ are already constructed, we put

$$m_k = \sup_{x \in \mathscr{X}} \mathsf{P}\{g(x_1, \omega) \in B, \dots, g(x_k, \omega) \in B, g(x, \omega) \notin B\}.$$

If $m_k=0$, then the corresponding sequence is already constructed. If $m_k>0$, let x_{k+1} be a point such that

$$\mathsf{P}\{g(x_1,\omega)\in B,\dots,g(x_k,\omega)\in B,g(x_{k+1},\omega)\notin B\}\geqslant\frac{m_k}{2}.$$

Since the sets

$$L_k=\{\omega:g(x_i,\omega)\in B,\ i=1,2,\dots,k,\ g\{x_{k+1},\omega)\notin B\}$$

are disjoint,

$$1\geqslant\sum_{k=1}^{\infty}\mathsf{P}\{L_k\}\geqslant\tfrac{1}{2}\sum_{k=1}^{\infty}m_k.$$

Consequently, $m_k\to 0$ for $k\to\infty$. Therefore

$$\mathsf{P}\{g(x_k,\omega)\in B,\ k=1,2,\dots,g(x,\omega)\notin B\}\leqslant\lim m_k=0,$$

for any x, which proves Lemma 2. □

The following assertion can be easily deduced from the above:

Lemma 3. *Let $\mathfrak{M}_0$ be a countable class of sets, and $\mathfrak{M}$ a class consisting of intersections of all possible sequences of sets in $\mathfrak{M}_0$. There exists a finite or countable sequence of points $x_1,x_2,\dots,x_n,\dots$ and a set $N(x)$, for each x, such that*

$$\mathsf{P}\{N(x)\}=0$$

and

$$\{\omega:g(x_n,\omega)\in B,\ n=1,2,\dots,g(x,\omega)\notin B\}\subset N(x)$$

for any $B\in\mathfrak{M}$.

To prove the lemma we proceed as follows: Let I be a countable set of points in $\mathscr{X}$ which is a sum of sequences $\{x_n,n=1,2,\dots\}$ constructed for each $B\in\mathfrak{M}_0$ as indicated in Lemma 2 and let $N(x)=\bigcup_{B\in\mathfrak{M}_0}N(x,B)$. If $B'\in\mathfrak{M}$ and $B\supset B'$, $B\in\mathfrak{M}_0$ then

$$\{\omega:g(x_n,\omega)\in B',\ x_n\in I,\ g(x,\omega)\notin B\}\subset$$
$$\subset\{\omega:g(x_n,\omega)\in B,\ x_n\in I,\ g(x,\omega)\notin B\}\subset N(x,B)\subset N(x).$$

Moreover, if $B'=\bigcap_{k=1}^{\infty}B_k\in\mathfrak{M}$, then

$$\{\omega:g(x_n,\omega)\in B',\ x_n\in I,\ g(x,\omega)\notin B'\}\subset$$
$$\subset\bigcup_{k=1}^{\infty}\{\omega:g(x_n,\omega)\in B',\ x_n\in I,\ g(x,\omega)\notin B_k\}\subset$$
$$\subset\bigcup_{k=1}^{\infty}N(x,B_k)\subset N(x),$$

which proves the lemma. □

It will now be easy to prove the following theorem:

Theorem 1. (J. L. Doob) *Let $\mathcal{X}$ and $\mathcal{Y}$ be metric spaces, $\mathcal{X}$ be separable, $\mathcal{Y}$ be compact. An arbitrary random function $g(x, \omega)$, $x \in \mathcal{X}$ with values in $\mathcal{Y}$ is stochastically equivalent to a certain separable random function.*

Proof. We fix a certain everywhere dense set of points L in $\mathcal{Y}$ and let $\mathfrak{M}_0$ be the class of sets which are complements of the spheres of rational radii with centers at points of L. Then $\mathfrak{M}$ being the class of all intersections of the sets in $\mathfrak{M}_0$ contains all the closed sets of the space $\mathcal{Y}$. Next for each $S \in V$ we consider a random function $g(x, \omega)$ as defined only for $x \in S$ and construct a sequence $I = I(S)$ and the sets $N(x) = N_S(x)$ as indicated in Lemma 3. Let

$$J = \bigcup_{S \in V} I(S), \quad N_x = \bigcup_{S \in V} N_S(x).$$

$$\tilde{g}(x, \omega) = g(x, \omega),$$

if $x \in I$ or $\omega \notin N_x$; if however, $\omega \in N_x$, $x \notin I$ then we define $\tilde{g}(x, \omega)$ in an arbitrary manner provided only that $\tilde{g}(x, \omega) \in A(x, \omega)$. Since for the points $x \in I$ the values of the functions $\tilde{g}(x, \omega)$ and $g(x, \omega)$ coincide, the sets $A(x, \omega)$ constructed for the functions $\tilde{g}(x, \omega)$ and $g(x, \omega)$ also coincide. It follows from the definition of $\tilde{g}(x, \omega)$ that

$$\tilde{g}(x, \omega) \in A(x, \omega)$$

for arbitrary x and ω. Since $\{\omega : g(x, \omega) \neq \tilde{g}(x, \omega)\} \subset N_x$, $\mathsf{P}\{\tilde{g}(x, \omega) = g(x, \omega)\} = 1$, which completes the proof of the theorem. □

Theorem 1 can be directly generalized to the case of random functions with values in separable locally compact spaces.

Theorem 2. *Let $\mathcal{Y}$ be a separable locally-compact space and $\mathcal{X}$ be an arbitrary metric separable space. For an arbitrary random function $g(x, \omega)$ defined on $\mathcal{X}$ with values in $\mathcal{Y}$, there exists a stochastically equivalent separable random function $\tilde{g}(x, \omega)$ taking on values in a certain compact extension $\tilde{\mathcal{Y}}$ of the space $\mathcal{Y}$, $\tilde{\mathcal{Y}} \supset \mathcal{Y}$.*

The proof follows from the fact that every locally compact separable space $\mathcal{Y}$ can be considered as a subset of a certain compact $\tilde{\mathcal{Y}}$. For example if $g(x, \omega)$ is a random function with values in a finite-dimensional space $\mathcal{Y}$, then by adjoining to $\mathcal{Y}$ a single point "at infinity" ∞, it is easy to obtain a new compact space $\tilde{\mathcal{Y}} = \mathcal{Y} \cup \{\infty\}$ with a new metric such that every closed set $F \subset \mathcal{Y}$ (relative to the topology of space $\mathcal{Y}$) will also be closed in $\tilde{\mathcal{Y}}$ (with respect to the new metric). When constructing a separable realization of a random function, it may be necessary to assign to this function the additional value ∞, but clearly for a fixed x the probability of this is zero. □

Stochastic continuity. In many problems it is important to know which set I may serve as a separability set. Before answering this question we introduce one important notion and present related simple theorems.

Definition 2. A random function $g(x, \omega)$ with values in $\mathscr{Y}$ is called *stochastically continuous* at point x_0, $x_0 \in \mathscr{X}$ if for any $\varepsilon > 0$

$$\mathsf{P}\{\varrho(g(x_0, \omega), g(x, \omega)) > \varepsilon\} \to 0 \quad \text{as} \quad r(x, x_0) \to 0. \tag{4}$$

If $g(x, \omega)$ is stochastically continuous in every point of a certain set $B \subset \mathscr{X}$, then it is called stochastically continuous on B.

Note that the condition of stochastic continuity is a condition imposed on the "two-dimensional" distributions of a random function, i.e. on the joint distribution of the random elements, $g(x_1, \omega)$ and $g(x_2, \omega)$, $x_1, x_2 \in \mathscr{X}$. In particular this notion is applicable to wide-sense random functions.

The requirement of stochastic continuity at point x_0 means that $\zeta(x) = g(x, \omega)$ converges in probability to $\zeta(x_0)$ as $x \to x_0$.

Definition 3. If there exists a point $y \in \mathscr{Y}$ such that for $K \to \infty$

$$\sup_{x \in B} \mathsf{P}\{\varrho[g(x, \omega), y] > K\} \to 0, \tag{5}$$

then the random function $g(x, \omega)$ is called *stochastically bounded* in B.

Theorem 3. *A random function $g(x, \omega)$ which is stochastically continuous on a compact set $\mathscr{X}$ is also stochastically bounded on $\mathscr{X}$.*

Proof. Let $\varepsilon > 0$ be an arbitrary number given in advance. For each point x we construct a sphere S_x with the center at x, such that

$$\mathsf{P}\{\varrho(g(x, \omega), g(x', \omega)) > 1\} < \frac{\varepsilon}{2}$$

for any point $x' \in S_x$. From the totality of the spheres S_x we select a sequence $S_{x_1}, S_{x_2}, \ldots, S_{x_n}$ which forms a finite cover of $\mathscr{X}$. Then for any y

$$\varrho\{g(x, \omega), y\} \leqslant \varrho(g(x_1, \omega), y) + \\ + \max_{i=2,\ldots,n} \varrho(g(x_1, \omega), g(x_i, \omega)) + \varrho(g(x_j, \omega), g(x, \omega)),$$

where x_j denotes the center of one of those spheres S_{x_k} $(k = 1, 2, \ldots, n)$ in the interior of which the point x is located. The summands in the r.h.s. of the equality are finite random variables. Therefore for N sufficiently large

$$\mathsf{P}\{\varrho(g(x_1, \omega), y) + \max_{i=2,\ldots,n} \varrho(g(x_1, \omega), g(x_i, \omega)) > N\} < \frac{\varepsilon}{2}.$$

If we assume that $N>1$, then for any $x\in\mathscr{X}$

$$\mathsf{P}\{\varrho(g(x,\omega),y)>2N\}\leqslant\mathsf{P}\{\varrho(g(x_j,\omega),g(x,\omega))>1\}+$$
$$+\mathsf{P}\{\varrho(g(x_1,\omega),y)+\max_{i=2,\dots,n}\varrho(g(x_1,\omega),g(x_i,\omega))>N\}<\varepsilon;$$

from here it follows that

$$\sup_{x\in\mathscr{X}}\mathsf{P}\{\varrho(g(x,\omega),y)>2N\}<\varepsilon.\quad\square$$

Definition 4. A random function $g(x,\omega)$ is called *uniformly stochastically continuous* on $\mathscr{X}$, if for arbitrary positive ε and ε_1 as small as desired, a $\delta>0$ can be found such that

$$\mathsf{P}\{\varrho(g(x,\omega),g(x',\omega))>\varepsilon\}<\varepsilon_1, \tag{6}$$

as long as $r(x,x')<\delta$.

Theorem 4. *If $g(x,\omega)$ is stochastically continuous on the compact set $\mathscr{X}$, then $g(x,\omega)$ is uniformly stochastically continuous.*

Indeed, suppose the assertion is not true, then one can find a pair of positive numbers ε and ε_1 and for any $\delta_n>0$ a pair of points x_n and x_n' such that $r(x_n,x_n')<\delta_n$ and

$$\mathsf{P}\{\varrho(g(x_n,\omega),g(x_n',\omega))>\varepsilon\}>\varepsilon_1,$$

It may be assumed that $\delta_n\to0$ and $x_n\to x_0$, then $x_n'\to x_0$ and

$$\varepsilon_1<\mathsf{P}\{\varrho(g(x_n,\omega),g(x_n',\omega))>\varepsilon\}\leqslant\mathsf{P}\left\{\varrho(g(x_n,\omega),g(x_0,\omega))>\frac{\varepsilon}{2}\right\}+$$
$$+\mathsf{P}\left\{\varrho(g(x_0,\omega),g(x_n',\omega))>\frac{\varepsilon}{2}\right\}.$$

But this inequality contradicts the condition of stochastic continuity.

Theorem 5. *Let $\mathscr{X}$ be a separable space, $\mathscr{Y}$ an arbitrary metric space and $g(x,\omega)$ a separable stochastically continuous random function with values in $\mathscr{Y}$. Then any countable everywhere dense set of points in $\mathscr{X}$ may serve as a set of separability of the random function $g(x,\omega)$.*

Proof. Let $V=\{S\}$ be a countable set of spheres in $\mathscr{X}$ introduced above, $I=\{x_k, k=1,2,\dots,n,\dots\}$ be the set of separability of the random function $g(x,\omega)$, N be the exceptional set of values ω, appearing in the definition of separability and J be an arbitrary everywhere dense set of points in $\mathscr{X}$. Let $B(S,\omega)$ denote the closure of the set of values $g(x_k',\omega)$ as the point x_k' runs through the set $J\cap S$, and $N(S,k)$ be the event that $g(x_k,\omega)\notin B(S,\omega)$ provided $x_k\in S$. The events $N(S,k)$ have probability 0.

Indeed, let x'_r, $r=1, 2, \ldots, n, \ldots$, be an arbitrary sequence of points in $J \cap S$ converging to x_k. Then

$$\mathsf{P}\{g(x_k, \omega) \notin B(S, \omega)\} \leqslant \mathsf{P}\left\{\varliminf_{r\to\infty} \varrho(g(x_k, \omega), g(x'_r, \omega)) > 0\right\} \leqslant$$

$$\leqslant \lim_{n\to\infty} \mathsf{P}\left\{\varliminf_{r\to\infty} \varrho(g(x_k, \omega), g(x'_r, \omega)) > \frac{1}{n}\right\} \leqslant$$

$$\leqslant \lim_{n\to\infty} \varliminf_{r\to\infty} \mathsf{P}\left\{\varrho(g(x_k, \omega), g(x'_r, \omega)) > \frac{1}{n}\right\} = 0.$$

Let $N' = \bigcup_S \bigcup_{x_k \in S} N(S, k)$, then $\mathsf{P}\{N'\} = 0$. If $\omega \notin N \cup N'$ and $g(x, \omega) \in F$ for all $x \in J \cap G$, where G is an open set and $F \subset \mathscr{Y}$ is a closed set, then for every $x_k \in G$ and for S such that $x_k \in S \subset G$, we have

$$g(x_k, \omega) \in B(S, \omega) \subset F.$$

It follows from the definition of the set $\{x_k\}$ that $g(x, \omega) \in F$ for all $x \in G$ and $\omega \notin N \cup N'$. Thus the set J satisfies the condition appearing in the definition of the set of separability of a random function. □

§3. Measurable Random Functions

Let $\mathscr{X}$ and $\mathscr{Y}$ denote metric spaces as before with metrics $r(x_1, x_2)$ and $\varrho(y_1, y_2)$ respectively; let $g(x, \omega)$ be a random function with values in $\mathscr{Y}$ and the domain of definition $\mathscr{X}$, and let ω be an elementary event in the probability space $\{\Omega, \mathfrak{S}, \mathsf{P}\}$.

Assume that a σ-algebra of sets $\mathfrak{A}$ is defined on $\mathscr{X}$ containing Borel sets and a certain complete measure μ is defined on $\mathfrak{A}$. Denote by $\sigma\{\mathfrak{A} \times \mathfrak{S}\}$ the smallest σ-algebra generated in $\mathscr{X} \times \Omega$ by the product of σ-algebras $\mathfrak{A}$ and $\mathfrak{S}$ and by $\tilde{\sigma}\{\mathfrak{A} \times \mathfrak{S}\}$ its completion relative to measure $\mu \times \mathsf{P}$.

Definition 1. The random function $g(x, \omega)$ is called *measurable* if it is measurable with respect to $\tilde{\sigma}\{\mathfrak{A} \times \mathfrak{S}\}$.

Denote by $\mathfrak{B}$ the σ-algebra of Borel sets of the space $\mathscr{Y}$. Recall that in the general case it follows from the definition of a random function that for any $B \in \mathfrak{B}$ and fixed x

$$\{\omega : g(x, \omega) \in B\} \in \mathfrak{S}.$$

If, however, the random function $g(x, \omega)$ is measurable, then

$$\{(x, \omega) : g(x, \omega) \in B\} \in \tilde{\sigma}\{\mathfrak{A} \times \mathfrak{S}\}.$$

It follows from here and from Fubini's theorem that $g(x, \omega)$, when considered as a function of x is $\mathfrak{A}$-measurable with probability 1.

Consider the problem of existence for a given random function of a stochastically equivalent measurable and separable function.

Theorem 1. *Let $\mathscr{X}$ be a complete separable metric space, $\mathscr{Y}$ a separable and locally compact space and let measure μ be σ-finite. If for μ-almost all x the random function $g(x, \omega)$ is stochastically continuous, then there exists a measurable separable function $g^*(x, \omega)$ which is stochastically equivalent to function $g(x, \omega)$.*

Proof. First assume that $\mathscr{X}$ and $\mathscr{Y}$ are both compact and $\mu(\mathscr{X})<\infty$. It follows from Theorem 1 of Section 2 that there exists a separable random function $\tilde{g}(x, \omega)$ stochastically equivalent to function $g(x, \omega)$. Let I be the set of separability of the function $\tilde{g}(x, \omega)$. I is everywhere dense in $\mathscr{X}$. Arrange the points of I in a certain sequence $\{x_1, x_2, \ldots, x_n, \ldots\}$ and set $r_n = \min\{r(x_k, x_s), k, s=1, \ldots, n\}$.

For each n we construct a finite cover of the set $\mathscr{X}$ by the spheres $S_1^{(n)}, \ldots, S_{m_n}^{(n)}$ whose radius is equal to $r_n/2$ with centers at the points $x_j^{(n)} \in I$. It is assumed here that $x_j^{(n)} = x_j$ for $j=1, 2, \ldots, n$ and the other points $x_j^{(n)}$ $(j=n+1, \ldots, m_n)$ are chosen arbitrarily from I, provided the spheres $S_j^{(n)}$ $(j=n+1, \ldots, m_n)$ form a cover of the set $\mathscr{X}$. Set $\bar{g}_n(x, \omega) = \tilde{g}(x_k, \omega)$ if $x \in S_k^{(n)}$, $k=1, 2, \ldots, n$ (these spheres do not intersect so that the definition is proper) and $\bar{g}_n(x, \omega) = \tilde{g}(x_j^{(n)}, \omega)$ if $x \in S_j^{(n)} \setminus \bigcup_{i=1}^{j-1} S_i^{(n)}$, $j = n+1, \ldots, m_n$ where $x_j^{(n)}$ is the center of the sphere $S_j^{(n)}$.

Note that $\bar{g}(x, \omega)$ are Borel functions of argument x for a fixed ω, $\sigma\{\mathfrak{A} \times \mathfrak{S}\}$ are measurable as functions of the pair (x, ω). Moreover, $r_n \to 0$ and

$$\varrho[\bar{g}_n(x, \omega), \tilde{g}(x, \omega)] = \varrho[\tilde{g}(x_k^{(n)}, \omega), \tilde{g}(x, \omega)] \quad \text{for} \quad r(x_k^{(n)}, x) < \frac{r_n}{2}. \tag{1}$$

If we let

$$G_{nm}(x) = \mathsf{P}\{\omega : \varrho[\bar{g}_n(x, \omega), \bar{g}_{n+m}(x, \omega)] > \varepsilon\},$$

then in view of the condition of the theorem, $G_{nm}(x) \to 0$ as $n \to \infty$ μ-almost for all x. Therefore

$$(\mu \times \mathsf{P})\{(x, \omega) : \varrho[\bar{g}_n(x, \omega), \bar{g}_{n+m}(x, \omega)] > \varepsilon\} = \int_{\mathscr{X}} G_{nm}(x)\, \mu(dx) \to 0$$

as $n \to \infty$, i.e. sequence $\bar{g}_n(x, \omega)$ is fundamental in measure $\mu \times \mathsf{P}$. A subsequence $\bar{g}_{n_k}(x, \omega)$ may be extracted from this sequence which converges $\mu \times \mathsf{P}$-almost everywhere to some $\sigma\{\mathfrak{A} \times \mathfrak{S}\}$-measurable function $\bar{g}(x, \omega)$. Denote by K the set of points (x, ω) on which this convergence does not take place. Since K has measure 0, μ-almost all its cross sections have P-measure 0. Denote by X_1 the set of all those x for which this measure is >0. By virtue of the preceding construction we may assume that $X_1 \cap I = \emptyset$ and $\bar{g}(x_n, \omega) = \tilde{g}(x_n, \omega)$. Let X_2 denote the set of those x for which the stochastic continuity does not hold. It follows from (1) that

$$\mathsf{P}\{\bar{g}(x, \omega) \neq \tilde{g}(x, \omega)\} = 0, \quad \text{if} \quad x \notin X_1 \cup X_2.$$

We now set $g^*(x, \omega) = \bar{g}(x, \omega)$ if $(x, \omega) \notin K$ and $x \notin X_1 \cup X_2$ and $g^*(x, \omega) = \tilde{g}(x, \omega)$ if $(x, \omega) \in K$ or $x \in X_1 \cup X_2$. Then $\mathsf{P}\{g(x, \omega) \neq \tilde{g}(x, \omega)\} = 0$ for all x so that $g^*(x, \omega)$ is stochastically equivalent to $\tilde{g}(x, \omega)$. The function $g^*(x, \omega)$ is $\bar{\sigma}\{\mathfrak{A} \times \mathfrak{S}\}$- measurable since it differs from a $\sigma(\mathfrak{A} \times \times \mathfrak{S}\}$-measurable function on a set of $\mu \times \mathsf{P}$-measure 0. It remains to show that $g^*(x, \omega)$ is separable. Let $A(G, \omega)$ denote (as in Section 2) the closure of the set of values $\tilde{g}(x, \omega)$ obtained as x runs through the set $G \cap I$, and $A(x, \omega)$ be the intersection of the sets $A(S, \omega)$ where S is an arbitrary sphere with the center at point x. Separability of the function $\tilde{g}(x, \omega)$ is equivalent to condition $\tilde{g}(x, \omega) \in A(x, \omega)$. Since $g^*(x, \omega) = \tilde{g}(x, \omega)$ for $x \in I$, the set $A(x, \omega)$ constructed for $g^*(x, \omega)$ coincides with $A(x, \omega)$. Next it follows from the definition of $\bar{g}_n(x, \omega)$ that $g^*(x, \omega) = \bar{g}(x, \omega) = \lim \bar{g}_n(x, \omega) \in A(x, \omega)$ for any $x \notin X_1 \cup X_2$ and $(x, \omega) \notin K$; moreover $g^*(x, \omega) = \tilde{g}(x, \omega) \in A(x, \omega)$, by definition, for any $x \in X_1 \cup X_2$ or $(x, \omega) \in K$. Thus $g^*(x, \omega)$ is a separable random function and the theorem is proved in the particular case under consideration. It is now easy to obtain the proof in the general case. The requirement of compactness of the space is required only in order to be able to refer to Theorem 1 of Section 2. Here, however, we may refer to Theorem 2 of Section 2. Moreover, the separable and measurable representation $g^*(x, \omega)$ of function $g(x, \omega)$ takes on, in general, values on some compact topological extension of space $\mathscr{Y}$. Next if $\mathscr{X}$ is a complete separable space and the measure μ is σ-finite, then $\mathscr{X}$ can be represented as a sum of a countable number of compacts $\{K_n, n = 1, 2, \ldots\}$ of a finite measure and of a set N of μ-measure 0. The latter follows from the fact that in a complete separable metric space every measurable set A of a finite measure can be approximated in the measure as closely as desired by a compact $K \subset A$. The foregoing arguments are applicable to each one of the compacts K_n. From here the general assertion of the theorem easily follows. □

Remark 1. Theorem 1 holds for Euclidean spaces $\mathscr{X}$ and $\mathscr{Y}$ in particular if the measure μ is the Lebesgue measure on $\mathscr{X}$.

Remark 2. The proof of Theorem 1 would have been simpler if the separability of the measurable representation of the given function was not required. In such a case it would not have been necessary to consider the set I and the points $x_k^{(n)}$ could have been chosen arbitrarily from the corresponding sets. The only property used would have been the completeness of the space $\mathscr{Y}$. Therefore if $\mathscr{Y}$ is complete, $\mathscr{X}$ is a complete and separable metric space and μ is a σ-finite measure; then the random function $g(x, \omega)$ with values in $\mathscr{Y}$, $x \in \mathscr{X}$, $\omega \in \Omega$, stochastically continuous for μ-almost all x, is stochastically equivalent to a measurable random function.

The next important result follows directly from Fubini's theorem.

Theorem 2. *Let $\xi(x) = g(x, \omega)$ be a measurable random function taking on real or complex values. If*

$$\int_{\mathscr{X}} \mathsf{E} \,|\, \xi(x) \,|\, \mu(dx) < \infty,$$

then for any $A \in \mathfrak{A}$

$$\int_A \mathsf{E}\xi(x)\, \mu(dx) = \mathsf{E} \int_A \xi(x)\, \mu(dx). \quad \square$$

The last equality indicates the permutability of the operations of taking the mathematical expectation of a random variable and the integration with respect to x.

§4. A Criterion for the Absence of Discontinuities of the Second Kind

Functions with no discontinuities of the second kind. Let $\xi(t)$, $t \in [a, b]$ be a random process with values in a complete metric space $\mathscr{Y}$.

Definition 1. If the sample functions of the process have for each $t \in (a, b)$ with probability 1 left-hand and right-hand limits, and possess at point $a(b)$ a right (left)-hand limit, then the process is referred to as *without discontinuities of the second kind* on the interval (a, b).

In the present section it will always be assumed that the process $\xi(t)$ is separable. The separability set of the process is denoted by J.

Definition 2. *The function $y = f(t)$, $y \in \mathscr{Y}$ possesses at least m ε-oscillations ($\varepsilon > 0$) on the interval $[a, b]$, if there exist points $t_0, \ldots, t_m$, $a \leqslant t_0 < t_1 < \ldots < t_m \leqslant b$, such that*

$$\varrho(f(t_{k-1}), f(t_k)) > \varepsilon, \qquad k = 1, 2, \ldots, m.$$

Lemma 1. *A function $y=f(t)$ has no discontinuities of the second kind on the interval $[a, b]$ if and only if for any $\varepsilon>0$ it has only a finite number of ε-oscillations on $[a, b]$.*

Proof. Necessity. Let the number of ε-oscillations be infinite. Then there exists a sequence $t_0, t_1, \ldots, t_n, \ldots$, such that $t_n \uparrow t_0$ or $t_n \downarrow t_0$ and $\varrho(f(t_n), f(t_{n+1}))>\varepsilon$. But this implies that $f(t_0-0)$ or $f(t_0+0)$ do not exist.

Sufficiency. Let the one-sided limit (say the left-hand one) be nonexistent at a certain point t_0. Then a sequence $t_n \uparrow t_0$ can be found such that for any n $\sup_{m>n} \varrho(f(t_m), f(t_n))>\varepsilon$, i.e. the number of ε-oscillations is infinite. □

Note that definition 2 is trivially carried over to the case of random functions defined on an arbitrary set of real-valued t.

Henceforth when dealing with functions with no discontinuities of the second kind we shall not distinguish between two functions having at each point $t \in [a, b]$ the same left-hand and right-hand limits. Therefore it is natural to choose a certain convention concerning the values of these functions at the discontinuity point. Denote by $D[a, b]= =D[a, b; \mathscr{Y}]$ the space of functions defined on $[a, b]$ with values in $\mathscr{Y}$ which do not possess discontinuity of the second kind and which are continuous from the left or from the right at each point $t \in [a, b]$. Set

$$\begin{aligned} \Delta_c(f) = \sup\{\min[\varrho(f(t'), f(t)), \varrho(f(t''), f(t))]; \\ t-c \leqslant t' < t < t'' \leqslant t+c,\ t', t, t'' \in [a, b]\} + \\ + \sup\{\varrho(f(t), f(a));\ a<t<a+c\} + \sup\{\varrho(f(t), f(b));\ b-c<t<b\}. \quad (1) \end{aligned}$$

Lemma 2. *A function $y=f(t)$ has no discontinuities of the second kind if and only if*

$$\lim_{c \to 0} \Delta_c(f)=0. \quad (2)$$

Proof. Necessity. It follows from the definition that for any function $f \in D[a, b]$ the last two terms in the r.h.s. of (1) tend to zero as $c \to 0$.

Let condition 2 not be satisfied. Then sequences t'_n, t_n, t''_n can be found such that $t'_n<t_n<t''_n$, $t''_n-t'_n \to 0$ and $\varrho(f(t'_n), f(t_n))>\varepsilon$, $\varrho(f(t''_n), f(t''_n))>\varepsilon$ for some $\varepsilon>0$. It may be assumed that t_n converges to some t_0 (if this is not the case we replace the sequence t_n by a certain convergent subsequence of it). At least two out of three sequences $\{t'_n\}$, $\{t_n\}$ and $\{t''_n\}$ possess infinitely many points located on one side of t_0. If, for example, $\{t'_n\}$ and $\{t_n\}$ are located to the left of t_0, then $f(t_n) \to f(t-0)$, $f(t'_n) \to f(t-0)$ which contradicts the condition $\varrho(f(t'_n), f(t_n))>\varepsilon$. The case for which $\{t_n\}$ and $\{t''_n\}$ possess infinitely many values located to the right of t_0 is dealt with analogously. All other cases may be reduced to these two.

Sufficiency. It follows from condition (2) that $f(t)$ is continuous from the right at point a and from the left at point b. If for some $t_0 \in (a, b)$, $f(t_0+0)$ does not exist, then a sequence $t_n \downarrow t$ and $\varepsilon > 0$ can be found such that $\varrho(f(t_n), f(t_{n+1})) > \varepsilon$ which contradicts (2). Therefore $f(t_0+0)$ exists for any $t_0 \in [a, b)$. Analogously the existence of $f(t_0-0)$ is obtained. It follows from relation (2) that either $f(t_0) = f(t_0-0)$ or $f(t_0) = f(t_0+0)$. The lemma is thus proved. □

Some inequalities. Lemma 3. *Let $\xi(t)$, $t \in [0, T]$ be a separable stochastically continuous process with values in $\mathscr{Y}$ and let there exist a non-negative monotonically increasing function $g(h)$ and a function $q(C, h) \geqslant 0$, $h > 0$ such that*

$$\mathsf{P}\{[\varrho(\xi(t), \xi(t-h) > Cg(h)] \cap [\varrho(\xi(t+h), \xi(t)) > Cg(h)\} \leqslant q(C, h) \tag{3}$$

and

$$G = \sum_{n=0}^{\infty} g(T2^{-n}) < \infty, \quad Q(C) = \sum_{n=1}^{\infty} 2^n q(C, T2^{-n}) < \infty. \tag{4}$$

Then

$$\mathsf{P}\{\sup_{t', t'' \in [0, T]} \varrho(\xi(t'), \xi(t'')) > N\} \leqslant$$

$$\leqslant \mathsf{P}\left\{\varrho(\xi(0), \xi(T)) > \frac{N}{2G}\right\} + Q\left(\frac{N}{2G}\right), \quad \forall N > 0.$$

Proof. Put

$$A_{nk} = \left\{\varrho\left(\xi\left(\frac{k+1}{2^n} T\right), \xi\left(\frac{k}{2^n} T\right)\right) \leqslant Cg(T2^{-n})\right\},$$

$$k = 0, 1, 2, \ldots, 2^n - 1, \quad n = 0, 1, 2, \ldots$$

$$B_{nk} = A_{nk-1} \cup A_{nk}, \quad D_n = \bigcap_{m=n}^{\infty} \bigcap_{k=1}^{2^m-1} B_{mk} \quad (n \geqslant 1),$$

$$D_0 = A_{00} \cap D_1.$$

In view of stochastic continuity, the separability set J of process $\xi(t)$ can be assumed to be the set of numbers of form $k/2^n$, $k = 0, 1, 2, \ldots$, $n = 0, 1, 2, \ldots$ (cf. Theorem 5, Section 2). We have

$$\mathsf{P}\{\bar{D}_n\} \leqslant \sum_{m=n}^{\infty} \sum_{k=1}^{2^m-1} \mathsf{P}\{\bar{B}_{mk}\} \leqslant \sum_{m=n}^{\infty} 2^m q(C, T2^{-m}) = Q(n, C), \tag{5}$$

where

$$Q(n, C) = \sum_{m=n}^{\infty} 2^m q(C, T2^{-m}).$$

It follows from D_0 that $\varrho(\xi(T), \xi(0)) \leqslant Cg(T)$ and that one of the following events takes place: either $\varrho(\xi(T/2), \xi(0)) \leqslant Cg(T2^{-1})$ or $\varrho(\xi(T), \xi(T/2)) \leqslant Cg(T2^{-1})$. In both cases

$$\varrho(\xi(0), \xi(T/2)) \leqslant Cg(T) + Cg(T2^{-1}),$$
$$\varrho(\xi(T/2), \xi(T)) \leqslant Cg(T) + Cg(T2^{-1}).$$

We now apply the induction method. Assume that inequality

$$\varrho\left(\xi\left(\frac{k}{2^m}T\right), \xi\left(\frac{j}{2^m}T\right)\right) < Cg(T) + 2C \sum_{s=1}^{m} g\left(\frac{T}{2^s}\right) \tag{6}$$

is proved for $m = n$ and for $k, j = 0, 1, \dots 2^n$ under the assumption that D_0 is valid. We prove that an analogous inequality holds also for $m = n+1$. Let k and j be odd numbers $k = 2k_1 + 1, j = 2j_1 - 1$. Since it follows from D_{n+1} that at least one of the inequalities

$$\varrho\left(\xi\left(\frac{k_1}{2^n}T\right), \xi\left(\frac{2k_1+1}{2^{n+1}}T\right)\right) \leqslant Cg\left(\frac{T}{2^{n+1}}\right),$$
$$\varrho\left(\xi\left(\frac{k_1+1}{2^n}T\right), \xi\left(\frac{2k_1+1}{2^{n+1}}T\right)\right) \leqslant Cg\left(\frac{T}{2^{n+1}}\right)$$

is satisfied, we obtain that

$$\varrho\left(\xi\left(\frac{k}{2^{n+1}}T\right), \xi\left(\frac{k'}{2^n}T\right)\right) \leqslant Cg(T2^{-(n+1)}),$$

where k' is either equal to k_1 or to $k_1 + 1$. Analogously an integer j' can be found such that

$$\varrho\left(\xi\left(\frac{j}{2^{n+1}}T\right), \xi\left(\frac{j'}{2^n}T\right)\right) \leqslant Cg(T2^{-(n+1)}).$$

Taking the induction assumption into account we obtain

$$\varrho\left(\xi\left(\frac{k}{2^{n+1}}, T\right), \xi\left(\frac{j}{2^{n+1}}T\right)\right) \leqslant Cg(T) + 2C \sum_{s=1}^{n+1} g(T2^{-s}).$$

The case when k or j are even is dealt with analogously. Therefore inequality (6) is proved for all $m \geqslant 1$. It follows from the separability of the process that if event D_0 occurs, then

$$\sup\{\varrho(\xi(t'), \xi(t'')), t', t'' \in [0, T]\} \leqslant 2CG$$

with probability 1. From here it follows that

$$\mathsf{P}\{\sup_{t', t'' \in [0, T]} \varrho(\xi(t'), \xi(t'')) > N\} \leqslant$$
$$\leqslant Q\left(\frac{N}{2G}\right) + \mathsf{P}\left\{\varrho(\xi(0), \xi(T)) > \frac{N}{2G}\right\}.$$

The lemma is thus proved. □

Lemma 4. *Let the conditions of the previous lemma be satisfied. Then*

$$\mathsf{P}\left\{\Delta_\varepsilon(\xi) > CG\left(\left[\lg_2 \frac{T}{2_\varepsilon}\right]\right)\right\} \leqslant Q\left(\left[\lg_2 \frac{T}{2\varepsilon}\right], C\right),$$

where

$$G(n) = \sum_{m=n}^{\infty} g(T2^{-m}), \quad Q(n, C) = \sum_{m=n}^{\infty} 2^m q(C, T2^{-m}).$$

Proof. We continue the argument of the previous lemma. Let event D_n occur. Using induction we shall prove that for any k and m an integer j_{nm} $(0 \leqslant j_{nm} < 2^{m+1})$ can be found such that

$$\max_{0 \leqslant j \leqslant j_{nm}} \varrho\left(\xi\left(\frac{k-1}{2^n} T\right), \xi\left(\left[\frac{k-1}{2^n} + \frac{j}{2^{n+m}}\right] T\right)\right) \leqslant C \sum_{s=n}^{n+m} g(T2^{-s}), \tag{7}$$

$$\max_{j_{nm}+1 \leqslant j \leqslant 2^{m+1}} \varrho\left(\xi\left(\left[\frac{k-1}{2^n} + \frac{j}{2^{n+m}}\right] T\right), \xi\left(\frac{k+1}{2^n} T\right)\right) \leqslant$$
$$\leqslant C \sum_{s=n}^{n+m} g(T2^{-s}). \tag{8}$$

Moreover quantity $j_{mn} 2^{-(n+m)}$ regarded as a function of m (for fixed n and k) is monotonically non-decreasing. For $m=0$ we choose $j_{n0}=0$ if $\varrho\left(\xi\left(\frac{k}{2^n}\right), \xi\left(\frac{k+1}{2^n}\right)\right) \leqslant Cg(T2^{-n})$ and $j_{n0}=1$ if $\varrho\left(\xi\left(\frac{k-1}{2^n}\right), \xi\left(\frac{k}{2^n}\right)\right) \leqslant$ $\leqslant Cg(T2^{-n})$. Under the assumption of occurrence of D_n, one of these inequalities will necessarily hold. Assume j_{nm} has been chosen. We then define $j_{nm+1} = 2j_{nm}$ if

$$\varrho\left(\xi\left(\left[\frac{k-1}{2^n} + \frac{2j_{nm}+1}{2^{n+m+1}}\right] T\right), \xi\left(\left[\frac{k-1}{2^n} + \frac{j_{nm}+1}{2^{n+m}}\right] T\right)\right) \leqslant$$
$$\leqslant Cg(T2^{-(n+m+1)}),$$

and $j_{nm+1} = 2j_{nm} + 1$, if

$$\varrho\left(\xi\left(\left[\frac{k-1}{2^n} + \frac{j_{nm}}{2^{n+m}}\right] T\right), \xi\left(\left[\frac{k-1}{2^n} + \frac{2j_{nm}+1}{2^{n+m+1}}\right] T\right)\right) \leqslant$$
$$\leqslant Cg(T2^{-(n+m+1)}).$$

Such a choice of j_{nm+1} is possible since one of these two inequalities necessarily holds if D_n occurs; in the case when both of these inequalities are satisfied the choice between the above stated values of j_{nm+1} is arbitrary.

Approaching to the limit in (7) and (8) as $m \to \infty$ we obtain that for any sample function for which D_n holds a number $\tau = \tau(\omega)$, $0 \leqslant \tau \leqslant \leqslant T2^{-(n-1)}$ can be found such that

$$\sup_{\substack{0<t<\tau \\ t \in J}} \varrho\left(\xi\left(\frac{k-1}{2^n} T\right), \xi\left(\frac{k-1}{2^n} T + t\right)\right) \leqslant CG(n),$$

and

$$\sup_{\substack{\tau<t<T2^{-(n-1)} \\ t \in J}} \varrho\left(\xi\left(\frac{k-1}{2^n} T + t\right), \xi\left(\frac{k+1}{2^n} T\right)\right) \leqslant CG(n).$$

Let $\varepsilon \in [2^{-(n+1)}T, 2\ \ T]$ and $0 < t'' - t' < \varepsilon$. Then a k can be found such that $(k-1)\,2^{-n}T \leqslant t' < t'' < (k+1)\,2^{-n}T$. If $t \in [t', t'']$ then either $(t', t) \in \in [(k-1)\,2^{-n}T, (k-1)\,2^{-n}T + \tau]$ or $(t, t'') \subset [(k-1)\,2^{-n}T + \tau, (k+1)\,2^{-n}T]$. If, moreover, the values t', t, and t'' are chosen from J, then at least one of the inequalities

$$\varrho(\xi(t'), \xi(t)) \leqslant 2CG(n), \quad \varrho(\xi(t), \xi(t'')) \leqslant 2CG(n)$$

is satisfied.

It follows from the separability of the process that one of these inequalities holds with probability 1 for any sample function of the process. Therefore it follows from D_n that with probability 1

$$\Delta_\varepsilon(\xi) \leqslant 2CG(n).$$

In view of inequality (5)

$$\mathsf{P}\{\Delta_\varepsilon(\xi) > 2CG(n)\} \leqslant \mathsf{P}(\bar{D}_n) \leqslant Q(n, C),$$

or taking into account that $\varepsilon \geqslant 2^{-(n+1)}T$ and the monotonicity of functions $g(h)$ and $q(h)$ we finally obtain

$$\mathsf{P}\left\{\Delta_\varepsilon(\xi) > CG\left(\left[\lg_2 \frac{T}{2\varepsilon}\right]\right)\right\} \leqslant Q\left(\left[\lg_2 \frac{T}{2\varepsilon}\right], C\right).$$

The lemma is thus proved. □

The conditions based on marginal distributions of the process for absence of discontinuities of the second kind.
From the preceding lemma the following is immediately obtained.

Theorem 1. *If $\xi(t)$, $t\in[0, T]$ is a separable stochastically continuous process with values in $\mathscr{Y}$ satisfying conditions*

$$\mathsf{P}\{[\varrho(\xi(t), \xi(t-h))\geqslant Cg(h)]\cap[\varrho(\xi(t+h), \xi(t))\geqslant Cg(h)]\}\leqslant q(C, h), \quad (9)$$

where

$$\sum_{n=1}^{\infty} g(T2^{-n})<\infty, \qquad \sum_{n=1}^{\infty} 2^n q(C, T2^{-n})<\infty, \quad (10)$$

then with probability 1 *$\xi(t)$ has no discontinuities of the second kind. If moreover*

$$Q(n, C)=\sum_{m=n}^{\infty} 2^m q(C, T2^{-m})\to 0 \quad (11)$$

for some n and $C\to\infty$ then for each sample function of the process with probability 1 a constant α can be found such that

$$\Delta_\varepsilon(\xi)\leqslant\alpha G\left(\left[\lg_2\frac{T}{2\varepsilon}\right]\right) \quad \text{for} \quad 0<\varepsilon<\varepsilon_0, \quad \text{where} \quad G(n)=\sum_{m=n}^{\infty} g(T2^{-m}).$$

Proof. Setting $C=1$ into inequality from Lemma 4 we observe that under the conditions of the theorem $\Delta_\varepsilon(\xi)\to 0$ in probability as $\varepsilon\to 0$. But $\Delta_\varepsilon(\xi)$ regarded as a function of ε is monotonically decreasing as $\varepsilon\downarrow 0$. Therefore $\lim \Delta_\varepsilon(\xi)$ as $\varepsilon\to 0$ exists with probability 1 and equals 0. This proves the first assertion of the lemma. The second assertion also follows from condition (11) and Lemma 4. □

As a particular case of Theorem 1 consider a separable stochastically continuous random process satisfying condition

$$\mathsf{E}[\varrho(\xi(t+h), \xi(t))\,\varrho(\xi(t), \xi(t-h))]^p\leqslant Kh^{1+r}, \quad (12)$$

where $p>0$ and $r>0$. Substituting $g(h)=h^{r'/2p}$ and utilizing Chebyshev's inequality we observe that relations (9), (10) and (11) are satisfied for $q(C, h)=\frac{K}{C^{2p}}h^{1+r-r'}$ and $0<r'<r$. We thus obtain

Corollary 1. *If a separable stochastically continuous random process satisfies condition (12) then its sample functions satisfy with probability* 1 *relation*

$$\Delta_\varepsilon(\xi)\leqslant\alpha\varepsilon^{r'/2p},$$

where $\alpha=\alpha(\omega)$ is a constant and r' is an arbitrary number from $(0, r)$.

We also state the following corollary of Theorem 1.

Corollary 2. *Let a wide-sense stochastically continuous random process be defined on* $[0, T]$ *with values in a complete separable locally compact space* $\mathscr{Y}$ *whose "three-dimensional" marginal distributions satisfy conditions* (9) *and* (10). *Then there exists a representation of this process without discontinuities of the second kind.*

The condition based on conditional probabilities for absence of discontinuities of the second kind.

In the previous theorem the condition for absence of discontinuities of the second kind was expressed in terms of properties of marginal ("three-dimensional") distributions of a random process.

We now present results of a somewhat different nature. They utilize the assumptions related to conditional probabilities and are applicable in the case when information on the properties of conditional distributions of the process is available.

Let $\{\mathfrak{F}_t, t\in[0, T]\}$ be a current of σ-algebras. We shall say that the process $\xi(t)$ *obeys (or is subordinated to) the current of* σ*-algebras* $\{\mathfrak{F}_t, t\in[0, T]\}$ if for each $t\in[0, T]$ the random element $\xi(t)$ is $\mathfrak{F}_t$-measurable.

We introduce quantity

$$\alpha(\varepsilon, \delta)=\inf \sup[\mathsf{P}\{\varrho(\xi(s), \xi(t))\geqslant\varepsilon \mid \mathfrak{F}_s\}; 0\leqslant s\leqslant t\leqslant s+\delta\leqslant T, \omega\in\Omega'], \tag{13}$$

where the inf is taken over all the subsets Ω' $(\Omega'\in\mathfrak{S})$ which have probability 1. It is easy to verify that there exists a Ω^0 such that $\mathsf{P}(\Omega^0)=1$, $\Omega^0\in\mathfrak{S}$ and such that the *inf* is attained on this set, so that

$$\alpha(\varepsilon, \delta)=\sup\{\mathsf{P}\{\varrho(\xi(s), \xi(t))\geqslant\varepsilon \mid \mathfrak{F}_s\}; 0\leqslant s\leqslant t\leqslant s+\delta\leqslant T, \omega\in\Omega^0\}.$$

We show that the condition that $\alpha(\varepsilon, \delta)\to 0$ as $\delta\to 0$ and any $\varepsilon>0$ assures the absence of discontinuities of the second kind for separable processes. Let $[c, d]$ be a fixed interval, $[c, d]\subset[0, T]$ and I be an arbitrary finite sequence of instants of time $t_1, t_2, \ldots, t_n$, $s\leqslant c\leqslant t_1<t_2\ldots<t_n\leqslant d$. Denote by $A(\varepsilon, Z)$ the event: the sample function of a random process $\xi(t)$ on $[c, d]\cap Z$ has at least one ε-oscillation.

Lemma 5. *With probability* 1

$$\mathsf{P}\{A(\varepsilon, I) \mid \mathfrak{F}_s\}\leqslant 2\alpha\left(\frac{\varepsilon}{4}, d-c\right). \tag{14}$$

Proof. We first note that since for $s<t, \mathfrak{F}_s\subset\mathfrak{F}_t$ it follows from the properties of conditional mathematical expectation that

$$\begin{aligned}\mathsf{P}\{\varrho(\xi(t), \xi(u))\geqslant\varepsilon \mid \mathfrak{F}_s\}&=\\ &=\mathsf{E}\{\mathsf{P}\{\varrho(\xi(t), \xi(u))\geqslant\varepsilon \mid \mathfrak{F}_t\} \mid \mathfrak{F}_s\}\leqslant\alpha(\varepsilon, u-s).\end{aligned} \tag{15}$$

for $s<t<u$.

We now introduce events

$$B_k=\left\{\varrho(\xi(c),\xi(t_i))<\frac{\varepsilon}{2},\right.$$

$$\left. i=1,2,\dots,k-1,\ \varrho(\xi(c),\xi(t_k))>\frac{\varepsilon}{2}\right\},$$

$$C_k=\left\{\varrho(\xi(t_k),\xi(d))\geqslant\frac{\varepsilon}{4}\right\},\quad D_k=B_k\cap C_k,\ k=1,2,\dots,n,$$

$$C_0=\left\{\varrho(\xi(c),\xi(d))\geqslant\frac{\varepsilon}{4}\right\}.$$

Events B_k are disjoint and if we put $D=\bigcup_{k=1}^{n} D_k$, then $A(\varepsilon, I)\subset C_0\cup D$. Indeed if $A(\varepsilon, I)$ holds then for some k inequality $\varrho(\xi(c),\xi(t_k))\geqslant\frac{\varepsilon}{2}$ is satisfied for the first time, i.e. one of the events $B_k (k=1,\dots,n)$ occurs. If, moreover D does not hold, i.e. if $\varrho(\xi(t_k),\xi(d))<\frac{\varepsilon}{4}$, then $\varrho(\xi(c),\xi(d))\geqslant$ $\geqslant\varrho(\xi(c),\xi(t_k))-\varrho(\xi(t_k),\xi(d))>\frac{\varepsilon}{4}$, i.e. the event C_0 occurs. Thus $A(\varepsilon, I)\subset$ $\subset C_0\cup D$. We now have with probability 1

$$\mathsf{P}\{D_k\mid\mathfrak{F}_s\}=\mathsf{E}\{\chi_{D_k}\mid\mathfrak{F}_s\}=\mathsf{E}\{\mathsf{E}\{\chi_{B_k}\chi_{C_k}\mid\mathfrak{F}_{t_k}\}\mid\mathfrak{F}_s\}=$$
$$=\mathsf{E}\{\chi_{B_k}\mathsf{P}\{C_k\mid\mathfrak{F}_{t_k}\}\mid\mathfrak{F}_s\}\leqslant\alpha\left(\frac{\varepsilon}{4},d-c\right)\mathsf{E}\{\chi_{B_k}\mid\mathfrak{F}_s\},$$

where χ_A denotes, as usual, the indicator of event A. From here it follows that

$$\mathsf{P}\{D\mid\mathfrak{F}_s\}=\sum_{k=1}^{n}\mathsf{P}\{D_k\mid\mathfrak{F}_s\}\leqslant\alpha\left(\frac{\varepsilon}{4},d-c\right)\mathsf{E}\left\{\sum_{k=1}^{n}\chi_{B_k}\mid\mathfrak{F}_s\right\}\leqslant$$
$$\leqslant\alpha\left(\frac{\varepsilon}{4},d-c\right)(\operatorname{mod}\mathsf{P}).$$

In view of (15) $\mathsf{P}\{C_0\mid\mathfrak{F}_s\}\leqslant\alpha\left(\frac{\varepsilon}{4},d-c\right)$. Therefore

$$\mathsf{P}\{A(\varepsilon,I)\mid\mathfrak{F}_s\}\leqslant\mathsf{P}\{D\mid\mathfrak{F}_s\}+\mathsf{P}\{C_0\mid\mathfrak{F}_s\}\leqslant2\alpha\left(\frac{\varepsilon}{4},d-c\right)(\operatorname{mod}\mathsf{P}),$$

which proves the lemma. □

Lemma 6. *Let $A^k(\varepsilon, I)$ denote the event: $\xi(t)$ has at least k ε-oscillations on I. Then*

$$\mathsf{P}\{A^k(\varepsilon, I) \mid \mathfrak{F}_s\} \leqslant \left[2\alpha\left(\frac{\varepsilon}{4}, d-c\right)\right]^k \pmod{\mathsf{P}}.$$

Proof. Let $B_r(\varepsilon, I)$ denote the event: the sample function of process $\xi(t)$ possesses at least $k-1$ ε-oscillations on the set $(t_1, \ldots, t_r)$, but the number of ε-oscillations on the set $(t_1, \ldots, t_{r-1})$ is less than $k-1$. The events $B_r(\varepsilon, I)$ $(r=1, \ldots, n)$ are disjoint and $\bigcup_{r=1}^{n} B_r(\varepsilon, I) = A^{k-1}(\varepsilon, I) \supset A^k(\varepsilon, I)$. On the other hand, it follows from $A^k(\varepsilon, I) \subset B_r(\varepsilon, I)$ that at least one ε-oscillation exists on the set $(t_r, t_{r+1}, \ldots, t_n)$. Consequently,

$$A^k(\varepsilon, I) \subset \bigcup_{r=1}^{n} (B_r(\varepsilon, I) \cap C_r(\varepsilon, I)),$$

where $C_r(\varepsilon, I)$ denotes the event that $\xi(t)$ has at least one ε-oscillation on $(t_r, t_{r+1}, \ldots, t_n)$. Therefore

$$\mathsf{P}\{A^k(\varepsilon, I) \mid \mathfrak{F}_s\} \leqslant \sum_{r=1}^{n} \mathsf{P}\{B_r(\varepsilon, I) \cap C_r(\varepsilon, I) \mid \mathfrak{F}_s\} \pmod{\mathsf{P}}. \tag{16}$$

Using the properties of conditional mathematical expectations we obtain

$$\begin{aligned}\mathsf{P}\{B_r(\varepsilon, I) \cap C_r(\varepsilon, I) \mid \mathfrak{F}_s\} &= \mathsf{E}\{\mathsf{E}\{\chi_{B_r(\varepsilon, I)}\chi_{C_r(\varepsilon, I)} \mid \mathfrak{F}_{t_r}\} \mid \mathfrak{F}_s\} \leqslant \\ &\leqslant \mathsf{E}\{\chi_{B_r(\varepsilon, I)} \mathsf{P}\{C_r(\varepsilon, I) \mid \mathfrak{F}_{t_r}\} \mid \mathfrak{F}_s\} \leqslant \\ &\leqslant 2\alpha\left(\frac{\varepsilon}{4}, d-c\right) \mathsf{P}\{B_r(\varepsilon, I) \mid \mathfrak{F}_s\} \pmod{\mathsf{P}}.\end{aligned}$$

It follows from the inequality obtained and from (16) that

$$\begin{aligned}\mathsf{P}\{A^k(\varepsilon, I) \mid \mathfrak{F}_s\} &\leqslant 2\alpha\left(\frac{\varepsilon}{4}, d-c\right) \sum_{r=1}^{n} \mathsf{P}\{B_r(\varepsilon, I) \mid \mathfrak{F}_s\} = \\ &= 2\alpha\left(\frac{\varepsilon}{4}, d-c\right) \mathsf{P}\{A^{k-1}(\varepsilon, I) \mid \mathfrak{F}_s\} \pmod{\mathsf{P}},\end{aligned}$$

which yields the required assertion. □

Theorem 2. *If $\xi(t)$ is a separable process and for any $\varepsilon > 0$*

$$\lim_{\delta \to 0} \alpha(\varepsilon, \delta) = 0, \tag{17}$$

then process $\xi(t)$ has no discontinuities of the second kind.

It is sufficient to prove that with probability 1 every sample function of

$\xi(t)$ possesses only a finite number of ε-oscillations. Let J be the separability set of process $\xi(t)$. We represent this set as $J = \bigcup_{n=1}^{\infty} I_n$, where $\{I_n\}$ is a monotonically increasing sequence of sets consisting of a finite number of elements. Let $\varepsilon > 0$ be given. Subdivide $[0, T]$ into m intervals Δ_r, $r = 1, \ldots, m$, of equal lengths such that

$$2\alpha\left(\frac{\varepsilon}{4}, \frac{T}{m}\right) = \beta < 1.$$

Then

$$\mathsf{P}\{A^{\infty}(\varepsilon, J \cap \Delta_r) \mid \mathfrak{F}_s\} \leqslant \mathsf{P}\{A^k(\varepsilon, J \cap \Delta_r) \mid \mathfrak{F}_s\} = \lim_{n \to \infty} \mathsf{P}\{A^k(\varepsilon, I_n \cap \Delta_r) \mid \mathfrak{F}_s\} \leqslant \beta^k,$$

hence

$$\mathsf{P}\{A^{\infty}(\varepsilon, J \cap \Delta_r) \mid \mathfrak{F}_s\} = 0 \pmod{\mathsf{P}} \quad \text{and} \quad \mathsf{P}\{A^{\infty}(\varepsilon, J \cap \Delta_r)\} = 0.$$

Consequently, $\mathsf{P}\{A^{\infty}(\varepsilon, J)\} = 0$. The theorem is thus proved. □

We present a number of important corollaries of the theorem just proved.

Theorem 3. *A separable stochastically continuous process $\xi(t)$ $t \in [0, T]$ with independent increments and with values in a linear normed space $\mathscr{Y}$ has no discontinuities of the second kind.*

Indeed, we have from the definition of a process with independent increments

$$\mathsf{P}\{|\xi(s) - \xi(t)| \geqslant \varepsilon \mid \mathfrak{F}_s\} = \mathsf{P}\{|\xi(s) - \xi(t)| \geqslant \varepsilon\} \pmod{\mathsf{P}}.$$

On the other hand, it follows from the property of uniform stochastic continuity (c.f. Theorem 4, Section 2) that

$$\alpha(\varepsilon, \delta) = \sup[\mathsf{P}\{|\xi(s) - \xi(t)| \geqslant \varepsilon\};\ 0 \leqslant s \leqslant t \leqslant s + \delta \leqslant T]$$

tends to zero as $\delta \to 0$ for any $\varepsilon > 0$. Therefore the conditions of Theorem 2 are satisfied. □

Theorem 2 implies some sharp results for Markov processes also.

Theorem 4. *If $\xi(t)$ $t \in [0, T]$ is a separable Markov process with values in a metric space $\mathscr{Y}$ and transition function $\mathsf{P}(t, x, s, A)$ satisfying condition*

$$\alpha(\varepsilon, \delta) = \sup[\mathsf{P}\{s, y, t, \bar{S}_{\varepsilon}(y)\};\ y \in \mathscr{Y},\ 0 \leqslant s \leqslant t \leqslant s + \delta \leqslant T] \to 0$$

as $\delta \to 0$, where $S_{\varepsilon}(y)$ is a sphere of radius ε with center at point y and $\bar{S}_{\varepsilon}(y)$ is its complement, then the process $\xi(t)$ has no discontinuities of the second kind.

The last assertion follows directly from Theorem 2 and the definition of a Markov process. □

Regularization of sample functions of a process without discontinuities of the second kind. As was mentioned previously, when considering functions without discontinuities of the second kind we identify functions which have the same right-hand and left-hand limits at each point.

Recall that if the process is separable then the values of the sample functions of $\xi(t)$ with probability 1 are limiting values of the sequences $\xi(t_i)$ for $t_i \to t$ and t_i belong to the separability set. If, moreover, the process does not have discontinuities of the second kind then, with probability 1, $\xi(t)$ is equal to $\xi(t-0)$ or $\xi(t+0)$ for each t.

Theorem 5. *If $\xi(t)$ is a stochastically continuous process without discontinuities of the second kind and with values in a metric space $\mathscr{Y}$, then there exists an equivalent process $\xi'(t)$ whose sample functions are continuous from the right* (mod P).

Proof. Define event A: $\lim\limits_{n\to\infty} \xi\left(t+\frac{1}{n}\right)$ exists for each $t \in [0.\ T)$ with probability 1. Set $\xi'(t) = \lim\limits_{n\to\infty} \xi\left(t+\frac{1}{n}\right)$ in the case of A and $\xi'(t) = \xi(t)$ in the case of $\bar{A}$. We have

$$\{\xi'(t) \neq \xi(t)\} = \bigcup_{m=1}^{\infty} \left\{\varrho(\xi(t), \xi'(t)) > \frac{1}{m}\right\} \cap A,$$

$$\mathsf{P}\{\xi'(t) \neq \xi(t)\} = \lim_{m\to\infty} \mathsf{P}\left\{\left[\varrho(\xi(t), \xi'(t)) > \frac{1}{m}\right] \cap A\right\}.$$

On the other hand

$$\mathsf{P}\left\{\varrho(\xi(t), \xi'(t)) > \frac{1}{m}\right\} =$$

$$= \mathsf{P}\left\{\bigcup_{k=1}^{\infty} \bigcap_{n=k}^{\infty} \left\{\varrho\left(\xi(t), \xi\left(t+\frac{1}{n}\right)\right) > \frac{1}{m}\right\}\right\} =$$

$$= \lim_{k\to\infty} \mathsf{P}\left\{\bigcap_{n=k}^{\infty} \left\{\varrho\left(\xi(t), \xi\left(t+\frac{1}{n}\right)\right) > \frac{1}{m}\right\}\right\} \leqslant$$

$$\leqslant \lim_{n\to\infty} \mathsf{P}\left\{\varrho\left(\xi(t), \xi\left(t+\frac{1}{n}\right)\right) > \frac{1}{m}\right\}.$$

Therefore $\mathsf{P}\{\xi'(t) \neq \xi(t)\} = 0$. It remains to observe that the function $\xi'(t)$ is continuous from the right on A. The theorem is thus proved. □

The existence of a stochastically equivalent process continuous from the left is proved analogously.

Martingales. Consider properties of sample functions of a separable semimartingale $\{\xi(t), \mathfrak{F}_t, t\in[0, T]\}$. The general definition of semimartingales and martingales was given earlier (Chapter II, Section 2). In that Chapter we obtained important properties of semimartingales with discrete arguments. Note that the inequalities obtained in Section 2 of Chapter II can easily be carried over to the case of separable submartingales. Indeed for a separable process the event $\sup\{\xi(t), t\in[0, T]\}\neq \neq\sup\{\xi(t), t\in I\}$ where I is the separability set of processes $\xi(t)$ has probability 0. Therefore the corresponding random variables have the same distribution. Furthermore, if the sample function of the process $\xi(t)$ on the interval $[0, T]$ upcrosses the half-interval $[a, b)$ n times and $\xi(\cdot)\notin N$, where N is the exceptional set appearing in the definition of separability, then $\xi(t)$, restricted to I, also upcrosses the half-interval $[a, b)$ n times. This means that the distribution of variables $\nu_{[0, T]}[a, b)$ and $\nu_I[a, b)$ is the same (the lower index at ν denotes the set to which the process is restricted).

Theorem 6. *A separable semimartingale on $[0, T]$ has no discontinuities of the second kind.*

To prove this theorem one should basically repeat the argument given in the proof of Theorem 1, Section 2, Chapter II. It follows from inequality $\mathsf{P}\{\sup\{\xi^+(t), t\in[0, T]\}>C\}\leqslant C^{-1}\,\mathsf{E}\xi^+(T)$ that $\sup\{\xi(t), t\in[0, T]\}<\infty$ with probability 1. Analogously it follows from inequality $\mathsf{E}\nu[a, b)\leqslant(b-a)^{-1}\ \mathsf{E}(\xi(T)-b)^+$ as $a\to-\infty$ that with probability 1 $\inf\{\xi(t), t\in[0, T]\}>-\infty$. Furthermore since $\nu[a, b)$ is integrable there exists a set $N_1\in\mathfrak{S}$ with $\mathsf{P}(N_1)=0$ such that for $\omega\notin N$, $\xi(t)$ crosses any half interval $[a, b)$ only a finite number of times and consequently $\xi(t)$ has no discontinuities of the second kind. We may choose N_1 to be the sum of all $N(a, b)$ where $N(a, b)$ is the set on which $\nu[a, b)=\infty$ when a and b run through all the rational numbers and $a<b$. The theorem is thus proved. □

Let $\mathfrak{F}_{t-0}$ denote the minimal σ-algebra containing $\mathfrak{F}_s$ for $s<t$ and $\mathfrak{F}_{t+0}$ the intersection of $\mathfrak{F}_s$ for $s>t$. Clearly $\mathfrak{F}_{t-0}\subset\mathfrak{F}_t\subset\mathfrak{F}_{t+0}$ and $\xi(t-0)$ is an $\mathfrak{F}_{t-0}$ measurable variable, while $\xi(t+0)$ is a $\mathfrak{F}_{t+0}$ measurable variable.

Theorem 7. *Let $\{\xi(t), \mathfrak{F}_t, t\in[0, T]\}$ be a separable submartingale. Then $\{\xi(t+0), \mathfrak{F}_{t+0}, t\in[0, T]\}$ $(\xi(T+0)=\xi(T))$ is also a submartingale whose sample functions are continuous from the right with probability* 1. *Moreover, $\mathsf{P}\{\xi(t)=\xi(t+0)\}=1$ at each point at which $\mathsf{E}\xi(t)$ is continuous and $\mathfrak{F}_t=\mathfrak{F}_{t+0}$.*

Proof. Note that $\max(\xi(t), a)$ is a submartingale (Chapter II, Section 2) and moreover a uniformly integrable family of random variable (Chapter II, Section 2). Therefore, for $s \leqslant t$

$$\int_A \max(\xi(s), a)\, d\mathsf{P} \leqslant \lim_{t' \downarrow t} \int_A \max(\xi(t'), a)\, d\mathsf{P} = \int_A \max(\xi(t+0), a)\, d\mathsf{P}$$

for any $A \in \mathfrak{F}_t$. Approaching $a \to -\infty$ we obtain

$$\int_A \xi(s)\, d\mathsf{P} \leqslant \int_A \xi(t+0)\, d\mathsf{P},$$

i.e.

$$\xi(s) \leqslant \mathsf{E}\{\xi(t+0) \mid \mathfrak{F}_s\} \quad \text{for} \quad s \leqslant t.$$

Hence

$$\xi(s+0) \leqslant \mathsf{E}\{\xi(t+0) \mid \mathfrak{F}_{s+0}\}, \qquad s \leqslant t.$$

This proves the first part of the theorem. It is easy to verify that the previous discussions also yield the following inequalities

$$\xi(t) \leqslant \mathsf{E}\{\xi(t+0) \mid \mathfrak{F}_t\} \leqslant \mathsf{E}\{\xi(t') \mid \mathfrak{F}_t\} \pmod{\mathsf{P}} \quad \text{for} \quad t' > t,$$

which implies that

$$\xi(t) \leqslant \xi(t+0) \leqslant \mathsf{E}\{\xi(t') \mid \mathfrak{F}_t\} \pmod{\mathsf{P}}.$$

at point t such that $\mathfrak{F}_t = \mathfrak{F}_{t+0}$.

Now if $\mathsf{E}\xi(t') \to \mathsf{E}\xi(t)$, then $\xi(t+0) = \xi(t) \pmod{\mathsf{P}}$. It is also easy to verify that the functions $\xi(t+0)$ are continuous from the right. The theorem is thus proved. □

§ 5. Continuous Processes

Conditions for continuity of processes without discontinuities of the second kind.

We shall assume here as before that $\mathscr{Y}$ is a complete metric space, and $\xi(t)$, $t \in [0, T]$ is a random process with values in $\mathscr{Y}$.

Definition 1. The process $\xi(t)$, $t \in [0, T]$ is called *continuous* if almost all of its sample functions are continuous on $[0, T]$.

For processes without discontinuities of the second kind one can formulate a rather simple sufficient condition for continuity.

Theorem 1. *Let* $\{t_{nk}, k = 0, 1, \ldots, m_n\}$, $n = 1, 2, \ldots$, *be a sequence of sub-*

divisions of the interval $[0, T]$, $0=t_{n0}<t_{n1}<\dots<t_{nm_n}=T$ *and* $\lambda_n=\max_{1\leqslant k\leqslant m_n}(t_{nk}-t_{nk-1})\to 0$ *as* $n\to\infty$. *If the separable process* $\xi(t)$ *has no discontinuities of the second kind, then condition*

$$\sum_{k=1}^{m_n} \mathsf{P}\{\varrho[\xi(t_{nk}), \xi(t_{nk-1})]>\varepsilon\}\to 0 \quad \text{as} \quad n\to\infty \tag{1}$$

for any $\varepsilon>0$ *is a sufficient condition for continuity of this process.*

Proof. Denote by $\nu_\varepsilon(0\leqslant\nu_\varepsilon\leqslant\infty)$ the number of values t such that $\varrho[\xi(t+0), \xi(t-0)]>2\varepsilon$, and by $\nu_\varepsilon^{(n)}$ the number of indices k such that $\varrho[\xi(t_{nk}), \xi(t_{nk-1})]>\varepsilon$. Clearly $\nu_\varepsilon\leqslant\varliminf_{n\to\infty}\nu_\varepsilon^{(n)}$. On the other hand

$$\mathsf{E}\nu_\varepsilon^{(n)}=\sum_{k=1}^{m_n}\mathsf{P}\{\varrho[\xi(t_{nk}), \xi(t_{nk-1})]>\varepsilon\}.$$

In view of the Fatou lemma $\mathsf{E}\nu_\varepsilon\leqslant\mathsf{E}\varliminf_{n\to\infty}\nu_\varepsilon^{(n)}\leqslant\varliminf_{n\to\infty}\mathsf{E}\nu_\varepsilon^{(n)}$. Hence $\mathsf{E}\nu_\varepsilon=0$, i.e. $\nu_\varepsilon=0$ with probability 1 for any $\varepsilon>0$. Consequently $\xi(t-0)=\xi(t+0)$ for any t with probability 1. In view of the separability of the process $\xi(t)=\xi(t-0)=\xi(t+0)$, i.e. the process is continuous. □

Corollary. *If* $\{\xi(t), \mathfrak{F}_t, t\in[0, T]\}$ *is a separable semi-martingale and*

$$\sum_{k=1}^{m_n}\mathsf{P}\{|\xi(t_{nk})-\xi(t_{nk-1})|>\varepsilon\}\to 0 \quad \text{as} \quad \lambda_n\to 0,$$

then $\xi(t)$ *is a continuous process.*

This corollary follows from the fact that a separable semi-martingale has no discontinuities of the second kind. □

We now apply Theorem 1 to the processes satisfying the conditions of Theorem 2 in Section 4. Let $\alpha(\varepsilon, \delta)$ be determined by relation (13) of Section 4.

Theorem 2. *If process* $\xi(t)$ *is separable and*

$$\lim_{\delta\to 0}\frac{\mathsf{E}\alpha(\varepsilon, \delta)}{\delta}=0 \tag{2}$$

for any $\varepsilon>0$, *then process* $\xi(t)$ *is continuous.*

Proof. Since process $\xi(t)$ has no discontinuities of the second kind provided condition (2) is satisfied it is sufficient to verify relation (1). Noting that

$$\mathsf{P}\{\varrho[\xi(t_{nk}), \xi(t_{nk-1})]>\varepsilon\}\leqslant\alpha(\varepsilon, \Delta t_{nk}), \quad \text{where} \quad \Delta t_{nk}=t_{nk}-t_{nk-1},$$

we obtain that

$$\sum_{k=1}^{m_n} \mathsf{P}\{\varrho[\xi(t_{nk}), \xi(t_{nk-1})] > \varepsilon\} \leqslant (b-a) \max_{1 \leqslant k \leqslant n} \frac{\mathsf{E}\alpha(\varepsilon, \Delta t_{nk})}{\Delta t_{nk}} \to 0$$

as $\lambda_n \to 0$. The theorem is thus proved. □

Applying Theorem 2 to Markov processes we obtain the following condition for continuity of a Markov process:

Theorem 3. *Let $\xi(t)$ be a separable Markov process and let*

$$\frac{1}{\delta} \mathsf{P}(s, y, t, \bar{S}_\varepsilon(y)) \to 0$$

for $\delta \to 0$ and for any fixed $\varepsilon > 0$ uniformly in y, s and t where $0 \leqslant t-s \leqslant \delta$, then the process $\xi(t)$ is continuous.

Here $\bar{S}_\varepsilon(x)$ denotes the complement of the sphere $S_\varepsilon(x)$ with the center at point x and radius ε.

Processes with independent increments. The theorem just proved gives only sufficient conditions for continuity of a random process. It turns out however that for the particular case of processes with independent increments the conditions of Theorem 1 are also necessary.

Theorem 4. *If the process $\xi(t)$ with independent increments is continuous, then condition* (1) *is satisfied for an arbitrary sequence $\{t_{nk}, k=0, \ldots, m_n\}$ $n=1, 2, \ldots$ of subdivisions of the interval $[0, T]$ for which $\lambda_n = \max_{1 \leqslant k \leqslant m_n} (t_{nk} - t_{nk-1}) \to 0$.*

Proof. Set $\Delta_h = \sup_{|t_1 - t_2| \leqslant h} \varrho[\xi(t_1), \xi(t_2)]$. In view of the continuity of the process $\xi(t)$, $\Delta_n \to 0$ for $h \to 0$ with probability 1. Therefore $\lim_{h \to 0} \mathsf{P}\{\Delta_h > \varepsilon\} = 0$. On the other hand if $\lambda_n < h$, then

$$\mathsf{P}\{\Delta_h > \varepsilon\} \geqslant \mathsf{P}\{\sup \varrho[\xi(t_{nk}), \xi(t_{nk-1})] > \varepsilon\} = \mathsf{P}\{\varrho[\xi(t_{n1}), \xi(t_{n0})] > \varepsilon\} +$$
$$+ \mathsf{P}\{\varrho[\xi(t_{n1}), \xi(t_{n0})] \leqslant \varepsilon\} \, \mathsf{P}\{\varrho[\xi(t_{n2}), \xi(t_{n1})] > \varepsilon\} + \ldots$$
$$\ldots + \prod_{k=1}^{m_n - 1} \mathsf{P}\{\varrho[\xi(t_{nk}), \xi(t_{nk-1})] \leqslant \varepsilon\} \, \mathsf{P}\{\varrho[\xi(t_{nm_n}), \xi(t_{nm_n-1})] > \varepsilon\} \geqslant$$
$$\geqslant \mathsf{P}\{\Delta_h \leqslant \varepsilon\} \left[\sum_{k=1}^{m_n} \mathsf{P}\{\varrho[\xi(t_{nk}), \xi(t_{nk-1})\right] > \varepsilon\};$$

hence $\sum_{k=1}^{m_n} \mathsf{P}\{\varrho[\xi(t_{nk}), \xi(t_{nk-1})] > \varepsilon\} \leqslant \frac{\mathsf{P}\{\Delta_h > \varepsilon\}}{\mathsf{P}\{\Delta_h \leqslant \varepsilon\}} \to 0$ as $h \to 0$ and any $\varepsilon > 0$.

The theorem is thus proved. □

It is now easy to present a complete description of continuous processes with independent increments and values in a finite-dimensional space.

Theorem 5. *A random process $\xi(t)$, $t\geqslant 0$, $\xi(0)=0$ with values in $\mathscr{R}^m$ and with independent increments is continuous if and only if $\xi(t)$ is a Gaussian process with continuous mean value $a(t)$ and continuous matrix correlation function $R(t,s)=\sigma^2(\min(t,s))$, where $\sigma^2(t)$ is a matrix function $\sigma^2(0)=0$ and $\sigma^2(t)-\sigma^2(s)$ for $s<t$ is a non-negative definite matrix.*

Proof. Let $\xi(t)$ be a continuous process with independent increments. We prove that $\xi(t)-\xi(s)$, $(s<t)$ has a normal distribution. Choose an arbitrary vector z, $z\in\mathscr{R}^m$. The scalar process $\eta(t)=(z,\xi(t))$ is also a continuous process with independent increments. If we show that $\eta(t)-\eta(s)$ has a normal distribution, then it will follow that $\xi(t)-\xi(s)$ has an m-dimensional normal distribution. Let t_{nk}, $k=1,\dots,m_n$ be the subdivision of the interval (s,t) into intervals of equal length such that (cf. Theorem 4)

$$\sum_{k=1}^{m_n}\mathsf{P}\left\{|\eta(t_{nk})-\eta(t_{nk-1})|>\frac{1}{n}\right\}<\frac{1}{n}. \tag{3}$$

Set $\eta'_{nk}=\eta(t_{nk})-\eta(t_{nk-1})=\Delta\eta_{nk}$, if $|\eta(t_{nk})-\eta(t_{nk-1})|\leqslant\frac{1}{n}$ and $\eta'_{nk}=0$ otherwise, and let $\eta'_n=\sum_k\eta'_{nk}$. It follows from inequality (3) that $\mathsf{P}\{\eta'_n\neq\eta(t)-\eta(s)\}<\frac{1}{n}$; hence the P-$\lim\eta'_n=\eta(t)-\eta(s)$. Let $a'_{nk}=\mathsf{E}\eta'_{nk}$, $\sigma^2_{nk}=\mathsf{V}\eta'_{nk}$, $a'_n=\sum_k a'_{nk}$, $\sigma^2_n=\sum_k\sigma^2_{nk}$. Consider the following two cases:

1) $\underline{\lim}\,\sigma^2_n<\infty$; 2) $\lim\sigma^2_n=\infty$. In the first case there exists a subsequence n_r such that $\lim\sigma^2_{n_r}=\sigma^2<\infty$. Since

$$\eta'_{n_r}=a'_{n_r}+\sum_k(\eta'_{n_rk}-a'_{n_rk}),$$

the central limit theorem is applicable to the sum in the r.h.s. of the last equation. The distribution of η'_{nr} thus converges weakly to the normal distribution with parameters $(0,\sigma^2)$. Since η'_{n_r} converges in probability to a limit, a'_{n_r} should also converge to a certain limit a. Thus $\eta(t)-\eta(s)=$ $=a+\eta$, where η is a Gaussian random variable.

In the second case for any $c>0$ a q_n can be found such that $\sum_{k=1}^{q_n}\sigma^2_{nk}\to c$. This follows from the fact that the quantities σ^2_{nk} are uniformly small $\left(\sigma^2_{nk}<\frac{1}{n^2}\right)$. The central limit theorem is applicable to the sum $\sum_{1}^{q_n}(\eta_{nk}-a_{nk})$. But then it follows from the equation

$$\mathsf{E}e^{iu\eta'_n}=e^{iua'_n}\prod_{k=1}^{m_n}\mathsf{E}e^{iu(\eta'_{nk}-a'_{nk})}$$

that

$$\overline{\lim}\,|\mathsf{E}e^{iu\eta'_n}|\leqslant\lim\left|\prod_{k=1}^{q_n}\mathsf{E}e^{iu(\eta'_{nk}-a_{nk})}\right|=e^{-(u^2c^2/2)},$$

where c is an arbitrary number. Therefore $\lim \mathsf{E}e^{iu\eta'_n}=0$, which contradicts the convergence of η'_n to $\eta(t)-\eta(s)$. Hence the second case can not hold. Thus we have shown that $\xi(t)-\xi(s)$ has a normal distribution. Let $a(t)=\mathsf{E}\xi(t)$, $\sigma^2(t)=\mathsf{E}(\xi(t)-a(t))\,(\xi(t)-a(t))^*$. If $t>s$, we have for the matrix correlation function

$$R(t,s)=\mathsf{E}(\xi(t)-a(t))\,(\xi(s)-a(s))^*=\sigma^2(s).$$

It follows from the continuity of $\xi(t)$ that the characteristic function

$$J(u,t)=\mathsf{E}e^{i(u,\xi(t))}=e^{i(a(t),u)-\frac{1}{2}(\sigma^2(t)u,u)}$$

is continuous in t. This is possible if and only if $a(t)$ and $\sigma^2(t)$ are continuous functions in t and $\sigma^2(t)$ satisfies the conditions of the theorem. The first part of the theorem is thus proved.

Now let $\xi(t)$ be a Gaussian process with mean $a(t)$ and matrix correlation $R(t,s)=\sigma^2(\min(t,s))$, where $a(t)$ and $\sigma^2(t)$ are continuous. Set $\xi'(t)=\xi(t)-a(t)$. Then if $t_1<t_2<t_3<t_4$,

$$\begin{aligned}\mathsf{E}(\xi'(t_4)-\xi'(t_3))\,(\xi'(t_2)-\xi'(t_1))^*&=\\&=R(t_4,t_2)-R(t_3,t_2)-R(t_4,t_1)+R(t_3,t_1)=\\&=\sigma^2(t_2)-\sigma^2(t_2)-\sigma^2(t_1)+\sigma^2(t_1)=0,\end{aligned}$$

i.e. the process $\xi(t)$ has independent increments. Next

$$\mathsf{E}(\xi'(t_2)-\xi'(t_1))\,(\xi'(t_2)-\xi'(t_1))=\sigma^2(t_2)-\sigma^2(t_1),$$

and from the well-known expression for the moments of a Gaussian distribution we have

$$\mathsf{E}|\xi'(t_2)-\xi'(t_1)|^4=3\,[\mathrm{Sp}\,\{\sigma^2(t_2)-\sigma^2(t_1)\}]^2.$$

Utilizing Chebyshev's inequality we obtain

$$\begin{aligned}\sum_{k=1}^{m_n}\mathsf{P}\,\{|\xi'(t_{nk})-\xi'(t_{nk-1})|>\varepsilon\}&\leqslant\\&\leqslant\sum_{k=1}^{m_n}\frac{3\,[\mathrm{Sp}\,\{\sigma^2(t_{nk})-\sigma^2(t_{nk-1})\}]^2}{\varepsilon^4}\leqslant\end{aligned}$$

$$\leqslant \frac{3 \max \operatorname{Sp}\{\sigma^2(t_{nk})-\sigma^2(t_{nk-1})\}}{\varepsilon^4} \operatorname{Sp}\sigma^2(T) \to 0$$

as $\max(t_{nk}-t_{nk-1}) \to 0$. In view of Theorem 1 the process $\xi'(t)$ and therefore the process $\xi(t)$ are continuous. The theorem is thus proved. □

Kolmogorov's conditions for continuity of a random process. We prove a convenient and direct sufficient condition for continuity of a random process, which does not utilize the assumption of absence of discontinuities of the second kind. This condition is based on the simplified version of Lemmas 3 and 4 in Section 4.

Lemma 1. *Let $\xi(t)$, $t \in [0, T]$ be a separable process satisfying the following condition: there exists a non-negative monotonically non-decreasing function $g(h)$ and a function $q(c, h)$, $h \geqslant 0$ such that*

$$\mathsf{P}\{\varrho(\xi(t+h), \xi(t)) > Cg(h)\} \leqslant q(C, h) \tag{4}$$

and

$$G = \sum_{n=0}^{\infty} g(2^{-n}T) < \infty, \quad Q(C) = \sum_{n=1}^{\infty} 2^n q(C, 2^{-n}T) < \infty. \tag{5}$$

Then

$$\mathsf{P}\{\sup_{0 \leqslant t' < t'' \leqslant T} \varrho(\xi(t'), \xi(t'')) > N\} \leqslant Q\left(\frac{N}{2G}\right) \tag{6}$$

and

$$\mathsf{P}\left\{\sup_{|t'-t''| \leqslant \varepsilon} \varrho(\xi(t'), \xi(t'')) > CG\left(\left[\lg_2 \frac{T}{2\varepsilon}\right]\right)\right\} \leqslant Q\left(\left[\lg_2 \frac{T}{2\varepsilon}\right], C\right), \tag{7}$$

where

$$G(m) = \sum_{n=m}^{\infty} g(2^{-n}T), \quad Q(m, C) = \sum_{n=m}^{\infty} 2^n q(C, 2^{-n}T). \tag{8}$$

To prove this lemma it is sufficient to repeat in simplified form the arguments presented in the proofs of Lemmas 3 and 4 of Section 4. We shall omit the details and present a brief outline of the argument. We introduce the events

$$A_{nk} = \left\{\varrho\left(\xi\left(\frac{k+1}{2^n} T\right), \xi\left(\frac{k}{2^n} T\right)\right) \leqslant Cg(2^{-n}T)\right\},$$

$$k = 0, 1, \ldots, 2^n - 1, \quad n = 0, 1, 2, \ldots$$

and set $D_n = \bigcap_{m=n}^{\infty} \bigcap_{k=0}^{2^n-1} A_{nk}$. Then

$$\mathsf{P}\{\bar{D}_n\} \leqslant Q(n, C).$$

It follows from D_n that for any t' and t'' belonging to J (as defined in Lemma 3, Section 4)

$$\varrho(\xi(t'), \xi(t'')) \leqslant 2CG;$$

if moreover, in addition to D_n the inequalities $0 \leqslant t''-t' \leqslant 2^{-n}$ are satisfied, then $\varrho(\xi(t'), \xi(t'')) \leqslant 2CG(n)$. Arguing in the same manner as at the conclusion of the proof of the above-mentioned lemmas we obtain the required assertion. One should also keep in mind that conditions (4) and (5) yield stochastic continuity of the process $\xi(t)$. □

Theorem 6. *Let the conditions of Lemma* 1 *be satisfied. Then the process* $\xi(t)$ *is continuous. If, moreover,* $Q(m, C) \to 0$ *for some* m *and* $C \to \infty$, *the process* $\xi(t)$ *then possesses the following property: with probability* 1 *there exists a constant* $\gamma = \gamma(\omega)$ *such that*

$$\sup_{|t'-t''|<\varepsilon} \varrho(\xi(t'), \xi(t'')) \leqslant \gamma G\left(\left[\lg_2 \frac{T}{2\varepsilon}\right]\right) \pmod{\mathsf{P}}. \tag{9}$$

The theorem follows from Lemma 1. □

As a particular case in which conditions (4) and (5) are fulfilled we consider the process satisfying

$$\mathsf{E}\varrho^p[\xi(t'), \xi(t'')] \leqslant L\,|t''-t'|^{1+r}, \tag{10}$$

where $p>0$, $r>0$. Set $g(h)=h^{r'/p}$, where $0<r'<r$. Then

$$G\left(\left[\lg_2 \frac{T}{2\varepsilon}\right]\right) \leqslant K_1 \varepsilon^{r'/p}, \quad \text{and} \quad Q\left(\left[\lg_2 \frac{T}{2\varepsilon}\right], C\right) \leqslant C^{-p} K_2 \varepsilon^{(r-r')},$$

where K_1 and K_2 are constants. From Theorem 6 now follows

Corollary 1. *If a separable random process* $\xi(t)$ *satisfies condition* (10) *then its sample function satisfies with probability one a Lipschitz condition* $\varrho[\xi(t'), \xi(t'')] \leqslant \gamma\,|t''-t'|^{r'/p}$, *where* $\gamma = \gamma(\omega)$ *is a constant and* r' *can be any number in* $(0, r)$.

Corollary 2. *Consider the Wiener process for which* $\mathsf{E}[\xi(t+h)-\xi(t)]=0$, $\mathsf{E}[\xi(t+h)-\xi(t)]^2=h$. *Since* $\mathsf{E}|\xi(t+h)-\xi(t)|^{2m}=(2m-1)!!|\,h|^m$ *for an arbitrary integer* m, *the sample function of the separable Wiener process satisfy with probability* 1 *a Lipschitz condition of order* $\frac{1}{2}-\varepsilon$, *where* ε *is an arbitrary positive number.*

Corollary 3. *If a separable process satisfies conditions* (4) *and* (5) *and a* $q^m G(m) \leqslant K$ *for all* m *and some* $q>1$, *then the sample functions of this process satisfy with probability* 1 *a Lipschitz condition*

$$\varrho(\xi(t'), \xi(t'')) \leqslant \gamma\,|t''-t'|^{\lg_2 q}.$$

Consider now another condition more general than (10) which assures the fulfillment of assumptions (4) and (5). Let

$$\mathsf{E}\varrho^p[\xi(t), \xi(t+h)] \leqslant \frac{L|h|}{|\lg_2|h||^{1+r}}, \quad p<r \tag{11}$$

If we set

$$g(h) = |\lg_2|h||^{-r'/p}, \quad \text{where} \quad p<r'<r,$$

then

$$G = \sum_{n=0}^{\infty} |\lg_2|2^{-n}T||^{-r'/p} < \infty,$$

$$Q(C) \leqslant \frac{LT}{\sum_{n=0}^{\infty} C^p |\lg_2|2^{-n}T||^{1+r-r'}} < \infty.$$

Corollary 4. *If a separable process $\xi(t)$ satisfies relation* (11) *then the process is continuous.*

Gaussian processes. We now apply the preceding results to a one-dimensional separable real Gaussian process $\xi(t)$, $t\in[0, T]$ with the correlation function $R(s, t)$ and the mean value 0. The difference $\xi(t+h)-\xi(t)$ has the variance

$$\sigma^2(t, h) = R(t+h, t+h) - 2R(t, t+h) + R(t, t).$$

Therefore

$$\mathsf{P}\{|\xi(t+h)-\xi(t)| > Cg(h)\} = \frac{2}{\sqrt{2\pi}} \int_{\alpha}^{\infty} e^{-t^2/2}\, dt,$$

where $\alpha = Cg(h)\,\sigma^{-1}(t, h)$. Utilizing inequality

$$\int_{\alpha}^{\infty} e^{-t^2/2}\, dt \leqslant \frac{1}{\alpha} e^{-\alpha^2/2}, \tag{12}$$

(which is easily verified using integration by parts), we obtain

$$\mathsf{P}\{|\xi(t+h)-\xi(t)| > Cg(h)\} \leqslant \frac{2}{\sqrt{2\pi}} \frac{\sigma(t, h)}{Cg(h)} e^{-C^2 g^2(h)/2\sigma^2(t, h)}. \tag{13}$$

Theorem 7. *If a Gaussian process satisfies condition*

$$\sigma^2(t, h) \leqslant \frac{K}{|\ln|h||^p}, \quad p>3, \tag{14}$$

then the process is continuous.

Proof. Set $g(h)=\left|\ln|h|\right|^{-p'}$, where p' is an arbitrary number satisfying inequality $1<p'<\frac{p-1}{2}$. We then can choose

$$q(C,h)=\frac{K'}{C\left|\ln|h|\right|^{(p/2)-p'}}\,e^{-(C^2/2K)\left|\ln|h|\right|^{p-2p'}},$$

and series (5) will be convergent. From here the assertion of the theorem follows. □

Moreover, it follows from the second part of Theorem 6 that for each sample function of the process with probability 1 a constant $\gamma=\gamma(\omega)$ can be found such that

$$|\xi(t+h)-\xi(t)|\leqslant\gamma\left[\ln\frac{T}{|h|}\right]^{1-p'}.$$

If we now assume that the correlation function of the process $\xi(t)$ is smoother, then the sample functions will also be smoother. Assume that

$$\sigma^2(t,h)\leqslant K|h|^p,\qquad p>0. \tag{15}$$

It follows from (13) that $q(C,h)$ can be chosen as

$$q(C,h)=\frac{K_1|h|^{p/2}}{Cg(h)}\,e^{-C^2g^2(h)/2K|h|^p}.$$

Set $g(h)=|h|^{p/2}\left|\ln|h|\right|^{1+\varepsilon}$, where $\varepsilon>0$. Then

$$Q(m,C)=\sum_{n=m}^{\infty}\frac{K_1}{C}\,\frac{1}{|n-\ln T|^{1+\varepsilon}}\,e^{-(C^2/2K')|n-\ln T|^{2+2\varepsilon}+n\ln 2}$$

tends to zero as $C\to\infty$. Next we have

$$G(m)\leqslant K_3m^{1+\varepsilon}2^{-mp/2}.$$

We have thus obtained:

Theorem 8. *If the correlation function of a Gaussian process satisfies* (15) *then its sample functions satisfy with probability* 1 *the following inequality*

$$|\xi(t+h)-\xi(t)|\leqslant\gamma|h|^{p/2}\left|\ln|h|\right|^{1+\varepsilon},$$

where ε is an arbitrary positive number and γ is a constant.

In particular, sample functions of a separable Wiener process satisfy with probability 1 the following inequality

$$|\xi(t+h)-\xi(t)|\leqslant\gamma\sqrt{|h|}\cdot\left|\ln|h|\right|^{1+\varepsilon},\ t,\ t+h\in[0,T]$$

for any $\varepsilon>0$. This result represents a refinement of Corollary 2 of Theorem 6.

Chapter IV

Linear Theory of Random Processes

§ 1. Correlation Functions

Positive definite kernels. The existence of an important and sufficiently wide class of problems whose solution requires a knowledge of only the very general properties of a random function and its first and second moment is indeed remarkable and non-trivial. A substantial part of the theory of linear transformations of random functions is devoted to these problems and they constitute the main topic of the present chapter. We shall therefore discuss here random functions with values in a linear space with finite moments of second order unless otherwise explicitly stipulated.

Let $\zeta(x)$, $x \in X$ be a complex-valued random function with finite moments of the second order. We call these random functions Hilbert random functions. A Hilbert random function can be interpreted as a function defined on X with values in the Hilbert space of random variables $\mathscr{L}_2$:

$$x \to \zeta(x) = f(x, \omega) \in \mathscr{L}_2 .$$

In particular if X is an interval (a, b) on the real line and $\zeta(x)$ is a curve in $\mathscr{L}_2$, then the notation $\zeta = \zeta(x)$, $x \in (a, b)$ is a parametric equation of this curve. In the present chapter mainly Hilbert random functions are considered, therefore the word Hilbert will often be omitted.

Set

$$a(x) = \mathsf{E}\zeta(x),$$

$$R(x, y) = \mathsf{E}(\zeta(x) - a(x)) \overline{(\zeta(y) - a(y))} . \tag{1}$$

The function $a(x)$ is called the *mean value* of $\zeta(x)$ and $R(x, y)$ is its *correlation function.* If we put $x = y$ then $R(x, x) = \mathsf{E}|\zeta(x) - a(x)|^2 = \sigma^2(x)$ gives us the variance of the complex-valued random variable $\zeta(x)$. The correlation function coincides with the previously introduced covariance of the set of random variables $\zeta(x) - a(x)$.

It is sometimes more advantageous to utilize the correlation function in place of the covariance since the correlation function has an important probabilistic interpretation – namely, it characterizes the degree of linear dependence between the values of a random function at two points.

On the other hand, the distinction between correlation function and covariance is inconsequential. While the correlation function is the covariance of the random function $\zeta(x)-a(x)$, the covariance, on the other hand, can be interpreted as the correlation function of $\zeta(x)\,e^{i\varphi}$, where φ is a random variable uniformly distributed on $(-\pi, \pi)$ independent of $\{\zeta(x), x\in X\}$. This means that the classes of correlation and covariance functions coincide. Henceforth we shall consider correlation as well as covariance functions.

The covariance $B(x_1, x_2)=\mathsf{E}\zeta(x_1)\overline{\zeta(x_2)}$ of a random function $\zeta(x)$ possesses the characteristic property of positive definiteness: Let X be an arbitrary set.

Definition 1. A complex-valued function $C(x_1, x_2)$ $(x_1, x_2)\in X^2$ is called *a positive definite kernel on* X^2 if for any $n\,(n=1, 2, \ldots)$, $x_k\in X$ and any complex numbers z_k $(k=1, 2, \ldots)$

$$\sum_{k,r=1}^{n} C(x_k, x_r)\, z_k \bar{z}_r \geqslant 0. \tag{2}$$

Covariance $B(x_1, x_2)$ is a positive definite kernel on X^2. Indeed

$$\sum_{k,r=1}^{n} B(x_k, x_r)\, z_k \bar{z}_r = \mathsf{E}\left|\sum_{1}^{n} \zeta(x_k)\, z_k\right|^2 \geqslant 0.$$

The following properties of a positive definite kernel are easily derived from its definition

1) $$C(x, x)\geqslant 0, \tag{3}$$

2) $$C(x_1, x_2)=\overline{C(x_2, x_1)}, \tag{4}$$

3) $$|C(x_1, x_2)|^2\leqslant C(x_1, x_1)\, C(x_2, x_2), \tag{5}$$

4) $$|C(x_1, x_3)-C(x_2, x_3)|^2\leqslant \leqslant C(x_3, x_3)\,[C(x_1, x_1)+C(x_2, x_2)-2\operatorname{Re} C(x_1, x_2)]. \tag{6}$$

These properties may be easily verified directly for covariance. To obtain inequalities (3)–(6) in the general case we first put $n=1$ into (2). We thus obtain $C(x_1, x_1)\,|z_1|^2\geqslant 0$, which yields (3). Next we assume $n=2$. We first note that $C(x_1, x_2)\, z_1\bar{z}_2+C(x_2, x_1)\,\bar{z}_1 z_2$ is real which implies (4). Inequality (5) is the condition of positive definiteness of a Hermetian quadratic form

$$\sum_{k,r=1}^{2} C(x_k, x_r)\, z_k \bar{z}_r .$$

To obtain (6) we set $n=3$, $z_1=z$, $z_2=-z$ in (2). Then

$$[C(x_1, x_1)+C(x_2, x_2)-2\,\mathrm{Re}\,C(x_1, x_2)]\,|z|^2+$$
$$+2\,\mathrm{Re}[C(x_1, x_3)-C(x_2, x_3)]\, z\bar{z}_3+C(x_3, x_3)\,|z_3|^2\geqslant 0,$$

which implies (6).

The cross correlation function characterizes the degree of linear dependence between the two random functions $\zeta_1(x)$ and $\zeta_2(x)$ with finite moments.

Definition 2. Let $\zeta_1(x)$, $\zeta_2(x)$ be Hilbert random functions, $\mathsf{E}\zeta_i(x)=a_i(x)$. Then

$$R_{\zeta_1\zeta_2}(x, y)=\mathsf{E}[\zeta_1(x)-a_1(x)]\,\overline{[\zeta_2(y)-a_2(y)]}$$

is called the *cross correlation function* of $\zeta_1(x)$ and $\zeta_2(x)$.

To describe the class of possible cross correlation functions and for solutions of many other problems, it is convenient to consider a sequence of Hilbert random functions $\zeta^{(1)}(x), \zeta^{(2)}(x), \dots \zeta^{(m)}(x)$, $x\in X$ as components of a single random vector function $\zeta(x)$ with values in $\mathscr{Z}^m$. Moreover as above, $\zeta(x)$ denotes a column-vector, and $\zeta^*(x)$ denotes a row-vector with components $\zeta_k(x)=\overline{\zeta^{(k)}(x)}$, $k=1, 2, \dots, m$. Set

$$a(x)=\mathsf{E}\zeta(x)=\{\mathsf{E}\zeta^{(1)}(x), \mathsf{E}\zeta^{(2)}(x), \dots, \mathsf{E}\zeta^{(m)}(x)\},$$
$$R(x, y)=|R_k^j(x, y)|=\mathsf{E}(\zeta(x)-a(x))\,(\zeta(y)-a(y))^*.$$

The vector function $a(x)=(a^{(1)}(x), \dots, a^{(m)}(x))$ is called the *mean value* and $R(x, y)$ is the *matrix correlation function* of $\zeta(x)$.

We also note that

$$R_k^j(x, y)=\mathsf{E}(\zeta^{(j)}(x)-a^{(j)}(x))\,\overline{(\zeta^{(k)}(y)-a^{(k)}(y))},$$
$$j, k=1, 2, \dots, m.$$

Definition 3. A matrix function $C(x, y)=\|C_k^j(x, y)\|$ $j, k=1, \dots, m$ is called a *matrix positive definite kernel on* X^2, if for an arbitrary n, an arbitrary sequence of complex-valued vectors z_k $(z_k\in\mathscr{Z}^m)$ and arbitrary points x_k $(x_k\in X)$ we have

$$\sum_{j,k=1}^{n} z_j^* C(x_j, x_k)\, z_k \geqslant 0. \tag{7}$$

A correlation matrix function is a matrix positive definite kernel.

Indeed,

$$\sum_{j,k=1}^{n} z_j^* R(x_j, x_k)\, z_k =$$

$$= \mathsf{E} \sum_{j,k=1}^{n} z_j^* (\zeta(x_j) - a(x_j))\, (\zeta(x_k) - a(x_k))^*\, z_k =$$

$$= \mathsf{E} \left| \sum_{k=1}^{n} (\zeta(x_k) - a(x_k))^*\, z^k \right|^2 \geqslant 0.$$

We note several properties of matrix positive definite kernels $C(x, y)$.

1. Matrix $C(x, x)$ is positive definite, i.e.

$$z^* C(x, x)\, z = \sum_{j,k=1}^{n} C_k^j(x, x)\, \bar{z}^j z^k \geqslant 0, \tag{8}$$

2. $$C^*(x, y) = C(y, x). \tag{9}$$

3. $$|C_k^j(x, y)| \leqslant C_j^i(x, x)\, C_k^k(y, y). \tag{10}$$

Property (8) coincides with (7) for $n=1$. Equality (9) follows from the fact that the matrix $z_1^* C(x, y)\, z_2 + z_2^* C(y, x)\, z_1$ is real for any complex-values vectors z_1 and $z_2\, (z_k \in \mathscr{Z}^m)$. Observe also that inequality (7) is equivalent to the requirement that for any n and any $x_1, x_2, \ldots$ the block matrix

$$\left\| \begin{matrix} C(x_1, x_1) & C(x_1, x_2) & \ldots & C(x_1, x_n) \\ C(x_2, x_1) & C(x_2, x_2) & \ldots & C(x_2, x_n) \\ C(x_n, x_1) & C(x_n, x_2) & \ldots & C(x_n, x_n) \end{matrix} \right\|$$

be positive definite. Utilizing this remark we obtain in the case $n=2$ inequality (10).

The property of positive definiteness is characteristic of the correlation (matrix) function.

Theorem 1. *In order that a function $R(x_1, x_2)$, $x_i \in X$ be a correlation function it is necessary and sufficient that it be a positive definite kernel.*

Proof. The necessity follows from the preceeding definitions. Sufficiency follows from the fact that given a positive-definite kernel $R(x_1, x_2)$ one can construct a complex Gaussian random function $\zeta(x)$ for which $R(x_1, x_2)$ is the correlation function. □

Remark. One can prove analogously that Theorem 1 is valid also for matrix correlation functions: in order that a matrix function $R(x_1, x_2)$ be a correlation function of the vector $\zeta(x)$, $x \in X$, it is necessary and sufficient that it be a positive-definite matrix kernel.

Let $X = \mathscr{X}$ be a metric space with metric ϱ.

Definition 4. A Hilbert random function $\{\zeta(x), x \in \mathscr{X}\}$ is called *continuous at point x_0 in the mean square* (briefly *m.s. continuous*) if

$$\mathsf{E}|\zeta(x)-\zeta(x_0)|^2 \to 0 \quad \text{as} \quad \varrho(x, x_0) \to 0.$$

From Lemma 3 of Section 1 in Chapter I we obtain the following

Theorem 2. *For the m.s. continuity of $\zeta(x)$ at point x_0 it is necessary and sufficient that the covariance $B(x_1, x_2) = \mathsf{E}\zeta(x_1)\,\zeta(x_2)$ be continuous at point (x_0, x_0).*

Remark 1. Stochastic continuity of $\zeta(x)$ at point x_0 follows from the m.s. continuity of $\zeta(x)$ at the same point. Indeed in view of Chebyshev's inequality

$$\mathsf{P}\{|\zeta(x)-\zeta(x_0)|>\varepsilon\} \leqslant \frac{\mathsf{E}|\zeta(x)-\zeta(x_0)|^2}{\varepsilon^2}.$$

Remark 2. If $\zeta(x)$ is m.s. continuous on $\mathscr{X}$ (i.e. at each point of x) it does not mean that the sample functions are continuous with probability 1 on $\mathscr{X}$. Indeed for the Poisson process we have $\mathsf{E}|\zeta(t+h)-\zeta(t)|^2 = \lambda h + (\lambda h)^2$, but the sample functions $\zeta(t)$ are discontinuous with a positive probability.

Wide-sense stationary processes. If we assume that the random function $\zeta(x)$ possesses certain invariant properties with respect to variable x, then the class of corresponding correlation functions also possesses a certain invariance and it becomes possible to describe this class in more detail. In this subsection we shall consider limitations of this kind and a description of the corresponding class of correlation functions will be given in the next.

We start with an important generalization of the notion of stationarity of a random process.

Let $\zeta(t) = \{\zeta^1(t), \zeta^2(t), \ldots, \zeta^m(t)\}$, $t \in (-\infty, \infty)$, be a stationary process with values in $\mathscr{Z}^m$. Then the variables

$$a(t) = \mathsf{E}\zeta(t), \quad R(t_0+t, t_0) = \mathsf{E}(\zeta(t_0+t) - a(t_0+t))\,(\zeta(t_0) - a(t_0))^*$$

are independent of t,

$$a(t) = a = \text{const}, \qquad R(t_1, t_2) = R(t_1 - t_2, 0) = R(t_1 - t_2). \tag{11}$$

Function $R(t) = R(t+t_0, t_0)$ is also called the *(matrix) correlation function* of a stationary process.

Clearly, even if for a certain random process the equalities (11) are satisfied, still it is not sufficient for the process to be stationary. However, in problems whose solution depends only on the values of the moments of the first two orders of the process, the stationarity condition is utilized only to the extent expressed in relations (11). Therefore it would seem natural to introduce the following important class of processes first investigated by A. Ja. Khinchin.

Definition 5. A Hilbert m.s. continuous random proces $\zeta(t)$, $-\infty<t<\infty$ with values in $\mathscr{Z}^m$ is called a *wide-sense stationary process* (or a *stationary process* in *Khinchin's sense*) if

$$\mathsf{E}\zeta(t)=a=\text{const}, \quad \mathsf{E}(\zeta(t_1)-a)(\zeta(t_2)-a)^*=R(t_1-t_2).$$

Let $\zeta(t)$ be a one-dimensional wide-sense stationary process. Since the correlation function $R(t_1-t_2)$ is a positive definite kernel it follows that

$$\sum_{j,k=1}^{n} R(t_j-t_k)\,\bar{z}_j z_k \geqslant 0$$

for any n, $t_i \in(-\infty,\infty)$ and z_j $(j=1,\dots,n)$. Positive definite kernels on the linear space $\mathscr{X}$ which depend on the difference of the arguments are important in various problems of analysis. These are called positive definite functions on $\mathscr{X}$.

Definition 6. Let $\mathscr{X}$ be a linear space. A complex-valued function $f(x)$, $x\in\mathscr{X}$ is called *positive definite* if for any n, $x_j\in\mathscr{X}$ and complex numbers z_j $(j=1,2,\dots)$

$$\sum_{j,k=1}^{n} f(x_j-x_k)\,\bar{z}_j z_k \geqslant 0.$$

A positive definite function has the following properties (cf. (3)–(6)):

1) $$f(0)\geqslant 0, \tag{12}$$

2) $$\overline{f(x)}=f(-x), \tag{13}$$

3) $$|f(x)|\leqslant f(0), \tag{14}$$

4) $$|f(x_1)-f(x_2)|^2\leqslant 2f(0)\,[f(0)-\operatorname{Re} f(x_2-x_1)]. \tag{15}$$

In particular a positive definite function is bounded on $\mathscr{X}$. Next if it is continuous at $x=0$ it is then uniformly continuous on the whole space $\mathscr{X}$.

We now return to wide sense stationary processes with values in $\mathscr{Z}^m$. Each component of such a process is a one-dimensional wide-sense stationary process and the cross correlation function of two components $\zeta^j(t)$ and $\zeta^k(t)$ of the process depends only on the difference of the arguments:

$$R_{\zeta^j\zeta^k}(t_1,t_2)=\mathsf{E}(\zeta^j(t_1)-a^j)\,(\overline{\zeta^k(t_2)-a^k})=R^j_k(t_1-t_2).$$

Definition 7. If $\zeta(t)$ and $\eta(t)$ are wide-sense stationary random processes and the compound process $\xi(t)=\{\zeta(t), \eta(t)\}$ is also wide-sense stationary then the processes $\zeta(t)$ and $\eta(t)$ are called *stationary correlated (in the wide sense).*

It follows from this definition that any group of components of a wide-sense stationary process regarded as a "self-contained" stationary process is stationary correlated with any other group of components of this process.

Definition 8. Let $\mathscr{X}$ be a linear space. The matrix function $C(x)=\|C_k^j(x)\|$, $x\in\mathscr{X}$, $j, k=1,\ldots, m$ is called *positive definite* if for any n, x_j, z_j, where $x_j\in\mathscr{X}$ and z_j are complex vectors in $\mathscr{Z}^m$,

$$\sum_{j,k=1}^{n} z_j^* C(x_j-x_k)\, z_k \geqslant 0.$$

Since $C(x_1-x_2)$ is a matrix positive definite kernel, $C(x)$ possesses the following properties (cf. (8)–(10)):

1) $C(0)$ is a positive definite matrix, (16)

2) $$C^*(x)=C(-x), \tag{17}$$

3) $$|C_j^i(x)|^2 \leqslant C_i^i(0)\, C_j^j(0). \tag{18}$$

It follows from Definition 5 that the matrix correlation function of a wide-sense stationary process is a positive-definite matrix function. In particular it possesses properties (16)–(18).

The definition of a wide-sense stationary process directly carries over to the case of random sequences $\{\zeta(n), n=0, \pm 1, \pm 2,\ldots\}$. In this case

$$\mathsf{E}\zeta(n)=a=\text{const}, \qquad \mathsf{E}(\zeta(k+n)-a)\,(\zeta(k)-a)^*=R(n).$$

The matrix correlation function here is a sequence of matrices.

We now present a number of examples of correlation functions of wide-sense stationary sequences.

Example 1. A standard non-correlated sequence of random vectors $\{\zeta(n), n=0, \pm 1, \pm 2,\ldots\}$ is a sequence which satisfies the following conditions

$$a=\mathsf{E}\zeta(n)=0, \qquad R(0)=I, \qquad R(n)=0 \quad \text{for} \quad n\neq 0,$$

where I is the unit matrix.

Example 2. *Markov stationary Gaussian sequence.* We shall confine ourselves to the case of a vector sequence with real components, zero mean vector and nondegenerate matrix $R(0)=\mathsf{E}\zeta(n)\,\zeta^*(n)$. It follows from the last assumption that the distribution of $\zeta(n)$ is not concentrated in a proper subspace of the space of values of $\zeta(n)$.

Since $\mathsf{E}(\zeta(n+1)-A\zeta(n))\,\zeta^*(n)=0$ if $A=R(1)\times R^{-1}(0)$, it follows from the Gaussian property of the process that $\zeta(n)$ and $\eta(n)=\zeta(n+1)-A\zeta(n)$ are independent. Therefore

$$\mathsf{E}\{\zeta(n+1)\mid\zeta(n)\}=\mathsf{E}\{A\zeta(n)+\eta(n)\mid\zeta(n)\}=A\zeta(n).$$

Let $\mathfrak{F}_n$ be a σ-algebra generated by the random variables $\zeta(s)$, $s\leqslant n$. Since the process is Markovian $\mathsf{E}\{\zeta(s+n)\mid\mathfrak{F}_n\}=\mathsf{E}\{\zeta(s+n)\mid\zeta(n)\}$. Therefore

$$\begin{aligned}\mathsf{E}\{\zeta(s+n)\mid\zeta(s)\}&=\mathsf{E}\{\mathsf{E}\{\zeta(s+n)\mid\mathfrak{F}_{s+n-1}\}\mid\zeta(s)\}=\\&=\mathsf{E}\{\mathsf{E}\{\zeta(s+n)\mid\zeta(s+n-1)\}\mid\zeta(s)\}=\\&=\mathsf{E}\{A\zeta(s+n-1)\mid\zeta(s)\}=A^n\zeta(s).\end{aligned}$$

Finally for $n\geqslant 0$

$$R(n)=\mathsf{E}\{\zeta(s+n)\,\zeta^*(s)\}=\mathsf{E}\{\mathsf{E}\{\zeta(s+n)\mid\zeta(s)\}\,\zeta^*(s)\}=A^nR(0).$$

Thus the correlation function of a stationary Markov Gaussian sequence is of the form

$$R(n)=A^nR(0),\qquad (n\geqslant 0).\tag{19}$$

For example in the one-dimensional case ($|a|\leqslant 1$)

$$R(n)=\sigma^2a^n\qquad (n\geqslant 0).\tag{20}$$

Example 3. *The process of moving averages.* Let $\{\xi(n),\ n=0,\ \pm 1,\dots\}$ be a standard non-correlated sequence of random vectors with values in $\mathscr{Z}^m$. Set

$$\zeta(n)=\sum_{k=0}^{\infty}A_k\xi(n-k),\tag{21}$$

where A_k, $k=0, 1, 2,\dots$ is a sequence of matrices (operators) which map $\mathscr{Z}^m$ into itself. The series in the r.h.s. of the last equality represents the sum of orthogonal vectors in $\mathscr{L}_2^{(m)}\{\Omega, \mathfrak{S}, \mathsf{P}\}$.

For the convergence of this series it is sufficient that

$$\mathsf{E}\sum_{k=0}^{\infty}|A_k\xi(n-k)|^2\leqslant\mathsf{E}\sum_{k=0}^{\infty}|A_k|^2\,|\xi(n-k)|^2=\sum_{k=0}^{\infty}|A_k|^2<\infty.$$

Here $|A|$ denotes the norm of the matrix A,

$$|A|=\sqrt{\mathrm{Sp}(AA^*)}=\sqrt{\sum_{j,k=1}^{m}|a_{jk}|^2}.$$

For the matrix correlation function we have for $n\geqslant 0$ the expression

$$R(n)=\sum_{k=0}^{\infty}A_{n+k}A_k^*.\tag{22}$$

Example 4. *Autoregression process.* Let $\xi(n)$ be a one-dimensional standard non-correlated sequence. Consider the finite-difference equation for determining sequence $\zeta(n)$:

$$\zeta(n)+b_1\zeta(n-1)+\ldots+b_s\zeta(n-s)=\\ =a_0\xi(n)+a_1\xi(n-1)+\ldots+a_s\xi(n-s). \tag{23}$$

Many applied problems lead to equations of type (23) which is called the autoregression equation. Clearly if the values $\zeta(0), \zeta(1), \ldots, \zeta(s-1)$ are given, equation (23) will enable us to express successively $\zeta(s), \zeta(s+1), \ldots$ in terms of the "initial values" $\zeta(0), \ldots, \zeta(s-1)$ and the values $\xi(0), \xi(1), \ldots$. Consider the problem of existence of the stationary solution of (23) in which $\zeta(n)$ is expressed in terms of $\xi(m)$, $m \leqslant n$. For this purpose we shall seek a solution of equation (23) in the form of a moving averages process

$$\zeta(n)=\sum_{k=0}^{\infty} c_k\xi(n-k). \tag{24}$$

Equation (23) is reduced to the system

$$\left.\begin{array}{l} c_0=a_0,\ c_1+b_1c_0=a_1, \ldots, \\ c_s+b_1c_{s-1}+\ldots+b_sc_0=a_s, \\ c_p+b_1c_{p-1}+\ldots+b_sc_{p-s}=0 \quad \text{for} \quad p>s. \end{array}\right\} \tag{25}$$

We introduce the generating functions $A(z), B(z), C(z)$ of the sequences $\{a_n\}, \{b_n\}$ and $\{c_n\}$,

$$A(z)=\sum_{n=0}^{s} a_nz^n, \quad B(z)=\sum_{n=0}^{s} b_nz^n, \quad C(z)=\sum_{n=0}^{\infty} c_nz^n,$$

where $b_0=1$. Multiplying equation (25) by $1, z, z^2, \ldots$ and summing them up, we obtain $C(z)\,B(z)=A(z)$, or

$$C(z)=\frac{A(z)}{B(z)}=a_0+\frac{zA_1(z)}{B(z)},$$

where $A_1(z)$ is a polynomial of degree at most $s-1$. Assume that all the roots of the polynomial $B(z)$ are simple. Then there exists a partial fraction expansion of the form

$$\frac{A_1(z)}{B(z)}=\frac{A_1}{z_1-z}+\frac{A_2}{z_2-z}+\ldots+\frac{A_s}{z_s-z},$$

hence

$$C(z)=a_0+\sum_{n=1}^{\infty}\left(\frac{A_1}{z_1^n}+\frac{A_1}{z_2^n}+\ldots+\frac{A_s}{z_s^n}\right)z^n$$

and

$$c_n = \sum_{k=1}^{s} A_k z_k^{-n}, \qquad n \geqslant 1. \tag{26}$$

Moreover if the roots of the polynomial $B(z)$ are located outside the circle $|z| \leqslant 1$, then the series is m.s. convergent. It is easy to see that this result remains valid also in the case when $B(z)$ has multiple roots. We have thus proved the following

Theorem 3. *The autoregression equation* (23) *possesses a stationary solution* (24), (26) *provided all the roots of the polynomial* $B(z)$ *are located outside the circle* $|z| \leqslant 1$. *The correlation function* $R(n)$ *of this process satisfies the difference equation*

$$R(n) + b_1 R(n-1) + \dots + b_s R(n-s) = 0, \qquad n > s,$$

$$R(n) + b_1 R(n-1) + \dots + b_s R(n-s) = \\ = a_n \bar{c}_0 + a_{n-1} \bar{c}_1 + \dots + a_s \bar{c}_{s-n} \quad \text{for} \quad 0 \leqslant n \leqslant s.$$

We now present several examples of correlation functions of wide-sense continuous-parameter stationary processes.

One is prompted to consider, as the simplest example, the process $\zeta(t)$ such that

$$\mathsf{E}\zeta(t) = 0, \quad \mathsf{E}|\zeta(t)|^2 = 1, \quad \mathsf{E}\zeta(t)\overline{\zeta(s)} = 0 \quad \text{for} \quad t \neq s.$$

The correlation function of this process is discontinuous; thus the process is not m.s. continuous and does not belong to the class of processes studied in this section. It can be shown that this process is not equivalent (stochastically) to a process with measurable sample functions. On the other hand, some processes similar to this example and having even more irregular behavior are studied in the theory of generalized random processes.

Example 5. *Random oscillations.* Oscillation processes occur in many physical and technical problems; these are represented in complex form by a function of the type

$$\zeta(t) = \sum_k \gamma_k e^{iu_k t}. \tag{27}$$

Each component of this sum represents a simple harmonic (periodic) oscillation with frequency $u_k/2\pi$ and power $|\gamma_k|^2$. The totality of the quantities $\{u_k\}$ is called the spectrum (or the frequency spectrum) of process $\zeta(t)$. Assume that γ_k are mutually orthogonal random variables

$$\mathsf{E}\gamma_k = 0, \quad \mathsf{E}|\gamma_k|^2 = c_k^2, \quad \mathsf{E}\gamma_k \bar{\gamma}_j = 0 \quad \text{for} \quad k \neq j.$$

Then the correlation function of the process $\zeta(t)$ is equal to

$$R(t_1, t_2) = \mathsf{E}\zeta(t_1)\overline{\zeta(t_2)} = \mathsf{E}\sum_{k,j}\gamma_k\bar{\gamma}_j e^{i(u_k t_1 - u_j t_2)} = \\ = \sum_k c_k^2 e^{iu_k(t_1 - t_2)},$$

i.e. $\zeta(t)$ is a wide-sense stationary process. Its correlation function is completely determined by the frequency spectrum and the averages (the mathematical expectations) of measures of power corresponding to each one of the simple harmonic oscillations appearing in process $\zeta(t)$. In connection with this power representation we introduce the important characteristic of a stationary process called the spectral function of a process.

The spectral function $F(u)$ of process (27) is determined by the relation

$$F(u) = \sum_{k,\, u_k < u} c_k^2 .$$

This means that $F(u)$ is equal to the average power carried by the harmonic components of the process $\zeta(t)$ with frequencies smaller than the given value u. The function $F(u)$ completely characterizes the mean power of each harmonic component of the process $\zeta(t)$ as well as the total average power of the harmonic components of the process with the frequencies lying within any given interval. Indeed,

$$c_k^2 = F(u_k + 0) - F(u_k), \quad \sum_{u_1 \leqslant u_k \leqslant u_2} c_k^2 = F(u_2) - F(u_1).$$

In terms of the spectral function the correlation function of the process $\zeta(t)$ can be written in the form

$$R(t) = \int_{-\infty}^{\infty} e^{itu}\, dF(u). \tag{28}$$

From the mathematical point of view the spectral function is a non-negative non-decreasing function continuous from the left, which is constant everywhere except at a finite number of points where it has jumps of size c_k^2. It turns out that the notion of a spectral function can be introduced for arbitrary wide-sense stationary processes. This problem as well as the problem of generalization of representation (28) for arbitrary random processes is considered in the following sections.

§ 2. Spectral Representations of Correlation Functions

Stationary sequences. Consider first a wide-sense stationary sequence of complex-valued random variables $\{\zeta(n), n=\ldots, -1, 0, 1, \ldots\}$ for which

$$\mathsf{E}\zeta(n)=0, \ \mathsf{E}\zeta(k+n)\overline{\zeta(k)}=R(n).$$

The sequence of numbers $R(n)$ is positive-definite, i.e. for any n and any complex numbers z_k, $k=0, 1, 2, \ldots, n$,

$$\sum_{j,k=0}^{n} R(j-k)\,\bar{z}_j z_k \geqslant 0.$$

Theorem 1. *The function $\{R(n), n=0, \pm 1, \pm 2, \ldots\}$ is a correlation function of a wide-sense stationary sequence of random variables iff it can be represented in the form*

$$R(n)=\int_{-\pi}^{\pi} e^{inu} F(du), \tag{1}$$

where $F(\cdot)$ is a finite measure on $[-\pi, \pi]$. The measure F is uniquely defined on the Borel sets of the interval $[-\pi, \pi]$.

Proof. Sufficiency. Sequence (1) is positive definite since

$$\sum_{j,k=0}^{n} R(j-k)\,\bar{z}_j z_k = \int_{-\pi}^{\pi} \left(\sum_{j=0}^{n} e^{iju}\bar{z}_j\right)\overline{\left(\sum_{k=0}^{n} e^{iku}\bar{z}_k\right)} F(du) =$$

$$= \int_{-\pi}^{\pi} \left|\sum_{j=0}^{n} e^{iju}\bar{z}_j\right|^2 F(du) \geqslant 0.$$

Therefore the sequence is a correlation function of a certain wide-sense stationary sequence.

Necessity. Let $R(n)$ be a correlation function of a certain wide-sense stationary sequence. Put

$$f(u,\varrho)=\sum_{n=0}^{\infty}\sum_{m=0}^{\infty} e^{-i(n-m)u} R(n-m)\,\varrho^{n+m}, \quad 0<\varrho<1. \tag{2}$$

The series in the r.h.s. of (2) is absolutely convergent since

$$\sum_{n=0}^{N}\sum_{m=0}^{N} |e^{-i(n-m)u} R(n-m)\,\varrho^{n+m}| \leqslant R(0)\left|\sum_{n=0}^{N}\varrho^n\right|^2 \leqslant \frac{R(0)}{(1-\varrho)^2}.$$

It follows from the positive-definiteness of $R(n)$ that $f(u,\varrho)\geqslant 0$. Changing the order of summation in (2) we obtain

$$f(u,\varrho)=\sum_{k=-\infty}^{\infty} e^{-iku}R(k)\sum_{j=0}^{\infty}\varrho^{|k|+2j}=\sum_{-\infty}^{\infty}\frac{\varrho^{|k|}}{1-\varrho^2}R(k)\,e^{-iku}.$$

The relation obtained shows that the quantities $\frac{\varrho^{|k|}}{1-\varrho^2}R(-k)$ are the Fourier coefficients of a positive function $f(u,\varrho)$. Hence

$$\frac{\varrho^{|n|}}{1-\varrho^2}R(n)=\frac{1}{2\pi}\int_{-\pi}^{\pi} e^{inu}f(u,\varrho)\,du$$

or

$$\varrho^{|n|}R(n)=\int_{-\pi}^{\pi} e^{inu}F_\varrho(du), \tag{3}$$

where

$$F_\varrho(A)=\frac{1-\varrho^2}{2\pi}\int_A f(u,\varrho)\,du,$$

and $F_\varrho[-\pi,\pi]=R(0)<\infty$. The family of measures $F_\varrho(\cdot)$ on $[-\pi,\pi]$ is weakly compact. Therefore a sequence $\varrho_k\uparrow 1$ can be found such that $F_{\varrho_k}(\cdot)$ converges weakly to a certain measure $F(\cdot)$. Approaching the limit in (3) as $\varrho=\varrho_k\to 1$ we obtain formula (1).

We now prove the uniqueness of measure F. Assume that there exist measures F_1 and F_2 defined on the Borel sets of the interval $[-\pi,\pi]$ such that $R(n)$ can be represented by (1). Denote by K the class of Borel functions $f(u)$ on $[-\pi,\pi]$ for which

$$\int_{-\pi}^{\pi} f(u)\,F_1(du)=\int_{-\pi}^{\pi} f(u)\,F_2(du).$$

The class K is linear and closed with respect to the operation of uniform limits and limits of bounded monotone sequences of functions. Since this class contains functions of the type e^{inu} $(n=0,\pm 1,\dots)$ it contains, in view of Weierstrass' approximation theorem for continuous functions, all continuous functions and hence all the bounded Borel functions. Putting $f(u)=\chi_A(u)$ where A is an arbitrary Borel set on $[-\pi,\pi]$ we obtained that $F_1(A)=F_2(A)$. The theorem is thus proved. □

The measure F is called the *spectral measure* of a stationary sequence and the corresponding distribution function $F(u)=F(-\infty, u)$ is called the *spectral function*. If $F(du)=f(u)\,du$, i.e. if measure F is absolutely continuous with respect to the Lebesgue measure, then $f(u)$ is called the *spectral density* of the sequence $\zeta(n)$. We note that condition

$$\sum_{-\infty}^{\infty} |R(n)| < \infty$$

assures the existence of a spectral density. Indeed in this case the Fourier series

$$2\pi f(u) = \sum_{n=-\infty}^{\infty} R(n)\, e^{-inu} \tag{4}$$

converges uniformly and absolutely. Therefore

$$R(n) = \int_{-\pi}^{\pi} e^{inu} f(u)\, du.$$

Homogeneous random fields. We shall generalize Theorem 1 to the case of continuous parameter wide-sense stationary fields.

Definition 1. A random function $\{\zeta(x),\ x\in\mathscr{R}^m\}$ is called a *homogeneous field* in $\mathscr{R}^m$ if

$$\left.\begin{aligned} &\mathsf{E}\zeta(x) = a = \text{const},\\ &R(x_1, x_2) = \mathsf{E}[\zeta(x_1)-a]\,\overline{[\zeta(x_2)-a]} = R(x_1-x_2). \end{aligned}\right\} \tag{5}$$

Thus a correlation function of a homogeneous random field $R(x_1, x_2)$ depends only on the vector which joins the points x_1 and x_2. The function $R(x)$ in the r.h.s. of equality (5) is also called a correlation function of a homogeneous field. The condition of positive definiteness of a correlation function is of the form

$$\sum_{j,k=1}^{n} R(x_j - x_k)\, \bar{z}_j z_k \geqslant 0.$$

It follows from relation

$$\mathsf{E}|\zeta(x+h)-\zeta(x)|^2 = 2[R(0) - \operatorname{Re} R(h)]$$

that if function $R(x)$ is continuous at $x=0$ then the field $\zeta(x)$ is m.s. continuous at each point $x\in\mathscr{R}^m$.

Theorem 2. *In order that function $R(x)$ $(x\in\mathscr{R}^m)$ be the correlation function of a homogeneous m.s. continuous random field $\{\zeta(x),\ x\in\mathscr{R}^m\}$ it is necessary* *and sufficient that it admit representation*

$$R(x)=\int_{\mathscr{R}^m} e^{i(x,u)} F(du), \tag{6}$$

where F is a finite measure on the Borel sets of $\mathscr{R}^m$. Moreover the measure F is uniquely determined on $\mathfrak{B}^m$.

Sufficiency. The function $R(x)$ determined by formula (6) is continuous and positive-definite:

$$\sum_{j,k=1}^{n} R(x_j-x_k)\,\bar{z}_j z_k=\int_{\mathscr{R}^m}\left(\sum_{j,k=1}^{n} e^{i(x_j-x_k,u)}\bar{z}_j z_k\right)F(du)=$$

$$=\int_{\mathscr{R}^m}\left|\sum_{k=1}^{n} e^{-i(x_k,u)}z_k\right|^2 F(du)\geqslant 0.$$

Therefore $R(n)$ is the correlation function of a certain m.s. continuous complex Gaussian field (cf. Section 1 of Chapter III). Moreover one can construct a very simple example of a homogeneous field with the correlation function given by (6). To do this we introduce a random vector ξ in $\mathscr{R}^m$ with the distribution

$$\mathsf{P}\{\xi\in A\}=\frac{1}{F_0}F(A), \qquad F_0=F(\mathscr{R}^m)$$

for an arbitrary Borel set $A\subset\mathscr{R}^m$. Set $\zeta(x)=\sqrt{F_0}\,e^{i[(\xi,x)+\varphi]}$, where φ is a random variable uniformly distributed on $(-\pi,\pi)$, and φ and ξ are mutually independent. Then

$$\mathsf{E}\zeta(x)=0, \qquad R(x,y)=\mathsf{E}\zeta(x)\,\overline{\zeta(y)}=F_0\mathsf{E}e^{i(\xi,x-y)}=\int_{\mathscr{R}^m} e^{i(x-y,u)}F(du).$$

Necessity. We show now that an arbitrary continuous positive definite function admits representation (6). It follows from the condition of positive definiteness that for an arbitrary function $g(x)$ integrable in $\mathscr{R}^m$ the inequality

$$\int_{\mathscr{R}^m}\int_{\mathscr{R}^m} R(x-y)\,\bar{g}(x)\,g(y)\,dx\,dy\geqslant 0$$

is valid. We set $g(x)=\exp\left\{-\dfrac{|x|^2}{2N}+i(x,z)\right\}$, where $N>0$, and $z\in\mathscr{R}^m$.

Then

$$\int_{\mathcal{R}^m}\int_{\mathcal{R}^m} R(x-y)\exp\left\{-\frac{|x|^2+|y|^2}{2N}-i(x-y,z)\right\}dx\,dy\geqslant 0.$$

Carrying out the following orthogonal transformation of the coordinates in $\mathcal{R}^m\times\mathcal{R}^m$:

$$x-y=\sqrt{2}u,\qquad x+y=\sqrt{2}v,$$

we obtain

$$0\leqslant\int_{\mathcal{R}^m}\int_{\mathcal{R}^m} R(u)\exp\left\{-\frac{|u|^2+|v|^2}{2N}-i(u,z)\right\}du\,dv=$$

$$=(2\pi N)^{m/2}\int_{\mathcal{R}^m} R(u)\exp\left\{-\frac{|u|^2}{2N}-i(u,z)\right\}du.$$

Thus the function

$$\tilde{R}_N(z)=\frac{1}{(2\pi)^{m/2}}\int_{\mathcal{R}^m} R(u)\,e^{-|u|^2/2N}\cdot e^{-i(u,z)}\,du$$

is non-negative. Moreover this function is the Fourier transform of an integrable continuous function $R(u)\,e^{-|u|^2/2N}$ and is also differentiable. We show that $\tilde{R}_N(z)$ is integrable. Since $\tilde{R}_N(z)$ and $e^{-\varepsilon|z|^2/2}$, $\varepsilon>0$, are the Fourier transforms of the functions

$$R(u)\,e^{-|u|^2/2N}\quad\text{and}\quad \varepsilon^{-m/2}\,e^{-|u|^2/2\varepsilon},$$

correspondingly, it follows from Parseval's equality that

$$\int_{\mathcal{R}^m}\tilde{R}_N(z)\,e^{-\varepsilon|z|^2/2}\,dz=\int_{\mathcal{R}^m} R(u)\,e^{-|u|^2/2N}\frac{1}{\varepsilon^{m/2}}\,e^{-|u|^2/2\varepsilon}\,du\leqslant$$

$$\leqslant R(0)\int_{\mathcal{R}^m}\frac{1}{\varepsilon^{m/2}}\,e^{-|u|^2/2\varepsilon}\,du=(2\pi)^{m/2}R(0).$$

Let $\varepsilon\to 0$. Utilizing Fatou's lemma we obtain

$$\int_{\mathcal{R}^m}\tilde{R}_N(z)\,dz\leqslant(2\pi)^{m/2}R(0).$$

From the integrability of $\tilde{R}_N(z)$ it follows that the inversion formula for

Fourier transforms is applicable to this function

$$R(u)\,e^{-|u|^2/2N}=\int\limits_{\mathscr{R}^m} e^{i(u,z)}\frac{1}{(2\pi)^{m/2}}\,\tilde{R}_N(z)\,dz=$$

$$=\int\limits_{\mathscr{R}^m} e^{i(u,z)}F_N(dz), \qquad (7)$$

where

$$F_N(A)=\int\limits_A \frac{1}{(2\pi)^{m/2}}\,\tilde{R}_N(z)\,dz.$$

Thus the function $\dfrac{R(u)}{R(0)}\,e^{-|u|^2/2N}$ is the characteristic function of a certain distribution in $\mathscr{R}^m$ and converges as $N\to\infty$ to a continuous function. Therefore (Chapter I, Section 1, Theorem 3) $R(u)/R(0)$ is also a characteristic function. The uniqueness of measure F in representation (6) follows from the theorem on the uniqueness of a distribution with a given characteristic function (Chapter I, Section 1, Theorem 2). The theorem is thus proved. □

As in the case of sequences, measure $F(\cdot)$ in representation (6) is called the *spectral measure*, and the corresponding distribution function $F(u)=F(I_u)$, where $I_u=\{x:x<u,\ x\in\mathscr{R}^m\}$ is called the *spectral function*.

If the spectral measure is absolutely continuous:

$$F(A)=\int\limits_A f(u)\,du,$$

then $f(u)$ is called the *spectral density* of a random field. If the spectral density exists, then the spectral representation of a correlation function becomes

$$R(x)=\int\limits_{-\infty}^{\infty} e^{i(x,u)}f(u)\,du.$$

We note the following criterion for the existence of the spectral density: *if $R(x)$ is an absolutely integrable function $(x\in\mathscr{R}^m)$, then the spectral density exists.*

To verify this criterion, we utilize the notation and relations obtained in the course of the proof of the preceding theorem. By virtue of Parseval's equality for Fourier integrals we have

$$\int\limits_K \tilde{R}_N(z)\,dz=\int\limits_{\mathscr{R}^m} \tilde{R}_N(z)\,\chi_K(z)\,dz=$$

$$= \int_{\mathscr{R}^m} R(u)\, e^{-|u|^2/2N} \frac{1}{(2\pi)^{m/2}} \prod \frac{e^{i(x^k+h^k)u^k} - e^{i(x^k-h^k)u^k}}{iu^k}\, du,$$

where $K = \{z : x^k - h^k < z^k < x^k + h^k\}$. Hence

$$\int_K \tilde{R}_N(z)\, dz \leqslant \frac{V(K)}{(2\pi)^{m/2}} \int_{\mathscr{R}^m} |R(u)|\, du,$$

where $V(K)$ is the volume of the parallepiped K. Hence measure $F(\cdot)$ is absolutely continuous with respect to Lebesgue's measure. □

Corollary 1. *A function $R(t)$, $t \in (-\infty, \infty)$ is the correlation function of a wide-sense stationary process if and only if*

$$R(t) = \int_{-\infty}^{\infty} e^{itu} F(du),$$

where $F(\cdot)$ is a finite measure on $\mathfrak{B}^1$.

Corollary 2. *A function $J(u)$, $u \in \mathscr{R}^m$, $J(0) = 1$ is the characteristic function of a distribution in $\mathscr{R}^m$ if and only if it is continuous and positive-definite.*

Homogeneous and isotropic fields. Formula (6) can be further specialized if certain additional assumptions are imposed on the random field. An important and at the same time quite general property is the isotropy of a random field. A random field is called *isotropic* if its correlation function $R(x_1, x_2)$ depends only on x_2 and on the distance between the points x_1 and x_2. If the field is, in addition, homogeneous then

$$R(x_1, x_2) = R(\varrho)$$

where ϱ is the distance between x_1 and x_2, $\varrho = \sqrt{\sum_{j=1}^{m} (x_1^j - x_2^j)^2}$.

We derive a representation of the correlation function of an m.s. continuous homogeneous and isotropic random field. Since the field is homogeneous its correlation function is of form (6). Integrating both sides of this formula over the surface of the sphere S_ϱ of radius ϱ we obtain

$$R(\varrho) = \frac{\Gamma(m/2)}{2\pi^{m/2} \varrho^{m-1}} \int_{\mathscr{R}^m} \left\{ \int_{S_\varrho} e^{i(x, u)} s(dx) \right\} F(du),$$

where $s(dx)$ in the inner integral denotes the integration over the surface of the sphere S_ϱ. Note that if V_ϱ denotes a sphere of radius ϱ with the center at the origin, then

$$\int_{S_\varrho} f(x)\, s(dx) = \frac{d}{d\varrho} \int_{V_\varrho} f(x)\, dx .$$

On the other hand

$$\int_{V_\varrho} e^{i(x,u)} dx = \left(\frac{2\pi\varrho}{|u|}\right)^{m/2} I_{m/2}(\varrho|u|),$$

where $I_\nu(x)$ is the Bessel function of the first kind. It follows from here that

$$\int_{S_\varrho} e^{i(x,u)} s(dx) = \left(\frac{2\pi\varrho}{|u|}\right)^{m/2} |u|\, I_{(m-2)/2}(\varrho|u|). \tag{8}$$

We introduce a measure g on the semi-axis $[0, \infty)$ by putting $g([a, b)) = F\{V_b \backslash V_a\}$, $0 \leqslant a \leqslant b$, where V_ϱ denotes the open sphere of radius ϱ. Then

$$R(\varrho) = 2^{(m-2)/2} \Gamma\left(\frac{m}{2}\right) \int_0^\infty \frac{I_{(m-2)/2}(\lambda\varrho)}{(\lambda\varrho)^{(m-2)/2}}\, g(d\lambda), \tag{9}$$

and $g([0, \infty)) = F(\mathcal{R}^m) = R(0)$.

We have thus proved the following theorem.

Theorem 3. *In order that $R(\varrho)$ be a correlation function of a homogeneous and isotropic m.s. continuous m-dimensional random field, it is necessary and sufficient that this function admit representation* (9), *where g is a finite measure on $[0, \infty)$.*

For $n=2$ formula (9) becomes

$$R(\varrho) = \int_0^\infty I_0(\lambda\varrho)\, g(d\lambda), \tag{10}$$

and for $n=3$

$$R(\varrho) = 2 \int_0^\infty \frac{\sin \lambda\varrho}{\lambda\varrho}\, g(d\lambda). \tag{11}$$

The same argument shows that the m.s. continuous random field $\xi(t, x)$, $\infty - < t < \infty$, $x \in \mathcal{R}^m$ will be homogeneous in variables (t, x) and isotropic in "the spatial" variables x, i.e. its correlation function will depend only on t and ϱ:

$$\mathsf{E}\xi(t+s, x)\overline{\xi(s, y)} = R(t, \varrho),$$

where ϱ is the distance between x and y, if and only if this correlation function is of the form

$$R(t, \varrho) = \int_{-\infty}^{\infty}\int_{0}^{\infty} e^{itv}\Omega_m(\varrho\lambda)\, g(dv \times d\lambda), \tag{12}$$

where

$$\Omega_m(x) = \left(\frac{2}{x}\right)^{(m-2)/2} \Gamma\left(\frac{m}{2}\right) I_{(m-2)/2}(x) \tag{13}$$

and g is a measure on the half-plane (λ, v), $\lambda \in [0, \infty)$, $v \in (-\infty, \infty)$.

We now obtain the general form of the correlation function of an m.s. continuous homogeneous isotropic field in a Hilbert space. If $R(\varrho)$ is such a correlation function then for any m the function $R(\varrho)$, $\varrho^2 = \sum_{k=1}^{m} (x_k)^2$ is a correlation function of an m.s. continuous homogeneous field in $\mathscr{R}^m$. We note that the function $e^{-\lambda^2\varrho^2/2}$ possesses this property for any λ. Indeed, for any m

$$e^{-(x_1^2+\dots+x_m^2)\lambda^2/2} = \frac{1}{(\sqrt{2\pi}\lambda)^m} \int_{-\infty}^{\infty}\!\!\dots\!\int e^{i\sum_1^m x_k y_k}\, e^{-(1/2\lambda^2)\sum_1^m y_k^2} dy_1 \dots dy_m,$$

i.e. the function $e^{-\lambda^2\varrho^2/2}$, $\varrho^2 = \sum_{k=1}^{m} x_k^2$, is the Fourier transform of a positive function and therefore is positive-definite. From here it follows that the function

$$R(\varrho) = \int_0^{\infty} e^{-\lambda^2\varrho^2/2}\, g(d\lambda) \tag{14}$$

is also positive-definite for any finite measure g on $[0, \infty)$ and for any m if $\varrho^2 = \sum_{k=1}^{m} x_k^2$. We show that formula (14) exhausts all possible positive definite continuous functions in a Hilbert space which depend only on ϱ.

Theorem 4. *In order that function $R(\varrho)$ be the correlation function of an* *m.s. continuous homogeneous and isotropic random field in a Hilbert space it is necessary and sufficient that it be of the form* (14).

The sufficiency follows from the arguments above. To prove necessity we note that, in view of Theorem 3 and the discussion following this theorem, we have for each m

$$R(\varrho)=\int_0^\infty \Omega_m(\lambda\varrho)\, g_m(d\lambda), \qquad g_m[0,\infty)=R(0),$$

$$\Omega_m(x)=\Gamma\left(\frac{m}{2}\right)\left(\frac{2}{x}\right)^{(m-2)/2} I_{(m-2)/2}(x)=$$
$$=1-\frac{x^2}{2m}+\frac{x^4}{2\cdot 4\cdot m(m+2)}-\frac{x^6}{2\cdot 4\cdot 6m(m+2)(m+4)}+\dots.$$

Moreover $\Omega_m(x\sqrt{m})\to e^{-x^2/2}$ for $m\to\infty$ uniformly on each finite interval $|x|\leqslant N$. Therefore it is sufficient to prove the uniform boundedness of the family of functions $\Omega_m(x)$ for $x\in[0,\infty)$ and the weak compactness of the family of distributions $g_m(\sqrt{m}\,u)$. With this in mind we observe that (8) yields the following equation

$$\Omega_{m+2}(\varrho)=\frac{\int_{S_\varrho} e^{i(u,z)}\, s(du)}{V(S_\varrho)},$$

where S_ϱ is the sphere $|u|=\varrho$ in the space $\mathscr{R}^m$, $V(S_\varrho)$ is its surface area and $|z|=1$. Therefore

$$|\Omega_m(x)|\leqslant 1.$$

To prove the weak compactness of the sequence of distribution functions $\bar g_m(u)=g_m(\sqrt{m}\,u)$ we multiply the relationship

$$R(0)-R(\varrho)=\int_0^\infty (1-\Omega_m(\varrho u\sqrt{m}))\,\bar g_m(du)\geqslant$$
$$\geqslant\int_{2/a}^\infty (1-\Omega_m(\varrho u\sqrt{m}))\,\bar g_m(du)$$

by ϱ and integrate it from 0 to a. We thus obtain

$$\frac{2}{a^2}\int_0^a [R(0)-R(\varrho)]\,\varrho\,d\varrho \geq \int_{2/a}^{\infty}\left(1-\frac{2}{a^2}\int_0^a \Omega_m(\varrho u\sqrt{m})\,\varrho\,d\varrho\right)\bar{g}_m(du).$$

It follows from the formula

$$\frac{d}{dz}\Omega_m(z) = -\frac{1}{m}\,z\Omega_{m+2}(z)$$

that for $m \geq 3$ and $u \geq \frac{2}{a}$

$$\frac{2}{a^2}\int_0^a \Omega_m(\varrho u\sqrt{m})\,\varrho\,d\varrho = \frac{2(m-2)}{a^2u^2m}\,[1-\Omega_{m-2}(au\sqrt{m})] \leq \tfrac{1}{2},$$

hence

$$\frac{2}{a^2}\int_0^a [R(0)-R(\varrho)]\,\varrho\,d\varrho \geq \tfrac{1}{2}\bar{g}_m\left(\left[\frac{2}{a},\infty\right)\right).$$

The compactness of the measures $\bar{g}_m$ follows from the fact that the left-hand side of the last inequality tends to zero as $a\to\infty$ (Chapter I, Section 1, Theorem 1). The theorem is thus proved. □

Vector-valued homogeneous fields. Let $\{\zeta(x);\, x\in\mathscr{R}^m\}$ be a vector-valued random field with values in $\mathscr{Z}^s$. This field is called *homogeneous* if $\mathsf{E}\zeta(x) = a = \text{const}$ (in the sequel we shall assume $a=0$) and if

$$R(x_1, x_2) = \mathsf{E}\zeta(x_1)\,\zeta(x_2)^* = R(x_1 - x_2).$$

A matrix countably-additive set function $F(A) = \{F_{kj}(A)\}$, $k, j = 1, \ldots, s$, $A\in\mathfrak{B}^m$ is called *positive definite* if the matrix $F(A)$ is positive definite for any $A\in\mathfrak{B}^m$, i.e. if for any $c\in\mathscr{Z}^s$ the set function $\mu_c(A) = c^*F(A)\,c$ is a finite measure on $\mathfrak{B}^m$. Applying Theorem 2 we obtain the following result.

Theorem 5. *In order that function $R(x)$ be the matrix correlation function of an m.s. continuous homogeneous vector field $\zeta(x)$ it is necessary and sufficient that*

$$R(x) = \int_{\mathscr{R}^m} e^{i(x,u)}\,F(du), \tag{15}$$

where F is a positive definite matrix countably additive set function on $\{\mathscr{R}^m, \mathfrak{B}^m\}$.

Proof. Let $R(x)$ be the matrix correlation function of an m.s. continuous homogeneous field $\zeta(x)$. For any $c \in \mathscr{Z}^s$ we introduce the scalar field $\zeta_c(x) = (\zeta(x), c)$. This field is clearly m.s. continuous and homogeneous,

$$\mathsf{E}\zeta_c(x) = 0, \quad R_c(x) = \mathsf{E}\zeta_c(x + x_0)\,\overline{\zeta_c(x_0)} = c^* R(x)\, c. \tag{16}$$

In view of Theorem 2 the correlation function $R_c(x)$ can be represented in the form

$$R_c(x) = \int_{\mathscr{R}^m} e^{i(x,u)} F_c(du), \tag{17}$$

where F_c is a finite measure in $\{\mathscr{R}^m, \mathfrak{B}^m\}$. Let $e_k \in \mathscr{Z}^s$, $e_k^j = \delta_k^j$, $R(x) = \{R_k^j(x)\}$, $k, j = 1, \ldots, s$. Then $R_{e_k}(x) = R_{kk}(x)$. Set $e_{kj} = e_k + e_j$, $\tilde{e}_{kj} = ie_k + e_j$. It is easy to obtain that

$$2R_k^j(x) = [R_{e_{kj}}(x) - R_{e_k}(x) - R_{e_j}(x)] - i[R_{\tilde{e}_{kj}}(x) - R_{e_k}(x) - R_{e_j}(x)].$$

If we put

$$F_k^k(A) = F_{e_k}(A), \quad F_k^j(A) = [F_{e_{kj}}(A) - F_{kk}(A) - F_{jj}(A)] - \\ - i[F_{\tilde{e}_{kj}}(A) - F_{kk}(A) - F_{jj}(A)],$$

then it will follow from (16) that

$$R_k^j(x) = \int_{\mathscr{R}^m} e^{i(x,u)} F_k^j(du),$$

and moreover $F_k^j(A)$ are countably additive (complex-valued) finite set functions on $\{\mathscr{R}^m, \mathfrak{B}^m\}$. In view of the uniqueness of representation (17) $c^* F(A)\, c = F_c(A)$, which implies that the matrix $F(A) = \{F_k^j(A)\}$, k, $j = 1, \ldots, s$ is positive definite. The necessity is thus proved.

To prove the sufficiency one must show that the function $R(x)$ defined by formula (15) where F satisfies the conditions of the theorem is a continuous positive-definite matrix function. Its continuity is obvious. Moreover, for any $z_p \in \mathscr{Z}^s$

$$\sum_{p,q=1}^{n} z_p^* R(x_p - x_q)\, z_q = \int_{\mathscr{R}^m} w^* F(du)\, w = F_w(\mathscr{R}^m) \geqslant 0,$$

where $w = \sum_{p=1}^{n} e^{-i(x_p, u)} z_p$. The theorem is thus proved. □

Theorem 1 admits an analogous generalization.

Theorem 6. *A sequence of matrices* $R(n) = \{R_k^j(n)\}$ $n = 0, \pm 1, \pm 2, \ldots$ *is the matrix correlation function of a wide-sense stationary vector-valued* *sequence* $\{\zeta(n), n=0, \pm 1, \pm 2, \ldots\}$ *if and only if it can be represented in the form*

$$R(n)=\int_{-\pi}^{\pi} e^{inu} F(du),$$

where $F(A)$ *is a matric positive definite countably-additive set function defined on the Borel sets of the interval* $[-\pi, \pi]$.

§ 3. A Basic Analysis of Hilbert Random Functions

The study of Hilbert random functions is formally the study of functions in the ordinary sense with values in a Hilbert space. However, since, when analysing Hilbert random functions, we utilize the notion of co-variance and other specific probabilistic notions and investigate various types of convergences, the problems dealing with random functions have certain specific features.

Integration. Let $\{\mathscr{X}, \mathfrak{A}, m\}$ be a complete separable metric space with a σ-finite complete measure and let $\{\zeta(x), x \in \mathscr{X}\}$ be a Hilbert random function. Assume that $\zeta(x)=\zeta(x, \omega)$ is a measurable and separable random function. As it is known from the above (Chapter III, Section 3), if the covariance $B(x, y)$ is continuous at point (x, x) m-almost for all x, then for any $\zeta(x)$ there exists a stochastically equivalent measurable and separable random function. This observation shows how restrictive the above assumption is. Theorem 2 (Chapter III, Section 3) yields the following corollary:

Theorem 1. *If*

$$\int_{\mathscr{X}} B(x, x)\, m(dx)<\infty, \tag{1}$$

then with probability 1

$$\int_{\mathscr{X}} |\zeta(x)|^2\, m(dx)<\infty$$

and

$$\mathsf{E} \int_{\mathscr{X}} |\zeta(x)|^2\, m(dx)=\int_{\mathscr{X}} B(x, x)\, m(dx). \tag{2}$$

Corollary. *Let* $f_i(x)$, $i=1, 2$ *be functions belonging to* $\mathscr{L}_2(\mathscr{X}, \mathfrak{A}, m)$ *and let condition* (1) *be satisfied. Then with probability* 1 *the following integrals* *exist:*

$$\eta_i = \int_{\mathscr{X}} f_i(x)\,\zeta(x)\,m(dx),$$

and moreover in view of Fubini's theorem

$$\mathsf{E}\eta_1\bar{\eta}_2 = \mathsf{E}\int_{\mathscr{X}}\int_{\mathscr{X}} f_1(x)\overline{f_2(y)}\,\zeta(x)\,\overline{\zeta(y)}\,m(dx)\,m(dy) =$$

$$= \int_{\mathscr{X}}\int_{\mathscr{X}} f_1(x)\,B(x,y)\overline{f_2(y)}\,m(dx)\,m(dy).$$

A few remarks concerning the definition of integrals of random functions would seem in order.

Remark 1. Let condition (1) be satisfied and let $m(\mathscr{X}) < \infty$. Then the integral

$$\int_{\mathscr{X}} \zeta(x)\,m(dx) \tag{3}$$

where $\zeta(x)$ is a measurable random function, is defined and finite with probability 1 for each realization of $\zeta(x)$. However a different approach can be taken for the definition of integral (3). Firstly, integral (3) can be defined as the m.s. limit of the Lebesgue integral sums of $\zeta(x)$. It is easy to verify that this definition coincides with the usual one. To prove this it is sufficient to consider non-negative random variables. By definition integral (3) is the limit as $n \to \infty$ of

$$\int_{\mathscr{X}} \zeta_n(x)\,m(dx),$$

where $\zeta_n(x)$ is a monotonically non-decreasing sequence of random functions taking on a finite number of values and such that $\lim \zeta_n(x) = \zeta(x)$ with probability 1. Since $|\zeta(x) - \zeta_n(x)| \leqslant |\zeta(x)|$, and in view of Lebesgue's theorem on bounded convergence we obtain that

$$\mathsf{E}\left|\int_{\mathscr{X}} \zeta(x)\,m(dx) - \int_{\mathscr{X}} \zeta_n(dx)\,m(dx)\right|^2 \leqslant \mathsf{E}\int_{\mathscr{X}} |\zeta(x) - \zeta_n(x)|^2\,m(dx)\,m(\mathscr{X}) \to 0$$

as $n\to\infty$, so that

$$\int_{\mathscr{X}} \zeta(x)\, m(dx) = \text{l.i.m.} \int_{\mathscr{X}} \zeta_n(x)\, m(dx).$$

Remark 2. Consider a random process $\{\zeta(t), t\in[a, b]\}$. The integral

$$\int_a^b \zeta(t)\, dt$$

is often defined as the m.s. limit of the integral sums

$$\sum_{k=1}^{n} \zeta(t_{nk})\, \Delta t_{nk},$$

$$\Delta t_{nk} = t_{nk} - t_{nk-1}, \qquad a = t_{n0} < t_{n1} < \dots < t_{nn} = b.$$

In view of Lemma 3 (Chapter I, Section 1), it is necessary and sufficient for the existence of the m.s. limit of these sums that the limit of

$$\mathsf{E} \sum_{k=1}^{n} \zeta(t_{nk})\, \Delta t_{nk} \sum_{k=1}^{m} \overline{\zeta(t_{mk})}\, \Delta t_{mk} = \sum_{k=1}^{n} \sum_{r=1}^{m} B(t_{nk}, t_{mr})\, \Delta t_{nk} \Delta t_{mr}$$

exist as $n, m\to\infty$, i.e. that the function $B(t, s)$ $(a\leqslant t, s\leqslant b)$ be Riemann integrable. Thus the given definition of the integral is more restrictive than the original one but has the advantage of not being dependent on the notion of measurability of the process. It is easy to verify that the latter definition of the integral – when applicable – coincides (mod P) with the initial definition.

Indeed

$$\mathsf{E}\left|\int_a^b \zeta(t)\, dt - \sum_{k=1}^{n} \zeta(t_{nk})\, \Delta t_{nk}\right|^2 =$$

$$= \sum_{k=1}^{n} \sum_{r=1}^{n} \int_{t_{nk-1}}^{t_{nk}} \int_{t_{nr-1}}^{t_{nr}} [B(t, s) - B(t, t_{nr}) - B(t_{nk}, s) +$$

$$+ B(t_{nk}, t_{nr})]\, dt\, ds \leqslant 2 \sum_{k=1}^{n} \sum_{r=1}^{n} \Omega_{nkr} \to 0,$$

where Ω_{nkr} is the oscillation of the function $B(t, s)$ in the rectangle $t_{nk-1}\leqslant t\leqslant t_{nk}$, $t_{nr-1}\leqslant s\leqslant t_{nr}$.

Remark 3. The improper m.s. integral

$$\int_{-\infty}^{\infty} \zeta(t)\,dt \quad \left(\text{or} \quad \int_{a}^{\infty} \zeta(t)\,dt\right) \tag{4}$$

is defined as the limit

$$\underset{N\to\infty}{\text{l.i.m.}} \int_{-N}^{N} \zeta(t)\,dt \quad \left(\underset{N\to\infty}{\text{l.i.m.}} \int_{a}^{N} \zeta(t)\,dt\right).$$

In view of Lemma 3 (Chapter I, Section 1) a necessary and sufficient condition for the existence of these integrals is the existence of the limits

$$\lim_{N,N'\to\infty} \int_{-N}^{N}\int_{-N'}^{N'} B(t,s)\,dt\,ds \quad \left(\lim_{N,N'\to\infty} \int_{a}^{N}\int_{a}^{N'} B(t,s)\,dt\,ds\right).$$

This definition of improper integrals is, in certain cases, less restrictive than the interpretation of the integrals (4) as Lebesgue integrals of function $\zeta(t)$ for fixed ω.

The law of large numbers. Let $\{\zeta(t), t\geqslant 0\}$ be a measurable Hilbert process with integrable covariance in each finite interval. We say that $\{\zeta(t), t\geqslant 0\}$ satisfies the *law* of *large numbers* if

$$\frac{1}{T}\int_{0}^{T} \zeta(t)\,dt$$

approaches a constant c in a certain sense as $T\to\infty$.

It follows from Lemma 3 (Chapter 1, Section 1) that in order that the mean

$$\underset{T\to\infty}{\text{l.i.m.}} \frac{1}{T}\int_{0}^{T} \zeta(t)\,dt$$

exist, it is necessary and sufficient that the limit

$$\lim_{T,T'\to\infty} \mathsf{E}\frac{1}{T}\int_{0}^{T} \zeta(t)\,dt\,\frac{1}{T'}\int_{0}^{T'} \overline{\zeta(t)}\,dt = \lim_{T,T'\to\infty} \frac{1}{TT'}\int_{0}^{T}\int_{0}^{T'} B(t,s)\,dt\,ds$$

exist. Furthermore for the validity of the equality

$$\operatorname*{l.i.m.}_{T\to\infty}\left\{\frac{1}{T}\int_0^T \zeta(t)\,dt-\frac{1}{T}\int_0^T \mathsf{E}\zeta(t)\,dt\right\}=0$$

it is necessary and sufficient that the relation

$$\lim_{T,T'\to\infty}\frac{1}{TT'}\int_0^T\int_0^{T'} R(t,s)\,dt\,ds=0 \tag{5}$$

be valid, where $R(t, s)$ is the correlation function of the process.

It is easy to observe that

$$\left|\int_0^T\int_0^{T'} R(t,s)\,dt\,ds\right|^2 \leqslant \int_0^T\int_0^T R(t,s)\,dt\,ds \int_0^{T'}\int_0^{T'} R(t,s)\,dt\,ds.$$

Therefore equality (5) holds if and only if

$$\lim_{T\to 0}\frac{1}{T^2}\int_0^T\int_0^T R(t,s)\,dt\,ds=0. \tag{6}$$

For a wide-sense stationary process $R(t, s)=R(t-s)$. Since

$$\frac{1}{T^2}\int_0^T\int_0^T R(t-s)\,dt\,ds=\frac{1}{T}\int_{-T}^T R(t)\left(1-\frac{|t|}{T}\right)dt,$$

we obtain the following result.

Theorem 2. *If $\zeta(t)$ is a wide-sense stationary process, then for the equality*

$$\operatorname*{l.i.m.}_{T\to\infty}\frac{1}{T}\int_0^T \zeta(t)\,dt=\mathsf{E}\zeta(t) \tag{7}$$

it is necessary and sufficient that

$$\lim_{T\to\infty}\frac{1}{T}\int_{-T}^T R(t)\left(1-\frac{|t|}{T}\right)dt=0. \tag{8}$$

In particular, conditions (8) of the theorem are satisfied if the mean value of the correlation function is zero:

$$\lim_{T\to\infty} \frac{1}{2T} \int_{-T}^{T} R(s)\,ds = 0.$$

We express condition (8) in terms of the spectral function of the process. We have

$$\frac{1}{T}\int_{-T}^{T} R(t)\left(1-\frac{|t|}{T}\right)dt = \int_{-\infty}^{\infty} F(du)\frac{1}{T}\int_{-T}^{T} e^{itu}\left(1-\frac{|t|}{T}\right)dt,$$

hence

$$\frac{1}{T}\int_{-T}^{T} R(t)\left(1-\frac{|t|}{T}\right)dt = \int_{-\infty}^{\infty} \frac{2(1-\cos Tu)}{T^2u^2}\,F(du) =$$

$$= F(\{0\}) + \int_{-\infty}^{\infty} \frac{2(1-\cos Tu)}{T^2u^2}\,\tilde{F}(du),$$

where $\tilde{F}(A) = F(A\backslash\{0\})$, where $\{0\}$ is the singleton containing the point $u=0$. It is easily verified that the last integral tends to 0 as $T\to\infty$. Therefore

$$\lim_{T\to\infty} \frac{1}{T}\int_{-T}^{T} R(t)\left(1-\frac{|t|}{T}\right)dt = F(\{O\}). \tag{9}$$

Thus the following theorem is valid.

Theorem 3. *For a wide-sense stationary process, the equality* (7) *holds if and only if its spectral function is continuous at the point $u=0$.*

Differentiation. Let $\{\zeta(t),\ t\in(a,b)\}$ $-\infty \leqslant a < b \leqslant +\infty$ be a Hilbert random process.

Definition 1. A random process $\zeta(t)$, $t\in(a,b)$ is *m.s. differentiable at t_0 (mean square differentiable)* if there exists the limit

$$\zeta'(t_0) = \underset{h\to 0}{\text{l.i.m.}}\, \frac{\zeta(t_0+h)-\zeta(t_0)}{h}, \qquad t_0,\ t_0+h\in(a,b).$$

The random variable $\zeta'(t_0)$ is called the *m.s. (mean square) derivative* of the random process at point t_0.

It is easy to obtain the necessary and sufficient conditions for the m.s. differentiability of a random process. Since

$$\mathsf{E}\,\frac{\zeta(t_0+h)-\zeta(t_0)}{h}\cdot\frac{\overline{\zeta(t_0+h_1)-\zeta(t_0)}}{h_1}=\frac{1}{h_1 h}\{B(t_0+h,\,t_0+h_1)-$$
$$-B(t_0,\,t_0+h_1)-B(t_0+h,\,t_0)+B(t_0,\,t_0)\}, \tag{10}$$

it follows from Lemma 3 (Chapter I, Section 1) that for the m.s. differentiability of the process $\zeta(t)$ at t_0 it is necessary and sufficient that the generalized mixed derivative

$$\left.\frac{\partial^2 B(t,t')}{\partial t\,\partial t'}\right|_{t=t'=t_0}=$$
$$=\lim_{h,h_1\to 0}\frac{B(t_0+h,\,t_0+h_1)-B(t_0,\,t_0+h_1)-B(t_0+h,\,t_0)+B(t_0,\,t_0)}{hh_1}$$

exist.

It follows from the m.s. differentiability of the process at point t and the inequality

$$\left|\mathsf{E}\left(\zeta'(t)-\frac{\zeta(t+h)-\zeta(t)}{h}\right)\right|\leqslant\left\{\mathsf{E}\left|\zeta'(t)-\frac{\zeta(t+h)-\zeta(t)}{h}\right|^2\right\}^{1/2}$$

that

$$\mathsf{E}\zeta'(t)=\frac{d}{dt}\,\mathsf{E}\zeta(t), \tag{11}$$

and, moreover, the derivative on the right exists.

If the process is m.s. differentiable at each point $t\in(a,b)$, then the derivative $\zeta'(t)$ forms a Hilbert random process on (a,b).

Theorem 4. *Let $\{\zeta(t),\,t\in(a,b)\}$ be a Hilbert random process and let the generalized derivative*

$$\left.\frac{\partial^2 B(t,t')}{\partial t\,\partial t'}\right|_{t=t'}$$

exist for each value of $t\in(a,b)$. Then the process $\zeta(t)$ is m.s. differentiable on (a,b) and

$$B_{\zeta'\zeta'}(t,t')=\frac{\partial^2 B(t,t')}{\partial t\,\partial t'}, \tag{12}$$

$$B_{\zeta'\zeta}(t,t')=\frac{\partial B(t,t')}{\partial t}, \tag{13}$$

where $B_{\zeta'\zeta'}(t, t') = \mathsf{E}\zeta'(t)\overline{\zeta'(t')}$ *is the covariance of the process* $\zeta'(t)$, *and* $B_{\zeta'\zeta}(t, t') = \mathsf{E}\zeta'(t)\cdot\overline{\zeta(t')}$ *is the cross covariance of the processes* $\zeta'(t)$ *and* $\zeta(t)$.

Only formulas (12) and (13) require a proof. We have

$$B_{\zeta'\zeta}(t, t') = \mathsf{E}\overline{\zeta(t')}\,\zeta'(t) = \lim_{h\to 0} \mathsf{E}\overline{\zeta(t')}\,\frac{\zeta(t+h)-\zeta(t)}{h} = \lim_{h\to 0}\frac{B(t+h, t')-B(t, t')}{h}.$$

Consequently, the derivative $\dfrac{\partial B(t, t')}{\partial t}$ exists and the cross covariance of the processes $\zeta'(t)$ and $\zeta(t)$ is given by formula (13). Furthermore

$$B_{\zeta'\zeta'}(t, t') = \lim_{h, h'\to 0} \mathsf{E}\,\frac{\overline{\zeta(t'+h')-\zeta(t')}}{h'}\,\frac{\zeta(t+h)-\zeta(t)}{h} = \lim_{h, h'\to 0}\frac{B(t+h, t'+h')-B(t, t'+h')-B(t+h, t')+B(t, t')}{hh'}.$$

We thus obtain the existence of the generalized second derivative

$$\frac{\partial^2 B(t, t')}{\partial t\,\partial t'}$$

(its existence in the condition of the theorem was assumed only at $t=t'$) and the validity of (12) is verified. □

If the process $\zeta(t)$ is wide-sense stationary, then $B(t, t') = B(t-t')$ and Theorem 4 implies

Corollary 1. *In order that a wide-sense stationary process* $\zeta(t)$ $(t\in T)$ *be m.s. differentiable it is necessary and sufficient that the generalized second derivative of the correlation function* $R(t)$ *exist at* $t=0$. *If this condition is satisfied then the generalized derivative* $\dfrac{d^2R(t)}{dt^2}$ *exists and*

$$R_{\zeta'\zeta'}(t_0, t_0+t) = -\frac{d^2R(t)}{dt^2},$$

$$R_{\zeta'\zeta}(t_0+t, t_0) = R_{\zeta'\zeta}(t) = \frac{dR(t)}{dt},$$

Analogous results are valid for the m.s. derivatives of higher orders.

Corollary 2. *If* $\zeta(t)$ *is a wide-sense stationary process*, $t\in(-\infty, \infty)$ *and*

$$\int_{-\infty}^{\infty} u^2 F(du)\infty >,$$

where F is the spectral measure of the process, then $\zeta(t)$ is m.s. differentiable, $(\zeta'(t), \zeta(t))$ is a wide-sense stationary process and its matrix correlation function $R(t)$ is of the form

$$\left\| \begin{matrix} \int_{-\infty}^{\infty} e^{itu} u^2 F(du) & \int_{-\infty}^{\infty} e^{itu} iuF(du) \\ -\int_{-\infty}^{\infty} e^{itu} iuF(du) & \int_{-\infty}^{\infty} e^{itu} F(du) \end{matrix} \right\|$$

Decomposition of a random process into orthogonal series. Let $\{\zeta(t), t\in[a, b]\}$ be a measurable m.s. continuous Hilbert process. Its covariance $B(t_1, t_2)$ is a continuous non-negative definite kernel in the square $[a, b]\times[a, b]$. According to the theory of integral equations, the kernel $B(t_1, t_2)$ can be expanded into uniformly convergent series in terms of its eigenfunctions $\varphi_n(t)$:

$$B(t_1, t_2)=\sum_{n=1}^{\infty} \lambda_n \varphi_n(t_1)\overline{\varphi_n(t_2)},$$

where

$$\lambda_n \varphi_n(t)=\int_a^b B(t, \tau)\varphi_n(\tau)\,d\tau, \quad \int_a^b \varphi_n(t)\overline{\varphi_m(t)}\,dt=\delta_{nm},$$

moreover the eigen-values λ_n are positive.

Set

$$\xi_n=\int_a^b \zeta(t)\overline{\varphi_n(t)}\,dt.$$

This integral exists (Theorem 1) and, in view of the corollary to Theorem 1,

$$\mathsf{E}\xi_n\bar{\xi}_m=\int_a^b\int_a^b B(t, \tau)\overline{\varphi_n(t)}\varphi_m(\tau)\,dt\,d\tau=\lambda_n\delta_{nm},$$

i.e. the sequence of random variables ξ_n $(n=1, 2, \ldots)$ is orthogonal. Furthermore

$$\mathsf{E}\zeta(t)\bar{\xi}_n=\int^b B(t, \tau)\varphi_n(\tau)\,d\tau=\lambda_n\varphi_n(t).$$

It thus follows that

$$\mathsf{E}\left|\zeta(t)-\sum_{k=1}^{n}\xi_k\varphi_k(t)\right|^2=$$

$$=B(t,t)-2\sum_{k=1}^{n}\overline{\varphi_k(t)}\,\mathsf{E}\zeta(t)\,\bar{\xi}_k+\sum_{k=1}^{n}\lambda_k|\varphi_k(t)|^2=$$

$$=B(t,t)-\sum_{k=1}^{n}\lambda_k|\varphi_k(t)|^2\to 0$$

as $n\to\infty$ uniformly in t in view of Dini's theorem.

Theorem 5. *A measurable m.s. continuous Hilbert process $\zeta(t)$, $t\in[a,b]$, admits the series expansion*

$$\zeta(t)=\sum_{k=1}^{\infty}\xi_k\varphi_k(t), \tag{14}$$

which is convergent in $\mathscr{L}_2$ for each $t\in[a,b]$. In this expansion, ξ_k is an orthogonal sequence of random variables, $\mathsf{E}|\xi_k|^2=\lambda_k$, where λ_k are the eigenvalues and $\varphi_k(t)$ the eigenfunction of the covariance of the process.

Remark 1. If the process $\zeta(t)$ is Gaussian, then its m.s. derivative and integrals of the form $\int_a^b f(t)\,\zeta(t)\,dt$ are Gaussian random variables. Therefore if $\zeta(t)$ is a real Gaussian process and $\mathsf{E}\zeta(t)=0$, then the coefficients ξ_k in series (14) are independent Gaussian variables and series (14) is convergent with probability 1 for each t.

Indeed, the independence of the variables ξ_k follows from the fact that they are orthogonal and Gaussian. For convergence with probability 1 of series (14) it is sufficient that the series $\sum_{k=1}^{\infty}\mathsf{E}(\xi_k\varphi_k(t))^2=\sum_{k=1}^{\infty}\lambda_k|\varphi_k(t)|^2$ be convergent. However, as it has already been mentioned, this series is convergent (and its sum is equal to $B(t,t)$).

Theorem 6. *If*

$$\mathsf{E}|\zeta(t)-\zeta(t+h)|^2\leqslant\frac{L|h|}{|\lg|h||^{3+r}},\quad r>0,\quad a\leqslant t\leqslant b, \tag{15}$$

then for any $\varepsilon>0$

$$\mathsf{P}\left\{\sup_{a\leqslant t\leqslant b}\left|\zeta(t)-\sum_{1}^{n}\xi_k\varphi_k(t)\right|>\varepsilon\right\}\to 0\quad as\quad n\to\infty.$$

The proof is based on Lemma 1 of Section 5 in Chapter III. Set

$$\zeta_n(t)=\sum_{k=1}^{n}\xi_k\varphi_k(t),\qquad \zeta(t)-\zeta_n(t)=\zeta_n'(t),\qquad \gamma_n=\sup_{a\leqslant t\leqslant b}|\zeta(t)-\zeta_n(t)|.$$

Then

$$\mathsf{P}\{\gamma_n>\varepsilon\}\leqslant\mathsf{P}\left\{|\zeta_n'(0)|>\frac{\varepsilon}{2}\right\}+\mathsf{P}\left\{\sup_{a\leqslant t\leqslant b}|\zeta_n'(t)-\zeta_n'(0)|>\frac{\varepsilon}{2}\right\}\leqslant$$

$$\leqslant\frac{4\mathsf{E}|\zeta_n'(0)|^2}{\varepsilon^2}+Q\left(n,\frac{\varepsilon}{4G}\right),$$

where $Q(n, c)$ and G are as defined in the above-mentioned lemma. We have

$$\mathsf{P}\{|\zeta_n'(t+h)-\zeta_n'(t)|>Cg(h)\}\leqslant\frac{\sigma_n^2(t,h)}{C^2g^2(h)},$$

where

$$\sigma_n^2(t,h)=\mathsf{E}|\zeta_n'(t+h)-\zeta_n'(t)|^2=$$

$$=\mathsf{E}\left|\sum_{k=n+1}^{\infty}\xi_k[\varphi_k(t+h)-\varphi_k(t)]\right|^2=$$

$$=\sum_{k=n+1}^{\infty}\lambda_k|\varphi_k(t+h)-\varphi_k(t)|^2.$$

Taking (15) into account, we observe that the functions $|\lg|h||^{3+r'}\times\sigma_n^2(t,h)\cdot(L|h|)^{-1}$ $(0<r'<r)$ are continuous with respect to $t\in[a,b]$, $h\in[0,h_0]$ and are monotonically decreasing as n increases and approach 0 as $n\to\infty$. In view of Dini's theorem this convergence is uniform. Consequently,

$$\max\left\{\frac{|\lg|h||^{3+r'}\sigma_n^2(t,h)}{L|h|};\ t\in[a,b],\ h\in[0,h_0]\right\}=\delta_n\to 0$$

as $n\to\infty$.

Putting

$$g(h)=|\lg|h||^{-(1+r'')},\ 0<r''<\frac{r'}{2},$$

and

$$q_n(C,h)=\frac{L\delta_n|h|}{C^2g^2(h)\,|\lg|h||^{3+r'}},$$

(cf. Chapter III, Section 5) we obtain that

$$G<\infty,\quad Q(n,C)\leqslant\frac{K\delta_n}{C^2},$$

where K is a constant independent of n. Thus $Q\left(n, \frac{\varepsilon}{4G}\right) \to 0$ as $n \to \infty$. Moreover $\mathsf{E}|\zeta_n'(0)|^2 \to 0$ also. The theorem is thus proved. □

Consider, as an example, the expansion of a Brownian motion into an orthogonal series on the interval $[0, 1]$. Here $\zeta(0)=0$, $\mathsf{E}\zeta(t)=0$, $\mathsf{V}\zeta(t)=t$, $B(t, s)=\mathsf{E}\zeta(t)\zeta(s)=\min(t, s)$. The eigenvalues and eigenfunctions of the kernel $B(t, s)$ are easily obtained. From the equation

$$\lambda_n \varphi_n(t) = \int_0^1 \min(t, s)\, \varphi_n(s)\, ds = \int_0^t s\varphi_n(s)\, ds + \int_t^1 t\varphi_n(s)\, ds$$

we have firstly $\varphi_n(0)=0$. Differentiating with respect to t we obtain $\lambda_n \varphi_n'(t) = \int_t^1 \varphi_n(s)\, ds$, hence $\varphi_n'(1)=0$. Repeating the differentiation we obtain equation $\lambda_n \varphi_n''(t) = -\varphi_n(t)$. The normalized solutions of the last equation which satisfy the boundary conditions $\varphi_n(0)=0$, $\varphi_n'(1)=0$ are of the form

$$\varphi_n(t) = \sqrt{2}\, \sin\left(n+\tfrac{1}{2}\right)\pi t, \qquad \lambda_n^{-1} = \left(n+\tfrac{1}{2}\right)^2 \pi^2, \qquad n=0, 1, \ldots.$$

Thus

$$\zeta(t) = \sqrt{2} \sum_{n=0}^{\infty} \xi_n \frac{\sin\left(n+\frac{1}{2}\right)\pi t}{\left(n+\frac{1}{2}\right)\pi}, \tag{16}$$

where ξ_n is a sequence of independent Gaussian random variables with parameters (0, 1). For a fixed t this series is convergent with probability 1. Since $\zeta(t)$ is a Gaussian process and $\mathsf{E}|\zeta(t+h)-\zeta(t)|^2 = h$, it follows that

$$\sup_{0 \le t \le 1} \left| \zeta(t) - \sqrt{2} \sum_{k=1}^{n} \xi_k \frac{\sin\left(k+\frac{1}{2}\right)\pi t}{\left(k+\frac{1}{2}\right)\pi} \right| \to 0$$

in probability.

Another expansion of a Brownian motion process can be obtained as follows. Set $\xi(t) = \zeta(t) - t\zeta(1)$. Then $\xi(t)$ is a Gaussian process with covariance $B_1(t, s) = \min(t, s) - ts$ and $\mathsf{E}\xi(t)=0$. The eigenvalues and eigenfunctions of the kernel $B_1(t, s)$ are obtained in the same manner as in the previous case. We again arrive at equation $\lambda_n \varphi_n''(t) = -\varphi_n(t)$ with the boundary conditions $\varphi_n(0) = \varphi_n(1) = 0$. The solutions of this equation are of the form

$$\varphi_n(t) = \sqrt{2}\, \sin n\pi t, \qquad \lambda_n^{-1} = n^2\pi^2, \qquad n=1, 2, \ldots.$$

Thus

$$\xi(t)=\zeta(t)-t\zeta(1)=\sqrt{2}\sum_{n=1}^{\infty}\xi_n\frac{\sin n\pi t}{n\pi},$$

where $\xi_n (n=1, 2, \ldots)$ is a normalized sequence of independent Gaussian random variables and moreover

$$\xi_n=\sqrt{2}\int_0^1 \xi(t)\sin n\pi t\,dt.$$

Since

$$\mathsf{E}\zeta(1)=1,\ \mathsf{E}\zeta^2(1)=1,$$

$$\mathsf{E}\xi_n\zeta(1)=\sqrt{2}\int_0^1 \mathsf{E}(\zeta(t)-t\zeta(1))\,\zeta(1)\sin n\pi t\,dt=0,$$

putting $\xi_0=\zeta(1)$ we obtain

$$\zeta(t)=t\xi_0+\sqrt{2}\sum_{n=1}^{\infty}\xi_n\frac{\sin n\pi t}{n\pi},\tag{17}$$

where $\xi_0, \xi_1, \ldots, \xi_n, \ldots$ are independent and are normally distributed with parameters (0, 1). The convergence properties of series (17) are the same as those of (16).

§ 4. Stochastic Measures and Integrals

Integrals of the form

$$\int_a^b f(t)\,d\zeta(t),\tag{1}$$

play an important role in a number of problems. Here $f(t)$ is a given (*non-random*) function and $\zeta(t)$ is a random process. Generally realizations of the process $\zeta(t)$ are functions of unbounded variation and the integral (1) can not be interpreted as a Stieltjes or Lebesgue-Stieltjes integral, existing for almost all realizations of $\zeta(t)$. However, even in this case, integral (1) can be defined in a manner such that it will possess properties shared by the ordinary integrals.

In the present section we define and investigate properties of integrals in which the integration is taken with respect to a random measure. Such integrals are called *stochastic* integrals.

Let $\{\Omega, \mathfrak{S}, \mathsf{P}\}$ be a probability space, $\mathscr{L}_2=\mathscr{L}_2(\Omega, \mathfrak{S}, \mathsf{P}), E$ be a set and $\mathfrak{M}$ be the semi-ring of subsets of E. Assume that for each $\Delta\in\mathfrak{M}$ there corresponds a complex-valued random variable $\zeta(\Delta)$ satisfying the following conditions:

1) $$\zeta(\Delta)\in\mathscr{L}_2 \quad \zeta(\emptyset)=0:$$

2) $$\zeta(\Delta_1\cup\Delta_2)=\zeta(\Delta_1)+\zeta(\Delta_2)\,(\mathrm{mod}\,\mathsf{P}), \quad \text{if} \quad \Delta_1\cap\Delta_2=\emptyset;$$

3) $$\mathsf{E}\,\zeta(\Delta_1)\,\overline{\zeta(\Delta_2)}=m(\Delta_1\cap\Delta_2),$$

where $m(\Delta)$ is a set function on $\mathfrak{M}$.

Definition 1. The family of random variables $\{\zeta(\Delta), \Delta\in\mathfrak{M}\}$ satisfying conditions 1)–3) is called an *elementary orthogonal stochastic measure* and $m(\Delta)$ is its *structural function*.

The orthogonality property of stochastic measures is expressed by condition 3); if $\Delta_1\cap\Delta_2=\emptyset$, then the variables $\zeta(\Delta_1)$ and $\zeta(\Delta_2)$ are orthogonal.

It follows from the definition of $m(\Delta)$ that this function is non-negative:

$$m(\Delta)=\mathsf{E}\,|\zeta(\Delta)|^2\geqslant 0, \quad m(\emptyset)=0,$$

and additive: if $\Delta_1\cap\Delta_2=\emptyset$, then

$$m(\Delta_1\cup\Delta_2)=\mathsf{E}\,|\zeta(\Delta_1)+\zeta(\Delta_2)|^2= \\ =m(\Delta_1)+m(\Delta_2)+2m(\Delta_1\cap\Delta_2)=m(\Delta_1)+m(\Delta_2).$$

Thus $m(\Delta)$ is an *elementary measure* on* $\mathfrak{M}$. Denote by $\mathscr{L}_0(\mathfrak{M})$ the class of all simple functions $f(x)$:

$$f(x)=\sum_{k=1}^{n} c_k\chi_{\Delta_k}(x), \quad \Delta_k\in\mathfrak{M}, \quad k=1, 2, \ldots, n, \tag{2}$$

where n is an arbitrary number and $\chi_A(x)$ is the indicator of the set A.

We define the stochastic integral of a function $f(x)\in\mathscr{L}_0\{\mathfrak{M}\}$ with respect to the elementary stochastic measure ζ as follows

$$\eta=\int f(x)\,\zeta(dx)=\sum_{k=1}^{n} c_k\zeta(\Delta_k). \tag{3}$$

* a non-negative, additive set function defined on a semi-ring.

Since $\mathfrak{M}$ is a semi-ring, any pair of functions in $\mathscr{L}_0(\mathfrak{M})$ can be represented as a linear combination of indicators of the same sets in $\mathfrak{M}$. Therefore if f and $g \in \mathscr{L}_0(\mathfrak{M})$, we assume that $f(x)$ is given by formula (2) and $g(x) = \sum_{k=0}^{n} d_k \chi_{\Delta_k}(x)$, where $\Delta_k \cap \Delta_r = \emptyset$ for $k \neq r$.

It follows from the orthogonality of ζ that

$$\mathsf{E}\left(\int f(x)\,\zeta(dx) \int \overline{g(x)\,\zeta(dx)}\right) = \sum_{k=1}^{n} c_k \bar{d}_k m(\Delta_k). \tag{4}$$

Assume that the elementary measure m satisfies the semi-additivity condition and therefore may be extended to a complete measure $\{E, \mathfrak{L}, m\}$. Then $\mathscr{L}_0\{\mathfrak{M}\}$ is the linear subset of the Hilbert space $\mathscr{L}_2\{m\} = \mathscr{L}_2\{E, \mathfrak{L}, m\}$ and $\mathscr{L}_2\{\mathfrak{M}\}$ is the closure of $\mathscr{L}_0\{\mathfrak{M}\}$ in the topology generated by the scalar product

$$(f, g) = \int f(x)\,\overline{g(x)}\, m(dx). \tag{5}$$

Moreover relation (4) can be rewritten in the following form

$$\mathsf{E} \int f(x)\,\zeta(dx) \int \overline{g(x)\,\zeta(dx)} = \int f(x)\,\overline{g(x)}\, m(dx) \tag{6}$$

for any pair of functions $f(x)$ and $g(x) \in \mathscr{L}_0\{m\}$.

We now introduce the linear span $\mathscr{L}_0\{\zeta\}$ of the family of random variables $\{\zeta(\Delta), \Delta \in \mathfrak{M}\}$, i.e. the set of random variables which can be represented in form (3) and the space $\mathscr{L}_2\{\zeta\}$ which is the closure of $\mathscr{L}_0\{\zeta\}$ in the Hilbert space of random variables $\mathscr{L}_2\{\Omega, \mathfrak{S}, \mathsf{P}\}$. Note that relation (3) determines the isometric correspondence $\eta = \psi(f)$ between $\mathscr{L}_0\{\mathfrak{M}\}$ and $\mathscr{L}_0(\zeta)$. This correspondence may be extended to the isometric correspondence ψ between $\mathscr{L}_2\{\mathfrak{M}\}$ and $\mathscr{L}_2\{\zeta\}$. If $\eta = \psi(f)$, $f \in \mathscr{L}_2\{\mathfrak{M}\}$, we set

$$\eta = \psi(f) = \int f(x)\,\zeta(dx) \tag{7}$$

by definition and call the random variable η the stochastic integral of function $f(x)$ with respect to measure ζ. We thus have the following

Theorem 1. a) *For a simple function* (2) *the value of a stochastic integral is given by formula* (3);

b) *For any $f(x)$ and $g(x)$ in $\mathscr{L}_2\{E, \mathfrak{L}, m\}$ equation* (6) *is valid*;

c) $$\int [\alpha f(x) + \beta g(x)]\,\zeta(dx) = \alpha \int f(x)\,\zeta(dx) + \beta \int g(x)\,\zeta(dx);$$

d) *for an arbitrary sequence of functions* $f^{(n)}(x) \in \mathscr{L}_2\{E, \mathfrak{L}, m\}$ *such that*

$$\int |f(x) - f^{(n)}(x)|^2 \, m(dx) \to 0, \quad n \to \infty, \tag{8}$$

the following relation

$$\int f(x)\, \zeta(dx) = \text{l.i.m.} \int f^{(n)}(x)\, \zeta(dx)$$

is satisfied.

Remark. In particular if $f^{(n)}(x)$ are simple functions

$$f^{(n)}(x) = \sum_{k=1}^{m_n} c_k^{(n)} \chi_{\Delta_k^{(n)}}(x), \quad \Delta_k^{(n)} \in \mathfrak{M}, \quad n = 1, 2, \ldots,$$

and (8) is satisfied then

$$\int f(x)\, \zeta(dx) = \text{l.i.m.} \sum_{k=1}^{m_n} c_k^{(n)} \zeta(\Delta_k^{(n)}).$$

The existence of a sequence of simple functions which approximate an arbitrary function $f(x) \in \mathscr{L}_2\{E, \mathfrak{L}, m\}$ follows from the general theorems of measure theory. Therefore a stochastic integral can be considered as the m.s. limit of the corresponding integral sums.

Denote by L_0 the class of all sets $A \in \mathfrak{L}$, for which $m(A) < \infty$. Define the random set function $\tilde{\zeta}(A)$ by

$$\tilde{\zeta}(A) = \int \chi_A(x)\, \zeta(dx) = \int_A \zeta(dx). \tag{9}$$

This function possesses the following properties:

a) $\tilde{\zeta}(A)$ is defined on the class of sets L_0;

b) if $A_n \in L_0$; $n = 0, 1, 2, \ldots$, $A_0 = \bigcup_{k=1}^{\infty} A_n$, $A_k \cap A_r = \emptyset$ for $k \neq r$, $k > 0$, $r > 0$, then $\tilde{\zeta}(A_0) = \sum_{n=1}^{\infty} \tilde{\zeta}(A_n)$ in the sense of m.s. convergence.

c) $\mathsf{E}\tilde{\zeta}(A)\overline{\tilde{\zeta}(B)} = m(A \cap B)$, $A, B \in L_0$;

d) $\tilde{\zeta}(\Delta) = \zeta(\Delta)$ for $\Delta \in \mathfrak{M}$.

Definition 2. A random set function $\tilde{\zeta}$ satisfying conditions a), b) and c) is called a *stochastic orthogonal measure.*

Property d) signifies that $\tilde{\zeta}$ is an extension of an elementary stochastic measure ζ. We thus have the following

Theorem 2. *If the structure function of an elementary stochastic measure ζ is semi-additive then ζ may be extended to a stochastic measure $\tilde{\zeta}$.*

Remark. Since $\mathscr{L}_2\{\zeta\} = \mathscr{L}_2\{\tilde{\zeta}\}$, we have

$$\int f(x)\,\zeta(dx) = \int f(x)\,\tilde{\zeta}(dx).$$

In accordance with this inequality, we shall henceforth identify the stochastic integral relative to an elementary orthogonal measure ζ, whose structure function is semi-additive, with the stochastic integral relative to the stochastic measure $\tilde{\zeta}$ defined by relation (9).

A few remarks would seem in order concerning the definition of a stochastic integral on a line. Let $\xi(t)$ $(a \leqslant t < b)$ be a process with orthogonal increments, i.e.

$$\mathsf{E}(\xi(t_2) - \xi(t_1))\,\overline{(\xi(t_4) - \xi(t_3))} = 0$$

for any $t_i \in [a, b)$, $t_1 < t_2 < t_3 < t_4$, m.s. continuous from the left:

$$\mathsf{E}|\xi(t) - \xi(s)|^2 \to 0 \quad \text{as} \quad s \uparrow t.$$

Set

$$F(t) = \mathsf{E}|\xi(t) - \xi(a)|^2.$$

In view of the orthogonality of the increments of the process $\xi(t)$ we have for $t_2 > t_1$

$$F(t_2) = \mathsf{E}|\xi(t_2) - \xi(t_1) + \xi(t_1) - \xi(a)|^2 = F(t_1) + \mathsf{E}|\xi(t_2) - \xi(t_1)|^2;$$

hence $F(t_2) \geqslant F(t_1)$ and $F(t) = \lim_{s \uparrow t} F(s)$. Therefore $F(t)$ is a monotonically non-decreasing function continuous from the left. Let $\mathfrak{M}$ be the class of all half-intervals $\Delta = [t_1, t_2)$, $a \leqslant t_1 \leqslant t_2 \leqslant b$, $\zeta([t_1, t_2)) = \xi(t_2) - \xi(t_1)$, $m([t_1, t_2)) = F(t_2) - F(t_1)$. Then $\mathfrak{M}$ is a semi-ring of sets,

$$\mathsf{E}\zeta(\Delta_1)\,\overline{\zeta(\Delta_2)} = m(\Delta_1 \cap \Delta_2),$$

and $\zeta(\Delta)$ is an elementary orthogonal stochastic measure with a structural function admitting extension to a measure. Therefore one can define a stochastic Stieltjes integral by means of the equality

$$\int_a^b f(t)\,d\xi(t) = \int_a^b f(t)\,\zeta(dt),$$

where $\xi(t)$ is a process with orthogonal increments. This integral exists for an arbitrary Borel function $f(t)$, $t \in [a, b)$ for which

$$\int_a^b |f(t)|^2 \, F(dt) < \infty,$$

where $F(A)$ is a measure corresponding to a monotonic function $F(t)$. A stochastic integral on the whole real line $(-\infty, \infty)$ is defined analogously.

We now prove several propositions concerning stochastic integrals.

Let ζ be an orthogonal stochastic measure with structural function m which is a complete measure on $\{E, \mathfrak{L}\}$ and let $g(x) \in \mathscr{L}_2\{m\}$. Set

$$\lambda(A) = \int \chi_A(x) \, g(x) \, \zeta(dx), \qquad A \in \mathfrak{L}.$$

Then

$$\mathsf{E}\lambda(A)\,\overline{\lambda(B)} = \int \chi_A(x)\, \chi_B(x)\, |g(x)|^2 \, m(dx) = \\ = \int_{A \cap B} |g(x)|^2 \, m(dx).$$

Introducing a new measure on $\mathfrak{L}$

$$l(A) = \int_A |g(x)|^2 \, m(dx),$$

we see that $\lambda(A)$ is an orthogonal stochastic measure with structural function $l(A)$, $A \in \mathfrak{L}$.

Lemma 1. *If $f(x) \in \mathscr{L}_2\{l\}$, then $f(x)\, g(x) \in \mathscr{L}_2\{m\}$ and*

$$\int f(x) \, \lambda(dx) = \int f(x) \, g(x) \, \zeta(dx).$$

Proof. The assertion of the lemma is obvious for simple functions $f(x)$, $f(x) = \sum_k c_k \chi_{A_k}(x)$, $A_k \in \mathfrak{L}$. Next, if $f_k(x)$ is a fundamental sequence of simple functions in $\mathscr{L}_2\{l\}$, then

$$\left| \int f_n(x) \, \lambda(dx) - \int f_{n+m}(x) \, \lambda(dx) \right|^2 = \\ = \int |f_n(x) - f_{n+m}(x)|^2 \, l(dx) =$$

$$=\int |f_n(x)-f_{n+m}(x)|^2 \, |g(x)|^2 \, m(dx),$$

i.e. $f_n(x)\, g(x)$ is a fundamental sequence in $\mathscr{L}_2\{m\}$. Approaching the limit in the equality

$$\int f_n(x)\, \lambda(dx)=\int f_n(x)\, g(x)\, \zeta(x)$$

as $n\to\infty$ we obtain the assertion of the lemma in the general case. □

Lemma 2. *If $A\in L_0$, then*

$$\zeta(A)=\int \frac{\chi_A(x)}{g(x)}\, \lambda(dx).$$

Proof. We first note that $g(x)=0$ on the set of l-measure 0; therefore $[g(x)]^{-1}\neq\infty \pmod{l}$. Next

$$\int \frac{\chi_A(x)}{|g(x)|^2}\, l(dx)=\int_A \frac{1}{|g(x)|^2}\, |g(x)|^2\, m(dx)=m(A)<\infty.$$

Consequently we may utilize Lemma 1 and get:

$$\int \frac{1}{g(x)}\, \chi_A(x)\, \lambda(dx)=\int \frac{1}{g(x)}\, \chi_A(x)\, g(x)\, \zeta(dx)=\zeta(A).$$

The proof of lemma 2 is thus completed. □

Let T be a finite or infinite segment on the line, $\mathfrak{B}$ be the σ-algebra of the Lebesgue measurable subsets of T, and let l be the Lebesgue measure.

Assume that the function $g(t, x)$ is $\mathfrak{B}\times\mathfrak{L}$ measurable, $g(t, x)\in\mathscr{L}_2\{l\times m\}$ and $g(t, x)\in\mathscr{L}_2\{m\}$ for an arbitrary $t\in T$. Consider the stochastic integral

$$\xi(t)=\int g(t, x)\, \zeta(dx). \tag{10}$$

This integral is defined with probability 1 for each t.

Lemma 3. *The stochastic integral* (10) *can be defined as a function of t in such a manner that the process $\xi(t)$ will be measurable.*

Proof. If

$$g(t, x)=\sum c_k \chi_{B_k}(t)\, \chi_{A_k}(x), \tag{11}$$

$B_k\in\mathfrak{B}$, $A_k\in\mathfrak{L}$, then $\xi(t)=\sum c_k \chi_{B_k}(t)\, \zeta(A_k)$ is a $\mathfrak{B}\times\mathfrak{S}$-measurable function in variables (t, ω), $t\in T$, $\omega\in\Omega$. In the general case a sequence of simple functions $g_n(t, x)$ of type (11) can be constructed such that

$$\iint |g(t, x)-g_n(t, x)|^2\, m(dx)\, dt\to 0 \quad \text{as} \quad n\to\infty .$$

Let $\xi_n(t)$ be a sequence of processes constructed in accordance with formula (10) for $g=g_n$. Then there exists a process $\tilde{\xi}(t)$ such that

$$\int \mathsf{E}|\tilde{\xi}(t)-\xi_n(t)|^2\, dt\to 0 \quad \text{as} \quad n\to\infty$$

and $\tilde{\xi}(t)$ is a $\mathfrak{B}\times\mathfrak{S}$-measurable function of (t, ω). On the other hand,

$$\int \mathsf{E}|\xi(t)-\xi_n(t)|^2\, dt=\iint |g(t, x)-g_n(t, x)|^2\, m(dx)\, dt\to 0,$$

so that $\mathsf{E}|\xi(t)-\tilde{\xi}(t)|^2=0$ for almost all t.

Set

$$\xi'(t)=\begin{cases}\tilde{\xi}(t), & \text{if} \quad \mathsf{P}\{\xi(t)\neq\tilde{\xi}(t)\}=0,\\ \xi(t), & \text{if} \quad \mathsf{P}\{\xi(t)\neq\tilde{\xi}(t)\}>0.\end{cases}$$

The process $\xi'(t)$ is measurable (since $\xi'(t)$ differs from a $\mathfrak{B}\times\mathfrak{S}$ measurable function $\tilde{\xi}(t)$ on a set of measure 0), and is stochastically equivalent to $\xi(t)$. The lemma is thus proved. □

We shall henceforth assume that the processes determined by stochastic integrals of type (10) and satisfying the conditions enumerated above are measurable.

Lemma 4. *If $g(t, s)$ and $h(t)$ are Borel functions,*

$$\int_a^b \int_{-\infty}^{\infty} |g(t, s)|^2\, dt\, m(ds)<\infty, \qquad \int_a^b |h(t)|^2\, dt<\infty, \tag{12}$$

and ζ is an orthogonal stochastic measure on $\{R^1, \mathfrak{B}^1\}$ then

$$\int_a^b h(t) \int_{-\infty}^{\infty} g(t, s)\, \zeta(ds)\, dt=\int_{-\infty}^{\infty} g_1(s)\, \zeta(ds), \tag{13}$$

where

$$g_1(s)=\int^b h(t)\, g(t, s)\, dt .$$

Proof. The mathematical expectation of the square of the absolute value of the integral in the l.h.s. of (13) is equal to

$$\int_a^b \int_a^b h(t_1)\overline{h(t_2)}\left(\int_{-\infty}^{\infty} g(t_1, s)\overline{g(t_2, s)}\, m(ds)\right) dt_1\, dt_2 =$$

$$= \int_{-\infty}^{\infty} \left|\int_a^b h(t)\, g(t, s)\, dt\right|^2 m(ds) \leqslant$$

$$\leqslant \int_a^b |h(t)|^2\, dt \cdot \int_{-\infty}^{\infty} \int_a^b |g(t, s)|^2\, dt\, m(ds).$$

The mathematical expectation of the square of the absolute value of the integral in the r.h.s. of (13) satisfies the inequality indicated in the second line of the last relationship. Consequently the r.h. and l.h. sides of equality (13) are continuous with respect to the limit transition over the sequences $g_n(t, s)$ converging in $\mathscr{L}_2\{\Phi\}$ where Φ is the direct product of the Lebesgue measure and the measure m in the strip $[a, b] \times (-\infty, \infty)$. Furthermore the set of functions $g(t, s)$ for which (13) is valid is linear and contains all the functions of the type $\sum c_k \chi_{A_k}(t)\, \chi_{B_k}(\tau)$. Consequently, this set contains all the functions belonging to $\mathscr{L}_2\{\Phi\}$. □

Remark. If the conditions of Lemma 4 are satisfied on each finite interval (a, b) and if the integral

$$\int_{-\infty}^{\infty} h(t)\, g(t, s)\, dt = \lim_{\substack{a \to -\infty \\ b \to +\infty}} \int_a^b h(t)\, g(t, s)\, dt$$

exists in the sense of $\mathscr{L}_2\{m\}$-convergence, then

$$\int_{-\infty}^{\infty} h(t) \int_{-\infty}^{\infty} g(t, s)\, \zeta(ds)\, dt = \int_{-\infty}^{\infty} f_1(s)\, \zeta(ds), \tag{14}$$

where

$$f_1(s) = \int_{-\infty}^{\infty} h(t)\, g(t, s)\, dt.$$

The proof follows directly from the fact that the l.h.s. of equation (14) is an m.s. limit of the l.h. side of equation (13) noting that the passage to the limit under the sign of the stochastic integral is permissible in the r.h.s. of formula (13). □

Consider now the generalization of the previous results to the case of vector-valued stochastic measures. Here we shall confine ourselves to the simplest case of integration of scalar functions which differs only little from integration with respect to real-valued stochastic measures.

Let $\mathscr{Z}^p$ denote a complex vector space of dimension p. For simplicity we shall assume that a certain basis in this space is fixed. Let there correspond to each $\Delta \in \mathfrak{M}$ a vector-valued random variable $\zeta(\Delta)$ with values in $\mathscr{Z}^p$, $\zeta(\Delta)=\{\zeta^1(\Delta), \zeta^2(\Delta), \ldots, \zeta^p(\Delta)\}$. Denote by $|\zeta(\Delta)|$ the norm of vector $\zeta(\Delta)$,

$$|\zeta(\Delta)|^2 = \sum_{k=1}^{p} |\zeta^k(\Delta)|^2 .$$

Assume that

1) $\mathsf{E}|\zeta(\Delta)|^2 < \infty$, $\zeta(\emptyset)=0$;
2) $\zeta(\Delta_1 \cup \Delta_2)=\zeta(\Delta_1)+\zeta(\Delta_2)$ (mod P), if $\Delta_1 \cap \Delta_2=\emptyset$;
3) $\mathsf{E}\zeta^k(\Delta_1)\,\overline{\zeta^j(\Delta_2)}=m_j^k(\Delta_1 \cap \Delta_2)$, $\Delta_i \in \mathfrak{M}$, $i=1,2$; $k,j=1,\ldots,p$.

The family of random vectors $\{\zeta(\Delta), \Delta \in \mathfrak{M}\}$ is called the *elementary vector-valued stochastic (orthogonal) measure* and the matrix $m(\Delta)= =\{m_j^k(\Delta)\}=\mathsf{E}\zeta(\Delta)\,\zeta^*(\Delta)$ is called *the structure matrix*.

Note that the matrix $m(\Delta_1 \cap \Delta_2)$, regarded as a function of Δ_1 and Δ_2, possesses properties of the correlation matrix of a vector-valued random function (cf. Section 1). Also if $\Delta_1 \cap \Delta_2=\emptyset$, then

$$m(\Delta_1 \cup \Delta_2)=m(\Delta_1)+m(\Delta_2).$$

From here it follows that the diagonal elements of the matrix $m(\Delta)$ are elementary measures. Moreover it follows from inequality

$$|m_j^k(\Delta)| \leqslant \sqrt{m_k^k(\Delta)\, m_j^j(\Delta)} \tag{15}$$

that

$$\sum_r |m_j^k(\Delta_r)| \leqslant \{\sum_r m_k^k(\Delta_r) \sum_r m_j^j(\Delta_r)\}^{1/2}, \tag{16}$$

and hence the set-functions m_j^k $(k, j=1,\ldots,p)$ are of bounded variation on Δ.

Set $m_0(\Delta)=\operatorname{Sp} m(\Delta)=\sum_{k=1}^{p} m_k^k(\Delta)$. It follows from (16) that if $\sum_{r=1}^{m_N} m_0(\Delta_r^N) \to 0$ as $N \to \infty$, then $\sum_{r=1}^{m_N} |m_j^k(\Delta_r^N)| \to 0$, also. We thus obtain that the functions $m_j^k(\Delta)$ can be extended to countably-additive set functions on $\mathfrak{B}$ provided $m_0(\Delta)$ is semi-additive on $\mathfrak{M}$.

Henceforth matrix functions obtained by means of such an extension measure will be called *positive definite matrix measures*.
from the structure function of an elementary orthogonal stochastic

In the above, $\mathscr{L}$ denoted the completion of $\sigma\{\mathfrak{M}\}$ with respect to the extended elementary measure $m_0(\Delta)$. For simplicity we shall retain the initial notations for the extensions of the functions m_j^k, m_0 and the matrix m on $\mathfrak{B}$ and shall assume in what follows that $m_0(\Delta)$ is semi-additive on $\mathfrak{M}$

Using the formula

$$\eta = \int f(x)\,\zeta(dx) = \sum_{k=1}^{n} c_k \zeta(\Delta_k), \tag{17}$$

we define on $\mathscr{L}_0\{\mathfrak{M}\}$ the stochastic integral, where

$$f(x) = \sum_{k=1}^{n} c_k \chi_{\Delta_k}(x), \quad \Delta_k \in \mathfrak{M} \quad (k=1,\dots,n).$$

The value of this integral is a random (column) vector with values in $\mathscr{Z}^p$. Denote by $\mathscr{L}_0^p\{\zeta\}$ the collection of all random vectors η of form (17). If $g(x) = \sum_{k=1}^{n} d_k \chi_{\Delta_k}(x)$, then

$$\mathsf{E}\left(\int f(x)\,\zeta(dx)\left(\int g(x)\,\zeta(dx)\right)^*\right) = \sum_{k=1}^{n} c_k \bar{d}_k m(\Delta_k);$$

this can be rewritten in the form

$$\mathsf{E}\left(\int f(x)\,\zeta(dx)\left(\int g(x)\,\zeta(dx)\right)^*\right) = \int f(x)\,\overline{g(x)}\,m(dx). \tag{18}$$

We thus obtain equation

$$\mathsf{E}\left|\int f(x)\,\zeta(dx)\right|^2 = \int |f(x)|^2\, m_0(dx). \tag{19}$$

Introduce the scalar product on $\mathscr{L}_0\{\mathfrak{M}\}$:

$$(f, g) = \int f(x)\,\overline{g(x)}\, m_0(dx).$$

Formula (17) establishes the isometric mapping $\eta = \psi(f)$ of the space $\mathscr{L}_0\{\mathfrak{M}\}$ into $\mathscr{L}_0^p\{\zeta\}$ provided we define in $\mathscr{L}_0^p\{\zeta\}$ the scalar product of the elements η_1 and η_2 by $\mathsf{E}\eta_2^*\eta_1$. The closure of the space of random functions $\mathscr{L}_0^p\{\zeta\}$ is denoted by $\mathscr{L}_2^p\{\zeta\}$ and the completion of $\mathscr{L}_0\{\mathfrak{M}\}$ by $\mathscr{L}_2\{\mathfrak{M}\}$.

Inequality

$$\int |f(x)|\, |m_j^k|\, (dx) \leq \left\{ \int |f(x)|\, m_k^k(dx) \int |f(x)|\, m_j^j(dx) \right\}^{1/2}, \tag{20}$$

is derived analogously to inequality (16) (here $|m_j^k|\,(A)$ is the absolute variation of the function m_j^k) first for simple functions and then using the limiting transition for arbitrary $\mathfrak{B}$-measurable functions. Inequality (20) yields the existence and continuity of the integral

$$\int f(x)\, \overline{g(x)}\, m_j^k(dx)$$

as a functional of f and g in $\mathscr{L}_2\{m_0\}$.

From here, an isometric correspondence $\eta = \psi(f)$ of the space $\mathscr{L}_0\{\mathfrak{M}\}$ into $\mathscr{L}_0^p(\zeta)$ can be extended up to the isometric correspondence of $\mathscr{L}_2\{\mathfrak{M}\}$ into $\mathscr{L}_2^p(\zeta)$.

The random vector η is called the *stochastic integral* and denoted as

$$\eta = \int f(x)\, \zeta(dx),$$

where $f(x) \in \mathscr{L}_2(m_0)$.

A *vector-valued stochastic measure* $\zeta(A)$ is defined analogously to the notion of a stochastic measure in the scalar case.

§5. Integral Representation of Random Functions

Utilizing the results of the previous section one can obtain different representations of random functions using stochastic integrals.

We first assume that a p-dimensional vector random function $\xi(x)$, $x \in \mathscr{X}$ can be represented in the form

$$\xi(x) = \int g(x, u)\, \zeta(du), \tag{1}$$

where ζ is a stochastic measure on a measurable space $\{\mathscr{U}, \mathfrak{B}\}$ with values in $\mathscr{Z}^p$ and structure matrix $m(A)$ (here the notations used in the previous section are retained) $g(x, u)$ is a scalar function and moreover for each $x \in \mathscr{X}$

$$g(x, u) \in \mathscr{L}_2\{m_0\} = \mathscr{L}_2\{\mathscr{U}, \mathfrak{B}, m_0\},\ m_0(A) = \operatorname{Sp} m(A).$$

In view of formula (18) in Section 4 the covariance matrix of the random function $\xi(x)$ is of the form

$$B(x_1, x_2) = \mathsf{E}\xi(x_1)\,\xi^*(x_2) = \int g(x_1, u)\,\overline{g(x_2, u)}\, m(du), \tag{2}$$

and it follows from (19) in Section 4 that

$$\mathsf{E}\xi^*(x_2)\,\xi(x_1) = \int g(x_1, u)\,\overline{g(x_2, u)}\, m_0(du). \tag{3}$$

Recall that $\{\mathscr{U}, \mathfrak{B}, m_0\}$ – is a space with a complete measure, $\mathscr{L}_2\{m_0\}$ is a Hilbert space of $\mathfrak{B}$-measurable complex-valued square m_0-integrable functions.

Denote by $\mathscr{L}_2\{g\}$ the closure in $\mathscr{L}_2\{m_0\}$ of the linear span generated by the system of functions $\{g(x, u), x\in\mathscr{X}\}$. Then $\mathscr{L}_2\{g\}$ becomes a linear closed subspace of $\mathscr{L}_2\{m_0\}$. If $\mathscr{L}_2\{g\} = \mathscr{L}_2\{m_0\}$, the system of functions $\{g(x, u), x\in\mathscr{X}\}$ is called complete in $\mathscr{L}_2\{m_0\}$.

Let $\{\xi(x), x\in\mathscr{X}\}$ be a Hilbert random function with values in $\mathscr{Z}^p$, $\mathscr{L}_0\{\xi\}$ be the set of all random vectors

$$\eta = \sum_{k=1}^{n} c_k \xi(x_k), \quad n = 1, 2, \ldots, \quad x_k \in \mathscr{X}, \tag{4}$$

where c_k are arbitrary complex numbers, and $\mathscr{L}_2\{\xi\}$ is the closure of $\mathscr{L}_0\{\xi\}$ in the sense of mean square convergence of random vectors.

Definition 1. The family of random vectors $\{\eta_\alpha, \alpha\in A\}$, $\eta_\alpha \in \mathscr{L}_2\{\Omega, \mathfrak{S}, \mathsf{P}\}$ is called *subordinated to a random function* $\{\xi(x), x\in\mathscr{X}\}$ *if* $\eta_\alpha \in \mathscr{L}_2\{\xi\}$, $\alpha\in A$.

Theorem 1. *Let the covariance matrix of a random function* $\{\xi(x), x\in\mathscr{X}\}$ *admit representation* (2), *where* m *is a positive definite matrix measure on* $\{\mathscr{U}, \mathfrak{B}\}$, $g(x, u)\in\mathscr{L}_2\{m_0\}$ $x\in\mathscr{X}$ *and the family* $\{g(x, u), x\in\mathscr{X}\}$ *is complete in* $\mathscr{L}_2\{\mathscr{U}, \mathfrak{B}, m_0\}$. *Then* $\xi(x)$ *can be represented by formula* (1), *where* $\{\zeta(B), B\in\mathfrak{B}\}$ *is a stochastic orthogonal vector-valued measure subordinated to a random function* $\xi(x)$ *with structural function* $m(\cdot)$ *and equality* (1) *is satisfied with probability* 1 *for each* x.

Proof. We associate by means of relation (4) a random vector η, $\eta = \psi(f)$ with each linear combination

$$f(u) = \sum_{k=1}^{n} c_k g(x_k, u), \quad x_k \in \mathscr{X}. \tag{5}$$

Denote by $\mathscr{L}_0\{g\}$ the set of functions (5). Define in $\mathscr{L}_0\{g\}$ the scalar product by means of relation

$$(f_1, f_2) = \int f_1(u)\,\overline{f_2(u)}\, m_0(du). \tag{6}$$

The correspondence $\eta=\psi(f)$ is an isometric mapping of $\mathscr{L}_0\{g\}$ into $\mathscr{L}_0\{\xi\}$. Hence it can be extended to an isometric mapping of $\mathscr{L}_2\{g\}$ into $\mathscr{L}_2\{\xi\}$. In view of the completeness of the family of functions $\{g(x, u), x\in\mathscr{X}\}$, if $B\in\mathfrak{B}$, then $\chi_B(x)\in\mathscr{L}_2\{m_0\}=\mathscr{L}_2\{g\}$. Set $\zeta(A)=\psi(\chi_A)$. Then $\zeta(A)$ is a vector-valued stochastic measure and its structure function coincides with m:

$$\mathsf{E}\zeta(A_1)\,\zeta^*(A_2)=\int \chi_{A_1}(x)\,\overline{\chi_{A_2}(x)}\,m(dx)=m(A_1\cap A_2).$$

Now define a random function $\tilde{\xi}(x)$ by means of the stochastic integral

$$\tilde{\xi}(x)=\int g(x, u)\,\zeta(du).$$

Since

$$\mathsf{E}\xi(x)\,\zeta^*(A)=\int g(x, u)\,\chi_A(u)\,m(du),$$

it follows from the isometry of the correspondence $\eta=\psi(f)$ that

$$\mathsf{E}\xi(x)\,\tilde{\xi}^*(x)=\int g(x, u)\,\overline{g(x, u)}\,m(du).$$

We thus obtain

$$\begin{aligned}&\mathsf{E}|\xi(x)-\tilde{\xi}(x)|^2=\\&=\mathsf{E}\xi^*(x)\,\xi(x)-\mathsf{E}\tilde{\xi}^*(x)\,\xi(x)-\mathsf{E}\xi^*(x)\,\tilde{\xi}(x)+\mathsf{E}\tilde{\xi}^*(x)\,\tilde{\xi}(x)=0,\end{aligned}$$

which proves the theorem. □

We now present several applications of the theorem just proved. For brevity we shall refer to "a wide-sense stationary process" as simply a "stationary process" in the remainder of this section.

In view of Theorem 2 of Section 2 in Chapter IV the correlation matrix of a stationary and m.s. continuous process can be represented in the form

$$R(t_1, t_2)=R(t_1-t_2)=\int_{-\infty}^{\infty} e^{iu(t_1-t_2)}\,F(du), \tag{7}$$

where $F(\cdot)$ is a non-negative definite matrix measure (the spectral matrix of the process). Expression (7) is a particular case of (2) in which functions $g(x, u)$ are replaced by e^{iut}, $x\leftrightarrow t$ and the collection of functions $\{e^{iut}, -\infty<u<\infty\}$ is complete in $\mathscr{L}_2\{m_0\}$ where m_0 is an arbitrary bounded measure on the line. Thus Theorem 1 is applicable and the following result is obtained:

Theorem 2. *A vector-valued m.s. continuous random process* $\xi(t)$ $(-\infty<t<\infty)$ *with* $\mathsf{E}\xi(t)=0$ *admits representation*

$$\xi(t)=\int_{-\infty}^{\infty} e^{itu}\zeta(du), \tag{8}$$

where $\zeta(A)$ *is a vector-valued orthogonal stochastic measure on* $\mathfrak{B}$ *subordinated to* $\xi(t)$. *There exists an isometric correspondence between* $\mathscr{L}_2\{\xi\}$ *and* $\mathscr{L}_2\{F_0\}$ *where* $F_0(\cdot)=\operatorname{Sp}F(\cdot)$ *such that*

a) $\xi(t)\leftrightarrow e^{itu}$, $\zeta(A)\leftrightarrow\chi_A(u)$;

b) *if* $\eta_i\leftrightarrow g_i(u)$ $(i=1,2)$, *then*

$$\eta_i=\int g_i(u)\,\zeta(du)$$

and

$$\mathsf{E}\eta_1\eta_2^*=\int g_1(u)\,\overline{g_2(u)}\,F(du).$$

Formula 8 is called the *spectral decomposition* (*representation*) of a stationary process, and the measure $\zeta(A)$ is the *stochastic spectral measure* of the process. It follows from Theorem 2, that

$$\mathsf{E}\zeta(A_1)\,\zeta^*(A_2)=\int_{A_1\cap A_2} F(du)=F(A_1\cap A_2), \tag{9}$$

i.e. $F(\cdot)$ is the structure function of a vector-valued stochastic measure $\zeta(\cdot)$.

Remark 1. We have $\mathsf{E}\eta=0$ for any $\eta\in\mathscr{L}_2\{\xi\}$. In particular for any $A\in\mathfrak{B}$ we have $\mathsf{E}\zeta(A)=0$.

Remark 2. If $\mathsf{E}\xi(t)=a\neq0$, then the previous theorem is applicable to the process $\xi(t)-a$. On the other hand representation (8) can be retained in the general case also if we add to $\zeta(A)$ the measure of value equal to a, concentrated at point $u=0$.

As an example of an application of Theorem 2 we shall derive the Kotel'nikov-Shannon theorem for a one-dimensional random process whose spectral measure is concentrated on the finite interval $[-B,B]$. We expand the function e^{iut} into a Fourier series on the interval $[-B,B]$. We have

$$e^{iut}=\sum_{n=-\infty}^{\infty}\frac{\sin(Bt-\pi n)}{Bt-\pi n}e^{i(\pi n/B)u}.$$

The series in the r.h. side of the last formula converges uniformly in u in an arbitrary segment $[-B', B']$, $B' < B$ and its partial sums are bounded; hence the series is convergent also in $\mathscr{L}_2\{m_0\}$. In view of the isomorphism between the spaces $\mathscr{L}_2\{m_0\}$ and $\mathscr{L}_2\{\xi\}$ we have (in the sense of m.s. convergence)

$$\xi(t) = \sum_{n=-\infty}^{\infty} \frac{\sin(Bt - \pi n)}{Bt - \pi n} \xi\left(\frac{\pi n}{B}\right). \tag{10}$$

Thus, the value of a random function $\xi(t)$ at any instant t is uniquely recovered by means of its values in the equidistant instants of time $\pi n/B$, $n=0, \pm 1, \pm 2$.

For stationary vector-valued sequences ξ_n, $n=0, \pm 1, \pm 2, \ldots$ one can formulate a theorem completely analogous to Theorem 2. The only difference is that the spectral measure of the sequence is concentrated on the half-interval $[-\pi, \pi)$ rather than on the whole real line as in the case of a continuous parameter process (cf. Theorem 1, Section 2).

Theorems 1 and 2 of Section 2 yield the following generalization of Theorem 2 on spectral decomposition of a homogeneous m.s. continuous field.

Theorem 3. *A vector-valued homogeneous m.s. continuous field $\xi(x)$, $x \in \mathscr{R}^m$ can be represented in the form*

$$\xi(x) = a + \int_{\mathscr{R}^m} e^{i(x,u)} \zeta(du), \quad a = \mathsf{E}\xi(x),$$

where ζ is a vector orthogonal measure on $\mathfrak{B}^m$ subordinated to the field $\xi(x)$. There exists an isometric correspondence between $\mathscr{L}_2\{\xi\}$ and $\mathscr{L}_2\{F_0\}$, where $F_0(\cdot) = \operatorname{Sp} F(\cdot)$, such that

a) $\zeta(x) \leftrightarrow e^{i(x,u)}$;

b) *if* $\eta_i \leftrightarrow g_i(u)$, $\eta_i \in \mathscr{L}_2\{\xi\}$, $g_i(u) \in \mathscr{L}_2\{F_0\}$, $i=1, 2$, *then*

$$\eta_i = \int_{\mathscr{R}^m} g_i(u) \zeta(du),$$

and

$$\mathsf{E}\eta_1\eta_2^* = \int_{\mathscr{R}^m} g_1(u) \overline{g_2(u)} F(du).$$

Corollary. If a homogeneous (scalar) field $\xi(x)$ (*with* $\mathsf{E}\xi(x)=0$) *has a* *bounded spectrum, i.e.*

$$R(x)=\int_{-B_1}^{B_1}\cdots\int_{-B_m}^{B_m} e^{i(x,u)}F(du),$$

then this field is uniquely determined by the values at the points of the lattice $\left\{x_n=\left(\frac{\pi n^1}{B_1},\frac{\pi n^2}{B_2},\ldots,\frac{\pi n^m}{B_m}\right),\ n^k=0,\pm1,\pm2,\ldots\right\}$ *by the formula:*

$$\xi(x)=\sum_{n=(n^1,\ldots,n^m)}\prod_{k=1}^{m}\frac{\sin(B_k x^k-\pi n^k)}{B_k x^k-\pi n^k}\,\xi\left(\frac{\pi n^1}{B_1},\frac{\pi n^2}{B_2},\ldots,\frac{\pi n^m}{B_m}\right). \tag{11}$$

In this formula the summation is carried out over all possible integer-valued vectors n and the series in the r.h.s. of the formula is convergent in the m.s. for each x. □

Consider also the spectral decomposition of an m.s. continuous isotropic two-dimensional random field. In view of formula (10) in Section 2, the correlation function of the field is of the form

$$R(x_1,x_2)=R(\varrho)=\int_0^{\infty} J_0(u\varrho)\,g(du), \tag{12}$$

where x_1 and x_2 are points in the plane and ϱ is the distance between these points. If (r_i,θ_i) are the polar coordinates of the point x_i $(i=1,2)$, then

$$\varrho=\sqrt{r_1^2+r_2^2-2r_1r_2\cos(\theta_1-\theta_2)}.$$

Applying the addition formula to function J_0

$$J_0(u\varrho)=\sum_{k=-\infty}^{\infty} J_k(ur_1)\,J_k(ur_2)\,e^{ik(\theta_1-\theta_2)},$$

we rewrite formula (12) in the form

$$R(\varrho)=\int_0^{\infty}\int_{-\infty}^{\infty} J_\nu(ur_1)\,e^{i\nu\theta_1}\,J_\nu(ur_2)\,e^{i\nu\theta_2}\,g(du)\,\varepsilon(d\nu),$$

where $\varepsilon(d\nu)$ is the measure concentrated at the points $k=0,\pm1,\pm2,\ldots$ and $\varepsilon(\{k\})=1$. In view of Theorem 1, the planar, isotropic, homogeneous and m.s. continuous field $\xi(x)$, $x=re^{i\theta}$, $(\mathsf{E}\xi(x)=0)$ admits representation in the form

$$\xi(x)=\sum_{k=-\infty}^{\infty} e^{ik\theta}\int_{0}^{\infty} J_k(ur)\,\zeta_k(du), \tag{13}$$

where ζ_k is a sequence of mutually orthogonal stochastic measures on the line $[0, \infty)$.

§6. Linear Transformations

Let a system Σ (an apparatus or a device) be designed to transform time-dependent signals (functions) $x(t)$. The function to be transformed is called the function at the input (input function) of the system and the transformed function is called the function at the output (output function) or the response on the input function. Mathematically any system is defined by a class D of "admissible" functions at the input and the relations of the form

$$z(t)=T(x\mid t),$$

where $x=x(s)$ $(-\infty<s<\infty)$ is an input function, $x(s)\in D$, and $z(t)$ is the value of the output function at time t. The system Σ is called *linear* if

a) the class of admissible functions D is linear

b) the operator T satisfies the additivity principle

$$T(\alpha x_1+\beta x_2\mid t)=\alpha T(x_1\mid t)+\beta T(x_2\mid t).$$

We introduce the time-shift operation S_τ $(-\infty<\tau<\infty)$ by means of the relationship

$$x_\tau(t)=S_\tau(x\mid t)=x(t+\tau).$$

This operation is defined on the set of all functions in variable t $(-\infty<t<\infty)$ and is linear. The system Σ is called *homogeneous* in time (or simply homogeneous) if the class of admissible functions D is invariant with respect to the shift operation S_τ, $S_\tau D=D$, and

$$T(x_\tau\mid t)=T(x\mid t+\tau) \quad \text{or} \quad T(S_\tau x\mid t)=S_\tau T(x\mid t),$$

i.e. if the transformation T is commutative with the shift operation S_τ $(-\infty<\tau<\infty)$.

The simplest example of a linear transformation is the transformation of the form

$$z(t)=\int_{-\infty}^{\infty} h(t, s)\, x(s)\, ds, \tag{1}$$

here the class of admissible functions D depends on the properties of the function $h(t, s)$. Let the input function be δ_{x-s}, where δ_x is the Dirac function. Then $z(t)=h(t, s)$ for $t>s$ and $z(t)=0$ for $t<s$. Hence the function $h(t, s)$ should be interpreted as the response of the system on the δ-function at the instant s. Accordingly, $h(t, s)$ is called the *impulse transition function* of the system. If the system Σ is homogeneous in time then, formally,

$$h(t, a-c)=T(\delta_{a-c} \mid t)=T(S_c\delta_a \mid t)=S_cT(\delta_a \mid t)=h(t+c, a),$$

or replacing a by c and t by $t-c$, we have

$$h(t-c, 0)=h(t, c).$$

The function $h(t)=h(t+c, c)$ is called the *impulse transition function* of a homogeneous system.

Thus for a homogeneous system equation 1 becomes

$$z(t)=\int_{-\infty}^{\infty} h(t-s)\, x(s)\, ds. \tag{2}$$

The operation in the r.h.s. of relation (2) is called the convolution of functions $h(t)$ and $x(t)$.

If the function at the input differs from the function at the output by the scalar factor only (i.e. the transformation T does not distort the form of the signal)

$$T(f \mid t)=\lambda f(t) \quad (-\infty<t<\infty),$$

then $f(t)$ is called the *eigenfunction*, and λ is called the *eigenvalue* of the transformation T. The functions e^{itu} (where u is an arbitrary number) are eigenfunctions for time homogeneous systems with an integrable impulse transition function. Indeed, all the bounded measurable functions are admissible and

$$\int_{-\infty}^{\infty} h(t-s)\, e^{ius}\, ds=\int_{-\infty}^{\infty} h(s)\, e^{iu(t-s)}\, ds=H(iu)\, e^{iut},$$

where

$$H(iu)=\int_{-\infty}^{\infty} h(s)\, e^{-isu}\, ds \tag{3}$$

– the Fourier transform of an impulse transition function – is an eigenvalue of the transformation.

Therefore, the ratio of the system's response on a simple harmonic function e^{iut} to this function,

$$H(iu)=\frac{T(e^{isu}\mid t)}{e^{iut}},$$

is independent of time. The function $H(iu)$ is called the frequency characteristic of the system or the transmission coefficient.

One can give a somewhat different interpretation of the frequency characteristic of system (2) by considering a different class of admissible functions. Let $x(t)$ be integrable. In view of Fubini's theorem,

$$\int_{-\infty}^{\infty}|z(t)|\,dt\leqslant\int_{-\infty}^{\infty}\int_{-\infty}^{\infty}|h(t-s)|\,|x(s)|\,ds\,dt=$$

$$=\int_{-\infty}^{\infty}|x(s)|\,ds\int_{-\infty}^{\infty}|h(t)|\,dt\times\infty,$$

i.e. the function $z(t)$ is also integrable. Consider the Fourier transform of function $z(t)$. Applying Fubini's theorem we obtain

$$\tilde{z}(u)=\int_{-\infty}^{\infty}e^{-itu}z(t)\,dt=$$

$$=\int_{-\infty}^{\infty}\int_{-\infty}^{\infty}e^{-iu(t-s)}h(t-s)\,e^{-ius}x(s)\,ds\,dt=H(iu)\,\tilde{x}(u),$$

$$\tilde{x}(u)=\int_{-\infty}^{\infty}e^{-ius}x(s)\,ds.$$

Consequently, the ratio of the Fourier transform of the function at the output to the Fourier transform of the function at the input does not depend on the function at the input and is equal to its frequency characteristic

$$H(iu)=\frac{\tilde{z}(u)}{\tilde{x}(u)}.$$

The response at time t as given in formula (1) depends on the values of the input function at times $s<t$ as well as at times $s>t$. However, in physical devices there is no possibility to anticipate the future. Therefore in this case

$$h(t, s)=0 \quad \text{for} \quad t<s \tag{4}$$

Relation (4) is called the *condition of physical realization of the system.* For systems which satisfy condition (4) formula (1) becomes

$$z(t)=\int_{-\infty}^{t} h(t, s)\, x(s)\, ds, \tag{5}$$

and if the system is homogeneous, then

$$z(t)=\int_{-\infty}^{t} h(t-s)\, x(s)\, ds=\int_{0}^{\infty} h(s)\, x(t-s)\, ds. \tag{6}$$

If the input functions satisfies $x(s)=0$ for $s<0$, then

$$z(t)=\int_{0}^{t} h(t-s)\, x(s)\, ds. \tag{7}$$

For these systems it is more convenient to use the Laplace transforms

$$\tilde{z}(p)=\int_{0}^{\infty} e^{-pt} z(t)\, dt. \tag{8}$$

(rather than the Fourier).

It follows from formula (7) that

$$\tilde{z}(p)=H(p)\, \tilde{x}(p), \quad \tilde{x}(p)=\int_{0}^{\infty} e^{-pt} x(t)\, dt \tag{9}$$

for $\operatorname{Re} p \geqslant \alpha$ if the functions $e^{-\alpha t} h(t)$ and $e^{-\alpha t} x(t)$ are absolutely integrable.

We now proceed to the basic topic of this section–linear transformations of random processes. We consider here transformations of random processes which are homogeneous in time. We shall also briefly comment on a more general case.

Let $\xi(t)$ be a measurable Hilbert process $(-\infty<t<\infty)$ with covariance $B(t, s)$ where $B(t, t)$ is integrable with respect to t in each finite interval and also let the function $|h(t, s)|^2$ be integrable for fixed t. Then with probability 1 there exists for each a and b the integral

$$\zeta(t)=\int_a^b h(t, s)\,\xi(s)\,ds.$$

Define the improper integral from $-\infty$ to $+\infty$ as the m.s. limit of integrals over finite intervals of integration:

$$\int_{-\infty}^{\infty} h(t, s)\,\xi(s)\,ds=\underset{\substack{a\to-\infty\\ b\to+\infty}}{\text{l.i.m.}}\int_a^b h(t, s)\,\xi(s)\,ds.$$

In order that this limit exist it is necessary and sufficient that the integral

$$\int_{-\infty}^{\infty}\int_{-\infty}^{\infty} h(t, s_1)\,B(s_1, s_2)\,\overline{h(t, s_2)}\,ds_1\,ds_2$$

exists as an improper Cauchy integral on the plane. If this integral exists for $t\in T$, then $\zeta(t)$ becomes a Hilbert random process on T with the covariance

$$B_\zeta(t_1, t_2)=\int_{-\infty}^{\infty}\int_{-\infty}^{\infty} h(t_1, s_1)\,B(s_1, s_2)\,\overline{h(t_2, s_2)}\,ds_1\,ds_2. \tag{10}$$

Assume now that $\xi(t)$ is a wide-sense stationary process with the spectral measure $F(du)$ and $\mathsf{E}\xi(t)=0$. This assumption remains valid until the end of this section. The integral

$$\eta(t)=\int_{-\infty}^{\infty} h(t-s)\,\xi(s)\,ds \tag{11}$$

exists (in the sense defined above) if and only if the integral

$$\int_{-\infty}^{\infty}\int_{-\infty}^{\infty} h(t-s_1)\,R(s_1-s_2)\,\overline{h(t-s_2)}\,ds_1\,ds_2=$$

$$=\int_{-\infty}^{\infty}\int_{-\infty}^{\infty} h(s_1)\,R(s_2-s_1)\,\overline{h(s_2)}\,ds_1\,ds_2$$

exists where $R(t)$ is the correlation function of the process. For this to hold it is, in turn, sufficient that the function $h(t)$ be absolutely integrable on $(-\infty, \infty)$. In this case, using the spectral representation of the correlation function $R(t)$ we obtain the following expression for the correlation function $R_\eta(t, t)$ of the process $\eta(t)$:

$$R_\eta(t_1, t_2) = \int_{-\infty}^{\infty}\int_{-\infty}^{\infty} h(t_1 - s_1)\, R(s_1 - s_2)\, \overline{h(t_2 - s_2)}\, ds_1\, ds_2 =$$

$$= \int_{-\infty}^{\infty}\int_{-\infty}^{\infty}\int_{-\infty}^{\infty} h(t_1 - s_1)\, e^{iu(s_1 - s_2)}\, \overline{h(t_2 - s_2)}\, ds_1\, ds_2 F(du) =$$

$$= \int_{-\infty}^{\infty} e^{i(t_1 - t_2)u} |H(iu)|^2\, F(du) = R_\eta(t_1 - t_2).$$

Thus the process $\eta(t)$ is also a wide-sense stationary process.

Definition 1. For a given process $\xi(t)$, the transformation T is called an *admissible filter* (or simply a *filter*) if it is defined by formula (11), where $h(t)$ is an absolutely integrable function on $(-\infty, \infty)$ and square integrable on any finite interval, or if T is the m.s. limit of sequences of such transformations (in $\mathscr{L}_2\{\xi\}$).

The following relationship stipulates the condition for convergence of transformations (11) $\eta_n(t) = T_n(\xi \mid t)$ with impulse transition functions $h_n(t)$ and frequency characteristics $H_n(iu)$:

$$\mathsf{E}|\eta_n(t) - \eta_m(t)|^2 = \int_{-\infty}^{\infty} |H_n(iu) - H_m(iu)|^2\, F(du) \to 0 \tag{12}$$

as $n, m \to \infty$.

This means that the sequence $H_n(iu)$ is fundamental in $\mathscr{L}_2\{F\}$. In this case, however, the limit $H(iu) = \text{l.i.m.}\, H_n(iu)$ (in $\mathscr{L}_2\{F\}$) exists which is called the *frequency characteristic of the limiting filter* and if $\eta(t) = \text{l.i.m.}\, \eta_n(t)$, then

$$R_\eta(t) = \int_{-\infty}^{\infty} e^{itu} |H(iu)|^2\, F(du). \tag{13}$$

Conversely, any function $H(iu) \in \mathscr{L}_2\{F\}$ can be approximated in the sense of convergence in $\mathscr{L}_2\{F\}$ by means of functions which are Fourier transforms of absolutely integrable functions. Thus it is convenient to define filters by means of their frequency characteristics.

Theorem 1. *In order that the function $H(iu)$ be the frequency characteristic of an admissible filter for the process $\xi(t)$ with spectral measure F it is necessary and sufficient that $H(iu) \in \mathscr{L}_2\{F\}$. The correlation function of the process at the output of the filter with frequency characteristic $H(iu)$ is given by formula* (13).

Recalling the power interpretation of the spectral function we observe from (13) that $|H(iu)|^2$ represents the amount of increase in the power of simple harmonic components of the process – when passing through the filter – with frequencies in the interval $(u, u+du)$.

Theorem 2. *If the process $\xi(t)$ at the input of a filter with spectral characteristic $H(iu)$ possesses the spectral representation*

$$\xi(t) = \int_{-\infty}^{\infty} e^{iut}\, \zeta(du), \tag{14}$$

then the process $\eta(t)$ at the output of the filter will be of the form

$$\eta(t) = \int_{-\infty}^{\infty} e^{iut}\, H(iu)\, \zeta(du). \tag{15}$$

Indeed if the filter possesses an absolutely integrable impulse transfer function then

$$\eta(t) = \int_{-\infty}^{\infty} h(t-s)\, \xi(s)\, ds = \int_{-\infty}^{\infty} e^{iut}\, H(iu)\, \zeta(du).$$

The proof in the general case is obtained by means of the limiting transition over the sequences $H_n(iu)$ converging to $H(iu)$ in $\mathscr{L}_2\{F\}$. □

Let $\eta_k(t)$ be the process at the output of the filter with the frequency characteristic $H_k(iu)$, $\mathsf{E}\eta_k(t) = 0$ $(k = 1, 2)$. We find the mutual correlation function of the processes $\eta_1(t)$ and $\eta_2(t)$. It follows directly from the isomorphisms of the spaces $\mathscr{L}_2\{\zeta\}$ and $\mathscr{L}_2\{F\}$ that

$$R_{12}(t) = \mathsf{E}\eta_1(t+s)\, \overline{\eta_2(s)} = \int_{-\infty}^{\infty} e^{iut}\, H_1(iu)\, \overline{H_2(iu)}\, F(du). \tag{16}$$

We now present several examples of filters and their frequency characteristics.

1. A band-pass filter passes through only those harmonic components of a process with frequencies in the given range (a, b). The frequency characteristic of the process is equal to $H(iu)=\chi_{(a,b)}(u)$ and the filter is admissible for an arbitrary process. The impulse transfer function is obtained by means of the Fourier formula

$$h(t)=\frac{1}{2\pi}\int_a^b e^{itu}\,du=\frac{e^{ibt}-e^{iat}}{2\pi it}.$$

2. A high-pass filter which supresses the low frequencies without altering the high ones. Its frequency characteristic $H(iu)=\chi_{(|u|>a)}(u)$ and the impulse transfer function does not exist.

3. Consider the operation of the m.s. differentiation of a wide-sense stationary process. In order that the m.s. derivative of the process $\xi(t)$ exist, it is sufficient to require the existence of $R''(0)$ (Section 3, Corollary 1). This condition is equivalent to requirement

$$\int_{-\infty}^{\infty} u^2 F(du)<\infty. \tag{17}$$

(cf. Theorem 4, Section 1, Chapter I).

On the other hand, if this condition is satisfied, then for $h\to 0$

$$\frac{e^{ihu}-1}{h}\to iu \quad (\text{in } \mathscr{L}_2\{F\})$$

and it is permissible to pass to the limit as $h\to 0$ under the sign of the stochastic integral in relation

$$\frac{\xi(t+h)-\xi(t)}{h}=\int_{-\infty}^{\infty} e^{itu}\frac{e^{ihu}-1}{h}\zeta(du).$$

Consequently,

$$\xi'(t)=\int_{-\infty}^{\infty} e^{itu}\, iu\zeta(du). \tag{18}$$

Thus a filter with the frequency characteristic iu, which is admissible for all the stationary processes satisfying condition (17) corresponds to the differentiation operation. The impulse transfer function does not exist, but the filter can be considered as a limiting filter (as $\varepsilon\to 0$) of a collection of filters with impulse transition functions $h_\varepsilon(t)=0$ for $|t|\geqslant\varepsilon$

and $h_\varepsilon(t) = -\frac{\operatorname{sgn} t}{\varepsilon^2}$ for $|t| < \varepsilon$, with the corresponding frequency characteristics $-\frac{4\sin^2\frac{u\varepsilon}{2}}{iu\varepsilon^2}$.

4. The shift operation. Since

$$\xi(t+s) = \int_{-\infty}^{\infty} e^{iut} e^{ius} \zeta(du),$$

it follows that the frequency characteristic $H(iu) = e^{ius}$ corresponds to the shift operation T_s, $T_s(\xi \mid t) = \xi(t+s)$. The impulse transition function does not exist.

5. Differential equations. Consider a filter defined by a linear differential equation with constant coefficients

$$L\eta = M\xi \tag{19}$$

where

$$L = a_0 \frac{d^n}{dt^n} + a_1 \frac{d^{n-1}}{dt^{n-1}} + \ldots + a_n,$$

$$M = b_0 \frac{d^m}{dt^m} + b_1 \frac{d^{m-1}}{dt^{m-1}} + \ldots + b_m.$$

Equation (19) is meaningful only if process $\xi(t)$ is m times m.s. differentiable. We then seek an n times m.s. differentiable stationary process $\eta(t)$, satisfying (19). Assume that (19) admits a stationary solution. It can be represented in the form

$$\eta(t) = \int_{-\infty}^{\infty} e^{iut} H(iu)\, \zeta(du).$$

Applying operations M and L to processes $\xi(t)$ and $\eta(t)$ respectively, we obtain:

$$\int_{-\infty}^{\infty} e^{iut} L(iu)\, H(iu)\, \zeta(du) = \int_{-\infty}^{\infty} e^{iut} M(iu)\, \zeta(du),$$

where $L(iu) = \sum_{k=0}^{n} a_k (iu)^{n-k}$, $M(iu) = \sum_{k=0}^{\infty} b_k (iu)^{m-k}$. Therefore, if $L(iu)$ has no real roots, we have

$$H(iu)=\frac{M(iu)}{L(iu)}. \tag{20}$$

Conversely, if process $\xi(t)$ is m times m.s. differentiable $M(iu)\in\mathscr{L}_2(F)$, $L(iu)\neq 0\ (-\infty<u<\infty)$, then process

$$\eta(t)=\int_{-\infty}^{\infty} e^{iut}\frac{M(iu)}{L(iu)}\zeta(du)$$

is n times m.s. differentiable and satisfies equation (19). Therefore under the condition $M(iu)\in\mathscr{L}_2(F)$, $L(iu)\neq 0$ there exists a unique filter corresponding to the differential equation (19). Note, however, that the solution of equation (19) can be determined in more general cases. Assume that $L(iu)$ has no real roots. The filter with the frequency characteristic $M(iu)/L(iu)$ exists even if $M(iu)\notin\mathscr{L}_2\{F\}$; it is sufficient to require here only that $\frac{M(iu)}{L(iu)}\in\mathscr{L}_2\{F\}$. The latter holds always if the degree n of the polynomial is greater or equal to m. Therefore for $n\geqslant m$ a filter with the frequency characteristic (20) with a non-vanishing denominator for real u is admissible for an arbitrary process at the input, while the process at the output is identified with the stationary solution of equation (19). Considering, as before, only those differential equations for which polynomial $L(x)$ has no purely imaginary roots, we extract from the rational function $M(x)/L(x)$ its integral part ($\mathsf{P}(x)$ which is non-zero if $m\geqslant n$) and expand the remainder into a partial fraction expansion. We thus obtain

$$\frac{M(iu)}{L(iu)}=P(iu)+\sum_{k=1}^{n'}\sum_{s=1}^{l'_k}\frac{c'_{ks}}{(iu-p'_k)^s}+\sum_{k=1}^{n''}\sum_{s=1}^{l''_k}\frac{c''_{ks}}{(iu-p''_k)^s},$$

where $P(iu)=\sum_{k=0}^{m-n} a_k(iu)^k \quad (m\geqslant n)$ and $P(iu)=0 \quad (m<n)$,

$\operatorname{Re} p'_k<0$ and $\operatorname{Re} p''_k>0$, p'_k and p''_k are the roots of the polynomials $L(x)=0$. Since

$$\frac{1}{(iu-p)^s}=\frac{1}{(s-1)!}\frac{d^{s-1}}{dp^{s-1}}\int_0^{\infty} e^{pt}e^{-iut}\,dt=\int_0^{\infty}\frac{t^{s-1}}{(s-1)!}e^{pt}e^{-iut}\,dt$$

$$(\operatorname{Re} p<0)$$

and

$$\frac{1}{(iu-p)^s} = -\int_{-\infty}^{0} \frac{t^{s-1}}{(s-1)!} e^{pt} e^{-iut} dt \quad (\operatorname{Re} p > 0),$$

the process at the output of the filter can be represented as

$$\eta(t) = \sum_{k=0}^{m-n} a_k \xi^{(k)}(t) + \int_0^\infty \xi(t-\tau) G_1(\tau) d\tau + \int_0^\infty \xi(t+\tau) G_2(-\tau) d\tau,$$

where

$$G_1(t) = \sum_{k=1}^{n'} \left(\sum_{s=1}^{l'_k} \frac{c'_{ks} t^s}{(s-1)!} \right) e^{p'_k t} \quad (t>0),$$

$$G_2(t) = -\sum_{k=1}^{n''} \left(\sum_{s=1}^{l''_k} \frac{c''_{ks} t^s}{(s-1)!} \right) e^{p''_k t} \quad (t<0).$$

We note that if the polynomial $L(x)$ has roots with a positive real part, then the corresponding filter can not be physically realized in practice.

§7. Physically Realizable Filters

In the present section the following problem is investigated: what are the spectral functions that can be obtained at the output of a physically realizable filter? Here at the input of the filter the simplest (in a certain sense) random process is considered.

The processes discussed in this section are always assumed to be one-dimensional and wide-stationary. Therefore the word "stationary" will sometimes be omitted, while the word "wide" is always omitted.

We start with stationary sequences. We shall not restate all the definitions and heuristic considerations given for continuous-parameter processes for the sequences, but we shall utilize the corresponding terminology. Consider a system such that the states at the input and output register only at the integer-valued instants of time $t=0, \pm 1, \pm 2, \ldots$.

Let a unit impulse enter the system at time 0. The system's response at time t is denoted by a_t. If the system does not anticipate the future, $a_t = 0$ for $t<0$. If the system is homogeneous in time, its response to the unit impulse applied at time s is equal to a_{t-s}. The response at time t of a linear, homogeneous and physically realizable system to the sequence of impulses $\xi(n)$ $(-\infty<n<\infty)$ will be

$$\eta(t)=\sum_{n=-\infty}^{t} a_{t-n}\xi(n)=\sum_{n=0}^{\infty} a_n\xi(t-n), \tag{1}$$

i.e. is a process with moving averages.

Assume that $\xi(n)$ is a standard uncorrelated sequence

$$\mathsf{E}\xi(n)=0,\ \mathsf{E}\xi(n)\,\overline{\xi(m)}=\delta_{nm}$$
$$(-\infty<n,\ m<\infty).$$

This sequence has a constant spectral density.

In order that series (1) be m.s. convergent, it is necessary and sufficient that

$$\sum_{n=0}^{\infty} |a_n|^2<\infty. \tag{2}$$

If the condition is satisfied then the process $\eta(t)$ is also a wide-stationary one and

$$\mathsf{E}\eta(t)=0, \qquad R_\eta(t)=\sum_{n=0}^{\infty} a_{n+t}\bar{a}_n. \tag{3}$$

What kind of sequences can be obtained in this manner?

Lemma 1. *In order that a stationary sequence $\eta(n)$ be a response of a physically realizable filter to an uncorrelated sequence, it is necessary and sufficient that the sequence $\eta(n)$ have an absolutely continuous spectral measure and its spectral density $f(u)$ admit representation*

$$f(u)=|g(e^{iu})|^2,\ g(e^{iu})=\sum_{n=0}^{\infty} b_n e^{inu},\ \sum_{n=0}^{\infty} |b_n|^2<\infty. \tag{4}$$

Proof. Necessity. Let the sequence be represented by (1). Set

$$g(e^{iu})=\frac{1}{\sqrt{2\pi}}\sum_{n=0}^{\infty} \bar{a}_n e^{inu}. \tag{5}$$

In view of Parseval's equality

$$R_\eta(t)=\sum_{n=0}^{\infty} a_{n+t}\bar{a}_n=\int_{-\pi}^{\pi} e^{itu}|g(e^{iu})|^2\,du,$$

i.e. the sequence $\eta(n)$ possesses an absolutely continuous spectrum with density $f(u)=|g(e^{iu})|^2$.

Sufficiency. Let $\eta(n)$ be a sequence with the correlation function

$$R_\eta(t)=\int_{-\pi}^{\pi} e^{itu} f(u)\, du$$

and $f(u)=|g(e^{iu})|^2$, where $g(e^{iu})$ is determined by relation (4). The sequence $\eta(n)$ has the spectral representation

$$\eta(n)=\int_{-\pi}^{\pi} e^{inu}\, \zeta(du).$$

We construct the following stochastic measure on the σ-algebra of Borel subsets of the interval $[-\pi, \pi)$:

$$\xi(A)=\int_{-\pi}^{\pi} \frac{1}{\sqrt{2\pi}\, g(e^{iu})}\, \chi_A(u)\, \zeta(du).$$

Then

$$\mathsf{E}\xi(A)\,\overline{\xi(B)}=\int_{-\pi}^{\pi} \chi_A(u)\, \chi_B(u)\, \frac{1}{2\pi|g(e^{iu})|^2} f(u)\, du=\frac{1}{2\pi}\int_{A\cap B} du,$$

i.e. $\xi(A)$ is an orthogonal measure with structure function $l(A\cap B)$, where l is the Lebesgue measure. Using Lemma 2 and 1 of Section 4, we obtain

$$\eta(n)=\int_{-\pi}^{\pi} e^{inu}\, \zeta(du)=\int_{-\pi}^{\pi} e^{inu}\, \sqrt{2\pi}\, g(e^{iu})\, \xi(du)=$$

$$=\sum_{k=0}^{\infty} \sqrt{2\pi}\, b_k \int_{-\pi}^{\pi} e^{i(n-k)u}\, \xi(du)=\sum_{k=0}^{\infty} a_k \xi(n-k),$$

where

$$a_n=\sqrt{2\pi}\, b_n, \quad \xi(n)=\int_{-\pi}^{\pi} e^{iun}\, \xi(du) \quad \text{and}$$

$$\mathsf{E}\xi(n)\,\overline{\xi(m)}=\frac{1}{2\pi}\int_{-\pi}^{\pi} e^{i(n-m)u}\, du=\delta_{nm}.$$

Thus $\xi(n)$ is a standard uncorrelated sequence. □

The lemma just proved gives us a simple answer to the question posed above. But this answer is not sufficiently effective in the general case, since it is still unclear when the spectral density can be represented by formula (4).

We now obtain the conditions on $f(u)$ required to admit such a representation. Denote by H_2 the set of all functions $f(z)$ analytical in the circle $D=\{z:|z|<1\}$ and such that

$$\|f(z)\|^2=\lim_{r\uparrow 1}\int_{-\pi}^{\pi}|f(re^{i\theta})|^2\,d\theta<\infty.$$

If $f(z)=\sum_{n=0}^{\infty}a_n z^n$, then $f(re^{i\theta})=\sum_{n=0}^{\infty}a_n r^n e^{in\theta}$, i.e. $a_n r^n$ are the Fourier coefficients of the function $f(re^{i\theta})$. In view of Parseval's theorem

$$\int_{-\pi}^{\pi}|f(re^{i\theta})|^2\,d\theta=2\pi\sum_{n=0}^{\infty}|a_n|^2\,r^{2n}.$$

It thus follows that $f(z)\in H_2$ if and only if

$$\sum_{n=0}^{\infty}|a_n|^2<\infty.$$

Consequently, one can define for each function $f(z)\in H_2$ a series $f(e^{i\theta})=$ $=\sum_{n=0}^{\infty}a_n\,e^{in\theta}$ which is convergent in $\mathscr{L}_2(l)$ where l is the Lebesgue measure on $[-\pi,\pi)$. The function $f(z)$ $(|z|<1)$ can be determined from the function $f(e^{i\theta})$ by means of Poisson's formula

$$f(re^{i\theta})=\frac{1}{2\pi}\int_{-\pi}^{\pi}f(e^{iu})\,P(r,\theta,u)\,du, \tag{6}$$

where

$$P(r,\theta,u)=\frac{1-r^2}{1-2r\cos(\theta-u)+r^2}=\sum_{n=-\infty}^{\infty}r^{|n|}\,e^{in(\theta-u)}.$$

The proof of this assertion follows directly from Parseval's theorem. □

It is proved in the theory of complex functions (cf. Privalov [46])*, that if in formula (6) the function $f(e^{i\theta})$ is Lebesgue integrable, then for

* See also P. L. Duren, *Theory of H^p Spaces*, Academic Press, N.Y., 1970 (p. 5, Corollary 2). *Translator's Remark.*

almost all θ there exists the limit

$$\lim_{r\uparrow 1} f(re^{i\theta}) = f(e^{i\theta}).$$

The function $f(e^{i\theta})$ is called the *boundary value* of function $f(z)$ ($|z|<1$).

Theorem 1. *Let $f(u)$ be a non-negative and Lebesgue integrable function on $[-\pi, \pi)$. For the existence of the function $g(z)\in H_2$ such that*

$$f(u) = |g(e^{iu})|^2, \tag{7}$$

it is necessary and sufficient that

$$\int_{-\pi}^{\pi} |\ln f(u)|\, du < \infty. \tag{8}$$

Proof. Necessity: Let $g(z) = \sum_{n=0}^{\infty} a_n z^n \in H_2$ and let (7) be valid. It can be assumed that $g(0)\neq 0$ (otherwise the function $z^{-m}g(z)$ may be taken in place of $g(z)$, where m is the multiplicity of the zero at $z=0$ of the function $g(z)$) and that $g(0)=1$. Let $0<r<1$ and $A=\{u: |g(re^{iu})|\leqslant 1\}$, $B=\{u: |g(re^{iu})|>1\}$. Then

$$\int_{-\pi}^{\pi} |\ln|g(re^{iu})||\, du = \int_B \ln|g(re^{iu})|\, du -$$

$$-\int_A \ln|g(re^{iu})|\, du = 2\int_B \ln|g(re^{iu})|\, du - \int_{-\pi}^{\pi} \ln|g(re^{iu})|\, du.$$

It follows from Jensen's formula that

$$\frac{1}{2\pi}\int_{-\pi}^{\pi} \ln|f(re^{iu})|\, du = \ln \prod_{k=1}^{n} \frac{r}{|z_k|} \geqslant 0,$$

where z_k are the zeros of function $f(z)$ in the interior of circle $|z|<r$ and $|f(0)|=1$. Consequently,

$$\int_{-\pi}^{\pi} |\ln|g(re^{iu})||\, du \leqslant 2\int_B \ln|g(re^{iu})|\, du \leqslant \int_B |g(re^{iu})|^2\, du \leqslant$$

$$\leqslant \int_{-\pi}^{\pi} |g(re^{iu})|^2 \, du \leqslant 2\pi \sum_{n=0}^{\infty} |a_n|^2.$$

Applying Fatou's lemma, we obtain

$$\int_{-\pi}^{\pi} |\ln|g(e^{iu})|| \, du = \int_{-\pi}^{\pi} \varliminf_{r\uparrow 1} |\ln|g(re^{iu})|| \, du \leqslant$$

$$\leqslant \varliminf_{r\uparrow 1} \int_{-\pi}^{\pi} |\ln|g(re^{iu})|| \, du \leqslant 2\pi \sum_{n=0}^{\infty} |a_n|^2,$$

which proves the necessity of condition (8).

Sufficiency. Let condition (8) be satisfied. The function

$$u(r, \theta) = \frac{1}{2\pi} \int_{-\pi}^{\pi} \ln f(u) \, P(r, \theta, u) \, du$$

is harmonic in the circle $D = \{z : |z| < 1\}$. We note that the Jensen inequality yields

$$u(r, \theta) \leqslant \ln \left\{ \frac{1}{2\pi} \int_{-\pi}^{\pi} f(u) \, P(r, \theta, u) \, du \right\}.$$

Denote the analytic function in D with the real part $u(r, \theta)$ by $\varphi(z)$. Set

$$g(z) = e^{(1/2)\varphi(z)}.$$

Then

$$|g(re^{i\theta})|^2 = e^{\operatorname{Re}\varphi(z)} = e^{u(r,\theta)} \leqslant \frac{1}{2\pi} \int_{-\pi}^{\pi} f(u) \, P(r, \theta, u) \, du$$

and

$$\int_{-\pi}^{\pi} |g(re^{i\theta})|^2 \, d\theta \leqslant \int_{-\pi}^{\pi} |f(u)| \, du < \infty.$$

Thus $g(z) \in H_2$ and $\lim_{r\uparrow 1} |g(re^{i\theta})|^2 = e^{\lim_{r\uparrow 1} u(r,\theta)} = f(\theta)$ almost everywhere. The theorem is proved. □

Remark 1. It follows from the proof of the theorem that function $g(z)$ can be chosen in such a manner that it will be positive for $z=0$ and have no zeros in D.

Remark 2. Function $g(z)$ whose existence was established in Theorem 1 is not uniquely determined. However, if $g(z)$ satisfies conditions

$$\text{a) } g(z)\neq 0, \qquad z\in D, \qquad \text{b) } g(0)>0,$$

then this function is unique and therefore coincides with the one obtained in the theorem.

Indeed, if $g_i(z)$ $(i=1,2)$ be two such functions, then $\psi(z)=\dfrac{g_1(z)}{g_2(z)}$ is analytic in D and is non-vanishing and its absolute value is one on the boundary of D. The function $\ln\psi(z)$ will be analytic in D and its real part will be zero on the boundary of D. Therefore $\ln\psi(z)=ik$, where k is a real number. Since $\ln\varphi(0)$ is real it follows that $\ln\psi(z)=0$. □

Combining Lemma 1 with Theorem 1 we obtain the following assertion.

Theorem 2. *In order that the sequence $\eta(t)$ admit representation*

$$\eta(t)=\sum_{n=0}^{\infty} a_n\xi(t-n), \qquad \sum_{n=0}^{\infty} |a_n|^2<\infty,$$

where $\xi(n)$ is an uncorrelated sequence it is necessary and sufficient that $\eta(t)$ possess an absolutely continuous spectral measure and its spectral density satisfy condition

$$\int_{-\pi}^{\pi} \ln f(u)\,du>-\infty.$$

Let $\zeta_1(x)$, $\zeta_2(x)$, $x\in\mathscr{X}$ be two Hilbert random functions. Denote by $\mathscr{L}_2\{\zeta_i\}$ the closed linear span of the system of random variables $\{\zeta_i(x), x\in\mathscr{X}\}$ in $\mathscr{L}_2$.

Definition 1. If $\mathscr{L}_2(\zeta_1)\subset\mathscr{L}_2(\zeta_2)$, then the random function $\zeta_1(x)$ is called *subordinated* to $\zeta_2(x)$. If, however, $\mathscr{L}_2(\zeta_1)=\mathscr{L}_2(\zeta_2)$, then $\zeta_1(x)$ and $\zeta_2(x)$ are called *equivalent*.

Remark 1. It follows from the proof of Lemma 1 that the sequences $\xi(n)$ and $\eta(n)$ are equivalent.

We now show how to express the coefficients a_n in the operation of moving averages in terms of the spectral density $f(u)$ of the sequence $\eta(t)$.

The function $\varphi(z)$ introduced in the course of the proof of Theorem 1 is analytic in D and its real part has the boundary value $\ln f(u)$. Hence using Schwarz's formula

$$\varphi(z)=\frac{1}{2\pi}\int_{-\pi}^{\pi}\ln f(u)\frac{e^{iu}+z}{e^{iu}-z}\,du. \tag{9}$$

Expanding the function $g(z)=\exp\{\frac{1}{2}\varphi(z)\}$ into the power series $g(z)=\sum_{n=0}^{\infty} b_n z^n$ we obtain the following values for the coefficients

$$a_n=\sqrt{2\pi}\,\bar{b}_n.$$

On the other hand the expression for $g(z)$ can be transformed as follows. Since

$$\frac{e^{iu}+z}{e^{iu}-z}=1+\frac{2ze^{-iu}}{1-ze^{-iu}}=1+2\sum_{k=1}^{\infty} z^k e^{-iku},$$

it follows that

$$\overline{g(z)}=\exp\left\{\frac{1}{4\pi}\int_{-\pi}^{\pi}\ln f(u)\,du+\frac{1}{2\pi}\sum_{k=1}^{\infty} d_k\bar{z}^k\right\},$$

where

$$d_k=\int_{-\pi}^{\pi} e^{iku}\ln f(u)\,du.$$

Setting

$$P=\exp\left\{\frac{1}{4\pi}\int_{-\pi}^{\pi}\ln f(u)\,du\right\},\quad \exp\left\{\frac{1}{2\pi}\sum_{k=1}^{\infty} d_k z^k\right\}=\sum_{k=0}^{\infty} c_k z^k \quad (c_0=1),$$

we obtain

$$\overline{g(z)}=P\sum_{k=0}^{\infty} c_k\bar{z}^k.$$

Hence

$$a_n=\sqrt{2\pi}\,Pc_n. \tag{10}$$

We now proceed to continuous-parameter processes. The operation which corresponds to the random process $\xi(t)$, a process $\eta(t)$, $t\in(-\infty,\infty)$

determined by the formula

$$\eta(t)=\int_0^\infty a(s)\,d\xi(t-s). \tag{11}$$

may serve as a generalization of the operation of moving averages to the case of continuous-parameter random processes.

A process with orthogonal increments $\xi(t)$ will be called *standard* if

$$\mathsf{E}\xi(t)=0, \quad \mathsf{E}|\xi(t+h)-\xi(t)|^2=h.$$

In view of the remarks in Section 4 concerning Stieltjes' stochastic integrals, a certain stochastic orthogonal measure $\xi(A)$, defined on the σ-algebra of Lebesgue measurable sets, is associated with process $\xi(t)$. This measure is also called the *standard stochastic measure.* A necessary and sufficient condition for the existence of integral (11) is that $a(t)$ be Lebesgue measurable and

$$\int_0^\infty |a(t)|^2\,dt<\infty.$$

Note that the standard process $\xi(t)$ is not m.s. differentiable. However the quotients

$$\xi'_\Delta(t_k)=\frac{\xi(t_{k+1}+\Delta)-\xi(t_k)}{\Delta}, \quad \Delta=t_{k+1}-t_k,$$

are orthogonal for all t_k and for Δ arbitrarily small. Therefore the fictitious derivative $\xi'(t)$ should be regarded as a process whose values in any two instants of time are orthogonal and their variance is infinite. This fictitious process is often utilized in arguments and is called "*white noise*". A precise definition of white noise is given in the theory of generalized random processes (Gel'fand and Vilenkin [10]). Symbolically formula (11) can be written as

$$\eta(t)=\int_{-\infty}^{t} a(t-s)\,\xi'(s)\,ds,$$

where $\eta(t)$ is interpreted as the response of a physically realizable filter to "white noise". The impulse transfer function of this filter equals zero for $t<0$ and equals $a(t)$ for $t>0$. Note that the admissible physically realizable filters for the process $\xi'(t)$ are all given by formula (11). Indeed, any admissible physically realizable filter is by definition either of the form (11) or is a limit of filters of this form. The condition for m.s. convergence of filters of type (11) with the impulse transition functions $a_n(t)$ is as follows:

$$\int_0^\infty |a_n(s) - a_{n'}(s)|^2 \, ds \to 0, \quad \text{as} \quad n, n' \to \infty.$$

If, however, this condition is satisfied, $\text{l.i.m.}\, a_n(t) = a(t)$ (with respect to the Lebesgue measure on $(0, \infty)$) exists and

$$\text{l.i.m.}\, \eta_n(t) = \text{l.i.m.} \int_0^\infty a_n(s) \, d\xi(t-s) = \int_0^\infty a(s) \, d\xi(t-s).$$

Thus, the limiting transition in filters of type (11) does not extend the class of filters.

Formula (11) can be rewritten as follows:

$$\eta(t) = \int_{-\infty}^\infty a(t-s) \, d\xi(s), \quad a(t) = 0 \quad \text{for} \quad t < 0.$$

Hence the correlation function of the process $\eta(t)$ is equal to

$$R(t) = \int_{-\infty}^\infty a(t+s-u) \overline{a(s-u)} \, du$$

or

$$R(t) = \int_0^\infty a(t+s) \overline{a(s)} \, ds. \tag{12}$$

Lemma 2. *In order that a wide-sense stationary process $\eta(t)$ be a response of a physically realizable filter to "white noise" subordinated to this process it is necessary and sufficient that the process $\eta(t)$ possess an absolutely continuous spectral measure and that its spectral density $f(u)$ admit representation*

$$f(u) = |h(iu)|^2, \tag{13}$$

where

$$h(iu) = \int_0^\infty b(s) \, e^{-ius} \, ds, \quad \int_0^\infty |b(s)|^2 \, ds < \infty. \tag{14}$$

Proof. Necessity. Let the process $\eta(t)$ admit representation (11). Set

$$h(iu)=\frac{1}{\sqrt{2\pi}}\int_0^\infty a(s)\,e^{-isu}\,ds.$$

In view of Parseval's equality

$$R(t)=\int_0^\infty a(t+s)\,\overline{a(s)}\,ds=\int_{-\infty}^\infty e^{iut}|h(iu)|^2\,du,$$

i.e. the spectrum of the process is absolutely continuous and the spectral density is of the form (13), (14).

Sufficiency. Let the conditions of the lemma be satisfied. Consider the spectral representation of the process $\eta(t)$:

$$\eta(t)=\int_{-\infty}^\infty e^{iut}\zeta(du),$$

and stochastic measure

$$\mu(A)=\int_{-\infty}^\infty \frac{\chi_A(u)}{h(iu)}\,\zeta(du). \tag{15}$$

The stochastic integral (15) is meaningful for an arbitrary bounded Borel set A since $\frac{\chi_A(u)}{h(iu)}\in\mathscr{L}_2\{F\}$, where F is the spectral measure of the process, $F(A)=\int_A |h(iu)|^2\,du$.

It is easy to verify that $\mu(A)$ is an orthogonal measure and moreover

$$\mathsf{E}\mu(A)\,\overline{\mu(B)}=\int_{A\cap B} du.$$

Set

$$\xi(t_2)-\xi(t_1)=\frac{1}{\sqrt{2\pi}}\int_{-\infty}^\infty \frac{e^{-iut_2}-e^{-iut_1}}{-iu}\,\mu(du). \tag{16}$$

Obviously the stochastic integral (16) exists. The random function of the interval $\xi(\Delta)=\xi(t_2)-\xi(t_1)$, $\Delta=[t_1, t_2)$ is an elementary measure which corresponds to the standard process. Indeed, $\mathsf{E}\xi(\Delta)=0$. Next, utilizing Parseval's equality for Fourier integrals, we obtain

$$\mathsf{E}\xi(\Delta_1)\overline{\xi(\Delta_2)}=\frac{1}{2\pi}\int_{-\infty}^{\infty}\frac{e^{-iut_2}-e^{-iut_1}}{-iu}\cdot\frac{e^{iut_4}-e^{iut_3}}{iu}\,du=$$

$$=\int_{-\infty}^{\infty}\chi_{\Delta_1}(t)\,\chi_{\Delta_2}(t)\,dt=l(\Delta_1\cap\Delta_2),$$

where $\Delta_1=[t_1, t_2)$, $\Delta_2=[t_3, t_4)$, and l is the Lebesgue measure on the real line. In view of Lemma 1 in Section 4 and formula (15) we obtain

$$\eta(t)=\int_{-\infty}^{\infty}e^{iut}h(iu)\,\mu(du). \tag{17}$$

Now we note that if

$$h(iu)=\frac{1}{\sqrt{2\pi}}\int_{-\infty}^{\infty}a(s)\,e^{ius}\,ds, \quad \text{where} \quad \int_{-\infty}^{\infty}|a(s)|^2\,ds<\infty, \tag{18}$$

then

$$\int_{-\infty}^{\infty}h(iu)\,\mu(du)=\int_{-\infty}^{\infty}a(-s)\,\xi(ds).$$

Indeed, since the spaces $\mathscr{L}_2(\mu)$ and $\mathscr{L}_2\{\xi\}$ are isomorphic to the space $\mathscr{L}_2\{l\}$, where l is the Lebesgue measure on the whole real line $(-\infty, \infty)$ and the Fourier transform leaves the scalar product invariant in $\mathscr{L}_2\{l\}$, it is sufficient to verify formula (18) for simple functions. Let $a(t)=$ $=\sum c_k\chi_{\Delta_k}(t)$, where Δ_k is the interval (or semi-interval) (a_k, b_k). Then

$$\int_{-\infty}^{\infty}a(-s)\,\xi(ds)=\frac{1}{\sqrt{2\pi}}\int_{-\infty}^{\infty}\sum_k c_k\frac{e^{iub_k}-e^{iua_k}}{iu}\,\mu(du),$$

which is a particular case of (18). Thus formula (18) is established. It follows from (18) that

$$\int_{-\infty}^{\infty}e^{iut}\,h(iu)\,\mu(du)=\int_{-\infty}^{\infty}a(s)\,d\xi(t-s), \tag{19}$$

since, in view of formula (16), the multiplication of measure ξ by e^{iut} yields the shift in the argument of function ξ by the amount t. We obtain from (17) and (19) that

$$\eta(t)=\int_0^\infty a(s)\,d\xi(t-s), \quad \text{where} \quad a(t)=\frac{1}{\sqrt{2\pi}}\int_{-\infty}^{\infty} h(iu)\,e^{-iut}\,du.$$

The lemma is thus proved. □

Let the spectral density $f(u)$ of a process $\eta(t)$ be given. The following questions arise. When does the spectral density admit representation (13) (14), (or as it is commonly called factorization)? How can we find the function $h(iu)$, given the function $f(u)$ (or analogously, how can one find function $a(t)$)? Answers to these problems can be obtained by redusing them to the case of factorizations of functions on a circle – a problem which has already been solved. We introduce the transformation $w=\dfrac{1+z}{1-z}$ which maps the circle $D=\{z:|z|<1\}$ into the right-hand half-plane $\Pi^+=\{w:\operatorname{Re} w>0\}$. On the boundary of the corresponding regions $(w=iu,\ z=e^{i\theta})$ this transformation is of the form $u=\operatorname{ctg}\theta/2$. Let $f(u)$ admit factorization (13), (14). Set

$$\begin{aligned} g(z)&=(1+w)\,h(w)=\frac{2}{1-z}\,h\left(\frac{1+z}{1-z}\right),\\ \tilde{f}(\theta)&=f(u)\,(1+u^2). \end{aligned} \tag{20}$$

The function $\tilde{f}(\cdot)$ admits factorization $|\tilde{f}(\theta)|=|g(e^{i\theta})|^2$ where $g(z)$ is analytic in D and integrable in $(-\pi,\pi)$,

$$\int_{-\pi}^{\pi} \tilde{f}(\theta)\,d\theta=2\int_{-\infty}^{\infty} f(u)\,du<\infty,$$

i.e. $g(z)\in H_2$. In view of Theorem 1

$$-\infty<\int_{-\pi}^{\pi} \ln\tilde{f}(\theta)\,d\theta=2\int_{-\infty}^{\infty}\frac{\ln f(u)+\ln(1+u^2)}{1+u^2}\,du,$$

hence

$$\int_{-\infty}^{\infty}\frac{\ln f(u)}{1+u^2}\,du>-\infty. \tag{21}$$

Assume the converse. Let $f(u)$ be non-negative, integrable and satisfy relation (21). Define $\tilde{f}(\cdot)$ by means of (20). Then $\tilde{f}(\cdot)$ is integrable and

$$\int_{-\pi}^{\pi} \ln \tilde{f}(\theta)\, d\theta > -\infty .$$

It follows from Theorem 1 that $\tilde{f}(\cdot)$ admits factorization

$$\tilde{f}(\theta) = |g(e^{i\theta})|^2, \quad g(z) = \sum_0^\infty a_n z^n, \quad \sum_{n=0}^\infty |a_n|^2 < \infty .$$

Set

$$h(w) = \frac{1}{1+w} \sum_{n=0}^\infty a_n \left(\frac{1-w}{1+w}\right)^n .$$

The function $h(\cdot)$ is analytic in the right-hand half-plane and $f(u) = |h(iu)|^2$,

$$h(iu) = \sum_{n=0}^\infty a_n \frac{(1-iu)^n}{(1+iu)^{n+1}} . \tag{22}$$

Recalling that functions $\frac{1}{\sqrt{2\pi}} e^{in\theta}$ form a complete orthonormal sequence in $\mathscr{L}_2(-\pi, \pi)$ it is easy to verify that the sequence $\frac{1}{\sqrt{2\pi}} \frac{(1-iu)^n}{(1+iu)^{n+1}}$ is a complete orthonormal sequence in $\mathscr{L}_2(-\infty, \infty)$ with respect to the Lebesgue measure. Therefore the series for $h(iu)$ written above (eq. (22)) is m.s. convergent. We now note that

$$\frac{(1-iu)^n}{(1+iu)^{n+1}} = \sum_{k=1}^{n+1} \frac{A_k}{(1+iu)^k} = \sum_{k=1}^{n+1} \frac{A_k}{(k-1)!} \int_0^\infty e^{-(1+iu)t} t^{k-1}\, dt =$$

$$= \int_0^\infty e^{-iut} B_n(t)\, dt ,$$

so that

$$h(iu) = \int_0^\infty e^{-iut} b(t)\, dt, \quad \text{where} \quad b(t) = \sum_{n=0}^\infty a_n B_n(t) .$$

One should keep in mind that the partial sums of the series (22) are

Fourier transforms (up to a multiplicative factor) of functions equal to $\sum_1^N a_n B_n(t)$ for $t \geqslant 0$ and equal to zero for $t<0$. Since the Fourier transform does not change the norm of functions in $\mathscr{L}_2(-\infty, \infty)$ the m.s. convergence of the series for $b(t)$ and the fact that $\int_0^\infty |b(t)|^2\,dt<\infty$ follow from the m.s. convergence of series (22). As far as the uniqueness of the factorization obtained of the function $f(u)$ is concerned, the situation is analogous to the case of factorization of functions on a circle. The expression for $h(w)$ will be obtained from formula (9) by writing w in place of z and performing the corresponding change of variables under the integral sign:

$$h(w)=\exp\left\{\frac{1}{2\pi}\int_{-\infty}^{\infty}\frac{\ln f(u)}{1+u^2}\,\frac{i+uw}{u+iw}\,du\right\}. \tag{23}$$

Theorem 3. *In order that a non-negative integrable function* $f(u)$ $(-\infty<u<\infty)$ *admit factorization* (13), (14) *it is necessary and sufficient that*

$$\int_{-\infty}^{\infty}\frac{\ln f(u)}{1+u^2}\,du>-\infty. \tag{24}$$

Under additional assumptions that $h(w)\neq 0$ $(\mathrm{Re}\,w>0)$, $h(1)>0$, the function $h(w)$ is unique and is defined by formula (23).

Theorem 4. *In order that a stationary process* $\eta(t)$ $(-\infty<t<\infty)$ *admit representation* (11) *it is necessary and sufficient that it possess an absolutely continuous spectrum and that its spectral density satisfies condition* (24).

§8. Forecasting and Filtering Stationary Processes

One of the important problems in the theory of random processes which has numerous practical applications is as follows: it is required to estimate in the best possible manner the value of a random variable ζ by observing a certain set of random variables $\{\xi_\alpha, \alpha\in A\}$. Thus it is required to find a function $f(\xi_\alpha \mid \alpha\in A)$ of variables ξ_α, $\alpha\in A$ which satisfies the approximate equality

$$\zeta \approx \hat{\zeta} = f(\xi_\alpha \mid \alpha \in A). \tag{1}$$

with the least possible error.

An example of such a problem is the forecasting (or extrapolation) of a random process. Here it is required to predict the value of a random process at time t^* by means of its values on a set of instants of time preceding t^*.

Another example is the problem of filtering a random process. The problem is as follows: at times $t' \in T' \subset T$ the process $\xi(t) = \eta(t) + \zeta(t)$ consisting of the "useful" signal $\zeta(t)$ and the additive "noise" $\eta(t)$ was observed. It is required to separate the noise from the signal, i.e. it is required to obtain for some $t^* \subset T$ the best approximation to $\zeta(t)$ of the form

$$\zeta(t^*) \approx \hat{\zeta} = f(\xi(t') \mid t' \in T').$$

The problem stated is not completely defined since the meaning of "the best approximation" is not clear. Evidently the optimality criterion depends on the practical nature of the problem under consideration. As far as the mathematical theory is concerned the most advanced solutions of the problem are based on choosing the mean square deviation as the measure of precision of the approximate equality (1).

Quantity

$$\delta = \{\mathsf{E}[\zeta - f(\xi_\alpha \mid \alpha \in A)]^2\}^{1/2} \tag{2}$$

is called the *mean square error* of the approximate formula (1). The problem is to determine a function f such that (2) admits the minimal value. In the case when A is a finite set, $f(\xi_\alpha \mid \alpha \in A)$ represents a Borel measurable function of arguments ξ_α, $\alpha \in A$. If, however, A is infinite, then this symbol represents a random variable measurable with respect to the σ-algebra $\mathfrak{F} = \sigma\{\xi_\alpha, \alpha \in A\}$ generated by the set of random variables $\{\xi_\alpha, \alpha \in A\}$.

Henceforth it will be assumed that both ζ and $f(\xi_\alpha, \alpha \in A)$ possess moments of the second order.

Set

$$\gamma = \mathsf{E}(\zeta \mid \mathfrak{F}). \tag{3}$$

Then

$$\delta^2 = \mathsf{E}\{\zeta - f(\xi_\alpha \mid \alpha \in A)\}^2 =$$
$$= \mathsf{E}(\zeta - \gamma)^2 + 2\mathsf{E}(\zeta - \gamma)(\gamma - f(\xi_\alpha \mid \alpha \in A)) + \mathsf{E}(\gamma - f(\xi_\alpha \mid \alpha \in A))^2.$$

Since $\gamma - f(\xi_\alpha \mid \alpha \in A)$ is $\mathfrak{F}$-measurable,

$$\mathsf{E}(\zeta - \gamma)(\gamma - f(\xi_\alpha \mid \alpha \in A)) =$$

$$= \mathsf{E}\,\mathsf{E}\{(\zeta-\gamma)(\gamma-f(\xi_\alpha \mid \alpha\in A)) \mid \mathfrak{F}\} =$$
$$= \mathsf{E}(\gamma-f(\xi_\alpha \mid \alpha\in A))\, \mathsf{E}\{(\zeta-\gamma) \mid \mathfrak{F}\} = 0.$$

Hence,

$$\delta^2 = \mathsf{E}(\zeta-\gamma)^2 + \mathsf{E}(\gamma - f(\xi_\alpha \mid \alpha\in A))^2.$$

We thus obtain

Theorem 1. *An approximation to a random variable ζ possessing finite second order moments – with minimal mean square error – by means of a $\mathfrak{F}=\sigma\{\xi_\alpha, \alpha\in A\}$-measurable random variable is unique* (mod P) *and is given by formula*

$$\gamma = \mathsf{E}\{\zeta \mid \mathfrak{F}\}.$$

Remark. The estimator $\hat{\zeta}=\gamma$ of the random variable ζ is unbiased, i.e.

$$\mathsf{E}\gamma = \mathsf{E}\,\mathsf{E}\{\zeta \mid \mathfrak{F}\} = \mathsf{E}\zeta$$

and the variables $\zeta-\gamma$ and ξ_α are uncorrelated for any $\alpha\in A$:

$$\mathsf{E}(\zeta-\gamma)\,\xi_\alpha = \mathsf{E}\,\mathsf{E}\{(\zeta-\gamma)\,\xi_\alpha \mid \mathfrak{F}\} = \mathsf{E}\xi_\alpha \mathsf{E}\{(\zeta-\gamma) \mid \mathfrak{F}\} = 0.$$

Unfortunately practical application of Theorem 1 is often very difficult. In case of Gaussian random variables one may proceed, however, one step further. First we note that a simpler problem leading in many cases to complete and analytically accessible solutions, is that of obtaining an optimal approximation not in the class of all measurable functions of given random variables but in a narrower class of linear functions. More precisely this means the following. Let $\{\Omega, \mathfrak{S}, \mathsf{P}\}$ be the basic probability space. Assume that the variables ξ_α and ζ possess finite moments of the second order. We introduce the subspace $\mathscr{L}_2\{\xi_\alpha, \alpha\in A\}$ of the Hilbert space $\mathscr{L}_2\{\Omega, \mathfrak{S}, \mathsf{P}\}$ which is a closed linear span of the variables ξ_α, $\alpha\in A$ and a constant. One may regard the subspace $\mathscr{L}_2\{\xi_\alpha, \alpha\in A\}$ as the set of all linear (nonhomogeneous) functions of ξ_α with finite variances. The best linear approximation $\tilde{\zeta}$ of random variable ζ is the element of $\mathscr{L}_2\{\xi_\alpha, \alpha\in A\}$ which is closest to ζ, i.e.

$$\delta^2 = \mathsf{E}|\tilde{\zeta}-\zeta|^2 \leqslant \mathsf{E}|\zeta'-\zeta|^2$$

for any $\zeta'\in\mathscr{L}_2\{\xi_\alpha, \alpha\in A\}$. We know from the theory of Hilbert spaces that the problem of determining an element $\tilde{\zeta}$ belonging to subspace H_0 which is closest to the given element ζ always has a unique solution. Namely $\tilde{\zeta}$ is the projection of ζ onto H_0. The element $\tilde{\zeta}$ can always be determined and moreover uniquely determined from the system of equations $(\zeta-\tilde{\zeta}, \zeta'')=0$ for any $\zeta''\in\mathscr{L}_2\{\xi_\alpha, \alpha\in A\}$. In our case this system of equations reduces to equation

$$\mathsf{E}(\tilde{\zeta}\bar{\xi}_\alpha) = \mathsf{E}(\zeta\bar{\xi}_\alpha), \tag{4}$$

and since the variable $\zeta \equiv 1$ belongs to $\mathscr{L}_2\{\xi_\alpha, \alpha \in A\}$

$$\mathsf{E}\tilde{\zeta} = \mathsf{E}\zeta,$$

so that the optimal linear estimators $\tilde{\zeta}$ are necessarily unbiased. We may assume that $\mathsf{E}\xi_\alpha = 0$ for any α. Therefore in what follows we shall confine ourselves to the subspace of the random variables in $\mathscr{L}_2\{\Omega, \mathfrak{S}, \mathsf{P}\}$ with zero mathematical expectations.

Obviously, linear estimators of ζ may not always be suitable. For example, if $\xi(n) = e^{i(vn+\varphi)}$ where v is uniformly distributed on $(-\pi, \pi)$, then $\mathsf{E}\{\xi(n)\overline{\xi(m)}\} = 0$ (for $n \neq m$) and the best linear estimator of the variable $\xi(m)$ in terms of the values of all $\xi(n)$ $(n \neq m)$ is of the form $\tilde{\xi}(m) = 0$, i.e. it does not utilize the values of the variables $\xi(n)$, while an arbitrary pair of observations $\xi(k)$ and $\xi(k+1)$ is sufficient to determine exactly the whole sequence $\xi(n)$, namely $\xi(n) = \left(\frac{\xi(k+1)}{\xi(k)}\right)^{n-k} \xi(k)$.

We now assume that the finite dimensional distributions of the system $\{\zeta, \xi_\alpha, \alpha \in A\}$ are normal and $\mathsf{E}\xi_\alpha = 0$, $\mathsf{E}\zeta = 0$. In this case the fact that variables $\zeta - \tilde{\zeta}$ and ξ_α are uncorrelated implies that they are independent. Therefore $\zeta - \tilde{\zeta}$ does not depend on the σ-algebra $\mathfrak{F}$ and

$$\mathsf{E}\{\zeta \mid \mathfrak{F}\} = \mathsf{E}\{\zeta - \tilde{\zeta} + \tilde{\zeta} \mid \mathfrak{F}\} = \mathsf{E}(\zeta - \tilde{\zeta}) + \tilde{\zeta} = \tilde{\zeta}.$$

Theorem 2. *Given a system of Gaussian random variables $\{\zeta, \xi_\alpha, \alpha \in A\}$, the best estimator (in the mean-square sense) of variable ζ by means of a $\sigma\{\xi_\alpha, \alpha \in A\}$-measurable function coincides with the best linear estimator in $\mathscr{L}_2\{\xi_\alpha, \alpha \in A\}$.*

Below a number of specific problems of construction of optimal linear estimators are examined.

A) The number of random variables ξ_α is finite ($\alpha = 1, 2, \ldots, n$). The solution of this problem is simple and well known from linear algebra. Assuming that ξ_α are linearly independent, one may represent the projection $\tilde{\zeta}$ of the variable ζ onto the finite dimensional space H_0 spanned by the variables ξ_α ($\alpha = 1, \ldots, n$) by means of formula

$$\tilde{\zeta} = \frac{1}{\Gamma} \begin{vmatrix} (\xi_1, \xi_1) & \cdots & (\xi_n, \xi_1) & \xi_1 \\ \cdots & \cdots & \cdots & \cdots \\ (\xi_n, \xi_1) & \cdots & (\xi_n, \xi_n) & \xi_n \\ (\zeta, \xi_1) & \cdots & (\zeta, \xi_n) & 0 \end{vmatrix},$$

where $\Gamma = \Gamma(\xi_1, \xi_2, \ldots, \xi_n)$ is the determinant of the Gram matrix constructed from the vectors $\xi_1, \xi_2, \ldots, \xi_n$,

$$\Gamma(\xi_1, \xi_2, \ldots, \xi_n) = \begin{vmatrix} (\xi_1, \xi_1) & \ldots & (\xi_1, \xi_1) \\ \cdots & \cdots & \cdots \\ (\xi_n, \xi_1) & \ldots & (\xi_n, \xi_1) \end{vmatrix}$$

and $(\xi, \eta) = \mathsf{E}(\xi\bar{\eta})$. The mean square error δ of the approximate equality $\zeta \approx \tilde{\zeta}$ is equal to the length of the perpendicular dropped from the end of the vector ξ onto the space H_0 and is given by the formula

$$\delta^2 = \frac{\Gamma(\xi_1, \xi_2, \ldots, \xi_n, \zeta)}{\Gamma(\xi_1, \xi_2, \ldots, \xi_n)}.$$

B) Consider the problem of estimating the random variable ζ by means of the results of observations on an m.s. continuous random process $\xi(t)$ during a finite time-interval $T = [a, b]$. Let $R(t, s)$ be the correlation function of the process $\xi(t)$. In view of Theorem 5 in Section 3 the process $\xi(t)$ is expandable into the series

$$\xi(t) = \sum_{k=1}^{\infty} \sqrt{\lambda_k}\, \varphi_k(t)\, \xi_k,$$

where $\varphi_k(t)$ is an orthonormal sequence of eigenfunctions and λ_k are the eigenvalues of the correlation function on (a, b):

$$\lambda_k \varphi_k(t) = \int_a^b R(t, s)\, \varphi_k(s)\, ds,$$

and ξ_k is a normalized uncorrelated sequence

$$\mathsf{E}\xi_k \bar{\xi}_r = \delta_{kr}.$$

Clearly $\{\xi_k\}$ $k = 1, 2, \ldots$ form a basis in $\mathscr{L}_2\{\xi(t); t \in (a, b)\}$. Therefore

$$\tilde{\zeta} = \sum_{n=1}^{\infty} c_n \xi_n,$$

where $\sqrt{\lambda_n}\, \xi_n = \int_a^b \xi(t)\, \overline{\varphi_n(t)}\, dt$, $n = 1, 2, \ldots$ and

$$c_n = \mathsf{E}\zeta\bar{\xi}_n = \int_a^b R_{\zeta\xi}(t)\, \varphi_n(t)\, dt, \qquad R_{\zeta\xi}(t) = \mathsf{E}\zeta\bar{\xi}(t).$$

The mean square error δ of the estimator is calculated from the formula

$$\delta^2 = \mathsf{E}|\zeta|^2 - \mathsf{E}|\tilde{\zeta}|^2 = \mathsf{E}|\zeta|^2 - \sum_{n=0}^{\infty} \left| \int_a^b R_{\zeta\xi}(t)\, \varphi_n(t)\, dt \right|^2.$$

The practical applicability of this method is hindered by the complexity in computing eigenfunctions and eigenvalues of the kernel $R(t, s)$.

Wiener's method. Let $\xi(t)$ and $\zeta(t)$ $t \in T$ be two Hilbert random functions. Assume that the process $\xi(t)$ is observed on a certain set T^* of values of t. The problem is to determine the optimal estimate of the value of $\zeta(t_0)$, $t_0 \in T$ in terms of the observed values of $\xi(t)$, $t \in T^*$. If we assume that the required estimate is of the form

$$\tilde{\zeta}(t_0) = \int_{T^*} c(s)\, \xi(s)\, m(ds), \tag{5}$$

where m is a measure on T^* and the conditions are satisfied which assures that the integral is meaningful, then equation (4) becomes

$$\int_{T^*} c(s)\, R_{\xi\xi}(s, t)\, m(ds) = R_{\zeta\xi}(t_0, t), \qquad t \in T^*, \tag{6}$$

where $R_{\xi\xi}$ is the correlation function of $\xi(t)$ and $R_{\zeta\xi}$ is the cross correlation function of $\zeta(t)$ and $\xi(t)$. Equation (6) is a Fredholm's integral equation of the first kind with a symmetric (Hermitian) kernal. A solution for this equation doesn't always exist. However, if

$$\int_T \mathsf{E}|\xi(t)|^2\, m(dt) < \infty,$$

then the integral equation (6) possesses a solution $c(s) \in \mathscr{L}_2\{m\}$ if and only if the optimal linear estimate $\tilde{\zeta}(t_0)$ of the value $\zeta(t)$ is of the form (5).

Let T be the real line, $T^* = (a, b)$ and let the processes $\xi(t)$ and $\zeta(t)$ be stationary and stationary correlated (in the wide sense) and let m be the Lebesgue measure. Then equation (6) becomes

$$\int_a^b c(s)\, R_{\xi\xi}(s-t)\, ds = R_{\zeta\xi}(t_0 - t), \qquad t \in (a, b). \tag{7}$$

If, however, $\zeta(t) = \xi(t)$ $(-\infty < t < \infty)$ and $t_0 > b$, i.e. if the problem is to estimate the value of $\xi(t_0)$ in terms of the values of $\xi(t)$ in the past, the problem then is one of *pure forecasting*.

We shall discuss in some detail the problem of forecasting the values of $\zeta(t+q)$ knowing some values of the process $\xi(s)$ before the time t, $t \geq s$. It will be assumed here that the processes $\xi(t)$ and $\zeta(t)$ are stationary and stationary correlated (in the wide sense). The forecasting variable $\tilde{\zeta}(t)$ will be regarded as a function of t for a fixed q. It is easy to observe that the variable $\tilde{\zeta}(t)$ defined by equation (7) is a stationary process. Indeed, equation (7) is of the form

$$\int_{-\infty}^{t} c_t(s)\, R_{\xi\xi}(s-u)\, ds = R_{\zeta\xi}(t+q-u), \qquad u \leq t.$$

Transforming the variables $t-u=v$ and $t-s=\tau$ we get the following form of (7) in terms of v and τ:

$$\int_{0}^{\infty} c_t(t-\tau)\, R_{\xi\xi}(v-\tau)\, d\tau = R_{\zeta\xi}(q+v), \qquad v \geq 0. \tag{8}$$

We thus observe that the function $c_t(t-\tau)$ is independent of t. Set $c(\tau) = c_t(t-\tau)$. Equation (8) becomes

$$\int_{0}^{\infty} c(s)\, R_{\xi\xi}(t-s)\, ds = R_{\zeta\xi}(q+t), \qquad t \geq 0, \tag{9}$$

and formula (5) for the forecasting function is of the form

$$\tilde{\zeta}(t) = \int_{-\infty}^{t} c(t-s)\, \xi(s)\, ds = \int_{0}^{\infty} c(s)\, \xi(t-s)\, ds. \tag{10}$$

Thus the process $\tilde{\zeta}(t) = \tilde{\zeta}_q(t)$ is stationary. It follows from (10) that $c(t)$ is an impulse transition function of a physically realizable filter which transforms the observed process into an optimal estimator of the variable $\zeta(t+q)$.

It is easy to derive the expression for the mean square error δ of the forecasting function $\tilde{\zeta}(t)$. Since δ^2 is the square of the length of the perpendicular dropped from the end of vector $\zeta(t+q)$ onto $\mathscr{L}_2\{\xi(s), s \leq t\}$,

$$\delta^2 = \mathsf{E}|\zeta(t+q)|^2 - \mathsf{E}|\tilde{\zeta}(t)|^2 =$$

$$= R_{\zeta\zeta}(0) - \int_{0}^{\infty}\int_{0}^{\infty} \overline{c(t)}\, R_{\xi\xi}(t-s)\, c(s)\, ds\, dt. \tag{11}$$

Setting $R_{\zeta\zeta}(0)=\sigma_\zeta^2$ and using the spectral representation of the function $R_{\xi\xi}(t)$ we obtain

$$\delta^2=\sigma_\zeta^2-\int_{-\infty}^{\infty}|c(iu)|^2\,dF_{\xi\xi}(u), \tag{12}$$

where $F_{\xi\xi}(u)$ is the spectral function of the process $\xi(t)$ and

$$c(iu)=\int_0^{\infty}c(t)\,e^{-iut}\,dt.$$

We shall briefly describe the method of solving equation (9) proposed by N. Wiener. Assume that the spectrum of the process $\xi(t)$ is absolutely continuous and the spectral density $f_{\xi\xi}(u)$ admits factorization (cf. Theorem 3, Section 7)

$$f_{\xi\xi}(u)=|h(iu)|^2,\quad h(z)=\frac{1}{\sqrt{2\pi}}\int_0^{\infty}a(t)\,e^{-zt}\,dt,\quad \operatorname{Re}z\geqslant 0.$$

It follows from Parseval's equality for the Fourier transform that

$$R_{\xi\xi}(t)=\int_{-\infty}^{\infty}e^{itu}\,|h(iu)|^2\,du=\int_0^{\infty}a(t+s)\overline{a(s)}\,ds.$$

Assume also that the cross spectral function of the processes $\zeta(t)$ and $\xi(t)$ is absolutely continuous and its density $f_{\zeta\xi}(u)$ satisfies the condition

$$\frac{f_{\zeta\xi}(u)}{h(iu)}=k(iu)\in\mathscr{L}_2. \tag{13}$$

Then

$$R_{\zeta\xi}(t)=\int_{-\infty}^{\infty}e^{itu}\,f_{\zeta\xi}(u)\,du=\int_{-\infty}^{\infty}e^{itu}\,k(iu)\,\overline{h(iu)}\,du=\int_0^{\infty}b(t+s)\,\overline{a(s)}\,ds,$$

where

$$b(t)=\frac{1}{\sqrt{2\pi}}\int_{-\infty}^{\infty}k(iu)\,e^{itu}\,du.$$

Using the above expressions, equation (9) can be rewritten as

$$\int_0^\infty \left[b(q+t+s) - \int_0^\infty c(\tau)\, a(t-\tau+s)\, d\tau\right] \overline{a(s)}\, ds = 0, \qquad t>0. \tag{14}$$

In order that (14) hold it is sufficient that the function $c(t)$ satisfy equation

$$b(q+x) = \int_0^\infty c(\tau)\, a(x-\tau)\, d\tau, \qquad x>0. \tag{15}$$

Equation (15) is of the same type as equation (9), the only difference being that the function $a(t)$ vanishes for negative values of t. Writing (15) in the form

$$b(q+x) = \int_0^x c(\tau)\, a(x-\tau)\, d\tau, \qquad x>0, \tag{16}$$

we can solve this equation directly using the Laplace transform. Multiplying equality (16) by e^{-zx} and integrating from 0 to ∞, we obtain

$$B_q(z) = C(z)\, h(z),$$

where

$$B_q(z) = \frac{1}{\sqrt{2\pi}} \int_0^\infty b(q+x)\, e^{-zx}\, dx,$$

$$C(z) = \frac{1}{\sqrt{2\pi}} \int_0^\infty c(t)\, e^{-zt}\, dt.$$

Thus

$$C(z) = \frac{B_q(z)}{h(z)}, \qquad c(t) = \frac{1}{\sqrt{2\pi}} \int_{-\infty}^\infty e^{iut} \frac{B_q(iu)}{h(iu)}\, du, \tag{17}$$

where the expression for $B_q(z)$, $\operatorname{Re}(z)>0$ can be presented as

$$B_q(z) = \frac{1}{\sqrt{2\pi}} \int_{-\infty}^\infty e^{iqu} \frac{f_{\zeta\xi}(u)}{h(iu)} \frac{du}{(z-iu)}. \tag{18}$$

The statement of the assumptions required for the validity of formulas (17) and (18) is very cumbersome. It is simpler in each particular case, to check directly the admissibility of the transformations leading to the solution of the problem at hand.

Yaglom's method. In contrast to Wiener's method, Yaglom's procedure determines the frequency characteristics of an optimal filter rather than the impulse transition function which may not even exist. The general formulas for the solution of the problem are not given, only the method of choosing the function sought, based on the conditions it should satisfy, is presented. In many important cases this choice can be easily made.

Let a two-dimensional stationary process $(\xi(t), \zeta(t))$ admit the spectral representation

$$\xi(t)=\int_{-\infty}^{\infty} e^{iut}\, v_1(du), \qquad \zeta(t)=\int_{-\infty}^{\infty} e^{iut}\, v_2(du)$$

with the matrix of spectral densities given by

$$\begin{pmatrix} f_{\xi\xi}(u) & f_{\xi\zeta}(u) \\ f_{\zeta\xi}(u) & f_{\zeta\zeta}(u) \end{pmatrix}.$$

As before we shall consider the problem of an optimal estimate of the variable $\zeta(t+q)$ given some values of the process $\xi(s)$ for $s\leqslant t$. The forecasting process $\tilde{\zeta}(t)$ is subordinated to $\xi(t)$. Therefore

$$\tilde{\zeta}(t)=\int_{-\infty}^{\infty} e^{iut}\, c(iu)\, v_1(du), \qquad \int_{-\infty}^{\infty} |c(iu)|^2\, f_{\xi\xi}(u)\, du<\infty. \tag{19}$$

The equation

$$\mathsf{E}\zeta(t+q)\,\overline{\xi(s)}=\mathsf{E}\tilde{\zeta}(t)\,\overline{\xi(s)}, \qquad s\leqslant t,$$

which determines the process $\tilde{\zeta}(t)$ becomes

$$\int_{-\infty}^{\infty} e^{ius}\{e^{iuq}\, f_{\zeta\xi}(u)-c(iu)\, f_{\xi\xi}(u)\}\, du=0, \qquad s>0. \tag{20}$$

In addition to conditions (19) and (20) we also have the requirement that $c(iu)$ be the frequency characteristic of a physically realizable filter. These conditions will be satisfied if:

a) the function $f_{\xi\xi}(u)$ is bounded;
b) $c(iu)$ is the limiting value of the function $c(z) \in \mathscr{H}_2^+$;
c) $\psi(iu) = e^{iuq} f_{\zeta\xi}(u) - c(iu)\, f_{\xi\xi}(u)$ is the limiting value of the function $\psi(z)$ in $\mathscr{H}_2^-$.

Here $\mathscr{H}_2^+$ $(\mathscr{H}_2^-)$ denotes the space of functions $h(z)$ analytic in the right-hand (left-hand) half-plane for which the integral

$$\int_{-\infty}^{\infty} |h(x+iu)|^2 \, du$$

is uniformly bounded for $x>0$ $(x<0)$.

Indeed it follows from b) that $\int_{-\infty}^{\infty} |c(iu)|^2 \, du < \infty$ and this in conjunction with a) assures that condition (19) will be satisfied. Moreover, it follows from b) that $c(iu)$ is the frequency characteristic of a physically realizable filter. It follows from condition c) that $e^{iuq} f_{\zeta\xi}(u) - c(iu)\, f_{\xi\xi}(u)$ is the Fourier transform of a function which vanishes for the positive values of the argument. Thus, the relation (20) is proved.

Note that condition b) rejects all the filters with frequency characteristics increasing at infinity. Such frequency characteristics correspond to operations connected with differentiation of the process $\xi(t)$ and are often encountered in the course of construction of optimal filters. Therefore it may be desired to substitute condition a) by a less restrictive one. Assume that $c(z)$ is a function analytic in the right-hand half-plane and let $|c(z)| \to \infty$ as $|z| \to \infty$ but not faster than a certain power of z (say the r-th). The function

$$c_n(z) = \frac{c(z)}{\left(1+\frac{z}{n}\right)^{r+1}} \in \mathscr{H}_2^+ .$$

Since $|c_n(z)| \leqslant |c(z)|$, we have

$$\lim_{n\to\infty} \int_{-\infty}^{\infty} |c_n(iu) - c(iu)|^2 \, f_{\xi\xi}(u)\, du = 0,$$

provided condition (19) is satisfied. Thus $c(iu)$ is the limit in $\mathscr{L}_2(F_{\xi\xi})$ of the frequency characteristics of physically realizable filters and therefore $c(iu)$ is also a frequency characteristic of such a filter. We have thus obtained

Theorem 3. *If the spectral density $f_{\xi\xi}(u)$ of the process $\xi(t)$ is bounded, then the conditions*

$$\text{a)} \int_{-\infty}^{\infty} |c(iu)|^2 f_{\xi\xi}(u)\, du < \infty,$$

b) *$c(iu)$ is the limiting value of the function $c(z)$ analytic in the right-hand half-plane and increasing as $|z|\to\infty$ not faster than a certain power of $|z|$,*

c) *$\psi(iu) = e^{iuq} f_{\zeta\xi}(u) - c(iu) f_{\xi\xi}(u)$ is the limit value of the function $\psi(z)$ belonging to $\mathscr{H}_2^-$*
determine the frequency characteristic $c(iu)$ of an optimal filter estimating the variable $\zeta(t+q)$.

The mean square error δ of the optimal estimator is equal to

$$\delta = \{\mathsf{E}|\zeta(t+q)|^2 - \mathsf{E}|\tilde{\zeta}(t)|^2\}^{1/2} =$$

$$= \left\{\sigma_\zeta^2 - \int_{-\infty}^{\infty} |c(iu)|^2 f_{\xi\xi}(u)\, du\right\}^{1/2}. \tag{21}$$

Example 1. Consider the problem of pure forecasting of the process $\xi(t)$, $(\xi(t) = \zeta(t))$ with correlation function $R(t) = \sigma^2 e^{-\alpha|t|}$ $(\alpha > 0)$. The spectral density is easily found to be $f_{\xi\xi}(u) = \frac{\sigma^2\alpha}{\pi} \frac{1}{u^2+\alpha^2}$. The analytical continuation of the function $\psi(iu)$ is of the form

$$\psi(z) = \frac{c(z) - e^{zq}}{(z+\alpha)(z-\alpha)} \frac{\sigma^2\alpha}{\pi}.$$

The function $\psi(z)$ possesses a unique pole at the left-hand half-plane $z = -\alpha$. To neutralize this pole by means of function $c(z)$ analytic in the right-hand half-plane it is sufficient to take $c(z) = \text{const} = e^{-\alpha q}$. With this choice of $c(z)$, condition a) of Theorem 3 is satisfied. Thus

$$c(iu) = e^{-\alpha q}, \quad \tilde{\xi}(t) = \int_{-\infty}^{\infty} e^{iut} e^{-\alpha q} \nu(du),$$

i.e. the best formula for the optimal forecasting of the variable $\xi(t+q)$ is the following formula:

$$\xi(t+q) \approx e^{-\alpha q} \xi(t),$$

which depends only on the value of $\xi(t)$ at the last observed instant of time. The mean square error of extrapolation is equal to

$$\delta=\sigma\sqrt{1-e^{-2\alpha q}}.$$

Example 2. Consider again the problem of pure forecasting of the process $\xi(t)$, i.e. the estimate of $\xi(t+q)$ by means of the observed values of $\xi(s)$, $s<t$. If the spectrum of the process $\xi(t)$ is absolutely continuous and condition (24) of Section 7 is satisfied, then the spectral density of the process admits factorization $f_{\xi\xi}(u)=|h(iu)|^2$, where $h(z)\in\mathscr{H}_2^+$ and has no zeros in the right-hand half-plane.

Consider the important practical case when $h(z)=\dfrac{P(z)}{Q(z)}$ where $P(z)$ is a polynomial of degree m and $Q(z)$ is a polynomial of degree $n(m<n)$. Assume also that the spectral density $f_{\xi\xi}(u)$ is bounded and does not vanish. Then the zeros of the polynomials $P(z)$ and $Q(z)$ lie in the left-hand half-plane. Let

$$P(z)=A\prod_{j=1}^{p}(z-z_j)^{\alpha_j},\qquad Q(z)=B\prod_{j=1}^{r}(z-\tilde{z}_j)^{\beta_j},$$

$$\sum_{j=1}^{p}\alpha_j=m,\qquad \sum_{j=1}^{r}\beta_j=n.$$

Set

$$P_1(z)=(-1)^m\bar{A}\prod_{j=1}^{p}(z+\bar{z}_j)^{\alpha_j},\qquad Q_1(z)=(-1)^n\bar{B}\prod_{j=1}^{r}(z+\bar{\tilde{z}}_j)^{\beta_j}.$$

The analytic continuation of the function $\psi(iu)$ is of the form:

$$\psi(z)=(e^{zq}-c(z))\frac{P(z)\,P_1(z)}{Q(z)\,Q_1(z)}.$$

The function $c(z)$ should be analytic in the right-hand half-plane, and $\psi(z)$ in the left-hand one. Therefore $c(z)$ should be analytic in the whole complex plane and may have poles at the zeros of the polynomial $P(z)$ where the order of the pole does not exceed the order of the corresponding zero of $P(z)$ Therefore

$$c(z)=\frac{M(z)}{P(z)},$$

where $M(z)$ is analytic in the z-plane and has no singularities for finite z. Since the growth of $c(z)$ is at most exponential, $M(z)$ is a polynomial. In view of the integrability of the square of the absolute value of the function

$$c(iu)\frac{P(iu)}{Q(iu)}=\frac{M(iu)}{Q(iu)}$$

the degree m_1 of the polynomial $M(iu)$ cannot exceed $n-1$, $m_1 \leqslant n-1$.

On the other hand, the choice of function $c(z)$ as given above guarantees that conditions a) and b) of Theorem 3 will be satisfied. It remains to choose polynomial $M(z)$ in such a manner that function

$$\psi(z)=\frac{[e^{zq}P(z)-M(z)]}{Q(z)}\frac{P_1(z)}{Q_1(z)},$$

or equivalently function

$$\psi_1(z)=\frac{e^{zq}P(z)-M(z)}{Q(z)}$$

will have no poles in the left-hand half-plane. A necessary and sufficient condition for this is the fulfillment of equalities

$$\left.\frac{d^j M(z)}{dz^j}\right|_{z=\tilde{z}_k}=\left.\frac{d^j(e^{zq}P(z))}{dz^j}\right|_{z=\tilde{z}_k},$$

$$j=0,1,\ldots,\beta_k-1, \qquad k=1,\ldots,r. \tag{22}$$

The problem of constructing a polynomial $M(z)$ satisfying condition (22) is a standard problem in interpolation theory and always has a unique solution in the class of polynomials of degree $n-1$. If we find polynomial $M(z)$ we thus obtain at the same time the frequency characteristic of the optimal forecasting filter

$$c(iu)=\frac{M(iu)}{P(iu)}.$$

The following method of determining function $c(z)$ may be used. Expand the functions $P(z)\,Q^{-1}(z)$ and $M(z)\,Q^{-1}(z)$ into partial fractions. Let

$$\frac{P(z)}{Q(z)}=\sum_{k=1}^{r}\sum_{j=1}^{\beta_k}\frac{c_{kj}}{(z-\tilde{z}_k)^j}, \qquad \frac{M(z)}{Q(z)}=\sum_{k=1}^{r}\sum_{j=1}^{\beta_k}\frac{\gamma_{kj}}{(z-\tilde{z}_k)^j}.$$

In order that the function $\psi_1(z)$ have no poles at points $\tilde{z}_k$, $k=1,\ldots,r$, it is necessary and sufficient that

$$\frac{d^j}{dx^j}(z-\tilde{z}_k)^{\beta_k}\psi_1(z)\,|_{z=\tilde{z}_k}=0, \qquad j=0,1,\ldots,\beta_k-1,$$

and moreover

$$\psi_1(z)=\sum_{k=1}^{r}\sum_{j=1}^{\beta_k}\frac{c_{kj}\,e^{zq}-\gamma_{kj}}{(z-\tilde{z}_k)^j}.$$

Simple calculations show that

$$\gamma_{kj}=\left[c_{kj}+\frac{q}{1!}c_{kj+1}+\frac{q^2}{2!}c_{kj+2}+\ldots+\frac{q^{\beta_k-1}}{(\beta_k-j)!}c_{k\beta_k}\right]e^{\tilde{z}_k q},$$

$k=1,\ldots,r$.

Knowing the coefficients γ_{kj} we can write down the expression for $c(iu)$:

$$c(iu)=\frac{1}{h(iu)}\sum_{k=1}^{r}\sum_{j=1}^{\beta_k}\frac{\gamma_{kj}}{(z-\tilde{z}_k)^j}=\frac{\sum_{k=1}^{r}\sum_{j=1}^{\beta_k}\frac{\gamma_{kj}}{(z-\tilde{z}_k)^j}}{\sum_{k=1}^{r}\sum_{j=1}^{\beta_k}\frac{c_{kj}}{(z-\tilde{z}_k)^j}}.$$

Example 3. Assume that the process $\zeta(s)$ $(s\leqslant t)$ is observed, but the results of the observations are distorted by various interferences (noise) so that the observed values yield a certain function $\xi(s)$ $s\leqslant t$ which is different from $\zeta(s)$. Assume that the value of the interference (noise) $\eta(t)=$ $=\xi(t)-\zeta(t)$ is a stationary process with mean value 0. It is required to estimate the value of $\zeta(t+q)$ using the results of observations on $\xi(s)=$ $=\zeta(s)+\eta(s)$, $s\leqslant t$.

Such problems are called filtering or smoothing (here we are required to filter out the noise $\eta(t)$ or to smooth the process $\xi(t)$, i.e. the non-regular noise is to be extracted). Moreover if $q>0$ we have filtering with forecasting and if $q<0$ the problem is filtering with delay.

Assume that the noise $\eta(t)$ and the process $\zeta(t)$ are uncorrelated and possess spectral densities $f_{\zeta\zeta}(u)$ and $f_{\eta\eta}(u)$. Then

$$R_{\xi\xi}(t)=R_{\eta\eta}(t)+R_{\zeta\zeta}(t),\qquad f_{\xi\xi}(u)=f_{\eta\eta}(u)+f_{\zeta\zeta}(u).$$

Since $R_{\zeta\xi}(t)=R_{\zeta\zeta}(t)$, there exists a cross spectral density of the processes $\zeta(t)$ and $\xi(t)$ and $f_{\zeta\xi}(u)=f_{\zeta\zeta}(u)$.

Let

$$f_{\zeta\zeta}(u)=\frac{c_1}{u^2+\alpha^2},\qquad f_{\eta\eta}(u)=\frac{c_2}{u^2+\beta^2}.$$

Then

$$f_{\xi\xi}(u)=\frac{c_3(u^2+\gamma^2)}{(u^2+\alpha^2)(u^2+\beta^2)},\qquad c_3=c_1+c_2,\qquad \gamma^2=\frac{c_2\alpha^2+c_1\beta^2}{c_1+c_2}.$$

For the function $\psi(z)$ the following expression is obtained

$$\psi(z) = \frac{-c_1 e^{zq}(z^2-\beta^2)+c_3 c(z)(z^2-\gamma^2)}{(z^2-\alpha^2)(z^2-\beta^2)}.$$

Let $q>0$. The function $\psi(z)$ should be analytic in the left-hand half-plane and must belong to $\mathscr{H}_2^-$. In order that this be satisfied it is required that the numerator vanish at points $z=-\alpha$ and $z=-\beta$. This leads to equations

$$c(-\beta)=0, \quad c(-\alpha)=\frac{c_1}{c_3}\frac{e^{-\alpha q}(\alpha^2-\beta^2)}{\alpha^2-\gamma^2}. \tag{23}$$

Moreover $c(z)$ is analytic in the left-hand half-plane (and also in the right-hand in view of condition b) except at the point $z=-\gamma$, where it has a simple pole. Therefore

$$c(z)=\frac{\varphi(z)}{z+\gamma},$$

where $\varphi(z)$ is an entire function. It follows from the finiteness of the integral

$$\int_{-\infty}^{\infty} |c(iu)|^2 f_{\xi\xi}(u)\, du$$

that $\varphi(z)$ is a linear function, $\varphi(z)=Az+B$.

From (23) we obtain

$$c(z)=A\frac{z+\beta}{z+\gamma}, \quad A=\frac{c_1}{c_3}\frac{\beta+\alpha}{\gamma+\alpha}e^{-\alpha q}.$$

Therefore the formula of optimal smoothing with forecasting has the form

$$\zeta_q(t)=A\int_{-\infty}^{\infty} e^{iut}\frac{iu+\beta}{iu+\gamma}\nu_1(du).$$

Recalling that $(iu+\gamma)^{-1}$ is the frequency characteristic of a physically realizable filter with impulse transfer function $e^{-\gamma t}$, we obtain

$$\zeta_q(t)=\frac{c_1}{c_3}\frac{\beta+\alpha}{\gamma+\alpha}e^{-\alpha q}\{\xi(t)+(\beta-\gamma)\int_{-\infty}^{t} e^{-\gamma(t-s)}\xi(s)\,ds\}. \tag{24}$$

For $q<0$ formula (23) is not valid. Formally it is due to the fact that the function $\psi(z)$ is not bounded in this case in the left-hand half-plane. Function $\psi(z)$ for $q<0$ can be determined from the following considerations. Let $\psi_1(z)=-c_1 e^{zq}(z^2-\beta^2)+c_3 c(z)(z^2-\gamma^2)$. Then $c(z)$ should be analytic in the left-hand half-plane except for the point $z=-\gamma$ and $\psi_1(-\alpha)=\psi_1(-\beta)=0$. Since

$$c(z)=\frac{\psi_1(z)+c_1 e^{zq}(z^2-\beta^2)}{c_3(z^2-\gamma^2)}$$

and $c(z)$ is analytic in the right-hand half-plane, $\psi_1(z)$ is an entire function and

$$\psi_1(\gamma)=-c_1 e^{\gamma q}(\gamma^2-\beta^2). \tag{25}$$

Set

$$\psi_1(z)=A(z)(z+\alpha)(z+\beta).$$

The function $A(z)$ should be entire. It then follows from condition a) of Theorem 3 that $A(z)=\text{const}=A$. The value A is determined from equation (25)

$$A=c_1 e^{\gamma q}\frac{-\gamma+\beta}{\alpha+\gamma}.$$

Hence

$$c(iu)=\frac{c_1}{c_3}\frac{(\alpha+\gamma)(u^2+\beta^2)e^{iuq}-e^{\gamma q}(-\gamma+\beta)(iu+\alpha)(iu+\beta)}{(\alpha+\gamma)(u^2+\gamma^2)}. \quad \square \tag{26}$$

The methods of forecasting and filtering analogous to those described for continuous parameter processes are applicable for stationary sequences. The general solution of forecasting of stationary sequences is presented in the next section. Here we confine ourselves to one example.

Example 4. Consider a stationary sequence $\xi(t)$ which satisfies the simplest autoregression equation

$$a_0\xi(t)+a_1\xi(t-1)+\ldots+a_p\xi(t-p)=\eta(t), \tag{27}$$

where $\eta(t)$ is a standard uncorrelated sequence and $\xi(t)$ is subordinated to $\eta(t)$. Let

$$\eta(t)=\int_{-\pi}^{\pi} e^{itu}\, d\zeta(u)$$

be the spectral representation of the sequence $\eta(t)$, $\zeta(t)$ be a process with uncorrelated increments and structure function $\frac{1}{2\pi} l(A\cap B)$, where l

is the Lebesgue measure. The spectral representation of the sequence $\xi(t)$ should be of the form

$$\xi(t)=\int_{-\pi}^{\pi} e^{itu}\,\varphi(u)\,d\zeta(u),\quad \text{where} \quad \int_{-\pi}^{\pi} |\varphi(u)|^2\,du<\infty. \tag{28}$$

Substituting (28) in (27), we obtain

$$\int_{-\pi}^{\pi} e^{itu}\,\overline{P(e^{iu})}\,\varphi(u)\,d\zeta(u)=\int_{-\pi}^{\pi} e^{itu}\,d\zeta(u),$$

where $P(z)=\sum_{k=0}^{n} \bar{a}_k z^k$. Hence

$$\varphi(u)=\frac{1}{\overline{P(e^{iu})}}\ (\mathrm{mod}\, l).$$

Assume that $P(z)$ has no zeros in the closed circle $|z|\leqslant 1$. Then $\frac{1}{P(z)}\in H_2$. If

$$\frac{1}{P(z)}=\sum_{k=0}^{\infty} \bar{b}_k z^k \quad \left(b_0=\frac{1}{a_0}\right),$$

then

$$\xi(t)=\sum_{n=0}^{\infty} b_n \eta(t-n),$$

and we have obtained the representation of the sequence $\xi(t)$ in the form of a response of a physically realizable filter on an uncorrelated sequence $\eta(t)$. Since

$$\xi(t)=-\frac{1}{a_0}\,[a_1\xi(t-1)+\ldots+a_p\xi(t-p)+\eta(t)], \tag{29}$$

the optimal forecast based on the given $\xi(t-n)$ $(n=1,2,\ldots)$ is of the form

$$\tilde{\zeta}(t)=-\frac{1}{a_0}\,[a_1\xi(t-1)+a_2\xi(t-2)+\ldots+a_p\xi(t-p)].$$

The minimal mean square error of the forecast is equal to

$$\delta(t)=\left\{\mathsf{E}\,\frac{|\eta(t)|^2}{|a_0|^2}\right\}^{1/2}=\frac{1}{|a_0|}.$$

Repeated application of formula (29) allows us to obtain an optimal forecast for a number of steps forward.

§9. General Theorems on Forecasting Stationary Processes

In this section certain general theorems on forecasting stationary sequences and processes from the infinite past are discussed. As before we refer to a wide-sense stationary process with expectation 0 as simply a stationary process.

Forecasting stationary sequences. Let $\{\xi(t),\ t=0,\ \pm 1,\ \pm 2, \ldots\}$ be a stationary sequence. Denote by $\mathscr{H}_\xi$ the closed linear span in $\mathscr{L}_2$ generated by all the variables $\xi(t)$ and by $\mathscr{H}_\xi(t)$ the closed linear span generated by the variables $\xi(n)$, $n \leqslant t$. Clearly, $\mathscr{H}_\xi(t) \subset \mathscr{H}_\xi(t+1)$ and $\mathscr{H}_\xi$ is the closure of $\bigcup_{t=-\infty}^{\infty} \mathscr{H}_\xi(t)$. Consider the shift operation S in $\mathscr{H}_\xi$. This operation is defined by the equality

$$S\eta = \sum c_k \xi(t_k+1).$$

provided $\eta \in \mathscr{H}_\xi$ is of the form $\eta = \sum c_k \xi(t_k)$. The operation S possesses the inverse S^{-1}

$$S^{-1}\eta = \sum c_k \xi(t_k - 1)$$

and preserves the scalar product:

$$\begin{aligned} &\mathsf{E}\big(S\big(\sum c_k \xi(t_k)\big)\, \overline{S\big(\sum d_k \xi(\tau_k)\big)}\big) = \\ &= \sum_k \sum_r c_k \bar{d}_r \mathsf{E}\big(\xi(t_k+1)\, \overline{\xi(\tau_r+1)}\big) = \\ &= \sum_k \sum_r c_k \bar{d}_r \mathsf{E}\big(\xi(t_k)\, \overline{\xi(\tau_r)}\big) = \mathsf{E}\big(\sum_k c_k \xi(t_k)\, \overline{\sum_r d_r \xi(\tau_r)}\big). \end{aligned}$$

Hence S can be extended by continuity to the whole of $\mathscr{H}_\xi$. Moreover it becomes a unitary operation in $\mathscr{H}_\xi$.

We introduce the spectral representation of the sequence $\xi(t)$

$$\xi(t) = \int_{-\pi}^{\pi} e^{iut}\, v(du),$$

where v is the spectral stochastic measure with the structure function F. Henceforth we shall not distinguish between measure $F(A)$ and the spectral function of the sequence $F(u) = F[-\pi, u)$ which generates this measure $F(A)$.

Recall that the random variable η belongs to $\mathscr{H}_\xi$ if and only if

$$\eta=\int_{-\pi}^{\pi} \varphi(u)\, \nu(du), \quad \text{where} \quad \varphi\in\mathscr{L}_2\{F\}.$$

Consider the sequence of random variables

$$\eta(t)=S^t\eta \qquad (t=0, \pm 1, \pm 2, \ldots).$$

Lemma 1. *The sequence $\eta(t)$ is stationary and has spectral representation*

$$\eta(t)=\int_{-\pi}^{\pi} e^{itu}\varphi(u)\, \nu(du). \tag{1}$$

The stationarity follows from the fact that the operator S is unitary

$$\mathsf{E}\eta(t+s)\,\overline{\eta(s)}=(\eta(t+s), \eta(s))=$$
$$=(S^{t+s}\eta, S^s\eta)=(S^t\eta, \eta)=\mathsf{E}\eta(t)\,\overline{\eta(0)}.$$

Finally the spectral representation (1) can be easily verified for the elements η of the form $\eta=\sum a_k\xi(t_k)$ $(\varphi(u)=\sum a_k\, e^{iut_k})$, and this representation is obtained for arbitrary η by means of the limiting transition. □

We note the following additional properties of the operator S:

a) $S\mathscr{H}_\xi(t)=\mathscr{H}_\xi(t+1)$;

b) if $\xi^{(p)}(t)$ is the projection of $\xi(t)$ on $\mathscr{H}_\xi(t-p)$, then

$$S\xi^{(1)}(t)=\xi^{(1)}(t+1), \qquad S^q\xi^{(p)}(t)=\xi^{(p)}(t+q).$$

Since

$$\mathsf{E}\,|\xi^{(p)}(t+q)|^2=\mathsf{E}\,|S^q\xi^{(p)}(t)|^2=\mathsf{E}\,|\xi^{(p)}(t)|^2,$$

the quantity $\mathsf{E}\,|\xi^{(p)}(t)|^2$ does not depend on t. Therefore the following quantity

$$\delta^2(p)=\mathsf{E}\,|\xi(t)-\xi^{(p)}(t)|^2=\mathsf{E}\,|\xi(t)|^2-\mathsf{E}\,|\xi^{(p)}(t)|^2,$$

which equals the square of the minimal mean square error of forecasting $\xi(t)$ by means of $\xi(n)$, $n\leqslant t-p$, is also independent of t.

Clearly,

$$\delta^2(1)\leqslant\delta^2(2)\leqslant\ldots\leqslant\sigma^2=\mathsf{E}\,|\xi(t)|^2.$$

The equality $\delta^2(n)=\sigma^2$ means that $\xi(t)$ is uncorrelated with all the variables $\xi(k)$, $k\leqslant t-n$ for all t, so that the knowledge of these terms does not help as far as the forecasting of $\xi(t)$ is concerned. If $\delta(1)=0$, then

$\xi(t) \in \mathscr{H}_\xi(t-1)$, hence $\mathscr{H}_\xi(t-1) = \mathscr{H}_\xi(t)$ and in general $\mathscr{H}_\xi(n) = \mathscr{H}_\xi(t)$ for any t and $n < t$. Set $\mathscr{H}^s_\xi = \bigcap_t \mathscr{H}_\xi(t)$. In our case $\mathscr{H}^s_\xi = \mathscr{H}_\xi$. This means that if the sequence of values of the process $\xi(n)$, $n \leqslant t$, is known, then all the succeeding terms of this sequence can with probability 1 be linearly expressed in terms of those observed. In a certain sense the opposite case is that in which $\mathscr{H}^s_\xi = 0$ (where 0 denotes the trivial subspace of $\mathscr{H}_\xi$ consisting of the singleton 0). Here the knowledge of the terms of the sequence $\xi(n)$ $(n \leqslant s)$ add little as far as the forecasting of the variable $\xi(s+t)$ is concerned for large t, since $\lim_{t\to\infty} \mathsf{E}|\xi^{(t)}(n)|^2 = 0$ and $\lim_{t\to\infty} \delta^2(t) = \sigma^2$.

Definition 1. If $\mathscr{H}^s_\xi = \mathscr{H}_\xi$, then the process $\xi(t)$ is called *singular* (or *determinate**); if $\delta(1) > 0$ the process $\xi(t)$ is called *undeterminate*; if $\mathscr{H}^s_\xi = 0$ the process is called *regular* (or *completely undeterminate*).

Definition 2. Let $\xi_i(t)$, $i = 1, 2$, be Hilbert random processes, $t \in T$, T be an arbitrary set of real numbers and $\mathscr{H}_\xi(t) = \mathscr{L}_2\{\xi(s);\ s \leqslant t,\ s \in T\}$. We say that $\xi_1(t)$ is *completely subordinate to the process* $\xi_2(t)$ if $\mathscr{H}_{\xi_1}(t) \subset \subset \mathscr{H}_{\xi_2}(t)$ for all $t \in T$.

Theorem 1. *An arbitrary stationary sequence admits representation of the form*

$$\xi(t) = \xi_s(t) + \eta(t), \tag{2}$$

where $\xi_s(t)$ and $\eta(t)$ are mutually uncorrelated sequences, completely subordinate to $\xi(t)$, $\xi_s(t)$ is singular and $\eta(t)$ is regular. The representation (2) *is unique.*

Proof. Clearly $S\mathscr{H}^s_\xi = \mathscr{H}^s_\xi$. Since S is unitary, the orthogonal complement of $\mathscr{H}^s_\xi$ is invariant under S, i.e. S maps one-to-one the subspace $\mathscr{H}^r_\xi = \mathscr{H}_\xi \ominus \mathscr{H}^s_\xi$ on itself (here $\mathscr{H}^r_\xi$ is the orthogonal complement of $\mathscr{H}^s_\xi$ in the space $\mathscr{H}_\xi$).

Let $\xi_s(0)$ be the projections of $\xi(0)$ on $\mathscr{H}^s_\xi$, $\eta(0)$ be the projection of $\xi(0)$ on $\mathscr{H}^r_\xi$ and $\xi_s(t) = S^t \xi_s(0)$, $\eta(t) = S^t \eta(0)$, $t = 0, \pm 1, \pm 2, \ldots$. Since $\xi(0) = \xi_s(0) + \eta(0)$, then $\xi(t) = S^t \xi(0) = \xi_s(t) + \eta(t)$, where the sequences $\eta(t)$ and $\xi_s(t)$ are stationary, mutually uncorrelated and subordinate to $\xi(t)$ (Lemma 1).

Next since in relation (2) $\xi_s(t) \in \mathscr{H}^s_\xi$ and $\eta(t) \in \mathscr{H}^r_\xi$, then $\mathscr{H}_\xi(t) \cap \mathscr{H}^s_\xi \subset \subset \mathscr{H}_{\xi_s}(t)$. Therefore $\mathscr{H}^s_\xi \subset \mathscr{H}^s_{\xi_s}$. On the other hand $\xi_s(t) \in \mathscr{H}^s_\xi$ yields $\mathscr{H}_\xi(t) \subset \mathscr{H}^s_\xi$. Therefore for any $t = \mathscr{H}_{\xi_s}(t) = \mathscr{H}^s_\xi = \mathscr{H}^s_{\xi_s}$ i.e., the sequence $\xi_s(t)$ is singular. Next it follows from the equality $\eta(t) = \xi(t) - \xi_s(t)$ that $\eta(t) \in \mathscr{H}_\xi(t)$. Therefore $\mathscr{H}^s_\eta = \bigcap \mathscr{H}_\eta(t) \subset \mathscr{H}^s_\xi$. On the other hand $\mathscr{H}_\eta(t)$ is

* deterministic using Doob's terminology. *Translator's Remark.*

orthogonal to $\mathcal{H}_\xi^s$ by definition. Therefore $\mathcal{H}_\eta^s=0$, the process $\eta(t)$ is regular.

The uniqueness of representation (2) follows from the fact that under the conditions of the theorem the projection of $\eta(t)$ on $\mathcal{H}_\xi^s$ is zero, $\mathcal{H}_\xi^s= =\mathcal{H}_{\xi_s}^s=\mathcal{H}_{\xi_s}$ and consequently, $\xi_s(t)$ is the projection of $\xi(t)$ on $\mathcal{H}_\xi^s$. The theorem is thus proved. □

Sequences $\eta(t)$ and $\xi_s(t)$ are called the *regular* and *singular* components of the process $\xi(t)$ respectively.

Theorem 2. *The regular component $\eta(t)$ of a stationary sequence can be represented in the form*

$$\eta(t)=\sum_{n=0}^{\infty} a(n)\,\zeta(t-n), \tag{3}$$

where $\zeta(t)$ $(t=0, \pm 1,\ldots)$ is a standard uncorrelated sequence, $\mathcal{H}_\zeta(t)= =\mathcal{H}_\eta(t)$ and $\sum_{n=0}^{\infty}|a(n)|^2<\infty$.

Proof. We introduce the subspace $G(t)=\mathcal{H}_\eta(t)\ominus\mathcal{H}_\eta(t-1)$. This space is one dimensional (if it were 0-dimensional, then $\delta_\eta^2(1)=0$ and $\eta(t)$ would be a singular sequence). We choose in $G(0)$ a unit vector $\zeta(0)$. Then the sequence $\zeta(t)=S^t\zeta(0)$ is orthonormal ($\zeta(t)\in\mathcal{H}_\eta(t)\ominus\mathcal{H}_\eta(t-1)$ therefore $\zeta(t)$ is orthogonal to $\mathcal{H}_\eta(t-1)$, and moreover $\zeta(k)\in\mathcal{H}_\eta(t-1)$ for $k<t$),

$$\mathcal{H}_\zeta(t)\subset\mathcal{H}_\eta(t), \qquad \bigcap_t \mathcal{H}_\zeta(t)\subset\bigcap_t \mathcal{H}_\eta(t)=0.$$

This means that the sequence $\zeta(t)$ forms a basis in $\mathcal{H}_\zeta$. Expanding $\eta(0)$ in terms of this basis we obtain

$$\eta(0)=\sum_{n=0}^{\infty} a(n)\,\zeta(-n), \quad \text{where} \quad \sum_{n=0}^{\infty}|a(n)|^2=\mathsf{E}\,|\eta(0)|^2<\infty.$$

Applying operator S^t to the given expansion of the variable $\eta(0)$ we obtain equality (3). Relation $\mathcal{H}_\eta(t)\subset\mathcal{H}_\xi(t)$ follows directly from (3) and the inclusion in the other direction follows from the definition of $\zeta(t)$. The theorem is thus proved. □

Remark 1. We may assume without loss of generality that $a(0)$ is positive.

Lemma 2. *Let the spectral function $F(u)$ of a stationary process $\xi(t)$ be equal to $F_1(u)+F_2(u)$ where $F_i(u)$ are non-negative monotonically non-decreasing functions and the measures $F_i(A)$ which correspond to functions $F_i(u)$ are singular. Then a decomposition $\xi(t)=\xi_1(t)+\xi_2(t)$ exists where the processes $\xi_i(t)$ are subordinate to $\xi(t)$, are orthogonal and have spectral functions $F_i(u)$ $(i=1, 2)$.*

To prove this assertion we represent the interval as the sum of disjoint sets P_1 and P_2 such that $F_2(P_1)=F_1(P_2)=0$. Set

$$\xi_1(t)=\int_{-\pi}^{\pi} e^{itu}\,\chi_{P_1}(u)\,v(du), \quad \xi_2(t)=\int_{-\pi}^{\pi} e^{itu}\,\chi_{P_2}(u)\,v(du),$$

where v is the stochastic spectral measure of the process $\xi(t)$, $\chi_{P_j}(u)$ is the indicator of the set P_i. Then

$$\xi_1(t)+\xi_2(t)=\int_{-\pi}^{\pi} e^{itu}\,v(du)=\xi(t),$$

$$\mathsf{E}\xi_1(t_1)\,\overline{\xi_2(t_2)}=\int_{-\pi}^{\pi} e^{i(t_1-t_2)u}\chi_{P_1}(u)\,\chi_{P_2}(u)dF(u)=0,$$

$$\mathsf{E}\xi_j(t_1)\,\overline{\xi_j(t_2)}=\int_{-\pi}^{\pi} e^{i(t_1-t_2)u}\,\chi_{P_j}(u)dF(u)=$$

$$=\int_{-\pi}^{\pi} e^{i(t_1-t_2)u}\,dF_j(u), \qquad j=1,2,$$

which proves the lemma. □

Theorem 3. *On order that a sequence $\xi(t)$ be non-determinate it is necessary and sufficient that*

$$\int_{-\pi}^{\pi} \ln f(u)\,du > -\infty, \tag{4}$$

where $f(u)$ is the derivative of absolutely continuous component of $F(A)$ (with respect to the Lebesgue measure).

Proof. Necessity. Let $F_r(u)$, $F_s(u)$ be the spectral functions of the sequences $\eta(t)$ and $\xi_s(t)$. Since $\eta(t)$ and $\xi_s(t)$ are uncorrelated, it follows that

$$F(u)=F_r(u)+F_s(u).$$

In view of Theorem 2 and Theorem 2 of Section 7, $F_r(u)$ is absolutely continuous and for $f_r(u)=F_r'(u)$ the condition $\int_{-\pi}^{\pi} \ln f_r(u)\,du > -\infty$ is satisfied. Decomposing the measures $F(A)$ and $F_s(A)$ into absolutely continuous and singular components with respect to the Lebesgue measure, we obtain

$$F(A)=\int_A f(u)\,du+F^*(A),\quad F_s(A)=\int_A f_s(u)\,du+F_s^*(A).$$

It thus follows that

$$f(u)=f_r(u)+f_s(u)$$

and

$$\int_{-\infty}^{\infty} \ln f(u)\,du \geqslant \int_{-\infty}^{\infty} \ln f_r(u)\,du > -\infty .$$

Hence, if the process is non-determinate, (4) is valid.

Sufficiency. Assume that the process $\xi(t)$ is singular. In this case the decomposition of $\xi(t)$ into two uncorrelated components $\xi_1(t)$ and $\xi_2(t)$ subordinate to $\xi(t)$ corresponds to the decomposition of $F(A)=F_s(A)=$ $=\int_A f_s(u)\,du+F_s^*(A)$ (cf. Lemma 2). Let $\int \ln f(u)\,du=\int \ln f_s(u)\,du>-\infty$.

Then in view of Theorem 2, Section 7, $\xi_1(t)=\sum_{n=0}^{\infty} a'(n)\,\xi'(t-n)$, where ξ' is an uncorrelated sequence. Since

$$\mathscr{H}_\xi(t)\subset\mathscr{H}_{\xi_1}(t)\oplus\mathscr{H}_{\xi_2}(t)\quad\text{and}\quad \cap\mathscr{H}_{\xi_1}(t)=0,$$

we have $\bigcap_t \mathscr{H}_\xi(t)\subset\bigcap_t \mathscr{H}_{\xi_2}(t)\subset\mathscr{H}_{\xi_2}$, which contradicts the relation $\mathscr{H}_\xi=\mathscr{H}_{\xi_1}\oplus\mathscr{H}_{\xi_2}$ so that the process $\xi(t)$ cannot be singular. Therefore

$$\int_{-\pi}^{\pi} \ln f(u)\,du=-\infty .$$

The theorem is thus proved. □

We now consider the problem of forecasting non-determinate processes. Utilizing Theorems 1 and 2 we write

$$\xi(t)=\xi_s(t)+\eta(t),\quad \eta(t)=\sum_{n=0}^{\infty} a_n\zeta(t-n).$$

Since $\xi_s(t)$ is completely predicted "from the past" it is sufficient to consider the forecasting of the regular component $\eta(t)$ of the process $\xi(t)$. It follows from Theorem 2 that the projection of $\eta(t)$ on $\mathscr{H}_\eta(t-q)$ coincides with the projection on $\mathscr{H}_\zeta(t-q)$. Consequently

$$\eta^{(q)}(t)=\sum_{n=q}^{\infty} a_n\zeta(t-n). \tag{5}$$

The value of the mean square error is determined from the equalities

$$\delta^2(q)=\sum_{n=0}^{q-1} |a_n|^2. \tag{6}$$

We now obtain a formula for optimal forecast which does not involve the sequence $\zeta(n)$. Since $\zeta(0)\in\mathscr{H}_\eta$,

$$\zeta(0)=\int_{-\pi}^{\pi} \varphi(u)\,\tilde{v}(du), \quad \int_{-\pi}^{\pi} |\varphi(u)|^2\, dF_r(u)<\infty,$$

where $\tilde{v}$ is the spectral stochastic measure of the process $\eta(t)$ and $F_r'(u)=|g(e^{iu})|^2$, $g(e^{iu})=\dfrac{1}{\sqrt{2\pi}}\sum_{n=0}^{\infty} \bar{a}_n e^{iun}$ (Lemma 1, Section 7).

Consequently (Lemma 1),

$$\zeta(t)=S^t\zeta(0)=\int_{-\pi}^{\pi} e^{itu}\,\varphi(u)\,\tilde{v}(du).$$

We utilize formula (3) to obtain function $\varphi(u)$. We have

$$\eta(t)=\sum_{n=0}^{\infty} a_n\zeta(t-n)=\int_{-\pi}^{\pi} e^{itu}\,\varphi(u)\sum_{n=0}^{\infty} a_n\, e^{-inu}\,\tilde{v}(du).$$

Comparing this equality with

$$\eta(t)=\int_{-\pi}^{\pi} e^{itu}\,\tilde{v}(du),$$

we obtain

$$\varphi(u)=\left(\sum_{n=0}^{\infty} a_n\, e^{-inu}\right)^{-1}=(\sqrt{2\pi}\,g(e^{iu}))^{-1}.$$

Now we have

$$\eta^{(q)}(t)=\int_{-\pi}^{\pi}\left(\sum_{n=q}^{\infty} a_n e^{-inu}\right)\varphi(u)\, e^{itu}\, \tilde{v}(du),$$

hence

$$\eta^{(q)}(t)=\int_{-\pi}^{\pi} e^{itu}\left[1-\frac{\overline{g_q(e^{iu})}}{\overline{g(e^{iu})}}\right]\tilde{v}(du), \tag{7}$$

where

$$g_q(e^{iu})=\frac{1}{\sqrt{2\pi}}\sum_{n=0}^{q-1}\bar{a}_n\, e^{inu}. \tag{8}$$

We now present a method to determine the function $g(z)=\sum_{n=0}^{\infty} b_n z^n$, where $b_n=\frac{1}{\sqrt{2\pi}}\,\bar{a}_n$. This will give us a general solution of the problem of forecasting a stationary sequence as well as the formula for computing a mean square error of the forecast. The function $g(z)\in H_2$, $g(0)=\frac{a_0}{\sqrt{2\pi}}$ is real (cf. Remark 1 after Theorem 2). The spectral density of the sequence $\eta(t)$ is factorized in terms of the function $g(z)$, namely we have $f_r(u)=|g(e^{iu})|^2$. In view of Remark 2 after Theorem 1 Section 7, function $g(z)$ is determined uniquely by means of $f_r(u)$ provided it has no zeros in the circle $|z|<1$ and $g(0)>0$. Therefore if function $g(z)$ – constructed in accordance with Theorem 1 of Section 7 – does not vanish for $|z|<1$, it is then identical with function $g(z)$ obtained in the course of the proof of Theorem 1 in Section 7.

Lemma 3. *Function $g(z)$ does not vanish in $|z|<1$.*

Proof. First we note that if $f_\eta(u)=|h(e^{iu})|^2$, $h(z)=\sum_{n=0}^{\infty} c_n z^n$, $\sum_{n=0}^{\infty}|c_n|^2<\infty$, then $\delta^2(1)\geqslant 2\pi\,|c_0|^2$. Indeed,

$$\mathsf{E}\left|\eta(0)-\sum_{k=1}^{N} d_k\eta(-k)\right|^2=\int_{-\pi}^{\pi}\left|\left(1-\sum_{k=1}^{N} d_k\, e^{-iku}\right)\left(\sum_{k=0}^{\infty}\bar{c}_k\, e^{-iku}\right)\right|^2 du\geqslant 2\pi|c_0|^2.$$

Since this inequality is valid for any d_k and N,

$$\delta^2(1)\geqslant 2\pi|c_0|^2. \tag{9}$$

Assume now that $g(z_0)=0$, $|z_0|<1$. The function $\tilde{g}(z)=\frac{1}{\sqrt{2\pi}}\sum_{n=0}^{\infty} a_n z^n$ vanishes at point $\bar{z}_0$. Set

$$\tilde{g}(z)=(z-\bar{z}_0)\sum_{n=0}^{\infty} b'_n z^n, \quad \text{where} \quad b'_0=-\frac{a_0}{\sqrt{2\pi}\,\bar{z}_0}.$$

Then

$$|g(e^{iu})|=\left|\sum_{n=0}^{\infty}\frac{1}{\sqrt{2\pi}}a_n e^{-inu}\right|=|\tilde{g}(e^{-iu})|=$$

$$=\left|\frac{1-e^{-iu}z_0}{e^{-iu}-\bar{z}_0}\right||e^{-iu}-\bar{z}_0|\left|\sum_{n=0}^{\infty} b'_n e^{-inu}\right|=$$

$$=\left|\sum_{n=0}^{\infty} b''_n e^{-inu}\right|=\left|\sum_{n=0}^{\infty}\bar{b}''_n e^{inu}\right|,$$

where $b''_0=b'_0=-\frac{a_0}{\sqrt{2\pi}\,\bar{z}_0}$. It follows from (9) that

$$\delta^2(1)\geqslant 2\pi|\bar{b}''_0|^2=\left|\frac{a_0}{z_0}\right|^2,$$

which is impossible in view of (6) if $|z_0|<1$. The lemma is thus proved. □

Corollary. *In the formula of the optimal forecasting* (7) *the function* $g(z)\in H_2$ *is uniquely determined (provided that* $g(0)$ *is positive) and coincides with the function obtained in Theorem* 1 *of Section* 7.

We have thus solved the problem of forecasting for the regular part of a non-determinate sequence. The following points should now be clarified: how to express the spectral density of the sequence $\eta(t)$ in terms of the spectral function of the process $\xi(t)$. What is the form of the forecasting formula for the sequence $\xi(t)$ expressed in terms of the characteristic quantities of $\xi(t)$?

Lemma 4. *Let a non-determinate process* $\xi(t)$ *be represented as* $\xi(t)=\eta(t)+\xi_s(t)$, *where* $\eta(t)$ *and* $\xi_s(t)$ *are uncorrelated,* $\xi_s(t)$ *is a singular process and* $\eta(t)$ *is a regular process,* $F(u)$, $F_r(u)$ *and* $F_s(u)$ *are spectral functions of the sequences* $\xi(t)$, $\eta(t)$ *and* $\xi_s(t)$.

Then equality

$$F(u)=F_r(u)+F_s(u) \tag{10}$$

is a decomposition of function $F(u)$ *into absolutely continuous* $F_r(u)$ *and singular* $F_s(u)$ *components with respect to the Lebesgue measure.*

Proof. Formula (10) follows from the fact that the sequences $\eta(t)$ and $\xi_s(t)$ are uncorrelated. We introduce a spectral representation of an uncorrelated sequence $\zeta(t)$ appearing in representation (3),

$$\zeta(t)=\int_{-\pi}^{\pi} e^{itu}\,\zeta(du), \tag{11}$$

where $\zeta(A)$ is a stochastic measure with a structure function $\frac{1}{2\pi}\,l(A)$, where l is the Lebesgue measure. Substituting (11) into (3) we obtain

$$\eta(t)=\int_{-\pi}^{\pi} e^{itu}\sqrt{2\pi}\,\overline{g(e^{iu})}\,\zeta(du).$$

Let

$$\xi_s(t)=\int_{-\pi}^{\pi} e^{itu}\,v_s(du) \tag{12}$$

be the spectral representation of the sequence $\xi_s(t)$. Then

$$\xi(t)=\int_{-\pi}^{\pi} e^{itu}\,v(du)=\int_{-\pi}^{\pi} e^{itu}\left[\sqrt{2\pi}\,\overline{g(e^{iu})}\,\zeta(du)+v_s(du)\right].$$

It follows from the last equality that

$$\int_{-\pi}^{\pi}\varphi(u)\,v(du)=\int_{-\pi}^{\pi}\varphi(u)\left[\sqrt{2\pi}\,\overline{g(e^{iu})}\,\zeta(du)+v_s(du)\right] \tag{13}$$

for any function $\varphi(u)\in\mathscr{L}_2\{F\}$.

Another spectral representation may be given for function $\xi(t)$. Since $\xi_s(0)\in\mathscr{H}_\xi$, it follows that

$$\xi_s(0)=\int_{-\pi}^{\pi}\varphi_s(u)\,v(du),$$

hence

$$\xi_s(t)=S^t\xi_s(0)=\int_{-\pi}^{\pi} e^{itu}\,\varphi_s(u)\,v(du).$$

Taking (13) into account we obtain

$$\xi_s(t)=\int_{-\pi}^{\pi} e^{itu}\,\varphi_s(u)\,[\sqrt{2\pi}\,\overline{g(e^{iu})}\,\zeta(du)+v_s(du)]. \tag{14}$$

Comparing (14) with (12) we see that

$$-\int_{-\pi}^{\pi} e^{itu}(\varphi_s(u)-1)\,v_s(du)=\int_{-\pi}^{\pi} e^{itu}\,\varphi_s(u)\,\sqrt{2\pi}\,\overline{g(e^{iu})}\,\zeta(du).$$

The elements appearing in the different parts of this equality belong to mutually orthogonal subspaces. Therefore they are equal to zero. Consequently,

$$\varphi_s(u)=1\,(\operatorname{mod} F_s),\qquad \varphi_s(u)\,g(e^{iu})=0\,(\operatorname{mod} l).$$

Since $g(e^{iu})$ may equal zero only on the set of l-measure 0, $\varphi_s(u)$ is zero almost everywhere. Let S be the set on which $\varphi_s(u)=1$. Then $l(S)=0$. Therefore

$$F_s(A)=\int_A |\varphi_s(u)|^2\,dF(u)=F(A\cap S),$$

$$F_r(A)=\int_A 2\pi\,|g(e^{iu})|^2\,du.$$

The lemma is thus proved. □

Lemma 5. *Let $\varphi_1(u)$, $\varphi_2(u)$ and $\varphi_3(u)$ be such that*

$$\int_{-\pi}^{\pi}\varphi_1(u)\,v(du),\quad \int_{-\pi}^{\pi}\varphi_2(u)\,\tilde{v}(du),\quad \int_{-\pi}^{\pi}\varphi_3(u)\,v_s(du)$$

are projections of the variables $\xi(t)$, $\eta(t)$ and $\xi_s(t)$ on the spaces $\mathscr{H}_\xi(t-q)$, $\mathscr{H}_\eta(t-q)$ and $\mathscr{H}_{\xi_s}(t-q)$ respectively.

Then

$$\varphi_1(u)=\varphi_2(u)=\varphi_3(u)=e^{itu}\left(1-\frac{\overline{g_q(e^{iu})}}{\overline{g(e^{iu})}}\right)(\operatorname{mod} F).$$

In view of formula (7) it is sufficient to prove that $\varphi_1(u)=\varphi_2(u)=\varphi_3(u)$.

It follows from equality

$$\xi(t)-\int_{-\pi}^{\pi}\varphi_1(u)\,v(du)=$$

$$=\left[\eta(t)-\int_{-\pi}^{\pi}\varphi_1(u)\,\tilde{v}(du)\right]+\left[\xi_s(t)-\int_{-\pi}^{\pi}\varphi_1(u)\,v_s(du)\right] \quad (15)$$

and from the orthogonality of the summands appearing in the brackets of the r.h.s. of equation (15) that

$$\delta_\xi^2(q)=\mathsf{E}\eta(t)-\int_{-\pi}^{\pi}\varphi_1(u)\,\tilde{v}(du)|^2+$$

$$+\mathsf{E}\left|\xi_s(t)-\int_{-\pi}^{\pi}\varphi_1(u)\,v_s(du)\right|^2\geqslant\delta_\eta^2(q),$$

where the equality is attained if and only if

$$\varphi_1(u)=\varphi_2(u)\ (\mathrm{mod}\,F_r),\qquad \varphi_1(u)=\varphi_3(u)\ (\mathrm{mod}\,F_s),$$

$$\xi_s(t)=\int_{-\pi}^{\pi}\varphi_1(u)\,v_s(du).$$

On the other hand, in view of the definition of $\xi_s(t)$ $\delta_\xi^2(q)=\delta_\eta^2(q)$. The lemma is thus proved. □

The results obtained may be formulated as follows:

Theorem 4. *If $\xi(t)$ is a non-determinate stationary sequence, then the optimal forecast $\xi^{(q)}(t)$ of the variable $\xi(t)$ based on the observations of $\xi(s)$, $s\leqslant t-q$ is given by the formula*

$$\xi^q(t)=\int_{-\pi}^{\pi}e^{itu}\left[1-\frac{\overline{g_q(e^{iu})}}{\overline{g(e^{iu})}}\right]v(du),$$

where v is the spectral stochastic measure of the sequence $\xi(t)$,

$$g(z)=\sum_{n=0}^{\infty}b_nz^n,\qquad g_q(z)=\sum_{n=0}^{q-1}b_nz^n;$$

moreover, the function $g(z)\in H_2$ does not vanish in the circle $|z|<1$, $g(0)$

is positive and $|g(e^{iu})|^2 = f(u)$, *where* $f(u)$ *is the derivative of an absolutely continuous component of the spectral function of the sequence* $\xi(t)$. *The square of the mean square error in forecasting is given by*

$$\delta^2(q) = 2\pi\, e^{\frac{1}{4\pi}\int_{-\pi}^{\pi} \ln f(u)\, du} \sum_{n=0}^{q-1} |c_n|^2,$$

where c_n *is determined from*

$$\exp\left\{\frac{1}{2\pi}\sum_{n=1}^{\infty} z^n \int_{-\pi}^{\pi} e^{inu} \ln f(u)\, du\right\} = \sum_{n=0}^{\infty} c_n z^n.$$

In particular

$$\delta^2(1) = 2\pi\, e^{\frac{1}{4\pi}\int_{-\pi}^{\pi} \ln f(u)\, du} \tag{16}$$

The theorem follows directly from Lemmas 4 and 5, formula (7) of the present section, and from Theorem 1 and Remark 2 in Section 7. □

Forecast of continuous parameter processes. Let $\xi(t)$ $(-\infty < t < \infty)$ be a stationary process.

$$\xi(t) = \int_{-\infty}^{\infty} e^{iut}\, \nu(du),$$

where ν is an orthogonal measure on the line $(-\infty < u < \infty)$,

$$\mathsf{E}\xi(t) = 0, \quad R(t) = \mathsf{E}\xi(t+s)\,\overline{\xi(s)} = \int_{-\infty}^{\infty} e^{itu}\, dF(u), \quad F(+\infty) = \sigma^2.$$

We introduce the Hilbert space

$$\mathscr{H}_\xi = \mathscr{H}\{\xi(t), -\infty < t < \infty\}$$

and its subspaces $\mathscr{H}_\xi(t) = \mathscr{H}\{\xi(s), -\infty < s \leqslant t\}$. Define in $\mathscr{H}_\xi$ the group of shift operators S^n $(-\infty < h < \infty)$, by putting

$$S^h\Big(\sum_k c_k \xi(t_k)\Big) = \sum_k c_k \xi(t_k + h)$$

and extend by continuity the definition of S^h over the whole $\mathscr{H}_\xi$. Then S^h forms a group of unitary transformations of $\mathscr{H}_\xi$. This group possesses the same properties (with obvious modifications) as the group of transformation S^n in the discrete parameter case. The problem of optimal linear forecast of processes $\xi(t)$ is to find a random variable $\xi_T(t)$ such that

$$\mathsf{E}|\xi(t)-\xi_T(t)|^2 \leqslant \mathsf{E}|\xi(t)-\eta|^2$$

for any $\eta \in \mathscr{H}_\xi(t-T)$. This problem has a unique solution: the variable $\xi_T(t)$ is the projection of $\xi(t)$ on $\mathscr{H}_\xi(t-T)$. Set

$$\delta_\xi(T)=\delta(T)=\sqrt{\mathsf{E}|\xi(t)-\xi_T(t)|^2}.$$

The quantity $\delta(T)$ – the mean square error of the forecast – is a monotone non-increasing function of T and $0 \leqslant \delta(T) \leqslant \sigma$. If $\lim \delta(T)=\sigma$, then the process is called regular (purely non-determinate). If $\delta(T_0)=0$ from some T_0, then $\mathscr{H}_\xi(t) \subset \mathscr{H}_\xi(t-T_0)$ for any t. Consequently

$$\mathscr{H}_\xi(t) \subset \bigcap_{k=1}^{\infty} \mathscr{H}_\xi(t-kT_0)$$

for any t and $\delta(T)=0$ for all $T>0$. In this case the process is called singular (determinate). Non-singular processes will be called non-determinate.

The proof of Theorem 1 is directly carried over to continuous-time processes: an arbitrary stationary process admits decomposition

$$\xi(t)=\eta(t)+\xi_s(t),$$

where $\eta(t)$ is a regular and $\xi_s(t)$ a singular stationary process, $\eta(t)$ and $\xi_s(t)$ are uncorrelated and are subordinate to $\xi(t)$. □

The following theorem is a continuous analogue of Theorem 2.

Theorem 5. *In order that a stationary process be regular it is necessary and sufficient that it be represented as*

$$\eta(t)=\int_{-\infty}^{t} a(t-s)\,\zeta(ds), \tag{17}$$

where $\zeta(s)$ is the standard process with orthogonal increments and

$$\mathscr{H}_\zeta(t)=\mathscr{H}_\eta(t), \qquad \int_{-\infty}^{\infty} |a(t)|^2\,dt<\infty.$$

In view of Theorem 4 of Section 7 this theorem is equivalent to the following

Theorem 6. *A stationary process is regular if and only if it possesses the*

spectral density $f(u)$ and

$$\int_{-\infty}^{\infty} \frac{\ln f(u)}{1+u^2}\, du > -\infty. \tag{18}$$

We first show that a process admitting representation (17) is regular. For this purpose we introduce projection $\eta_T(t)$ of the random variable $\eta(t)$ on $\mathscr{H}_\eta(t-T)$. Since $\mathscr{H}_\eta(t-T)=\mathscr{H}_\zeta(t-T)$, the random variable $\eta_T(t)$ can be written as

$$\eta_T(t)=\int_{-\infty}^{t-T} \varphi(s)\,\zeta(ds).$$

On the other hand, the difference $\eta(t)-\eta_T(t)$ should be orthogonal to any variable $\varphi\in\mathscr{H}_\zeta(t-T)$ and in particular, to the variable $\psi=\zeta(A)$, where A is an arbitrary measurable set contained in $(-\infty, t-T)$. Since

$$\mathsf{E}\,(\eta(t)-\eta_T(t)\,\overline{\zeta(A)}=\int_{-\infty}^{t} a(t-s)\,\chi_A(s)\,ds-$$

$$-\int_{-\infty}^{t-T} \varphi(s)\,\chi_A(s)\,(ds)=\int_A [a(t-s)-\varphi(s)]\,ds,$$

it follows that $\varphi(s)=a(t-s)$, $s\leqslant t-T$. Thus

$$\eta_T(t)=\int_{-\infty}^{t-T} a(t-s)\,\zeta(ds)$$

and

$$\|\eta_T(t)\|^2=\mathsf{E}\,|\eta_T(t)|^2=\int_{-\infty}^{t-T} |a(t-s)|^2\,ds=\int_T^{\infty} |a(s)|^2\,ds.$$

Therefore $\|\eta_T(t)\|\to 0$ as $T\to\infty$ which shows that the process $\eta(t)$ is regular. The converse assertion is more profound; it states that *every regular process may be represented by means of formula* (17) *or, equivalently, may possess the spectral density $f(u)$ satisfying condition* (18).

To prove this assertion we shall utilize the results analogous to those obtained for discrete-parameter processes.

Let $\xi(t)$ be an arbitrary stationary process and let

$$\xi(t)=\int_{-\infty}^{\infty} e^{itu}\,\nu(du)$$

be its spectral representation. Using the transformation $u=\tan\frac{\theta}{2}$ we transfer the measure ν from the whole real line $(-\infty, \infty)$ onto the interval $(-\pi, \pi)$ and denote the transformed measure by $\tilde{\nu}$. Now let the stationary sequence $\tilde{\xi}(\cdot)=K(\xi)$ with

$$\tilde{\xi}(n)=\int_{-\pi}^{\pi} e^{in\theta}\,\tilde{\nu}(d\theta). \tag{19}$$

correspond to the process $\xi(t)$. We now utilize the following statement to be proved below: the process $\xi(t)$ is regular if and only if the process $\tilde{\xi}(n)$ is regular (Lemma 7). Therefore if $\eta(t)$ is a regular process then $\tilde{\eta}(\cdot)=K(\eta)$ is a regular sequence. If $\tilde{f}(\theta)$ is its spectral density then in view of Theorem 2 of Section 7

$$\int_{-\pi}^{\pi} \ln \tilde{f}(\theta)\,d\theta>-\infty. \tag{20}$$

But then the process $\eta(t)$ also has the spectral density $f(u)$ and moreover $(1+u^2)\,f(u)=\tilde{f}(\theta)$, $\theta=2\arctan u$. Therefore (20) implies (18). The theorem is thus proved. $\square$

We now prove the assertion utilized in the proof of the theorem.

Let $\xi(t)$ be an arbitrary stationary process and let $\tilde{\xi}(n)$ be defined by means of the correspondence K as given above (the equation above equation 19).

Lemma 6. *The equality* $\mathscr{H}_{\xi}(0)=\mathscr{H}_{\tilde{\xi}}(0)$ *is valid.*

Proof. We show that $\tilde{\xi}(-n)\in\mathscr{H}_{\xi}(0)$, $n>0$ (for $n=0$ this is evident). Note that $e^{i\theta}=\dfrac{1+iu}{1-iu}$. Therefore,

$$\tilde{\xi}(-n)=\int_{-\pi}^{\pi} e^{-in\theta}\,\tilde{\nu}(d\theta)=\int_{-\infty}^{\infty}\left(\frac{1-iu}{1+iu}\right)^{n}\nu(du).$$

On the other hand,

$$\frac{1-iu}{1+iu}=-1+\frac{2}{1+iu}=-1+2\int_{-\infty}^{0} e^{ius}\, e^{s}\, ds.$$

Hence, function $\left(\frac{1-iu}{1+iu}\right)^n$ can be approximated by a bounded sequence of functions of the form $\sum_k a_k e^{ius_k}$, $s_k<0$, uniformly convergent on an arbitrary finite interval $(-A, A)$. It thus follows that $\tilde{\xi}(-n)\in\mathscr{H}_{\xi}(0)$. We now show that $\xi(t)$ belongs to $\mathscr{H}_{\tilde{\xi}}(0)$ for $t<0$. We have

$$\xi(t)=\int_{-\infty}^{\infty} e^{itu}\, v(du)=\int_{-\pi}^{\pi}\exp\left\{t\,\frac{e^{i\theta}-1}{e^{i\theta}+1}\right\}\tilde{v}(d\theta)=$$

$$=\lim_{\varrho\uparrow 1}\int_{-\pi}^{\pi}\exp\left\{t\,\frac{1-\varrho\, e^{-i\theta}}{1+\varrho\, e^{-i\theta}}\right\}\tilde{v}(d\theta),$$

since the integrand in the last integral is uniformly bounded for $0<\varrho<1$ and $t<0$.

On the other hand it follows from the equality

$$\frac{1-\varrho\, e^{-i\theta}}{1+\varrho\, e^{-i\theta}}=(1-\varrho\, e^{-i\theta})\sum_{n=0}^{\infty}(-1)^n\varrho^n\, e^{-in\theta}$$

that this integrand can be uniformly approximated (for fixed ϱ) by the functions of the form $\sum_{k=0}^{N} c_k\, e^{-ik\theta}$. Therefore $\xi(t)\in\mathscr{H}_{\tilde{\xi}}(0)$ for $t<0$. The lemma is thus proved. □

Lemma 7. *If $\xi(t)=\xi_s(t)+\eta(t)$ is a decomposition of the process $\xi(t)$ into singular and regular components, then equality $\tilde{\xi}(\cdot)=K(\xi_s)+K(\eta)$ is the decomposition of $\tilde{\xi}(n)$ into the same components.*

Proof. Note that $\mathscr{H}_{\xi_1}(t)\subset\mathscr{H}_{\xi_2}(t)$, $t\in(-\infty,\infty)$ iff $\mathscr{H}_{\tilde{\xi}_1}(n)\subset\mathscr{H}_{\tilde{\xi}_2}(n)$, $n=0, \pm 1,\ldots$. Indeed if $\mathscr{H}_{\xi_1}(0)\subset\mathscr{H}_{\xi_2}(0)$, then $\mathscr{H}_{\tilde{\xi}_1}(0)=\mathscr{H}_{\xi_1}(0)\subset\mathscr{H}_{\xi_2}(0)=$ $=\mathscr{H}_{\tilde{\xi}_2}(0)$, hence $\mathscr{H}_{\tilde{\xi}_1}(t)=S^t\mathscr{H}_{\tilde{\xi}_1}(0)\subset S^t\mathscr{H}_{\tilde{\xi}_2}(0)=\mathscr{H}_{\tilde{\xi}_2}(t)$. Moreover in these relations the roles of the processes $(\xi_1(t), \xi_2(t))$ and $(\tilde{\xi}_1(n), \tilde{\xi}_2(n))$ can be interchanged. Note also that since the measure v is subordinate to the process $\{\xi(t), -\infty<t<\infty\}$, it follows that $\mathscr{H}_{\xi}=\mathscr{H}_{\tilde{\xi}}$. Next let $\xi(t)$ be a singular process. Then $\mathscr{H}_{\tilde{\xi}}=\mathscr{H}_{\xi}=\mathscr{H}_{\xi}(0)=\mathscr{H}_{\tilde{\xi}}(0)$. This means that $\tilde{\xi}(n)$ is a singular process. Analogously we obtain that the singularity of $\tilde{\xi}(n)$ implies the singularity of $\xi(t)$. Let $\xi(t)$ be regular. If $\tilde{\xi}(n)$ were not regular we would have the equality $\tilde{\xi}(n)=\tilde{\xi}_1(n)+\tilde{\eta}_1(n)$, where the sequence $\tilde{\xi}_1(n)$ is singular and completely subordinate to $\xi(n)$. The decomposition $\xi(t)=$ $=\xi_1(t)+\eta_1(t)$ will then correspond to this decomposition, where $\xi_1(t)$ is as it has been shown, a singular process completely subordinate to $\xi(t)$. However this implies that $\bigcap_t \mathscr{H}_\xi(t)\supset\mathscr{H}_{\xi_1}(t)=\mathscr{H}_{\xi_1}(0)\neq 0$ which contradicts the regularity of the process $\xi(t)$. Thus $\tilde{\xi}(n)$ is regular provided $\xi(t)$ is such. The converse is analogously proved. The assertion of the lemma follows from here. □

Results obtained for forecasting stationary sequences can now be carried over with certain modifications in the statement of theorems and the proofs to the case of continuous parameter processes. Here one uses the spectral representation of continuous parameter stationary processes and refers to the results of Lemma 2 and Theorems 3 and 4 of Section 7.

For example, the statement of Lemma 4 is carried over verbatim to the case of continuous-parameter processes and only trivial modifications in the proof are required. From this analog of Lemma 4, we deduce the following

Theorem 7. *In order that the process $\xi(t)$ be non-determinate, it is necessary and sufficient that*

$$\int_{-\infty}^{\infty}\frac{\ln f(u)\,du}{1+u^2}>-\infty,$$

where $f(u)$ is the derivative of the absolutely continuous component of the spectral measure F of the process $\xi(t)$.

If $\xi(t)=\eta(t)+\xi_s(t)$ is the decomposition of the process $\xi(t)$ into regular and singular components, and in view of Theorem 5

$$\eta(t)=\int_{-\infty}^{t} a(t-s)\,\zeta(ds),$$

then

$$\xi_T(t)=\int_{-\infty}^{t-T} a(t-s)\,\zeta(ds)+\xi_s(t).$$

Moreover, the optimal mean square error of the forecast is determined from the relation

$$\delta^2(T)=\int_0^T |a(s)|^2\,ds.$$

An alternative expression of the optimal forecast is given by

$$\xi_T(t) = \int_{-\infty}^{\infty} e^{iut} \left[1 - \frac{h_T(iu)}{h(iu)}\right] \nu(du),$$

where ν is the stochastic spectral measure of the process $\xi(t)$,

$$h(iu) = \frac{1}{\sqrt{2\pi}} \int_0^{\infty} a(s)\, e^{-ius}\, ds, \quad h_T(iu) = \frac{1}{\sqrt{2\pi}} \int_0^{T} a(s)\, e^{-ius}\, ds.$$

The function $h(iu)$ is determined from the spectral density $f(u)$ by means of formula (23) of Section 7.

Chapter V

Probability Measures on Functional Spaces

§1. Measures Associated with Random Processes

Kolmogorov's theorem on the construction of a probability space from finite-dimensional distributions of a random process with values in a metric space $\mathscr{X}$ shows, in particular, how to construct a measure μ on a measurable space $(\mathscr{F}, \mathfrak{B})$ – where $\mathscr{F}$ is the space of all the functions with values in $\mathscr{X}$ and $\mathfrak{B}$ is the minimal σ-algebra containing all the cylinders in $\mathscr{F}$, such that, for any cylinder C, the value $\mu(C)$ coincides with the probability that the sample function of the random process belongs to C. This measure is called the measure associated with (or corresponding to) the random process $\xi(t)$ and it can always be constructed irrespective of the probability space on which the process $\xi(t)$ is defined. If the process $\xi(t, \omega)$ is defined on the probability space $\{\Omega, \mathfrak{S}, \mathsf{P}\}$, T is the mapping of Ω into $\mathscr{F}$ determined by the relation $\omega \xrightarrow{T} \xi(\cdot, \omega)$ and $\mathfrak{S}_0$ is a subalgebra of $\mathfrak{S}$ consisting of the sets of the form $T^{-1}C$, where $C \in \mathfrak{B}$, then the measure μ is the image of the measure $\tilde{\mathsf{P}}$, which is the contraction of the measure P on $\mathfrak{S}_0$ under the mapping T, i.e.

$$\mu(C) = \mathsf{P}(T^{-1}C). \tag{1}$$

The measurable space $(\mathscr{F}, \mathfrak{B})$ and the measure μ defined on it can be conveniently utilized in the investigation of random processes for which only finite dimensional distributions are given. Using these two notions one can investigate the existence of processes with given finite dimensional distributions – with sample functions satisfying certain regularity conditions – as well as to study various functionals on sample functions of the process, transformation of random processes and so on.

Consider measurable functionals of sample functions of a random process. We refer to an arbitrary random variable defined on the probability space $\{\mathscr{F}, \mathfrak{B}, \mu\}$ as the *functional* on the random process $\xi(t)$. Sometimes it may be convenient to consider a somewhat wider class of random variables: i.e. random variables defined on $\{\mathscr{F}, \bar{\mathfrak{B}}, \bar{\mu}\}$ where

$\{\mathfrak{B}, \bar{\mu}\}$ is the completion of the measure $\{\mathfrak{B}, \mu\}$. Since for any random variable $\bar{\xi}$ defined on $\{\mathscr{F}, \mathfrak{B}, \bar{\mu}\}$ a variable ξ on $\{\mathscr{F}, \mathfrak{B}, \mu\}$ can be found such that $\bar{\xi}=\xi\,(\mathrm{mod}\,\bar{\mu})$, the distinction is not a basic one. However, when constructing various specific functionals we may get $\bar{\mathfrak{B}}$-measurable functionals also. Obviously this can be avoided using a more elaborate construction – but this we shall not do.

Any functional on a sample function of a process is determined by the values of the process. We show that this is true for the class of functionals introduced above. We call a functional $f(x(\cdot))$ *cylindrical* if a Borel function $f_m(x_1, \ldots, x_m)$ in $\mathscr{X}^m$ and points $t_1, \ldots, t_m$ exist such that $f(x(\cdot))=f_m(x(t_1), \ldots, x(t_m))$. If f_m is continuous then the cylindrical function is also called *continuous*. Clearly every cylindrical functional is $\mathfrak{B}$-measurable and thus determines a random function on $\{\mathscr{F}, \mathfrak{B}, \mu\}$. Moreover, the value of the functional is determined by the sampling function of the process

$$f(\xi(\cdot))=f_m(\xi(t_1), \ldots, \xi(t_m)).$$

The distribution of the variable $f_m(\xi(t_1), \ldots, \xi(t_m))$ coincides with the distribution of the variable $f(x(\cdot))$ on $\{\mathscr{F}, \mathfrak{B}, \mu\}$. It is natural to consider, as a functional on sample functions of a process $\xi(t)$, a random variable η for which a sequence of cylindrical functionals $f^{(m)}(x(\cdot))$ can be found such that $f^{(m)}(\xi(\cdot))\to\eta$ in probability as $m\to\infty$. In this case the sequence of functionals $f^{(m)}(x(\cdot))$ converges in measure μ to a certain $\bar{\mathfrak{B}}$-measurable functional. This follows from the relation

$$\begin{aligned}\mu(\{x(\cdot): |f^{(m)}(x(\cdot))-f^{(n)}(x(\cdot))|>\varepsilon\})= \\ =\mathsf{P}\{|f^{(m)}(\xi(\cdot))-f^{(n)}(\xi(\cdot))|>\varepsilon\}.\end{aligned}$$

(this equality is a particular case of equation (1)) and the convergence in probability of $f^{(m)}(\xi(\cdot))$. We now show that for any $\bar{\mathfrak{B}}$-measurable functional f there exists a sequence of cylindrical functionals $f^{(k)}$ which converges in measure $\bar{\mu}$ to f. To prove this it is sufficient to show that for any $\mathfrak{B}$-measurable set A there exists a sequence of cylindrical sets C_n such that $\chi_{C_n}(x)\xrightarrow{\mu}\chi_A(x)$. If $\mathfrak{B}_0$ is a collection of such sets then 1) $\mathfrak{B}_0$ is an algebra, 2) it is a monotone class, 3) it contains all the cylinders, namely $\mathfrak{B}_0$ is a σ-algebra which coincides with $\mathfrak{B}$.

Note that if the process $\xi(t)$ is defined on a probability space $\{\Omega, \mathfrak{S}, \mathsf{P}\}$ and $\mathfrak{S}_0$ is the σ-algebra introduced above, then all the $\mathfrak{S}_0$-measurable random variables are functionals of $\xi(\cdot)$. If $\eta(\omega)$ is a $\mathfrak{S}_0$-measurable variable, then $\eta(\omega)=f(T\omega)$, where f is a certain $\mathfrak{B}$-measurable functional. We shall write $\eta(\omega)=f(\xi(\cdot, \omega))$.

It follows from the formula of change of measures in an integral that

$$\mathsf{E} f(\xi(\cdot, \omega)) = \int f(x)\, \mu(dx) \tag{2}$$

for any functional f provided the integral on the right is meaningful.

We now consider the problem of feasibility of constructing a measure on a functional space which is smaller than the space of all the functions. Clearly one can take an arbitrary $\mathfrak{B}$-measurable set $\mathscr{F}_0$ such that $\mu(\mathscr{F}_0)=1$ and consider a measure on $\mathscr{F}_0$. However all the interesting sets of functions are not $\mathfrak{B}$-measurable since every $\mathfrak{B}$-measurable set of functions is determined by the behavior of the functions on at most a countable number of points and this does not determine such properties as continuity, differentiability, absence of discontinuities of the second kind, measurability, etc.

Therefore the following approach to the construction of a measure on a functional space $\mathscr{F}_0$ would seem appropriate. Assume that the space $\mathscr{F}_0$ is such that for any cylinder $C \in \mathscr{F}_0$, the set $C \cap \mathscr{F}_0$ is not empty. Then one can consider in $\mathscr{F}_0$ the minimal σ-algebra $\mathfrak{B}_0$ which contains all the sets of the form $C \cap \mathscr{F}_0$ (we shall call them the cylindrical sets in $\mathscr{F}_0$). Define on the cylindrical sets C_0 in $\mathscr{F}_0$ an additive set function $\mu_0(C_0)= =\mu(C)$, where $C_0 = C \cap \mathscr{F}_0$. Note that this definition is unique: if $C_0 = C \cap \mathscr{F}_0$ and $C_0 = C_1 \cap \mathscr{F}_0$, then the intersection $[(C-C_1) \cup (C_1 - -C)] \cap \mathscr{F}_0$ is not void, which is impossible if $C \neq C_1$. Clearly μ_0 is an additive non-negative set function. A necessary and sufficient condition for extending μ_0 to a measure on $\mathfrak{B}_0$ is the condition that for any sequence C_0^n of cylindrical sets in $\mathscr{F}_0$ such that $\bigcup_n C_0^n = \mathscr{F}_0$ *the inequality* $\sum_n \mu_0(C_0^n) \geqslant \mu_0(\mathscr{F}_0)=1$ is satisfied. This requirement is equivalent to the following: $\sum_n \mu(C^n) \geqslant 1$ for any sequence of cylindrical sets $C^n \in \mathscr{F}$ such that $\mu_0(C_0^n)$

$$\bigcup_n C^n \supset \mathscr{F}_0 .$$

We define the outer measure μ^* in terms of a measure μ as follows: for any set A

$$\mu^*(A) = \inf\{\sum_n \mu(C^n);\ \cup C^n \supset A\}.$$

Then to construct a measure μ_0 on $\mathfrak{B}_0$ we must have: $\mu^*(\mathscr{F}_0)=1$. The measure μ_0 is then of the form $\mu_0(A)=\mu^*(A)$ for $A \in \mathfrak{B}_0$. To prove this we note that $\mu(S)=0$ for any $\mathfrak{B}$-measurable set S such that $S \cap \mathscr{F}_0 = \emptyset$. Indeed otherwise $\mathscr{F} - S \supset \mathscr{F}_0$ and hence

$$\mu^*(\mathscr{F}_0) \leqslant \mu^*(\mathscr{F} - S) = \mu(\mathscr{F} - S) = 1 - \mu(S) < 1. \quad \square$$

Clearly, the σ-algebra $\mathfrak{B}_0$ consists of the sets of the form $A \cap \mathscr{F}_0$, where A is a $\mathfrak{B}$-measurable set.

Let $A_0 \in \mathfrak{B}_0$, and $A_0 = A \cap \mathscr{F}_0$. Set $\bar{\mu}_0(A_0) = \mu(A)$. The definition is unique since if $A_0 = A \cap \mathscr{F}_0 = A' \cap \mathscr{F}_0$, then $(A - A') \cup (A' - A) \in \mathscr{F} - \mathscr{F}_0$, which implies that $\mu(A) = \mu(A')$. Note that $\bar{\mu}_0$ is a countably-additive measure on $\mathfrak{B}_0$: if A_0^k are not pairwise disjoint and $A_0^k = A^k \cap \mathscr{F}_0$, then $A^k \cap A^j \subset \mathscr{F} - \mathscr{F}_0$ for $k \neq j$ and $\mu(A^k \cap A^j) = 0$, i.e.

$$\bar{\mu}_0(\cup A_0^k) = \mu(\cup A^k) = \sum_n \mu(A^k) = \sum_n \bar{\mu}_0(A_0^k).$$

Moreover for any cylindrical set C_0, $\bar{\mu}_0(C_0) = \mu_0(C_0)$. Hence, $\bar{\mu}_0 = \mu_0$. On the other hand

$$\mu^*(A_0) = \inf\{\mu(A'), \quad A' \in \mathfrak{B}, \quad A' \supset A_0\} = \mu(A),$$

if $A_0 = A \cap \mathscr{F}_0$.

Thus a measure associated with a random process can be considered on any set of functions $\mathscr{F}_0$ with outer measure 1 and this measure coincides on this set with the outer measure.

What then are the measurable functionals on the space $\{\mathscr{F}_0, \mathfrak{B}_0, \mu_0\}$? We show that for any $\mathfrak{B}_0$-measurable functional $f_0(x)$ there exists a $\mathfrak{B}$-measurable functional $f(x)$ such that $f(x) = f_0(x)$ for $x \in \mathscr{F}_0$. Let $E_0^{k,n}$ be the set in $\mathfrak{B}_0$ defined by

$$E_0^{k,n} = \left\{x : \frac{k}{2^n} \leqslant f_0(x) < \frac{k+1}{2^n}\right\}, \qquad n > 0, \qquad k = 0, \pm 1, \pm 2, \ldots$$

Denote by $E^{k,n}$ the $\mathfrak{B}$-measurable set such that $E_0^{k,n} = E^{k,n} \cap \mathscr{F}_0$. The set $E^{k,n}$ can always be chosen in such a manner that the following conditions are satisfied:

1) $E^{k,n}$ are pairwise disjoint for a fixed n (otherwise we take the sets $\tilde{E}^{k,n} = E^{k,n} - \bigcup_{j=-\infty}^{k-1} E^{j,n}$ in place of $E^{k,n}$),

2) $E^{k,n} = E^{2k,n+1} \cup E^{2k+1,n+1}$, (otherwise we put $\tilde{E}^{2k,n+1} = E^{k,n} \cap E^{2k,n+1}$, $\tilde{E}^{2k+1,n+1} = E^{k,n} - \tilde{E}^{2k,n+1}$). Define now the function $f^{(n)}(x)$ equal to $k/2^n$ on the set $E^{k,n}$ and equal to $+\infty$ if $x \notin \bigcup_k E^{k,n}$. With this definition $f^{(n)}(x) \leqslant f^{(n+1)}(x) \leqslant f^{(n)}(x) + 1/2^n$ and hence $f^{(n)}(x)$ is uniformly convergent to a measurable function $f(x)$. Moreover $|f^{(n)}(x) - f_0(x)| < 1/2^n$ if $x \in \mathscr{F}_0$. Therefore $f^{(n)}(x) \to f_0(x)$ for $x \in \mathscr{F}_0$. Hence $f_0(x) = f(x)$ for $x \in \mathscr{F}_0$. Let another $\mathfrak{B}$-measurable function $f'(x)$ exist which coincides with $f_0(x)$ on $\mathscr{F}_0$. Then the $\mathfrak{B}$-measurable set $\{x : f(x) \neq f'(x)\}$ is disjoint from $\mathscr{F}_0$ and consequently has μ-measure 0. Thus each $\mathfrak{B}_0$-measurable function can be uniquely extended $(\operatorname{mod} \mu)$ to a $\mathfrak{B}$-measurable function.

The last consideration shows that in the study of functionals of random processes the transfer of a measure to a more restrictive space $\mathscr{F}_0$ has no significant implications. However, functionals defined on $\mathscr{F}_0$ often present a clearer picture. For example, if $\mathscr{F}_0$ is the space of continuous functions then the functional

$$f_0(x) = \sup_t x(t)$$

will be measurable on $\mathscr{F}_0$. This functional can be extended to $\mathscr{F}$ by means of the formula

$$f(x) = \sup_{t \in N} x(t),$$

where N is a countable everywhere dense set of the values of the argument. Clearly the form of the first functional is more natural.

The following observation plays an important role when one studies measures and the feasibility of their transfer to $\mathscr{F}_0$: in many cases one can find a countable set N of values of the argument such that the completion of the σ-algebra $\mathfrak{B}^N$ – the minimal σ-algebra generated by cylindrical sets determined by the values $x(t)$ for $t \in N$ – coincides with $\bar{\mathfrak{B}}$. For example, if the process is stochastically continuous then any everywhere dense set of values of the argument can be chosen as N. The same is true in the case of a one-side stochastically continuous process. A necessary and sufficient condition for the existence of such a set can easily be given:

Lemma. *In order that a set N exist such that $\bar{\mathfrak{B}}^N = \bar{\mathfrak{B}}$ it is necessary and sufficient that the Hilbert space $\mathscr{L}_2(\mu)$ of $\mathfrak{B}$-measurable functions square-integrable with respect to measure μ be separable.*

Proof. If such a set N exists then the separability of $\mathscr{L}_2(\mu)$ follows from the fact that $\mathscr{L}_2(\mu)$ coincides with $\mathscr{L}_2(\mu^N)$, where μ^N is the restriction of measure μ on $\mathfrak{B}^N$. (The separability of $\mathscr{L}_2(\mu^N)$ follows from the fact that bounded cylindrical functions are dense on it and, in turn, that continuous cylindrical functions are dense in the set of bounded cylindrical functions.)

Now let $\mathscr{L}_2(\mu)$ be a separable space and $f_1, f_2, \ldots$ be a basis in $\mathscr{L}_2(\mu)$ For each $\mathfrak{B}$-measurable function f_k there exists a countable set N_k such that f_k is measurable with respect to $\mathfrak{B}^{N_k}$. This follows from the possibility of approximating f_k in terms of cylindrical functions. In this case the union $\bigcup_k N_k$ can be used for N. □

Note that the existence of an N such that $\bar{\mathfrak{B}}^N = \bar{\mathfrak{B}}$ should not be confused with the existence of an N such that the process possesses an N-separable equivalent. The construction of a separable equivalent corresponds to the transfer of the measure μ onto the set $\mathscr{F}_0$ of all N-separable functions.

However to verify the continuity of the process it is sufficient to consider the values of the process on the set N not only in the case when the process is N-separable but also in the case when $\mathfrak{B}^N=\mathfrak{B}$. The first case is discussed in Chapter III, Section 5. As far as the second is concerned, we note that for the evaluation of the outer measure of the set $\mathscr{F}_0$ it is sufficient to utilize cylindrical sets in $\mathfrak{B}^N$.

In the conclusion of this section we state some general conditions which assure the construction of a measure on the sets $\mathscr{C}\subset\mathscr{F}$ and $\mathscr{D}\subset\mathscr{F}$, where $\mathscr{C}$ is the set of continuous functions and $\mathscr{D}$ is the set of functions without discontinuities of the second kind. Clearly, in the first case the process should be stochastically continuous, while in the second it should have no more than a countable number of stochastic discontinuities. In both cases it is easy to find N such that $\mathfrak{B}^N=\mathfrak{B}$ (N is a countable everywhere dense set of values of the argument). The process itself is assumed to be defined on a certain compactum K ($t\in K$) in the first case and on a closed interval in the second. The evaluation of the outer measures of the corresponding sets is simplified by the fact that a minimal $\mathfrak{B}^N$-measurable set can be found which contains the sets $\mathscr{C}$ and $\mathscr{D}$.

Condition for the existence of a measure on $\mathscr{C}$. *In order that a measure μ be transferable onto $\mathscr{C}$ it is necessary and sufficient that relation*

$$\mu\left(\bigcap_{r=1}^{\infty}\bigcup_{l=1}^{\infty}\bigcap_{\substack{|t-s|<1/l\\ t\in N,\, s\in N}}\left\{x(\cdot):|x(t)-x(s)|<\frac{1}{r}\right\}\right)=1.$$

be satisfied.

It is easy to verify that the set in the round brackets appearing under the sign of the measure μ is the minimal $\mathfrak{B}^N$-measurable set containing $\mathscr{C}$ (to do this it is sufficient to consider functions $x(\cdot)$ defined only on N).

Condition for the existence of a measure on $\mathscr{D}$. *In order that a measure μ be transferable onto $\mathscr{D}$ it is necessary and sufficient that*

$$\mu\left(\bigcap_{r=1}^{\infty}\bigcup_{l=1}^{\infty}\bigcap_{\substack{u,s,t\in N\\ s<t<u<s+1/l}}\left\{x(\cdot):|x(t)-x(s)|<\frac{1}{r}\right\}\cup\right.$$

$$\left.\cup\left\{x(\cdot):|x(u)-x(t)|<\frac{1}{r}\right\}\right)=1.$$

The fact that the set in the round brackets appearing under the sign μ is the minimal $\mathfrak{B}^N$-measurable set containing $\mathscr{D}$ follows from the fact that the functions $x(t)$ have no discontinuities of the second kind (cf. Chapter III, Section 4).

§2. Measures in Metric Spaces

In the previous section we pointed out the feasibility of transfering measure μ associated with a random process from the space of all functions $\mathscr{F}$ to a certain smaller functional space $\mathscr{F}_0$. In the present section we shall concern ourselves with the case when $\mathscr{F}_0$ is a separable metric space and σ-algebra $\mathfrak{B}_0$ coincides with the σ-algebra of all Borel sets in $\mathscr{F}_0$. To show that we will be dealing with an interesting situation, consider the case when $\mathscr{F}_0$ coincides with the space $\mathscr{C}$ of all real continuous functions. Clearly $\mathscr{C}$ is a metric space with the metric $\varrho(x, y) = \sup_t |x(t) - -y(t)|$. If we consider processes defined on a compactum, then $\mathscr{C}$ will be separable. We show that $\mathfrak{B}_0$ coincides with the σ-algebra of Borel sets. First we note that the cylindrical set $\{x(\cdot): x(t) \in A\} \cap \mathscr{C}$ – where A is a Borel set – is a Borel set in $\mathscr{C}$. Therefore all the sets in $\mathfrak{B}_0$ are Borel sets in $\mathscr{C}$. To show that all Borel sets in $\mathscr{C}$ belong to σ-algebra $\mathfrak{B}_0$ it is sufficient to show that an arbitrary closed sphere in $\mathscr{C}$ belongs to $\mathfrak{B}_0$. Let

$$S = \{x(\cdot): \sup_t |x(t) - y(t)| \leqslant \varrho\}$$

be a sphere with the center at $y(\cdot) \in \mathscr{C}$ of radius ϱ. Then

$$S = \mathscr{C} \cap \left[\bigcap_{t \in N} \{x(\cdot): |x(t) - y(t)| \leqslant \varrho\}\right],$$

where N is an arbitrary countable everywhere dense set in the domain of definition of the values of the argument.

As we shall see in the sequel, one can introduce a metric in the space of functions with discontinuities of the second kind $\mathscr{D}$ such that $\mathfrak{B}_0$ will coincide with the σ-algebra of Borel sets in $\mathscr{D}$. For future discussions the specific form of the space is irrelevant. We shall consider an abstract separable metric space $\mathscr{X}$ with elements $x, y, \ldots$ and metric $\varrho(x, y)$. Denote by $\mathfrak{B}$ the σ-algebra of Borel sets in $\mathscr{X}$ and let the measure μ be defined on $\mathfrak{B}$. If K is a subset of $\mathscr{X}$, we denote by $\mathscr{C}_K$ the space of all continuous bounded functions defined on K. We denote the space $\mathscr{C}_{\mathscr{X}}$ simply by $\mathscr{C}$. A natural metric in $\mathscr{C}_K$ is

$$\varrho_K(f, g) = \sup_{x \in K} |f(x) - g(x)|.$$

Continous functions on $\mathscr{X}$ form a simpler but at the same time sufficiently broad class of functions; all the $\mathfrak{B}$-measurable functions can be obtained from these functions by means of the limiting operation. Therefore the measure μ is completely determined by the values of the integrals $\int f(x)\,\mu(dx)$ for $f \in \mathscr{C}$: taking a sequence $f_n \in \mathscr{C}$ such that $f^n \xrightarrow{\mu} \chi_A$ where χ_A is the indicator of the set A in $\mathfrak{B}$ one can determine $\mu(A)$. In many cases the measure μ is not given and it is unknown whether it exists. Only the values of the integrals $L(f) = \int f(x)\,\mu(dx)$ are assigned. The question is under what conditions does the functional $L(f)$ defined on $\mathscr{C}$ admit representations in the form of the integral with respect to some finite measure? The answer in the case of a complete metric space is given by the following

Theorem 1. *In order that the functional $L(f)$, defined on the space $\mathscr{C}$ of continuous bounded functions given on a complete metric separable space $\mathscr{X}$ admit the representation*

$$L(f) = \int f(x)\,\mu(dx), \tag{1}$$

where μ is a finite measure on $\mathfrak{B}$, it is necessary and sufficient that the following conditions be satisfied

1) $L(f) \geqslant 0$ *for all* $f \geqslant 0$;
2) $L(c_1 f_1 + c_2 f_2) = c_1 L(f_1) + c_2 L(f_2)$;
3) *for any $\varepsilon > 0$ there exists a compactum K_ε such that for any function $f(x)$ for which $f(x) = 0$ for $x \in K_\varepsilon$, the inequality*

$$|L(f)| \leqslant \varepsilon \|f\|,$$

is fulfilled where $\|f\| = \sup_x |f(x)|$.

Proof. Necessity. The necessity of conditions 1) and 2) is obvious. Since for f satisfying 3) the inequality

$$\left|\int f(x)\,\mu(dx)\right| = \left|\int_{\mathscr{X} - K_\varepsilon} f(x)\,\mu(dx)\right| \leqslant \|f\|\,\mu(\mathscr{X} - K_\varepsilon)$$

holds, to prove the necessity of condition 3) we show that for any $\varepsilon > 0$ there exists a compactum K_ε such that $\mu(\mathscr{X} - K_\varepsilon) \leqslant \varepsilon$. Let $\{x_k, k = 1, 2, \ldots\}$ be a sequence everywhere dense in $\mathscr{X}$ and let $S_r(x)$ be the closed sphere with center at x of radius r. For each r, an N_r can be found such that

$$\mu(\mathscr{X}) - \mu\left(\bigcup_{k=1}^{N_r} S_r(x_k)\right) < r\varepsilon.$$

Set

$$K_\varepsilon = \bigcap_{n=1}^{\infty} \bigcup_{k=1}^{N_{2^{-n}}} S_{2^{-n}}(x_k),$$

then K_ε is a closed set and for each n it possesses a finite 2^{-n}-net. Thus K_ε is a compactum. Next we have

$$\mu(\mathscr{X} - K_\varepsilon) \leqslant \sum_{n=1}^{\infty} \mu\left(\mathscr{X} - \bigcup_{k=1}^{N_{2^{-n}}} S_{2^{-n}}(x_k)\right) \leqslant \sum_{n=1}^{\infty} \varepsilon 2^{-n} = \varepsilon.$$

The necessity of the conditions of the theorem is thus proved.

Sufficiency. Let F be a set. Set $\bar{\mu}(F) = \inf L(f)$, where the infimum is taken over all $f \geqslant 0$ such that $f(x) \geqslant 1$ for $x \in F$. We include F in class $\mathfrak{B}_0$ provided $\bar{\mu}(F') = 0$, where F' is the boundary of F. We show that $\mathfrak{B}_0$ forms an algebra of sets and that $\bar{\mu}$ is an additive function on $\mathfrak{B}_0$. To do this we note that $\bar{\mu}(A \cup B) \leqslant \bar{\mu}(A) + \bar{\mu}(B)$ and $\bar{\mu}(A) \leqslant \bar{\mu}(B)$ for $A \subset B$, as it easily follows from the definition of $\bar{\mu}$. Next $\bar{\mu}((\mathscr{X} - F)') = \bar{\mu}(F')$, and $\bar{\mu}((F_1 \cup F_2)') \leqslant \bar{\mu}(F_1' \cup F_2') \leqslant \bar{\mu}(F_1') + \bar{\mu}(F_2')$, and thus the sets F such that $\mu(F') = 0$ form an algebra. We now prove the additivity of $\bar{\mu}$ on $\mathfrak{B}_0$. Let F_1 and F_2 be two disjoint sets on $\mathfrak{B}_0$. We first show that

$$\bar{\mu}(F_1 \cup F_2) \geqslant \bar{\mu}(F_1) + \bar{\mu}(F_2).$$

Taking an arbitrary $\varepsilon > 0$ we can find functions f and φ, $f \geqslant 0$, $1 \geqslant \varphi \geqslant 0$ such that $L(\varphi) \leqslant \varepsilon$, $L(f) \leqslant \bar{\mu}(F_1 \cup F_2) + \varepsilon$, $\varphi(x) = 1$ for $x \in [F_1] \cap [F_2]$ and $f(x) \geqslant 1$ for all $x \in F_1 \cup F_2$. Now set

$$f_1(x) = \begin{cases} 1, & x \in [F_1], \\ \varphi(x), & x \in [F_2] \end{cases}$$

and extend by continuity the definition of $f_1(x)$ over the whole space $\mathscr{X}$ in such a manner that $0 \leqslant f_1(x) \leqslant f(x) + \varphi(x)$ (if $g(x)$ is an arbitrary continuous non-negative extension then $f_1(x) = \min[g(x), f(x) + \varphi(x)]$).

Let $f_2(x) = f(x) + \varphi(x) - f_1(x)$. Clearly, f_2 is a non-negative continuous function and $f_2(x) = f(x) = 1$ for $x \in F_2$. Therefore

$$\bar{\mu}(F_1) + \bar{\mu}(F_2) \leqslant L(f_1) + L(f_2) = L(f) + L(\varphi) \leqslant \bar{\mu}(F_1 \cup F_2) + 2\varepsilon,$$

and since ε is arbitrary

$$\bar{\mu}(F_1) + \bar{\mu}(F_2) \leqslant \bar{\mu}(F_1 \cup F_2).$$

The additivity of $\bar{\mu}$ on $\mathfrak{B}_0$ follows from this relation and the semi-additivity of $\bar{\mu}$.

Note that in the case when $[F_1] \cap [F_2] = \emptyset$, the function φ can be chosen to be zero and therefore for such sets the relation

$$\bar{\mu}(F_1 \cup F_2) = \bar{\mu}(F_1) + \bar{\mu}(F_2).$$

holds even if they do not belong to $\mathfrak{B}_0$.

We now show that $\bar{\mu}$ can be extended as a measure on $\sigma(\mathfrak{B}_0)$. For this purpose it is sufficient to prove that for any sequence $\mathcal{G}_n$ of decreasing sets in $\mathfrak{B}_0$ such that $\bigcap \mathcal{G}_n = \emptyset$, $\bar{\mu}(\mathcal{G}_n) \to 0$ as $n \to \infty$. Assuming that this is not the case one can find a sequence of sets $\mathcal{G}_n \in \mathfrak{B}_0$ such that $\bar{\mu}(\mathcal{G}_n) > \delta > 0$ and $\bigcap \mathcal{G}_n = \emptyset$. We note that for any $\mathcal{G} \in \mathfrak{B}_0$ and any $\varepsilon > 0$ there exists a closed set $F \in \mathfrak{B}_0$ such that $F \subset \mathcal{G}$ and $\bar{\mu}(\mathcal{G}) \leqslant \bar{\mu}(F) + \varepsilon$.

Indeed, let $f(x)$ be a continuous function and let $F_c = \{x : f(x) = c\}$. Then for all c, except possibly a finite number of them, $\bar{\mu}(F_c) = 0$ since for different $c_1, \ldots, c_l$ the sets $F_{c_1}, \ldots, F_{c_l}$ are closed and are pairwise disjoint and hence $\sum \bar{\mu}(F_{c_i}) = \bar{\mu}(\bigcup_i F_{c_i}) \leqslant \bar{\mu}(\mathscr{X})$.

Let $f(x)$ be the function defined as $f(x) = 1$ for $x \in [\mathscr{X} - \mathcal{G}]$, $f(x) < 1$ for $x \notin [\mathscr{X} - \mathcal{G}]$ and $\bar{\mu}(\mathscr{X} - \mathcal{G}) \geqslant L(f) - \frac{\varepsilon}{2}$.

Let $\lambda < 1$ be a number such that $\bar{\mu}(F_\lambda) = 0$. Denote by S the set $\{x : f(x) > \lambda\}$. Then

$$\bar{\mu}(S) \leqslant L\left(\frac{1}{\lambda} f\right) = \frac{1}{\lambda} L(f) \leqslant \frac{1}{\lambda} \bar{\mu}(\mathscr{X} - \mathcal{G}) + \frac{\varepsilon}{2\lambda}.$$

Since $S \in \mathfrak{B}_0$, the set $\mathscr{X} - S = \{x : f(x) \leqslant \lambda\}$ is a closed set belonging to $\mathfrak{B}_0$ and

$$\bar{\mu}(\mathscr{X} - S) = \bar{\mu}(S) \geqslant \bar{\mu}(\mathscr{X}) - \frac{1}{\lambda} \bar{\mu}(\mathscr{X} - \mathcal{G}) - \frac{\varepsilon}{2\lambda} = \bar{\mu}(\mathcal{G}) - \frac{1-\lambda}{\lambda} \bar{\mu}(\mathscr{X} - \mathcal{G}) - \frac{\varepsilon}{2\lambda}.$$

It remains to choose λ close to 1 in such a manner that $\frac{1-\lambda}{\lambda} \bar{\mu}(\mathscr{X} - \mathcal{G}) + \frac{\varepsilon}{2\lambda}$ be less than ε.

Now let $\tilde{F}_k$ be closed sets in $\mathfrak{B}_0$ such that $\tilde{F}_k \subset \mathcal{G}_k$ and $\bar{\mu}(\mathcal{G}_k) \leqslant \bar{\mu}(\tilde{F}_k) + \frac{\delta}{2^{k+1}}$. Set $F_n = \bigcap_{k=1}^{n} \tilde{F}_k$. Then

$$\bar{\mu}(F_n) \geqslant \bar{\mu}(\mathcal{G}_n) - \sum_{k=1}^{n} \bar{\mu}(\mathcal{G}_k - \tilde{F}_k) \geqslant \delta - \sum_{k=1}^{n} \delta \frac{1}{2^{k+1}} > \frac{\delta}{2}.$$

Therefore a decreasing sequence F_n of closed sets belonging to $\mathfrak{B}_0$ has been constructed such that $\bar{\mu}(F_n) \geqslant \frac{\delta}{2}$ and $\bigcap_{n=1}^{\infty} F_n = \emptyset$. Using condition 3) we now choose a compactum K such that for all $f(x)$, satisfying $f(x) = 0$

for $x \in K$, $L(f) \leqslant \frac{\delta}{4}\|f\|$. We show that the intersection of F_n with K is non-void for all n. Indeed, if $F_n \cap K = \emptyset$, then a continuous function $g(x)$ can be constructed such that $0 \leqslant g(x) \leqslant 1$, $g(x)=1$ for $x \in F_n$ $g(x)=0$ for $x \in K$. Then $\bar{\mu}(F_n) \leqslant L(g) \leqslant \|g\| \frac{\delta}{4} = \frac{\delta}{4}$ which contradicts the construction of F_n. Hence the sequence of non-empty compacts sets $K_n = F_n \cap K$ satisfies conditions $K_n \supset K_{n+1}$ *and* $\cap K_n = \emptyset$ which is impossible. From the contradiction thus obtained the countable additivity of $\bar{\mu}$ on $\mathfrak{B}_0$ follows as well as the feasibility of its extension over $\sigma(\mathfrak{B}_0)$. Denote the measure obtained on $\sigma(\mathfrak{B}_0)$ by μ. We now recall that for almost all c the set $\{x: f(x) < c\} \in \mathfrak{B}_0$ for an arbitrary continuous function f. Therefore $\{x: f(x) < c\} \in \sigma(\mathfrak{B}_0)$ for all c provided $f \in \mathscr{C}$. We thus obtain that $\sigma(\mathfrak{B}_0)$ contains $\mathfrak{B}$. Finally we show that equality (1) is satisfied. Let $0 \leqslant f(x) \leqslant 1$ and $c_0 < 0 < c_1 < \ldots$ $\ldots < c_{n-1} < 1 < c_n$ be such that the sets $\{x: f(x) = c_i\} \in \mathfrak{B}_0$. Then for any $\varepsilon > 0$ continuous functions $\varphi_k(x) \geqslant 0$ and $\varphi_k(x) = 1$ for $x \in E_k = \{x: c_k < f(x) < c_{k+1}\}$, $k = 0, \ldots, n-1$, can be found such that $L(\varphi_k) < \mu(E_k) + \frac{\varepsilon}{n}$. Therefore

$$L(f) \leqslant L\left(\sum_{k=0}^{n-1} c_{k+1}\varphi_k\right) < \sum_{k=1}^{n-1} c_{k+1}\mu(E_k) + \varepsilon \leqslant$$

$$\leqslant \int f(x)\,\mu(dx) + \varepsilon + \max_k (c_{k+1} - c_k).$$

Since $\varepsilon + \max_k (c_{k+1} - c_k)$ can be chosen arbitrarily small, it follows that

$$L(f) \leqslant \int f(x)\,\mu(dx).$$

Analogously,

$$L(1) - L(f) = L(1-f) \leqslant \mu(\mathscr{X}) - \int f(x)\,\mu(dx).$$

Since $L(1) = \mu(\mathscr{X})$, we have $-L(f) \leqslant -\int f(x)\,\mu(dx)$.

Thus equality (1) and the theorem are proved. □

Remark 1. It follows from condition 3) of the theorem that for any finite measure μ on $\mathfrak{B}$ and for any $\varepsilon > 0$, one can find a compactum K such that $\mu(\mathscr{X} - K) < \varepsilon$.

Remark 2. It is easy to see that the completeness was not used in the proof of the sufficiency part of the theorem. However completeness is essential in the proof of necessity. In the case when the space $\mathscr{X}$ is a Borel set in its completion $\bar{\mathscr{X}}$ the conditions of the theorem are also necessary. Then for any $\varepsilon>0$ a compactum K can be constructed in $\bar{\mathscr{X}}$ such that $\mathscr{X}\supset K$ and $\mu(\bar{\mathscr{X}}-K)<\varepsilon$.

On the other hand if $\mathscr{X}$ is not complete then the proof of necessity of condition 3) as given above implies the existence of a compact set K in $\bar{\mathscr{X}}$ (or completely bounded K in $\mathscr{X}$) satisfying $\mu(\mathscr{X}-K)<\varepsilon$. Considering the functional $L(\varphi)$ only on those functions φ which can be extended by continuity to the whole space $\bar{\mathscr{X}}$ (these are the functions uniformly continuous on each completely bounded set K in $\mathscr{X}$) we can construct a measure $\bar{\mu}$ on $(\bar{\mathscr{X}}, \mathfrak{B})$ where $\mathfrak{B}$ is a σ-algebra of Borel sets in the space $\mathscr{X}$. If it turns out that $\mathscr{X}$ viewed as a subset of $\bar{\mathscr{X}}$ has an outer measure which coincides with the measure of the space $\bar{\mathscr{X}}$ (i.e. with $L(1)$), then the measure $\bar{\mu}$ can be transferred onto $\bar{\mathscr{X}}$ as was indicated in the previous section. In order that the outer measure of the space $\mathscr{X}$ be equal to $L(1)$ it is sufficient that for any sequence of non-negative continuous functions φ_n such that $\sum_1^\infty \varphi_n(x)\geqslant 1$, and for each x the inequality $\sum_1^\infty L(\varphi_n)\geqslant L(1)$ be satisfied. The latter assertion follows from the fact that the outer measure of any set A in $\bar{\mathscr{X}}$ can be defined as $\inf \sum_n L(\varphi_n)$, where the infimum is taken over all sequences of non-negative continuous functions on $\bar{\mathscr{X}}$ such that

$$\sum \varphi_n(x)\geqslant 1, \quad x\in A.$$

The stated condition is equivalent to the following: for any monotonic sequence of non-negative continuous functions φ_n in $\mathscr{C}$ such that $\varphi_n(x)\downarrow 0$ as $n\to\infty$ for all x, we have $\lim_{n\to\infty} L(\varphi_n)=0$. It follows from the Lebesgue theorem on monotone convergence that this condition is also necessary. Therefore we have for the case of an incomplete space the following

Theorem 2. *In order that a functional $L(f)$ defined on the space $\mathscr{C}$ of continuous functions given on a metric separable space $\mathscr{X}$ admit representation* (1), *where μ is a finite measure on $\mathscr{X}$ it is necessary and sufficient that conditions*

1) $L(f)\geqslant 0$ *for all* $f\geqslant 0$;

2) $L(c_1 f_1+c_2 f_2)=c_1 L(f_1)+c_2 L(f_2)$ *for all real c_1 and c_2 and f_1, $f_2\in\mathscr{C}$*;

3) *for any decreasing sequence of non-negative functions $\varphi_n\in\mathscr{C}$ such that $\varphi_n(x)\to 0$ for all x, $L(\varphi_n)\to 0$,*

4) *for any $\varepsilon>0$ one can find a completely bounded set K such that $|L(f)|\leqslant\varepsilon\,\|f\|$ for all $f\in\mathscr{C}$ satisfying $f(x)=0$ for $x\in K$.*

In the conclusion of this section we consider the problem of determining integrals of continuous functions in the case when $\mathscr{X}$ is the space of continuous functions on $[a, b]$ with the metric $\varrho(x, y) = \sup_t |x(t) - y(t)|$ and the measure μ is a measure associated with a certain random process. We shall assume that partial distributions of this random process are known and let $F_{t_1, \ldots, t_k}(dx_1, \ldots, dx_k)$ be the joint distribution of the values of the process at points $t_1, \ldots, t_k$. Denote by α a certain subdivision $\{a = t_0^{(\alpha)} < t_1^{(\alpha)} < \ldots < t_n^{(\alpha)} = b\}$ of the segment $[a, b]$, $|\alpha| = \max_k |t_{k+1}^{(\alpha)} - t_k^{(\alpha)}|$. Let $x(\cdot) \in \mathscr{X}$. Set

$$x_\alpha(t) = x(t_k^{(\alpha)}) + \frac{t - t_k^{(\alpha)}}{t_{k+1}^{(\alpha)} - t_k^{(\alpha)}} [x(t_{k+1}^{(\alpha)}) - x(t_k^{(\alpha)})]$$

for

$$t_k^{(\alpha)} \leqslant t \leqslant t_{k+1}^{(\alpha)}.$$

Clearly, $x_\alpha(t)$ is a piecewise-linear function which coincides with $x(t)$ at the points of subdivision α. If $x(\cdot) \in \mathscr{X}$, then $\varrho(x(\cdot), x_\alpha(\cdot)) \to 0$ as $|\alpha| \to 0$. Let f be a certain continuous functional. Then for all $x \in \mathscr{X}$

$$f(x) = \lim_{|\alpha| \to 0} f(x_\alpha).$$

Denote $f(x_\alpha) = f_\alpha(x)$. The functional $f_\alpha(x)$ is a continuous cylindrical functional. If $\|f\| < \infty$, then $\|f_\alpha\| \leqslant \|f\|$ and in view of Lebesgue's theorem on bounded convergence

$$\int f(x)\, \mu(dx) = \lim_{|\alpha| \to 0} \int f_\alpha(x)\, \mu(dx). \tag{2}$$

Functional $f_\alpha(x)$ is of the form $\varphi_\alpha(x(t_0^{(\alpha)}), \ldots, x(t_n^{(\alpha)}))$. Therefore the integral in the r.h.s. of (2) can be evaluated by means of finite dimensional distributions:

$$\int f_\alpha(x)\, \mu(dx) = \int \varphi_\alpha(x_0, \ldots, x_n)\, F_{t_0^{(\alpha)}, \ldots, t_n^{(\alpha)}}(dx_0, \ldots, dx_n). \tag{3}$$

Formulas (2) and (3) allow us to define integrals of continuous functionals.

§3. Measures on Linear Spaces. Characteristic Functionals

Let $\mathscr{X}$ be the real line. Then the space $\mathscr{F}$ of all functions $x(t)$ defined on a certain set T and taking on values in $\mathscr{X}$ is a linear real space. Denote by L the space of all linear functionals l on $\mathscr{F}$ of the form

$$l(x) = \sum_{k=1}^{n} c_k x(t_k), \tag{1}$$

where n is an arbitrary positive integer, $\{t_1, \ldots, t_n\}$ is a set of points in the domain of the definition of the process, c_k are real numbers and the σ-algebra $\mathfrak{B}$ defined in Section 1 coincides with the minimal σ-algebra with respect to which all the functionals in $\mathscr{L}$ are measurable. The measure μ on $\mathfrak{B}$ is completely determined by its values on the sets of the form $\{x : l(x) < \alpha\}$ for all possible l. This follows from the fact that knowing the values of measure μ on these sets one can evaluate the integral

$$\int e^{il(x)} \mu(dx) = \mathsf{E} \exp\left\{i \sum_{k=1}^{n} c_k \xi(t_k)\right\}, \tag{2}$$

where $\xi(t)$ is the random process associated with measure μ; consequently, one can compute the joint characteristic function of the variables $\xi(t_1), \ldots, \xi(t_n)$ which will allow us to obtain the joint distribution of $\xi(t_1), \ldots, \xi(t_n)$ for any selection of the values of the argument. Thus if integral (2) is known, one can determine the marginal distributions of the process $\xi(t)$, which in turn completely determine the measure μ.

In the case when measure μ is transferred from $\mathscr{F}$ into a smaller space $\mathscr{F}_0$, the space $\mathscr{F}_0$ often turns out to be a linear space. At least linear functionals of form (1) are defined on this space and the σ-algebra of $\mathfrak{B}_0$-measurable sets in $\mathscr{F}_0$ will also coincide with the minimal σ-algebra with respect to which all the functionals of form (1) are measurable. It follows from the above that in order to define a measure in such a case, it is sufficient to know the distribution of all linear functionals on $\mathscr{F}_0$. The specific form of the space is often irrelevant when studying measures on various linear functional spaces. Therefore the following scheme will be adopted. Let $\mathscr{X}$ be an arbitrary linear space (over the field of reals) and $\mathscr{L}$ be a linear set of linear functionals $l(x)$ defined on $\mathscr{X}$. Denote by $\mathfrak{B}$ the minimal σ-algebra with respect to which all the functions $l(x) \in \mathscr{L}$ are measurable. We shall consider probability measures μ on $\mathscr{X}$. *The measure μ is completely determined by its characteristic functional*

$$\chi(l) = \int e^{il(x)} \mu(dx). \tag{3}$$

We now prove this assertion. We shall call any set of the form

$$\{x : l_1(x) \in A_1, \ldots, l_n(x) \in A_n\},$$

where n is a positive integer, $l_1, \ldots, l_n$ are functionals in $\mathscr{L}$ and $A_1, \ldots, A_n$ are Borel sets on the line, a *cylindrical set* in $\mathscr{X}$. Let $\mathfrak{S}_0$ be the algebra of all cylindrical sets. Clearly, every functional $l \in \mathscr{L}$ is measurable with respect to $\mathfrak{S}_0$ so that $\sigma(\mathfrak{S}_0) = \mathfrak{B}$. Therefore it is sufficient to define the measure μ on $\mathfrak{S}_0$. If the functional $\varphi(x)$ defined on $\mathscr{X}$ is of the form $\varphi(x) = g(l_1(x), \ldots, l_n(x))$ where n is an integer, $l_1, l_2, \ldots, l_n \in \mathscr{L}$ and $g(s_1, \ldots, s_n)$

is a Borel function of n variables, then we call φ a cylindrical function and term it continuous cylindrical if φ is continuous. To define measure μ on $\mathfrak{S}_0$ it is sufficient to know the integrals $\int \varphi(x)\,\mu(dx)$ for continuous bounded cylindrical functions φ, but these functions are limits of an everywhere convergent jointly bounded sequence of trigonometric polynomials of the form

$$T(x)=\sum_{k=1}^{N} c_k \exp\left\{i \sum_{j=1}^{N} \lambda_{kj} l_j(x)\right\}. \tag{4}$$

It remains to remark that formula (3) determines the values of the integrals of function $T(x)$ of the form (4)

$$\int T(x)\,\mu(dx)=\sum_{k=1}^{n} c_k \chi\left(\sum_{j=1}^{n} \lambda_{kj} l_j\right). \quad \square$$

We now investigate the degree of arbitrariness of a characteristic functional $\chi(l)$:

1) A characteristic functional should be positive definite: for any $l_1, \ldots, l_n$ belonging to $\mathscr{L}$ and any complex numbers $\alpha_1, \ldots, \alpha_n$

$$\sum_{k,j=1}^{n} \chi(l_k-l_j)\,\alpha_k \bar{\alpha}_j \geqslant 0, \tag{5}$$

since

$$\sum_{k,j=1}^{n} \chi(l_k-l_j)\,\alpha_k \bar{\alpha}_j=\int \left|\sum_{k=1}^{n} \alpha_k\, e^{il(x)}\right|^2 \mu(dx).$$

2) Moreover, the functional $\chi(l)$ must be continuous in the following sense: define $l_n \to l$ if $l_n(x) \to l(x)$ for all $x \in \mathscr{X}$, then $\chi(l_n) \to \chi(l)$ as $l_n \to l$.

Now let a functional $\chi(l)$, which is positive-definite and continuous in the above sense, be defined on $\mathscr{L}$. Are these properties sufficient for the existence of a measure μ such that formula (3) is satisfied? Note that for any $l_1, l_2, \ldots, l_n \in \mathscr{L}$ the function $\varphi(s_1, \ldots, s_n)=\chi\left(\sum_{k=1}^{n} s_k l_k\right)$ is a characteristic functional in arguments $s_1, s_2, \ldots, s_n$. Therefore a distribution $\mathsf{P}_{l_1, \ldots, l_n}(du_1, \ldots, du_n)$ in the n-dimensional space exists such that

$$\varphi(s_1, \ldots, s_n)=\int e^{i\sum s_k u_k}\, \mathsf{P}_{l_1, \ldots, l_n}(du_1, \ldots, du_n).$$

Define a set function by the relation

$$\mu(\{x: l_1(x) \in A_1, \ldots, l_n(x) \in A_n\})=\int_{A_1} \cdots \int_{A_n} \mathsf{P}_{l_1, \ldots, l_n}(du_1, \ldots, du_n).$$

It is easy to show that when the same cylindrical set is expressed in several different ways we obtain the same expression for the function μ. The function μ will be additive on $\mathfrak{S}_0$ and can be extended to a countably additive function on each of the σ-algebras $\mathfrak{S}_0^{l_1,\ldots,l_n}$ – the minimal σ-algebra with respect to which the functionals $l_1,\ldots,l_n$ are measurable. Therefore, for any bounded Borel function $g(u_1,\ldots,u_n)$ and any $l_1,\ldots,l_n$ one can define the integral

$$\int g(l_1(x),\ldots,l_n(x)\,\mu(dx).$$

In particular,

$$\int e^{il(x)}\,\mu(dx)=\chi(l).$$

Therefore, one can always construct, given $\chi(l)$, a finite-additive set function μ which is countably additive on each σ-algebra $\mathfrak{S}_0^{l_1,\ldots,l_n}$ such that equality (3) will be satisfied. Simple examples (to be given in Section 6) show that μ is not always countably-additive on and hence it cannot always be extended to a measure defined on $\mathfrak{B}$. However, one can always construct a certain extension $\tilde{\mathscr{X}}$ of the space $\mathscr{X}$ in which such a measure μ exists; moreover $\tilde{\mathscr{X}}$ will also be linear and functionals in $\mathscr{L}$ can be extended on $\tilde{\mathscr{X}}$ in such a manner that they will be linear on $\tilde{\mathscr{X}}$. We shall show how this can be accomplished.

Let $\mathscr{F}_{\mathscr{L}}$ denote the space of all numerical functions $\varphi(l)$ defined on $\mathscr{L}$ (these functions may admit also infinite values but of one definite (fixed) sign). Define a real random function $\xi(l)$ on $\mathscr{L}$ such that for any choice of $l_1,\ldots,l_n$ in $\mathscr{L}$ the joint distribution of the variables $\xi(l_1),\ldots,\xi(l_n)$ will be given by the following characteristic function:

$$\mathsf{E}\exp\left\{i\sum_{k=1}^{n}\lambda_k\xi(l_k)\right\}=\chi\left(\sum_{k=1}^{n}\lambda_k l_k\right).$$

It is easy to verify the compatability of the corresponding distributions so that the existence of the random function $\xi(l)$ follows from Theorem 2, Section 4, Chapter I. Let $\bar{\mu}$ be a measure on $\mathscr{F}_{\mathscr{L}}$, corresponding to $\xi(l)$. We transfer the measure $\bar{\mu}$ to a smaller space.

Denote by $A_{\mathscr{L}}$ the set of all linear functions $\lambda(l)$ on $\mathscr{L}$ such that

$$\lambda(c_1l_1+c_2l_2)=c_1\lambda(l_1)+c_2\lambda(l_2)$$

for all l_1, $l_2\in\mathscr{L}$ and all real c_1 and c_2.

We show that the outer measure of the set $A_{\mathscr{L}}$ is equal to 1.

Let S_n be an arbitrary decreasing sequence of cylindrical sets in $\mathscr{F}_{\mathscr{L}}$

for which $\bigcap_{n=1}^{\infty} S_n \cap A_{\mathscr{L}}$ is void. Without loss of generality one can assume that the sets S_n are determined by the values of the function φ at the points $l_1, \ldots, l_n$, where $\{l_k, k=1, 2, \ldots\}$ is a sequence of functionals. We shall write in order all the linear relations satisfied by the functionals:

$$\sum_{k=1}^{n} c_{nk} l_k = 0$$

(if l_n is linearly independent of $l_1, \ldots, l_{n-1}$, then the coefficient $c_{nn}=0$, otherwise $c_{nn} \neq 0$).

Let $D_n = \{\varphi: \sum c_{nk} \varphi(l_k) = 0\}$. Then

$$\emptyset = \bigcap_{n=1}^{\infty} S_n \cap A_{\mathscr{L}} = \bigcap_{n=1}^{\infty} (S_n \cap D_1 \cap \ldots \cap D_n).$$

Since $S_n \cap D_1 \cap \ldots \cap D_n$ is a decreasing sequence of cylindrical sets it follows from the relation

$$\emptyset = \bigcap_{n=1}^{\infty} [S_n \cap D_1 \cap \ldots \cap D_n]$$

that: $S_n \cap D_1 \cap \ldots \cap D_n \to \emptyset$ as $n \to \infty$. Finally note that $\tilde{\mu}(D_n) = 1$ for all n and therefore $\tilde{\mu}(S_n \cap D_1 \cap \ldots \cap D_n) = \tilde{\mu}(S_n)$ and hence $\tilde{\mu}(S_n) \to 0$. This means that the outer measure of $A_{\mathscr{L}}$ is 1. Hence the measure $\tilde{\mu}$ can be transferred onto $A_{\mathscr{L}}$. Next let X_0 be a linear manifold in $\mathscr{X}$, for which $l(x)=0$ for all $x \in X_0$, $l \in \mathscr{L}$ and X^1 be the quotient group of $\mathscr{X}$ by X_0. Each element $x^1 \in X^1$ can be considered as a linear functional on $\mathscr{L}: x^1(l) = l(x)$ where x is any representative of the residue class of x^1 modulo X_0. Denote by $\tilde{\mathscr{X}}$ the set of pairs $\tilde{x} = (x; \lambda)$ where $x \in X_0$, $\lambda \in A_{\mathscr{L}}$. Let P be a linear operator which maps $\mathscr{X}$ into X_0 such that $Px = x$ for all $x \in X_0$, and $x^1(x)$ denotes the residue class in X^1 to which x belongs. Then there exists a natural embedding of $\mathscr{X}$ into $\tilde{\mathscr{X}}$:

$$x \to (Px, x^1(x)).$$

Define on $\tilde{\mathscr{X}}$ a σ-algebra $\tilde{\mathfrak{B}}$ of sets of the form

$$\tilde{\mathscr{G}} = \{\tilde{x} = (x; \lambda): \lambda \in \mathscr{G}\},$$

where $\mathscr{G}$ is an arbitrary subset in $\mathfrak{A}$ and $\mathfrak{A}$ is the σ-algebra of subsets in $A_{\mathscr{L}}$ on which the measure $\tilde{\mu}$ is defined. We next set $\mu(\tilde{\mathscr{G}}) = \tilde{\mu}(\mathscr{G})$ and show that this is the required measure. Note that functionals l can be defined on $\tilde{\mathscr{X}}$ by the formula

$$l(\tilde{x}) = l((x; \lambda)) = \lambda(l).$$

This functional is linear and it coincides with $l(x)$ on $\mathscr{X}$ viewed as a subset of $\tilde{\mathscr{X}}$: $l(x)=l((Px;x^1(x)))$, since $x^1(x)$ as an element of $A_{\mathscr{L}}$ is determined by the formula $x^1(x)(l)=l(x)$. Finally consider the integral (3). From the construction of measure $\tilde{\mu}$ on cylindrical sets we have:

$$\int e^{il(\tilde{x})}\,\mu(d\tilde{x})=\int e^{i\lambda(l)}\,\tilde{\mu}(d\lambda)=\mathsf{E}\,e^{i\xi(l)}=\chi(l).$$

Thus μ is the required measure.

It is especially simple to construct $\tilde{\mathscr{X}}$ in the case when X_0 is the singleton $\{0\}$. This will be the case if the set of functionals $\mathscr{L}$ is so large that for each pair $x_1\neq x_2$ in $\mathscr{X}$ a functional l can be found such that $l(x_1)\neq$ $\neq l(x_2)$. In this case the space $A_{\mathscr{L}}$ can be chosen for $\mathscr{X}$, where each element $x\in\mathscr{X}$ determines an element in $A_{\mathscr{L}}$ by means of the formula

$$x(l)=l(x).$$

Clearly, every measure on $\mathscr{X}$ determines a characteristic functional $\chi(l)$ on $\mathscr{L}$ however the functional is not necessarily continuous in the sense indicated in condition 2). In order that condition 2) be satisfied it is necessary and sufficient that measure $\tilde{\mu}$ possess the following property: for any sequence l_n such that $l_n(x)\to 0$ for all $x\in\mathscr{X}$, $l_n(\tilde{x})\to 0$ in measure $\tilde{\mu}$. Note also that if $\mathscr{X}$ and $\mathscr{L}$ are chosen in such a manner that the space $A_{\mathscr{L}}$ coincides with $\mathscr{X}$ then the constructed measure will be a measure on $\mathscr{X}$.

Now let $\mathscr{X}$ be a linear normed complete separable space. It is natural to choose for $\mathscr{L}$ the space of all continuous linear functionals $\mathscr{X}^*$ (the elements of which will be denoted by x^*). The minimal σ-algebra with respect to which all the functionals $x^*(x)$ are measurable, coincides with the σ-algebra $\mathfrak{B}$ of all Borel sets in $\mathscr{X}$. Every probability measure μ on $\mathfrak{B}$ is determined by its characteristic functional

$$\chi(x^*)=\int e^{ix^*(x)}\,\mu(dx).$$

This characteristic functional will be positive definite and weakly continuous on $\mathscr{X}^*$. If a functional $\chi(x^*)$ possessing these properties is given, one can then construct a finite-additive measure μ on the algebra $\mathfrak{S}_0$ of all cylindrical sets. In the case when $\mathscr{X}^*$ is a separable space we shall find a necessary and sufficient condition for this measure to be extended to a countably-additive one on $\mathfrak{B}$. The requirement is that $\mu(S_n)\to 0$ for any sequence S_n of cylindrical sets satisfying the condition $\bigcap_n S_n=\emptyset$ and $S_n\supset S_{n+1}$. Let the set S_n be of the form $\{x:(l_1(x),\dots,l_n(x))\in A^n\}$, where A^n is a Borel set in $\mathscr{R}^n$. We call S_n closed if A^n is closed in $\mathscr{R}^n$. It turns out that it is sufficient to verify the continuity only on closed S_n since for any

$\varepsilon_n > 0$ one can find a closed $F^n \subset A^n$ such that

$$\mu(S_n) - \mu(\{x: (l_1(x), \ldots, l_n(x)) \in F^n\}) < \varepsilon_n.$$

Utilizing this observation we find the condition for the existence of measure μ on $\mathfrak{B}$ with a given characteristic functional.

Let $\{x_k^*(x), k=1,2,\ldots\}$ be an everywhere dense set on the unit sphere of the space $\mathscr{X}^*$. Since $|x| = \sup_k x_k^*(x)$, the relation

$$0 = \lim_{N\to\infty} \mu(\{x: |x| > N\}) = \lim_{N\to\infty} \lim_{n\to\infty} \mu(\{x: \sup_{k\leqslant n} x_k^*(x) > N\})$$

is satisfied, provided μ is a countably-additive measure. Therefore in order that μ be countably-additive it is necessary that condition

$$\lim_{N\to\infty} \lim_{n\to\infty} \mu(\{x: \sup_{k\leqslant n} x_k^*(x) > N\}) = 0 \tag{6}$$

be satisfied.

We now show that this condition is also sufficient. Let μ be a additive measure on cylindrical sets constructed by means of χ. It follows from (6) that for any $\varepsilon > 0$ an N can be found such that for all n

$$\mu(\{x: x_k^*(x) \leqslant N, k=1, \ldots, n\}) \geqslant 1 - \varepsilon.$$

Assume that for a certain sequence of closed cylindrical sets $S_n = \{x: (\tilde{x}_1^*(x), \ldots, \tilde{x}_n^*(x)) \in F^n\}$, $S_n \supset S_{n+1}$, the relation $\mu(S_n) \geqslant 2\varepsilon$ is satisfied. We show that $\cap S_n$ is not void. Let

$$K_N = \{x: |x| \leqslant N\}, \quad K_N^n = \{x: x_k^*(x) \leqslant N, k=1, \ldots, n\}.$$

The intersection $S_n \cap K_N$ is not void. Indeed if $S_n \cap K_N$ is void then $d = \inf_{x\in S_n} |x| > N$ (since the infimum is achieved). Set

$$g(u_1, \ldots, u_n) = \inf\{|x|: \tilde{x}_1^*(x) = u_1, \ldots, \tilde{x}_n^*(x) = u_n\}.$$

Then the set S_n is contained in the set $\{x: g(\tilde{x}_1^*(x), \ldots, \tilde{x}_n^*(x)) \geqslant d\}$. The set $\{(u_1, \ldots, u_n): N < g(u_1, \ldots, u_n) < d\}$ is open in $\mathscr{R}^n$ and is the difference of two simply connected regions. Therefore a polyhedron exists with faces of the form $\left\{(u_1, \ldots, u_n): \sum_1^n r_{ij} u_i = b_j\right\}, j=1, \ldots, m$, such that the set $\{(u_1, \ldots, u_n): g(u_1, \ldots, u_n) \leqslant N\}$ is totally contained within the polyhedron and the set $\{(u_1, \ldots, u_n): g(u_1, \ldots, u_n) \geqslant d\}$ outside of this polyhedron. Then the set S_n is contained entirely in the union of the sets

$$\bigcup_{j=1}^m \left\{x: \sum_1^n r_{ij} \tilde{x}_i^*(x) > b_j\right\},$$

and each one of the summands is disjoint of K_N. Denote

$$y_j^*(x)=\sum_1^n r_{ij}\tilde{x}_i^*(x).$$

The set $\{x: y_j^*(x)\geqslant b_j\}$ is disjoint of K_N if and only if $b_j>\|y_j^*\|\,N$.

Let $\frac{y_j^*}{\|y_j^*\|}=z_j^*$. Then S_n is entirely contained in the set $\bigcup_{j=1}^m \{x: z_j^*(x)\geqslant \geqslant N+\delta\}$, where $\delta=\inf_j \frac{1}{\|y_j^*\|}\, b_j-N>0$.

It follows from the continuity of $\chi(x^*)$ that for almost all $\alpha_1,\dots,\alpha_m$

$$\lim_{k\to\infty} \mu(\{x: z_{1,k}^*(x)<\alpha_1,\dots, z_{m,k}^*(x)<\alpha_m\})=$$
$$=\mu(\{z_1^*(x)<\alpha_1,\dots, z_m^*(x)<\alpha_m\}),$$

provided only $\|z_{j,k}^*-z_j^*\|\to 0$. We choose in the set $\{x_k^*, k=1,2,\dots\}$ a sequence $x_{j,k}^*, j=1,\dots,m$ such that $\|x_{j,k}^*-z_j^*\|\to 0$. Then

$$\mu(S_n)\leqslant\mu\left(\bigcup_{j=1}^m \{x: z_j^*(x)\geqslant N+\delta\}\right)=$$
$$=1-\mu(\{x:\sup_{j\leqslant m} z_j^*(x)<N+\delta\})\leqslant 1-\lim_{k\to\infty}\mu(\{x:\sup_{j\leqslant m} x_{j,k}^*(x)\leqslant N\})\leqslant\varepsilon.$$

This contradicts the inequality $\mu(S_n)\geqslant 2\varepsilon$. Thus $S_n\cap K_N$ is not void. Therefore the sequence of imbedded weakly closed sets $S_n\cap K_N$ belongs to the weakly compact set K_N and therefore $\bigcap_n \{S_n\cap K_N\}$ is non void and hence $\bigcap_n S_n$ is not void. Consequently, if S_n is a decreasing sequence of cylindrical sets such that $\bigcap_n S_n=\emptyset$, then $\lim_{n\to\infty}\mu(S_n)=0$ i.e. μ is countably additive. We have thus proved the following

Theorem. *In order that a continuous positive definite functional $\chi(x^*)$ be the characteristic function of a measure on $(\mathfrak{X},\mathfrak{B})$, where $\mathfrak{X}$ is a Banach space such that $\mathfrak{X}^*$ is separable, it is necessary and sufficient that a countably-additive measure μ generated by the functional $\chi(x^*)$ satisfy condition* (6) *for a certain set $\{x_k^*, k=1,\dots,n,\dots\}$ which is everywhere dense on the unit sphere of the space $\mathfrak{X}^*$.*

Remark. Condition (6) can be replaced by the following: let $h_n(x)$ be a sequence of continuous cylindrical functions such that $\lim_{n\to\infty} h_n(x)=|x|$. Then

$$\lim_{N\to\infty}\lim_{n\to\infty}\mu(\{x: h_n(x)>N\})=0. \tag{7}$$

Condition (7) becomes (6) if $h_n(x)=\sup_{k\leqslant n} x_k^*(x)$. Moreover (6) can be expressed in terms of $\chi(x^*)$ using the inversion formula.

§4. Measures in $\mathscr{L}_p$ Spaces

Spaces $\mathscr{L}_p[a, b]$ of real measurable functions $x(t)$ defined on $[a, b]$ such that

$$\int_a^b |x(t)|^p\, dt<\infty$$

serve as an important class of linear normed spaces. We shall consider only the case when $p\geqslant 1$. Let a certain probability space $\{\Omega, \mathfrak{S}, \mathsf{P}\}$ be fixed. We shall study the conditions under which a measure in $\mathscr{L}_p$ is associated with a given numerical process $\xi(t, \omega)$ defined on $[a, b]$. Assume that $\xi(t, \omega)$ is a measurable process. Then in view of Fubini's theorem $\xi(t, \omega)$ as a function of t is measurable with probability 1. Therefore with probability 1 the integral

$$\int_a^b |\xi(t, \omega)|^p\, dt$$

is defined (this integral may also admit infinite values). Moreover the integral is also a measurable function of ω.

Assume that

$$\mathsf{P}\left\{\int_a^b |\xi(t, \omega)|^p\, dt<\infty\right\}=1. \tag{1}$$

We show that under this condition a measure μ can be constructed in space $\mathscr{L}_p$ which is associated with the process $\xi(t, \omega)$, i.e. a measure μ such that for any Borel set B of the space $\mathscr{L}_p$

$$\mu(B)=\mathsf{P}(\{\omega:\xi(\cdot, \omega)\in B\}). \tag{2}$$

Relation (2) can be taken as the definition of measure μ provided we prove that $\{\omega:\xi(\cdot, \omega)\in B\}\in\mathfrak{S}$ for $B\in\mathfrak{B}$. To show this consider the class L of functionals defined on $\mathscr{L}_p$ of the form

$$l(x)=\int^b l(t)\, x(t)\, dt,$$

where $l(t)$ is a bounded measurable function. Functionals in L are defined on $\mathscr{L}_p$ for any $p \geqslant 1$ and L is dense in the space of linear continuous functionals for any p. Denote by $\mathfrak{B}_0$ the set of all those $B \in \mathfrak{B}$ such that $\{\omega : \xi(\cdot, \omega) \in B\} \in \mathfrak{S}$. Clearly $\mathfrak{B}_0$ is a σ-algebra. Since $l(\xi(\cdot, \omega))$ is measurable with respect to $\mathfrak{S}$ for any l in L, then for each continuous linear functional l on $\mathscr{L}_p$ the variable $l(\xi(\cdot, \omega))$ is $\mathfrak{S}$-measurable. Therefore $\mathfrak{B}_0$ coincides with the minimal σ-algebra with respect to which all continuous linear functionals $l(x)$ on $\mathscr{L}_p$ are measurable and this σ-algebra coincides with $\mathfrak{B}$. Thus relation (2) indeed defines a certain measure μ. Since for any continuous functional l on $\mathscr{L}_p$ the variable $l(\xi(\cdot))$ is measurable with respect to $\mathfrak{S}$, the measure μ can be given by means of the characteristic functional

$$\chi(l) = \int e^{il(x)} \mu(dx) = \mathsf{E}\, e^{il(\xi(\cdot, \omega))}. \tag{3}$$

This characteristic functional uniquely determines measure μ.

The question arises – how is the constructed measure μ connected with the marginal distributions of the process $\xi(t, \omega)$? In other words, given marginal distributions can one construct a measure μ and conversely given the measure μ can one determine the marginal distributions? We now show that for stochastically continuous processes the answer is positive.

Let $\xi(t, \omega)$ be a stochastically continuous measurable process (it follows from Theorem 1 of Section 3 in Chapter III that for a stochastically continuous process there always exists an equivalent measurable process). Set

$$\xi_N(t, \omega) = g_N(\xi(t, \omega)); \qquad g_N(x) = \begin{cases} x, & |x| \leqslant N, \\ N \operatorname{sign} x, & |x| > N. \end{cases}$$

Then

$$\int_a^b |\xi_N(t, \omega)|\, dt \to \int_a^b |\xi(t, \omega)|\, dt$$

as $N \to \infty$ almost for all ω. The process $\xi_N(t, \omega)$ is also stochastically continuous.

We show that

$$\int_a^b \xi_N(t, \omega)\, dt = \lim_{\lambda \to 0} \sum_{k=0}^{n-1} \xi_N(t_k, \omega)\, \Delta t_k, \tag{4}$$

where $a = t_0 < t_1 < \ldots < t_n = b$, $\Delta t_k = t_{k+1} - t_k$, $\lambda = \max_k \Delta t_k$, and the limit is taken in the sense of convergence in probability. We have

$$\mathsf{E}\left|\int_a^b \xi_N(t,\omega)\,dt-\sum_{k=0}^{n-1}\xi_N(t_k,\omega)\,\Delta t_k\right|\leqslant$$

$$\leqslant\sum_{k=0}^{n-1}\int_{t_k}^{t_{k+1}}\mathsf{E}\,|\xi_N(t,\omega)-\xi_N(t_k,\omega)|\,dt\leqslant$$

$$\leqslant\varepsilon(b-a)+\sum_{k=0}^{n-1}2N\int_{t_k}^{t_{k+1}}\mathsf{P}\{|\xi_N(t,\omega)-\xi_N(t_k,\omega)|>\varepsilon\}\,dt\leqslant$$

$$\leqslant\varepsilon(b-a)+2N(b-a)\sup\left[\mathsf{P}\{|\xi_N(t,\omega)-\xi_N(s,\omega)|>\varepsilon\};\,|t-s|\leqslant\lambda\right]$$

for any $\varepsilon>0$. Approaching the limit as $\lambda\to 0$ and taking the stochastic continuity of $\xi_N(t)$ into account (and hence the uniform stochastic continuity) as well as the fact that $\varepsilon>0$ is arbitrary we obtain the validity of equality (4).

Similarly,

$$\int_a^b |\xi(t,\omega)|^p\,dt=\lim_{N\to\infty}\lim_{\lambda\to 0}\sum_{k=0}^{n-1}|\xi_N(t_k,\omega)|^p\,\Delta t_k \tag{5}$$

holds and hence the fulfilment of condition (1) can be checked by means of the marginal distributions of the process $\xi(t,\omega)$. If condition (1) is satisfied, then, using equality

$$\int_a^b l(t)\,\xi(t,\omega)\,dt=\lim_{N\to\infty}\lim_{\lambda\to 0}\sum_{k=0}^{n-1}l(t_k)\,\xi_N(t_k,\omega)\,\Delta t_k, \tag{6}$$

which is valid for any continuous function $l(t)$ on $[a,b]$, we can define

$$\chi(l)=\lim_{N\to\infty}\lim_{\lambda\to 0}\exp\left\{i\sum_{k=0}^{n-1}l(t_k)\,\xi_N(t_k,\omega)\,\Delta t_k\right\} \tag{7}$$

for all continuous $l(t)$. In view of the continuity of $\chi(l)$ (7) determines the value of $\chi(l)$ on the closure of the set L of all the functionals

$$l(x)=\int^b l(t)\,x(t)\,dt$$

with continuous $l(t)$. And since L is everywhere dense in the space of all continuous functionals on $\mathscr{L}_p$ it means that $\chi(l)$ is completely determined by relation (7).

Assume now that for a measurable stochastically continuous process $\xi(t, \omega)$ on $[a, b]$ condition (1) is satisfied and the measure μ on $\mathscr{L}_p$ is defined or equivalently the characteristic functional $\chi(l)$ is defined. Let $l(t)$ be a certain continuous function, $N>0$ and n be a positive integer Set

$$t_{nk}=a+\frac{k}{n}(b-a),$$

$$I_{n,N}(l)=\sum_{k=0}^{n-1}\frac{b-a}{n}\,l(t_{nk})\,g_N\left(\frac{1}{t_{nk+1}-t_{nk}}\int_{t_{nk}}^{t_{nk+1}}\xi(t,\omega)\,dt\right).$$

For almost all ω

$$\lim_{n\to\infty} I_{n,N}(l)=\int_a^b l(t)\,\xi_N(t,\omega)\,dt.$$

Therefore for a characteristic functional $\chi_N(l)$ of the process $\xi_N(t, \omega)$ in $\mathscr{L}_p$ the following relation is valid

$$\chi_N(l)=\mathsf{E}\exp\left\{i\int_a^b l(t)\,\xi_N(t,\omega)\,dt\right\}=\lim_{n\to\infty}\int e^{iI_{n,N}(l,x)}\,\mu(dx),$$

where $l(t)$ is an arbitrary continuous function and

$$I_{n,N}(l,x)=\sum_{k=0}^{n-1}\frac{b-a}{n}\,l(t_{n,k})\,g_N\left(\frac{n}{b-a}\int_{t_{nk}}^{t_{nk+1}}x(t)\,dt\right)$$

is a continuous and therefore $\mathfrak{B}$-measurable functional on $\mathscr{L}_p$. By continuity the functional $\chi_N(l)$ can be extended on the whole L. We now show that

$$\frac{1}{h}\int_t^{t+h}\xi_N(t,\omega)\,dt\to\xi_N(t,\omega) \tag{8}$$

in probability as $h\to 0$. Indeed

$$\mathsf{E}\left|\frac{1}{h}\int_t^{t+h}(\xi_N(s,\omega)-\xi_N(t,\omega))\,dt\right| \leqslant$$

$$\leqslant \varepsilon + 2N\sup\left[\mathsf{P}\{|\xi_N(s,\omega)-\xi_N(t,\omega)|>\varepsilon\};\, t<s<t+h\right].$$

The last expression can be made arbitrarily small by the proper choice of $\varepsilon>0$ and $h>0$.

Denote by $l_{t_k,h}(x)$ the functional on $\mathscr{L}_p$ defined by the equality

$$l_{t_k,h}(x)=\frac{1}{h}\int_{t_k}^{t_k+h} x(t)\,dt.$$

Then it follows from (8) that for all real u_k and all points $t_1<\ldots<t_n$ in $[a,b)$

$$\mathsf{E}\exp\left\{i\sum_{k=1}^{n} u_k\xi_N(t_k,\omega)\right\}=\lim_{h\to 0}\chi_N\left(\sum_{k=1}^{n} u_k l_{t_k,h}(\xi_N(\cdot))\right). \tag{9}$$

Relation (9) determines the marginal distributions of the process $\xi_N(t,\omega)$. Approximating the limit as $N\to\infty$ we thus obtain the marginal distributions of the process $\xi(t,\omega)$. Formulas (5) and (7) are inconvenient because the truncated process $\xi_N(t,\omega)$ rather than the original process $\xi(t,\omega)$ appears in these formulas. Moreover, it is desirable to obtain a condition sufficient for the fulfilment of relation (1) in terms of the probabilistic characteristics of the random variables rather than in terms of the limit of random variables themselves. To obtain simpler statements regarding the integrability in the p-th degree of a process we need the following

Lemma. *Let $\xi(t)$ be a measurable stochastically continuous non-negative process defined on $[a,b]$. For any sequence of subdivisions of the interval $[a,b]$ $a=t_{n0}<\ldots<t_{nn}=b$ for which $\lambda_n=\max_k(t_{nk+1}-t_{nk})\to 0$ and any random variables τ_{nk} independent of $\xi(t)$ and uniformly distributed on the intervals $[t_{nk},t_{nk+1}]$ respectively, $k=0,1,\ldots,n-1$ the relation*

$$\sum_{k=0}^{n-1}\xi(\tau_{nk})\,\Delta t_{nk}\to\int_a^b\xi(t)\,dt$$

is satisfied in probability as $n\to\infty$ (the integral on the right may take the value $+\infty$).

Proof. Since the process $\xi(t)$ can be considered separately on each one of the sets

$$\left\{\omega: \int_a^b \xi(t)\,dt < \infty\right\} \quad \text{and} \quad \left\{\omega: \int_a^b \xi(t)\,dt = +\infty\right\},$$

it is sufficient to consider the two cases when only one of these conditions is satisfied with probability 1. First let

$$\mathsf{P}\left\{\int_a^b \xi(t)\,dt = +\infty\right\} = 1.$$

Since as it was shown above

$$\sum_{k=0}^{n-1} \xi_N(\tau_{nk})\,\Delta t_{nk} \to \int_a^b \xi_N(t)\,dt$$

in probability, and hence for each $c > 0$ the relation

$$\lim_{n\to\infty} \mathsf{P}\left\{\sum_{k=0}^{n-1} \xi_N(\tau_{nk})\Delta t_{nk} \geqslant c\right\} \geqslant$$

$$\geqslant \lim_{n\to\infty} \mathsf{P}\left\{\sum_{k=0}^{n-1} \xi_N(\tau_{nk})\,\Delta t_{nk} \geqslant c\right\} \geqslant \mathsf{P}\left\{\int_a^b \xi_N(t)\,dt > c\right\}$$

is satisfied.

Approaching the limit as $N \to \infty$ we verify that $\sum_{k=0}^{n-1} \xi(\tau_{nk})\,\Delta t_{nk}$ converges in probability to $+\infty$.

Now let

$$\mathsf{P}\left\{\int_a^b \xi(t)\,dt < \infty\right\} = 1.$$

Set

$$\xi^m(t) = \xi(t), \quad a \leqslant t \leqslant b, \quad \text{if} \quad \int_a^b \xi(t)\,dt \leqslant m,$$

$$\xi^m(t) = 0, \quad a \leqslant t \leqslant b, \quad \text{if} \quad \int \xi(t)\,dt > m.$$

The process $\xi^m(t)$ is non-negative, stochastically continuous and measurable.

Since

$$\int_a^b \xi^m(t)\,dt \leqslant m,$$

then in view of Fubini's theorem the expectation $\mathsf{E}\xi^m(t)$ exists for almost all t and $\int_a^b \mathsf{E}\xi^m(t)\,dt \leqslant m$. Set $\xi_N^m(t) = g_N(\xi^m(t))$. The process $\xi_N^m(t)$ is stochastically continuous and bounded by N. Hence for every $\varepsilon > 0$ an $h > 0$ can be found such that

$$\mathsf{E}|\xi_N^m(t) - \xi_N^m(s)| \leqslant \varepsilon$$

provided $|t-s| \leqslant h$.

Utilizing this fact, we obtain

$$\overline{\lim_{n\to\infty}}\, \mathsf{E}\left|\int_a^b \xi^m(t)\,dt - \sum_{k=0}^{n-1} \xi^m(\tau_{nk})\,\Delta t_{nk}\right| \leqslant$$

$$\leqslant \overline{\lim_{n\to\infty}} \sum_{k=0}^{n-1} \int_{t_{nk}}^{t_{nk+1}} \int_{t_{nk}}^{t_{nk+1}} \mathsf{E}|\xi^m(t) - \xi^m(s)|\,\frac{ds}{\Delta t_{nk}} \leqslant$$

$$\leqslant \overline{\lim_{n\to\infty}} \sum_{k=0}^{n-1} \frac{1}{\Delta t_{nk}} \int_{t_{nk}}^{t_{nk+1}} \int_{t_{nk}}^{t_{nk+1}} [\mathsf{E}|\xi_N^m(t) - \xi_N^m(s)| +$$

$$+ 2\mathsf{E}|\xi_N^m(t) - \xi^m(t)|]\,dt\,ds \leqslant$$

$$\leqslant \overline{\lim_{n\to\infty}} \sup_{|t-s|\leqslant\lambda_n} \mathsf{E}|\xi_N^m(t) - \xi_N^m(s)|\,(b-a) +$$

$$+ 2\int_a^b \mathsf{E}|\xi_N^m(t) - \xi^m(t)|\,dt = 2\int_a^b \mathsf{E}|\xi_N^m(t) - \xi^m(t)|\,dt.$$

The last expression tends to 0 as $N \to \infty$. Finally

$$\mathsf{P}\left\{\left|\int_a^b \xi(t)\,dt - \sum_{k=0}^{n-1} \xi(\tau_{nk})\,\Delta t_{nk}\right| > \varepsilon\right\} \leqslant$$

$$\leqslant \mathsf{P}\left\{\int_a^b \xi(t)\,dt > m\right\} + \frac{1}{\varepsilon}\,\mathsf{E}\left|\int_a^b \xi^m(t)\,dt - \sum_{k=0}^{n-1} \xi^m(\tau_{nk})\,\Delta t_{nk}\right|.$$

Approaching the limit first as $n\to\infty$ and then as $m\to\infty$ we obtain the proof of the lemma. □

Corollary. *Under the conditions of the lemma the non-random points* $s_{nk}\in[t_{nk}, t_{nk+1}]$ *exist, such that*

$$\sum_{k=0}^{n-1} \xi(s_{nk})\,\Delta t_{nk} \to \int_a^b \xi(t)\,dt$$

in probability as $n\to\infty$.

Remark. If for some sequence of subdivisions of the interval $[a, b]$ of the form given in the lemma and for some choice of points $s_{nk}\in[t_{nk}, t_{nk+1}]$, independent of $\xi(t)$ the quantity $\sum_{k=0}^{n-1} \xi(s_{nk})\,\Delta t_{nk}$ $(n\to\infty)$ is bounded in probability, then

$$\mathsf{P}\left\{\int_a^b \xi(t)\,dt < \infty\right\} = 1.$$

Indeed, for any $\varepsilon>0$ and $c>0$

$$\mathsf{P}\left\{\int_a^b \xi_N(t)\,dt > c\right\} \leqslant \varlimsup_{n\to\infty} \mathsf{P}\left\{\sum_{k=0}^{n-1} \xi_N(s_{nk})\,\Delta t_{nk} > c-\varepsilon\right\} \leqslant$$

$$\leqslant \varlimsup_{n\to\infty} \mathsf{P}\left\{\sum_{k=0}^{n-1} \xi(s_{nk})\,\Delta t_{nk} > c-\varepsilon\right\}.$$

Hence also

$$\mathsf{P}\left\{\int_a^b \xi(t)\,dt > c\right\} \leqslant \varlimsup_{n\to\infty} \mathsf{P}\left\{\sum_{k=0}^{n-1} \xi(s_{nk})\,\Delta t_{nk} > c-\varepsilon\right\}. \quad \square$$

We now present a condition necessary and sufficient for finiteness of the integral $\int_a^b |\xi(t)|^\alpha\,dt$ with $\alpha\in(0, 2]$ expressed in terms of the characteristic functional of the process. Since we don't know as yet in which space the process can be considered we utilize the characteristic functional determined by the relation

$$\chi_0(g) = \mathsf{E} \exp\left\{i \int_a^b \xi(t)\, dg(t)\right\}$$

for any step function $g(t)$ defined on $[a, b]$. Clearly defining $\chi_0(g)$ is equivalent to defining the marginal distributions of the process.

We construct a random function $v_n^\alpha(t)$ defined on $[a, b]$ as follows. Let $t_{nk} = a + \frac{k}{n}(b-a)$, $k = 0, \ldots, n$, $\eta_0, \eta_1, \ldots, \eta_n, \ldots$ be random variables independent of $\xi(t)$ each one of which is uniformly distributed on $[0, 1]$ (otherwise, the joint distribution of η_k can be arbitrary). Finally let the variables $\zeta_0, \zeta_1, \ldots$ which depend neither on $\xi(t)$ nor on $\eta_0, \eta_1, \ldots$, be independent and identically distributed and moreover $\mathsf{E}\, e^{is\zeta_0} = e^{-|s|^\alpha}$, i.e. ζ_k has a symmetric stable distribution with index α. Set $v_n^\alpha(a) = 0$, $v_n^\alpha(t)$ is constant for $(t-a)\, n \in [(j+\eta_j)(b-a), (j+1+\eta_{j+1})(b-a)]$ and

$$v_n^\alpha\left(a + \frac{j+\eta_j}{n}(b-a) + 0\right) - v_n^\alpha\left(a + \frac{j+\eta_j}{n}(b-a) - 0\right) = \zeta_j / n^{1/\alpha}.$$

These conditions uniquely determine $v_n^\alpha(t)$ (except at the discontinuity points). Moreover $v_n^\alpha(t)$ is a step function with probability 1 since expression $\chi_0(v_n^\alpha)$ is determined with probability 1.

Theorem. *In order that the integral* $\int_a^b |\xi(t)|^\alpha\, dt$ *(for some* $\alpha \in (0, 2]$*) be finite with probability* 1 *where* $\xi(t)$ *is a stochastically continuous measurable process it is necessary and sufficient that for all* $\lambda > 0$ *the limit*

$$\psi(\lambda) = \lim_{n \to \infty} \mathsf{E} \chi_0(\lambda v_n^\alpha),$$

exists satisfying condition $\psi(0+) = 1$. *Moreover*

$$\psi(\lambda) = \mathsf{E} \exp\left\{-\frac{\lambda^\alpha}{b-a} \int_a^b |\xi(t)|^\alpha\, dt\right\}.$$

Proof. Denote by $\mathfrak{A}$ the σ-algebra generated by the variables $\xi(t)$, $t \in [a, b]$ and η_k, $k = 0, \ldots$. Under the conditions of the theorem, ζ_k are independent of this σ-algebra. Therefore

$$\mathsf{E}(\chi_0(\lambda v_n^\alpha) \mid \mathfrak{A}) = \mathsf{E}\left(\exp\left\{\frac{i\lambda}{n^{1/\alpha}} \sum_{k=0}^{n-1} \zeta_k \xi\left(a + \frac{k+\eta_k}{n}(b-a)\right)\right\} \Bigg| \mathfrak{A}\right) =$$

$$=\exp\left\{-\frac{|\lambda|^\alpha}{n}\sum_{k=0}^{n-1}\left|\xi\left(a+\frac{k+\eta_k}{n}(b-a)\right)\right|^\alpha\right\}.$$

It follows from the lemma that

$$\frac{b-a}{n}\sum_{k=0}^{n-1}\left|\xi\left(a+\frac{k+\eta_k}{n}(b-a)\right)\right|^\alpha\to\int_a^b|\xi(t)|^\alpha\,dt$$

in probability. Therefore

$$\exp\left\{-\frac{|\lambda|^\alpha}{n}\sum_{k=0}^{n-1}\left|\xi\left(a+\frac{k+\eta_k}{n}(b-a)\right)\right|^\alpha\right\}\to\exp\left\{-\frac{|\lambda|^\alpha}{b-a}\int_a^b|\xi(t)|^\alpha\,dt\right\}$$

also in probability (we define $e^{-\infty}=0$). Since the quantities under consideration do not exceed 1, it follows that

$$\lim_{n\to\infty}\mathsf{E}\exp\left\{-\frac{|\lambda|^\alpha}{n}\sum_{k=0}^{n-1}\left|\xi\left(a+\frac{k+\eta_k}{n}(b-a)\right)\right|^\alpha\right\}=\mathsf{E}\exp\left\{-\frac{|\lambda|^\alpha}{b-a}\int_a^b|\xi(t)|^\alpha\,dt\right\}$$

also.

Moreover since

$$\lim_{n\to\infty}\mathsf{E}\exp\left\{-\frac{|\lambda|^\alpha}{n}\sum_{k=0}^{n-1}\left|\xi\left(a+\frac{k+\eta_k}{n}(b-a)\right)\right|^\alpha\right\}=$$

$$=\lim_{n\to\infty}\mathsf{E}(\chi(\lambda v_n^\alpha)\mid\mathfrak{A})=\lim_{n\to\infty}\mathsf{E}\chi(\lambda v_n^\alpha),$$

we have

$$\psi(\lambda)=\mathsf{E}\exp\left\{-\frac{|\lambda|^\alpha}{b-a}\int_a^b|\xi(t)|^\alpha\,dt\right\}.$$

Clearly,

$$\psi(0+)=\mathsf{P}\left\{\int_a^b|\xi(t)|^\alpha\,dt<\infty\right\}.$$

The proof of the theorem follows from the last relation. □

§5. Measures in Hilbert Spaces

The space $\mathscr{L}_2$ is the most interesting among the spaces $\mathscr{L}_p$ discussed in the previous sections. It is a separable Hilbert space. Since all the separable Hilbert spaces are isometric, it is more convenient to consider the abstract separable Hilbert space $\mathscr{X}$. The results obtained for such a space can easily be restated for various specific Hilbert spaces, for example, for the space of measurable functions on an arbitrary measurable space with the measure taking on values in a separable Banach space and square integrable in the norm.

Denote by $\mathfrak{B}$ the σ-algebra of Borel sets $\mathscr{X}$. The pair $(\mathscr{X}, \mathfrak{B})$ is called a *measurable* Hilbert space. Measures μ defined on a measurable Hilbert space $(\mathscr{X}, \mathfrak{B})$ are the main subject of study in this section. As before we are interested in probability measures, but since the results of this section are applicable for any finite measures, condition $\mu(\mathscr{X})=1$ is not imposed.

The scalar product in $\mathscr{X}$ will be denoted by (x, y) and we denote the norm of x by $|x|=\sqrt{(x, x)}$. A measure on $(\mathscr{X}, \mathfrak{B})$ – as in any linear space-- can be defined by means of a characteristic functional. Every continuous linear functional $l(x)$ defined on $\mathscr{X}$ is of the form $l(x)=(x, z)$ where z is an arbitrary element in $\mathscr{X}$. A function $\varphi(z)$ defined by equality

$$\varphi(z)=\int e^{i(z, x)}\, \mu(dx) \tag{1}$$

for all $z\in\mathscr{X}$ is called the *characteristic functional* of measure μ on $(\mathscr{X}, \mathfrak{B})$.

Let $\mathscr{L}$ be an arbitrary finite dimensional subspace of the space $\mathscr{X}$, and $\mathfrak{B}_{\mathscr{L}}$ be the σ-algebra of Borel subsets of $\mathscr{L}$. A set of the form

$$P_{\mathscr{L}}^{-1} A_{\mathscr{L}}=\{x: P_{\mathscr{L}} x\in A_{\mathscr{L}}\},$$

where $A_{\mathscr{L}}\in\mathfrak{B}_{\mathscr{L}}$ and $P_{\mathscr{L}}$ is the projector into $\mathscr{L}$ is called a *cylindrical* set with base in $\mathscr{L}$. The totality $\mathfrak{B}^{\mathscr{L}}$ of all cylindrical sets with bases in $\mathscr{L}$ is also a σ-algebra. The sets belonging to $\mathfrak{B}^{\mathscr{L}}$ for some $\mathscr{L}$ are called *cylindrical* sets and functions measurable relative to $\mathfrak{B}^{\mathscr{L}}$ for some $\mathscr{L}$ are called *cylindrical* functions.

Given a measure μ on $(\mathscr{X}, \mathfrak{B})$ one can associate with it a set of its finite dimensional projections (finite-dimensional distributions) $\{\mu_{\mathscr{L}}\}$ defined by the equality

$$\mu_{\mathscr{L}}(A_{\mathscr{L}})=\mu(P_{\mathscr{L}}^{-1} A_{\mathscr{L}}) \quad \text{for all} \quad A_{\mathscr{L}}\in\mathfrak{B}_{\mathscr{L}}.$$

Measures $\mu_{\mathscr{L}}$ are sufficient for evaluating integrals of cylindrical functions: For each $\mathfrak{B}_{\mathscr{L}}$-measurable bounded function $h(x)$

$$\int h(x)\, \mu_{\mathscr{L}}(dx)=\int h(P_{\mathscr{L}} x)\, \mu(dx). \tag{2}$$

We note that any cylindrical function is of the form $h(P_{\mathscr{L}}x)$ for some $\mathscr{L}$ where h is $\mathfrak{B}_{\mathscr{L}}$-measurable. Measures $\mu_{\mathscr{L}}$ for various $\mathscr{L}$ are coordinated in the following manner: if $\mathscr{L}\subset\mathscr{L}'$, then

$$\mu_{\mathscr{L}}(A_{\mathscr{L}})=\mu_{\mathscr{L}'}(\mathscr{L}'\cap P_{\mathscr{L}}^{-1} A_{\mathscr{L}}). \tag{3}$$

This relation follows from (2) and the feasibility of representing function $h(P_{\mathscr{L}}x)$ in the form $h'(P_{\mathscr{L}'}x)$ where h' is $\mathfrak{B}_{\mathscr{L}'}$-measurable. Condition (3) will be referred to in what follows as the *condition* of *compatability* and the family of measures $\{\mu_{\mathscr{L}}\}$ defined on all the finite-dimensional subspaces $\mathscr{L}$ and satisfying condition (3) is called a *compatible* family of finite-dimensional distributions.

To define measure μ it is sufficient to know only its one-dimensional projections. This follows from formula

$$\varphi(z)=\int e^{i(z,x)}\,\mu_{\mathscr{L}_z}(dx),$$

where $\mathscr{L}_z$ is a one-dimensional subspace generated by vector z. Conversely, given characteristic functional $\varphi(z)$ one can easily determine all the measures $\mu_{\mathscr{L}}$ by means of their characteristic functionals as follows: for $z\in\mathscr{L}$

$$\varphi_{\mathscr{L}}(z)=\int e^{i(z,x)}\,\mu_{\mathscr{L}}(dx),\qquad z\in\mathscr{L}.$$

Moreover, it turns out that the existence of a function $\varphi(z)$ satisfying the equality

$$\varphi(z)=\int e^{i(z,x)}\,\mu_{\mathscr{L}}(dx),\qquad z\in\mathscr{L},$$

for any finite-dimensional subspace $\mathscr{L}$ is a necessary and sufficient condition for compatability of the family of finite dimensional distributions $\{\mu_{\mathscr{L}}\}$. We now prove this assertion. Let the family $\{\mu_{\mathscr{L}}\}$ satisfy (3). Set

$$\varphi_{\mathscr{L}}(z)=\int e^{i(z,x)}\,\mu_{\mathscr{L}}(dx),\qquad z\in\mathscr{L}.$$

If $\mathscr{L}\subset\mathscr{L}'$ and $z\in\mathscr{L}$, then in view of (3)

$$\varphi_{\mathscr{L}'}(z)=\int e^{i(z,x)}\,\mu_{\mathscr{L}'}(dx)=\int e^{i(P_{\mathscr{L}}z,x)}\mu_{\mathscr{L}'}(dx)=$$

$$=\int e^{i(z,P_{\mathscr{L}}x)}\mu_{\mathscr{L}'}(dx)=\int e^{i(z,x)}\,\mu_{\mathscr{L}}(z)=\varphi_{\mathscr{L}}(z).$$

Set $\varphi(z)=\varphi_{\mathscr{L}_z}(z)$, where $\mathscr{L}_z$ is a one-dimensional subspace generated by the vector z. Since for $z\in\mathscr{L}$ and $\mathscr{L}_z\subset\mathscr{L}$ we have $\varphi_{\mathscr{L}}(z)=\varphi_{\mathscr{L}_z}(z)=\varphi(z)$. Conversely, if $\varphi(z)$ is a function such that $\varphi(z)=\varphi_{\mathscr{L}}(z)$ for $z\in\mathscr{L}$, then for $\mathscr{L}\subset\mathscr{L}'$ and $z\in\mathscr{L}$ we have the relation

$$\int e^{i(z,x)}\,\mu_{\mathscr{L}}(dx)=\int e^{i(z,x)}\,\mu_{\mathscr{L}'}(dx)=$$

$$=\int e^{i(z,P_{\mathscr{L}}x)}\,\mu_{\mathscr{L}'}(dx)=\int e^{i(z,x)}\,\mu_{\mathscr{L}'}(P_{\mathscr{L}}^{-1}\,dx),$$

which implies that the measures $\mu_{\mathscr{L}}(dx)$ and $\mu_{\mathscr{L}'}(P_{\mathscr{L}}^{-1}\,dx)$ coincide (since their characteristic functionals coincide). Thus condition (3) is fulfilled.

Moment forms. Important characteristics of measure μ on $(\mathscr{X},\mathfrak{B})$ are the moment forms of this measure. A *moment form* of order k of measure μ is determined by the relation

$$m_k(z_1,\dots,z_k)=\int(x,z_1)\dots(x,z_k)\,\mu(dx)$$

under the condition that the integral on the right is well defined (and finite) for all choices of $z_1,\dots,z_k\in\mathscr{X}$. Clearly for the existence of a moment form of order k it is necessary and sufficient that for all z the relation

$$\int|(x,z)|^k\,\mu(dx)<\infty \tag{4}$$

be satisfied.

The function $m_k(z_1,\dots,z_k)$ is a symmetric function in its arguments and moreover is continuous and homogeneous in each one of them. We show that the moment form of order k (under the condition that it is well defined) is a continuous symmetric k-linear form. For this purpose it is sufficient to show that

$$\sup_{|z|\leqslant 1}\int|(z,x|^k\,\mu(dx)<\infty \tag{5}$$

is valid.

We introduce the functions

$$m_n(z)=\int\frac{n|(x,z)|^k}{n+|x|^k}\,\mu(dx),\qquad m(z)=\int|(x,z)|^k\,\mu(dx).$$

Functions $m_n(z)$ are weakly continuous in z and $m_n(z)\uparrow m(z)$ for all z as $n\to\infty$. We set

$$K_{n,l}=\{z:m_n(z)\geqslant l\}\cap\{z:|z|\leqslant 1\}.$$

The set $K_{n,l}$ is weakly closed and weakly compact (since it is bounded). To prove (5) it is sufficient to show that $K^l=\bigcap_n K_{n,l}$ is void for some l (then $\sup_{|z|\leqslant 1} m(z)\leqslant l$). The sets K^l are also weakly closed and weakly compact. If all K^l are non-empty, then the intersection $\bigcap_l K^l$ will also be non-empty. But $m_n(z)\to\infty$ for $z\in\bigcap_l K^l$ which is impossible. Our assertion is thus proved. □

The first two moment forms are the most often used. The form $m_1(z)$ is a continuous linear functional with respect to z provided it is well defined. Hence there exists a vector $a\in\mathscr{X}$ such that

$$\int (x, z)\,\mu(dx)=m_1(z)=(a, z).$$

This vector is called the *mean value* of the measure μ. If the form $m_2(z_1, z_2)$ is defined (in this case the form $m_1(z)$ will also be defined), then the expression

$$m_2(z_1, z_2)-m_1(z_1)\,m_1(z_2)$$

will be a continuous symmetrical bilinear functional. Consequently a symmetrical bounded linear operator B exists such that

$$m_2(z_1, z_2)-m_1(z_1)\,m_1(z_2)=(Bz_1, z_2).$$

This operator is called the *correlation operator* of the measure μ. It follows from relation

$$0\leqslant\int (x-a, z)^2\,\mu(dx)=\int (x, z)^2\,\mu(dx)-(a, z)^2=$$
$$=m_2(z, z)-(m_1(z))^2=(Bz, z)$$

that B is a non-negative operator.

We note one important property of the correlation operator. Recall that a symmetric non-negative operator B is called a *kernel* (or *nuclear*) operator if it is completely continuous and the series $\sum\lambda_k$ of its eigenvalues converges (each value appears in the sum as many times as is its multiplicity). A symmetric non-negative operator B is a kernel operator if in some orthonormal basis $\{e_k\}$ of the space $\mathscr{X}$ the series $\sum(Be_k, e_k)$ is convergent. In this case this series will be convergent for any choice of the basis and its sum will be independent of the choice of the basis. This sum is called the *trace (spur) of the operator* and is denoted by $\operatorname{Sp} B$.

Lemma. *The correlation operator B of a measure μ is a kernel operator if and only if the following condition is satisfied*

$$\int |x|^2 \, \mu(dx) < \infty.$$

Moreover

$$\operatorname{Sp} B = \int |x|^2 \, \mu(dx) - |a|^2,$$

where a is the mean value of μ.

The proof follows from the equality

$$\sum_{k=1}^{n} (Be_k, e_k) = \int \sum_{k=1}^{n} (x, e_k)^2 \, \mu(dx) - \sum_{k=1}^{n} (a, e_k)^2,$$

which is valid for any choice of $e_1, \ldots, e_n$. Taking the vectors from an orthonormal basis and approaching the limit as $n \to \infty$ (this limiting approach under the sign of the integral is justified in view of the monotonicity of the sequence $\sum_{k=1}^{n} (x, e_k)^2$ in n) we obtain the assertion of the lemma. □

The Minlos-Sazonov theorem. As was mentioned above, a measure μ on a measurable Hilbert space $(\mathscr{X}, \mathfrak{B})$ can be defined by either its finite dimensional projections or its characteristic functional. As it turns out the two methods do not differ significantly.

Let a compatible family of finite dimensional distributions $\{\mu_{\mathscr{L}}\}$ be given. Under what conditions does a measure μ exist on $(\mathscr{X}, \mathfrak{B})$ such that $\{\mu_{\mathscr{L}}\}$ are its projections. Since $\mu_{\mathscr{L}}$ enable us to construct a functional $\varphi(z)$ which coincides for $z \in \mathscr{L}$ with the characteristic functional of the measure $\mu_{\mathscr{L}}$, the problem posed reduces to the following: under what conditions is $\varphi(z)$ a characteristic functional of a certain measure μ on $(\mathscr{X}, \mathfrak{B})$? The answer to this last question is given by Minlos-Sazonov's theorem.

Theorem 1. *In order that a complex-valued continuous positive definite function $\varphi(z)$ defined for $z \in \mathscr{X}$ be a characteristic functional of a certain measure μ on $(\mathscr{X}, \mathfrak{B})$ it is necessary and sufficient that for any $\varepsilon > 0$ a kernel operator A_ε be found such that* $\operatorname{Re}(\varphi(0) - \varphi(z)) < \varepsilon$ *as long as* $(A_\varepsilon z, z) \leq 1$.

Proof. Necessity. Let $\varphi(z)$ be a characteristic operator of the measure μ. Then

$$\operatorname{Re}(\varphi(0)-\varphi(z))=\int(1-\cos(z,x))\,\mu(dx)\leqslant$$

$$\leqslant\int_{|x|\leqslant c}2\sin^2\frac{(x,z)}{2}\,\mu(dx)+2\int_{|x|\leqslant c}\mu(dx)\leqslant$$

$$\leqslant\tfrac{1}{2}\int_{|x|\leqslant c}(x,z)^2\,\mu(dx)+2\mu(\{x:|x|>c\}).$$

The expression $\int_{|x|\leqslant c}(x,z)^2\,\mu(dx)$ is for each c a quadratic functional relative to z and admits representation in the form $(B_c z, z)$ where B_c is a kernel operator (in view of the lemma above) since

$$\int_{|x|\leqslant c}|x|^2\,\mu(dx)\leqslant\mu(\mathscr{X})\,c^2.$$

We choose c in such a manner that the inequality $\mu(\{x:|x|>c\})<\varepsilon/4$ is satisfied. Then taking $A_\varepsilon=\frac{1}{\varepsilon}B_c$, we obtain for $(A_\varepsilon z, z)\leqslant 1$

$$\operatorname{Re}(\varphi(0)-\varphi(z))<\frac{\varepsilon}{2}+\frac{1}{2}(B_c z, z)=\frac{\varepsilon}{2}+\frac{\varepsilon}{2}(A_\varepsilon z, z)\leqslant\varepsilon.$$

The necessity of the condition of the theorem is thus proved.

Sufficiency. Let $\{\mu_{\mathscr{L}}\}$ be a compatible family of finite dimensional distributions constructed using $\varphi(z)$. It follows from the theorem in Section 3 and the remark in this theorem that it is sufficient to prove the existence of N such that for any $\varepsilon>0$ and all the finite-dimensional subspaces $\mathscr{L}$ the inequality

$$\mu_{\mathscr{L}}(\{x:|x|>N\})<\varepsilon \tag{6}$$

is fulfilled. Indeed, in place of functions $h_n(x)$ appearing in the remark in Section 3 we can use functions $|P_{\mathscr{L}_n}x|$, where $\mathscr{L}_n$ is an increasing sequence of finite dimensional subspaces for which $\bigcup\mathscr{L}_n$ is dense in $\mathscr{X}$, and $P_{\mathscr{L}_n}$ is the projector into $\mathscr{L}_n$.

To prove formula (6) we utilize Chebyshev's inequality, which yields

$$\mu_{\mathscr{L}}(\{x:|x|>N\})\leqslant(1-e^{-\lambda N^2/2})^{-1}\int_{\mathscr{L}}(1-e^{-(\lambda/2)|x|^2})\,\mu_{\mathscr{L}}(dx)=$$

$$=(1-e^{-\lambda N^2/2})^{-1}\int (2\pi\lambda)^{-\frac{r_{\mathscr{L}}}{2}}\int_{\mathscr{L}} (1-e^{i(x,z)})\, e^{-(1/2\lambda)|z|^2}\, m_{\mathscr{L}}(dz)\,\mu(dx),$$

where $m_{\mathscr{L}}(dz)$ is the Lebesgue measure on $\mathscr{L}$, $r_{\mathscr{L}}$ is the dimension of $\mathscr{L}$. Interchanging the order of integration we obtain

$$(1-e^{-\lambda N^2/2})\,\mu_{\mathscr{L}}(\{x:|x|>N\})\leqslant$$

$$\leqslant(2\pi\lambda)^{-\frac{r_{\mathscr{L}}}{2}}\int_{\mathscr{L}} (\varphi(0)-\varphi(z))\, e^{-(1/2\lambda)|z|^2}\, m_{\mathscr{L}}(dz).$$

Next we choose a kernel operator A such that $\mathrm{Re}(\varphi(0)-\varphi(z))\leqslant\frac{\varepsilon}{2}$ for $(Az, z)\leqslant 1$. Then

$$(1-e^{-\lambda N^2/2})\,\mu_{\mathscr{L}}(\{x:|x|>N\})\leqslant$$

$$\leqslant\frac{\varepsilon}{2}+(2\pi\lambda)^{-\frac{r_{\mathscr{L}}}{2}}\times\int_{(Az,z)>1} 2e^{-(1/2\lambda)|z|^2}\, m_{\mathscr{L}}(dz)\leqslant$$

$$\leqslant\frac{\varepsilon}{2}+2(2\pi\lambda)^{-\frac{r_{\mathscr{L}}}{2}}\int (Az, z)\, e^{-(1/2\lambda)|z|^2}\, m_{\mathscr{L}}(dz)\leqslant\frac{\varepsilon}{2}+2\lambda\,\mathrm{Sp}\,A,$$

since

$$(2\pi\lambda)^{-\frac{r_{\mathscr{L}}}{2}}\int (Az, z)\, e^{-(1/2\lambda)|z|^2}\, m_{\mathscr{L}}(dz)=$$

$$=\sum_{i,j=1}^{r_{\mathscr{L}}}(2\pi\lambda)^{-\frac{r_{\mathscr{L}}}{2}}\int (Ae_i, e_j)\,(z, e_i)\,(z, e_j)\prod_{1}^{r_{\mathscr{L}}} e^{-(z,e_i)^2/2\lambda}\, m_{\mathscr{L}}(dz)=$$

$$=\sum_{i=1}^{r_{\mathscr{L}}}(Ae_i, e_i)\frac{1}{\sqrt{2\pi\lambda}}\int t^2 e^{-t^2/2\lambda}\,dt=\lambda\sum_{i=1}^{r_{\mathscr{L}}}(Ae_i, e_i)\leqslant\lambda\,\mathrm{Sp}\,A.$$

Thus

$$\mu_{\mathscr{L}}\{x:|x|>N\}\leqslant\left(\frac{\varepsilon}{2}+2\lambda\,\mathrm{Sp}\,A\right)(1-e^{-\lambda N^2/2})^{-1}. \tag{7}$$

Clearly, one can choose λ and N in such a manner that the right-hand-side of (7) will be less than ε. The theorem is thus proved. □

Generalized measures on a Hilbert space. In Section 3 a procedure was described which enables us to construct, for each positive definite function $\varphi(x^*)$ defined on $\mathscr{X}^*$ (the conjugate of the space $\mathscr{X}$) an extension $\tilde{\mathscr{X}}$ of the space $\mathscr{X}$ and the measure μ in $\tilde{\mathscr{X}}$ such that $\varphi(x^*)$ becomes the characteristic functional for this measure. Let $\varphi(z)$ be a positive definite function defined on the Hilbert space $\mathscr{X}$. Utilizing the above-mentioned result, one can construct an extension of the space $\mathscr{X}$ and a measure on that extension such that $\varphi(z)$ will be the characteristic functional for this measure. However, the procedure presented in Section 3 results in a space $\tilde{\mathscr{X}}$ which is too wide. In the case where $\mathscr{X}$ is a Hilbert space, $\tilde{\mathscr{X}}$ can also be constructed as a Hilbert space obtained by completion of $\mathscr{X}$ in a certain scalar product depending on the continuity conditions of $\varphi(z)$. We shall now consider this construction due to Yu. L. Daletskiĭ.

Let B be a bounded symmetric positive linear operator. Introduce in $\mathscr{X}$ a new scalar product

$$(x, y)_- = (Bx, y), \quad |x|_-^2 = (Bx, x). \tag{8}$$

The space $\mathscr{X}$ is in general incomplete in the metric generated by this scalar product. Denote the completion of $\mathscr{X}$ in the norm $|\cdot|_-$ by $\mathscr{X}_-^B$ (this set can be regarded as an extension of $\mathscr{X}$), $\mathscr{X}$ is an everywhere dense set in $\mathscr{X}_-^B$; $\mathscr{X}_-^B$ and $\mathscr{X}$ coincide if B^{-1} is a bounded operator. Denote by $\mathscr{X}_+^B$ the Hilbert space obtained from the domain of definition of the operator $B^{-1/2}$ (which is dense in $\mathscr{X}$) by introducing the scalar product

$$(x, y)_+ = (B^{-1/2}x, B^{-1/2}y) = (B^{-1}x, y). \tag{9}$$

The second equality in formula (9) requires certain clarification. Note that for any $x \in \mathscr{X}_+^B$ the scalar product (x, z) defined for z on $\mathscr{X}$ can be extended by continuity in the metric of $\mathscr{X}_-^B$ to the whole space $\mathscr{X}_-^B$. Indeed, let $x = B^{1/2}x_0$, $x_0 \in \mathscr{X}$. Then

$$\begin{aligned}|(x, z_n - z_m)| &= |(B^{1/2}x_0, z_n - z_m)| = |(x_0, B^{1/2}(z_n - z_m))| \leqslant \\ &\leqslant |x_0|\,(B^{1/2}(z_n - z_m), B^{1/2}(z_n - z_m))^{1/2} = |x_0|\,|z_n - z_m|_- .\end{aligned}$$

Therefore the linear functional (x, z) on $\mathscr{X}$ is continuous in the metric of $\mathscr{X}_-^B$ and hence it can be extended by continuity (in z) to the space $\mathscr{X}_-^B$. In what follows, the expression (x, z) where $x \in \mathscr{X}_+^B$ and $z \in \mathscr{X}_-^B$ is understood to be this extension. The operator B can also be extended by continuity to $\mathscr{X}_-^B$ since

$$\begin{aligned}|Bx|_- &= \sqrt{(Bx, Bx)_-} = \sqrt{(B^2x, Bx)} = \\ &= \sqrt{(B^2B^{1/2}x, B^{1/2}x)} \leqslant \sqrt{\|B^2\|\,(B^{1/2}x, B^{1/2}x)} \leqslant \|B\|\,|x|_- .\end{aligned}$$

In what follows B will be regarded as extended to $\mathscr{X}_-^B$. Moreover the following relations are satisfied

$$B^{1/2}\mathscr{X}_-^B = \mathscr{X}, \quad B^{1/2}\mathscr{X} = \mathscr{X}_+^B, \quad B\mathscr{X}_-^B = \mathscr{X}_+^B .$$

The third relation is a consequence of the first two. The second follows from the definition of $\mathscr{X}^B_+$. We now prove the first assertion. Let z be an arbitrary element in $\mathscr{X}^B_-$, $z_n \in \mathscr{X}$ and $|z_n - z|_- \to 0$. This means that

$$(B(z_n - z_m), z_n - z_m) = |B^{1/2} z_n - B^{1/2} z_m|^2 \to 0, \quad n, m \to \infty.$$

However $B^{1/2} z_n \in \mathscr{X}$, therefore $B^{1/2} z \in \mathscr{X}$ also. We now return to equality (9). From the above, operator $B^{-1} x$ is (well) defined for $x \in \mathscr{X}^B_+$ and belongs to $\mathscr{X}^B_-$; since $y \in \mathscr{X}^B_+$, the scalar product $(B^{-1}x, y)$ is also well defined.

Now let a certain measure μ be defined on $\mathscr{X}^B_-$. Then in terms of this measure one can construct the characteristic functional $\varphi_-(z) = \int e^{i(x,z)^-} \mu(dx)$, defined for $z \in \mathscr{X}^B_-$. Since $(x, z)_- = (Bz, x)$ and $Bz \in \mathscr{X}^B_+$, the measure μ can also be defined by means of the characteristic functional $\varphi(z) = \int e^{i(z,x)} \mu(dx)$ where $z \in \mathscr{X}^B_+$. Note that

$$\varphi_-(z) = \varphi(Bz), \quad \varphi(z) = \varphi_-(B^{-1} z).$$

It follows from Theorem 1 that $\varphi(z)$ is the characteristic functional of a measure on $\mathscr{X}^B_-$ if and only if a kernel operator S exists on $\mathscr{X}^B_-$ such that for each $\varepsilon > 0$, $\operatorname{Re}(\varphi_-(0) - \varphi_-(z)) \leqslant \varepsilon$, provided $(Sz, z)_- \leqslant 1$. We utilize this result to construct an extension of $\mathscr{X}$ in order that the given positive-definite functional $\varphi(z)$ be a characteristic functional on this extension.

Theorem 2. *Let $\varphi(z)$ be a continuous positive definite functional defined on $\mathscr{X}$. Then for any kernel operator B, $\varphi(z)$ is a characteristic function of a certain measure on $\mathscr{X}^B_-$.*

Proof. It follows from the continuity of $\varphi(z)$ that for any $\varepsilon > 0$ a $\delta > 0$ can be found such that $\operatorname{Re}(\varphi(0) - \varphi(z)) \leqslant \varepsilon$ provided $(z, z) \leqslant \delta$. The $\operatorname{Re}(\varphi(0) - \varphi(Bz)) \leqslant \varepsilon$ for $z \in \mathscr{X}^B_-$ provided $(Bz, Bz) \leqslant \delta$, i.e. $\operatorname{Re}(\varphi_-(0) - \varphi_-(z)) \leqslant \varepsilon$ if $\left(\frac{1}{\delta} Bz, z\right)_- \leqslant 1$.

We show that the operator $\frac{1}{\delta} B$ defined on $\mathscr{X}^B_-$ is a kernel operator. It is sufficient to show that B is such an operator. But

1) $(Bx, y)_- = (B^2 x, y) = (Bx, By) = (x, By)_-$;

2) $(Bx, x)_- = (Bx, Bx) \geqslant 0$.

Finally we show that

3) $\operatorname{Sp}_- B = \sum_{k=1}^{\infty} (Be_k, e_k)_- < \infty$,

where $\{e_k\}$ is an orthonormal basis in $\mathscr{X}_-^B$. Indeed, set $e_k = f_k/\sqrt{\lambda_k}$ where $\{f_k\}$ is the basis consisting of the eigenvectors of the operator B in $\mathscr{X}$, $\lambda_k = (Bf_k, f_k)$, $(f_k, f_k) = 1$. Then

$$\mathrm{Sp}_-(B) = \sum_{k=1}^{\infty} (B^2 e_k, e_k) = \sum_{k=1}^{\infty} \frac{(B^2 f_k, f_k)}{\lambda_k} = \sum_{k=1}^{\infty} \lambda_k = \mathrm{Sp}\, B.$$

The theorem is thus proved. □

Remark 1. Let a positive definite function $\varphi(z)$ satisfy the condition: for any $\varepsilon > 0$ there exists $\delta > 0$ such that $\mathrm{Re}(\varphi(0) - \varphi(z)) < \varepsilon$, provided only $(Vz, z) < \delta$ where V is a bounded symmetric positive operator. Consider the space $\mathscr{X}_-^S$ where S is a certain symmetric positive operator commuting with V. We now obtain the conditions that must be imposed on S in order that a measure exist in $\mathscr{X}_-^S$ with characteristic functional $\varphi(z)$. Since

$$\mathrm{Re}(\varphi_-(0) - \varphi_-(z)) = \mathrm{Re}(\varphi(0) - \varphi(Sz))$$

for $(VSz, Sz) = (VS, z)_- < \delta$ and $\mathrm{Sp}_-\, VS = \mathrm{Sp}\, VS$, such a measure exists, provided $\mathrm{Sp}\, VS < \infty$. This assertion is valid also in the case when $\varphi(z)$ is defined on a linear manifold which is dense in $\mathscr{X}$ and V is an unbounded operator.

Measures defined on $\mathscr{X}_-^S$, for some S, whose characteristic functionals in a scalar product in $\mathscr{X}$ are defined on an everywhere dense set in $\mathscr{X}$ are called generalized measures on $\mathscr{X}$. Theorem 2 shows that a generalized measure is not uniquely constructed by means of its characteristic functional defined on $\mathscr{X}$. Let $\mathscr{X}'$ and $\mathscr{X}''$ be two extensions of the space $\mathscr{X}$ in which the measures μ' and μ'' are defined which correspond to the same characteristic functional $\varphi(z)$. Then one can find an extension $\mathscr{X}'''$ which is included in each one of the extensions $\mathscr{X}'$ and $\mathscr{X}''$ such that $\mu'(\mathscr{X}' - \mathscr{X}''') = 0$ and $\mu''(\mathscr{X}' - \mathscr{X}'') = 0$ and μ' coincides with μ'' on $\mathscr{X}'''$. This extension $\mathscr{X}'''$ is easily constructed as follows: if

$$\mathscr{X}' = \mathscr{X}_-^{S_1}, \quad \mathscr{X}'' = \mathscr{X}_-^{S_2}, \quad \text{then} \quad \mathscr{X}''' = \mathscr{X}_-^{S_1 + S_2}.$$

Therefore a generalized measure is, in a certain sense, uniquely constructed.

It is more convenient to formulate the conditions of Theorem 1 in the spaces $\mathscr{X}_-^B$.

Remark 2. In order that the conditions of Theorem 1 be satisfied it is necessary and sufficient that a kernel operator B exist, such that the functional $\varphi(z)$ be continuous in the metric of $\mathscr{X}_-^B$ and hence extendable on $\mathscr{X}_-^B$. Indeed if $\varphi(z)$ is continuous in metric $\mathscr{X}_-^B$ then for any $\varepsilon > 0$ a $\delta > 0$

can be found such that $\operatorname{Re}(\varphi(0)-\varphi(z))\leqslant\varepsilon$, provided that $|z|_-\leqslant\delta$, i.e. $\left(\frac{1}{\delta}Bz, z\right)\leqslant 1$, where $\frac{1}{\delta}B$ is a kernel operator.

Conversely we show that the existence of operator B follows from the condition of Theorem 1. We choose a sequence $\varepsilon_n\downarrow 0$ and let the operators A_n satisfy the condition of Theorem 1 for $\varepsilon=\varepsilon_n$. Let c_n be a sequence ($c_n\downarrow 0$) such that

$$\sum_{n=1}^{\infty} c_n \operatorname{Sp} A_n<\infty.$$

Then the operator $B=\sum_{n=1}^{\infty} c_n A_n$ is the required kernel operator. Indeed, for any $\varepsilon>0$ an n can be found such that $\varepsilon_n<\varepsilon$. Then for $(Bz, z)<c_n$ we have: $(A_n z, z)<1$ and hence

$$\operatorname{Re}(\varphi(0)-\varphi(z))\leqslant\varepsilon_n<\varepsilon.$$

Next

$$|\varphi(z_1)-\varphi(z_2)|\leqslant\int |e^{i(z_1,x)}-e^{i(z_2,x)}|\,\mu(dx)\leqslant$$
$$\leqslant\left(\int |e^{i(z_1-z_2,x)}-1|^2\,\mu(dx)\right)^{1/2}=\sqrt{2\operatorname{Re}(\varphi(0)-\varphi(z_1-z_2))}.$$

It follows from this inequality that $|\varphi(z_1)-\varphi(z_2)|\to 0$ as

$$(B(z_1-z_2), z_1-z_2)\to 0.$$

From the last remark we deduce, in particular, that the function $e^{-|z|^2}$ which is a continuous and positive definite function cannot be a characteristic functional of a measure on $\mathscr{X}$ since it cannot be extended continuously onto $\mathscr{X}_-^B$ with the kernel operator B.

§6. Gaussian Measures in a Hilbert Space

Let μ be a probability measure on a measurable Hilbert space $(\mathscr{X}, \mathfrak{B})$. Then $(\mathscr{X}, \mathfrak{B}, \mu)$ is a probability space and any $\mathfrak{B}$-measurable function $g(x)$ is a random variable on this space. The measure μ is called *Gaussian* if every continuous linear functional $l_z(x)=(z, x)$ is a normally distributed random variable. Let

$$\alpha_z=\mathsf{E}(z, x)=\int (z, x)\,\mu(dx),$$

$$\beta_z = \mathsf{E}(z, x)^2 - \alpha_z^2 = \int (z, x)^2 \mu(dx) - \alpha_z^2.$$

Since the distribution of (z, x) is normal these variables are defined for each z and hence, as it was established in Section 5 in the course of the study of polylinear moment forms, a vector a and a bounded symmetric non-negative linear operator B exist such that

$$\alpha_z = (a, z), \quad \beta_z = (Bz, z).$$

Since (z, x) is normally distributed, we have

$$\mathsf{E}\, e^{i(z, x)} = \int e^{i(z, x)} \mu(dx) = \exp\{i(a, z) - \tfrac{1}{2}(Bz, z)\}.$$

Thus any Gaussian measure possesses the mean value a and the correlation operator B, and moreover the characteristic functional of this measure is of the form

$$\varphi(z) = \exp\{i(a, z) - \tfrac{1}{2}(Bz, z)\}. \tag{1}$$

Conversely, if the characteristic functional of the measure μ is of the form (1), then

$$\int e^{it(z, x)} \mu(dx) = \varphi(tz) = \exp\left\{it(a, z) - \frac{t^2}{2}(Bz, z)\right\}$$

and hence the variable (z, x) is normally distributed with the mean (a, z) and the variance (Bz, z). Hence the necessary and sufficient condition for a measure μ to be Gaussian is that the characteristic functional of this measure admit representation (1).

It follows from formula (1) that for any finite set of vectors $z_1, \ldots, z_n$ the joint distribution of the variables $(z_1, x), \ldots, (z_n, x)$ is also Gaussian. Indeed,

$$\mathsf{E} \exp\{i \sum t_k (z_k, x)\} = \varphi(\sum t_k z_k) =$$
$$= \exp\{i \sum t_k (a, z_k) - \tfrac{1}{2} \sum t_k t_j (Bz_k, z_j)\}.$$

How arbitrary is the choice of quantities a and B in formula (1)? If B is a positive definite operator, then the function $\varphi(z)$ defined by (1) is positive definite. Other restrictions on a and B are imposed by Minlos-Sazonov's theorem. Let A be a kernel operator for a given $\varepsilon > 0$, such that

$$\mathrm{Re}(1 - \varphi(z)) < \varepsilon \quad \text{for} \quad (Az, z) < 1.$$

Then for $(Az, z) < 1$ the following inequalities are satisfied

$$\tfrac{1}{2}(Bz, z) < \exp\{\tfrac{1}{2}(Bz, z)\} - 1 <$$

$$<[1-\exp\{-\tfrac{1}{2}(Bz, z)\}]\,[1-(1-\exp\{-\tfrac{1}{2}(Bz, z)\})]^{-1}<$$
$$<[1-\exp\{-\tfrac{1}{2}(Bz, z)\}\cos(a, z)]\times$$
$$\times[1-(1-\exp\{-\tfrac{1}{2}(Bz, z)\}\cos(a, z))]^{-1}<\frac{\varepsilon}{1-\varepsilon}$$

(if $\varepsilon<1$, then $\cos(a, z)>0$). Therefore

$$(Bz, z)<\frac{2\varepsilon}{1-\varepsilon}(Az, z)$$

and

$$\operatorname{Sp} B<\frac{2\varepsilon}{1-\varepsilon}\operatorname{Sp} A.$$

Consequently, condition $\operatorname{Sp} B<\infty$ is a necessary condition for formula (1) to determine a characteristic functional of a measure on $(\mathscr{X}, \mathfrak{B})$. We now show that this condition is also sufficient. Since

$$|1-\varphi(z)|<\tfrac{1}{2}(Bz, z)+|(a, z)|,$$

it follows that $|1-\varphi(z)|<\varepsilon$, provided only $\frac{1}{\varepsilon}(Bz, z)+\frac{4}{\varepsilon^2}(a, z)^2<1$. Setting

$$A_\varepsilon=\frac{1}{\varepsilon}B+\frac{4}{\varepsilon^2}P_a, \quad \text{where} \quad P_a z=(a, z)\,a,$$

we observe that the conditions of Minlos-Sazonov's theorem are satisfied, since

$$\operatorname{Sp} A_\varepsilon=\frac{1}{\varepsilon}\operatorname{Sp} B+\frac{4}{\varepsilon^2}|a|^2<\infty.$$

Therefore the following result is obtained:

Theorem 1. *A measure μ is a Gaussian measure on $(\mathscr{X}, \mathfrak{B})$ if and only if its characteristic functional $\varphi(z)$ admits representation* (1) *where a is an arbitrary vector in $\mathscr{X}$ and B is a kernel operator. Moreover, a is the mean value of the measure μ and B is its correlation operator.*

As it follows from the lemma in Section 5, for any Gaussian measure

$$\int |x|^2\,\mu(dx)<\infty.$$

Let $e_1, e_2, \ldots$ be a orthonormal basis of the eigenvectors of B, whose existence follows from the fact that B is completely continuous. If λ_k is an eigenvalue of B corresponding to e_k, then

$$(Be_i, e_j) = \lambda_i \delta_{ij}.$$

Therefore the random variables (x, e_k), $k=1,\ldots,n$ possess the joint characteristic function

$$\mathsf{E}\, e^{i\sum t_k(x, e_k)} = \exp\{i\sum t_k(a, e_k) - \tfrac{1}{2}\sum \lambda_k t_k^2\}.$$

The last formula shows that the variables (x, e_k), $k=1,\ldots,n$, are jointly independent. If $\lambda_k \neq 0$, then the variable $\dfrac{(x-a, e_k)}{\sqrt{\lambda_k}} = \xi_k$ is normally distributed with the mean 0 and variance 1. Regarding x as a random element on the probability space $(\mathscr{X}, \mathfrak{B}, \mu)$ one can write

$$x = a + \sum \sqrt{\lambda_k}\, \xi_k e_k, \tag{2}$$

where ξ_k are independent identically distributed Gaussian random variables on $(\mathscr{X}, \mathfrak{B}, \mu)$ defined for all k such that $\lambda_k > 0$, $\mathsf{E}\xi_k = 0$, $\operatorname{Var}\xi_k = 1$. The representation (2) can be utilized for various calculations. We consider an example of an application of formula (2). We shall evaluate the Laplace transform of the variable $|x|^2$. Since

$$|x|^2 = \sum \lambda_k \xi_k^2 + 2\sum \sqrt{\lambda_k}\, a_k \xi_k + |a|^2, \quad \text{where} \quad a_k = (a, e_k),$$

it follows that

$$\int e^{s|x|^2} \mu(dx) = e^{s|a|^2} \prod_{k=1}^{\infty} \mathsf{E} \exp\{s\lambda_k \xi_k^2 + 2s\sqrt{\lambda_k}\, \alpha_k \xi_k\} =$$

$$= e^{s|a|^2} \prod_{k=1}^{\infty} \int \frac{1}{\sqrt{2\pi}} e^{-\frac{1}{2}t^2 + s\lambda_k t^2 + 2s\sqrt{\lambda_k}\,\alpha_k t}\, dt =$$

$$= e^{s|\alpha|^2} \prod_{k=1}^{\infty} \frac{\exp\left\{\dfrac{2\lambda_k \alpha_k^2 s^2}{1 - 2\lambda_k s}\right\}}{\sqrt{1 - 2s\lambda_k}}.$$

The last infinite product converges for $\operatorname{Re} s < \dfrac{1}{2\|B\|}$; this follows from the convergence of the series $\sum \lambda_k = \operatorname{Sp} B$. The infinite product obtained can be expressed by means of operator

$$R_s(B) = (I - 2sB)^{-1}, \tag{3}$$

which is easily expressed in terms of the resolvent of operator B.

Indeed

$$\prod_{k=1}^{\infty} \frac{1}{\sqrt{1-2s\lambda_k}} = \exp\left\{-\tfrac{1}{2}\sum_{k=1}^{\infty} \ln(1-2s\lambda_k)\right\} =$$

$$= \exp\left\{\int_0^s \sum_{k=1}^{\infty} \frac{\lambda_k dt}{1-2t\lambda_k}\right\} = \exp\left\{\int_0^s \operatorname{Sp} BR_t(B)\, dt\right\},$$

$$\sum_{k=1}^{\infty} \frac{2\lambda_k \alpha_k^2}{1-2s\lambda_k} = 2(BR_s(B)\, a, a).$$

Thus for all $s < \dfrac{1}{2\|B\|}$ the following formula is valid:

$$\int e^{s|x|^2} \mu(dx) = \exp\left\{2s^2 (BR_s(B)\, a, a) + \int_0^s \operatorname{Sp} BR_t(B)\, dt + s|a|^2\right\}. \tag{4}$$

This formula can also be used for the determination of the Laplace transform of the variable (Vx, x) on the probability space $(\mathscr{X}, \mathfrak{B}, \mu)$ provided only that V is a non-negative symmetric operator. Let $V = U^2$, where U is also a non-negative operator. In this case $(Vx, x) = |Ux|^2$ where Ux is a random element, with values in $\mathscr{X}$ on the probability space $(\mathscr{X}, \mathfrak{B}, \mu)$. The characteristic functional of the variable Ux is of the form

$$\int e^{i(Ux, z)} \mu(dx) = \int e^{i(x, Uz)} \mu(dx) = \exp\{i(Ua, z) - \tfrac{1}{2}(UBUz, z)\}.$$

Consequently, Ux has a Gaussian distribution with the mean Ua and correlation operator UBU. Hence we have, in view of formula (4);

$$\int e^{s(Vx, x)} \mu(dx) = \exp\left\{2s^2 (UBUR_s(UBU)\, Ua, Ua) + \int_0^s \operatorname{Sp} UBUR_t(UBU)\, dt + s|Ua|^2\right\}. \tag{5}$$

Linear and quadratic functionals. Let μ be a Gaussian measure on $(\mathscr{X}, \mathfrak{B})$. Any measurable function $g(x)$ admitting representation as the limit in measure μ of a sequence of continuous linear functionals

$$g(x) = \lim_{n\to\infty} (x, z_n)$$

is called a *measurable linear functional* with respect to measure μ. Since the variables (x, z_n), $n = 1, 2, \ldots$, have a joint Gaussian distribution it follows from the convergence of (x, z_n) in measure μ to $g(x)$ that $g(x)$ will also have a normal distribution and also that (x, z_n) converges to $g(x)$ in the mean square. Hence

$$\lim_{n,m\to\infty} \int [(x, z_n)-(x, z_m)]^2 \mu(dx) = \lim_{n,m\to\infty} [(a, z_n - z_m)^2 + \\ +(B(z_n - z_m), z_n - z_m)].$$

Set $A = B + P_a$, where $P_a z = (a, z)\, a$. A is then a kernel operator. We introduce scalar product $(x, y)_- = (Ax, y)$.

Let $\mathscr{X}^A_-$ be the completion of $\mathscr{X}$ in this scalar product. If

$$\int [(z_n, x)-(z_m, x)]^2 \mu(dx) \to 0, \quad (z_n - z_m, z_n - z_m)_- \to 0,$$

i.e. the sequence z_n is fundamental (Cauchy) in $\mathscr{X}^A_-$. It is natural to associate with function $g(x)$, which is the limit (in measure μ) of the sequence (x, z_n), the element in $\mathscr{X}^A_-$ which is the limit of z_n. If $\lim z_n = z^*$, we denote $g(x) = (x, z^*)$. It is easy to see that this correspondence between the measurable linear functionals and $\mathscr{X}^A_-$ is one-to-one. In what follows we shall identify the space of linear functionals with $\mathscr{X}^A_-$. The space of measurable linear functionals is a Hilbert space with scalar product $(x, y)_-$. For any set of $z_1^*, \dots, z_n^*$ belonging to $\mathscr{X}^A_-$, the functionals (x, z_1^*), $\dots, (x, z_n^*)$ have a joint normal distribution and moreover

$$\mathsf{E} \exp\{i \sum t_k (z_k^*, x)\} = \exp\left\{i \sum_k t_k (z_k^*, a) - \tfrac{1}{2} \sum_{k,j} t_k t_j (B z_k^*, z_j^*)\right\},$$

where (z^*, a) is defined as the limit $\lim\limits_{n\to\infty} (z_n, a)$, $z_n \in \mathscr{X}$, $z_n \to z^*$ in $\mathscr{X}^A_-$ and $(B z_k^*, z_j^*)$ is defined as the limit $\lim\limits_{n\to\infty} (B z_k^n, z_j^n)$, $z_k^n \to z_k^*$, $z_j^n \to z_j^*$ in $\mathscr{X}^A_-$. The existence of both limits follows from the inequalities

$$(z_n - z_m, a)^2 \leqslant (z_n - z_m, z_n - z_m)_-,$$
$$(B(z_n - z_m), z_n - z_m) \leqslant (z_n - z_m, z_n - z_m)_-.$$

Since every element $z^* \in \mathscr{X}^A_-$ can be represented as $B^{-1/2} z$, $z \in \mathscr{X}$, we use in place of (z^*, x) the notation $(z, B^{-1/2} x)$ where $z \in \mathscr{X}$.

We now derive the representation of measurable linear functionals which uses decomposition (2). For any z

$$(z, x) = (z, a) + \sum \xi_k \sqrt{\lambda_k} (e_k, z),$$

where ξ_k is a sequence of independent identically distributed random variables with $\mathsf{E}\xi_k = 0$ and $\operatorname{Var} \xi_k = 1$. If $(z_n, x) \to (z^*, x)$ in the mean square, then $(z_n, a) \to (z^*, a)$, and the limit of (z_n, e_k) will exist for those k which satisfy $\lambda_k > 0$. We denote this limit by (z^*, e_k). Then

$$(z^*, x) = (z^*, a) + \sum \xi_k \sqrt{\lambda_k}(z^*, e_k). \tag{6}$$

Conversely, formula (6) defines a measurable linear functional for any sequence of numbers (z^*, e_k) such that the series

$$\sum \lambda_k (z^*, e_k)^2 \quad \text{and} \quad \sum (z^*, e_k)(a, e_k) = (z^*, a),$$

are convergent.

We shall now study measurable quadratic functionals for measures μ with mean zero. We shall distinguish between measurable quadratic functionals and measurable quadratic *centered* functionals.

A random variable $g(x)$ on a probability space $(\mathscr{X}, \mathfrak{B}, \mu)$ is called a measurable quadratic functional if a sequence of symmetric linear bounded operators A_n exists such that

$$g(x) = \lim_{n\to\infty} (A_n x, x)$$

in measure μ. A random variable $g(x)$ is called a *measurable centered quadratic functional* if a sequence of symmetric linear bounded operators A_n and constants c_n exist such that

$$g(x) = \lim_{n\to\infty} [(A_n x, x) + c_n]$$

in measure μ.

Assume that operator B is nondegenerate (otherwise one can consider a measure on the closure of the range of the values of operator B). Next let $\alpha_k^n = (A_n e_i, e_k)$ where e_k are eigenvalues of operator B. Utilizing decomposition (2) one can write

$$(A_n x, x) = \sum_{i,k} \sqrt{\lambda_i \lambda_k}\, \alpha_{ik}^n \xi_i \xi_k.$$

Since $\sqrt{\lambda_i \lambda_k}\, \alpha_{ik}^n = (B^{1/2} A_n B^{1/2} e_i, e_k)$ one can formally represent $(A_n x, x)$ in the form

$$(A_n x, x) = (B^{1/2} A_n B^{1/2} y, y),$$

where $y = \sum \xi_k e_k$ is a certain generalized random element in $\mathscr{X}$, i.e. a random element whose distribution is a generalized measure in $\mathscr{X}$ (cf. Section 5). We note that for any $z \in \mathscr{X}$ the scalar product

$$(z, y) = \sum \xi_k (z, e_k)$$

is defined, since ξ_k are independent with $\mathsf{E}\xi_k = 0$ and

$$\mathsf{V}(z, e_k)\,\xi_k = (z, e_k)^2, \quad \sum_{k=1}^{\infty} \mathsf{V}(z, e_k)\,\xi_k = |z|^2.$$

Hence if v is a generalized measure which is the distribution of the element y, then its characteristic functional $\varphi_v(z)$ is equal to

$$\varphi_v(z)=e^{-\frac{1}{2}|z|^2}$$

Let $f_1, f_2, \ldots$ be an orthonormal basis of eigenvectors of the operator $B^{1/2}A_nB^{1/2}$ (it is completely continuous). Then one can set

$$y=\sum \eta_k f_k,$$

where $\eta_k=(y,f_k)=\sum_j (e_j,f_k)\,\xi_j$ is a sequence of random variables on the probability space $(\mathscr{X}, \mathfrak{B}, \mu)$. It follows from the relation

$$e^{-\frac{1}{2}|z|^2}=\mathsf{E}\exp\{i\sum \eta_k(z,f_k)\}=\exp\{-\tfrac{1}{2}\sum(z,f_k)^2\}$$

that η_k are also independent Gaussian random variables with $\mathsf{E}\eta_k=0$ and $\mathsf{V}\eta_k=1$. Let c_k^n be eigenvalues of the operator $B^{1/2}A_nB^{1/2}$ corresponding to f_k. Then

$$(A_n x, x)=\sum_k c_k^n\eta_k^2.$$

Lemma. *Let for each n a sequence η_{nk} of independent Gaussian variables be given with $\mathsf{E}\eta_{nk}=0$ and $\mathsf{V}\eta_{nk}=1$. If constants d_n exist such that for $n\to\infty$ $\sum c_k^n\eta_{nk}^2+d_n\to 0$ in probability then*

$$\sum_k c_k^n+d_n\to 0 \quad \text{and} \quad \sum_k (c_k^n)^2\to 0.$$

Proof. First note that under the assumptions of the lemma $\sup_k|c_k^n|\to 0$, since for each k

$$\mathsf{P}\left\{\left|\sum_i c_j^n\eta_{nj}^2+d_n\right|\leqslant\varepsilon\right\}\leqslant\sup_t \mathsf{P}\{|c_k^n\eta_{nk}^2+t|\leqslant\varepsilon\}=$$

$$=\mathsf{P}\left\{|\eta_{nk}|\leqslant\sqrt{\frac{\varepsilon}{|c_k^n|}}\right\}\leqslant\frac{2}{\sqrt{2\pi}}\sqrt{\frac{\varepsilon}{|c_k^n|}}$$

and hence for each $\varepsilon>0$

$$\sup_k|c_k^n|\leqslant\frac{2}{\pi}\varepsilon\left(\mathsf{P}\left\{\left|\sum_j c_j^n\eta_{nj}^2+d_n\right|\leqslant\varepsilon\right\}\right)^{-2}.$$

Now if the inequality $\sum_k (c_k^n)^2>\delta$ were satisfied for some sequence of indices n then the variable $\sum_k c_k^n\eta_{nk}^2$ would, in view of the central limit theorem be asymptotically normal and therefore for each $\varepsilon>0$ we would have

$$1=\varliminf_{n\to\infty} \mathsf{P}\left\{\left|\sum_k c_k^n\eta_{nk}^2+d_n\right|\leqslant\varepsilon\right\}\leqslant \varliminf_{n\to\infty} \sup_t \mathsf{P}\left\{\left|\sum_k c_k^n\eta_{nk}^2+t\right|\leqslant\varepsilon\right\}\leqslant\frac{2\varepsilon}{\sqrt{2\pi\delta}},$$

which is impossible. Finally utilizing relation

$$\sum_j c_j^n\eta_{nj}^2+d_n=\sum_j c_j^n(\eta_{nj}^2-1)+\sum_j c_j^n+d_n$$

and the fact that

$$\mathsf{E}\Big(\sum_j c_j^n(\eta_{nj}^2-1)\Big)^2=2\sum_j (c_j^n)^2\to 0,$$

we observe that $\sum_j c_j^n+d_n\to 0$. The lemma is thus proved. □

It follows from the lemma that convergence of $(A_nx, x)+d_n$ in measure to a certain limit implies the convergence of $(A_nx, x)+d_n$ to the same limit in the mean square. It is easy to verify the calculations:

$$\begin{aligned}\int [(A_nx, x)+d_n]^2\,\mu(dx)&=2\sum_j (c_j^n)^2+\Big(\sum_j c_j^n+d_n\Big)^2=\\ &=2\,\mathrm{Sp}(B^{1/2}A_nBA_nB^{1/2})+(\mathrm{Sp}(B^{1/2}A_nB^{1/2})+d_n)^2=\\ &=2\,\mathrm{Sp}(A_nB)^2+(\mathrm{Sp}\,A_nB+d_n)^2.\end{aligned} \tag{7}$$

Therefore the following assertion is valid:

Corollary. *In order that the limit in measure μ of the expressions $(A_nx, x)+d_n$ exist, (for a certain choice of d_n) it is necessary and sufficient that the equality*

$$\lim_{n,\,m\to\infty} \mathrm{Sp}([A_n-A_m]\,B)^2=0$$

be satisfied. Moreover we can choose d_n equal to $-\mathrm{Sp}\,A_nB$. If the limit $\lim_{n\to\infty}\mathrm{Sp}\,A_nB$ exists then d_n can be chosen equal to zero.

We use these arguments to obtain a general form of the quadratic functional. Since

$$(A_nx, x)=\sum_{k,j}\sqrt{\lambda_k}\,\alpha_{kj}^n\sqrt{\lambda_j}\,(\xi_k\xi_j-\delta_{kj})+\sum_k\lambda_k\alpha_{kk}^n,$$

$$\mathrm{Sp}([A_n-A_m]\,B)^2=\sum_{k,j}\big(\sqrt{\lambda_k}\,\alpha_{kj}^n\sqrt{\lambda_j}-\sqrt{\lambda_k}\,\alpha_{kj}^m\sqrt{\lambda_j}\big)^2,$$

it follows that if the limit $\lim_{n\to\infty}(A_nx, x)+d_n$ in measure exists the limits

$$\lim_{n\to\infty}\sqrt{\lambda_k}\,\alpha_{kj}^n\sqrt{\lambda_j}=\beta_{kj},\qquad \sum_{k,j}\big(\sqrt{\lambda_k}\,\alpha_{kj}^n\sqrt{\lambda_j}-\beta_{kj}\big)^2\to 0.$$

also exist. Moreover,

$$\lim_{n\to\infty} \sum_{k,j} \sqrt{\lambda_k}\, \alpha_{kj}^n \sqrt{\lambda_j}(\xi_k\xi_j - \delta_{kj}) = \sum_{k,j} \beta_{kj}(\xi_k\xi_j - \delta_{kj}).$$

Functionals which are the limits of expressions

$$\lim_{n\to\infty} [(A_n x, x) - \operatorname{Sp} A_n B]$$

will henceforth be termed centered quadratic functionals. The general form of a centered functional is given by formula

$$g(x) = \sum_{k,j} \beta_{kj} \left(\frac{(x, e_k)(x, e_j)}{\sqrt{\lambda_k \lambda_j}} - \delta_{kj} \right), \tag{8}$$

where β_{kj} are arbitrary numbers satisfying $\sum_{k,j} \beta_{kj}^2 < \infty$. If an arbitrary constant is added to a centered functional we obtain a general form of a measurable quadratic functional. A centered functional is obtained from an arbitrary one by subtracting the mathematical expectation.

Linear and quadratic functionals of stationary Gaussian processes. Let $\xi(t)$ be a real stationary Gaussian process with mean 0 and correlation function $R(t)$ and spectral function $F(\lambda)$:

$$R(t) = \int e^{it\lambda}\, dF(\lambda).$$

Next let $y(\lambda)$ be a complex-valued Gaussian process with orthogonal increments such that

$$\xi(t) = \int e^{i\lambda t}\, dy(\lambda).$$

We consider $\xi(t)$ on the interval $[-T, T]$. A probability measure on Hilbert space $\mathscr{L}_2[-T, T]$ of real square-integrable functions on $[-T, T]$ is associated with this process. We now utilize the preceding results for obtaining linear and quadratic functionals on the process $\xi(t)$.

An arbitrary random variable η admitting representation as the mean square limit of the variables

$$\eta_n = \int_{-T}^{T} \xi(t)\, x_n(t)\, dt,$$

where $x_n(t)$ is a sequence of continuous functions defined on $[-T, T]$ is called a *linear functional* on the process $\xi(t)$. We now obtain the general form of a linear functional on $\xi(t)$. Since

$$\int_{-T}^{T}\int e^{i\lambda t}\, dy(\lambda)\, x_n(t)\, dt = \int\left[\int_{-T}^{T} e^{i\lambda t}\, x_n(t)\, dt\right] dy(\lambda),$$

it follows that

$$\mathsf{E}\eta_n^2 = \int |\varphi_n(\lambda)|^2\, dF(\lambda),$$

where

$$\varphi_n(\lambda) = \int_{-T}^{T} e^{i\lambda t}\, x_n(t)\, dt. \tag{9}$$

Denote by $\mathscr{W}_T(F)$ the Hilbert space of functions containing all functions of the form (9) and completed in the scalar product

$$(\varphi_1, \varphi_2) = \int \varphi_1(\lambda)\, \overline{\varphi_2(\lambda)}\, dF(\lambda).$$

It then follows from the convergence of η_n to a certain limit that the functions $\varphi_n(\lambda)$ defined by equation (9) converge to a certain limit φ belonging to $\mathscr{W}_T(F)$, and moreover

$$\eta = \int \varphi(\lambda)\, dy(\lambda). \tag{10}$$

Clearly formula (10) represents for $\varphi \in \mathscr{W}_T(F)$ the general form of a functional on a process $\xi(t)$ defined on $[-T, T]$.

An arbitrary random variable ζ which is the mean-square limit of the variables

$$\zeta_n = \int_{-T}^{T}\int_{-T}^{T} g_n(t, s)\,[\xi(t)\,\xi(s) - R(t-s)]\, dt\, ds,$$

where $g_n(t, s)$ is a sequence of continuous real symmetric functions defined for all t and $s \in [-T, T]$ is called a centered quadratic functional on $\xi(t)$. It can be easily calculated that

$$\mathsf{E}|\zeta_n|^2 = \mathsf{E}\int_{-T}^{T}\int_{-T}^{T}\int_{-T}^{T}\int_{-T}^{T} g_n(t, s)\, g_n(u, v)\, \xi(t)\, \xi(s)\, \xi(u)\, \xi(v)\, dt\, ds\, du\, dv =$$

$$=\int_{-T}^{T}\int_{-T}^{T}\int_{-T}^{T}\int_{-T}^{T} g_n(t,s)\, g_n(u,v)\, R(t-s)\, R(u-v)\, dt\, ds\, du\, dv=$$

$$=2\iint |\varphi_n(\lambda,\mu)|^2\, dF(\lambda)\, dF(\mu),$$

where

$$\varphi_n(\lambda,\mu)=\int_{-T}^{T}\int_{-T}^{T} g_n(t,s)\, e^{i\lambda t-i\mu s}\, dt\, ds. \tag{11}$$

Denote by $\mathscr{W}_T^2(F)$ the Hilbert space of functions containing all functions of form (11) and completed in the scalar product

$$(\varphi_1,\varphi_2)=\iint \varphi_1(\lambda,\mu)\,\overline{\varphi_2(\lambda,\mu)}\, dF(\lambda)\, dF(\mu).$$

Then, if the variables ζ_n converge to a certain limit, φ_n converge to a certain function φ belonging to $\mathscr{W}_T^2(F)$. To express ζ in terms of φ we introduce the double stochastic integral

$$\iint \varphi(\lambda,\mu)\; dy(\lambda)\,\overline{dy(\mu)}. \tag{12}$$

We define this integral as an integral over a random measure with orthogonal values (cf. Chapter IV, Section 4). Let the measure ν on $\mathscr{R}^2$ be defined on the rectangles by relation

$$\nu([\lambda_1,\lambda_2]\times[\mu_1,\mu_2])=$$
$$=y([\lambda_1,\lambda_2])\,\overline{y([\mu_1,\mu_2])}-F([\lambda_1,\lambda_2]\cap[\mu_1,\mu_2]), \tag{13}$$

where

$$y([\lambda_1,\lambda_2])=y(\lambda_2)-y(\lambda_1),\qquad F([\lambda_1,\lambda_2])=F(\lambda_2)-F(\lambda_1).$$

The measure ν is a measure with orthogonal values for which

$$\mathsf{E}\;|\nu([\lambda_1,\lambda_2]\times[\mu_1,\mu_2])|^2=F([\lambda_1,\lambda_2])\,F([\mu_1,\mu_2]).$$

Therefore the integral

$$\iint \varphi(\lambda,\mu)\, dy(\lambda)\,\overline{dy(\mu)}=\iint \varphi(\lambda,\mu)\,\nu(d\lambda\times d\mu)$$

is defined for all φ such that

$$\iint |\varphi(\lambda, \mu)|^2 \, dF(\lambda) \, dF(\mu) < \infty$$

and in particular for $\varphi \in \mathscr{W}_T^2(F)$. We show that the integral (12) with $\varphi \in \mathscr{W}_T^2(F)$ represents the general form of a centered quadratic functional. For this purpose it is sufficient to verify that

$$\int_{-T}^{T} \int_{-T}^{T} g_n(t, s) \, [\xi(t) \, \xi(s) - R(t-s)] \, dt \, ds =$$

$$= \iint \varphi_n(\lambda, \mu) \, dy(\lambda) \, \overline{dy(\mu)}, \qquad (14)$$

where φ_n is connected with g_n by formula (11). Let

$$g_n(t, s) = g(t) \, k(s), \quad \tilde{g}(\lambda) = \int_{-T}^{T} e^{i\lambda t} g(t) \, dt \quad \text{and} \quad \tilde{k}(\lambda) = \int_{-T}^{T} e^{i\lambda t} k(t) \, dt.$$

Then

$$\int_{-T}^{T} \int_{-T}^{T} g(t) \, k(s) \, [\xi(t) \, \xi(s) - R(t-s)] \, dt \, ds =$$

$$= \int \tilde{g}(\lambda) \, dy(\lambda) \int \overline{\tilde{k}(\mu) \, dy(\mu)} - \int \tilde{g}(\lambda) \, \overline{\tilde{k}(\lambda)} \, dF(\lambda) =$$

$$= \iint \tilde{g}(\lambda) \, \overline{\tilde{k}(\mu)} \, dy(\lambda) \, \overline{dy(\mu)}$$

(the last equality follows from formula (13) for step functions, therefore it is valid for all continuous functions). Hence (14) is also valid for the linear combinations of the form

$$\sum_j g_j(t) \, k_j(s),$$

and hence for all $g_n(t, s)$. □

As a corollary from the above we obtain the following formula: if $\varphi(\lambda, \mu) = \sum_k c_k \varphi_k(\lambda) \, \varphi_k(\mu)$, then

$$\iint \varphi(\lambda, \mu) \, dy(\lambda) \, \overline{dy(\mu)} =$$

$$=\sum_k c_k\left(\left|\int \varphi_k(\lambda)\,dy(\lambda)\right|^2-\int|\varphi_k(\lambda)|^2\,dF(\lambda)\right). \tag{15}$$

This formula will be utilized in succeeding chapters.

Chapter VI

Limit Theorems for Random Processes

§ 1. Weak Convergences of Measures in Metric Spaces

Let $\mathscr{X}$ be a metric space with metric $\varrho(x, y)$, $\mathfrak{B}$ be the σ-algebra of its Borel subsets, $\mathscr{C}_{\mathscr{X}}$ be the space of all bounded continuous functions defined on $\mathscr{X}$ with the norm $\|f\|_{\mathscr{X}} = \sup_{\mathscr{X}} |f(x)|$. A sequence of measures μ_n defined on $\mathfrak{B}$ is called *weakly convergent to measure* μ if for any function f in $\mathscr{C}_{\mathscr{X}}$ the relation

$$\lim_{n\to\infty} \int f(x)\, \mu_n(dx) = \int f(x)\, \mu(dx)$$

is satisfied. The set M of measures $\{\mu\}$ defined on $\mathfrak{B}$ is called weakly compact if from any sequence of measures μ_n in M a weakly convergent subsequence can be extracted.

Theorem 1. *Let $\mathscr{X}$ be a complete separable space. In order that the set M of measures defined on $\mathfrak{B}$ be weakly compact it is necessary and sufficient that the following conditions be satisfied:*

a) $\sup\{\mu(\mathscr{X});\ \mu \in M\} < \infty$;

b) *for every $\varepsilon > 0$ there exists a compactum K such that*

$$\sup\{\mu(\mathscr{X} \setminus K);\ \mu \in M\} < \varepsilon.$$

Proof. Necessity. Since the set M is compact it follows that the set of numbers

$$\left\{\int f(x)\, \mu(dx);\ \mu \in M\right\}$$

is also compact for any continuous bounded function f and hence this set is bounded. Choosing $f = 1$ we obtain the necessity of condition a). We now prove the necessity of condition b). Denote by K_δ the set of x such that $\varrho(x, K) < \delta$, where $\varrho(x, K) = \inf_{y \in K} \varrho(x, y)$. We show that for any $\varepsilon > 0$ and $\delta > 0$ there exists a compactum K such that $\mu(\mathscr{X} \setminus K_\delta) \leqslant \varepsilon$ for all $\mu \in M$.

Assume the contrary, i.e. that such a compactum does not exist for the given $\varepsilon>0$ and $\delta>0$. We consider an arbitrary measure $\mu_1\in M$ and let $K^{(1)}$ be a compactum such that $\mu_1(\mathscr{X}\backslash K^{(1)})<\varepsilon$. Since $\sup_\mu \mu(\mathscr{X}\backslash K_\delta^{(1)})>\varepsilon$ a measure $\mu_2\in M$ can be found such that $\mu_2(\mathscr{X}\backslash K_\delta^{(1)})>\varepsilon$. Hence a compact set $K^{(2)}\subset\mathscr{X}\backslash K_\delta^{(1)}$ can be found such that $\mu_2(K^{(2)})>\varepsilon$. In view of the above assumption

$$\sup_\mu \mu(\mathscr{X}\backslash K_\delta^{(1)}\backslash K_\delta^{(2)})=\sup_\mu \mu(\mathscr{X}\backslash[K^{(1)}\cup K^{(2)}]_\delta)>\varepsilon.$$

Therefore one can find a measure $\mu_3\in M$ such that $\mu_3(\mathscr{X}\backslash K_\delta^{(1)}\backslash K_\delta^{(2)})>\varepsilon$ and a compactum $K^{(3)}\subset\mathscr{X}\backslash K_\delta^{(1)}\backslash K_\delta^{(2)}$ such that $\mu_3(K^{(3)}>\varepsilon$. Continuing this process we construct a sequence of measures μ_n and compacta $K^{(n)}$ such that $\mu_n(K^{(n)})>\varepsilon$, $K^{(n)}\subset\mathscr{X}\backslash K_\delta^{(1)}\backslash\ldots\backslash K_\delta^{(n-1)}$. Let

$$\chi_i(x)=1-\frac{2}{\delta}\varrho(x, K^{(i)})$$

for $x\in K_{\delta/2}^{(i)}$ and $\chi_i(x)=0$ for $x\notin K_{\delta/2}^{(i)}$. Since the distance between any two sets $K^{(n)}$ and $K^{(m)}$ is greater than δ, $\chi_n(x)\,\chi_m(x)=0$. Therefore the series $g_p(x)=\sum_{i=p}^{\infty}\chi_i(x)$ is convergent for each $x\in\mathscr{X}$ and the function $g_p(x)$ is continuous and bounded by 1. Since a weakly convergent subsequence can be chosen from the sequence $\{\mu_n\}$ we can assume without loss of generality that the sequence $\{\mu_n\}$ is weakly convergent to a certain measure μ. Then

$$\lim_{n\to\infty}\int g_p(x)\,\mu_n(dx)=\int g_p(x)\,\mu(dx).$$

Since

$$\int g_p(x)\,\mu_n(dx)\geqslant\int \chi_n(x)\,\mu_n(dx)\geqslant\varepsilon \quad \text{for} \quad n>p,$$

the inequality $\int g_p(x)\,\mu(dx)\geqslant\varepsilon$ is satisfied for all p. This is, however, impossible since $g_p(x)\to 0$ for all x as $p\to\infty$ and $0\leqslant g_p(x)\leqslant 1$, so that in view of Lebesgue's theorem $\lim_{p\to\infty}\int g_p(x)\,\mu(dx)=0$. We have thus proved the existence – for each $\varepsilon>0$ and $\delta>0$ – of a compactum K such that $\mu(\mathscr{X}\backslash K_\delta)\leqslant\varepsilon$ for all $\mu\in M$. Fixing $\varepsilon>0$ we construct a compactum $K^{(r)}$ such that

$$\sup_\mu \mu(\mathscr{X}\backslash K_{1/2^r}^{(r)})\leqslant\frac{\varepsilon}{2^r}.$$

Then $K=\bigcap_{r=1}^{\infty} K_{1/2^r}^{(r)}$ will be a compactum and

$$\mu(\mathscr{X}-K)\leqslant\sum_{r=1}^{\infty}\mu(\mathscr{X}\backslash K_{1/2^r}^{(r)})\leqslant\sum_{r=1}^{\infty}\varepsilon 2^{-r}=\varepsilon.$$

The necessity of condition b) is verified.

Sufficiency. Let the conditions of the theorem be satisfied. It follows from condition b) that a sequence of compacta K^n can be constructed such that

$$K^n\subset K^{n+1},\quad \mu\Big(\bigcup_n K^n\Big)=\mu(\mathscr{X})$$

for all $\mu\in M$ and $\mu(\mathscr{X}\backslash K^n)\leqslant\varepsilon_n$, $\varepsilon_n\downarrow 0$.

Let F be a countable set of functions $f_n\in\mathscr{C}_{\mathscr{X}}$, such that for all m the functions $\{f_n\}$ restricted to K^m are everywhere dense in $\mathscr{C}_{K^m}$. The existence of such a countable set follows from the separability of the spaces $\mathscr{C}_{K^m}$ and the feasibility of extending any function in $\mathscr{C}_{K^m}$ to a function in $\mathscr{C}_{\mathscr{X}}$. Let μ_n be an arbitrary sequence of measures in M. We choose a subsequence n_k such that for all $f\in F$ the limit

$$\lim_{k\to\infty}\int f(x)\,\mu_{n_k}(dx)=L(f)$$

exists. We now show that this limit exists for all $\varphi\in\mathscr{C}_{\mathscr{X}}$. Indeed, for any φ, $\|\varphi\|_{\mathscr{X}}\leqslant 1$, and $\varepsilon>0$, $\delta>0$ one can find a function $f\in F$ such that $\sup\{|f(x)-\varphi(x)|;\,x\in K^m\}\leqslant\delta$, $\mu(\mathscr{X}\backslash K^m)\leqslant\varepsilon$ and $\|f\|\leqslant 1$. Therefore

$$\left|\int f(x)\,\mu_{n_k}(dx)-\int\varphi(x)\,\mu_{n_k}(dx)\right|\leqslant\delta\sup_{\mu}\mu(\mathscr{X})+2\varepsilon,$$

and hence

$$\left|\overline{\lim_{k\to\infty}}\int\varphi(x)\,\mu_{n_k}(dx)-\underline{\lim_{k\to\infty}}\int\varphi(x)\,\mu_{n_k}(dx)\right|\leqslant 4\varepsilon+2\delta\sup_{\mu}\mu(\mathscr{X}).$$

Since $\varepsilon>0$ and $\delta>0$ are arbitrary, it follows that

$$\overline{\lim_{k\to\infty}}\int\varphi(x)\,\mu_{n_k}(dx)=\underline{\lim_{k\to\infty}}\int\varphi(x)\,\mu_{n_k}(dx)=$$

$$=\lim_{k\to\infty}\int\varphi(x)\,\mu_{n_k}(dx).$$

Hence for all $\varphi\in\mathscr{C}_{\mathscr{X}}$ the limit $\lim_{k\to\infty}\int\varphi(x)\,\mu_{n_k}(dx)$ exists.

We shall denote this limit by $L(\varphi)$. Clearly $L(\varphi)$ satisfies conditions 1) and 2) of Theorem 1, Section 2, Chapter V. Next if $\varphi=0$ for $x\in K^m$, then

$$|L(\varphi)|=\left|\lim_{k\to\infty}\int\varphi(x)\,\mu_{n_k}(dx)\right|\leqslant\|\varphi\|_{\mathscr{X}}\,\varepsilon_m.$$

Therefore condition 3) of this theorem is also satisfied. Hence the measure μ exists such that $L(\varphi)=\int\varphi(x)\,\mu(dx)$. The sufficiency of conditions of the theorem is thus established. □

Remark. The completeness of the space $\mathscr{X}$ was not utilized in the course of the proof of the sufficiency part. (It was also not utilized in the proof of the sufficiency part of Theorem 1, Section 2, Chapter V.)

Corollary. *If $\mathscr{X}$ is a complete metric separable space and the sequence of measures μ_n is such that for all $\varphi\in\mathscr{C}_{\mathscr{X}}$ the limit*

$$L(\varphi)=\lim_{n\to\infty}\int\varphi(x)\,\mu_n(dx)$$

exists, then there exists a measure μ such that

$$L(\varphi)=\int\varphi(x)\,\mu(dx),$$

i.e. the sequence of measures μ_n is necessarily weakly convergent.

Proof. We first show that the set $\{\mu_n\}$ is weakly compact. Condition a) of Theorem 1 is satisfied for this set. Assume that condition b) is not satisfied. In the same manner as in the proof of the necessity of condition b) of Theorem 1, we can construct for some $\varepsilon>0$ a subsequence μ_{n_k} and compacta $K^{(k)}$ located at a distance at least δ from each other and such that $\mu_{n_k}(K^{(k)})\geqslant\varepsilon$. Let the functions $\chi_i(x)$ be defined in the same manner as in the proof of Theorem 1. Define for each prime p the function

$$\psi_p(x)=\sum_{m=1}^{\infty}\chi_{p^m}(x).$$

Functions $\psi_p(x)$ satisfy relations $0\leqslant\psi_p(x)\leqslant1$ and $\psi_p(x)\cdot\psi_{p'}(x)=0$ for $p\neq p'$; $\psi_p\in\mathscr{C}_{\mathscr{X}}$ and hence the limit

$$L(\psi_p)=\lim_{k\to\infty}\int\psi_p(x)\,\mu_{n_k}(dx)\ \text{exists.}$$

Note that for $k=p^m$

$$\int \psi_p(x)\,\mu_{n_k}(dx) \geqslant \int \chi_k(x)\,\mu_{n_k}(dx).$$

Therefore

$$L(\psi_p)=\lim_{m\to\infty}\int \psi_p(x)\,\mu_{n_{p^m}}(dx)\geqslant \varepsilon.$$

Hence for any N

$$L(1)\geqslant L\Big(\sum_{j=1}^{N}\psi_{p_j}\Big)\geqslant N\varepsilon$$

(here $p_1,\ldots,p_N$ are distinct primes), since $L(\varphi)\leqslant L(f)$ for $\varphi\leqslant f$. This contradicts the fact that $L(1)$ is finite. Hence condition b) of the theorem is fulfilled. Let μ_{n_k} be a weakly convergent subsequence and let μ be its limit. Then

$$L(\varphi)=\lim_{k\to\infty}\int \varphi(x)\,\mu_{n_k}(dx)=\int \varphi(x)\,\mu(dx).$$

The assertion of the corollary is thus established.

We now consider the relation between the weak convergence of measures and the convergence of the values of measures on individual sets.

Definition. *Let μ be a finite measure on $\mathfrak{B}$. The set $A\in\mathfrak{B}$ is called the set of continuity of the measure μ if $\mu(A')=0$, where A' is the boundary of the set A. Hereafter, we shall use the notation $[A]$ to denote the closure of A and* Int A *for the set of interior points of A.*

Theorem 2. *In order that the sequence of measures μ_n converge weakly to measure μ it is necessary and sufficient that for each set A which is the set of continuity of measure μ the relation $\mu_n(A)\to\mu(A)$ as $n\to\infty$ be satisfied.*

Proof. We first establish the necessity of the condition stated in the theorem. Let μ_n converge weakly to μ and let A be an arbitrary set in $\mathfrak{B}$. We set

$$g_m(x)=\exp\{-m\varrho(x,A)\},$$

where $\varrho(x,A)$ is the distance from the point x to the set A. Since $g_m(x)=1$ for $x\in[A]$ and $g_m(x)\to 0$ for $x\notin[A]$, we have

$$\lim_{m\to\infty}\int g_m(x)\,\mu(dx)=\mu([A]),$$

and hence for every $\varepsilon>0$ an m can be found such that

$$\int g_m(x)\,\mu(dx)\leqslant\mu([A])+\varepsilon.$$

Therefore

$$\mu([A])\geqslant\int g_m(x)\,\mu(dx)-\varepsilon=\lim_{n\to\infty}\int g_m(x)\,\mu_n(dx)-\varepsilon\geqslant\overline{\lim_{n\to\infty}}\,\mu_n(A)-\varepsilon,$$

and hence since $\varepsilon>0$ is arbitrary

$$\overline{\lim_{n\to\infty}}\,\mu_n(A)\leqslant\mu([A]).$$

Thus

$$\mu([\mathscr{X}\setminus A])\geqslant\overline{\lim_{n\to\infty}}\,\mu_n(\mathscr{X}\setminus A)=\lim_{n\to\infty}\mu_n(\mathscr{X})-\underline{\lim_{n\to\infty}}\,\mu_n(A)=\mu(\mathscr{X})-\underline{\lim_{n\to\infty}}\,\mu_n(A).$$

Taking into account that $\mu([\mathscr{X}\setminus A])=\mu(\mathscr{X})-\mu(\operatorname{Int}A)$, we obtain

$$\mu(\operatorname{Int}A)\leqslant\underline{\lim_{n\to\infty}}\,\mu_n(A)\leqslant\overline{\lim_{n\to\infty}}\,\mu_n(A)\leqslant\mu([A]).\tag{1}$$

Since $\mu(\operatorname{Int}A)=\mu([A])$ for the set of continuity of measure μ the necessity of the condition of the theorem follows from (1).

To prove the sufficiency we take an arbitrary function f in $\mathscr{C}_{\mathscr{X}}$. The boundary of the set $\{x:a\leqslant f(x)<b\}$ belongs to the set $\{x:f(x)=a\}\cup\cup\{x:f(x)=b\}$. The sets $A_c=\{x:f(x)=c\}$ are disjoint for different values of c. Therefore there exists at most a countable number of c satisfying $\mu(A_c)>0$. We choose a sequence of numbers a_k, $k=1,\dots,N$ such that $\mu(A_{a_k})=0$, $a_k<a_{k+1}<a_k+\varepsilon$, $a_1<-\|f\|$, $a_N>\|f\|$. Denote by E_k the set $\{x:a_k\leqslant f(x)<a_{k+1}\}$. The sets E_k are the sets of continuity of measure μ. Therefore $\mu_n(E_k)\to\mu(E_k)$. Hence

$$\begin{aligned}\overline{\lim_{n\to\infty}}&\left|\int f(x)\,\mu_n(dx)-\int f(x)\,\mu(dx)\right|\leqslant\\&\leqslant\overline{\lim_{n\to\infty}}\left|\int f(x)\,\mu_n(dx)-\sum_{k=1}^{N-1}a_k\mu_n(E_k)\right|+\\&+\left|\int f(x)\,\mu(dx)-\sum_{k=1}^{N-1}a_k\mu(E_k)\right|\leqslant2\varepsilon\sum_{k=1}^{N-1}\mu(E_k)=2\varepsilon\mu(\mathscr{X}).\end{aligned}$$

Since $\varepsilon > 0$ is arbitrary we obtain the proof of the sufficiency condition. The theorem is thus proved. □

We now present theorems dealing with the weak convergence of measures which utilize the condition of weak compactness. In all these theorems the following single fact is utilized: a weakly compact sequence which possesses a unique limit point is weakly convergent.

We say that sequence $f_n \in \mathscr{C}_{\mathscr{X}}$ is *weakly convergent to* f if the functions f_n are jointly bounded and $f_n(x)$ converges to $f(x)$ for all $x \in \mathscr{X}$. Using this notion we naturally define a *weakly closed set of functions* and a *weak closure* of a *set* of *functions*.

Theorem 3. *In order that a sequence μ_n be weakly convergent to a certain measure μ it is necessary and sufficient that it be weakly compact and that for some set of functions $F_0 \subset \mathscr{C}_{\mathscr{X}}$ whose weak closure coincides with $\mathscr{C}_{\mathscr{X}}$ for all $f \in F_0$ the following relation be satisfied:*

$$\lim_{n \to \infty} \int f(x)\, \mu_n(dx) = \int f(x)\, \mu(dx).$$

Proof. We show that all weakly convergent subsequences of sequence μ_n converge in measure to μ. Indeed if for all $f \in \mathscr{C}_{\mathscr{X}}$

$$\lim_{k \to \infty} \int f(x)\, \mu_{n_k}(dx) = \int f(x)\, \bar{\mu}(dx)$$

then for $f \in F_0$ the equality

$$\int f(x)\, \bar{\mu}(dx) = \int f(x)\, \mu(dx)$$

is satisfied. However it follows from the Lebesgue theorem on bounded convergence that the set of $f \in \mathscr{C}_{\mathscr{X}}$ such that

$$\int f(x)\, \bar{\mu}(dx) = \int f(x)\, \mu(dx)$$

is weakly closed. Hence this relation is valid for all f belonging to the weak closure of F_0, i.e. for all $f \in \mathscr{C}_{\mathscr{X}}$. The sufficiency of the conditions of the theorem is thus established while the necessity is obvious. □

In the case when measures correspond to random processes, it is convenient to apply theorems in which the convergence of marginal distributions is postulated. We now prove a general theorem which enables us to deduce corollaries of this kind.

Theorem 4. *Let μ_n be a weakly compact sequence, μ be a certain measure, $\mathfrak{A}_0$ be a class of open sets closed under finite unions and intersections and satisfying conditions:*

1) *σ-closure of $\mathfrak{A}_0$ contains all the open sets,*

2) *all the sets belonging to $\mathfrak{A}_0$ are sets of continuity of measure μ. If $\lim_{n\to\infty} \mu_n(A)=\mu(A)$ for all $A\in\mathfrak{A}_0$, then μ_n is weakly convergent to μ.*

Proof. Let μ_{n_k} be a sequence weakly convergent to the measure $\bar{\mu}$. It follows from (1) that for all $A\in\mathfrak{A}_0$ the relations

$$\bar{\mu}(A)=\bar{\mu}(\operatorname{Int} A)\leqslant \varliminf_{k\to\infty} \mu_{n_k}(A)=\mu(A)$$

are satisfied. Therefore for all $A\in\mathfrak{A}_0$ the inequality $\bar{\mu}(A)\leqslant\mu(A)$ is valid. Clearly this relation is fulfilled for a monotone class of sets and this class contains all the sets belonging to $\mathfrak{A}_0$. Therefore this inequality is satisfied for every open set as well. And since each closed set is an intersection of a decreasing sequence of open sets, we have $\bar{\mu}(F)\leqslant\mu(F)$ also for every closed set F. Therefore for all sets $A\in\mathfrak{A}_0$ we have $\bar{\mu}(A')\leqslant\mu(A')=0$. But then $\bar{\mu}(A)=\lim_{k\to\infty}\mu_{n_k}(A)=\mu(A)$. Since the measures μ and $\bar{\mu}$ coincide on $\mathfrak{A}_0$ they also coincide on $\mathfrak{B}$. Thus all the limit points of the sequence μ_n coincide with μ. The theorem is thus proved. □

Remark. In the case of measures on different functional spaces the class $\mathfrak{A}_0$ is usually chosen to be the class of all open cylindrical sets of continuity of measure μ.

Given the weak convergence of measures we may establish the convergence of integrals for some discontinuous functions. Here we are utilizing the fact that the set of points of discontinuity of a $\mathfrak{B}$-measurable function is also a $\mathfrak{B}$-measurable set.

Lemma. *If μ_n is weakly convergent to μ, then*

$$\lim_{n\to\infty}\int f(x)\,\mu_n(dx)=\int f(x)\,\mu(dx)$$

for every $\mathfrak{B}$-measurable μ-almost everywhere continuous and bounded function $f(x)$.

Proof. Let Λ be the set of points of discontinuity of function $f(x)$. Set $\mathscr{G}_\alpha=\{x:f(x)<\alpha\}$ and let $\mathscr{G}'_\alpha$ be the boundary of the set $\mathscr{G}_\alpha$. For $\alpha<\beta$ the set $\mathscr{G}'_\alpha\cap\mathscr{G}'_\beta$ is contained in the intersection of the sets $[\mathscr{G}_\alpha]\cap[X\setminus\mathscr{G}_\beta]$ and therefore for $x\in\mathscr{G}'_\alpha\cap\mathscr{G}'_\beta$ the inequalities $\liminf_{y\to x} f(y)\leqslant\alpha$, $\limsup_{y\to x} f(y)\geqslant\beta$ are satisfied. Thus $\mathscr{G}'_\alpha\cap\mathscr{G}'_\beta\subset\Lambda$ and the sets $\mathscr{G}'_\alpha\setminus\Lambda$ with different α are disjoint. Therefore at most a countable number of α exists such that $\mu(\mathscr{G}'_\alpha)=\mu(\mathscr{G}'_\alpha\setminus\Lambda)>0$, i.e. all the sets $\mathscr{G}_\alpha$ except possibly a finite number of them are sets of continuity of measure μ. Therefore for all α except possibly a countable number of them

$$\lim_{n\to\infty} \mu_n(\{x: f(x)<\alpha\}) = \mu(\{x: f(x)<\alpha\}). \tag{2}$$

We obtain from the formula of change of variables in integrals that

$$\int f(x)\,\mu_n(dx) = \int \alpha d_\alpha \mu_n(\{x: f(x)<\alpha\}), \tag{3}$$

$$\int f(x)\,\mu(dx) = \int \alpha d_\alpha \mu(\{x: f(x)<\alpha\}). \tag{4}$$

The assertion of the lemma now follows from equalities (2)–(4). □

Consider finally conditions of weak convergence of measures in linear normed spaces. Let $\mathscr{X}$ be a separable Banach space, L be a linear set of linear functionals on $\mathscr{X}$ such that the minimal σ-algebra with respect to which all the functionals l and L are measurable coincides with the σ-algebra of all Borel sets $\mathfrak{B}$ of the space $\mathscr{X}$. Denote by $\chi_n(l)$ and $\chi(l)$ the characteristic functionals of measures μ_n and μ respectively:

$$\chi_n(l) = \int e^{il(x)}\,\mu_n(dx), \qquad \chi(l) = \int e^{il(x)}\,\mu(dx).$$

Theorem 5. *In order that a sequence of measures μ_n converge weakly to measure μ it is necessary and sufficient that the sequence μ_n be weakly compact and that for all $l \in L$ the equality*

$$\lim_{n\to\infty} \chi_n(l) = \chi(l) \tag{5}$$

be satisfied.

Proof. The necessity of conditions of the theorem is obvious. To prove sufficiency we can show, as before, that each limit point of sequence μ_n coincides with μ. Let $\bar{\mu}$ be such a limit point, then for all $l \in L$

$$\chi(l) = \bar{\chi}(l) = \int e^{il(x)}\,\bar{\mu}(dx).$$

The equality $\mu = \bar{\mu}$ thus follows from the fact that the characteristic functionals coincide. □

§ 2. Conditions for Weak Convergence of Measures in Hilbert Spaces

In this section $\mathscr{X}$ denotes a separable Hilbert space and $\mathfrak{B}$ denotes a σ-algebra of Borel sets in $\mathscr{X}$. Measures on $\mathfrak{B}$ are considered and the conditions for weak compactness and weak convergence of these measures are studied. As it is seen from the results of the preceding section, the basic difficulty is to obtain conditions for weak compactness of the family of measures. It turns out that in the case of a Hilbert space one can state necessary and sufficient conditions for the weak compactness of a family of measures in terms of characteristic functionals.

Hereafter the following notation will be used for certain families of linear operators in $\mathscr{X}$: T_c will denote the set of all symmetric nonnegative completely continuous operators, S – the set of all kernel operators and S_a – the subset of S consisting of operators with the trace at most a.

First we prove a convenient criterion of compactness of a set in $\mathscr{X}$.

Lemma 1. *For any $A \in T_c$ the set $\{x:|A^{-1}x| \leqslant 1\}$ is compact. For each compactum $K \subset \mathscr{X}$ one can find an operator $A \in T_c$ such that $K \subset \{x:|A^{-1}x| \leqslant 1\}$.*

Proof. The set $\{x:|A^{-1}x| \leqslant 1\}$ is the image of a unit sphere under transformation A and is compact in view of the complete continuity of operator A. Let K be a compact set and $\{e_k\}$ an arbitrary orthonormal basis in $\mathscr{X}$. Set $x^k = (x, e_k)$. It follows from the compactness of K that a sequence $c_n \downarrow 0$ exists such that $\sum_{k \geqslant n} (x^k)^2 \leqslant c_n$ for all $x \in K$. We now choose numbers $d_n \downarrow 0$ such that the relations $\sum_n d_n = +\infty$ and $\sum_n d_n c_n < \infty$ will be satisfied. Then

$$\sum_{k=1}^{\infty} (x^k)^2 \sum_{j=1}^{k} d_j = \sum_{k=1}^{\infty} d_k \sum_{j \geqslant k} (x^j)^2 \leqslant \sum_{k=1}^{\infty} d_k c_k < \infty .$$

We define an operator A for which e_k are eigenvectors and

$$A e_k = \sqrt{\frac{\sum_{j=1}^{\infty} d_j c_j}{\sum_{j=1}^{k} d_j}}\, e_k .$$

Then

$$|A^{-1}x|^2 = \sum_{k=1}^{\infty} (x^k)^2 \frac{\sum_{j=1}^{k} d_j}{\sum_{j=1}^{\infty} d_j c_j} \leqslant 1$$

for all $x \in K$. The fact that A is a completely continuous operator follows from condition

$$\lim_{k\to\infty} \sum_{j=1}^{\infty} d_j c_j \left(\sum_{j=1}^{k} d_j\right)^{-1} = 0.$$

The lemma is thus proved. □

Theorem 1. *Let M be a family of measures, $\chi_\mu(z)$ with $z \in \mathscr{X}$, be the characteristic functional of the measure $\mu \in M$. For weak compactness of the set M it is necessary and sufficient that:* a) *$\chi_\mu(0)$ be bounded for $\mu \in M$;* b) *for any $\varepsilon > 0$ one can find an operator B, $B \in T_c$ and for each $\mu \in M$ one can find an operator $A_\mu \in S_1$, such that* $\operatorname{Re}[\chi_\mu(0) - \chi_\mu(z)] \leqslant \varepsilon$ *provided only* $(BA_\mu Bz, z) \leqslant 1$.

Proof. The necessity of condition a) follows from the fact that for a weak compact set M condition a) of Theorem 1 in Section 1 is satisfied. We now establish the necessity of condition b). We can assume without loss of generality that $\chi_\mu(0) = 1$. In view of Theorem 1, Section 1, one can find for a weakly compact set M a compactum $K \in \mathscr{X}$ such that for all $\mu \in M$ the inequality $\mu(\mathscr{X} - K) < \varepsilon/2$ is satisfied. Then

$$\operatorname{Re}[\chi_\mu(0) - \chi_\mu(z)] = \\ = \int [1 - \cos(x, z)]\, \mu(dx) \leqslant \frac{\varepsilon}{2} + \frac{1}{2} \int_K (z, x)^2\, \mu(dx). \tag{1}$$

Let B be an operator in T_c such that $K \subset \{x : |B^{-1}x| \leqslant \sqrt{\varepsilon}\}$. Such an operator exists in view of Lemma 1. Let A_μ be a nonnegative symmetric operator for which

$$(BA_\mu Bz, z) = \frac{1}{\varepsilon} \int_{|B^{-1}x| \leqslant \sqrt{\varepsilon}} (x, z)^2\, \mu(dx). \tag{2}$$

Then

$$(A_\mu z, z) = \frac{1}{\varepsilon} \int_{|B^{-1}x| \leqslant \sqrt{\varepsilon}} (x, B^{-1}z)^2\, \mu(dx) = \frac{1}{\varepsilon} \int_{|B^{-1}x| \leqslant \sqrt{\varepsilon}} (B^{-1}x, z)^2\, \mu(dx)$$

and

$$\operatorname{Sp} A_\mu = \sum_k (A_\mu e_k, e_k) = \frac{1}{\varepsilon} \int_{|B^{-1}x| \leqslant \sqrt{\varepsilon}} |B^{-1}x|^2\, \mu(dx) \leqslant 1,$$

i.e. $A_\mu \in S_1$. Inequality

$$\operatorname{Re}[\chi_\mu(0) - \chi_\mu(z)] \leqslant \frac{\varepsilon}{2} + \frac{\varepsilon}{2} (BA_\mu Bz, z) \tag{3}$$

follows from relations (1) and (2). The necessity of condition b) is thus proved.

We now prove the sufficiency of the theorem's conditions. The boundedness of $\mu(\mathscr{X})$ follows from condition a). It follows from Theorem 1 of Section 1 that it is sufficient to show the existence, for each $\varepsilon>0$, of a compactum K such that $\mu(\mathscr{X}-K)\leqslant\varepsilon$ for all $\mu\in M$. Let $B\in T_c$ be an operator such that for all $\mu\in M$, the inequality $\operatorname{Re}[\chi_\mu(0)-\chi_\mu(z)]\leqslant\varepsilon/2$ is satisfied for $(BA_\mu Bz, z)\leqslant 1$, where $A_\mu\in S_1$. Then

$$\operatorname{Re}[\chi_\mu(0)-\chi_\mu(z)]\leqslant\frac{\varepsilon}{2}+2(BA_\mu Bz, z). \tag{4}$$

We now bound the integral

$$\int\left[1-\exp\left\{-\frac{\lambda}{2}(B^{-2}x, x)\right\}\right]\mu(dx)$$

where $\lambda>0$. Let $e_1, e_2, \ldots$ be a complete orthonormal sequence of eigenvectors of operator B and let β_k be the eigenvalue corresponding to e_k. Then

$$(B^{-2}x, x)=\sum_{k=1}^{\infty}\frac{(x^k)^2}{\beta_k^2},$$

where $x^k=(x, e_k)$. Note that

$$\exp\left\{-\frac{\lambda}{2}\sum_{k=1}^{n}\frac{(x^k)^2}{\beta_k^2}\right\}=$$

$$=(2\pi\lambda)^{-n/2}\prod_{k=1}^{n}\beta_k\int\ldots\int\exp\left\{i\sum_{k=1}^{n}x^k z^k-\frac{1}{2\lambda}\sum_{k=1}^{n}\beta_k^2(z^k)^2\right\}dz^1\ldots dz^n,$$

where $z^1,\ldots, z^n$ are real variables. Therefore

$$\int\left[1-\exp\left\{-\frac{\lambda}{2}(B^{-2}x, x)\right\}\right]\mu(dx)=$$

$$=\lim_{n\to\infty}\int(2\pi\lambda)^{-n/2}\prod_{k=1}^{n}\beta_k\int\ldots\int\left[1-e^{i\sum_{k=1}^{n}x^k z^k}\right]\times$$

$$\times\exp\left\{-\frac{1}{2\lambda}\sum_{k=1}^{n}\beta_k^2(z^k)^2\right\}dz^1\ldots dz^n\mu(dx)=$$

$$=\lim_{n\to\infty}(2\pi\lambda)^{-n/2}\prod_{k=1}^{n}\beta_k\int\ldots\int\operatorname{Re}\left[1-\chi_\mu\left(\sum_{k=1}^{n}z^k e_k\right)\right]\times$$

$$\times \exp\left\{-\frac{1}{2\lambda}\sum_{k=1}^{n}\beta_k^2(z^k)^2\right\}dz^1\ldots dz^n \leqslant \lim_{n\to\infty}(2\pi\lambda)^{-n/2}\prod_{k=1}^{n}\beta_k\times$$

$$\times\int\ldots\int\left[\frac{\varepsilon}{2}+2\left(BA_\mu B\sum_{k=1}^{n}z^k e_k, \sum_{k=1}^{n}z^k e_k\right)\right]\times$$

$$\times\exp\left\{-\frac{1}{2\lambda}\sum_{k=1}^{n}\beta_k^2(z^k)^2\right\}dz^1\ldots dz^n=$$

$$=\frac{\varepsilon}{2}+2\lambda\lim_{n\to\infty}\sum_{k=1}^{n}(A_\mu e_k, e_k)=\frac{\varepsilon}{2}+2\lambda\,\mathrm{Sp}\,A_\mu\leqslant\frac{\varepsilon}{2}+2\lambda.$$

Hence,

$$\left(1-\exp\left\{-\frac{\lambda C}{2}\right\}\right)\int_{(B^{-2}x,\,x)>C}\mu(dx)\leqslant$$

$$\leqslant\int_{(B^{-2}x,\,x)>C}\left[1-\exp\left\{-\frac{\lambda}{2}(B^{-2}x, x)\right\}\right]\mu(dx)\leqslant\frac{\varepsilon}{2}+2\lambda.$$

We choose $\lambda>0$ and $C>0$ in such a manner that $\dfrac{\varepsilon+4\lambda}{2-2\exp\left\{-\dfrac{\lambda C}{2}\right\}}<\varepsilon.$

Then for all $\mu\in M$ the inequality $\mu(\{x:|B^{-1}x|>C\})<\varepsilon$ is satisfied and hence $K=\{x:|B^{-1}x|\leqslant C\}$ is the required compactum. The theorem is thus proved. □

Remark 1. A simple example due to Yu. V. Prokhorov and V. V. Sazonov shows that for a weakly compact family of measures one cannot always find, for each $\varepsilon>0$, an operator $A\in S$ (the same for all measures $\mu\in M$) such that $\mathrm{Re}(\chi_\mu(0)-\chi_\mu(z))\leqslant\varepsilon$ for $(Az, z)\leqslant 1$. Let K be a compactum of the form $K=\{x:|B^{-1}x|\leqslant 1\}$, where $B\in T_c$ and $\mathrm{Sp}\,B^2=+\infty$. We define the measures μ_x by the equalities $\mu(\{x\})=\mu(\{-x\})=\frac{1}{2}$, and $\mu(\mathscr{X}\backslash\{x\}\backslash\{-x\})=0$ (here $\{x\}$ is the singleton containing x). Consider the family of measures $M=\{\mu_x, x\in K\}$. Since $\mu(\mathscr{X}\backslash K)=0$ for all $\mu\in M$, it follows that M is a weakly compact set. Next we have

$$\mathrm{Re}(\chi_{\mu_x}(0)-\chi_{\mu_x}(z))=1-\cos(x, z)=2\sin^2\frac{(x, z)}{2},$$

$$\sup_{\mu\in M}\mathrm{Re}(\chi_{\mu_x}(0)-\chi_{\mu_x}(z))=\sup_{|B^{-1}x|\leqslant 1}2\sin^2\frac{(B^{-1}x, Bz)}{2}=$$

$$= \sup_{|y|\leqslant 1} 2\sin^2\frac{(y, Bz)}{2} = \begin{cases} 2, & |Bz|\geqslant\pi, \\ 2\sin^2\dfrac{|Bz|}{2}, & |Bz|<\pi. \end{cases}$$

Let A be an operator such that $\operatorname{Re}(\chi_\mu(0)-\chi_\mu(z))\leqslant 1$ for $(Az, z)\leqslant 1$. Then $|Bz|<\pi$ for $(Az, z)\leqslant 1$. Hence $|BA^{-1/2}z|<\pi$ for $|z|\leqslant 1$, i.e. $BA^{-1/2}$ is a bounded operator. Therefore $B=CA^{1/2}$, where C is a bounded operator. Since $B=B^*=A^{1/2}C^*$, then $B^2=A^{1/2}C^*CA^{1/2}$ and for an orthonormal sequence $\{e_k\}$ of eigenvectors of operator A we obtain

$$\sum_k (B^2e_k, e_k)=\sum_k (C^*CA^{1/2}e_k, A^{1/2}e_k)\leqslant \|C^*C\| \sum_k (Ae_k, e_k).$$

Hence we necessarily have $\operatorname{Sp} A=+\infty$.

Condition b) of the theorem appears to be somewhat cumbersome. We therefore present certain modifications of this condition.

Lemma 2. *In order that a family of operators C_μ belonging to S admit representation in the form $C_\mu=BA_\mu B$ where $B\in T_c$, $A_\mu\in S_1$ it is necessary that in each orthonormal basis $\{e_k\}$ the series*

$$\operatorname{Sp} C_\mu=\sum_{k=1}^{\infty} (C_\mu e_k, e_k) \tag{5}$$

be convergent uniformly in μ and it is sufficient that the series be uniformly convergent in at least one basis.

Proof. Sufficiency. Let $\{e_k\}$ be the orthonormal basis in which series (5) converges uniformly. Set

$$\varrho_n=\sup_\mu \sum_{k=n}^{\infty} (C_\mu e_k, e_k)$$

and choose a sequence $\alpha_n>0$ such that $\sum \alpha_n=+\infty$ and $\sum \alpha_n\varrho_n<\infty$. Then

$$\sum_{n=1}^{\infty}\sum_{k=1}^{n} \alpha_k(C_\mu e_n, e_n)=\sum_{k=1}^{\infty}\alpha_k\sum_{n=k}^{\infty}(C_\mu e_n, e_n)\leqslant\sum_{k=1}^{\infty}\alpha_k\varrho_k.$$

Next let B be a symmetric operator such that $Be_k=\lambda_k e_k$ where $\lambda_k= =\left(\sum_{n=1}^{\infty}\alpha_n\varrho_n/\sum_{n=1}^{k}\alpha_n\right)^{1/2}$; since $\lambda_k\to 0$, $B\in T_c$. Set $A_\mu=B^{-1}C_\mu B^{-1}$. Then

$$\operatorname{Sp} A_\mu=\sum_{k=1}^{\infty}(B^{-1}C_\mu B^{-1}e_k, e_k)=\sum_{k=1}^{\infty}\frac{1}{\lambda_k^2}(C_\mu e_k, e_k)=$$

$$=\sum_{k=1}^{\infty}\sum_{n=1}^{k}\alpha_n(C_\mu e_k, e_k)/\sum_{n=1}^{\infty}\alpha_n\varrho_n\leqslant 1.$$

The sufficiency of the lemma's condition is thus shown.

Necessity. Let $C_\mu = BA_\mu B$, $B \in T_c$, $A_\mu \in S_1$ and $\{f_k\}$ be any orthonormal basis. Denote by P_N the projector on the linear subspace spanned by the vectors $f_N, f_{N+1}, \ldots$, and let $B_N = BP_N$. Then

$$\sum_{k=N}^{\infty} (BA_\mu Bf_k, f_k) = \sum_{k=1}^{\infty} P_N BA_\mu BP_N f_k, f_k) = \mathrm{Sp}(B_N^* A_\mu B_N) =$$
$$= \mathrm{Sp}(B_N^{*2} A_\mu) \leqslant \|B_N^{*2}\| \, \mathrm{Sp}\, A_\mu \leqslant \|B_N^*\|^2,$$

since $\mathrm{Sp}\, AB = \mathrm{Sp}\, B^* A^*$ and $\mathrm{Sp}\, AB \leqslant \mathrm{Sp}\, B \|A\|$, if $B \in S$.

To complete the proof we observe that $\|B_N^*\| \to 0$ as $N \to \infty$. Indeed, since the set K of vectors of the form Bx for $|x| \leqslant 1$ is compact and the functions $|P_N y|$ are continuous and monotonically decreasing to zero, it follows that $\sup\{|P_N y|, y \in K\} \to 0$ also as $N \to \infty$. But $\sup\{|P_N y|, y \in K\} = \sup_{|x| \leqslant 1} |P_N Bx| = \|B_N^*\|$. The lemma is thus proved. □

Denote by S^* the set of operators D such that the sum $\sum_k |(De_k, e_k)|$ is finite and bounded for all orthogonal bases $\{e_k\}$. Denote by $\mathrm{Sp}\,|D|$ the supremum of this sum. Denote by S_ε^* the set of all operators in S^* such that $\mathrm{Sp}\,|D| \leqslant \varepsilon$.

Corollary. *Let the family of operators C_μ admit representation for any $\varepsilon > 0$ in the form $C_\mu = B^{(\varepsilon)} A_\mu^{(\varepsilon)} B^{(\varepsilon)} + D^{(\varepsilon)}$, where $B^{(\varepsilon)} \in T_c$, $A_\mu^{(\varepsilon)} \in S_1$, $D^{(\varepsilon)} \in S_\varepsilon^*$. Then an operator $B \in T_c$ exists such that $C_\mu = BA'_\mu B$ and $A'_\mu \in S_1$.*

Indeed, it follows from Lemma 2 that it is sufficient to show that for some orthonormal basis $\{e_k\}$ the series $\sum_k (C_\mu e_k, e_k)$ converges uniformly in μ. But for every $\varepsilon > 0$

$$\sum_{k \geqslant N} (C_\mu e_k, e_k) \leqslant \sum_{k \geqslant N} (B^{(\varepsilon)} A_\mu^{(\varepsilon)} B^{(\varepsilon)} e_k, e_k) + \varepsilon$$

and as it follows from Lemma 2 by choosing N sufficiently large the sum

$$\sum_{k \geqslant N} (B^{(\varepsilon)} A_\mu^{(\varepsilon)} B^{(\varepsilon)} e_k, e_k)$$

becomes less than ε simultaneously for all N.

Denote by $\mathscr{H}_{\mathscr{X}}$ the Hilbert space of linear Hilbert-Schmidt operators defined on $\mathscr{X}$ (i.e. operators C such that $\mathrm{Sp}\, CC^* < \infty$) with the scalar product

$$(A, B) = \mathrm{Sp}\, AB^*.$$

Lemma 3. 1) *If $B \in T_c$, $A_\mu \in S_1$ and $B_\mu = BA_\mu B$ then the set of operators $B_\mu^{1/2}$ is compact in $\mathscr{H}_{\mathscr{X}}$.*

2) *For any compact set in $\mathscr{H}_{\mathscr{X}}$ of operators C_μ one can find an operator $B \in T_c$ such that $B^{-1} C_\mu^2 B^{-1} \in S_1$.*

Proof. 1) Let $\{e_k\}$ be a basis of eigenvectors of the operator B, let P_N be the projector on the subspace spanned by the vectors $e_1, \dots, e_N$. Then

$$\overline{\lim_{N\to\infty}} \sup_\mu \operatorname{Sp}(B_\mu^{1/2} - P_N B_\mu^{1/2} P_N)^2 =$$
$$= \overline{\lim_{N\to\infty}} \sup_\mu [\operatorname{Sp} B_\mu - \operatorname{Sp} P_N B_\mu^{1/2} P_N B_\mu^{1/2}] =$$
$$= \overline{\lim_{N\to\infty}} \sup_\mu [\operatorname{Sp}(I-P_N) B_\mu + \operatorname{Sp} P_N B_\mu^{1/2} (I-P_N) B_\mu^{1/2}] \leqslant$$
$$\leqslant \overline{\lim_{N\to\infty}} \sup_\mu [\operatorname{Sp}(I-P_N) B_\mu + \operatorname{Sp} B_\mu^{1/2} (I-P_N) B_\mu^{1/2}] =$$
$$= 2 \overline{\lim_{N\to\infty}} \sup_\mu \operatorname{Sp}(I-P_N) B_\mu = 2 \overline{\lim_{N\to\infty}} \sup_\mu \sum_{k>N} (B_\mu e_k, e_k) = 0$$

in view of Lemma 2. Hence the set $\{P_N B_\mu^{1/2} P_N\}$ is a ε-net for the set $\{B_\mu^{1/2}\}$ for N sufficiently large. The compactness of the set $\{P_N B_\mu^{1/2} P_N\}$ follows from the fact that it is a bounded set in the space of N^2 dimensions.

Assertion 1 of the lemma is thus proved.

2) Let $C_1, C_2, \dots, C_N$ be an ε-net in the set $\{C_\mu\}$. Denote by C'_μ the operator C_k with the smallest index k for which $\operatorname{Sp}(C_\mu - C_k)(C_\mu - C_k) \leqslant \varepsilon^2$. Then $C_\mu^2 = C_\mu'^2 + D_\mu$ where

$$D_\mu = C'_\mu (C_\mu - C'_\mu) + (C_\mu - C'_\mu) C'_\mu + (C_\mu - C'_\mu)^2.$$

It is easy to see that $D_\mu \in S^*$ and

$$\operatorname{Sp}|D_\mu| \leqslant 2\sqrt{\operatorname{Sp} C_\mu'^2 \operatorname{Sp}(C_\mu - C'_\mu)^2} + \varepsilon^2 = O(\varepsilon).$$

Now note that $C_\mu'^2$ takes on, for distinct μ, only a finite number of values and for each μ the series $\sum_k (C_\mu'^2 e_k, e_k)$ converges for any orthonormal basis $\{e_k\}$, therefore this convergence is uniform in μ. To complete the proof of the lemma we use the corollary of Lemma 2. □

The lemmas proved above enable us to find a more efficient condition of compactness of measures (as compared with that given in Theorem 1).

Theorem 2. *Let $M = \{\mu\}$ be a family of finite measures on $\mathfrak{B}$. In order that M be weakly compact it is necessary and sufficient that*

1) *for every $\varepsilon > 0$ there exists c such that $\mu\{x : |x| > c\} < \varepsilon$ for all $\mu \in M$;*

2) *for each c the family of operators B_μ^c defined by the relations*

$$\int_{|x| \leqslant c} (z, x)^2 \mu(dx) = |B_\mu^c z|^2,$$

is a compact set in $\mathscr{H}_\mathfrak{X}$. Condition 2) can be replaced by the following:

2′) *the series*

$$\sum_{k=1}^{\infty} |B_\mu^c e_k|^2$$

is convergent uniformly in μ for each $c>0$ in some basis (and hence in every basis).

Proof. Without loss of generality we shall assume that $\mu(\mathscr{X})=1$. The sufficiency follows from Theorem 1 since

$$\operatorname{Re}(1-\chi_\mu(z))\leqslant\tfrac{1}{2}\int\limits_{|x|\leqslant c}(x,z)^2\,\mu(dx)+$$

$$+\mu(\{x:|x|>c\})\leqslant\tfrac{1}{2}\,|B_\mu^c z|^2+\frac{\varepsilon}{2},$$

if c is sufficiently large, and in view of assertion 2 of Lemma 3 $(B_\mu^c)^2=$ $=BA_\mu^2 B$, where $B\in T_c$ and $\operatorname{Sp}A_\mu^2<1$. We now show the *necessity* of conditions of the theorem. Let M be a compact set and K be a compactum such that $\mu(\mathscr{X}-K)<\varepsilon$ for all $\mu\in M$. If c is such that $|x|\leqslant c$ for $x\in K$ it follows that $\mu(\{x:|x|>c\})<\varepsilon$. The necessity of condition 1 is proved. Next putting $V_c=\{x:|x|\leqslant c\}$ we obtain

$$|B_\mu^c z|^2=\int\limits_{K\cap V_c}(z,x)^2\,\mu(dx)+\int\limits_{V_c\setminus K}(z,x)^2\,\mu(dx).$$

Let $K=\{x:|B^{-1}x|\leqslant 1\}$, where $B\in T_c$. Then

$$\int\limits_{K\cap V_c}(z,x)^2\,\mu(dx)\leqslant\int\limits_{|B^{-1}x|\leqslant 1}(B^{-1}x,Bz)^2\,\mu(dx)=|A_\mu Bz|^2,$$

where operator A_μ is defined by the relation

$$|A_\mu z|^2=\int\limits_{|B^{-1}x|\leqslant 1}(B^{-1}x,z)^2\,\mu(dx).$$

Therefore

$$\operatorname{Sp}A_\mu^2=\int\limits_{|B^{-1}x|\leqslant 1}|B^{-1}x|^2\,\mu(dx)\leqslant 1.$$

Thus

$$\int\limits_{K\cap V_c}(z,x)^2\,\mu(dx)\leqslant|A_\mu Bz|^2,$$

where $A_\mu^2\in S_1$, and $B\in T_c$.

On the other hand if D is defined by equality

$$\int_{V_c\setminus K} (z, x)^2 \, \mu(dx) = (Dz, z),$$

then

$$\operatorname{Sp} D = \int_{V_c\setminus K} |x|^2 \, \mu(dx) \leqslant c^2 \mu(\mathscr{X}\setminus K).$$

By choosing compactum K in an appropriate manner, we can make this quantity arbitrarily small uniformly for all μ. Therefore in view of the corollary to Lemma 2 and assertion 1 of Lemma 3 the totality of operators $\{B_\mu^c\}$ is compact in $\mathscr{H}_{\mathscr{X}}$. The necessity of condition 2) is thus proved. The necessity of condition 2′) follows from Lemma 2). The theorem is thus proved. □

Corollary 1. *Let the correlation operators*

$$(A_\mu z, z) = \int (z, x)^2 \, \mu(dx)$$

exist for measures $\mu \in M$ and let $A_\mu^{1/2} \in \mathscr{H}_{\mathscr{X}}$. Then in order for the family of measures M to be compact, it is sufficient that the family of operators $\{A_\mu^{1/2}\}$ be compact in $\mathscr{H}_{\mathscr{X}}$. This condition is also necessary if one can find $c > 0$ such that $\mu(\{x : |x| > c\}) = 0$ for all $\mu \in M$.

Corollary 2. *Let operator $\tilde{A}_\mu$ be defined by equality*

$$(\tilde{A}_\mu z, z) = \int \frac{(z, x)^2}{1 + |x|^2} \, \mu(dx). \tag{6}$$

Then in order that a family of measures M be compact it is necessary and sufficient that:

1) *the set of operators $\{\tilde{A}_\mu^{1/2}\}$ be compact in $\mathscr{H}_{\mathscr{X}}$.*
2) $\lim\limits_{c \to \infty} \sup\limits_\mu \mu(\{x : |x| > c\}) = 0$.

We now state a convenient condition for weak convergence of measures.

Theorem 3. *In order that a sequence of measures μ_n converge weakly to measure μ it is necessary and sufficient that:*

1) *characteristic functionals $\chi_n(z)$ of measures μ_n for all $z \in \mathscr{X}$ converge to the characteristic functional $\chi(z)$ of measure μ.*

2) *the family of operators* $\{\tilde{A}_\mu^{1/2}\}$ *defined by equation* (6) *be a compactum in* $\mathscr{H}_\mathfrak{X}$.

Proof. The necessity of the conditions of the theorem follows from Corollary 2. In view of Corollary 2 and Theorem 5 of Section 1, in order to prove sufficiency one is required only to show that

$$\lim_{c\to\infty}\overline{\lim_{n\to\infty}}\,\mu_n(\{x:|x|>c\})=0. \tag{7}$$

Let

$$\nu_n(A)=\mu_n\left(\left\{x:\frac{x}{\sqrt{1+|x|^2}}\in A\right\}\right).$$

Since

$$\int (z,x)^2\,\nu_n(dx)=\int\frac{(z,x)^2}{1+|x|^2}\,\mu_n(dx)=(\tilde{A}_\mu z, z),$$

the family of measures ν_n is compact. Relation (7) is equivalent to the following:

$$\lim_{\varepsilon\to 0}\overline{\lim_{n\to\infty}}\,\nu_n(\{x:|x|>1-\varepsilon\})=0. \tag{8}$$

Assume that (8) is not satisfied. One can find a weakly convergent subsequence ν_{n_k} such that its limit $\bar{\nu}$ satisfies condition $\bar{\nu}(\{x:|x|=1\})>0$. Consequently for some z, $|z|=1$ and $\delta>0$ we have

$$\bar{\nu}(\{x:|x|=1,\ |(x,z)|>\delta\})>\delta.$$

Then for all $\varepsilon>0$ and for n_k sufficiently large,

$$\nu_{n_k}(\{x:|x|>1-\varepsilon;\ |(x,z)|>\delta\})>\delta,$$

and hence

$$\mu_{n_k}\left(\left\{x:\frac{|x|}{\sqrt{1+|x|^2}}>1-\varepsilon;\ \frac{|(x,z)|}{\sqrt{1+|x|^2}}>\delta\right\}\right)>\delta.$$

Therefore for each $\varepsilon>0$

$$\underline{\lim}_{k\to 0}\,\mu_{n_k}\left(\left\{x:|(x,z)|>\frac{\delta}{2\varepsilon}\right\}\right)>\delta.$$

On the other hand for each z

$$\lim_{c\to\infty}\overline{\lim_{n\to\infty}}\,\mu_n(\{x:|(x,z)|>c\})\leqslant$$

$$\leqslant \lim_{c\to\infty} \overline{\lim_{n\to\infty}} \frac{\pi}{\pi-1} \int \left(1 - \frac{\sin\frac{\pi}{c}(x,z)}{\frac{\pi}{c}(x,z)}\right) \mu_n(dx) =$$

$$= \lim_{c\to\infty} \overline{\lim_{n\to\infty}} \frac{\pi}{\pi-1} \frac{c}{2\pi} \int_{-\pi/c}^{\pi/c} (1-\chi_n(tz))\, dt =$$

$$= \lim_{c\to\infty} \frac{c}{2\pi-2} \int_{-\pi/c}^{\pi/c} (1-\chi(tz))\, dt = 0.$$

The contradiction obtained proves the theorem. □

Remark. If a sequence of measures μ_n is weakly convergent to measure μ, then $\chi_n(z) \to \chi(z)$ uniformly for $|z| \leqslant c$ for any $c > 0$. Let B be an operator in T_c such that operators $A_n \in S_1$ can be found with the property that $\operatorname{Re}(1-\chi_n(z)) < \varepsilon^2/8$ for $(BA_nBz, z) \leqslant 1$. If $|Bz| < 1$, then $(A_nBz, Bz) < 1$ all the more. Therefore

$$|\chi_n(z_1) - \chi_n(z_2)|^2 \leqslant \int |1 - e^{i(z_1 - z_2, x)}|^2 \mu_n(dx) =$$

$$= 2 \int (1 - \cos(z_1 - z_2, x))\, \mu_n(dx) = 2 \operatorname{Re}(1-\chi_n(z)) < \frac{\varepsilon^2}{4},$$

only if $|Bz_1 - Bz_2| < 1$. Since the set $\{Bz : |z| \leqslant c\}$ is a compactum, there exists a finite collection of points $z_1, \ldots, z_m$ such that $\inf_k |Bz - Bz_k| < 1$ for all z, $|z| \leqslant c$. Then

$$\overline{\lim_{n\to\infty}} \sup_{|z|\leqslant c} |\chi_n(z) - \chi(z)| \leqslant \overline{\lim_{n\to\infty}} \sup_k |\chi_n(z_k) - \chi(z_k)| +$$

$$+ 2 \overline{\lim_{n\to\infty}} \sup_k \sup \{|\chi_n(z) - \chi_n(z_k)|;\ |B(z - z_k)| \leqslant 1\} \leqslant \varepsilon.$$

Our assertion is thus proved. □

§3. Sums of Independent Random Variables with Values in a Hilbert Space

In this section we shall consider, in addition to probability measures on Hilbert spaces, random variables with values in Hilbert spaces having these measures as their distribution functions. Let $\{\Omega, \mathfrak{A}, \mathsf{P}\}$ be a probability space and $\{\mathscr{X}, \mathfrak{B}\}$ be a Hilbert space with a σ-algebra of Borel sets. A function $\xi(\omega)$ defined on Ω with values in $\mathscr{X}$ such that $\{\omega:\xi(\omega)\in B\}\in \in \mathfrak{A}$ for all $B\in\mathfrak{B}$ is called a random variable with values in $\mathscr{X}$. Hereafter we shall write ξ in place of $\xi(\omega)$. The distribution of the variable ξ is the measure

$$\mu_\xi(B)=\mathsf{P}\{\xi\in B\}=\mathsf{P}\{\omega:\xi(\omega)\in B\}.$$

With each random variable ξ one associates a subalgebra $\mathfrak{A}_\xi$ of algebra $\mathfrak{A}$ of events of the form $\{\omega:\xi(\omega)\in B\}$ where B is an arbitrary set in $\mathfrak{B}$. The random variables $\xi_1, \xi_2, \ldots, \xi_n, \ldots$ are called *independent* if the σ-algebras of events $\mathfrak{A}_{\xi_1}, \mathfrak{A}_{\xi_2}, \ldots, \mathfrak{A}_{\xi_n}, \ldots$ are independent, i.e. for any events $A_i\in\mathfrak{A}_{\xi_i}$

$$\mathsf{P}\left\{\bigcap_i A_i\right\}=\prod_i \mathsf{P}\{A_i\}.$$

We shall find expressions for characteristics of the sum of independent random variables in terms of characteristics of the summands. The function

$$\chi_\xi(z)=\mathsf{E}\, e^{i(z,\xi)}=\int e^{i(z,x)}\,\mu_\xi(dx),$$

is called the *characteristic functional* of a random variable ξ i.e. this is the characteristic functional of the distribution of the variable ξ.

If $\xi_1, \xi_2, \ldots, \xi_n$ are independent and $\chi_k(z)$ is the characteristic functional of the variable ξ_k, then

$$\mathsf{E}\exp\left\{i\left(z, \sum_{k=1}^n \xi_k\right)\right\}=\prod_{k=1}^n \chi_k(z). \tag{1}$$

Thus when adding independent random variables, the corresponding characteristic functionals are multiplied. To obtain the expression for the distribution of a sum of independent random variables, consider the case of two summands. Let $\xi=\xi_1+\xi_2$ and $\mu_\xi, \mu_{\xi_1}, \mu_{\xi_2}$, be the distributions of the variables ξ, ξ_1 and ξ_2 respectively. Then

$$\begin{aligned}\mu_\xi(B)=\mathsf{P}\{\xi_1+\xi_2\in B\}&=\mathsf{E}\mathsf{P}\{\xi_1+\xi_2\in B\mid \mathfrak{A}_{\xi_1}\}=\\&=\mathsf{E}\mathsf{P}\{\xi_2\in B-\xi_1\mid \mathfrak{A}_{\xi_1}\}=\mathsf{E}\mu_{\xi_2}(B-\xi_1),\end{aligned}$$

where $B-x$ is the set of y's such that $x+y\in B$. Note that $\mu_{\xi_2}(B-x)$ is a $\mathfrak{B}$-measurable function. Therefore

$$\mathsf{E}\mu_{\xi_2}(B-\xi_1)=\int \mu_{\xi_2}(B-x)\,\mu_{\xi_1}(dx).$$

Thus we have the formula

$$\mu_\xi(B)=\int \mu_{\xi_2}(B-x)\,\mu_{\xi_1}(dx)=\int \mu_{\xi_1}(B-x)\,\mu_{\xi_2}(dx), \tag{2}$$

i.e. the distribution of the sum of two independent random variables is a convolution of the distributions of the summands.

Convergence of the series in independent random variables. We present a number of inequalities which extend Kolmogorov's inequality and its various generalizations to the case of variables with values in a Hilbert space.

Lemma 1. *Let $\xi_1, \xi_2, \ldots, \xi_n$ be independent random variables such that $\mathsf{E}\xi_k=0$, $\mathsf{E}|\xi_k|^2<\infty$ and $\zeta_k=\sum_{i=1}^{k} \xi_i$. Then*

$$\mathsf{P}\{\sup_{k\leqslant n}|\zeta_k|>\varepsilon\}\leqslant\frac{1}{\varepsilon^2}\,\mathsf{E}|\zeta_n|^2. \tag{3}$$

The proof follows from the fact that $|\zeta_k|^2$ is a semi-martingale and from inequality (16) in Section 2 of Chapter II. □

Lemma 2. *If $\xi_1, \ldots, \xi_n$ are independent and $|\xi_i|\leqslant c$, then for any positive integer l and positive α,*

$$\mathsf{P}\{\sup_{k\leqslant n}|\zeta_k|>l\alpha+(l-1)\,c\}\leqslant\left(\mathsf{P}\left\{\sup_{k\leqslant n}|\zeta_k|>\frac{\alpha}{2}\right\}\right)^l.$$

Proof. Let $\chi_k=1$ if $|\zeta_k|>(l-1)\,\alpha+(l-2)\,c$ and $|\zeta_i|\leqslant(l-1)\,\alpha+(1-2)\,c$ for $i<k$ and let $\chi_k=0$ otherwise. Then

$$\begin{aligned}
&\mathsf{P}(\sup_{k\leqslant n}|\zeta_k|>l\alpha+(l-1)\,c\}=\\
&\quad=\sum_{i=1}^{n}\mathsf{P}\{\sup_{k\leqslant n}|\zeta_k|>l\alpha+(l-1)\,c\mid\chi_i=1\}\\
&\quad\leqslant\sum_{i=1}^{n}\mathsf{P}\{\sup_{i<k\leqslant n}|\zeta_k-\zeta_i|>\alpha\}\,\mathsf{P}\{\chi_i=1\}\leqslant\mathsf{P}\{\chi_i=1\}\leqslant\\
&\quad\leqslant\sup_{1\leqslant i\leqslant n}\mathsf{P}\{\sup_{i<k\leqslant n}|\zeta_k-\zeta_i|>\alpha\}\sum_{i=1}^{n}\mathsf{P}\{\chi_i=1\}\leqslant\\
&\quad\leqslant\mathsf{P}\{\sup_{1\leqslant i<k\leqslant n}|\zeta_k-\zeta_i|>\alpha\}\,\mathsf{P}\{\sup_{1\leqslant k\leqslant n}|\zeta_k|>(l-1)\,\alpha+(l-2)\,c\}.
\end{aligned}$$

Finally we note that

$$\mathsf{P}\{\sup_{1 \leqslant i < k \leqslant n} |\zeta_k - \zeta_i| > \alpha\} \leqslant \mathsf{P}\left\{\sup_{k \leqslant n} |\zeta_k| > \frac{\alpha}{2}\right\}.$$

The lemma is thus proved. □

A variable ξ is called *symmetric* if ξ and $-\xi$ are identically distributed.

Lemma 3. *If $\xi_1, \dots, \xi_n$ are symmetric independent random variables then*

$$\mathsf{P}\{\sup_{k \leqslant n} |\zeta_k| > \varepsilon\} \leqslant 2\mathsf{P}\{|\zeta_n| > \varepsilon\}.$$

Proof. Let $\chi_k = 1$ if $|\xi_k| > \varepsilon$ and $|\xi_i| \leqslant \varepsilon$ for $i < k$ and let $\chi_k = 0$ otherwise. Then

$$\mathsf{P}\{|\zeta_n| > \varepsilon, \chi_k = 1\} \geqslant \mathsf{P}\{(\zeta_n - \zeta_k, \zeta_k) \geqslant 0, \chi_k = 1\} =$$

$$= \mathsf{P}\{(\zeta_n - \zeta_k, \zeta_k) \geqslant 0 \mid \chi_k = 1\} \, \mathsf{P}\{\chi_k = 1\} \leqslant \tfrac{1}{2}\mathsf{E}\chi_k.$$

Therefore

$$\mathsf{P}\{|\zeta_n| > \varepsilon\} = \sum_{k=1}^{n} \mathsf{P}\{|\zeta_n| > \varepsilon, \chi_k = 1\} \geqslant \tfrac{1}{2} \sum_{k=1}^{n} \mathsf{E}\chi_k = \tfrac{1}{2}\,\mathsf{P}\{\sup_{k \leqslant n} |\zeta_k| > \varepsilon\}.$$

The lemma is proved. □

Lemma 4. *Let $\xi_1, \dots, \xi_n$ be independent random variables such that for all $k \leqslant n$*

$$\mathsf{P}\left\{\left|\sum_{j=k}^{n} \xi_j\right| > c\right\} < \alpha.$$

Then

$$\mathsf{P}\{\sup_{k \leqslant n} |\zeta_k| > a + c\} \leqslant \frac{1}{1-\alpha}\,\mathsf{P}\{|\xi_n| > a\}. \tag{4}$$

The proof of this assertion is analogous to the proof of Theorem 6, Section 3, Chapter II. □

Utilizing these lemmas we now prove Kolmogorov's three-series theorem in the case of a Hilbert space.

Theorem 1. *Let $\xi_1, \dots, \xi_n, \dots$ be a sequence of independent random variables with values in $\mathscr{X}$. Then in order that the series $\sum_{i=1}^{\infty} \xi_i$ converge, it is necessary that for any $c \geqslant 0$ the following series*

1) $\sum_{i=1}^{\infty} a_i, \quad a_i = \int_{|x| \leqslant c} x \mu_{\xi_i}(dx),$

2) $\sum_{i=1}^{\infty} \int_{|x| \leqslant c} |x-a_i|^2 \mu_{\xi_i}(dx)$,

3) $\sum_{i=1}^{\infty} \mathsf{P}\{|\xi_i|>c\}$,

be convergent and it is sufficient that the series converge for at least one $c>0$.

The sufficiency of the conditions of the theorem is proved in exactly the same manner as in the one-dimensional case (cf. Theorem 5, Section 3, Chapter II).

The necessity of condition 3) follows from the fact that for each $c>0$ only a finite number of events $\{|\xi_i|>c\}$ occur and from the Borel-Cantelli lemma. We shall prove only the necessity of conditions 1) and 2). Let $\xi'_k=\xi_k$ for $|\xi_k|\leqslant c$ and $\xi'_k=0$ for $|\xi_k|>c$. Since condition 3) yields that only a finite number of variables $\xi_k-\xi'_k$ do not vanish, the series $\sum_{k=1}^{\infty} \xi'_k$ is convergent as long as the series $\sum_{k=1}^{\infty} \xi_k$ is convergent. Therefore the variable $\sup_{n,p}\left|\sum_{k=n}^{n+p} \xi'_k\right|$ is bounded. In view of Lemma 2 for all positive integers l

$$\mathsf{P}\left\{\sup_n\left|\sum_{k=0}^{n} \xi'_k\right|>l(c+\alpha)\right\}\leqslant\left(\mathsf{P}\left\{\sup_{n,p}\left|\sum_{k=n}^{n+p} \xi'_k\right|>\frac{\alpha}{2}\right\}\right)^l.$$

We choose α such that

$$\mathsf{P}\left\{\sup_{n,p}\left|\sum_{k=n}^{n+p} \xi'_k\right|>\frac{\alpha}{2}\right\}\leqslant e^{-1}.$$

Then for all n

$$\mathsf{P}\left\{\left|\sum_{k=1}^{n} \xi'_k\right|>t\right\}\leqslant Ke^{-\lambda t},$$

where $\lambda=\dfrac{1}{c+\alpha}$, $K=e^{c+\alpha}$. It follows from this inequality that $\mathsf{E}\left|\sum_{k=1}^{n} \xi'_k\right|^s$ are uniformly bounded for all s and hence in view of the theorem on the passage to the limit under the sign of an integral, the limit

$$\lim_{n\to\infty} \mathsf{E}\left|\sum_{k=1}^{n} \xi'_k\right|^s=\mathsf{E}\left|\sum_{k=1}^{\infty} \xi'_k\right|^s$$

exists. In particular for $s=2$ the limit

$$\lim_{n\to\infty} \mathsf{E}\left|\sum_{k=1}^{n} \xi_k'\right|^2 = \lim_{n\to\infty}\left(\sum_{k=1}^{n} \mathsf{E}|\xi_k'-a_k|^2+\left|\sum_{k=1}^{n} a_k\right|^2\right).$$

Consequently, the series $\sum_{k=1}^{\infty} \mathsf{E}|\xi_k'-a_k|^2$ is convergent. But now it follows from the sufficiency of the conditions of the theorem that the series $\sum_{k=1}^{\infty} (\xi_k'-a_k)$ is convergent and since the series $\sum_{k=1}^{\infty} \xi_k'$ is also convergent, so is series $\sum_{k=1}^{\infty} a_k$. The theorem is thus proved.

Corollary. *In order that the series $\sum_{k=1}^{\infty} \xi_k$ of independent random variables converge it is sufficient that the series $\sum_{k=1}^{\infty} \mathsf{E}\xi_k$ and $\sum_{k=1}^{\infty} \mathsf{E}|\xi_k-\mathsf{E}\xi_k|^2$ converge where $\mathsf{E}\xi_k=\int x\mu_{\xi_k}(dx)$ is a vector in $\mathscr{X}$ such that for all $z\in\mathscr{X}$ $(\mathsf{E}\xi_k, z)=\int (x, z)\,\mu_{\xi_k}(dx)$.*

The conditions of convergence of series of independent random variables can be expressed in terms of characteristic functionals as well.

Theorem 2. *Let $\xi_1,\dots,\xi_n,\dots$ be independent random variables and let $\chi_n(z)$ be their characteristic functionals. For convergence of the series $\sum_{k=1}^{\infty} \xi_k$ it is necessary and sufficient that the product $\prod_{k=1}^{\infty} \chi_k(z)$ converge uniformly in each region $\{z:|z|\leqslant c\}$ to a certain characteristic functional $\chi(z)$.*

Proof. Necessity. Let $\zeta_n=\sum_{k=1}^{n} \xi_k$, $\zeta=\lim_{n\to\infty} \zeta_n$. Then for each $\delta>0$

$$\left|\prod_{k=1}^{\infty} \chi_k(z)-\prod_{k=1}^{n} \chi_k(z)\right|=|\mathsf{E}\, e^{i(z,\zeta)}-\mathsf{E}\, e^{i(z,\zeta_n)}|\leqslant$$

$$\leqslant \mathsf{E}|e^{i(\zeta-\zeta_n, z)}-1|\leqslant 2\mathsf{P}\{|\zeta-\zeta_n|>\delta\}+\delta|z|.$$

Since $\lim_{n\to\infty} \mathsf{P}\{|\zeta-\zeta_n|>\delta\}\to 0$ the necessity of the theorem's conditions is proved.

Sufficiency. We introduce mutually independent random variables ξ_k' which are independent of ξ_k having the same distribution as ξ_k. Set $\eta_k=\xi_k-\xi_k'$. We first prove the convergence of the series in η_k. Clearly,

$$\mathsf{E}\, e^{i(z,\eta_k)} = |\chi_k(z)|^2 .$$

It follows from inequality

$$1 - \prod_{k=1}^{n} |\chi_k(z)|^2 \leqslant 1 - \prod_{k=1}^{\infty} |\chi_k(z)|^2 = 1 - |\chi(z)|^2$$

and the fact that $|\chi(z)|^2$ is a characteristic functional of a certain measure, that the distribution of variables $\sum_{k=1}^{n} \eta_k$ forms a compact family of measures. Therefore

$$\lim_{c\to\infty} \sup_n \mathsf{P}\left\{\left|\sum_{k=1}^{n} \eta_k\right| > c\right\} = 0.$$

But then in view of Lemma 3

$$\lim_{c\to\infty} \mathsf{P}\left\{\sup_{n,p}\left|\sum_{k=n}^{n+p} \eta_k\right| > c\right\} \leqslant \lim_{c\to\infty} \mathsf{P}\left\{\sup_{n}\left|\sum_{k=1}^{n} \eta_k\right| > \frac{c}{2}\right\} \leqslant$$

$$\leqslant 2 \lim_{c\to\infty} \sup_n \mathsf{P}\left\{\left|\sum_{k=1}^{n} \eta_k\right| > \frac{c}{2}\right\} = 0.$$

Therefore the variable $\sup_{n,p}\left|\sum_{k=n}^{n+p} \eta_k\right|$ is bounded. In particular, the variable $\sup_k |\eta_k|$ is bounded. Since for c sufficiently large

$$0 < \mathsf{P}\{\sup_k |\eta_k| \leqslant c\} = \prod_{k=1}^{\infty} (1 - \mathsf{P}\{|\eta_k| > c\}) \leqslant \exp\left\{-\sum_{k=1}^{\infty} \mathsf{P}\{|\eta_k| > c\}\right\},$$

it follows that the series $\sum_{k=1}^{\infty} \mathsf{P}\{|\eta_k| > c\}$ is convergent. Let $\eta_k' = \eta_k$ for $|\eta_k| \leqslant c$ and $\eta_k' = 0$ for $|\eta_k| > c$. Then the variable

$$\sup_{n,p}\left|\sum_{k=n}^{n+p} \eta_k'\right|$$

is also bounded because $\eta_k' = \eta_k$ except possibly for a finite number of indices k. From this fact the convergence series $\sum_{k=1}^{\infty} \mathsf{E}|\eta_k'|^2$ follows (since $\mathsf{E}\eta_k' = 0$) analogously as was the case in Theorem 1. Hence in view of Theorem 1 the series

$$\sum_{k=1}^{\infty} \eta_k = \sum_{k=1}^{\infty} (\xi_k - \xi_k')$$

is convergent with probability 1. Therefore a sequence of vectors $x_1, \ldots, x_n \ldots$ belonging to $\mathscr{X}$ (which are the possible values of ξ'_k) can be found such that the series $\sum_{k=1}^{\infty} (\xi_k - x_k)$ will converge with probability 1. To complete the proof we must establish the convergence of the series $\sum_{k=1}^{\infty} x_k$. In view of the necessity of the conditions of the theorem which were proved the following infinite product

$$\prod_{k=1}^{\infty} e^{-i(z, x_k)} \chi_k(z)$$

is uniformly convergent for $|z| \leqslant c$. Since a δ can be found such that $|\chi(z)| > \frac{1}{2}$ for $|z| \leqslant \delta$, there exists the uniform limit

$$\lim_{n \to \infty} \prod_{k=1}^{n} e^{-i(z, x_k)} = \lim_{n \to \infty} \frac{1}{\chi(z)} \prod_{k=1}^{n} e^{-i(z, x_k)} \chi_k(z).$$

Therefore there exists the limit $\lim_{n \to \infty} \left(z, \sum_{k=1}^{n} x_k\right)$ uniformly in z for $|z| \leqslant \delta$ and hence uniformly in z for $|z| \leqslant c$ for any $c > 0$. This implies that $\sum_{k=1}^{n} x_k$ possesses the weak limit x:

$$\lim_{n \to \infty} \left(z, \sum_{k=1}^{n} x_k\right) = (z, x)$$

and that moreover $\left|\sum_{k=1}^{n} x_k\right|$ are jointly bounded. It follows from the uniform convergence that

$$\lim_{n \to \infty} \left(\sum_{k=1}^{n} x_k, \sum_{k=1}^{n} x_k\right) = \lim_{n \to \infty} \left(\sum_{k=1}^{n} x_k, x\right) = (x, x).$$

Therefore $\sum_{k=1}^{n} x_k$ converges weakly to x and $\left|\sum_{k=1}^{n} x_k\right| \to |x|$. Thus $\sum_{k=1}^{n} x_k \to$ $\to x$. The theorem is proved. □

Corollary. *If the series* $\sum_{k=1}^{\infty} \xi_k$ *converges in probability it then converges with probability* 1 *also.*

Indeed, in the proof of the necessity of conditions of Theorem 2 only convergence in probability of the series was utilized.

Infinitely divisible distributions in a Hilbert space. A distribution (measure) μ is called *infinitely divisible* if its characteristic functional $\chi(z)$ satisfies the condition: for any positive integer n there exists a characteristic functional $\chi_n(z)$ of a certain distribution such that $\chi(z)=(\chi_n(z))^n$.

We derive the general form of a characteristic functional of an infinitely divisible distribution.

Let ξ be a random variable with values in $\mathscr{X}$ such that $\mathsf{E}\, e^{i(z,\xi)}=\chi(z)$ and let $\xi_{n1},\dots,\xi_{nn}$ be independent identically distributed random variables such that $\mathsf{E}\, e^{i(z,\xi_{nk})}=\chi_n(z)$ and $\xi=\sum_{k=1}^{n}\xi_{nk}$. We show that, given an arbitrary $\varepsilon>0$ one can find c such that for all $k\leqslant n$,

$$\mathsf{P}\left\{\left|\sum_{j=k}^{n}\xi_{nj}\right|>c\right\}<\varepsilon. \tag{5}$$

Let S be a kernel operator such that

$$1-\operatorname{Re}\chi(z)\leqslant\frac{\varepsilon}{2}\quad\left(\varepsilon<\frac{1}{4}\right)\quad\text{for}\quad (Sz,z)\leqslant 1.$$

Then

$$|\operatorname{Im}\chi(z)|\leqslant\sqrt{1-(\operatorname{Re}\chi(z))^2}<\sqrt{\varepsilon}.$$

Therefore
$$|\arg\chi(z)|\leqslant\operatorname{arctg}\frac{\sqrt{\varepsilon}}{1-\frac{\varepsilon}{2}}<\frac{\pi}{4},$$

$$1-\operatorname{Re}(\chi_n(z))^{n-k}=1-|\chi(z)|^{\frac{n-k}{n}}\cos\left[\frac{n-k}{n}\arg\chi(z)\right]<$$
$$<1-|\chi(z)|\cdot\cos\arg\chi(z)\leqslant\frac{\varepsilon}{2}.$$

Utilizing inequality (7) Section 5 of Chapter V, we obtain

$$\mathsf{P}\left\{\left|\sum_{j=k}^{n}\xi_{nj}\right|>c\right\}\leqslant\left(\frac{\varepsilon}{2}+2\lambda\operatorname{Sp}S\right)(1-e^{-\lambda c^2/2})^{-1}.$$

In view of the latter inequality, one can choose λ and c such that (5) is fulfilled. We now obtain from Lemma 4 that

$$\mathsf{P}\left\{\sup_{k\leqslant n}\left|\sum_{j=1}^{k}\xi_{nj}\right|>2c\right\}\leqslant\frac{1}{1-\varepsilon}\,\mathsf{P}\left\{\left|\sum_{j=1}^{n}\xi_{nj}\right|>c\right\}\leqslant\frac{\varepsilon}{1-\varepsilon}. \tag{6}$$

Finally

$$\mathsf{P}\{\sup_{k\leqslant n}|\xi_{nk}|>4c\}\leqslant\mathsf{P}\left\{\sup_{k\leqslant n}\left|\sum_{j=1}^{k}\xi_{nj}\right|>2c\right\}\leqslant\frac{\varepsilon}{1-\varepsilon}.$$

Hence

$$\prod_{k=1}^{n}\mathsf{P}\{|\xi_{nk}|\leqslant 4c\}\geqslant 1-\frac{\varepsilon}{1-\varepsilon}$$

and

$$\exp\left\{-\sum_{k=1}^{n}\mathsf{P}\{|\xi_{nk}|>4c\}\right\}\geqslant\prod_{k=1}^{n}\mathsf{P}\{|\xi_{nk}|\leqslant 4c\}\geqslant\frac{1-2\varepsilon}{1-\varepsilon},$$

$$\sum_{k=1}^{n}\mathsf{P}\{|\xi_{nk}|>4c\}\leqslant\log\frac{1-\varepsilon}{1-2\varepsilon}.$$

The last inequality yields the following

Lemma 5. *For all c sufficiently large*

$$\sup_n n\mathsf{P}\{|\xi_{n1}|>c\}<\infty \tag{7}$$

and

$$\lim_{c\to\infty}\sup_n n\mathsf{P}\{|\xi_{n1}|>c\}=0. \quad \square \tag{8}$$

Define $\xi'_{ni}=\xi_{ni}$ for $|\xi_{ni}|\leqslant c$ and $\xi'_{ni}=0$ for $|\xi_{ni}|>c$, where c is such that

$$\sup_n n\mathsf{P}\{|\xi_{n1}|>c\}<\tfrac{1}{2}.$$

Then

$$\mathsf{P}\left\{\sup_{1\leqslant k\leqslant n}\left|\sum_{j=1}^{k}\xi'_{nj}\right|>\alpha\right\}\leqslant\mathsf{P}\left\{\sup_{1\leqslant k\leqslant n}\left|\sum_{j=1}^{k}\xi_{nj}\right|>\alpha\right\}+n\mathsf{P}\{|\xi_{n1}|>c\}.$$

It follows from (6) and the choice of c that for α sufficiently large

$$\mathsf{P}\left\{\sup_{1\leqslant k\leqslant n}\left|\sum_{j=1}^{k}\xi'_{nj}\right|>\alpha\right\}<\frac{1}{2}$$

holds for all n. Therefore in view of Lemma 2 the following quantities are uniformly bounded in n:

$$\mathsf{E}\left|\sum_{j=1}^{n}\xi'_{nj}\right|,\qquad\left|\mathsf{E}\sum_{j=1}^{n}\xi'_{nj}\right|\quad\text{and}\quad\mathsf{E}\left|\sum_{j=1}^{n}\xi'_{nj}\right|^{2}.$$

This in particular yields that

$$|\mathsf{E}\xi'_{n1}|=O\left(\frac{1}{n}\right). \tag{9}$$

Denote by μ_n the measure representing the distribution of variable ξ_{ni} and let π_n be the measure defined by relation

$$\pi_n(A)=n\int_A \frac{|x|^2}{1+|x|^2}\,\mu_n(dx).$$

Measures $\pi_n(A)$ are uniformly bounded:

$$\pi_n(\mathscr{X})\leqslant \mathsf{E}\left|\sum_{j=1}^{n}\xi'_{nj}\right|^2+n\mathsf{P}\{|\xi_{nj}|>c\}.$$

It follows from (8) that measures π_n satisfy

$$\lim_{c\to\infty}\sup_n \pi_n(\{x:|x|>c\})=0. \tag{10}$$

We now show that measures π_n are compact. To do this it is sufficient to show that for each $\varepsilon>0$ a kernel operator S exists such that

$$\pi_n(\mathscr{X})-\mathrm{Re}\int e^{i(z,x)}\,\pi_n(dx)\leqslant\varepsilon,$$

as long as $(Sz, z)\leqslant 1$. But

$$\pi_n(\mathscr{X})-\mathrm{Re}\int e^{i(z,x)}\,\pi_n(dx)=$$

$$=n\int(1-\cos(z,x))\frac{|x|^2}{1+|x|^2}\,\mu_n(dx)\leqslant n[1-\mathrm{Re}\,\chi_n(z)]=$$

$$=n\left[1-|\chi(z)|^{1/n}\cos\left(\frac{1}{n}\arg\chi(z)\right)\right]\leqslant n[1-|\chi(z)|^{1/n}]+$$

$$+n|\chi(z)|^{1/n}\left(1-\cos\left[\frac{1}{n}\arg\chi(z)\right]\right)\leqslant\frac{1-|\chi(z)|}{|\chi(z)|}+\frac{1}{2n}[\arg\chi(z)]^2.$$

Assume that $1-\mathrm{Re}\,\chi(z)\leqslant\frac{\varepsilon}{2}$ $(\varepsilon<1)$; then $|\mathrm{Im}\,\chi(z)|<\sqrt{\varepsilon}$. Hence in every connected region in which this assumption is satisfied, we have

$$|\arg\chi(z)|<\mathrm{arctg}\frac{\sqrt{\varepsilon}}{1-\frac{\varepsilon}{2}}<\frac{\pi}{4},\qquad 1-|\chi(z)|\leqslant\frac{\varepsilon}{2}.$$

If S is a kernel operator such that $1-\operatorname{Re}\chi(z)\leqslant\frac{\varepsilon}{2}$ for $(Sz, z)\leqslant 1$, we have for these z

$$\pi_n(\mathscr{X})-\operatorname{Re}\int e^{i(z,x)}\,\pi_n(dx)\leqslant\frac{\varepsilon/2}{1-\frac{\varepsilon}{2}}+\frac{1}{2n}\frac{\varepsilon}{\left(1-\frac{\varepsilon}{2}\right)^2}.$$

For $n>1$ and ε sufficiently small the r.h.s. is less than ε. The compactness of measures π_n is thus proved. □

Let a_n be defined by relation:

$$(a_n, z)=n\int\frac{(z, x)}{1+|x|^2}\,\mu_n(dx)=n\mathsf{E}\frac{(z, \xi_{ni})}{1+|\xi_{ni}|^2}.$$

It follows from (9) that a_n are jointly bounded. Finally we define symmetric operators V_n by the equality

$$(V_n z, z)=n\int\frac{(z, x)^2}{1+|x|^2}\,\mu_n(dx).$$

Note that

$$\operatorname{Sp} V_n=n\int\frac{|x|^2}{1+|x|^2}\,\mu_n(dx)=\pi_n(\mathscr{X})$$

and hence $\operatorname{Sp} V_n$ are uniformly bounded.

We choose a subsequence n' such that: 1) $\pi_{n'}$ is weakly convergent to π', 2) $a_{n'}$ is weakly convergent to some vector a and 3) for all z there exists the limit

$$\lim_{n'\to\infty}(V_{n'}z, z)=(Vz, z). \tag{11}$$

The last condition is attainable since in view of the uniform boundedness ($\|V_n\|\leqslant\operatorname{Sp} V_n$) it is sufficient that (11) be satisfied on a certain countable everywhere dense set in $\mathscr{X}$. Clearly V is also a kernel operator since

$$\operatorname{Sp} V\leqslant\varliminf_{n'\to\infty}\operatorname{Sp} V_{n'}.$$

Next set $\pi(A)=\pi'(A)$ if $0\notin A$ and $\pi(\{0\})=0$. We then have

$$\chi(z)=[\chi_{n'}(z)]^{n'}=\lim_{n'\to\infty}\Bigg[1+\frac{1}{n'}\Bigg\{i(a_{n'}, z)-\tfrac{1}{2}(V_{n'}z, z)+$$

$$+\int\left(e^{i(z,x)}-1-\frac{i(z, x)}{1+|x|^2}+\tfrac{1}{2}\frac{(z, x)^2}{1+|x|^2}\right)\frac{1+|x|^2}{|x|^2}\,\pi_{n'}(dx)\Bigg\}\Bigg]^{n'}=$$

$$=\exp\left\{i(a, z)-\tfrac{1}{2}(Vz, z)+\right.$$

$$\left.+\int\left(e^{i(z,x)}-1-\frac{i(z, x)}{1+|x|^2}+\frac{1}{2}\frac{(z, x)^2}{1+|x|^2}\right)\frac{1+|x|^2}{|x|^2}\pi'(dx)\right\}.$$

The function $\left(e^{i(z,x)}-1-\frac{i(z, x)}{1+|x|^2}+\frac{1}{2}\frac{(z, x)^2}{1+|x|^2}\right)\frac{1+|x|^2}{|x|^2}$ is defined by continuity at $x=0$ to equal 0. Hence

$$\chi(z)=\exp\left\{i(a, z)-\tfrac{1}{2}(Bz, z)+\right.$$

$$\left.+\int\left(e^{i(z,x)}-1-\frac{i(z, x)}{1+|x|^2}\right)\frac{1+|x|^2}{|x|^2}\pi(dx)\right\}, \qquad (12)$$

where

$$(Bz, z)=(Vz, z)-\int\frac{(z, x)^2}{|x|^2}\pi(dx).$$

Since

$$(Bz, z)=\lim_{n'\to\infty}\int\frac{(z, x)^2}{|x|^2}\pi_{n'}(dx)-\int\frac{(z, x)^2}{|x|^2}\pi(dx),$$

and for almost all $\varepsilon>0$ (such that $\pi(\{x:|x|=\varepsilon\})=0$)

$$\lim_{n'\to\infty}\int\limits_{|x|>\varepsilon}\frac{(z, x)^2}{|x|^2}\pi_{n'}(dx)=\int\limits_{|x|>\varepsilon}\frac{(z, x)^2}{|x|^2}\pi(dx),$$

and moreover the integral $\int\limits_{|x|\leqslant\varepsilon}\frac{(z, x)^2}{|x|^2}\pi(dx)$ becomes arbitrarily small for a suitable choice of $\varepsilon>0$, it follows that $(Bz, z)\geqslant 0$ for all z. Thus for any infinitely divisible distribution vectors $a\in\mathscr{X}$, a kernel operator B and a finite measure π, with $\pi(\{0\})=0$, can be found such that the characteristic functional $\chi(z)$ of this distribution is of the form (12).

We now show the converse, i.e. that formula (12) determines the characteristic functional of a certain distribution. The fact that $\chi(z)$ is positive definite follows from the observation that $\chi(Pz)$ – where P is the projector on a finite-dimensional subspace $\mathscr{L}$ with z varying in this subspace – is the characteristic functional of a certain infinitely divisible distribution in $\mathscr{L}$.

Next utilizing relations

$$1-|\chi(z)|\leqslant\tfrac{1}{2}(Bz, z)+\int(1-\cos(z, x))\,\pi(dx)+\frac{1}{2}\int\frac{(z, x)^2}{|x|^2}\,\pi(dx),$$

$$\arg\chi(z)=(a, z)+\int\sin(z, x)\,\pi(dx)+\int\frac{\sin(z, x)-(z, x)}{|x|^2}\,\pi(dx),$$

$$|\sin t-t|\leqslant\frac{t^2}{2},\qquad \sin^2 t\leqslant 4(1-\cos t),$$

we verify that for some C

$$1-\operatorname{Re}\chi(z)\leqslant 1-|\chi(z)|+\tfrac{1}{2}(\arg\chi(z))^2\leqslant$$

$$\leqslant C\left[(a, z)^2+(Bz, z)+\int(1-\cos(z, x))\,\pi(dx)+\right.$$

$$\left.+\int\frac{(z, x)^2}{|x|^2}\,\pi(dx)+\left(\int\frac{(z, x)^2}{|x|^2}\,\pi(dx)\right)^2\right].$$

For measure π one can find a kernel operator S' such that for each $\varepsilon>0$

$$\int(1-\cos(z, x))\,\pi(dx)<\frac{\varepsilon}{2C}\quad\text{for}\quad (S'z, z)<1.$$

Setting

$$S=\frac{2C}{\varepsilon-\varepsilon^2}(B+U)+S',$$

where the kernel operator U is defined by the equality

$$(Uz, z)=(a, z)^2+\int\frac{(z, x)^2}{|x|^2}\,\pi(dx),\qquad \operatorname{Sp}U=|a|^2+\pi(\mathscr{X}),$$

we have: $1-\operatorname{Re}\chi(z)<\varepsilon$ for $(Sz, z)<1$. Hence $\chi(z)$ is a characteristic functional. We have thus proved the following theorem:

Theorem 3. *In order that a functional $\chi(z)$ be the characteristic functional of a certain infinitely divisible distribution, it is necessary and sufficient that there exist vectors $a\in\mathscr{X}$, a kernel operator B and a finite measure π on $\mathfrak{B}$ with $\pi(\{0\})=0$ such that $\chi(z)$ is representable by formula* (12).

Remark. The representation of $\chi(z)$ by means of formula (12) is unique. Indeed,

$$(Bz, z)=-2\lim_{t\to\infty}\frac{1}{t^2}\ln\chi(tz).$$

Therefore we shall assume hereafter that $B=0$. Let $\{e_k\}$ be a basis and the numbers $c_k>0$ be such that $\sum_k |c_k|<\infty$. Then the series

$$\sum_{k=1}^{\infty} c_k\left[1-\frac{e^{it(x,e_k)}+e^{-it(x,e_k)}}{2}\right]=\sum_{k=1}^{\infty} c_k(1-\cos t(e_k,x))$$

converges monotonically to a bounded function and hence series

$$\sum_{k=1}^{\infty} c_k\left[\ln\chi(z)-\frac{\ln\chi(z+te_k)-\ln\chi(z-te_k)}{2}\right]=$$

$$=\int e^{i(z,x)}\sum_{k=1}^{\infty} c_k[1-\cos t(e_k,x)]\frac{1+|x|^2}{|x|^2}\pi(dx) \qquad (13)$$

is convergent.

Therefore knowing $\chi(z)$ one can determine the expression

$$\int e^{i(z,x)}\sum_{k=1}^{\infty} c_k\left(1-\frac{\sin\delta(e_k,x)}{\delta(e_k,x)}\right)\frac{1+|x|^2}{|x|^2}\pi(dx), \qquad (14)$$

which is obtained if the r.h.s. of (13) is integrated with respect to t from $-\delta$ up to δ and then divided by 2δ. It thus follows that the measure

$$\tilde{\pi}(A)=\int_A \sum_{k=1}^{\infty} c_k\left(1-\frac{\sin\delta(e_k,x)}{\delta(e_k,x)}\right)\frac{1+|x|^2}{|x|^2}\pi(dx)$$

is uniquely determined since (14) is the characteristic functional of this measure. The measure π is completely determined by the conditions:

1) $\pi(\{0\})=0$,
2) If $0\notin A$, then

$$\pi(A)=\int_A \left[\sum_{k=1}^{\infty} c_k\left(1-\frac{\sin\delta(e_k,x)}{\delta(e_k,x)}\right)\right]^{-1}\frac{|x|^2}{1+|x|^2}\tilde{\pi}(dx).$$

Hence the measure π is determined by the values of $\chi(z)$. Therefore a is also uniquely determined by the values of $\chi(z)$.

A limit theorem for sums of independent random variables. Let $\xi_{n1},\ldots,\xi_{nk_n}$ be a double sequence of independent random variables and let $\zeta_n=\sum_{k=1}^{k_n}\xi_{nk}$. The variables ξ_{nk} are assumed to be infinitely small, i.e. $\lim_{n\to\infty}\sup_k \mathsf{P}\{|\xi_{nk}|>\varepsilon\}=0$ for each $\varepsilon>0$. We now find the conditions under which the distribution of ζ_n converges as $n\to\infty$ to a certain limiting distribution. Denote by μ_{nk} the distribution of the variable ξ_{nk} and the distribution of ζ_n by ν_n. Next let $a_{nk} \in X$ be determined by the equality

$$(a_{nk}, z) = \int \frac{(z, x)}{1+|x-a_{nk}|^2} \mu_{nk}(dx)$$

for all $z \in \mathscr{X}$. The existence of these a_{nk} and their uniqueness, provided they satisfy the inequality $|a_{nk}| < \delta < 1$ for n sufficiently large, follows from the relations

$$|Ta| \leqslant \int \frac{|x|}{1+|x-a|^2} \mu_{nk}(dx) \leqslant \varepsilon + |a| \, \mu_{nk}(\{x : |x| > \varepsilon\}),$$

where

$$(Ta, z) = \int \frac{(x, z)}{1+|x-a|^2} \mu_{nk}(dx),$$

$$|Ta - Tb| \leqslant |a-b| \int \frac{(2|x| + |a| + |b|)\,|x|}{(1+|x-a|^2)(1+|x-b|^2)} \mu_{nk}(dx) \leqslant$$

$$\leqslant |a-b| \, [2\varepsilon^2 + 2\delta\varepsilon + L\mu_{nk}(\{x : |x| > \varepsilon\})],$$

$$L = \sup\left\{\frac{(2|x| + |a| + |b|)\,|x|}{(1+|x-a|^2)(1+|x-b|^2)} ;\ |a| \leqslant \delta,\ |b| \leqslant \delta,\ x \in X\right\} \leqslant 2\frac{(1+\delta)^2}{(1-\delta)^2}.$$

It also follows from these inequalities that the operator T in the region $|a| < \delta < 1$ is a contraction and maps this region into itself. Therefore a_{nk} exist and are unique.

Set

$$a_n = \sum_{k=1}^{k_n} a_{nk}, \quad (V_n z, z) = \sum_{k=1}^{k_n} (V_{nk} z, z),$$

$$(V_{nk} z, z) = \int \frac{(z, x-a_{nk})^2}{1+|x-a_{nk}|^2} \mu_{nk}(dx),$$

and let the measure μ_n be determined by the equality

$$\int e^{i(z, x)} \mu_n(dx) = \sum_{k=1}^{k_n} \int e^{i(z, x-a_{nk})} \frac{|x-a_{nk}|^2}{1+|x-a_{nk}|^2} \mu_{nk}(dx).$$

Theorem 4. *In order that a sequence of measures ν_n converge weakly as $n \to \infty$ to a measure ν it is necessary and sufficient that the following conditions be satisfied:*

1) *μ_n converges weakly to a certain measure π';*
2) *the limit $a = \lim_{n\to\infty} a_n$ exists;*

3) *the sequence of operators* V_n *is such that* $\sum_{k=1}^{\infty}(V_n e_k, e_k)$ *converges uniformly in* n *and for each* z *the limit* $\lim_{n\to\infty}(V_n z, z)=(Vz, z)$ *is valid, where* V *is a kernel operator. Moreover, the characteristic functional of the limiting distribution is given by formula* (12) *with*

$$(Bz, z)=(Vz, z)-\int\frac{(x, z)^2}{|x|^2}\,\pi(dx),$$

$$\pi(A)=\pi'(A)\quad\text{for}\quad 0\notin A,\quad \pi(\{0\})=0,$$

Proof. Sufficiency. We have

$$\mathsf{E}\,e^{i(z,\zeta_n)}=e^{i(z,a_n)}\prod_{k=1}^{k_n}\int e^{i(z,x-a_{nk})}\,\mu_{nk}(dx)=$$

$$=e^{i(z,a_n)}\prod_{k=1}^{k_n}\Bigg\{\bigg(1-\tfrac{1}{2}(V_{nk}z, z)+\int\bigg[e^{i(z,x-a_{nk})}-1-$$

$$-\frac{i(z, x-a_{nk})}{1+|x-a_{nk}|^2}+\frac{1}{2}\frac{(x-a_{nk}, z)^2}{1+|x-a_{nk}|^2}\bigg]\bigg)\,\mu_{nk}(dx)\Bigg\}.$$

Note that

$$\bigg|-\tfrac{1}{2}(V_{nk}z, z)+\int\bigg[e^{i(z,x-a_{nk})}-1-\frac{i(z, x-a_{nk})}{1+|x-a_{nk}|^2}+$$

$$+\frac{1}{2}\frac{(z, x-a_{nk})^2}{1+|x-a_{nk}|^2}\bigg]\,\mu_{nk}(dx)\bigg|\leqslant\bigg|\int[\cos(z, x-a_{nk})-1]\,\mu_{nk}(dx)+$$

$$+i\int\bigg[\sin(z, x-a_{nk})-\frac{(z, x-a_{nk})}{1+|x-a_{nk}|^2}\bigg]\,\mu_{nk}(dx)\bigg|=$$

$$=O((V_{nk}z, z)+(1+|z|)\,\mu_{nk}(\{x:|x|>1\})).$$

It follows from the last inequality that

$$\ln\mathsf{E}\,e^{i(z,\zeta_n)}=i(z, a_n)-\tfrac{1}{2}(V_n z, z)+$$

$$+\sum_{k=1}^{k_n}\int\bigg[e^{i(z,x-a_{nk})}-1-\frac{i(z, x-a_{nk})}{1+|x-a_{nk}|^2}+$$

$$+\frac{1}{2}\frac{(z, x-a_{nk})^2}{1+|x-a_{nk}|^2}\bigg]\,\mu_{nk}(dx)+$$

$$+O\big[\sup_k(V_{nk}z, z)+(1+|z|)\sup_k\mu_{nk}\{x:|x|>1\}\big].$$

It is easy to verify that

$$\lim_{n\to\infty} \sup_k |a_{nk}| = 0.$$

Moreover

$$\operatorname{Sp} V_{nk} = \int \frac{|x-a_{nk}|^2}{1+|x-a_{nk}|^2}\, \mu_{nk}(dx) \leqslant 2\delta^2 + |a_{nk}|^2 + \mathsf{P}\{|\xi_{nk}| > \delta\},$$

and hence

$$\lim_{n\to\infty} \sup_k \operatorname{Sp} V_{nk} = 0.$$

Therefore, in view of the conditions of the theorem, we have for all z

$$\begin{aligned}\lim_{n=\infty} \ln \mathsf{E}\, e^{i(z,\zeta_n)} &= i(z,a) - \tfrac{1}{2}(Vz,z) + \\ &+ \lim_{n\to\infty} \sum_{k=1}^{k_n} \int \Bigg[e^{i(z,x-a_{nk})} - 1 - \frac{i(z,x-a_{nk})}{1+|x-a_{nk}|^2} + \\ &+ \frac{1}{2}\frac{(z,x-a_{nk})^2}{1+|x-a_{nk}|^2} \Bigg] \mu_{nk}(dx).\end{aligned}$$

Note that

$$\begin{aligned}&\lim_{n\to\infty} \sum_{k=1}^{k_n} \int \Bigg[e^{i(z,x-a_{nk})} - 1 - \frac{i(z,x-a_{nk})}{1+|x-a_{nk}|^2} + \\ &\quad + \frac{1}{2}\frac{(z,x-a_{nk})^2}{1+|x-a_{nk}|^2} \Bigg] \mu_{nk}(dx) = \\ &= \lim_{n\to\infty} \int \left(e^{i(z,x)} - 1 - \frac{i(z,x)}{1+|x|^2} + \frac{1}{2}\frac{(z,x)^2}{1+|x|^2} \right) \frac{1+|x|^2}{|x|^2}\, \mu_n(dx) = \\ &= \int \left(e^{i(z,x)} - 1 - \frac{i(z,x)}{1+|x|^2} + \frac{1}{2}\frac{(z,x)^2}{1+|x|^2} \right) \frac{1+|x|^2}{|x|^2}\, \pi'(dx) = \\ &= \int \left(e^{i(z,x)} - 1 - \frac{i(z,x)}{1+|x|^2} + \frac{1}{2}\frac{(z,x)^2}{1+|x|^2} \right) \frac{1+|x|^2}{|x|^2}\, \pi(dx),\end{aligned}$$

since if we define the function $\left(e^{i(z,x)} - 1 - \frac{i(z,x)}{1+|x|^2} + \frac{1}{2}\frac{(z,x)^2}{1+|x|^2} \right) \frac{1+|x|^2}{|x|^2}$ to be zero at $x=0$, this function will remain continuous. Hence

$$\lim_{n\to\infty} \ln \mathsf{E}\, e^{i(z,\zeta_n)} = i(z,a) - \tfrac{1}{2}(Bz,z) + \int \left(e^{i(z,x)} - 1 - \frac{i(z,x)}{1+|x|^2} \right) \frac{1+|x|^2}{|x|^2}\, \pi(dx).$$

To prove the weak convergence of measures ν_n it is sufficient to verify that they are weakly compact. However,

$$\left|1-\int e^{i(z,x)}\, \nu_n(dx)\right| = \left|1-e^{i(a_n,z)}\prod_{k=1}^{k_n}\int e^{i(z,x-a_{nk})}\, \mu_{nk}(dx)\right| \leqslant$$

$$\leqslant |(a_n, z)| + \sum_{k=1}^{k_n}\left|1-\int e^{i(z,x-a_{nk})}\, \mu_{nk}(dx)\right| \leqslant$$

$$\leqslant |(a_n, z)| + \sum_{k=1}^{k_n}\int (1-\cos(z, x-a_{nk}))\, \mu_{nk}(dx) +$$

$$+\sum_{k=1}^{k_n}\left|\int\left[\sin(z, x-a_{nk}) - \frac{(z, x-a_{nk})}{1+|x-a_{nk}|^2}\right]\mu_{nk}(dx)\right|.$$

We utilize the bounds

$$|(a_n, z)| \leqslant \delta + \frac{1}{\delta}(a_n, z)^2,$$

$$\int (1-\cos(z, x-a_{nk}))\, \mu_{nk}(dx) \leqslant$$

$$\leqslant \int_{|x-a_{nk}|\leqslant c} (x-a_{nk}, z)^2\, \mu_{nk}(dx) + 2\int_{|x-a_{nk}|>c} \mu_{nk}(dx),$$

$$\left|\int \frac{|x-a_{nk}|^2}{1+|x-a_{nk}|^2}\sin(z, x-a_{nk})\, \mu_{nk}(dx)\right| \leqslant$$

$$\leqslant \delta\int \frac{|x-a_{nk}|^2}{1+|x-a_{nk}|^2}\, \mu_{nk}(dx) + \frac{1}{\delta}\int \sin^2(z, x-a_{nk})\, \mu_{nk}(dx) \leqslant$$

$$\leqslant \delta\int \frac{|x-a_{nk}|^2}{1+|x-a_{nk}|^2}\, \mu_{nk}(dx) +$$

$$+\frac{1}{\delta}\int_{|x-a_{nk}|\leqslant c} (z, x-a_{nk})^2\, \mu_{nk}(dx) + \frac{1}{\delta}\int_{|x-a_{nk}|>c} \mu_{nk}(dx),$$

$$\int_{|x-a_{nk}|\leqslant c} (z, x-a_{nk})^2\, \mu_{nk}(dx) \leqslant (1+c^2)\,(V_{nk}z, z).$$

These bounds yield the following inequality

$$\left|1-\int e^{i(z,x)}\, \nu_n(dx)\right| \leqslant (1+c^2)\left(1+\frac{1}{\delta}\right)(V_n z, z) + \frac{1}{\delta}(a_n, z)^2 +$$

$$+\delta(\mu_n(\mathscr{X})+1) + \left(2+\frac{1}{\delta}\right)\mu_n(\{x: |x| > c - \sup_k |a_{nk}|\}).$$

Since by a suitable choice of δ and c the expression

$$\delta\mu_n(\mathscr{X})+\delta+\left(2+\frac{1}{\delta}\right)\mu_n(\{x:|x|>c-\sup_k|a_{nk}|\})$$

becomes arbitrarily small, to prove the compactness of v_n it is sufficient to show that the series

$$\sum_{k=1}^{\infty}(S_n e_k, e_k)=\sum_{k=1}^{\infty}(V_n e_k, e_k)+\sum_{k=1}^{\infty}(a_n, e_k)^2$$

where the operator S_n is defined by the relation

$$(S_n z, z)=(V_n z, z)+(a_n, z)^2$$

is uniformly convergent in n in any orthonormal basis $\{e_k\}$. The uniform convergence in n of the series $\sum_{k=1}^{\infty}(V_n e_k, e_k)$ follows from condition 3) and the uniform convergence of the series $\sum_{k=1}^{\infty}(a_n, e_k)^2$ follows from the fact that in view of condition (2)

$$\lim_{n\to\infty}\sum_{k=1}^{\infty}[(a_n, e_k)-(a, e_k)]^2=0.$$

The sufficiency of the conditions of the theorem is thus proved.

Necessity. Let $\xi'_{n1},\dots,\xi'_{nk_n}$ be random variables, taking on values in $\mathscr{X}$, which are mutually independent and also independent of $\xi_{n1},\dots,\xi_{nk_n}$ and let ξ_{nk} and ξ'_{nk} have the same distribution. Since

$$\mathsf{P}\left\{\left|\sum_{k=1}^{k_n}(\xi_{nk}-\xi'_{nk})\right|>2c\right\}\leqslant 2\mathsf{P}\left\{\left|\sum_{k=1}^{k_n}\xi_{nk}\right|>c\right\}$$

and the variables $\xi_{nk}-\xi'_{nk}$ are symmetric, it follows from Lemma 3 that

$$\mathsf{P}\left\{\sup_{j\leqslant k_n}\left|\sum_{k=1}^{j}(\xi_{nk}-\xi'_{nk})\right|>2c\right\}\leqslant 4\mathsf{P}\left\{\left|\sum_{k=1}^{k_n}\xi_{nk}\right|>c\right\}.$$

Hence

$$\mathsf{P}\{\sup_{k\leqslant k_n}|\xi_{nk}-\xi'_{nk}|>4c\}\leqslant 4\mathsf{P}\left\{\left|\sum_{k=1}^{k_n}\xi_{nk}\right|>c\right\}.$$

Choosing c sufficiently large so that $4\mathsf{P}\left\{\left|\sum_{k=1}^{k_n}\xi_{nk}\right|>c\right\}<1$, we have, in view of the inequality $x\leqslant-\ln(1-x)$,

$$\sum_{k=1}^{k_n} \mathsf{P}\{|\xi_{nk}-\xi'_{nk}|>4c\} \leqslant -\sum_{k=1}^{k_n} \ln \mathsf{P}\{|\xi_{nk}-\xi'_{nk}|\leqslant 4c\} =$$
$$= -\ln \mathsf{P}\{\sup_{k\leqslant k_n} |\xi_{nk}-\xi'_{nk}|\leqslant 4c\} \leqslant$$
$$\leqslant -\ln\left(1-4\mathsf{P}\left\{\left|\sum_{k=1}^{k_n} \xi_{nk}\right|>c\right\}\right).$$

Consequently, vectors b_{nk} can be found such that

$$\sum_{k=1}^{k_n} \mathsf{P}\{|\xi_{nk}-b_{nk}|>4c\} \leqslant -\ln\left(1-4\mathsf{P}\left\{\left|\sum_{k=1}^{k_n} \xi_{nk}\right|>c\right\}\right).$$

Since $\sup_k \mathsf{P}\{|\xi_{nk}|>\varepsilon\}\to 0$ for every $\varepsilon>0$, we have $|b_{nk}|\leqslant c+\varepsilon$, provided only

$$-\ln\left(1-4\mathsf{P}\left\{\left|\sum_{k=1}^{n} \xi_{nk}\right|>c\right\}\right)<1-\sup_k \mathsf{P}\{|\xi_{nk}|>\varepsilon\}.$$

Choosing $\varepsilon=c$, we obtain

$$\sum_{k=1}^{k_n} \mathsf{P}\{|\xi_{nk}|>6c\} \leqslant -\ln\left(1-4\mathsf{P}\left\{\left|\sum_{k=1}^{k_n} \xi_{nk}\right|>c\right\}\right). \tag{13}$$

Let $\psi_c(x)=1$ for $|x|\leqslant c$ and $\psi(x)=0$ for $|x|>c$. Then

$$\mathsf{P}\left\{\sup_{k\leqslant k_n}\left|\sum_{j=1}^{k} \psi_c(\xi_{nj}-\xi'_{nj})(\xi_{nj}-\xi'_{nj})\right|>\alpha\right\} \leqslant$$
$$\leqslant 2\mathsf{P}\left\{\left|\sum_{j=1}^{k_n} \psi_c(\xi_{nj}-\xi'_{nj})(\xi_{nj}-\xi'_{nj})\right|>\alpha\right\} \leqslant$$
$$\leqslant 2\mathsf{P}\left\{\left|\sum_{j=1}^{k_n} (\xi_{nj}-\xi'_{nj})\right|>\alpha\right\}+2\mathsf{P}\{\sup_j|\xi_{nj}-\xi'_{nj}|>c\} \leqslant$$
$$\leqslant 2\mathsf{P}\left\{\left|\sum_{j=1}^{k_n} (\xi_{nj}-\xi'_{nj})\right|>\alpha\right\}+4\mathsf{P}\left\{\left|\sum_{j=1}^{k_n} (\xi_{nj}-\xi'_{nj})\right|>\frac{c}{2}\right\} \leqslant$$
$$\leqslant 4\mathsf{P}\left\{\left|\sum_{j=1}^{k_n} \xi_{nj}\right|>\frac{\alpha}{2}\right\}+8\mathsf{P}\left\{\left|\sum_{j=1}^{k_n} \xi_{nj}\right|>\frac{c}{4}\right\}.$$

For α and c sufficiently large, the r.h.s. of this inequality becomes arbitrarily small; therefore, in view of Lemma 2, the quantities

$$\mathsf{E}\left|\sum_{j=1}^{k_n} \psi_c(\xi_{nj}-\xi'_{nj})(\xi_{nj}-\xi'_{nj})\right|^2$$

are uniformly bounded in n (provided c are sufficiently large). But for all $\delta>0$

$$\mathsf{E}\psi_c(\xi_{nj}-\xi'_{nj})\,|\xi_{nj}-\xi'_{nj}|^2 \geqslant \mathsf{P}\{|\xi'_{nj}|\leqslant\delta\} \inf_{|a|\leqslant\delta} \int\limits_{|x|<c-\delta} |x-a|^2\,\mu_{nj}(dx).$$

Denote by $\tilde{a}_{nj}$ the value of a for which the infimum is obtained. For n sufficiently large

$$\tilde{a}_{nj}=\int\limits_{|x|<c-\delta} x\mu_{nj}(dx).$$

Since

$$\inf_j \mathsf{P}\{|\xi'_{nj}|\leqslant\delta\}=\inf_j \mathsf{P}\{|\xi_{nj}|\leqslant\delta\}\to 1$$

as $n\to\infty$, it follows that

$$\sup_n \sum_{k=1}^{k_n} \int\limits_{|x|\leqslant c-\delta} |x-\tilde{a}_{nk}|^2\,\mu_{nk}(dx)<\infty.$$

From the last inequality and (13) we obtain

$$\sup_n \sum_{k=1}^{k_n} \int \frac{|x-\tilde{a}_{nk}|^2}{1+|x-a_{nk}|^2}\,\mu_{nk}(dx)<\infty. \tag{15}$$

However

$$|x-\tilde{a}_{nk}|^2=|x-a_{nk}|^2+2(x-a_{nk},\tilde{a}_{nk}-a_{nk})+|\tilde{a}_{nk}-a_{nk}|^2.$$

Since

$$\int \frac{(x-a_{nk},a_{nk}-\tilde{a}_{nk})}{1+|x-a_{nk}|^2}\,\mu_{nk}(dx)=0,$$

we have

$$\sup_n \sum_{k=1}^{k_n} \int \frac{|x-a_{nk}|^2+|a_{nk}-\tilde{a}_{nk}|^2}{1+|x-a_{nk}|^2}\,\mu_{nk}(dx)<\infty.$$

It thus follows that

$$\sup_n \sum_{k=1}^{k_n} \int \frac{|x-a_{nk}|^2}{1+|x-a_{nk}|^2}\,\mu_{nk}(dx)<\infty. \tag{16}$$

The following relation is obvious

$$\sup_{k\leqslant k_n} \int \frac{|x-a_{nk}|^2}{1+|x-a_{nk}|^2}\,\mu_{nk}(dx)=o(1).$$

Therefore utilizing the inequality

$$\left|\int\left[e^{i(z,x-a_{nk})}-1-\frac{i(z,x-a_{nk})}{1+|x-a_{nk}|^2}\right]\mu_{nk}(dx)\right|\leqslant$$

$$\leqslant\tfrac{1}{2}\int\frac{(z,x-a_{nk})^2}{1+|x-a_{nk}|^2}\mu_{nk}(dx)+$$

$$+\left|\int(e^{i(z,x-a_{nk})}-1)\frac{|x-a_{nk}|^2}{1+|x-a_{nk}|^2}\mu_{nk}(dx)\right|,$$

we obtain the following expansion of $\ln\chi_n(z)$:

$$\ln\chi_n(z)=\ln \mathsf{E}\, e^{i(z,\zeta_n)}=i(a_n,z)+$$

$$+\sum_{k=1}^{k_n}\int\left(e^{i(z,x-a_{nk})}-1-\frac{i(z,x-a_{nk})}{1+|x-a_{nk}|^2}\right)\mu_{nk}(dx)+$$

$$+(V_nz,z)\,[O(\sup_{k\leqslant k_n}(V_{nk}z,z))+o(1)]+o(1).$$

It follows from the compactness of the measures ν_n that for each $\varepsilon>0$ an operator $B\in T_c$ and operators $A_n\in S_1$ can be found such that $1-\operatorname{Re}\chi_n(z)\leqslant\leqslant\varepsilon$ for $(BA_nBz,z)\leqslant 1$. Thus $-\ln|\chi_n(z)|<2\varepsilon$ for ε sufficiently small provided $(BA_nBz,z)\leqslant 1$. Hence

$$\sum_{k=1}^{k_n}\int(1-\cos(z,x-a_{nk}))\,\mu_{nk}(dx)<2\varepsilon.$$

However in this case,

$$\int(1-\cos(z,x))\,\mu_n(dx)=\sum_{k=1}^{k_n}\int(1-\cos(z,x-a_{nk}))\frac{|x-a_{nk}|^2}{1+|x-a_{nk}|^2}\mu_{nk}(dx)\leqslant$$

$$\leqslant\sum_{k=1}^{k_n}\int(1-\cos(z,x-a_{nk}))\,\mu_{nk}(dx)<2\varepsilon.$$

We have thus shown that the measures μ_n are compact. Hence in view of Corollary 2 of Theorem 2 in Section 2, operators $V_n^{1/2}$ are compact in $\mathscr{H}_{\mathfrak{X}}$, and therefore, in view of Lemma 2 in Section 2, the series $\sum_{k=1}^{\infty}(V_ne_k,e_k)$ converges uniformly in each basis $\{e_k\}$. Clearly

$$\lim_{n\to\infty}\chi_n(z)=\lim_{n\to\infty}\exp\left\{i(a_n,z)-\tfrac{1}{2}(V_nz,z)+\right.$$

$$\left.+\int\left(e^{i(z,x)}-1-\frac{i(z,x)}{1+|x|^2}+\frac{1}{2}\frac{(z,x)^2}{1+|x|^2}\right)\frac{1+|x|^2}{|x|^2}\mu_n(dx)\right\}.$$

We choose a subsequence n' such that the measures $\mu_{n'}$ converge to a certain measure π' and $(V_{n'}z, z) \to (Vz, z)$. In this case the limit of $(a_{n'}, z)$ (which is equal to (a, z)) exists. Thus

$$\lim_{n' \to \infty} \chi_{n'}(z) = \exp\left\{ i(a, z) - \tfrac{1}{2}(Vz, z) + \right.$$
$$\left. + \int \left(e^{i(z, x)} - 1 - \frac{i(z, x)}{1+|x|^2} + \frac{1}{2}\frac{(z, x)^2}{1+|x|^2} \right) \frac{1+|x|^2}{|x|^2}\, \pi'(dx) \right\}.$$

The weak convergence of μ_n to π' and also the convergence of $(V_n z, z)$ to (Vz, z) follow from the uniqueness of the representation of a characteristic function. Moreover, this yields the weak convergence of a_n to a. To show that the strong convergence of a_n to a is also valid, we note that $\chi_n(z)$ is uniformly convergent to $\chi(z)$ for $|z| \leqslant c$ (cf. the remark for Theorem 3 in Section 2), $(V_n z, z) \to (Vz, z)$ *also* uniformly for all $|z| \leqslant c$ (this follows from condition 3 of the theorem) and finally one can show in exactly the same manner as is done in the remark for Theorem 3 of Section 2 that the series

$$\sum_{k=1}^{k_n} \int \left(e^{i(z, x-a_{nk})} - 1 - \frac{i(z, x-a_{nk})}{1+|x-a_{nk}|^2} + \frac{1}{2}\frac{(z, x-a_{nk})^2}{1+|x-a_{nk}|^2} \right) \mu_{nk}(dx)$$

converges uniformly to

$$\int \left(e^{i(z, x)} - 1 - \frac{i(z, x)}{1+|x|^2} + \frac{1}{2}\frac{(z, x)^2}{1+|x|^2} \right) \pi'(dx).$$

Therefore (a_n, z) also converges uniformly to (a, z) for $|z| \leqslant c$. The same argument as in Theorem 2 now yields that a_n converges (strongly) to a. The theorem is thus proved. □

Remark. If we define a_{nk} by means of the relation

$$(a_{nk}, z) = \int_{|x| \leqslant c} (x, z)\, \mu_{nk}(dx),$$

and the variables μ_n, V_{nk} and V_n are defined as in Theorem 4, then under the conditions of Theorem 4, the distribution of the variable ξ_n will converge weakly to an infinitely divisible distribution with the characteristic function

$$\chi(z) = \exp\left\{ i(a, z) - \tfrac{1}{2}(Bz, z) + \right.$$

$$+\int\limits_{|x|\leqslant c}(e^{i(z,x)}-1-i(z,x))\frac{1+|x|^2}{|x|^2}\pi(dx)+$$

$$+\int\limits_{|x|>c}(e^{i(z,x)}-1)\frac{1+|x|^2}{|x|^2}\pi(dx)\Big\},$$

provided only c is chosen in such a manner that $\pi(\{x:|x|=c\})=0$. The proof of the necessity and sufficiency of the conditions is identical to the proof given in Theorem 4.

§ 4. Limit Theorems for Continuous Random Processes

In this section general theorems on weak convergence of measures in metric spaces, presented in Section 1, are applied to the derivation of limit theorems for random processes continuous with probability 1.

Let $\xi_n(t)$ be a sequence of random processes defined on the interval $[a, b]$ taking on values on a certain separable complete metric space $\mathscr{X}$ and continuous on $[a, b]$ with probability 1. Denote by $\mathscr{C}_{[a,b]}(\mathscr{X})$ the set of continuous functions $x(t)$ defined on $[a, b]$ and taking on values in $\mathscr{X}$.

We introduce the following metric on $\mathscr{C}_{[a,b]}(\mathscr{X})$

$$r(x(\cdot), y(\cdot))=\sup_{a\leqslant t\leqslant b}\varrho(x(t), y(t)),$$

where ϱ is the distance in $\mathscr{X}$. In this metric the space $\mathscr{C}_{[a,b]}(\mathscr{X})$ becomes a complete metric separable space. Denote by $\mathfrak{B}_{[a,b]}(\mathscr{X})$ the σ-algebra of all Borel sets in $\mathscr{C}_{[a,b]}(\mathscr{X})$. This σ-algebra coincides with the smallest σ-algebra containing all cylindrical sets in $\mathscr{C}_{[a,b]}(\mathscr{X})$ (cf. Section 2 of Chapter V – the proof given in that section is for the case of a linear $\mathscr{X}$; the proof in our case is the same). Therefore one can associate with each process $\xi_n(t)$ a measure μ_n on $\mathfrak{B}_{[a,b]}(\mathscr{X})$ such that the values of this measure on cylindrical sets coincide with the finite dimensional distributions of the process $\xi_n(t)$.

What is the meaning of the weak convergence of measures μ_n for random processes $\xi_n(t)$?

Let μ_n converge weakly to μ where μ is the measure associated with the process $\xi(t)$. Then for each μ-almost everywhere continuous and bounded $\mathfrak{B}_{[a,b]}(\mathscr{X})$-measurable functional $\varphi(x)$ defined on $\mathscr{C}_{[a,b]}(\mathscr{X})$ we have

$$\lim_{n\to\infty}\int\varphi(x)\,\mu_n(dx)=\int\varphi(x)\,\mu(dx)$$

(cf. lemma of Section 1). Therefore for each μ-almost everywhere continuous $\mathfrak{B}_{[a,b]}(\mathscr{X})$-measurable functional $f(x)$

$$\lim_{n\to\infty}\int e^{i\lambda f(x)}\,\mu_n(dx)=\int e^{i\lambda f(x)}\,\mu(dx)$$

for all real λ.

We now note that

$$\int e^{i\lambda f(x)}\,\mu_n(dx)=\mathsf{E}\,e^{i\lambda f(\xi_n(\cdot))},$$

$$\int e^{i\lambda f(x)}\,\mu(dx)=\mathsf{E}\,e^{i\lambda f(\xi(\cdot))},$$

here $f(\xi_n(\cdot))$ and $f(\xi(\cdot))$ are random variables for each $\mathfrak{B}_{[a,b]}(\mathscr{X})$- measurable functional f and the last formulas are corollaries of formula (2) in Section 1 of Chapter V. The convergence of the distribution of the variable $f(\xi_n(\cdot))$ to the distribution of the variable $f(\xi(\cdot))$ follows from the convergence of the characteristic function of the variable $f(\xi_n(\cdot))$ to the characteristic function of $f(\xi(\cdot))$. Thus the weak convergence of measures μ_n to μ yields the convergence of the distribution of the variables $f(\xi_n(\cdot))$ to the distribution of $f(\xi(\cdot))$ for each μ-almost everywhere continuous $\mathfrak{B}_{[a,b]}(\mathscr{X})$-measurable functional $f(x)$. Conversely, if the distribution of $f(\xi_n(\cdot))$ converges to the distribution of $f(\xi(\cdot))$ for each μ-almost everywhere continuous $\mathfrak{B}_{[a,b]}(\mathscr{X})$-measurable functional, then $\mathsf{E}\varphi(\xi_n(\cdot))\to\mathsf{E}\varphi(\xi(\cdot))$ for each bounded μ-everywhere continuous $\mathfrak{B}_{[a,b]}(\mathscr{X})$-measurable functional φ, i.e.

$$\lim_{n\to\infty}\int\varphi(x)\,\mu_n(dx)=\int\varphi(x)\,\mu(dx).$$

Therefore the weak convergence of measures μ_n to μ is equivalent to the convergence of the distributions of $f(\xi_n(\cdot))$ to the distribution of $f(\xi(\cdot))$ for each μ-everywhere continuous $\mathfrak{B}_{[a,b]}(\mathscr{X})$-measurable functional f.

It is common when investigating limit theorems for stochastic processes to assume the weak convergence of the marginal distributions, i.e. the convergence of the measures $\mu_n(A)$ to $\mu(A)$ for all cylindrical sets A which are the sets of continuity for the measure μ. For each open sphere in $\mathscr{C}_{[a,b]}(\mathscr{X})$ of the form

$$\{x(\cdot):\varrho(\bar{x}(t)\,x(t))<\varepsilon,\ a\leqslant t\leqslant b\},$$

where $\bar{x}(t)$ is a given function on $\mathscr{C}_{[a,b]}(\mathscr{X})$ the following relation is valid

$$\{x(\cdot):\varrho(\bar{x}(t),x(t))<\varepsilon,\ a\leqslant t\leqslant b\}=$$

$$= \bigcup_{m=1}^{\infty} \bigcap_{N=1}^{\infty} \left\{ x(\cdot): \varrho(\bar{x}(t_k), x(t_k)) < \varepsilon - \frac{1}{m}, \quad k=1, \dots, N \right\},$$

where $\{t_1, t_2, \dots\}$ is an everywhere dense sequence on $[a, b]$. Consequently, the algebra $\mathfrak{A}_0$ of open cylindrical sets which are the sets of continuity for the measure μ satisfy the conditions of Theorem 4 in Section 1.

To apply this theorem one must find the general form of a compactum in the space $\mathscr{C}_{[a,b]}(\mathscr{X})$. In the case when $\mathscr{X}$ is a finite-dimensional Euclidean space, the general form of a compactum in $\mathscr{C}_{[a,b]}(\mathscr{X})$ is given by the well-known Arzelà's theorem.

An analogous result is also valid in our case. We state this result as the following lemma:

Let λ_δ be a positive monotonic and continuous function defined for $\delta > 0$ satisfying condition $\lambda_\delta \downarrow 0$ as $\delta \downarrow 0$ and let X_1 be a compactum in $\mathscr{X}$. Denote by $K(X_1, \lambda_\delta)$ the set of functions $x(t)$ belonging to $\mathscr{C}_{[a,b]}(\mathscr{X})$ satisfying conditions: a) $x(t) \in X_1$, $a \leqslant t \leqslant b$; b) $\varrho(x(t_1), x(t_2)) \leqslant \lambda_\delta$ for $|t_1 - t_2| \leqslant \delta$.

Lemma 1. *The set $K(X_1, \lambda_\delta)$ is compact in $\mathscr{C}_{[a,b]}(\mathscr{X})$. For each compactum K_1 in $\mathscr{C}_{[a,b]}(\mathscr{X})$ we can find a compactum X_1 in $\mathscr{X}$ and a function λ_δ which is positive, continuous and increasing and satisfying the condition $\lambda_{+0} = 0$, such that $K_1 \subset K(X_1, \lambda_\delta)$.*

Proof. To prove the compactness of the set $K(X_1, \lambda_\delta)$ we consider an arbitrary sequence $x_n(\cdot)$ belonging to this set and show that a convergence subsequence can be extracted from this sequence. Utilizing the compactness of the set of values of $x_n(t)$ for each t we can select, using the diagonal method, a subsequence $x_{n_k}(t)$ such that $x_{n_k}(t)$ will converge to a certain limit for all rational t in $[a, b]$. Denote $x_{n_k}(t)$ by $y_k(t)$ and show that the sequence $y_k(t)$ is convergent. Let $a \leqslant t_1 < \dots < t_N \leqslant b$ be rational points such that the length of each of the intervals $[a, t_1]$, $[t_1, t_2], \dots,$ $[t_N, b]$ does not exceed δ. Then

$$\sup_{a \leqslant t \leqslant b} \varrho(y_k(t), y_l(t)) \leqslant \sup_{1 \leqslant i \leqslant N} \varrho(y_k(t_i), y_l(t_i)) + \\ + \sup \{ (\varrho(y_k(t_i), y_k(t)) + \varrho(y_l(t_i), y_l(t))); \\ |t - t_i| \leqslant \delta, \quad i = 1, \dots, N \}.$$

Therefore

$$\overline{\lim_{k, l \to \infty}} \, r(y_k(\cdot), y_l(\cdot)) \leqslant 2\lambda_\delta.$$

Since $\delta > 0$ is arbitrary, it follows that $y_k(\cdot)$ is a fundamental sequence and thus is convergent to a limit. To prove the second assertion of the lemma we denote by X_t the set of values $x(t)$ with $x(\cdot) \in K_1$. We show that $X_1=\bigcup_t X_t$, $t\in[a,b]$ is compact. Let $x_n\in X_1$. Then $x_n=y(t_n)$, where $y_n(\cdot)\in K_1$. Choosing the subsequence n_k in such a manner that $t_{n_k}\to t_0$, and $\varrho(y_{n_k}(\cdot),y(\cdot))\to 0$ we observe that $x_{n_k}\to y(t_0)\in X_1$. Next, setting $\lambda_\delta(x(\cdot))=\sup\{\varrho(x(t_1),x(t_2));\,|t_1-t_2|<\delta\}$, we can easily verify that $\lambda_\delta(x(\cdot))$ is jointly continuous in the variables $\delta>0$ and $x(\cdot)$. Therefore it follows from the compactness of K_1 that the function $\sup\{\lambda_\delta(x(\cdot));\ x(\cdot)\in K_1\}=\lambda_\delta$ is continuous in the variable δ. The monotonicity of λ_δ follows from the relation $\lambda_{\delta_1}(x(\cdot))\leqslant\lambda_{\delta_2}(x(\cdot))$ for $\delta_1<\delta_2$. Since $\lambda_\delta(x(\cdot))$ approaches zero monotonically as $\delta\downarrow 0$, it follows from Dini's theorem that the convergence is uniform on each compactum. Therefore

$$\lim_{\delta\downarrow 0}\lambda_\delta=\lim_{\delta\downarrow 0}\sup\{\lambda_\delta(x(\cdot));\ x(\cdot)\in K_1\}=0.$$

Hence $K_1\subset K(X_1,\lambda_\delta)$ and the lemma is proved. □

Theorem 1. *Let the marginal distributions of the processes $\xi_n(t)$ converge to the marginal distributions of the process $\xi(t)$. In order that for all functionals f continuous on $\mathscr{C}_{[a,b]}(\mathscr{X})$ the distribution of $f(\xi_n(\cdot))$ converge to the distribution of $f(\xi(\cdot))$ it is necessary and sufficient that for every $\varrho>0$ the relation*

$$\lim_{h\to 0}\sup_n \mathsf{P}\{\sup_{|t_1-t_2\leqslant h}\varrho(\xi_n(t_1),\xi_n(t_2))>\varrho\}=0. \tag{1}$$

be satisfied.

Proof. Necessity. If the assertion of the theorem is satisfied, then the sequence of measures μ_n associated with the processes $\xi_n(t)$ is weakly compact so that condition b) of Theorem 1 in Section 1 is satisfied. Therefore for each $\varepsilon>0$ a compactum $K(X_1,\lambda_\delta)$ exists such that

$$\sup_n \mu_n(\mathscr{C}_{[a,b]}(\mathscr{X})-K(X_1,\lambda_\delta))\leqslant\varepsilon.$$

Then

$$\sup_n \mathsf{P}\{\sup_{|t_1-t_2|\leqslant h}\varrho(\xi_n(t_1),\xi_n(t_2))>\lambda_h\}\leqslant\varepsilon.$$

If h is sufficiently small, then $\lambda_h<\varrho$ and

$$\overline{\lim_{h\to 0}}\,\mathsf{P}\{\sup_{|t_1-t_2|\leqslant h}\varrho(\xi_n(t_1),\xi_n(t_2))>\varrho\}\leqslant\varepsilon.$$

Since $\varepsilon>0$ is arbitrary, we obtain equation (1).

Sufficiency. Taking into account the convergence of measures μ_n to μ on cylindric open sets of continuity of the measure μ (which follows from the convergence of the marginal distributions of $\xi_n(t)$ to the marginal distributions of $\xi(t)$) and Theorem 4 of Section 1 we observe that it is
Sufficiency. Taking into account the convergence of measures μ_n to μ on

sufficient to establish the compactness of measures μ_n. We denote by ν_{nt} the measure on $\mathscr{X}$ which represents the distribution of $\xi_n(t)$ and show that the set of measures $\{\nu_{nt}, n=1, 2, \ldots, t\in[a, b]\}$ is compact. Indeed, if ν_{nt_n} is a sequence of measures we can, by choosing a subsequence n_k such that $t_{n_k}\to t_0$ easily verify that for each bounded and continuous function $\varphi(x)$, defined on $\mathscr{X}$,

$$\lim_{k\to\infty}\int \varphi(x)\,\nu_{n_k t_{n_k}}(dx)=\lim_{k\to\infty}\mathsf{E}\varphi(\xi_{n_k}(t_{n_k}))=$$
$$=\lim_{k\to\infty}\mathsf{E}\varphi(\xi_{n_k}(t_0))+\lim_{k\to\infty}\mathsf{E}[\varphi(\xi_{n_k}(t_{n_k}))-\varphi(\xi_{n_k}(t_0))]=\mathsf{E}\varphi(\xi(t_0)).$$

This is because for each compactum X_1 and $\delta>0$

$$\overline{\lim_{k\to\infty}}\,\mathsf{E}|\varphi(\xi_{n_k}(t_{n_k}))-\varphi(\xi_{n_k}(t_0))|\leqslant$$
$$\leqslant 2\sup_x|\varphi(x)|\,\overline{\lim_{k\to\infty}}\,[\mathsf{P}\{\xi_{n_k}(t_0)\notin X_1\}+$$
$$+\mathsf{P}\{\varrho(\xi_{n_k}(t_{n_k}), \xi_{n_k}(t_0))>\delta\}]+$$
$$+\sup\{|\varphi(x)-\varphi(y)|;\ x\in X_1, \varrho(x, y)\leqslant\delta\},$$

and the r.h.s. may become arbitrarily small in view of the continuity of $\varphi(x)$, the compactness of $\nu_{n_k t_0}$ and condition (1).

We now choose the sequence $\{h_k\}$ according to the following condition

$$\sup_n\mathsf{P}\{\sup_{|t'-t''|\leqslant h_k}\varrho(\xi_n(t'), \xi_n(t''))>2^{-k}\}\leqslant 2^{-k}.$$

Let $X^{(k)}$ be a compactum such that $\nu_{nt}(\mathscr{X}-X^{(k)})\leqslant 2^{-k}\dfrac{h^k}{b-a}$ for all n, $t\in[a, b]$. Denote by $X_1^{(k)}$ the set of x such that $\varrho(x, X^{(k)})\leqslant 2^{-k}$. Then

$$\mathsf{P}\{\xi_n(t)\in X_1^{(k)}, a\leqslant t\leqslant b\}\geqslant\mathsf{P}\Big\{\xi_n(a+lh_k)\in X^{(k)};\ 1\leqslant l\leqslant\frac{b-a}{h_k},$$
$$\sup_{|t_1-t_2|\leqslant h_k}\varrho(\xi_n(t_1), \xi_n(t_2))\leqslant 2^{-k}\Big\}.$$

Therefore

$$1-\mathsf{P}\{\xi_n(t)\in X_1^{(k)}, a\leqslant l\leqslant b\}\leqslant\sum_{l\leqslant\frac{b-a}{h_k}}\mathsf{P}\{\xi_n(a+lh_k)\notin X^{(k)}\}+$$
$$+\mathsf{P}\{\sup_{|t_1-t_2|\leqslant h_k}\varrho(\xi_n(t_1), \xi_n(t_2))>2^{-k}\}\leqslant 2\cdot 2^{-k}.$$

Note that $\bigcap_{k=m}^{\infty}X_1^{(k)}$ is a compactum in $\mathscr{X}$. We now construct for each $\varepsilon>0$

a compactum $K(X_1, \lambda_\delta)$ such that $\mu_n(\mathscr{C}_{[a,b]}(\mathscr{X}) - K(X_1, \lambda_\delta)) < \varepsilon$ for all n. To do this we choose an m such that $2 \sum_{k \geq m} 2^{-k} < \frac{\varepsilon}{2}$ and set $X_1 = \bigcap_{k=m}^{\infty} X_1^{(k)}$. Consider the sequence $\lambda_r \downarrow 0$. For each r an h_r can be found such that $h_r < h_{r-1}$ and

$$\sup_n \mathsf{P}\{\sup_{|t_1 - t_2| \leq h_r} \varrho(\xi_n(t_1), \xi_n(t_2)) > \lambda_r\} \leq \frac{\varepsilon}{2^{r+1}}.$$

Let λ_δ be a nonnegative continuous non-increasing function such that $\lambda_{h_r} = \lambda_{r-1}$. Clearly, $\lambda_\delta \downarrow 0$ as $\delta \downarrow 0$. Moreover,

$$\mathsf{P}\{\xi_n(\cdot) \notin K(X_1, \lambda_\delta)\} \leq 1 - \mathsf{P}\{\xi_n(t) \in X_1, a \leq t \leq b\} +$$
$$+ \sum_{r=1}^{\infty} \mathsf{P}\{\sup_{|t_1 - t_2| \leq h_r} \varrho(\xi_n(t_1), \xi_n(t_2)) > \lambda_r\} < \frac{\varepsilon}{2} + \sum_{r=1}^{\infty} \frac{\varepsilon}{2^{r+1}} = \varepsilon.$$

The theorem is thus proved. □

Remark 1. Condition (1) may be substituted by the following condition

$$\lim_{h \to 0} \overline{\lim_{n \to \infty}} \mathsf{P}\{\sup_{|t_1 - t_2| \leq h} \varrho(\xi_n(t_1), \xi_n(t_2)) > \varepsilon\} = 0. \tag{2}$$

(The latter is often easier to verify.) Indeed it follows from (2) that for any $\eta > 0$, a $\delta > 0$ and an N exist such that for $n > N$ and for $h < \delta$ we have

$$\mathsf{P}\{\sup_{|t_1 - t_2| \leq h} \varrho(\xi_n(t_1), \xi_n(t_2)) > \varepsilon\} \leq \eta. \tag{3}$$

The uniform continuity of the processes $\xi_n(t)$ follows from their continuity. We thus have for each n

$$\lim_{h \to 0} \mathsf{P}\{\sup_{|t_1 - t_2| \leq h} \varrho(\xi_n(t_1), \xi_n(t_2)) > \varepsilon\} = 0.$$

Therefore a δ can be chosen such that for $h < \delta$ relation (3) is satisfied for all n.

The following theorem is occasionally more convenient in applications.

Theorem 2. *Let the marginal distribution of the processes $\xi_n(t)$ converge to the finite-dimensional distributions of the process $\xi(t)$ and let $\alpha > 0$, $\beta > 0$ and $H > 0$ exist such that for all $t_1, t_2 \in [a, b]$ and for all n*

$$\mathsf{E}[\varrho(\xi_n(t_1), \xi_n(t_2))]^\alpha \leq H |t_1 - t_2|^{1+\beta}. \tag{4}$$

Then for all functionals f continuous on $\mathscr{C}_{[a,b]}(\mathscr{X})$ the distribution of $f(\xi_n(\cdot))$ converges to the distribution of $f(\xi(\cdot))$.

Proof. We utilize Lemma 1 of Section 5 in Chapter III. Condition (4) of this lemma is fulfilled for the process $\xi_n(t)$ if we set $g(h)=h^\gamma$, where $0<\gamma<\frac{\beta}{\alpha}$, $q(C,h)=HC^{-\alpha}h^{1+\delta}$, where $\delta=\beta-\alpha\gamma$. Here the functions $G(m)$ and $Q(m, C)$ defined by equation (8) of Section 5 in Chapter III are given by

$$G(m)=T^\gamma\frac{2^{-m\gamma}}{1-2^{-\gamma}},\quad Q(m,C)=HC^{-\alpha}T^{1+\delta}\frac{2^{-m\delta}}{1-2^{-\delta}},\quad T=b-a.$$

Hence in view of relation (7) of Section 5 in Chapter III the following inequality

$$\mathsf{P}\{\sup_{|t_1-t_2|\leqslant h}\varrho(\xi_n(t_1),\xi_n(t_2))>\varepsilon\}\leqslant L\varepsilon^{-\alpha}h^\beta$$

is satisfied where L is a constant. The remainder of the proof follows from Theorem 1. □

Convergence of processes constructed from the sums of independent random variables. Let $\xi_{n1},\ldots,\xi_{nk_n}$ be a double sequence (a sequence of series) of numerical random variables, independent in each series and satisfying conditions

1) $$\mathsf{E}\xi_{ni}=0,\quad i=1,\ldots,k_n;$$

2) $$\mathsf{V}\xi_{ni}=b_{ni},\quad \sum_{i=1}^{k_n}b_{ni}=1.$$

We construct the random function $\xi_n(t)$, $t\in[0,1]$ as follows: set

$$S_{nk}=\sum_{i=1}^{k}\xi_{ni},\quad t_{nk}=\sum_{i=1}^{k}b_{ni},$$

$$\xi_n(t)=S_{nk}+\frac{t-t_{nk}}{t_{nk+1}-t_{nk}}[S_{nk+1}-S_{nk}]$$

for $t\in[t_{nk},t_{nk+1}]$, $S_{n0}=0$, $t_{n0}=0$. Then $\xi_n(t)$ is a random broken line, joining the points of the plane $(t;\xi)$, with coordinates $(t_{nk};S_{nk})$, $k=0,1,\ldots,k_n$.

We study the conditions under which the marginal distributions of processes $\xi_n(t)$ and the distribution of functionals on these processes converge to the marginal distributions of corresponding functionals of the Brownian motion process $w(t)$.

Theorem 3. *Let the random variables ξ_{ni} satisfy conditions* 1) *and* 2) *as well as the Lindeberg condition: if $F_{ni}(x)$ is the distribution function of the variable ξ_{ni} then for each $\varepsilon>0$*

$$\lim_{n\to\infty}\sum_{i=1}^{k_n}\int_{|u|>\varepsilon}u^2\,dF_{ni}(u)=0. \tag{5}$$

Under these conditions the finite-dimensional distributions of the processes $\xi_n(t)$ converge to finite dimensional distributions of the process $w(t)$ and the distribution of $f(\xi_n(\cdot))$ converges to the distribution of $f(w(\cdot))$ for each continuous functional f on $\mathscr{C}_{[0,1]}$.

Proof. The convergence of finite dimensional distributions of processes $\xi_n(t)$ to finite dimensional distributions of $w(t)$ follows from the central limit theorem. To prove the convergence of distributions of $f(\xi_n(\cdot))$ to the distributions of $f(w(\cdot))$ for all functionals f continuous on $\mathscr{C}_{[0,1]}$ we verify that for an arbitrary $\varepsilon>0$ condition

$$\lim_{h\to 0}\varlimsup_{n\to\infty} \mathsf{P}\{\sup_{|t_1-t_2|\leqslant h} |\xi_n(t_1)-\xi_n(t_2)|>\varepsilon\}=0 \tag{6}$$

is fulfilled and utilize Remark 1 following Theorem 1. Since

$$\begin{aligned}\sup_{|t_1-t_2|\leqslant h} |\xi_n(t_1)-\xi_n(t_2)| &\leqslant 2\sup_k \sup_{kh<t\leqslant (k+2)h} |\xi_n(t)-\xi_n(kh)| \leqslant \\ &\leqslant 4\sup_k \sup_{kh<t\leqslant (k+1)h} |\xi_n(t)-\xi_n(kh)|,\end{aligned}$$

it follows that

$$\begin{aligned}&\mathsf{P}\{\sup_{|t_1-t_2|\leqslant h} |\xi_n(t_1)-\xi_n(t_2)|>\varepsilon\}\leqslant \\ &\qquad\leqslant \sum_{kh<1} \mathsf{P}\left\{\sup_{kh<t\leqslant (k+1)h} |\xi_n(t)-\xi_n(kh)|>\frac{\varepsilon}{4}\right\}.\end{aligned}$$

Note that

$$\sup_{kh<t\leqslant (k+1)h} |\xi_n(t)-\xi_n(kh)|\leqslant 2 \sup_{j_{n,k}<r\leqslant j_{n,k+1}} \left|\sum_{j=j_{n,k}}^{r} \xi_{nj}\right|,$$

where $j_{n,k}$ is the maximal of the indices j such that t_{nj} does not exceed kh. Since for $j_{n,k}<s<j_{n,k+1}$

$$\varlimsup_{n\to\infty} \sup_s \mathsf{P}\left\{\left|\sum_{j=s}^{j_{n,k+1}} \xi_{nj}\right|>\frac{\varepsilon}{16}\right\}\leqslant \frac{256}{\varepsilon^2}\,h,$$

it follows that for h sufficiently small we have, in view of Theorem 6 of Section 3 in Chapter II, that

$$\begin{aligned}&\varlimsup_{n\to\infty} \mathsf{P}\left\{\sup_{kh<t\leqslant (k+1)h} |\xi_n(t)-\xi_n(kh)|>\frac{\varepsilon}{4}\right\}\leqslant \\ &\qquad\leqslant \frac{1}{1-\dfrac{256}{\varepsilon^2}h} \varlimsup_{n\to\infty} \mathsf{P}\left\{|\xi_n(t_{nj_{n,k+1}})-\xi_n(t_{nj_{n,k}})|>\frac{\varepsilon}{16}\right\}.\end{aligned}$$

It follows from the convergence of finite dimensional distributions of $\xi_n(t)$ to finite dimensional distributions of $w(t)$ that

$$\overline{\lim_{n\to\infty}} \mathsf{P}\left\{|\xi_n(t_{nj_n,k+1})-\xi_n(t_{nj_n,k})|>\frac{\varepsilon}{16}\right\}=\frac{1}{\sqrt{2\pi}}\int\limits_{|u|>\frac{\varepsilon}{16\sqrt{h}}} e^{-u^2/2}\,du.$$

Consequently,

$$\overline{\lim_{n\to\infty}} \mathsf{P}\{\sup_{|t_1-t_2|\leq h}|\xi_n(t_1)-\xi_n(t_2)|>\varepsilon\}=$$

$$=O\left(\sum_{kh<1}\int\limits_{|u|>\varepsilon/\frac{\varepsilon}{16\sqrt{h}}} e^{-u^2/2}\,du\right)=O\left(\frac{1}{h}\int\limits_{|u|>\varepsilon/\frac{\varepsilon}{16\sqrt{h}}} e^{-u^2/2}\,du\right).$$

Since

$$\lim_{h\to 0}\frac{1}{h}\int\limits_{|u|>\frac{\varepsilon}{16\sqrt{h}}} e^{-u^2/2}\,du=0,$$

we thus obtain condition (6). The theorem is thus proved. □

The following result is a corollary of Theorem 3.

Theorem 4. *Let $\xi_1, \xi_2, \ldots, \xi_n, \ldots$ be a sequence of independent identically distributed random variables such that $\mathsf{E}\xi_i=0$, $\operatorname{Var}\xi_i=1$. Denote by $\xi_n(t)$ the random broken line with the vertices $\frac{k}{n}, \frac{1}{\sqrt{n}}S_k$, where $S_0=0$, $S_k=\xi_1+\ldots+\xi_k$. Then for each functional f, which is μ_w-almost everywhere defined and continuous on $\mathscr{C}_{[0,1]}$, where μ_w is the measure associated with process $w(t)$, the distribution of $f(\xi_n(\cdot))$ will converge to the distribution of $f(w(\cdot))$.*

Corollary. *If the conditions of Theorem 4 are satisfied, then*

$$\lim_{n\to\infty}\mathsf{P}\{\max_{1\leq k\leq n}|S_k|<\alpha\sqrt{n}\}=\mathsf{P}\{\sup_{0\leq t\leq 1}|w(t)|<\alpha\}$$

for almost all α.

This follows from the continuity of the function $f(x(\cdot))=\sup_{0\leq t\leq 1}|x(t)|$.

Theorem 5. *Let function $\varphi(x)$ be defined for $x\in\mathscr{R}^1$ and be Riemann integrable in each finite segment and let the variables ξ_k satisfy the conditions of Theorem 4. Then*

$$\lim_{n\to\infty}\mathsf{P}\left\{\frac{1}{n}\sum_{k=1}^{n}\varphi\left(\frac{1}{\sqrt{n}}S_k\right)<\alpha\right\}=\mathsf{P}\left\{\int_0^1\varphi(w(t))\,dt<\alpha\right\}$$

for all α *such that*

$$\mathsf{P}\left\{\int_0^1 \varphi(w(t))\,dt=\alpha\right\}=0.$$

Proof. We show that functional

$$f(x(\cdot))=\int_0^1 \varphi(x(t))\,dt$$

is μ_w-almost everywhere continuous (in the metric of $\mathscr{C}_{[0,1]}$). Let $x_n(t)\to$ $\to x(t)$ uniformly on $[0, 1]$. Then $\varphi(x_n(t))\to\varphi(x(t))$ for all t such that $x(t)\notin\Lambda_\varphi$, where Λ_φ is the set of discontinuities of the function φ. Denote by $\chi_\varphi(x)$ the indicator of the set Λ_φ. Then the functional $f(x(\cdot))$ will be continuous at point $x(\cdot)\in\mathscr{C}_{[0,1]}$ if $x(t)\notin\Lambda_\varphi$ for almost all t, i.e. if

$$\int_0^1 \chi_\varphi(x(s))\,ds=0,$$

since in this case $\varphi(x_n(t))\to\varphi(x(t))$ for almost all t and $\varphi(x_n(t))$ are bounded by the same constant since $\sup\limits_{n,t}|x_n(t)|$ is finite and $\varphi(x)$ is bounded on each finite interval. Since φ is Riemann integrable Λ_φ has the Lebesgue measure 0. We observe that

$$\mathsf{E}\int_0^1 \chi_\varphi(w(t))\,dt=\int_0^1 \mathsf{E}\chi_\varphi(w(t))\,dt=$$

$$=\int_0^1\int_{\Lambda\varphi} e^{-x^2/2t}\frac{1}{\sqrt{2\pi t}}\,dx\,dt=0,$$

The quantity $\int_0^1 \chi_\varphi(w(t))\,dt$ is nonnegative, therefore

$$\mathsf{P}\left\{\int_0^1 \chi_\varphi(w(t))\,dt\neq 0\right\}=0.$$

If we denote by $A\subset\mathscr{C}_{[0,1]}$ the set of points of discontinuity of the functional f, it follows that

$$A \subset \left\{ x(\cdot): \int_0^1 \chi_\varphi(x(s))\, ds > 0 \right\}$$

and hence

$$\mu_w(A) \leqslant \mathsf{P}\left\{ \int_0^1 \chi_\varphi(w(t))\, dt \neq 0 \right\} = 0.$$

If $\xi_n(t)$ is the process introduced in Theorem 4, then, in view of Theorem 4,

$$\lim_{n\to\infty} \mathsf{P}\left\{ \int_0^1 \varphi(\xi_n(t))\, dt < \alpha \right\} = \mathsf{P}\left\{ \int_0^1 \varphi(w(t))\, dt < \alpha \right\},$$

provided only that

$$\mathsf{P}\left\{ \int_0^1 \varphi(w(t))\, dt = \alpha \right\} = 0.$$

Let $\varphi_\varepsilon^+(x)$ and $\varphi_\varepsilon^-(x)$ be two continuous functions satisfying relations $\varphi_\varepsilon^-(x) < \varphi(x) < \varphi_\varepsilon^+(x)$ and

$$\int_{-\infty}^{\infty} [\varphi_\varepsilon^+(x) - \varphi_\varepsilon^-(x)]\, dx < \varepsilon.$$

For any continuous function $\bar\varphi(x)$

$$\left| \int_0^1 \bar\varphi(\xi_n(t))\, dt - \frac{1}{n} \sum_{k=1}^n \bar\varphi\left(\frac{1}{\sqrt{n}} S_k \right) \right| \leqslant$$

$$\leqslant \sum_{k=1}^n \int_{\frac{k-1}{n}}^{k/n} \left| \bar\varphi(\xi_n(t)) - \bar\varphi\left(\xi_n\left(\frac{k}{n} \right) \right) \right| dt \leqslant$$

$$\leqslant \sup\{ |\bar\varphi(x) - \bar\varphi(y)|;\ |x-y| \leqslant \eta_n,\ |x| \leqslant \zeta_n \},$$

where

$$\eta_n = \sup_k \left| \xi_n\left(\frac{k}{n}\right) - \xi_n\left(\frac{k+1}{n}\right) \right| = \frac{1}{\sqrt{n}} \sup_{k \leqslant n} |\xi_k|,$$

$$\zeta_n = \frac{1}{\sqrt{n}} \sup_{k \leqslant n} |S_k|.$$

Therefore

$$\mathsf{P}\left\{\left|\int_0^1 \bar{\varphi}(\xi_n(t))\,dt - \frac{1}{n}\sum_{k=1}^n \bar{\varphi}\left(\frac{1}{\sqrt{n}} S_k\right)\right| > \varepsilon\right\} \leqslant \mathsf{P}\{\eta_n > \delta\} + \mathsf{P}\{\zeta_n > c\},$$

provided only that δ and c are chosen such that for $|x-y| \leqslant \delta$, $|x| \leqslant c$, we have $|\bar{\varphi}(x) - \bar{\varphi}(y)| < \varepsilon$. However $\eta_n \to 0$ in probability and the probability $\mathsf{P}\{\zeta_n > c\}$ becomes as small as desired for all n by choosing c sufficiently large. Hence

$$\left|\int_0^1 \bar{\varphi}(\xi_n(t))\,dt - \frac{1}{n}\sum_{k=1}^n \bar{\varphi}\left(\frac{1}{\sqrt{n}} S_k\right)\right| \to 0$$

in probability so that

$$\lim_{n\to\infty} \mathsf{P}\left\{\frac{1}{n}\sum_{k=1}^n \bar{\varphi}\left(\frac{1}{\sqrt{n}} S_k\right) < \alpha\right\} = \mathsf{P}\left\{\int_0^1 \bar{\varphi}(w(t))\,dt < \alpha\right\},$$

provided only that

$$\mathsf{P}\left\{\int_0^1 \bar{\varphi}(w(t))\,dt = \alpha\right\} = 0.$$

Since

$$\mathsf{P}\left\{\frac{1}{n}\sum_{k=1}^n \varphi_\varepsilon^+\left(\frac{1}{\sqrt{n}} S_k\right) < \alpha\right\} \leqslant \mathsf{P}\left\{\frac{1}{n}\sum_{k=1}^n \varphi\left(\frac{1}{\sqrt{n}} S_k\right) < \alpha\right\} \leqslant$$

$$\leqslant \mathsf{P}\left\{\frac{1}{n}\sum_{k=1}^n \varphi_\varepsilon^-\left(\frac{1}{\sqrt{n}} S_k\right) < \alpha\right\},$$

then approaching the limit as $n \to \infty$ in this relation, we obtain for each $h > 0$

$$\mathsf{P}\left\{\int_0^1 \varphi_\varepsilon^+(w(t))\,dt < \alpha - h\right\} \leqslant \varliminf_{n\to\infty} \mathsf{P}\left\{\frac{1}{n}\sum_{k=1}^n \varphi\left(\frac{1}{\sqrt{n}} S_k\right) < \alpha\right\} \leqslant$$

$$\leqslant \overline{\lim_{n\to\infty}} \mathsf{P}\left\{\frac{1}{n}\sum_{k=1}^{n} \varphi\left(\frac{1}{\sqrt{n}} S_k\right) < \alpha\right\} \leqslant \mathsf{P}\left\{\int_0^1 \varphi_\varepsilon^-(w(t))\, dt < \alpha + h\right\}.$$

However,

$$\mathsf{E}\left|\int_0^1 \varphi_\varepsilon^+(w(t))\, dt - \int_0^1 \varphi(w(t))\, dt\right| \leqslant$$

$$\leqslant \mathsf{E}\left[\int_0^1 \varphi_\varepsilon^+(w(t))\, dt - \int_0^1 \varphi_\varepsilon^-(w(t))\, dt\right] \leqslant$$

$$\leqslant \frac{1}{\sqrt{2\pi}} \int_0^1 \frac{dt}{\sqrt{t}} \int_{-\infty}^{\infty} [\varphi_\varepsilon^+(x) - \varphi_\varepsilon^-(x)]\, e^{-x^2/2t}\, dx \leqslant \frac{2\varepsilon}{\sqrt{2\pi}}.$$

Thus the distribution of $\int_0^1 \varphi_\varepsilon^+(w(t))\, dt$ converges to the distribution of $\int_0^1 \varphi(w(t))\, dt$ as $\varepsilon \to 0$. An analogous assertion is valid also for φ_ε^-. Approaching the limit as $\varepsilon \to 0$ we observe that for all $h > 0$

$$\mathsf{P}\left\{\int_0^1 \varphi(w(t))\, dt < \alpha + h\right\} \leqslant \underline{\lim_{n\to\infty}} \mathsf{P}\left\{\frac{1}{n}\sum_{k=1}^{n} \varphi\left(\frac{1}{\sqrt{n}} S_k\right) < \alpha\right\} \leqslant$$

$$\leqslant \overline{\lim_{n\to\infty}} \mathsf{P}\left\{\frac{1}{n}\sum_{k=1}^{n} \varphi\left(\frac{1}{\sqrt{n}} S_k\right) < \alpha\right\} \leqslant \mathsf{P}\left\{\int_0^1 \varphi(w(t))\, dt < \alpha + h\right\}.$$

Approaching the limit as $h \to 0$ and taking into account the fact that the function $\mathsf{P}\left\{\int_0^1 \varphi(w(t))\, dt < z\right\}$ is continuous at $z = \alpha$ provided only that $\mathsf{P}\left\{\int_0^1 \varphi(w(t))\, dt = \alpha\right\} = 0$, we obtain the proof of the theorem. □

Convergence of continuous processes with independent increments. Consider continuous processes with independent increments and values in a certain Banach space $\mathscr{X}$. If $\xi(t)$, $a \leqslant t \leqslant b$ is such a process, then for all $\varepsilon > 0$

$$\lim_{\lambda \to 0} \sum_{k=0}^{n-1} \mathsf{P}\{|\xi(t_{k+1}) - \xi(t_k)| > \varepsilon\} = 0, \tag{7}$$

where $a = t_0 < t_1 < \ldots < t_n = b$, $\lambda = \max_k (t_{k+1} - t_k)$ (cf. Theorems 1 and 4 in Section 5 of Chapter III).

Theorem 6. *Let $\xi_n(t)$, $n = 0, 1, \ldots$ be a sequence of continuous processes with independent increments defined on $[a, b]$ and taking on values in $\mathscr{X}$. In order that for each function $\varphi(x)$ continuous on $\mathscr{C}_{[a,b]}(\mathscr{X})$ the distribution of the variable $\varphi(\xi_n(\cdot))$ converge to the distribution of the variable $\varphi(\xi_0(\cdot))$ it is necessary and sufficient that the following conditions be satisfied:*

1) *marginal distributions of the processes $\xi_n(t)$ converge to the marginal distributions of $\xi_0(t)$;*

2) *for each $\varepsilon > 0$*

$$\lim_{h \to 0} \overline{\lim_{n \to \infty}} \sup_{|t_1 - t_2| \leqslant h} \mathsf{P}\{|\xi_n(t_2) - \xi_n(t_1)| > \varepsilon\} = 0.$$

Proof. The necessity of condition 1) follows from the convergence of the distribution of $g(\xi_n(t_1), \ldots, \xi_n(t_k))$ to the distribution of $g(\xi_0(t_1), \ldots, \xi_0(t_k))$ for each bounded continuous function $g(x_1, \ldots x_k)$ defined on $\mathscr{X}^k$ (the functional $\varphi(x(\cdot)) = g(x(t_1), \ldots, x(t_k))$ is continuous on $\mathscr{C}_{[a,b]}(\mathscr{X})$). The necessity of condition 2) follows from the Remark following Theorem 1 since the following inequality is valid:

$$\sup_{|t_1 - t_2| \leqslant h} \mathsf{P}\{|\xi_n(t_1) - \xi_n(t_2)| > \varepsilon\} \leqslant \mathsf{P}\{\sup_{|t_1 - t_2| \leqslant h} |\xi_n(t_2) - \xi_n(t_1)| > \varepsilon\}.$$

In view of the Remark following Theorem 1, in order to prove the sufficiency of the conditions of the theorem it is only necessary to show that condition 2) yields for all $\varepsilon > 0$ equality

$$\lim_{h \to 0} \overline{\lim_{n \to \infty}} \mathsf{P}\{\sup_{|t_1 - t_2| \leqslant h} |\xi_n(t_1) - \xi_n(t_2)| > \varepsilon\} = 0. \tag{8}$$

In the same manner as in the Remark following Theorem 1, we show that condition 2) yields equality

$$\lim_{h \to 0} \sup_n \sup_{|t_1 - t_2| \leqslant h} \mathsf{P}\{|\xi_n(t_1) - \xi_n(t_2)| > \varepsilon\} = 0 \tag{9}$$

for each $\varepsilon > 0$. Choose for a given $\varepsilon > 0$ an h so small that

$$\sup_n \sup_{|t_1 - t_2| \leqslant 2h} \mathsf{P}\left\{|\xi_n(t_1) - \xi_n(t_2)| > \frac{\varepsilon}{4}\right\} \leqslant \tfrac{1}{2}.$$

Then utilizing the continuity of $\xi_n(t)$ and Lemma 4 in Section 3 we obtain that

$$\mathsf{P}\left\{\sup_{s\leqslant t\leqslant s+2h}|\xi_n(t)-\xi_n(s)|>\frac{\varepsilon}{2}\right\}\leqslant 2\mathsf{P}\left\{|\xi_n(s+2h)-\xi_n(s)|>\frac{\varepsilon}{4}\right\}.$$

Therefore

$$\mathsf{P}\{\sup_{|t-s|\leqslant h}|\xi_n(t)-\xi_n(s)|>\varepsilon\}\leqslant$$

$$\leqslant \mathsf{P}\left\{\sup\left[|\xi_n(t)-\xi_n(a+kh)|;\, kh\leqslant t-a\leqslant (k+2)h,\ 0\leqslant k<\frac{b-a}{h}\right]>\frac{\varepsilon}{2}\right\}\leqslant$$

$$\leqslant \sum_{kh<b-a}\mathsf{P}\left\{\sup[|\xi_n(t)-\xi_n(a+kh)|;\, kh\leqslant t-a\leqslant (k+2)\,h]>\frac{\varepsilon}{2}\right\}\leqslant$$

$$\leqslant 2\sum_{kh<b-a}\mathsf{P}\left\{|\xi_n(a+kh+2h)-\xi_n(a+kh)|>\frac{\varepsilon}{4}\right\},$$

(if $t>b$ we set $\xi_n(t)=\xi_n(b)$). In view of condition 1) of the theorem

$$\overline{\lim_{n\to\infty}}\,\mathsf{P}\{\sup_{|t_1-t_2|\leqslant h}|\xi_n(t_1)-\xi_n(t_2)|>\varepsilon\}\leqslant$$

$$\leqslant 2\sum_{kh<b-a}\mathsf{P}\left\{|\xi_0(a+(k+2)\,h)-\xi_0(a+kh)|>\frac{\varepsilon}{4}\right\}\leqslant$$

$$\leqslant 4\sum_{kh<b-a}\mathsf{P}\left\{|\xi_0(a+(k+1)\,h)-\xi_0(a+kh)|>\frac{\varepsilon}{8}\right\}.$$

The fact that the last sum tends to zero as $h\to 0$ follows from condition (7). The theorem is thus proved. □

Convergence of continuous Markov processes. Consider a sequence of continuous Markov processes $\xi_n(t)$, $n=0, 1,\ldots$ defined on the interval $[a, b]$ and taking on values in a complete metric space $(\mathscr{X}, \varrho)$. Denote by $P_n(t, x, s, A)$ the transition probability for the process $\xi_n(t)$. Let

$$V_\varepsilon(x)=\{y:\varrho(x, y)>\varepsilon\},$$
$$\alpha_n(h, \varepsilon)=\sup\{P_n(t_1, x, t_2, V_\varepsilon(x));\, x\in\mathscr{X},\, |t_1-t_2|\leqslant h\}.$$

Theorem 7. *Let the marginal distributions of the processes $\xi_n(t)$ converge to the marginal distributions of the process $\xi_0(t)$ and let the following conditions be satisfied:*

1) *for each $\varepsilon>0$ $\lim_{n\to 0}\sup_n \alpha_n(h, \varepsilon)=0$;*
2) *if $a=t_0<t_1<\ldots<t_n=b$, $\lambda=\max_k(t_{k+1}-t_k)$, then for every $\varepsilon>0$*

$$\lim_{\lambda\to 0}\sum_{k=0}^{n-1} \mathsf{P}\{\varrho(\xi_0(t_k), \xi_0(t_{k+1}))>\varepsilon\}=0.$$

Then for each function φ *in* $\mathscr{C}_{[a,b]}(\mathscr{X})$ *the distribution of* $\varphi(\xi_n(\cdot))$ *will converge to the distribution of* $\varphi(\xi_0(\cdot))$.

As a preliminary, we prove the following lemma.

Lemma 2. *If for a separable Markov process* $\xi(t)$ *the quantity* $\alpha(h, \varepsilon/2)$ *– defined in the same manner as* $\alpha_n(h, \varepsilon/2)$ *is defined for* $\xi_n(t)$) *– is less than* 1, *then*

$$\mathsf{P}\{\sup[\varrho(\xi(t), \xi(s)); s\in[t, t+h]]\geqslant\varepsilon\}\leqslant\frac{\mathsf{P}\left\{\varrho(\xi(t), \xi(t+h))\geqslant\frac{\varepsilon}{2}\right\}}{1-\alpha(h, \varepsilon/2)}. \tag{10}$$

Proof. Taking into account the separability of the process, it is sufficient to prove (10) in the case when the sup under the sign of the probability is taken over any finite subset I of the interval $[t, t+h]$. Let $I=\{t=t_0, \dots, t_n=t+h\}$. Denote by B_k the event $\{\varrho(\xi(t_0), \xi(t_k)\geqslant\varepsilon\}$, $C_k=\left\{\varrho(\xi(t_k), \xi(t_n))\geqslant\frac{\varepsilon}{2}\right\}$. Then

$$C_0\supset\bigcup_{j=1}^{n}\{\bar{B}_1\cap\dots\cap\bar{B}_{j-1}\cap B_j\cap\bar{C}_j\}.$$

Therefore

$$\begin{aligned}\mathsf{P}\{C_0\}&\geqslant\sum_{j=1}^{n}\mathsf{P}\{\bar{C}_j|\bar{B}_1\cap\dots\cap\bar{B}_{j-1}\cap B_j\}\,\mathsf{P}\{\bar{B}_1\cap\dots\cap\bar{B}_{j-1}\cap B_j\}=\\ &=\sum_{j=1}^{n}(1-\mathsf{P}\{C_j|\bar{B}_1\cap\dots\cap\bar{B}_{j-1}\cap B_j\})\,\mathsf{P}\{\bar{B}_1\cap\dots\bar{B}_{j-1}\cap B_j\}\geqslant\\ &\geqslant(1-\alpha(h, \varepsilon/2))\sum_{j=1}^{n}\mathsf{P}\{\bar{B}_1\cap\dots\cap\bar{B}_{j-1}\cap B_j\}.\end{aligned}$$

It remains to show that

$$\sum_{j=1}^{n}\mathsf{P}\{\bar{B}_1\cap\dots\cap\bar{B}_{j-1}\cap B_j\}=\mathsf{P}\{\sup[\varrho(\xi(t), \xi(s)), s\in I]\geqslant\varepsilon\}.$$

The lemma is thus proved. □

Proof of Theorem 7. Choose h so small that $\sup_n \alpha_n(2h, \varepsilon/8)<\frac{1}{2}$. It then follows from Lemma 2 that

$$\mathsf{P}\left\{\sup\ [\varrho(\xi_n(t), \xi_n(s)); s\in[t, t+2h]]\geqslant\frac{\varepsilon}{2}\right\}\leqslant 2\mathsf{P}\left\{\varrho(\xi(t), \xi(t+2h))\geqslant\frac{\varepsilon}{4}\right\}.$$

From this inequality we obtain, in the same manner as in the previous theorem, that

$$\overline{\lim_{n\to\infty}} \mathsf{P}\{\sup_{|t_1-t_2|\leqslant h} \varrho(\xi_n(t_1), \xi_n(t_2))>\varepsilon\}\leqslant$$

$$\leqslant 4\sum_{kh<b-a} \mathsf{P}\left\{\varrho(\xi_0(a+(k+1)h), \xi_0(a+kh))\geqslant\frac{\varepsilon}{8}\right\}$$

(if $t>b$ we set $\xi(t)=\xi(b)$). The proof of the theorem then follows from this inequality and condition 2).

§5. Limit Theorems for Processes without Discontinuities of the Second Kind

Metrics in the space of functions without discontinuities of the second kind. To apply the results of Section 1 to processes without discontinuities of the second kind, it is necessary as a preliminary, to introduce a suitable metric in the space of functions without discontinuities. Denote by $\mathscr{D}_{[a,b]}(\mathscr{X})$ the set of functions $x(t)$ defined on $[a, b]$ taking on values in a complete metric space $\mathscr{X}$ and possessing limit values $x(t+0)$ for $a\leqslant t<b$ and $x(t-0)$ for $a<t\leqslant b$. Since any interval $[a, b]$ can be mapped continuously and in a one-to-one manner into an interval $[0, 1]$ we shall consider the space $\mathscr{D}_{[0,1]}(\mathscr{X})$. Functions which coincide at all points of continuity will not be distinguished, therefore it is natural to give a standard definition for the values of the function $x(t)$ at the discontinuity points. In what follows, we shall assume that for all functions in $\mathscr{D}_{[0,1]}(\mathscr{X})$ the following relations are satisfied

$$x(t)=x(t+0), \quad x(0)=x(+0), \quad x(1)=x(1-0). \tag{1}$$

The value $\varrho(x(t-0), x(t))$ is called the size of the jump of $x(t)$ at point t. It is necessary to introduce in $\mathscr{D}_{[0,1]}(\mathscr{X})$ a metric which will transform $\mathscr{D}_{[0,1]}(\mathscr{X})$ into a separable metric space with the property that the minimal σ-algebra containing all cylindrical sets coincides with the σ-algebra of Borel sets in this space. It is also desirable that the metric be sufficiently "strong" (i.e. that there will be as few as possible converging sequences and hence as many functionals as possible, continuous in this metric). The uniform metric

$$\varrho_u(x(\cdot), y(\cdot))=\sup_{0\leqslant t\leqslant 1} \varrho(x(t), y(t))$$

is not suitable for these purposes since in this metric $\mathscr{D}_{[0,1]}(\mathscr{X})$ is not a separable space. Indeed the set of functions

$$x_s(t)=\begin{cases} x_1, & t<s, \\ x_2, & t\geqslant s, \end{cases} \qquad \varrho(x_1, x_2)=\delta>0, \quad 0<s<1,$$

is uncountable, but the distance between any two elements of this set is δ. We introduce in space $\mathcal{D}_{[0,1]}(\mathcal{X})$ a metric which is somewhat weaker than the uniform metric.

Denote by Λ the totality of all continuous monotonically increasing numerical functions $\lambda(t)$ with $t\in[0,1]$ such that $\lambda(0)=0$, $\lambda(1)=1$ (i.e. $\lambda(t)$ maps the interval $[0,1]$ into itself continuously and in a one-to-one fashion).

Note that for all $\lambda\in\Lambda$ the inverse functions λ^{-1} exist which also belong to Λ. If λ_1 and $\lambda_2\in\Lambda$, then the composite function $\lambda_1(\lambda_2)$ belongs to Λ as well.

Define for each pair $x(t)$ and $y(t)$ in $\mathcal{D}_{[0,1]}(\mathcal{X})$ the quantity

$$r_{\mathcal{D}}(x, y)=\inf\{\sup_{0\leqslant t\leqslant 1} \varrho(x(t), y(\lambda(t)))+\sup_{0\leqslant t\leqslant 1} |t-\lambda(t)|;\ \lambda\in\Lambda\}. \tag{2}$$

We show that $r_{\mathcal{D}}$ determines a metric in $\mathcal{D}_{[0,1]}(\mathcal{X})$. To do so it is necessary to verify that $r_{\mathcal{D}}$ satisfies the three axioms: a) $r_{\mathcal{D}}(x, y)\geqslant 0$ and is zero if $x=y$, b) $r_{\mathcal{D}}(x, y)=r_{\mathcal{D}}(y, x)$; c) $r_{\mathcal{D}}(x, z)\leqslant r_{\mathcal{D}}(x, y)+r_{\mathcal{D}}(y, z)$ for all $x(\cdot)$, $y(\cdot)$ and $z(\cdot)$ in $\mathcal{D}_{[0,1]}(\mathcal{X})$.

Condition a) is obvious. Condition b) follows from relation

$$\begin{aligned} r_{\mathcal{D}}(y, x) &= \inf_{\lambda\in\Lambda}\{\sup_{0\leqslant t\leqslant 1} \varrho(y(t), x(\lambda(t)))+\sup_{0\leqslant t\leqslant 1} |t-\lambda(t)|\}= \\ &=\inf_{\lambda\in\Lambda}\{\sup_{0\leqslant t\leqslant 1} \varrho(y(\lambda^{-1}(t)), x(t))+ \\ &\quad+\sup_{0\leqslant t\leqslant 1} |\lambda^{-1}(t)-t|;\ \lambda\in\Lambda\}=r_{\mathcal{D}}(x, y). \end{aligned}$$

We now discuss condition c) – the triangular inequality – in some detail. Let $x(\cdot)$, $y(\cdot)$ and $z(\cdot)$ be functions in $\mathcal{D}_{[0,1]}(\mathcal{X})$. For each $\varepsilon>0$ we can find functions $\lambda_1(t)$ and $\lambda_2(t)$ such that the following conditions be satisfied

$$\left.\begin{aligned} r_{\mathcal{D}}(x, y) &\geqslant \sup_{0\leqslant t\leqslant 1} \varrho(x(t), y(\lambda_1(t)))+\sup_{0\leqslant t\leqslant 1} |t-\lambda_1(t)|-\varepsilon, \\ r_{\mathcal{D}}(y, z) &\geqslant \sup_{0\leqslant t\leqslant 1} \varrho(y(t), z(\lambda_2(t)))+\sup_{0\leqslant t\leqslant 1} |t-\lambda_2(t)|-\varepsilon. \end{aligned}\right\} \tag{3}$$

Then

$$\begin{aligned} r_{\mathcal{D}}(x, z) &\leqslant \sup_{0\leqslant t\leqslant 1} \varrho(x(t), z(\lambda_2(\lambda_1(t))))+ \\ &\quad+\sup_{0\leqslant t\leqslant 1} |t-\lambda_2(\lambda_1(t))|\leqslant \sup_{0\leqslant t\leqslant 1} \varrho(x(t), y(\lambda_1(t)))+ \end{aligned}$$

$$+\sup_{0\leqslant t\leqslant 1}|t-\lambda_1(t)|+\sup_{0\leqslant t\leqslant 1}\varrho(y(\lambda_1(t)), z(\lambda_2(\lambda_1(t))))+$$
$$+\sup_{0\leqslant t\leqslant 1}|\lambda_1(t)-\lambda_2(\lambda_1(t))|=$$
$$=\sup_{0\leqslant t\leqslant 1}\varrho(x(t), y(\lambda_1(t))+\sup_{0\leqslant t\leqslant 1}|t-\lambda_1(t)|+$$
$$+\sup_{0\leqslant t\leqslant 1}\varrho(y(t), z(\lambda_2(t)))+\sup_{0\leqslant t\leqslant 1}|t-\lambda_2(t)|,$$

since if t runs through the interval $[0, 1]$, $\lambda_1(t)$ runs through the same interval. Taking into account relation (3) we have

$$r_{\mathscr{D}}(x, z)\leqslant r_{\mathscr{D}}(x, y)+r_{\mathscr{D}}(y, z)+2\varepsilon,$$

Since $\varepsilon>0$ is arbitrary condition c) is verified.

Therefore $r_{\mathscr{D}}$ is a proper distance function on $\mathscr{D}_{[0,1]}(\mathscr{X})$.

The following auxiliary assertions will be required for the further investigation of the properties of the metric $r_{\mathscr{D}}$:

Define for each function $x(\cdot)$ in $\mathscr{D}_{[0,1]}(\mathscr{X})$

$$\Delta_c(x)=\sup\{\min[\varrho(x(t'), x(t)); \varrho(x(t), x(t''))]; t-c\leqslant t'\leqslant t\leqslant t''\leqslant t+c\}+$$
$$+\sup_{0\leqslant t\leqslant c}\varrho(x(0), x(t))+\sup\{\varrho(x(t), x(1)); 1-c\leqslant t\leqslant 1\}. \qquad (4)$$

Then in view of Lemma 1, Section 4 in Chapter III we have $\lim_{c\to 0}\Delta_c(x)=0$.

Lemma 1. *Let $x(\cdot)$ be a function in $\mathscr{D}_{[0,1]}(\mathscr{X})$ and let $[\alpha, \beta]\subset[0, 1]$. If $x(\cdot)$ has no jumps on $[\alpha, \beta]$ of size exceeding ε, then for $|t'-t''|<c$ with $t', t''\in[\alpha, \beta]$ we have*

$$\varrho(x(t'), x(t''))\leqslant 2\Delta_c(x)+\varepsilon.$$

Proof. We choose an arbitrary $\delta\in(0, \varepsilon)$ and a point τ in $[t', t'']$ with the property that for $t\in[t', \tau)$

$$\varrho(x(t'), x(t))<\Delta_c(x)+\delta, \qquad \varrho(x(t'), x(\tau))\geqslant\Delta_c(x)+\delta.$$

If there is no such point, then $\varrho(x(t'), x(t''))<\Delta_c(x)+\delta$ and hence the assertion of the lemma is satisfied. If a point τ exists, then, in view of the fact that

$$\min[\varrho(x(t'), x(\tau)); \varrho(x(\tau), x(t''))]\leqslant\Delta_c(x),$$

and $\varrho(x(t'), x(\tau))\geqslant\Delta_c(x)+\delta$, we have $\varrho(x(\tau), x(t''))\leqslant\Delta_c(x)$. Therefore

$$\varrho(x(t'), x(t''))\leqslant\varrho(x(t'), x(\tau-0))+$$
$$+\varrho(x(\tau-0), x(\tau))+\varrho(x(\tau), x(t''))\leqslant\Delta_c(x)+\delta+\varepsilon+\Delta_c(x).$$

Approaching the limit as $\delta\downarrow 0$ we obtain the proof of the lemma. □

Denote by Y_m the countable set of points $y_{mk} \in \mathscr{X}$ such that $\bigcup_k S_{1/m}(y_{mk}) = \mathscr{X}$, where $S_a(x)$ denotes an open sphere of radius a with the center at x. Denote by $H_{m,n}$ the collection of functions $x(\cdot) \in \mathscr{D}_{[0,1]}(\mathscr{X})$ which are constant on each one of the intervals $\left[\frac{k}{n}, \frac{k+1}{n}\right)$ and taking on values in Y_m.

Lemma 2. *For each function $x(\cdot)$ in $\mathscr{D}_{[0,1]}(\mathscr{X})$ there exists a function $x^*(\cdot)$ in $H_{m,n}$ such that*

$$r_{\mathscr{D}}(x, x^*) \leqslant \frac{1}{n} + \frac{1}{m} + 4\Delta_{2/n}(x).$$

Proof. There is at most one point in each one of the intervals $\left[\frac{k}{n}, \frac{k+1}{n}\right]$ with a jump of size exceeding $2\Delta_{2/n}(x)$. Indeed, let τ be such a point, then

$$\varrho(x(s), x(\tau-0)) = \min[\varrho(x(s), x(\tau-0)); \varrho(x(\tau-0), x(\tau))] \leqslant$$
$$\leqslant \Delta_{1/n}(x) \quad \text{for} \quad s \in \left[\frac{k}{n}, \tau\right);$$

$$\varrho(x(s), x(\tau)) \leqslant \Delta_{1/n}(x) \quad \text{for} \quad s \in \left(\tau, \frac{k+1}{n}\right]$$

and hence

$$\varrho(x(s-0), x(s)) \leqslant 2\Delta_{1/n}(x) \leqslant 2\Delta_{2/n}(x), \qquad s \neq \tau.$$

Let τ_k be the point in the interval $\left[\frac{k}{n}, \frac{k+1}{n}\right]$ such that

$$\varrho(x(\tau_k - 0), x(\tau_k)) \geqslant 2\Delta_{2/n}(x),$$

provided such a point exists in this interval. Denote by $\lambda(t)$ a function in Λ such that $\lambda\left(\frac{k+1}{n}\right) = \tau_k$ and $t - \frac{1}{n} \leqslant \lambda(t) \leqslant t$. (A piecewise-linear function defined by the equalities $\lambda(0)=0$, $\lambda\left(\frac{k+1}{n}\right) = \tau_k$ and $\lambda(1)=1$ is such a function.) Set $\bar{x}(t) = x(\lambda(t))$. The function $\bar{x}(t)$ has jumps of size exceeding $2\Delta_{2/n}(x)$ only at the points of the form k/n and moreover

$$r_{\mathscr{D}}(x, \bar{x}) \leqslant \sup_{0 \leqslant t \leqslant 1} \varrho(\bar{x}(t), x(\lambda(t))) + \sup_{0 \leqslant t \leqslant 1} |t - \lambda(t)| \leqslant \frac{1}{n}.$$

Next let $\bar{x}^*(t)$ be the function equal to $\bar{x}(k/n)$ for $t\in\left[\frac{k}{n},\frac{k+1}{n}\right)$, $k\leqslant n-1$, and let $\bar{x}^*(1)=\bar{x}\left(\frac{n-1}{n}\right)$. Then

$$r_{\mathscr{D}}(\bar{x},\bar{x}^*)\leqslant \sup_{0\leqslant t\leqslant 1}\varrho(\bar{x}(t),\bar{x}^*(t))\leqslant$$
$$\leqslant \sup_k \sup\left[\varrho\left(\bar{x}(t),\bar{x}\left(\frac{k}{n}\right)\right);\frac{k}{n}\leqslant t<\frac{k+1}{n}\right].$$

Since the jumps of $\bar{x}(t)$ exceeding $2\Delta_{2/n}(x)$, occur only at the points of the form k/n, there are no such jumps in the half-interval $\left[\frac{k}{n},\frac{k+1}{n}\right)$ and hence in view of Lemma 1

$$\varrho\left(\bar{x}\left(\frac{k}{n}\right),\bar{x}(t)\right)\leqslant 2\Delta_{1/n}(\bar{x})+2\Delta_{2/n}(x)\quad\text{for}\quad t\in\left[\frac{k}{n},\frac{k+1}{n}\right).$$

We estimate $\Delta_{1/n}(\bar{x})$:

$$\Delta_{1/n}(\bar{x})=\sup\left\{\min\left[\varrho(\bar{x}(t'),\bar{x}(t));\varrho(\bar{x}(t),\bar{x}(t''))\right];\ t-\frac{1}{n}\leqslant t'\leqslant t\leqslant t''\leqslant t+\frac{1}{n}\right\}+$$
$$+\sup\left\{\varrho(\bar{x}(0),\bar{x}(t));0\leqslant t\leqslant\frac{1}{n}\right\}+$$
$$+\sup\left\{\varrho(\bar{x}(t),\bar{x}(1));1-\frac{1}{n}\leqslant t\leqslant 1\right\}=$$
$$=\sup\left\{\varrho(x(0),x(\lambda(t)));0\leqslant t\leqslant\frac{1}{n}\right\}+$$
$$+\sup\left\{\varrho(x(\lambda(t)),x(1));1-\frac{1}{n}\leqslant t\leqslant 1\right\}+$$
$$+\sup\left\{\min\left[\varrho(x(\lambda(t')),x(\lambda(t)));\varrho(x(\lambda(t)),x(\lambda(t'')))\right];\right.$$
$$\left. t-\frac{1}{n}\leqslant t'\leqslant t\leqslant t''\leqslant t+\frac{1}{n}\right\}.$$

Note that $t_1-\frac{1}{n}<\lambda(t_1)<\lambda(t_2)\leqslant t_2<t_1+\frac{1}{n}$ for $t_1<t_2<t_1+\frac{1}{n}$ so that

$0 \leqslant \lambda(t_2) - \lambda(t_1) \leqslant \frac{2}{n}$. Therefore $\Delta_{1/n}(\bar{x}) \leqslant \Delta_{2/n}(x)$. Hence, $r_{\mathscr{D}}(\bar{x}, \bar{x}^*) \leqslant 4\Delta_{2/n}(x)$.

Finally we set $x^*(t) = y_{mk}$, where k is the smallest index such that $\varrho(\bar{x}^*, y_{mk}) < 1/m$.

Since $\varrho(\bar{x}^*(t), x^*(t)) \leqslant 1/m$ it follows that $r_{\mathscr{D}}(\bar{x}^*, x^*) \leqslant 1/m$,

$$r_{\mathscr{D}}(x, x^*) \leqslant r_{\mathscr{D}}(x, \bar{x}) + r_{\mathscr{D}}(\bar{x}, \bar{x}^*) + r_{\mathscr{D}}(\bar{x}, x^*) \leqslant$$
$$\leqslant \frac{1}{n} + 4\Delta_{2/n}(x) + \frac{1}{m}.$$

The lemma is thus proved. □

Corollary. *The space $\mathscr{D}_{[0,1]}(\mathscr{X})$ with the metric $r_{\mathscr{D}}$ is a separable space.*

This follows from the fact that in view of Lemma 2, the countable set $\bigcup_{m,n} H_{m,n}$ is everywhere dense in $\mathscr{D}_{[0,1]}(\mathscr{X})$. □

Let X_1 be a compactum in $\mathscr{X}$ and let λ_δ be an increasing continuous function defined for $\delta > 0$ and satisfying condition $\lambda_{+0} = 0$. Denote by $K_{\mathscr{D}}(x_1, \lambda_\delta)$ the set of functions in $\mathscr{D}_{[0,1]}(\mathscr{X})$ such that $x(t) \in X_1$ for $t \in [0, 1]$ and let $\Delta_c(x) \leqslant \lambda_c$ for all $c > 0$.

Theorem 1. 1) *The set $K_{\mathscr{D}}(X_1, \lambda_\delta)$ is a compactum in $\mathscr{D}_{[0,1]}(\mathscr{X})$;* 2) *for each compactum K_1 one can find a compactum $X_1 \subset \mathscr{X}$ and an increasing continuous function λ_δ with $\lambda_{+0} = 0$ such that $K_1 \subset K_{\mathscr{D}}(X_1, \lambda_\delta)$.*

Proof. 1) We show that $K_{\mathscr{D}}(X_1, \lambda_\delta)$ possesses a finite ε-net for each $\varepsilon > 0$. To do this we note that for each m there exists K_m such that

$$\bigcup_{k=1}^{k_m} S_{1/m}(y_{mk}) \supset X_1.$$

We choose m and n satisfying $\frac{1}{n} + \frac{1}{m} + 4\lambda_{2/n} < \varepsilon$. Then the set of functions $H_{m,n} \cap F[y_{m1}, \ldots, y_{mk_m}]$ where $F[y_1, \ldots, y_s]$ is the set of functions taking on only the values $y_1, \ldots, y_s$ is a finite ε-net in the set $K_{\mathscr{D}}(X_1, \lambda_\delta)$. Indeed, in view of Lemma 2, $H_{m,n}$ is a $\left(\frac{1}{n} + \frac{1}{m} + 4\lambda_{2/n}\right)$-net in $K_{\mathscr{D}}(X_1, \lambda_\delta)$ and moreover functions with values in the set $[y_{m1}, \ldots, y_{mk_m}]$ form such a net. The set $K_{\mathscr{D}}(X_1, \lambda_\delta)$ is closed. It is easy to verify the relationship

$$\Delta_c(x) \leqslant \Delta_{c + r_{\mathscr{D}}(x, y)}(y) + 3r_{\mathscr{D}}(x, y).$$

Therefore if $r_{\mathscr{D}}(x_n, \bar{x}) \to 0$, then for each $\alpha > 0$

$$\Delta_c(\bar{x}) \leqslant \varliminf_{n \to \infty} \Delta_{c+\alpha}(x_n) \leqslant \lambda_{c+\alpha}.$$

Hence in view of the continuity of λ, $\Delta_c(\bar{x}) \leq \lambda_c$. It is also clear that $\lim_{n\to\infty} x_n(t) \in X_1$ if $x_n(t) \in X_1$ for all n.

Consequently, the limit of a sequence belonging to $K_{\mathscr{D}}(X_1, \lambda_\delta)$ will also belong to $K_{\mathscr{D}}(X_1, \lambda_\delta)$. It remains to show that each fundamental sequence $x_n(\cdot)$ belonging to $K_{\mathscr{D}}(X_1, \lambda_\delta)$ will be convergent. Let $x_n(\cdot)$ be a sequence of functions in $K_{\mathscr{D}}(X_1, \lambda_\delta)$ for which $r_{\mathscr{D}}(x_n, x_m) \to 0$ as $n \to \infty$ and $m \to \infty$ (i.e. $x_n(\cdot)$ is a fundamental sequence). It is sufficient to show that some subsequence $x_{n_k}(\cdot)$ possesses the limit $\bar{x}(\cdot)$. Therefore it can be assumed that the sequence $x_n(\cdot)$ is such that $r_{\mathscr{D}}(x_n, x_{n+1}) < 2^{-n-1}$. Then there exists a sequence of functions λ_n belonging to Λ such that

$$\sup_{0\leq t\leq 1} |t - \lambda_{n+1}(t)| \leq \frac{1}{2^{n+1}},$$

$$\sup_{0\leq t\leq 1} \varrho(x_n(t), x_{n+1}(\lambda_{n+1}(t))) \leq \frac{1}{2^{n+1}}.$$

Set $\mu_1(t) = \lambda_1(t)$, $\mu_n(t) = \lambda_n(\mu_{n-1}(t))$. Since

$$\sup_{0\leq t\leq 1} |\mu_n(t) - \mu_{n-1}(t)| \leq \sup_{0\leq t\leq 1} |\lambda_n(t) - t| \leq \frac{1}{2^n},$$

$\mu_n(t)$ converges to a certain non-decreasing continuous function $\mu(t)$ satisfying the conditions $\mu(0)=0$, $\mu(1)=0$. Next

$$\sup_{0\leq t\leq 1} \varrho(x_n(\mu_n(t)), x_{n-1}(\mu_{n-1}(t))) = \sup_{0\leq t\leq 1} \varrho(x_n(\lambda_n(t)), x_{n-1}(t)) \leq \frac{1}{2^n}$$

Therefore $x_n(\mu_n(t))$ is uniformly convergent to a certain function $x^*(t)$ in $\mathscr{D}_{[0,1]}(\mathscr{X})$. We investigate the connection between the functions $x^*(t)$ and $\mu(t)$. Let $\mu(t)$ be a constant of a certain interval $[\alpha, \beta]$. If $x^*(\alpha) = x^*(\beta)$ then $x^*(t)$ is also a constant on $[\alpha, \beta]$; if, however $x^*(\alpha) \neq x^*(\beta)$ then a $\gamma \in [\alpha, \beta]$ exists such that $x^*(t) = x^*(\alpha)$ for $t \in [\alpha, \gamma)$, and also $x^*(t) = x(\beta)$, $t \in [\gamma, \beta]$. Indeed, otherwise points $t' < t'' < t'''$ could have been found belonging to $[\alpha, \beta]$ such that $x^*(t') \neq x^*(t'')$, $x^*(t'') \neq x^*(t''')$ and then

$$\lim_{n\to\infty} \min[\varrho(x_n(\mu_n(t'')), x_n(\mu_n(t'''))); \varrho(x_n(\mu_n(t'')), x_n(\mu_n(t')))] =$$

$$= \min[\varrho(x^*(t''), x^*(t''')), \varrho(x^*(t''), x^*(t'))] > 0,$$

while $\mu_n(t') < \mu_n(t'') < \mu_n(t''')$ and $\mu_n(t')$, $\mu_n(t'')$, $\mu_n(t''')$ tend to $\mu(\alpha)$. This would contradict the fact that the sequence $x_n(\cdot)$ belongs to $K_{\mathscr{D}}(X_1, \lambda_\delta)$.

Denote by $\bar{x}(\cdot)$ the function in $\mathscr{D}_{[0,1]}(\mathscr{X})$ defined by relations

$$\bar{x}(t) = x^*(\mu(t)), \tag{5}$$

satisfied for all t such that $\mu(s) > \mu(t)$ for all $s \in (t, 1]$. Relation (5) determines a unique function $\bar{x}(t) \in \mathscr{D}_{[0,1]}(\mathscr{X})$.

We show that this function $\bar{x}(\cdot)$ is the limit of a sequence $x_n(\cdot)$. For this purpose we construct auxiliary functions $\varphi_n \in \Lambda$. Let $\tau_1, \ldots, \tau_k$ be all the points in $[0, 1]$ at which $\bar{x}(\cdot)$ possesses jumps of size exceeding $1/n$. Denote by $[\alpha_i, \beta_i]$ the maximal interval on which $\mu(t)$ takes on the value τ_i (this interval may also be a singleton).

Let γ_i be a point in $[\alpha_i, \beta_i]$ such that $x^*(t) = \bar{x}(\tau_i - 0)$ for $t \in [\alpha_i, \gamma_i)$ and $x^*(t) = \bar{x}(\tau_i)$ for $t \in [\gamma_i, \beta_i]$. In particular if $\alpha_i = \gamma_i$, then $x^*(t)$ takes the unique value $\bar{x}(\tau_i)$ on the interval $[\alpha_i, \beta_i]$. We choose an ε_n not exceeding $1/n$ such that $\Delta_{\varepsilon_n}(\bar{x}) < 1/n$. Let $\varphi_n(t)$ be the function satisfying the relations: $\varphi_n(\gamma_i) = \tau_i$, $|\varphi_n(t) - \mu(t)| < \varepsilon_n$.

We bound the $\sup\{\varrho(x^*(t), \bar{x}(\varphi_n(t)); 0 \leqslant t \leqslant 1\}$. If t does not belong to any one of the intervals $[\alpha_i, \beta_i]$ then in view of lemma 1

$$\varrho(x^*(t), \bar{x}(\varphi_n(t))) = \varrho(\bar{x}(\mu(t)), \bar{x}(\varphi_n(t))) \leqslant 2\Delta_{\varepsilon_n}(\bar{x}) + \frac{1}{n}$$

since $x(t)$ has no jumps between $\mu(t)$ and $\varphi_n(t)$ of size exceeding $1/n$. If $t \in [\alpha_i, \gamma_i)$, then

$$\varrho(x^*(t), \bar{x}(\varphi_n(t))) \leqslant \sup\{\varrho(\bar{x}(\tau_i - 0), \bar{x}(s)); s \in [\tau_i - \varepsilon_n, \tau_i)\} \leqslant \Delta_{\varepsilon_n}(x),$$

because $\varrho(\bar{x}(\tau_i - 0), \bar{x}(\tau_i)) > \frac{1}{n} > \Delta_{\varepsilon_n}(\bar{x})$. Analogously we show that for $t \in [\gamma_i, \beta_i]$, $\varrho(x^*(t), \bar{x}(\varphi_n(t))) \leqslant \Delta_{\varepsilon_n}(\bar{x})$. Consequently,

$$\sup_{0 \leqslant t \leqslant 1} \varrho(x^*(t), \bar{x}(\varphi_n(t))) \leqslant \frac{1}{n} + 2\Delta_{\varepsilon_n}(\bar{x}) \leqslant \frac{3}{n}.$$

We now estimate $r_{\mathscr{D}}(x_n, \bar{x})$. We have

$$\begin{aligned} r_{\mathscr{D}}(x_n, \bar{x}) &\leqslant r_{\mathscr{D}}(x_n(\cdot), x^*(\mu_n^{-1}(\cdot))) + \\ &+ r_{\mathscr{D}}(x^*(\mu_n^{-1}(\cdot)), \bar{x}(\varphi_n(\mu_n^{-1}(\cdot)))) + r_{\mathscr{D}}(\bar{x}(\cdot), \bar{x}(\varphi_n(\mu_n^{-1}))) \leqslant \\ &\leqslant \sup_{0 \leqslant t \leqslant 1} \varrho(x_n(\mu_n(t)), x^*(t)) + \sup_{0 \leqslant t \leqslant 1} \varrho(x^*(t), \bar{x}(\varphi_n(t))) + \\ &+ \sup_{0 \leqslant t \leqslant 1} |t - \varphi_n(\mu_n^{-1}(t))| \leqslant \frac{1}{2^n} + \frac{3}{n} + \sup_{0 \leqslant t \leqslant 1} |\mu_n(t) - \varphi_n(t)| \leqslant \\ &\leqslant \frac{1}{2^n} + \frac{3}{n} + \frac{1}{2^n} + \varepsilon_n. \end{aligned}$$

Therefore $r_{\mathscr{D}}(x_n, \bar{x}) \to 0$, i.e. the sequence $x_n(\cdot)$ converges to the function $\bar{x}(\cdot)$. Assertion 1) is proved.

2) Denote by X_t the set of values $x(t)$ and $x(t-0)$ with $x(\cdot) \in K_1$. The proof that $\bigcup_t X_t$ is a compactum is identical to one given in Lemma 1 of Section 4.

We set $\Delta_c = \sup\{\Delta_c(x);\ x(\cdot)\in K_1\}$ Clearly Δ_c is a monotonically increasing function of c. We show that $\lim_{c\downarrow 0}\Delta_c = 0$. Assume the contrary. Then one can find a sequence of functions $x_n(\cdot)\in K_1$ and a sequence $c_n \to 0$ such that $\Delta_{c_n}(x_n) \geqslant \delta$ for some $\delta > 0$. Since K_1 is compact one can assume that $x_n(\cdot)\to x_0(\cdot)$. However for $r_{\mathscr{D}}(x, y)\leqslant \varepsilon$

$$\Delta_c(x) \leqslant \Delta_{c+\varepsilon}(y) + 3\varepsilon.$$

Therefore for each $c > 0$

$$\Delta_c(x_0) \geqslant \Delta_{c - r_{\mathscr{D}}(x_n, x_0)}(x_n) - 3r_{\mathscr{D}}(x_n, x_0) \geqslant \delta - 3r_{\mathscr{D}}(x_n, x_0),$$

as long as $c_n < c - r_{\mathscr{D}}(x_n, x_0)$. Hence $\Delta_c(x_0)\geqslant\delta$ for all $c>0$ and this contradicts the condition that $\lim\Delta_c(x_0)=0$. Thus $\lim\Delta_c=0$. Clearly a continuous monotonic function λ_c can be constructed such that $\Delta_c < \lambda_c$, $\lambda_{+0}=0$. Then $K_1 \subset K(X_1, \lambda_\delta)$. The theorem is thus proved. □

The basic limit theorem for processes without discontinuities of the second kind

Theorem 2. *Let $\xi_n(t)$, $0\leqslant t\leqslant 1$, $n=0, 1, \dots$, be a sequence of processes without discontinuities of the second kind with values in $\mathscr{X}$; moreover let the marginal distributions of $\xi_n(t)$ converge to the marginal distributions of $\xi_0(t)$. In order for every functional f defined on $\mathscr{D}_{[0,1]}(\mathscr{X})$ and continuous in metric $r_{\mathscr{D}}$ the distribution of $f(\xi_n(\cdot))$ to converge to the distribution of $f(\xi_0(\cdot))$, it is necessary and sufficient that for all $\varepsilon>0$ condition*

$$\lim_{c\to 0}\overline{\lim_{n\to\infty}}\, \mathsf{P}\{\Delta_c(\xi_n(\cdot))>\varepsilon\}=0. \tag{6}$$

be satisfied.

Proof. It follows from equality (6) that for all $\varepsilon>0$ the relation

$$\lim_{c\to 0}\sup_n \mathsf{P}\{\Delta_c(\xi_n(\cdot))>\varepsilon\}=0$$

is satisfied.

From here in the same manner as in the proof of Theorem 1 in Section 4 we verify the existence of a continuous monotone function λ_δ, such that $\lambda_{+0}=0$ and

$$\sup_n \mathsf{P}\{\Delta_c(\xi_n(\cdot))\leqslant\lambda_c,\ 0<c\leqslant 1\} > 1-\varepsilon/2. \tag{7}$$

Utilizing the convergence of marginal distributions of the processes $\xi_n(t)$ we can show – in the same manner as in Theorem 1 of Section 4 – that the family of measures $\{\nu_{nt}, n=1, 2, \dots;\ 0\leqslant t\leqslant 1\}$, where $\nu_{nt}(A) = \mathsf{P}\{\xi_n(t)\in A\}$ is compact. Hence for each k one can find a compactum $X^{(k)}$ such that $\nu_{nt}(X^{(k)}) \geqslant 1 - 2^{-2k}\frac{\varepsilon}{4}$ for all n and t. Denote by $\tilde{X}^{(k)}$ the set of y

such that $\varrho(y, X^{(k)}) \leqslant \lambda_{2^{-k}}$. Then $X_1 = \bigcap_k \tilde{X}^{(k)}$ is a compactum. Since relations $x\left(\frac{l}{2^k}\right) \in X^{(k)}$, $x\left(\frac{l+1}{2^k}\right) \in X^{(k)}$, $\Delta_{2^{-k}}(x) < \lambda_{2^{-k}}$ imply that $x(t) \in \tilde{X}^{(k)}$ for $\frac{l}{2^k} \leqslant t \leqslant \frac{l+1}{2^k}$, we have

$$\mathsf{P}\{\xi_n(\cdot) \notin K_{\mathscr{D}}(X_1, \lambda_\delta)\} \leqslant 1 - \mathsf{P}\{\Delta_c(\xi_n(\cdot)) \leqslant \lambda_c, 0 < c \leqslant 1\} +$$

$$+ \sum_{k=1}^{\infty} \sum_{l=0}^{2^k} \mathsf{P}\left\{\xi_n\left(\frac{l}{2^k}\right) \notin X^{(k)}\right\} \leqslant \frac{\varepsilon}{2} + \sum_{k=1}^{\infty} \sum_{l=0}^{2^k} 2^{-2k} \frac{\varepsilon}{4} < \varepsilon.$$

Thus for each $\varepsilon > 0$ a compactum $K_{\mathscr{D}}(X_1, \lambda_\delta)$, is constructed such that for all measures μ_n associated with the random processes $\xi_n(\cdot)$ on $\mathscr{D}_{[0,1]}(\mathscr{X})$ the inequality

$$\mu_n(K_{\mathscr{D}}(X_1, \lambda_\delta)) \leqslant 1 - \varepsilon.$$

is satisfied. It now remains to apply Theorem 1 of Section 1 (as it follows from the remark following Theorem 1, the conditions of the theorem are sufficient also in an incomplete space). Sufficiency is thus established.

To prove the necessity of condition (6) we introduce the functional

$$F_a(x(\cdot)) = \sup_{0 \leqslant t \leqslant 1} \varrho(x(0), x(t))\, e^{-at} + \sup_{0 \leqslant t \leqslant 1} \varrho(x(1), x(t))\, e^{-a(1-t)} +$$

$$+ \sup\{\min[\varrho(x(t), x(s))\, e^{-a(t-s)}; \varrho(x(t), x(u))\, e^{-a(u-t)}]; 0 \leqslant s \leqslant t \leqslant u \leqslant 1\}.$$

It is easy to verify that $F_a(x(\cdot))$ is a continuous functional on $\mathscr{D}_{[0,1]}(\mathscr{X})$. Therefore if the distribution $f(\xi_n(\cdot))$ converges to the distribution of $f(\xi_0(\cdot))$ for all continuous functions f, we then have for each $\varepsilon > 0$

$$\varlimsup_{n\to\infty} \mathsf{P}\{F_a(\xi_n(\cdot)) > \varepsilon\} \leqslant \mathsf{P}\{F_a(\xi_0(\cdot)) \geqslant \varepsilon\}.$$

Now note that

$$\Delta_c(x(\cdot)) \leqslant e^{ac} F_a(x(\cdot)),$$

$$F_a(x(\cdot)) \leqslant \Delta_c(x(\cdot)) + 5\, e^{-ac} \sup_{0 \leqslant t \leqslant 1} \varrho(x(0), x(t)).$$

Therefore

$$\varlimsup_{n\to\infty} \mathsf{P}\{\Delta_c(\xi_n(\cdot)) < \varepsilon\} \leqslant \varlimsup_{n\to\infty} \mathsf{P}\{F_{1/c}(\xi_n(\cdot)) > e^{-1}\varepsilon\} \leqslant$$

$$\leqslant \mathsf{P}\{F_{1/c}(\xi_0(\cdot)) \geqslant e^{-1}\varepsilon\} \leqslant \mathsf{P}\{\Delta_{\sqrt{c}}(\xi_0(\cdot)) \geqslant \tfrac{1}{2} e^{-1}\varepsilon\} +$$

$$+ \mathsf{P}\left\{\sup_{0 \leqslant t \leqslant 1} \varrho(\xi_0(0), \xi_0(t)) \geqslant \frac{\varepsilon}{10} \exp\left(-1 + \frac{1}{\sqrt{c}}\right)\right\}.$$

In view of Lemma 1, Section 4, Chapter III, the finiteness of $\sup \varrho(x(0), x(t))$ for all $x(\cdot)\in\mathscr{D}_{[0,1]}(\mathscr{X})$ and the fact that $\xi_0(\cdot)\in\mathscr{D}_{[0,1]}(\mathscr{X})$ with probability 1, the right-hand side of the last inequality tends to zero as $c\to 0$. The theorem is thus proved. □

Theorem 3. *Let $\xi_n(t)$, $n=0, 1,\ldots$ be a sequence of random processes belonging to $\mathscr{D}_{[0,1]}(\mathscr{X})$ with probability 1 such that the marginal distributions of $\xi_n(t)$ converge to the marginal distributions of $\xi_0(t)$ and let $\alpha>0$, $\beta>0$ and $H>0$ exist for which the inequality*

$$\mathsf{E}[\varrho(\xi_n(t_1), \xi_n(t_2))\,\varrho(\xi_n(t_2), \xi_n(t_3))]^\beta \leqslant H(t_3-t_1)^{1+\alpha}$$

is satisfied with $n\geqslant 0$ and $t_1<t_2<t_3$.

Then for each continuous functional f on $\mathscr{D}_{[0,1]}(\mathscr{X})$ the distribution of the variables $f(\xi_n(\cdot))$ converges to the distribution of the variable $f(\xi_0(\cdot))$.

Proof. We utilize Lemma 4 of Section 4 in Chapter III. If we set $g(h)=h^\gamma$, where $0<\gamma<\beta/\alpha$, and $q(h)=2^{1+\alpha}Hh^\delta$, where $\delta=\beta-\alpha\gamma$, then for all processes $\xi_n(t)$ the conditions of this lemma are satisfied with

$$G(m)=\frac{2^{-m\gamma}}{1-2^{-\gamma}},\qquad Q(m, C)=C^{-\beta}H2^{1+\alpha}\frac{2^{-m\delta}}{1-2^{-\delta}}.$$

Hence for some L

$$\mathsf{P}\{\Delta_c(\xi_n(\cdot))>\varepsilon\}\leqslant L\varepsilon^{-\beta}c^\alpha.$$

To complete the proof it remains to apply Theorem 2. □

Limit theorems for Markov processes. Let $\xi_n(t)$ be a sequence of Markov processes defined on $[0, 1]$ with sample functions belonging to $\mathscr{D}_{[0,1]}(\mathscr{X})$ with probability 1. Denote by $\mathsf{P}_n(t, x, s, A)$ the transition probabilities of the process $\xi_n(t)$. Next let $V_\varepsilon(x)=\{y:\varrho(x, y)>\varepsilon\}$.

Theorem 4. *If the marginal distributions of the processes $\xi_n(t)$ converge to the marginal distributions of $\xi_0(t)$ and if for each $\varepsilon>0$*

$$\lim_{h\downarrow 0}\varlimsup_{n\to\infty}\sup\{P_n(t, x, s, V_\varepsilon(x)); x\in\mathscr{X}, 0\leqslant s-t\leqslant h\}=0$$

is satisfied then for each continuous functional f on $\mathscr{D}_{[0,1]}(\mathscr{X})$ the distribution of $f(\zeta_n(\cdot))$ converges to the distribution of $f(\xi_0(\cdot))$.

The proof of this theorem is based on the following lemma.

Lemma 3. *Let $\xi_1, \xi_2,\ldots, \xi_n$ be a Markov chain such that for all $k<l$*

$$\mathsf{P}\{\varrho(\xi_k, \xi_l)\geqslant\varepsilon \mid \xi_k\}\leqslant\alpha<1$$

with probability 1. *Then*

$$\mathsf{P}\{\sup\{\min[\varrho(\xi_i,\xi_j);\ \varrho(\xi_j,\xi_l)]\ ;\ 1\leqslant i<j<l\leqslant n\}\geqslant 4\varepsilon\}\leqslant$$
$$\leqslant\frac{\alpha}{(1-\alpha)^2}\,\mathsf{P}\{\varrho(\xi_l,\xi_n)\geqslant\varepsilon\}.$$

Proof. The event

$$\{\sup\{\min[\varrho(\xi_i,\xi_j);\varrho(\xi_j,\xi_l)];1\leqslant i<j<l\leqslant n\}\geqslant 4\varepsilon\}$$

implies one of the events $A_r\cap B_r$, where

$$A_r=\{\varrho(\xi_1,\xi_j\}<2\varepsilon, j=1,\dots,r-1;\varrho(\xi_1,\xi_r)\geqslant 2\varepsilon\},$$
$$B_r=\{\sup_{k>r}\varrho(\xi_r,\xi_k)\geqslant 2\varepsilon\}.$$

Therefore

$$\mathsf{P}\{\sup\{\min[\varrho(\xi_i,\xi_j);\varrho(\xi_j,\xi_l)];1\leqslant i<j<l\leqslant n\}\geqslant 4\varepsilon\}\leqslant$$
$$\leqslant\sum_{r=1}^{n}\int_{A_r}\mathsf{P}\{B_r\mid\xi_1,\dots,\xi_r\}\,\mathsf{P}(d\omega)=\sum_{r=1}^{n}\int_{A_r}\mathsf{P}\{B_r|\,\xi_r\}\,\mathsf{P}(d\omega).$$

In view of Lemma 2 in Section 4 we have

$$\mathsf{P}\{B_r\mid\xi_r\}\leqslant\frac{\alpha}{1-\alpha}$$

and

$$\sum_{r=1}^{n}\mathsf{P}\{A_r\}=\mathsf{P}\{\sup_k\varrho(\xi_1,\xi_k)\geqslant 2\varepsilon\}\leqslant\frac{1}{1-\alpha}\,\mathsf{P}\{\varrho(\xi_1,\xi_n)\geqslant\varepsilon\}.$$

The result follows from these 2 inequalities. The lemma is thus proved. □

Corollary. *If* $\xi(t)$ *is a separable Markov process for which the transition probability* $P(t,x,s,A)$ *satisfies the inequality* $P(t,x,s,V_\varepsilon(x))\leqslant\alpha<1$ *for* $t_1\leqslant t<s<t_2$, *then*

$$\mathsf{P}\{\sup\{\min[\varrho(\xi(t'),\xi(t''));\varrho(\xi(t''),\xi(t'''))];t_1\leqslant t'<t''<t'''\leqslant t_2\}\geqslant 4\varepsilon\}\leqslant$$
$$\leqslant\frac{\alpha}{(1-\alpha)^2}\,\mathsf{P}\{\varrho(\xi(t_1),\xi(t_2))\geqslant\varepsilon\}.$$

We now proceed to the proof of Theorem 4. It is sufficient to show that for each $\varepsilon>0$

$$\lim_{c\to 0}\overline{\lim_{n\to\infty}}\,\mathsf{P}\{\Delta_c(\xi_n(\cdot))>\varepsilon\}=0.$$

We estimate this probability. Let c be a small number such that for all

n sufficiently large

$$\sup\{P_n(t, x, s, V_{\varepsilon/8}(x));\ x\in\mathscr{X},\ 0<s-t\leqslant 3c\}<\tfrac{1}{2}.$$

Then

$$\Delta_c(\xi_n(\cdot))\leqslant \sup_{0\leqslant t\leqslant c}\varrho(\xi_n(0), \xi_n(t))+\sup_{0\leqslant 1-t\leqslant c}\varrho(\xi_n(1), \xi_n(t))+$$
$$+\sup\left\{\min[\varrho(\xi_n(t), \xi_n(t')); \varrho(\xi_n(t'), \xi_n(t''))];\ kc\leqslant t'<t<t''\leqslant\right.$$
$$\left.\leqslant(k+3)\,c,\ k<\frac{1}{c}\right\}.$$

Therefore

$$\mathsf{P}\{\Delta_c(\xi_n(\cdot))\geqslant\varepsilon\}\leqslant \mathsf{P}\left\{\sup_{0\leqslant t\leqslant c}\varrho(\xi_n(0), \xi_n(t))\geqslant\frac{\varepsilon}{4}\right\}+$$
$$+\mathsf{P}\left\{\sup_{0\leqslant 1-t\leqslant c}\varrho(\xi_n(1), \xi_n(t))\geqslant\frac{\varepsilon}{4}\right\}+$$
$$+\sum_{k<\frac{1}{c}}\mathsf{P}\left\{\sup\{\min[\varrho(\xi_n(t'), \xi_n(t)); \varrho(\xi_n(t), \xi_n(t''))];\right.$$
$$\left.kc\leqslant t'<t<t''\leqslant(k+3)\,c\}\geqslant\frac{\varepsilon}{2}\right\}\leqslant$$
$$\leqslant\frac{2\alpha_n}{1-\alpha_n}+\frac{\alpha_n}{(1-\alpha_n)^2}\sum_{k<\frac{1}{c}}\mathsf{P}\left\{\varrho(\xi_n(kc), \xi_n(kc+3c))\geqslant\frac{\varepsilon}{8}\right\},$$

where

$$\alpha_n=\sup\{P_n(t, x, s, V_{\varepsilon/8}(x));\ x\in\mathscr{X},\ 0<s-t\leqslant 3c\}.$$

Since

$$\mathsf{P}\left\{\varrho(\xi_n(kc), \xi_n(kc+3c))\geqslant\frac{\varepsilon}{8}\right\}\leqslant$$
$$\leqslant\mathsf{P}\left\{\varrho(\xi_n(kc), \xi(kc+c))\geqslant\frac{\varepsilon}{24}\right\}+$$
$$+\mathsf{P}\left\{\varrho(\xi_n(kc+c), \xi_n(kc+2c))\geqslant\frac{\varepsilon}{24}\right\}+$$
$$+\mathsf{P}\left\{\varrho(\xi_n(kc+2c), \xi_n(kc+3c))\geqslant\frac{\varepsilon}{24}\right\},$$

and $\frac{1}{1-\alpha_n}\leqslant 2$ it follows that

$$\mathsf{P}\{\Delta_c(\xi_n\cdot))\geqslant\varepsilon\}\leqslant 4\alpha_n\left[1+3\sum_{k<\frac{1}{c}}\mathsf{P}\left\{\varrho(\xi_n(kc),\xi_n(kc+c))\geqslant\frac{\varepsilon}{24}\right\}\right].$$

Therefore, in view of the condition of the theorem and the equality

$$\lim_{n\to\infty}\sum_{k<1/c}\mathsf{P}\left\{\varrho(\xi_n(kc),\xi_n(kc+c))\geqslant\frac{\varepsilon}{24}\right\}=$$

$$=\sum_{k<1/c}\mathsf{P}\left\{\varrho(\xi_0(kc),\xi_0(kc+c)\geqslant\frac{\varepsilon}{24}\right\},$$

which is valid for almost all $\varepsilon>0$ it is sufficient to show that the sum in the r.h.s. of the last equality remains bounded as $c\to 0$. We choose h in such a manner that

$$P_0(t,x,s,V_{\varepsilon_1}(x))\leqslant\tfrac{1}{3}\quad\text{for}\quad s-t\leqslant h,\qquad \varepsilon_1=\frac{\varepsilon}{96}.$$

It is sufficient to show that the sum

$$\sum_{\bar t\leqslant kc<\bar t+h}\mathsf{P}\{\varrho(\xi_0(kc),\xi_0(kc+c))\geqslant 4\varepsilon_1\}$$

is bounded for any $\bar t$ with a given h.

Let $\eta_k=1$ if $\varrho(\xi_0(kc),\xi_0(kc+c))\geqslant 4\varepsilon_1$, and $\eta_k=0$ otherwise.

We are required to show that $\sum_{\bar t\leqslant kc<\bar t+h}\mathsf{E}\eta_k$ is uniformly bounded in $\bar t$ and c.

We estimate

$$\mathsf{P}\{\sum_{\bar t\leqslant kc<\bar t+h}\eta_k>l\}.$$

We shall consider only η_k with index k for which $\bar t\leqslant kc<\bar t+h$.

Let A_r be the event

$$\{\omega:\sum_{k\leqslant r}\eta_k=l;\,\eta_r=1\}.$$

Then

$$\mathsf{P}\{\sum\eta_k>l\}=\sum_r\mathsf{P}\{A_r\cap\{\omega:\sum_{k>r}\eta_k>0\}\}=$$

$$=\sum_r\int_{A_r}\mathsf{P}\{\sum_{k>r}\eta_k>0\,|\,\xi_0(rc+c)\}\,\mathsf{P}(d\omega)\leqslant$$

$$\leqslant \sum_r \int_{A_r} \mathsf{P}\{\sup_{k>r} \varrho(\xi_0(kc), \xi_0(rc+c)) \geqslant 2\varepsilon_1 \mid \xi_0(rc+c)\} P(d\omega) \leqslant$$

$$\leqslant \frac{\frac{1}{3}}{1-\frac{1}{3}} \sum_r \mathsf{P}\{A_r\} \leqslant \tfrac{1}{2}\mathsf{P}\{\sum \eta_k > l-1\}.$$

Therefore for all $\bar{t}$ and c,

$$\sum_{\bar{t} \leqslant kc < \bar{t}+h} \mathsf{E}\eta_k \leqslant \sum_{l=1}^{\infty} l(\tfrac{1}{2})^l = \tfrac{1}{4}.$$

The theorem is thus proved. □

Processes with independent increments in a complete linear normed space X are particular cases of Markov processes. We thus have the following theorem as a corollary of Theorem 4.

Theorem 5. *Let $\xi_n(t)$, $n=0, 1, \ldots$ be a sequence processes with independent increments defined on $[0, 1]$ with values $\mathscr{X}$ and belonging to $\mathscr{D}_{[0,1]}(\mathscr{X})$ with probability 1. If the marginal distributions of the process $\xi_n(t)$ converge to the marginal distributions of $\xi_0(t)$ and if for each $\varepsilon > 0$*

$$\lim_{h \to 0} \overline{\lim_{n \to \infty}} \sup_{|t-s| \leqslant h} \mathsf{P}\{|\xi_n(t) - \xi_n(s)| > \varepsilon\} = 0,$$

then for each continuous functional f on $\mathscr{D}_{[0,1]}(\mathscr{X})$ the distribution of $f(\xi_n(\cdot))$ converges to the distribution of $f(\xi_0(\cdot))$.

Remark. It is sufficient to require in Theorems 2–5 that the functional f be measurable and μ_0-almost everywhere continuous where μ_0 is the measure associated with the limiting process in $\mathscr{D}_{[0,1]}(\mathscr{X})$.

Applications to Statistics. We shall apply the limit theorems discussed above to the study of the asymptotic behavior of the empiric distribution function used in mathematical statistics.

Suppose that the results of a certain experiment represent a random variable with an unknown continuous distribution function $F(x)$. How is one to estimate the function $F(x)$ if n results $\xi_1, \xi_2, \ldots, \xi_n$ of independent outcomes of the experiment are available?

For this purpose an empirical distribution function $F_n^*(x)$ is utilized in mathematical statistics. This function is defined by relation

$$F_n^*(x) = \frac{\nu_n(x)}{n},$$

where $\nu_n(x)$ is the number of variables ξ_k which fall into the interval $(-\infty, x)$. It follows from Bernoulli's theorem that $F_n^*(x)$ converges in probability to $F(x)$. Thus the function $F_n^*(x)$ can be taken as an estimator of $F(x)$. Obviously we are interested in the error incurred by choosing this estimator. On the other hand, it is often convenient to have an analytic expression for the approximation of $F(x)$. In this case the following problem should be solved: can a given function $\Phi(x)$ serve as an approximation of $F(x)$ if the results of the experiment $\xi_1, \xi_2, \ldots, \xi_n$ are known. In the first as well as the second case it is important to determine the behavior of the difference between the empirical and theoretical distribution functions.

To study this difference we introduce the process

$$\eta_n(t) = \sqrt{n}(F_n^*(t) - F(t)).$$

Lemma 4. *The marginal distributions of processes $\eta_n(t)$ converge to the marginal distributions of a Gaussian process $\eta(t)$ with $\mathsf{E}\eta(t)=0$ and $\mathsf{E}\eta(t)\,\eta(\tau)=F(t)\,[1-F(\tau)]$ for $t<\tau$.*

Proof. We note that

$$\eta_n(t) = \frac{1}{\sqrt{n}} \sum_{k=1}^{n} [\varepsilon(\xi_k - t) - F(t)],$$

where $\varepsilon(t)=0$ for $t \geqslant 0$ and $\varepsilon(t)=1$ for $t<0$. Since

$$\mathsf{E}\varepsilon(\xi_k - t) = F(t),$$
$$\mathsf{E}\varepsilon(\xi_k - t)\,\varepsilon(\xi_k - \tau) = F(t) \quad \text{for} \quad t<\tau,$$

and the processes $\varepsilon(\xi_k - t) - F(t)$ are independent for distinct values of k, the rest of the proof follows from Theorem 1, Section 1 in Chapter III. □

Corollary. *Let $F^{-1}(t)$ be the inverse of $F(t)$. Set*

$$\xi_n(t) = \eta_n(F^{-1}(t)), \quad \xi(t) = \eta(F^{-1}(t)).$$

Then the marginal distributions of the processes $\xi_n(t)$ converge to the marginal distributions of a Gaussian process $\xi(t)$ defined for $t \in [0, 1]$ such that

$$\mathsf{E}\xi(t)=0, \quad \mathsf{E}\xi(t)\,\xi(s) = t(1-s) \quad \text{for} \quad 0 \leqslant t < s \leqslant 1.$$

Remark 1. The process $\xi_n(t)$ can be represented in the form

$$\xi_n(t) = \frac{1}{\sqrt{n}} \sum_{k=1}^{n} [\varepsilon(\eta_k - t) - t],$$

where $\eta_k = F(\xi_k)$ are independent variables uniformly distributed on $[0, 1]$.

Remark 2. Finite dimensional distributions of the process $\xi(t)$ coincide with the conditional finite-dimensional distributions of a Brownian motion process $w(t)$, $0 \leqslant t \leqslant 1$ under the condition $w(1)=0$. Since the conditional distributions of the process $w(t)$ given $w(1)=0$ are Gaussian, it is sufficient to show that

$$\mathsf{E}(w(t) \mid w(1))_{w(1)=0}=0,$$
$$\mathsf{E}(w(t)\, w(s) \mid w(1))_{w(1)=0}=t(1-s), \quad \text{for} \quad 0 \leqslant t<s \leqslant 1.$$

The variable $\bar{\xi}(t)=w(t)-tw(1)$ is uncorrelated with $w(1)$. Since the variables $\bar{\xi}(t)$ and $w(1)$ have the joint Gaussian distribution the process $\bar{\xi}(t)$ is independent of $w(1)$. Therefore

$$\mathsf{E}(\bar{\xi}(t) \mid w(1))=\mathsf{E}\bar{\xi}(t)=0,$$
$$\mathsf{E}(\bar{\xi}(t)\, \bar{\xi}(s) \mid w(1))=\mathsf{E}\bar{\xi}(t)\, \bar{\xi}(s).$$

Utilizing relation $\bar{\xi}(t)=w(t)-tw(1)$ and the previous formulas we obtain that

$$\mathsf{E}(w(t)/w(1))=tw(1),$$
$$\mathsf{E}(w(t)\, w(s)/w(1))=\mathsf{E}\bar{\xi}(t)\,\bar{\xi}(s)+ts(w(1))^2=\min[t;s]-ts+ts[w(1)]^2.$$

Setting $w(1)=0$ we establish the validity of the statement at the beginning of Remark 2. □

Theorem 6. *For any functional f continuous on $\mathscr{D}_{[0,1]}(\mathscr{R}^1)$ the distribution of $f(\xi_n(\cdot))$ converges to the distribution of $f(\xi(\cdot))$.*

Proof. We first note that the separable process $\xi(t)$ is continuous so that $\xi(t)$ belongs to $\mathscr{D}_{[0,1]}(\mathscr{R}^1)$ with probability 1. Indeed $\xi(t+h)-\xi(t)$ has a Gaussian distribution and moreover

$$\mathsf{E}(\xi(t+h)-\xi(t))^4=3(\mathsf{E}[\xi(t+h)-\xi(t)]^2)^2=O(h^2).$$

Therefore in view of Theorem 7, Section 5, Chapter III, the process $\xi(t)$ is continuous. The convergence of the marginal distributions of the processes $\xi_n(t)$ to the marginal distributions of $\xi(t)$ is thus established. □

In view of Theorem 2 it is sufficient to show that relation (6) is fulfilled. Since

$$\Delta_c(x) \leqslant \sup\{|x(t')-x(t'')|;\ |t'-t''| \leqslant c\},$$

the theorem is proved if for all $\varepsilon>0$ relation

$$\lim_{c\to 0} \overline{\lim_{n\to\infty}} \mathsf{P}\{\sup_{|t'-t''|\leqslant c} |\xi_n(t')-\xi_n(t'')|>\varepsilon\}=0 \tag{8}$$

is verified. The process $\xi_n(t)+\sqrt{n}t$ is monotonically increasing, hence for $t_1<t_2<t_3<t_4$

$$-\sqrt{n}(t_4-t_1) \leqslant \xi_n(t_3)-\xi_n(t_2) \leqslant \xi_n(t_4)-\xi_n(t_1)+\sqrt{n}(t_4-t_1).$$

Therefore

$$\sup_{|t'-t''|\leqslant c} |\xi_n(t')-\xi_n(t'')| \leqslant \frac{2\sqrt{n}}{2^m} + \sup_{|k_1-k_2|\leqslant c2^{m+2}} \left|\xi_n\left(\frac{k_1}{2^m}\right)-\xi_n\left(\frac{k_2}{2^m}\right)\right|.$$

Let m_n be chosen in such a manner that $\frac{\sqrt{n}}{2^{m_n}} \to 0$ as $n\to\infty$ and $n2^{-m_n}\geqslant 1$.

To prove (8) it is sufficient to show that for each $\varepsilon>0$

$$\lim_{c\to 0}\overline{\lim_{n\to\infty}} \mathsf{P}\{\sup_{|k_1-k_2|\leqslant c2^{m_n}} |\xi_n(k_1 2^{-m_n})-\xi_n(k_2 2^{-m_n})|>\varepsilon\}=0.$$

Note that

$$\sup_{|k_1-k_2|\leqslant c2^{m_n}} |\xi_n(k_1 2^{-m_n})-\xi_n(k_2 2^{-m_n})| \leqslant 2\sum_{r=m^{(c)}}^{m_n} \sup_i \left|\xi_n\left(\frac{i+1}{2^r}\right)-\xi_n\left(\frac{i}{2^r}\right)\right|,$$

where $m^{(c)}$ is the smallest integer satisfying relation $c2^{m^{(c)}}\geqslant 1$ (see the proof of Lemma 3, Section 4 in Chapter II concerning the last inequality).

Choose an $a<1$ such that $2a^4>1$. Then

$$\mathsf{P}\{\sup_{|k_1-k_2|\leqslant c2^{m_n}} |\xi_n(k_1 2^{-m_n})-\xi_n(k_2 2^{-m_n})|>\varepsilon\}\leqslant$$

$$\leqslant \sum_{r=m^{(c)}}^{m_n} \mathsf{P}\left\{\sup_i \left|\xi_n\left(\frac{i+1}{2^r}\right)-\xi_n\left(\frac{i}{2^r}\right)\right|>\frac{\varepsilon}{2}\frac{a^{r-m^{(c)}}}{1-a}\right\}\leqslant$$

$$\leqslant \sum_{r=m^{(c)}}^{m_n}\sum_{i=0}^{2^r-1} \mathsf{P}\left\{\left|\xi_n\left(\frac{i+1}{2^r}\right)-\xi_n\left(\frac{i}{2^r}\right)\right|>\frac{\varepsilon a^{r-m^{(c)}}}{2(1-a)}\right\}\leqslant$$

$$\leqslant \sum_{r=m^{(c)}}^{m_n}\sum_{i=0}^{2^r-1} \mathsf{E}\left|\xi_n\left(\frac{i+1}{2^r}\right)-\xi_n\left(\frac{i}{2^r}\right)\right|^4\left(\frac{2(1-a)}{\varepsilon a^{r-m^{(c)}}}\right)^4. \quad (9)$$

Let μ_n be the number of variables η_i taking values in the interval $[t, t+h]$. Then

$$\mathsf{P}\{\mu_n=k\}=C_n^k h^k(1-h)^{n-k}$$

and

$$\xi(t+h)-\xi(t)=\sqrt{n}\left(\frac{\mu_n}{n}-h\right).$$

Calculations show (cf. B.V. Gnedenko [37], Chapter 6, Section 34, equation 9) that

$$\mathsf{E}(\xi_n(t+h)-\xi_n(t))^4\leqslant 3h^2+\frac{h}{n}\leqslant 3h^2+h2^{-m_n}.$$

Hence we have for $h\geqslant 2^{-m_n}$

$$\mathsf{E}(\xi_n(t+h)-\xi_n(t))^4\leqslant 4h^2.$$

Substituting this bound into inequality (9) we obtain

$$\mathsf{P}\{\sup_{|k_1-k_2|\leqslant c2^{m_n}} |\xi_n(k_1 2^{-m_n})-\xi_n(k_2 2^{-m_n})|>\varepsilon\}\leqslant$$

$$\leqslant \sum_{r=m^{(c)}}^{m_n} \frac{2^4(1-a)^4}{\varepsilon^4 a^{4(r-m_c)}}\cdot 4\cdot 2^{-r} \leqslant L_\varepsilon \frac{1}{2^{m^{(c)}}},$$

where

$$L_\varepsilon=\frac{2^6(1-a)^4}{\varepsilon^4}\sum_{r=0}^{\infty}(2a^4)^{-r}.$$

The theorem is thus proved. □

Chapter VII

Absolute Continuity of Measures Associated with Random Processes

§1. General Theorems on Absolute Continuity

First we shall review certain definitions in measure theory.

Let two measures μ_1 and μ_2 be defined on a measurable space $(\mathscr{X}, \mathfrak{B})$. The measure μ_2 is called *absolutely continuous with respect to measure* μ_1* (denoted by $\mu_2 \ll \mu_1$) if $\mu_2(A)=0$ for all $A \in \mathfrak{B}$ such that $\mu_1(A)=0$. If $\mu_1 \ll \mu_2$ and $\mu_2 \ll \mu_1$ we write $\mu_1 \sim \mu_2$ and call these measures *equivalent*. Measures μ_1 and μ_2 are called mutually singular if there exists a set A such that $\mu_1(A)=0_2$ and $\mu_2(\mathscr{X}-A)=0$. Mutually singular measures are also called *orthogonal* and are denoted by $\mu_1 \perp \mu_2$. If measures μ_1 and μ_2 are finite then $\mu_2 = \nu_1 + \nu_2$ where $\nu_1 \ll \mu_1$ and $\nu_2 \perp \mu_1$. This representation is unique. Measures ν_1 and ν_2 are called respectively the *absolute continuous* and *singular components of measures* μ_2 with respect to measure μ_1.

For finite measures the *Radon-Nikodym* theorem is valid: $\mu_2 \ll \mu_1$ *iff there exist a* $\mathfrak{B}$*-measurable function* $\varrho(x)$ *such that for all* $A \in \mathfrak{B}$ *the equality*

$$\mu_2(A) = \int_A \varrho(x)\, \mu_1(dx)$$

is satisfied.

The function $\varrho(x)$ which is unique up to equivalence in measure μ_1 is called the *density* or the *derivative* of *measure* μ_2 *with respect* to *measure* μ_1 and is denoted by $\varrho(x) = \frac{d\mu_2}{d\mu_1}(x)$. If μ_2 is not absolutely continuous with respect to μ_1 then $\frac{d\mu_2}{d\mu_1}$ denotes the derivative of an absolutely continuous component of measure μ_2 with respect to μ_1. In particular if $\mu_1 \perp \mu_2$ then $\frac{d\mu_2}{d\mu_1} = 0$.

---

* the term μ_1-continuous is also used. *Translator's Remark.*

In this chapter we shall consider the case where μ_1 and μ_2 are probabilistic measures, i.e. $\mu_i(\mathscr{X})=1$. If $\mathscr{X}$ is a functional space then $\mathfrak{B}$ denotes a σ-algebra generated by cylindrical sets so that the measure μ_i can be viewed as a measure associated with a certain random process. The subject of this chapter is the study of conditions of absolute continuity, equivalence and singularity of such a measure as well as the evaluation of the density of one measure with respect to another.

When proving theorems on absolute continuity of probabilistic measures on a measurable space $(\mathscr{X}, \mathfrak{B})$ the following procedure is very often used. Let $\mathfrak{B}_n$ be an increasing sequence of σ-algebras such that $\sigma\{\bigcup_n \mathfrak{B}_n\}=\mathfrak{B}$ and μ_i^n be the restriction of measure μ_i on $\mathfrak{B}_n$. It is assumed that the structure of the σ-algebras in $\mathfrak{B}_n$ is such that the absolute continuity of the measure μ_2^n with respect to measure μ_1^n can be easily verified. If $\mathscr{X}$ is a functional space then $\mathfrak{B}_n$ usually signify σ-algebras generated by the cylindrical sets with bases in a fixed finite-dimensional subspace of $\mathscr{X}$. Let $\mu_2^n \ll \mu_1^n$ and

$$\varrho_n(x)=\frac{d\mu_2^n}{d\mu_1^n}(x).$$

The variables $\varrho_n(x)$ form a martingale on the probability space $(\mathscr{X}, \mathfrak{B}, \mu_1)$. Indeed for each $\mathfrak{B}_n$-measurable function $f(x)$

$$\int f(x)\,\varrho_{n+1}(x)\,\mu_1(dx)=\int f(x)\,\varrho_{n+1}(x)\,\mu_1^{n+1}(dx)=$$

$$=\int f(x)\,\mu_2^{n+1}(dx)=\int f(x)\,\mu_2^n(dx)=$$

$$=\int f(x)\,\varrho_n(x)\,\mu_1^n(dx)=\int f(x)\,\varrho_n(x)\,\mu_1(dx).$$

From here, in view of the definition of the conditional mathematical expectation on the probability space $(\mathscr{X}, \mathfrak{B}, \mu_1)$ we have

$$\mathsf{E}(\varrho_{n+1}(x)\mid\mathfrak{B}_n)=\varrho_n(x).$$

However, the variable $\varrho_n(x)$ is $\mathfrak{B}_n$-measurable and hence $\varrho_n(x)$ is a martingale. Since $\varrho_n(x)\geqslant 0$ *and* $\int \varrho_n(x)\,\mu_1(dx)=\mu_2^n(\mathscr{X})=1$, it follows from the theorem on the limit of martingales (cf. Theorem 1, Section 2, Chapter II) that the limit

$$\lim_{n\to\infty}\varrho_n(x)=\varrho(x) \tag{1}$$

exists μ_1-almost everywhere.

Theorem 1. *The function* $\varrho(x)$ *defined by relation* (1) *is a density of an absolutely continuous component of the measure* μ_2 *with respect to* μ_1, *i.e.*

$$\varrho(x)=\frac{d\mu_2}{d\mu_1}(x).$$

Proof. Let $\mu_2=a\mu'+b\mu''$, where $a+b=1$, and $\mu'\ll\mu_1$, $\mu''\perp\mu_1$ so that μ' and μ'' are probability measures. Denote by μ'^n and μ''^n the contractions of these measures on $\mathfrak{B}_n$. Then

$$\varrho_n(x)=a\varrho_n'(x)+b\varrho_n''(x),$$

where

$$\varrho_n'(x)=\frac{d\mu'^n}{d\mu_1^n}(x), \quad \varrho_n''(x)=\frac{d\mu''^n}{d\mu_1^n}(x).$$

To prove the theorem it is sufficient to show that $\varrho_n''(x)\to 0$ and $\varrho_n'(x)\to \to\frac{d\mu'}{d\mu_1}(x)$ μ_1-almost everywhere. For each $\mathfrak{B}_n$-measurable bounded function $f(x)$ the equality

$$\int f(x)\frac{d\mu'}{d\mu_1}(x)\,d\mu_1(x)=\int f(x)\,\mu_2'(dx)=\int f(x)\,\mu'^n(dx)=$$

$$=\int f(x)\,\varrho_n'(x)\,\mu_1(dx) \qquad \text{is satisfied.}$$

Therefore

$$\varrho_n'=\mathsf{E}\left(\frac{d\mu'}{d\mu_1}(x)\,\Big|\,\mathfrak{B}_n\right),$$

where the conditional mathematical expectation is taken over the probability space $(\mathscr{X}, \mathfrak{B}, \mu_1)$. In view of Theorem 4, Section 2, Chapter II, for each monotonically increasing sequence of σ-algebras $\mathfrak{B}_n$ we have with probability 1

$$\lim_{n\to\infty}\mathsf{E}(\xi\mid\mathfrak{B}_n)=\mathsf{E}(\xi\mid\sigma\{\bigcup_n\mathfrak{B}_n\}).$$

Hence

$$\lim_{n\to\infty}\varrho_n'(x)=\mathsf{E}\left(\frac{d\mu'}{d\mu_1}(x)\,\Big|\,\mathfrak{B}\right)=\frac{d\mu'}{d\mu_1}(x).$$

We now show that $\varrho_n''(x)\to 0$ almost everywhere in measure μ_1. Let $\lim_{n\to\infty}\varrho_n''(x)=\varrho''(x)$ (the existence of the limit follows from the fact that

$\varrho_n''(x)$ is a martingale). For each $\mathfrak{B}_n$-measurable nonnegative function $f(x)$ we have, in view of Fatou's lemma

$$\int f(x)\,\mu''(dx)=\int f(x)\,\mu''^n(dx)=\int f(x)\,\varrho_n''(x)\,\mu_1^n(dx)=$$

$$=\lim_{n\to\infty}\int f(x)\,\varrho_n''(x)\,\mu_1(dx)\geqslant\int f(x)\,\varrho''(x)\,\mu_1(dx),$$

(where $n>m$). Thus for $A\in\mathfrak{B}$ $\int_A \varrho''(x)\,\mu_1(dx)\leqslant\mu''(A)$. Let A be such that $\mu''(A)=0$, $\mu_1(A)=1$. Then $\int_A \varrho''(x)\,\mu_1(dx)=0$. Hence $\varrho''(x)=0$ μ_1-almost everywhere on the set A and since $\mu_1(A)=1$ it follows that $\varrho''(x)=0$ μ_1-almost everywhere. The theorem is thus proved. □

Corollary 1. *If $\varrho(x)$ defined by relation* (1) *is positive μ_1-almost everywhere then $\mu_1\ll\mu_2$ and*

$$\frac{d\mu_1}{d\mu_2}(x)=\begin{cases}\dfrac{1}{\varrho(x)}, & x\in S,\\ 0, & x\notin S,\end{cases}$$

where $S\in\mathfrak{B}$ is such that $\mu_2''(S)=0$, $\mu_1(S)=1$.

Indeed, for each $\mathfrak{B}$-measurable nonnegative function $f(x)$ the equality

$$\int_S f(x)\,\mu_2(dx)=\int_S f(x)\,\varrho(x)\,\mu_1(dx)$$

is valid.

Taking $g(x)/\varrho(x)$ in place of $f(x)$ we obtain

$$\int_S g(x)\,\mu_1(dx)=\int_S \frac{g(x)}{\varrho(x)}\,\mu_2(dx),$$

which yields our assertion.

Corollary 2. *In order that measure μ_2 be absolutely continuous with respect to μ_1 it is necessary and sufficient that the function $\varrho(x)$ defined by relation* (1) *satisfy condition*

$$\int \varrho(x)\,\mu_1(dx)=1. \tag{2}$$

Since

$$\int \varrho_n(x)\, \mu_1(dx) = 1, \tag{3}$$

in order that (2) be satisfied it is necessary and sufficient that the limit transition under the sign of the integral be permissible in (3), i.e. that the functions $\varrho_n(x)$ be integrable with respect to measure μ_1 uniformly in n.

Sometimes in place of measures μ_i^n which are contractions of measures μ_i on $\mathfrak{B}_n$ it seems more convenient to consider approximations of measures μ_i by means of certain measures μ_i^n for which the evaluation of $d\mu_2^n/d\mu_1^n$ is simpler. This state of affairs is analogous to a certain extent to the situation considered in Theorem 1 and its corollaries.

Theorem 2. *Let two sequences of probability measures μ_n^1 and μ_n^2 be defined on $(\mathscr{X}, \mathfrak{B})$ satisfying conditions:*

1) *on a certain algebra $\mathfrak{B}_0$ whose σ-closure coincides with $\mathfrak{B}$ the sequence μ_n^i converges to μ_0^i: $\lim_{n\to\infty} \mu_n^i(A) = \mu_0^i(A)$ for $A \in \mathfrak{B}_0$.*

2) *measures μ_n^2 are absolutely continuous with respect to μ_n^1 for $n \geq 1$;*

3) *functions $\varrho_n(x) = d\mu_n^2/d\mu_n^1(x)$ are uniformly integrable with respect to μ_n^1, i.e. for each $\varepsilon > 0$ an N can be found such that for all n,*

$$\int \varrho_n(x)\, \chi_{[N,\infty)}(\varrho_n(x))\, \mu_n^1(dx) < \varepsilon,$$

where $\chi_{[N,\infty)}(t)$ is the indicator of the interval $[N, \infty)$. Then $\mu_0^2 \ll \mu_0^1$.

Proof. For any $A \in \mathfrak{B}_0$ we have

$$\mu_0^2(A) = \lim_{n\to\infty} \mu_n^2(A) = \lim_{n\to\infty} \int_A \varrho_n(x)\, \mu_n^1(dx) \leq N \lim_{n\to\infty} \mu_n^1(A) +$$

$$+ \overline{\lim_{n\to\infty}} \int_A \varrho_n(x)\, \chi_{[N,\infty)}(\varrho_n(x))\, \mu_n^1(dx) \leq N\mu_0^1(A) + \varepsilon,$$

provided N and ε are chosen in such a manner that the inequality stated in condition 3) is satisfied. The class of sets A such that

$$\mu_0^2(A) \leq N\mu_0^1(A) + \varepsilon \tag{4}$$

is a monotone class containing algebra $\mathfrak{B}_0$, hence (4) is satisfied for all $A \in \mathfrak{B}$. It follows from (4) that $\mu_0^2(A) = 0$ provided only that $\mu_0^1(A) = 0$ since the $\varepsilon > 0$ can be chosen arbitrarily small.

Remark. In order that condition 3) of Theorem 2 be satisfied, it is sufficient that one of the following conditions be fulfilled:

1) for some $\alpha > 1$

$$\sup_n \int [\varrho_n(x)]^\alpha \mu_n^1(dx) < \infty,$$

2) a positive continuous function $\varphi(t)$ exists such that

$$\lim_{t\to\infty} \frac{t}{\varphi(t)} = 0 \quad \text{and} \quad \sup_n \int \varphi(\varrho_n(x)) \mu_n^1(dx) < \infty,$$

3) $\sup_n \int \log \varrho_n(x) \mu_n^{(2)}(dx) = \sup_n \int \varrho_n(x) \log \varrho_n(x) \mu_n^1(dx) < \infty,$

4) measures μ_n^1 are also absolutely continuous with respect to μ_n^2 and for every $\varepsilon > 0$ an N can be chosen such that

$$\overline{\lim_{n\to\infty}} \mu_n^2 \left\{ x : \frac{d\mu_n^1}{d\mu_n^2}(x) < \frac{1}{N} \right\} < \varepsilon. \tag{5}$$

Indeed 1) and 3) are particular cases of 2) if we set $\varphi(t) = t^\alpha$ and $\varphi(t) = t \log t + 1$. The sufficiency of condition 2) follows from inequality

$$\int \varrho_n(x) \chi_{[N,\infty)}(\varrho_n(x)) \mu_n^1(dx) \leqslant \sup_{t \geqslant N} \frac{t}{\varphi(t)} \int \varphi(\varrho_n(x)) \mu_n^1(dx).$$

To prove the sufficiency of condition 4) we note that

$$\int \varrho_n(x) \chi_{[N,\infty)}(\varrho_n(x)) \mu_n^1(dx) = \mu_n^2 \left(\left\{ x : \frac{d\mu_n^1}{d\mu_n^2}(x) < \frac{1}{N} \right\} \right).$$

is positive μ_n^2-almost everywhere in view of the fact that $\mu_n^2 \ll \mu_n^1$, we have for all n

$$\lim_{N\to\infty} \mu_n^2 \left(\left\{ x : \frac{d\mu_n^1}{d\mu_n^2}(x) < \frac{1}{N} \right\} \right) = 0.$$

From here and utilizing (5) we derive condition 3).

Theorem 2 does not enable us to calculate $\frac{d\mu_0^2}{d\mu_0^1}(x)$. Since the functions $\varrho_n(x)$ are defined, each one with respect to its own measure, it is meaningless to speak of a limit of $\varrho_n(x)$ as $n \to \infty$. However, in the particular case when all the measures μ_n^1 coincide with μ_0^1 it makes sense to consider the limit of $\varrho_n(x)$ with respect to measure μ_0^1. If this limit exists, then provided the conditions of Theorem 2 are satisfied this limit coincides with the derivative $d\mu_0^2/d\mu_0^1$. We now prove a more general theorem concerning the density $\frac{d\mu_0^2}{d\mu_0^1}(x)$.

Theorem 3. *Let the random variables ξ_n^i, $i=1, 2, n=0, 1, \ldots$ be defined on the probability space $\{\Omega, \mathfrak{S}, \mathsf{P}\}$ with values in a measurable space $(\mathscr{X}, \mathfrak{B})$ and let algebra $\mathfrak{B}_0$ exist whose σ-closure coincides with $\mathfrak{B}$ such that for all $A \in \mathfrak{B}_0$*

$$\chi_A(\xi_n^i) \to \chi_A(\xi_0^i)$$

in probability P. If the measures $\mu_n^i(A) = \mathsf{P}\{\xi_n^i \in A)$ on $(\mathscr{X}, \mathfrak{B})$ satisfy conditions 2) and 3) of Theorem 2 and if

$$\lim_{n\to\infty} \varrho_n(\xi_n^1) = \varrho$$

exist in the sense of convergence in probability, then

$$\varrho = \frac{d\mu_0^2}{d\mu_0^1}(\xi_0^1).$$

Proof. For all $A \in \mathfrak{B}_0$ the limit in probability

$$\lim_{n\to\infty} \varrho_n(\xi_n^1)\, \chi_A(\xi_n^1) = \varrho\chi_A(\xi_0^1)$$

exists. In view of condition 3) of Theorem 2 the functions $\varrho_n(\xi_n^1)\,\chi_A(\xi_n^1)$ are uniformly integrable in measure μ_n^1 and therefore the limiting transition under the sign of the mathematical expectation is permissible:

$$\mu_0^2(A) = \lim_{n\to\infty} \mathsf{E}\chi_A(\xi_n^2) = \lim_{n\to\infty} \mathsf{E}\chi_A(\xi_n^1)\,\varrho_n(\xi_n^1) = \mathsf{E}\chi_A(\xi_0^1)\,\varrho = \int_A \varrho(x)\,\mu_0^1(dx).$$

These relations can be extended in an obvious manner over the whole space A. The theorem is thus proved. □

Remark 1. If the variables $\frac{d\mu_n^1}{d\mu_n^2}(\xi_n^2)$ converge in probability to a certain limit $\bar{\varrho} > 0$, condition 3) of Theorem 2 is automatically satisfied since in this case condition 4) of the Remark for Theorem 2 is valid.

Remark 2. If we do not require condition 3) in Theorem 2, but instead, it is assumed that $\mathsf{E}\varrho = 1$, then the assertion of Theorem 3 remains valid. Indeed, it follows from Fatou's lemma that

$$\mathsf{E}\varrho\chi_A(\xi_0^1) \leqslant \lim_{n\to\infty} \mathsf{E}\varrho_n(\xi_n^1)\,\chi_A(\xi_n^1) = \lim_{n\to\infty} \mu_n^2(A) = \mu_0^2(A)$$

for all $A\in\mathfrak{B}_0$. Moreover, the collection of sets A for which relation

$$\mathsf{E}\varrho\chi_A(\xi_0^1)\leqslant\mu_0^2(A) \tag{6}$$

is satisfied forms a monotone class. Therefore (6) is satisfied for all $A\in\mathfrak{B}$. If for some A the inequality

$$\mathsf{E}\varrho\chi_A(\xi_0^1)<\mu_0^2(A)$$

is valid then

$$\mathsf{E}\varrho=\mathsf{E}\varrho\chi_A(\xi_0^1)+\mathsf{E}\varrho\chi_{(\mathscr{X}-A)}(\xi_0^1)<\mu_0^2(A)+\mu_0^2(\mathscr{X}-A)=1,$$

which contradicts the assumption that $\mathsf{E}\varrho=1$. Hence $\mathsf{E}\varrho\chi_A(\xi_0^1)=\mu_0^2(A)$ for all $A\in\mathfrak{B}$ or $\varrho=\frac{d\mu_0^2}{d\mu_0^1}(\xi_0^1)$. The case when the probability space coincides with $(\mathscr{X},\mathfrak{B},\mu)$ and the random variables are defined as measurable mappings of $\mathscr{X}$ into $\mathscr{X}$ is of special interest. It follows from Theorem 3 that:

Corollary 1. *Let:* 1) *two sequences of measurable mappings $T_n^1(x)$ and $T_n^2(x)$ of $\mathscr{X}$ into $\mathscr{X}$ be given and let the measures μ_n^i be defined by the equalities $\mu_n^i(A)=\mu(T_n^{i-1}(A))$, where $T_n^{i-1}(A)$ if the total inverse image of A under the mapping T_n^i,*

2) *$T_n^i(x)\to T_0^i(x)$ μ-almost for all x*

3) *$\mu_n^1\sim\mu_n^2$ and μ-almost for all x, the nonnegative limits*

$$\lim_{n\to\infty}\frac{d\mu_n^2}{d\mu_n^1}(T_n^1(x))=\varrho_1(x),$$

$$\lim_{n\to\infty}\frac{d\mu_n^1}{d\mu_n^2}(T_n^2(x))=\varrho_2(x)\quad \textit{exist}.$$

Then $\mu_0^1\sim\mu_0^2$ and

$$\frac{d\mu_n^1}{d\mu_0^1}(T_0^1(x))=\varrho_1(x),\qquad \frac{d\mu_0^1}{d\mu_0^2}(T_0^2(x))=\varrho_2(x).$$

Indeed the conditions of Theorem 3 and Remark 1 will be fulfilled in this case if we choose for algebra $\mathfrak{B}_0$ the continuity sets* of the measure $\mu_0^1+\mu_0^2$ (measures μ_n^i are weakly convergent to u_0^i).

Corollary 2. *Let conditions* 1) *and* 2) *of corollary* 1 *be satisfied and moreover, let* 3) $\mu_0^1=\mu_0^2=\nu$, 4) $\mu_n^2\ll\mu$, $\mu\ll\mu_n^1$ *and the following non-zero limits*

---

* Cf. Chapter VI, Section 1.

in measure μ

$$\lim_{n\to\infty} \frac{d\mu_n^2}{d\mu}(x) = \varrho_1(x), \qquad \lim_{n\to\infty} \frac{d\mu}{d\mu_n^1}(T_n^1(x)) = \varrho_2(x)$$

exist. 5) *μ-almost for all x the equality*

$$\varrho_1(T_0^1(x))\,\varrho_2(x) = 1$$

be satisfied. Then

$$\mu \sim \nu \text{ and } \varrho_1(x) = \frac{d\nu}{d\mu}(x), \qquad \varrho_2(x) = \frac{d\mu}{d\nu}(T_0^1(x)).$$

Indeed it follows from Theorem 3 and Remark 1 that $\mu \sim \nu$. Next, as was established in Remark 2,

$$\varrho_1(x) \leqslant \frac{d\nu}{d\mu}(x), \qquad \varrho_2(x) \leqslant \frac{d\mu}{d\gamma}(T_0^1(x)).$$

Hence

$$1 = \varrho_1(T_0^1(x))\,\varrho_2(x) \leqslant \frac{d\nu}{d\mu}(T_0^1(x)) \frac{d\mu}{d\nu}(T_0^1(x)) = 1,$$

which implies that $\varrho_2(x) = \frac{d\mu}{d\nu}(T_0^1(x))$ μ-almost everywhere and $\varrho_1(T_0^1(x)) = \frac{d\nu}{d\mu}(T_0^1(x))$ μ-almost everywhere. But then $\varrho_1(x) = \frac{d\nu}{d\mu}(x)$ ν-almost everywhere and since $\nu \sim \mu$ the latter holds also μ-almost everywhere. The assertion is thus proved. □

We now study the absolute continuity of measures in the case when the corresponding spaces are mapped. Let $(\mathscr{X}_1, \mathfrak{B}_1)$ and $(\mathscr{X}_2, \mathfrak{B}_2)$ be two measurable spaces. The mapping φ of $\mathscr{X}_1$ into $\mathscr{X}_2$ is called *measurable* if $\varphi^{-1}(A) \in \mathfrak{B}_1$ for all $A \in \mathfrak{B}_2$. Assume that two measures μ_1 and ν_1 are defined on $\mathfrak{B}_1$, and let the measures μ_2 and ν_2 be defined on $\mathfrak{B}_2$ by the equalities $\mu_2(A) = \mu_1(\varphi^{-1}(A))$, $\nu_2(A) = \nu_1(\varphi^{-1}(A))$.

Theorem 4. *If $\nu_1 \ll \mu_1$ then $\nu_2 \ll \mu_2$, and moreover*

$$\frac{d\nu_2}{d\mu_2}(\varphi^{-1}(x)) = \mathsf{E}\left(\frac{d\nu_1}{d\mu_1} \,\middle|\, \tilde{\mathfrak{B}}_1\right),$$

where $\tilde{\mathfrak{B}}_1$ is the σ-algebra of the sets of the form $\varphi^{-1}(A)$, $A \in \mathfrak{B}_2$ and the conditional mathematical expectation is taken over the probability space $(\mathscr{X}_1, \mathfrak{B}_1, \mu_1)$.

Proof. Every $\mathfrak{B}_2$-measurable function $f(x)$ can be represented in the form $g(\varphi^{-1}(x))$, where g is $\mathfrak{B}_1$-measurable. Therefore

$$\int f(x)\,\nu_2(dx)=\int g(\varphi^{-1}(x))\,\nu_2(dx)=\int g(x)\,\nu_1(dx)=$$
$$=\int g(x)\frac{d\nu_1}{d\mu_1}(x)\,\mu_1(dx)=\int g(x)\,\mathsf{E}\left(\frac{d\nu_1}{d\mu_1}(x)\,\Big|\,\mathfrak{B}_1\right)\mu_1(dx).$$

Let $\mathsf{E}\left(\frac{d\nu_1}{d\mu_1}(x)\,\Big|\,\mathfrak{B}_1\right)=\varrho(x)$. Since $\varrho(x)$ is $\mathfrak{B}_1$-measurable, it follows that $\varrho(\varphi^{-1}(x))$ is $\mathfrak{B}_2$-measurable. Hence

$$\int g(x)\,\varrho(x)\,\mu_1(dx)=\int g(\varphi(x))\,\varrho(\varphi^{-1}(x))\,\mu_2(dx)=$$
$$=\int f(x)\,\varrho(\varphi^{-1}(x))\,\mu_2(dx).$$

The proof of the theorem follows from the last equality. □

§2. Admissible Shifts in Hilbert Spaces

As we shall see in the next section, when studying absolute continuity of measures under various transformations, the absolute continuity and density of measures under the simplest transformations – the translation – play an important role. Let μ be a measure in $(\mathscr{X}, \mathfrak{B})$, where $\mathscr{X}$ is the Hilbert space and $\mathfrak{B}$ be the σ-algebra of Borel sets in this space. We introduce the translation operator $S_a x=x+a$. Denote by μ_a the measure defined by relation $\mu_a(A)=\mu(S_{-a}A)$. Note that if μ is the distribution of a random element ξ with values in $\mathscr{X}$, then μ_a is the distribution of the random element $\xi+a$. The measure μ_a is uniquely determined by the relation

$$\int f(x)\,\mu_a(dx)=\int f(x+a)\,\mu(dx)$$

valid for all those measurable functions for which the integral on the right exists.

We say that a is an *admissible shift** of measure μ if $\mu_a \ll \mu$. The set of admissible shifts of a measure is denoted by M_μ or simply by M if there is no ambiguity concerning the measure under consideration. If $a \in M_\mu$, we denote

$$\varrho(a, x)=\varrho_\mu(a, x)=\frac{d\mu_a}{d\mu}(x).$$

* also called the admissible mean value for μ. *Translator's Remark.*

In this section the structure of the set M_μ and the properties of the density $\varrho_\mu(a, x)$ are investigated. In what follows probability measures on $\mathscr{X}$ are considered.

Theorem 1. *The set M_μ is an additive semi-group, i.e. $a+b\in M_\mu$, provided $a\in M_\mu$ and $b\in M_\mu$; moreover,*

$$\varrho(a+b, x)=\varrho(a, x)\,\varrho(b, x-a).$$

Proof: We have

$$\int f(x)\,\mu_{a+b}(dx)=\int f(x+a+b)\,\mu(dx)=$$

$$=\int f(x+a)\,\varrho(b, x)\,\mu(dx)=\int f(x)\,\varrho(b, x-a)\,\mu_a(dx)=$$

$$=\int f(x)\,\varrho(b, x-a)\,\varrho(a, x)\,\mu(dx).$$

The proof follows from the fact that these equalities are valid for any bounded measurable function $f(x)$. □

The following theorem shows that there are not that many admissible shifts. In particular, it follows from the theorem that in any infinite dimensional subspace of $\mathscr{X}$ the set M_μ is of the first category.

Theorem 2. *Let $\varphi(z)$ be the characteristic functional of measure μ and a completely continuous nonnegative symmetric operator B be such that $\varphi(z)\to 1$ as $(Bz, z)\to 0$. Then for each $a\in M_\mu$, $b\in\mathscr{X}$ exists such that $a=B^{1/2}b$*, i.e. $M_\mu\subset B^{1/2}\mathscr{X}$.

Proof. Let $a\in M_\mu$ and $\varrho(a, x)=\dfrac{d\mu_a}{d\mu}(x)$. *Then the characteristic functional* of measure μ_a can be represented as

$$\varphi_a(z)=\int e^{i(x, z)}\,\mu_a(dx)=\int e^{i(x+a, z)}\,\mu(dx)=e^{i(a, z)}\,\varphi(z).$$

Moreover

$$\varphi_a(z)=\int e^{i(z, x)}\,\varrho(a, x)\,\mu(dx).$$

Hence

$$\varphi_a(z)-1=\int (e^{i(z, x)}-1)\,\varrho(a, x)\,\mu(dx).$$

We show that $\varphi_a(z)\to 1$ for $(Bz, z)\to 0$. Since the following inequality

$$|1-\psi(z)|^2 \leqslant 2\,\mathrm{Re}(1-\psi(z))$$

is valid for any characteristic functional $\psi(z)$, it is sufficient to show that

$$\mathrm{Re}\int (1-e^{i(z,x)})\,\varrho(a,x)\,\mu(dx)\to 0$$

as $(Bz, z)\to 0$. We set $\varrho_N(x)=\varrho(a,x)$ for $\varrho(a,x)\leqslant N$, and $\varrho_N(x)=0$ for $\varrho(a,x)>N$. Then

$$\mathrm{Re}\int (1-e^{i(z,x)})\,\varrho(a,x)\,\mu(dx)=\int (1-\cos(z,x))\,\varrho_N(x)\,\mu(dx)+$$

$$+\int (1-\cos(z,x))\,[\varrho(a,x)-\varrho_N(x)]\,\mu(dx)\leqslant$$

$$\leqslant N\,\mathrm{Re}(1-\varphi(z))+2\int [\varrho(a,x)-\varrho_N(x)]\,\mu(dx).$$

The second summand becomes arbitrarily small for all z if N is chosen sufficiently large while the first summand approaches zero as $(Bz, z)\to 0$ for any N. We thus proved that $\varphi(z)\,e^{i(a,z)}\to 1$ as $(Bz, z)\to 0$. Hence $e^{i(a,z)}\to 1$ as $(Bz, z)\to 0$ and therefore $(a, z)\to 0$ as $(Bz, z)\to 0$. Let $|(a, z)|<\varepsilon$ provided $(Bz, z)<\delta$. Then for all z the inequality

$$|(a,z)|^2<\frac{\varepsilon^2}{\delta}(Bz, z)$$

is satisfied.

We note that a belongs to the closure of the range of values of the operator B since for all y such that $By=0$, we also have $(a, y)=0$. Let λ_k be the eigenvalues and e_k be the corresponding eigenvectors of the operator B. Putting $C=\frac{\varepsilon^2}{\delta}$, $z=\sum_{k=1}^{n}\frac{(a,e_k)}{\lambda_k}e_k$ we have

$$(a,z)^2=\left(\sum_{k=1}^{n}\frac{(a,e_k)^2}{\lambda_k}\right)^2\leqslant C\sum_{k=1}^{n}\frac{(a,e_k)^2}{\lambda_k};$$

hence $\sum_{k=1}^{n}\frac{(a,e_k)^2}{\lambda_k}<C$. Approaching the limit as $n\to\infty$ we verify the existence of the vector $b=\sum_{k=1}^{\infty}\frac{(a,e_k)}{\sqrt{\lambda_k}}e_k$, which satisfies the relation $B^{1/2}b=a$. The theorem is thus proved. □

We now investigate the transformations of the set M_μ and the function $\varrho(a, x)$ under the simplest transformations of the measure μ.

Theorem 3. 1) *If* $\nu = \mu_c$, *then for any* $c \in X$, $M_\nu = M_\mu$,

$$\varrho_\nu(a, x) = \varrho_\mu(a, x-c).$$

2) *If* $\nu \ll \mu$, $f(x) = \frac{d\nu}{d\mu}(x)$ and $a \in M_\mu$, *then* $a \in M_\nu$ *if and only if the expression*

$$\varrho_\nu(a, x) = \frac{f(x-a)}{f(x)}\, \varrho_\mu(a, x) \tag{1}$$

is defined μ*-almost everywhere, i.e. if*

$$\mu(\{x : f(x) = 0\} - \{x : f(x-a)\, \varrho_\mu(a, x) = 0\}) = 0$$

(we define $\frac{0}{0} = 0$*). Moreover* $\varrho_\nu(a, x)$ *is determined by formula* (1).

3) *If* $\nu(A) = \mu(L^{-1}A)$, *where* L *is an invertible linear operator, then*

$$M_\nu = LM_\mu, \quad \varrho_\nu(a, x) = \varrho_\mu(L^{-1}a, L^{-1}x).$$

Proof. 1) follows from the equality

$$\int g(x)\, \nu_a(dx) = \int g(x+a+c)\, \mu(dx) =$$

$$= \int g(x+c)\, \varrho_\mu(a, x)\, \mu(dx) = \int g(x)\, \varrho_\mu(a, x-c)\, \mu_c(dx) =$$

$$= \int g(x)\, \varrho_\mu(a, x-c)\, \nu(dx).$$

2) Let $g(x)$ be a bounded measurable function. If $\frac{f(x-a)}{f(x)}\, \varrho_\mu(x, a)$ is defined μ-almost everywhere, we than have

$$\int g(x)\, \nu_a(dx) = \int g(x+a)\, \nu(dx) = \int g(x+a)\, f(x)\, \mu(dx) =$$

$$= \int g(x)\, f(x-a)\, \mu_a(dx) = \int g(x)\, \varrho_\mu(a, x)\, f(x-a)\, \mu(dx) =$$

$$= \int g(x) \frac{f(x-a)}{f(x)}\, \varrho_\mu(a, x)\, \nu(dx).$$

If, however $a \in M_\nu$, then for every bounded measurable function $g(x)$ the relation

$$\int g(x)\,\varrho_\nu(a,x)\,f(x)\,\mu(dx)=\int g(x)\,\varrho_\mu(a,x)\,f(x-a)\,\mu(dx)$$

is valid. It thus follows that $\varrho_\nu(a,x)\,f(x)=\varrho_\mu(a,x)\,f(x-a)$ μ-almost everywhere. Assertion 2) is proved.

3) The following is valid for a bounded measurable function $g(x)$ and for $a=Lb$, with $b\in M_\mu$:

$$\int g(x)\,\nu_a(dx)=\int g(x+a)\,\nu(dx)=\int g(Lx+a)\,\mu(dx)=$$

$$=\int g(L(x+b))\,\mu(dx)=\int g(Lx)\,\varrho_\mu(b,x)\,\mu(dx)=$$

$$=\int g(x)\,\varrho_\mu(b,L^{-1}x)\,\nu(dx).$$

It thus follows that $M_\nu\supset LM_\mu$ and $\varrho_\nu(a,x)=\varrho_\mu(b,L^{-1}x)$ for $a\in LM_\mu$. Utilizing the fact that the operator L is invertable we obtain that $M_\nu\subset \subset LM_\mu$. Hence $M_\nu=LM_\mu$. The theorem is thus proved. □

Remark. 1) It follows from 1) that if ν is the distribution of the variable $\xi+\eta$, where ξ and η are independent random variables with values in $\mathscr{X}$ and μ is the distribution of the variable ξ, then $M_\nu\supset M_\mu$ and for $a\in M_\mu$ the equality

$$\varrho_\nu(a,x)=\mathsf{E}(\varrho_\mu(a,\xi)/\xi+\eta)_{\xi+\eta=x}$$

holds.

This equality follows from the relations

$$\int g(x)\,\nu_a(dx)=\int g(x+a)\,\nu(dx)=$$

$$=\iint g(x+y+a)\,\mu(dx)\,\mathsf{P}\{\eta\in dy\}=$$

$$=\iint g(x+y)\,\varrho_\mu(a,x)\,\mu(dx)\,\mathsf{P}\{\eta\in dy\}=$$

$$=\mathsf{E}g(\xi+\eta)\,\varrho_\mu(a,\xi)=\mathsf{E}g(\xi+\eta)\,\mathsf{E}(\varrho_\mu(a,\xi)/\xi+n).$$

2) If under condition 3) L is a non-invertable operator, then $M_\nu\supset LM_\mu$ and

$$\varrho_\nu(a,x)=\mathsf{E}(\varrho_\mu(b,\xi)/L\xi)_{L\xi=x},$$

where b is an arbitrary vector satisfying the relation $Lb=a$. This formula follows from Theorem 4 of Section 1.

Let $|a|=1$. We study the conditions under which $\lambda a \in M_\mu$ for all $\lambda>0$. Let $F(t)=\mu(\{x:(a, x)<t\})$; $F(t)$ is the distribution function of a random variable (a, x) on the probability space $(\mathscr{X}, \mathfrak{B}, \mu)$. The distribution function $F(t-\lambda)$ of variable $(a, x)+\lambda=(a, x+\lambda a)$ on the same probability space is absolutely continuous, for all $\lambda>0$, with respect to the distribution of (a, x). The following lemma describes the available information concerning $F(t)$ under these conditions.

Lemma. *If the measure $\nu_\lambda(E)$ defined on the Borel sets of the line $\mathscr{R}^1$ by the relation*

$$\nu_\lambda(E)=\int dF(t-\lambda)$$

is absolutely continuous with respect to measure $\nu_0(E)$ for all $\lambda>0$, then $F(t)$ is absolutely continuous and its derivative $p(t)=\frac{dF}{dt}(t)$ is such that for some t_1 (possibly equal to $-\infty$) $p(t)=0$ almost for all $t<t_1$ and $p(t)>0$ almost for all $t>t_1$.

Proof. We represent $F(t)$ in the form $F(t)=F_1(t)+F_2(t)$ where $F_1(t)$ is the absolutely continuous and $F_2(t)$ is the singular component of F. Let

$$\nu_\lambda^i(E)=\int dF_i(t-\lambda).$$

Since $\nu_\lambda^2 \ll \nu_0^1+\nu_0^2$, $\nu_\lambda^2 \perp \nu_0^1$, it follows that $\nu_\lambda^2 \ll \nu_0^2$. Therefore the measure

$$\bar{\nu}(E)=\int_0^\infty \nu_\lambda^2(E)\, e^{-\lambda}\, d\lambda$$

will be absolutely continuous with respect to measure ν_0^2. On the other hand

$$\bar{\nu}(E)=\int_0^\infty \int_{t\in E} dF(t-\lambda)\, e^{-\lambda}\, d\lambda=\iint_{\substack{u-t<0\\ t\in E}} dF_2(u)\, e^{u-t}\, dt \leqslant \operatorname{mes} E,$$

where $\operatorname{mes} E$ is the Lebesgue measure of the set E. Since $\bar{\nu}$ is singular with respect to the Lebesgue measure it follows that $\bar{\nu}=0$ and hence $F_2(t)=0$.

Now let

$$\bar{F}(t)=\int_0^\infty F(t-\lambda)\,e^{-\lambda}\,d\lambda \quad\text{and}\quad \tilde{\nu}(E)=\int dF(t).$$

It is easy to verify that $\frac{d\bar{\nu}}{d\nu}(t)=\frac{1}{p(t)}\int_0^\infty p(t-\lambda)\,e^{-\lambda}\,d\lambda$, and moreover, the numerator of this fraction must vanish for almost all t on the set of arguments s such that $p(s)=0$. If the set $\{t: p(t)=0\}$ is of a positive Lebesgue measure, then one can find s such that the Lebesgue measure of the set $\{t: p(t)=0\}\cap\{s-\delta, s\}$ is positive for all $\delta>0$. Then the function

$$\int_0^\infty p(t-\lambda)\,e^{-\lambda}\,d\lambda$$

vanishes for some $t\in(s-\delta, s)$ for any $\delta>0$ and since this function is continuous it follows that $\int_0^\infty p(s-\lambda)\,e^{-\lambda}\,d\lambda=0$ and hence $p(t)=0$ for almost all $t<s$. If $p(t)$ is not identically zero, then a maximal s with the above properties can be found. Thus $p(t)=0$ for almost all $t<s$ and $p(t)>0$ for almost all $t>s$. The lemma is proved. □

Let Γ_t denote the hyperplane $(a, x)=t$. Define measures μ^t on the Borel sets of Γ_t by means of the equalities

$$\mu^t(A)=\mu^t(A\cap\Gamma_t), \qquad \mu(A)=\int \mu^t(A)\,dF(t).$$

Thus $\mu^t(A)$ is the conditional distribution of x under the condition $(a, x)=t$ on the probability space $(\mathscr{X}, \mathfrak{B}, \mu)$. We introduce the conditional distributions of the projection of x on Γ_0 given $(a, x)=t: \nu^t(A)=\mu^t(S_{ta}A)$. Finally let ν be the unconditional distribution of the projection of x on Γ_0:

$$\nu(A)=\int \nu^t(A)\,p(t)\,dt.$$

We introduce measure μ^* by means of the equalities

$$\mu^*(A)=\int \nu(S_{-ta}[A\cap\Gamma_t])\,p(t)\,dt. \tag{2}$$

We note that μ^* is the distribution of a random variable such that the variable (a, ξ) and the projection of ξ on Γ_0 are independent and their distributions coincide with the distributions of the variable (a, x) and the projection of x on Γ_0 in the probability space $(\mathscr{X}, \mathfrak{B}, \mu)$. We show that the measure μ is absolutely continuous with respect to μ^*. Note that

$$\mu_{\lambda a}(A)=\mu(S_{-\lambda a}A)=\int v^t(S_{-ta}[S_{-\lambda a}A\cap\Gamma_t]\, p(t)\, dt=$$

$$=\int v^t(S_{-(t+\lambda)a}[A\cap\Gamma_{t+\lambda}])\, p(t)\, dt=$$

$$=\int v^{t-\lambda}(S_{-at}[A\cap\Gamma_t])\, p(t-\lambda)\, dt. \tag{3}$$

Consequently for any bounded integrable function $k(\lambda)$ the representation

$$\int \mu_{\lambda a}(A)\, k(\lambda)\, d\lambda=\iint p(t-\lambda)\, k(\lambda)\, v^{t-\lambda}(S_{-ta}[A\cap\Gamma_t])\, dt\, d\lambda$$

is valid. It follows from the definition of the measure μ^* that

$$\mu^*(A)=\iint v^s(S_{-ta}[A\cap\Gamma_t])\, p(s)\, p(t)\, dt\, ds.$$

If $\mu^*(A)=0$,then $v^s(S_{-ta}[A\cap\Gamma_t]=0$ for almost all $t>t_1$ and $s>t_1$ – in the Lebesgue measure – where t_1 is defined in the lemma. We may assume without loss of generality that $v^t(A)=0$ for $t<t_1$. Moreover it is natural to consider only those sets A which belong for some $\delta>0$ to the set $\{x:(a, x)>t_1+\delta\}$ since

$$\mu^*(\{x:(a, x)\leqslant t_1\})=\mu(\{x:(a, x)\leqslant t_1\})=0.$$

Under these assumptions the condition $\mu^*(A)=0$ implies the equality

$$\int_{-\delta}^{0} \mu_{\lambda a}(A)\, d\lambda=\int_{-\delta}^{0} d\lambda \int p(t-\lambda)\, v^{t-\lambda}(S_{-ta}[A\cap\Gamma_t])\, dt=0,$$

and hence $\mu_{\lambda a}(A)$ for almost all $\lambda\in[-\delta, 0]$. Our assertion is proved since $\mu\ll\mu_{\lambda a}$ and hence $\mu(A)=0$. □

We note that $\mu^*_{\lambda a}\ll\mu^*$ for all $\lambda>0$. Indeed from the definition

$$\mu^*_{\lambda a}(A)=\int p(t-\lambda)\, v(S_{-ta}[A\cap\Gamma_t])\, dt=$$

$$=\int \frac{p(t-\lambda)}{p(t)}\, v(S_{-ta}[A\cap\Gamma_t])\, p(t)\, dt.$$

Hence
$$\frac{d\mu^*_{\lambda a}}{d\mu^*}(x)=\frac{p((a, x)-\lambda)}{p((a, x))}.$$

Therefore utilizing Theorem 3 part 2) we obtain the following

Theorem 4. *In order that $\lambda a \in M_\mu$ for all $\lambda > 0$, it is necessary and sufficient that the following conditions be satisfied:*

1) the function $F(t)=\mu(\{x:(a, x)<t\})$ be absolutely continuous and there exist t_1 (possibly equal to $-\infty$) such that $p(t)$, the derivative of $F(t)$, satisfies $p(t)=0$ for almost all $t<t_1$ and $p(t)>0$ for almost all $t>t_1$.

2) the measure μ be absolutely continuous with respect to measure μ^* defined by
$$\mu^*(\{x:\alpha<(a, x)<\beta\}\cap\{x:Px\in A\})=$$
$$=\mu(\{x:\alpha<(a, x)<\beta\})\,\mu(\{x:Px\in A\}),$$
where P is the projector on the subspace Γ_0 and its density $\varrho(x)=\frac{d\mu}{d\mu^*}(x)$ is such that the expression
$$\varrho_\mu(\lambda, a, x)=\frac{p((a, x)-\lambda)}{p((a, x))}\frac{\varrho(x-\lambda a)}{\varrho(x)} \tag{4}$$
is defined μ^*-almost everywhere provided we assume that the expression is zero as long as the numerator is zero.

Remark. If $\lambda a \in M_\mu$ for all real λ, then $p(t)>0$ for almost all t and the measures μ and μ^* are equivalent. Indeed μ^* is equivalent to the measure $\int k(\lambda)\,\mu_{\lambda a}d\lambda$ provided $k(\lambda)$ is a positive integrable function and the measure $\int k(\lambda)\,\mu_{\lambda a}d\lambda$ is absolutely continuous with respect to μ.

Corollary. *If $a_1, \dots, a_n$ ($|a_k|=1$) are mutually orthogonal vectors such that $\lambda a_k \in M_\mu$ for all real λ and $k=1,\dots, n$, then:* 1) *the functions $F_k(t)=$ $=\mu(\{x:(a_k, x)<t)\})$ are absolutely continuous and their derivatives $p_k(t)=$ $=\frac{d}{dt}F_k(t)$ are positive for almost all t;* 2) *the measure μ is equivalent to the measure $\bar{\mu}$ defined by the relation*
$$\bar{\mu}\left(\left[\bigcap_{k=1}^{n}\{x:\alpha_k<(a_k, x)<\beta_k\}\right]\cap\{x:P_n x\in C\}\right)=$$
$$=\mu(\{x:P_n x\in C\})\prod_{k=1}^{n}\mu(\{x:\alpha_k<(a_k, x)<\beta_k\}),$$

where α_k, β_k, $k=1,\dots,n$ are arbitrary real numbers, P_n is the projector on the subspace $\mathscr{X}^n=\{x:(x, a_k)=0,\ k=1,\dots,n\}$ and C is an arbitrary Borel set in this subspace.

The proof of this assertion follows from the fact that any admissible shift of measure μ^*, constructed in the proof of Theorem 4, orthogonal to a will be an admissible shift of measure ν defined in the course of the same proof. Note that the measure $\bar{\mu}$ defined above is the distribution of the variable

$$\xi=\sum_{k=1}^{n} \eta_k a_k+\xi^n,$$

where $\eta_1,\dots,\eta_n$ are independent random variables with values in $\mathscr{R}^1$ and densities $p_k(t)$ and ξ^n is the variable, independent of η_i, with values in $\mathscr{X}^n$, distributed as the projection of x on $\mathscr{X}^n$ is distributed on the probability space $(\mathscr{X}, \mathfrak{B}, \mu)$.

Assume that an orthonormal sequence of vectors a_k can be constructed such that M_u contains the linear span of these vectors. Is it possible by analogy with the corollary formulated above to assert that measure μ be equivalent to a measure which is the distribution of the random variable

$$\xi=\sum_{k=1}^{\infty} \eta_k a_k, \tag{5}$$

where η_k is a sequence of independent numerical random variables possessing positive densities? To answer this query we investigate admissible shifts of measure $\bar{\mu}$. Denote by Π^∞ the set of measures which are distributions of variables ξ of form (5).

Theorem 5. *If η_k are independent random variables with positive densities $p_k(t)$ and $\bar{\mu}$ is the distribution of the variable ξ defined by equality* (5), *then $\bar{\mu}_a \ll \bar{\mu}$ iff the series*

$$\sum_{k=1}^{\infty} [\log p_k(\eta_k-\alpha_k)-\log p_k(\eta_k)], \tag{6}$$

is convergent with probability 1, *where $\alpha_k=(a, a_k)$. If this series is divergent then $\bar{\mu}_a \perp \bar{\mu}$.*

Proof. Let $\bar{\mu}^n$ and $\bar{\mu}_a^n$ denote the projections of measures $\bar{\mu}$ and $\bar{\mu}_a$ on the subspace $\mathscr{X}_n$ spanned by $a_1,\dots,a_n$. Then

$$\frac{d\bar{\mu}_a^n}{d\bar{\mu}^n}(x)=\prod_{k=1}^{n} \frac{p_k((x, a_k)-\alpha_k)}{p_k((x, a_k))}.$$

If $\bar{\mu}_a$ is not orthogonal to $\bar{\mu}$, then the limit

$$\lim_{n\to\infty} \prod_{k=1}^{n} \frac{p_k((x, a_k)-\alpha_k)}{p_k((x, a_k))}$$

exist μ- almost everywhere. Moreover this limit is not identically zero. Hence the series (6) of independent random variables converges with positive probability; therefore this series converges with probability 1. (Note that the sequence (x, a_k) is distributed on the same probability space $(\mathscr{X}, \mathfrak{B}, \mu)$ as is the sequence η_k). Hence if $\bar{\mu}_a \ll \bar{\mu}$ series (6) is convergent and therefore

$$\frac{d\bar{\mu}_a}{d\bar{\mu}}(x)=\exp\left\{\sum_{k=1}^{\infty} \log \frac{p_k((x, a_k)-\alpha_k)}{p_k((x, a_k))}\right\}$$

is everywhere positive, so that $\bar{\mu}_a \sim \bar{\mu}$. Conversely, the existence of the non-zero limit

$$\lim_{n\to\infty} \frac{d\bar{\mu}_a^n}{d\bar{\mu}^n}(x)$$

follows from the convergence of the series (6). Therefore in view of Corollary 1 to Theorem 1 of Section 1 we have that $\bar{\mu} \ll \bar{\mu}_a$ and hence as it was proved above $\bar{\mu} \sim \bar{\mu}_a$. The theorem is thus proved. □

Corollary. *If $\mu \sim \bar{\mu}$ where $\bar{\mu} \in \Pi^\infty$, then μ_a and μ are either equivalent or orthogonal.*

Thus if we construct a measure μ such that $\lambda a_k \in M_\mu$ *for all* λ and such that an a exists for which the measures μ and μ_a are neither equivalent nor orthogonal, we then construct a measure for which an *equivalent measure* $\bar{\mu}$ belonging to Π^∞ does not exist. We shall construct an example of such a measure below.

Admissible shifts of weighted measures. First we consider admissible shifts of linear combinations of measures. If $\mu=\mu^1+\mu^2$ then it is natural to consider only the case when the measures μ^1 and μ^2 are orthogonal since under the condition $\mu^2 \ll \mu^1$ the measures μ and μ^1 are equivalent and hence their admissible shifts coincide; if however μ^2 has an absolutely continuous component with respect to μ^1, it can be adjoined to μ^1 and the problem is reduced to the case of singular measures. In what follows we shall assume that $\mu_a^1 \perp \mu^2$ for all a. (Note that without this assumption a connection between admissible shifts of measures μ^1 and μ^2 can hardly be established. Indeed consider the measures

$$\mu^k(A)=\int_A f_k(x)\,\mu(dx), \quad k=1, 2, \quad f_1(x)+f_2(x)=1,$$

$$f_1(x) f_2(x)=0;$$

here μ^1 and μ^2 may have no admissible shifts at all, while μ may have them). If this assumption is satisfied, then $M_\mu = M_{\mu^1} \cap M_{\mu^2}$. Indeed it follows from the relations $\mu_a^1 \ll \mu^1$ and $\mu_a^2 \ll \mu^2$ that $\mu_a = \mu_a^1 + \mu_a^2 \ll \mu^1 + \mu^2 = \mu$. Conversely let $\mathscr{X} = E_1 \cup E_2$ and for some a let $\mu^1(E_2) = 0$, $\mu_a^1(E_2) = 0$, $\mu^2(E_1) = 0$ and $\mu_a^2(E_1) = 0$. Then if $a \in M_\mu$, it follows from the condition $\mu^1(A) = 0$ that $\mu(A \cap E_1) = 0$ and hence $0 = \mu_a(A \cap E_1) = \mu_a^1(A \cap E_1)$. Thus $\mu_a^1 \ll \mu^1$. In the same manner $\mu_a^2 \ll \mu^2$. Hence $M_\mu = M_{\mu_1} \cap M_{\mu_2}$. We now find the expression of $\varrho_\mu(a, x)$ in terms of $\varrho^1(a, x) = \frac{d\mu_a^1}{d\mu^1}(x)$ and $\varrho^2(a, x) = \frac{d\mu_a^2}{d\mu^2}(x)$. If the sets E_1 and E_2 are as defined above, then

$$\mu_a(A) = \mu_a^1(A \cap E_1) + \mu_a^2(A \cap E_2) = \int_{A \cap E_2} \varrho^1(a, x)\, \mu^1(dx) + \int \varrho^2(a, x)\, \mu^2(dx) = \int_A [\varrho^1(a, x)\chi_{E_1}(x) + \varrho^2(a, x)_{E_2}(x)]\, \mu(dx),$$

where χ_{E_k} is the indicator of the set E_k. Therefore

$$\varrho_\mu(a, x) = \sum_{i=1}^{2} \varrho^i(a, x)\, \chi_{E_i}(x).$$

The result obtained can be easily extended to the case of a countable number of measures.

Theorem 6. *Let $\mu^1, \mu^2, \ldots$ be a sequence of pairwise orthogonal measures such that $\mu^i \perp \mu_a^k$ for $i \neq k$ for any $a \in \mathscr{X}$. If $\mu = \sum_k p_k \mu^k$ where $p_k > 0$ and $\sum_k p_k = 1$, then $M_\mu = \bigcap_{k=1}^{\infty} M_{\mu^k}$ and one can find pairwise disjoint sets E_k such that*

$$\varrho_\mu(a, x) = \sum_{k=1}^{\infty} \varrho^k(a, x)\, \chi_{E_k}(x), \tag{7}$$

where $\varrho^k(a, x) = \frac{d\mu_a^k}{d\mu_2}(x)$.

Proof. The inclusion $M_\mu \supset \bigcap_{k=1}^{\infty} M_{\mu^k}$ follows from the fact that $\sum p_k \mu_a^k \ll \ll \sum p_k \mu^k$ provided $\mu_a^k \ll \mu^k$ for all k. On the other hand if for two singular measures μ^k and $\sum_{l \neq k} p_l \mu^l$ the condition $\mu_a^k \perp \sum_{l \neq k} p_l \mu^l$ is satisfied, then the shift is admissible for their sum provided it is admissible for each one of the summands. Hence $M_\mu \subset M_{\mu^k}$ and $M_\mu = \bigcap_{k=1}^{\infty} M_{\mu^k}$. Let the set $\mathscr{G}_l$ be such that $\mu^l(\mathscr{G}_l) = 1$, $\mu_a^l(\mathscr{G}_l) = 1$, $\sum_{k \neq l} p_k \mu^k(\mathscr{G}_l) = 0$. The existence of such a set follows from the fact that the measures $\mu^l+\mu_a^l$ and $\sum_{k\neq l} p_k\mu^k$ are *singular*.

Now set $E_l=\mathscr{G}_l\setminus\bigcup_{k\neq l}\mathscr{G}_k$. Clearly, $\mu^k(E_l)=\delta_{kl}$ and E_l are pairwise disjoint. Next

$$\mu_a(A)=\sum p_k\mu_a^k(A)=\sum_k p_k\mu_a^k(A\cap E_k)=$$

$$=\sum_k p_k\int_{A\cap E_k}\varrho^k(a,x)\,\mu^k(dx)=$$

$$=\sum_k\int_{A\cap E_k}\varrho^k(a,x)\sum_l p_l\mu^l(dx)=\sum_k\int_A\chi_{E_k}(x)\,\varrho^k(a,x)\,\mu(dx).$$

This equality yields formula (7). The theorem is thus proved. □

Consider now shifts of measures represented by integrals (rather than sums) of families of measures depending on a continuously varying parameter. Let Θ be a complete metric space, $\mathfrak{S}$ be the σ-algebra of its Borel subsets. Consider the family of measures μ^θ on $(\mathscr{X},\mathfrak{B})$, $\theta\in\Theta$ satisfying the following condition: for any continuous bounded function $f(x)$ defined on $\mathscr{X}$ the function $\int f(x)\,\mu^\theta(dx)$ is continuous in θ. From here it follows that $\mu^\theta(A)$ is a $\mathfrak{S}$-measurable function of θ for all $A\in\mathfrak{B}$. Let $\sigma(d\theta)$ be a certain (probability) measure on $\mathfrak{S}$. Consider the measure

$$\mu(A)=\int\mu^\theta(A)\,\sigma(d\theta),\qquad A\in\mathfrak{B}.\tag{8}$$

Theorem 7. *Let M^θ be the set of admissible shifts of measure μ^θ; then $M_\mu\supset\bigcap_\theta M^\theta$. If, moreover, the measures μ^θ are mutually orthogonal and there exists a $\mathfrak{B}$-measurable function $\theta(x)$ with values in Θ such that*

$$\mu^\theta(\{x:\theta(x)=\theta\})=1,$$

then one can construct a function $\varrho^\theta(a,x)$, $a\in\bigcap_\theta M^\theta$, measurable in θ and x on $\mathfrak{S}\times\mathfrak{B}$ and such that $\varrho^\theta(a,x)=\frac{d\mu_a^\theta}{d\mu^\theta}(x)\ (\mathrm{mod}\,\mu^\theta)$ for all $\theta\in\Theta$ and

$$\varrho_\mu(a,x)=\varrho^{\theta(x)}(a,x).\tag{9}$$

Proof. If $a\in\bigcap_\theta M_\theta$, then $\mu_a^\theta\ll\mu^\theta$ for all θ. Let $\mu(A)=0$, then $\mu^\theta(A)=0$ σ-almost for all θ. But then $\mu_a^\theta(A)=0$ σ-almost for all θ. Hence $\mu_a(A)=0$ and $\mu_a\ll\mu$.

Let $\mathscr{X}_n$ be an increasing sequence of finite-dimensional subspaces of $\mathscr{X}$, $\mu_a^\theta(n, \cdot)$ and $\mu^\theta(n, \cdot)$ be the projections of measures μ_a^θ and μ^θ on these subspaces and $\rho_n^\theta(a, x) = \frac{d\mu_a^\theta(n, \cdot)}{d\mu^\theta(n, \cdot)}(x)$. Note that the function

$$\int e^{-l\|x-y\|} \mu^\theta(n, dx),$$

$$\int \varrho_n^\theta(a, x)\, e^{-l\|x-y\|} \mu^\theta(n, dx) = \int e^{-l\|x+a-y\|} \mu^\theta(n, dx)$$

are continuous jointly in θ and y, therefore the function

$$g_l(\theta, y) = \left[\int e^{-l\|x-y\|} \mu^\theta(n, dx)\right]^{-1} \int \varrho_n^\theta(a, x)\, e^{-l\|x-y\|} \mu^\theta(n, dx)$$

also possesses this property. The latter function converges, as $l \to \infty$, $\mu^0(n, \cdot)$-almost everywhere, to $\varrho_n^\theta(a, y)$. Hence the function $\bar{\varrho}_n^\theta(a, x) = \lim_{l\to\infty} g_l(\theta, x)$ – where the limit exists – is measurable on $\mathfrak{S} \times \mathfrak{B}$ and for each θ coincides with $\varrho_n^\theta(a, x)$ $\mu^\theta(n, \cdot)$-almost everywhere. Set $\varrho^\theta(a, x) = \lim_{n\to\infty} \bar{\varrho}_n^\theta(a, x)$ (where the limit exists); this function is that required. To derive formula (9) we write the following equality

$$\int f(x)\, \mu_a(dx) = \int f(x+a)\, \mu(dx) = \int f(x+a) \int \mu^\theta(dx)\, \sigma(d\theta) =$$

$$= \int \left[\int f(x)\, \varrho^{\theta(x)}(a, x)\, \mu^\theta(dx)\right] \sigma(d\theta) = \int f(x)\, \varrho^{\theta(x)}(a, x)\, \mu(dx).$$

utilizing the fact that $\varrho^{\theta(x)}(a, x) = \varrho^\theta(a, x)$ μ^θ-almost for all x. The theorem is thus proved. □

Remark. It is easy to write the expression for $\varrho_\mu(a, x)$ under the condition that all μ^θ are absolutely continuous with respect to a certain measure ν and the function

$$g(\theta, x) = \frac{d\mu^\theta}{d\nu}(x)$$

is measurable jointly in variables θ and x. In this case

$$\int f(x)\, \mu_a(dx) = \int f(x+a)\, \mu(dx) = \int\int f(x+a)\, \mu^\theta(dx)\, \sigma(d\theta) =$$

$$= \int\int f(x)\, \varrho^\theta(a, x)\, g(\theta, x)\, \nu(dx)\, \sigma(d\theta) =$$

$$=\int f(x)\frac{\int \varrho^{\theta}(a,x)\,g(\theta,x)\,\sigma(d\theta)}{\int g(\theta',x)\,\sigma(d\theta')}\,\mu(dx),$$

since for any bounded measurable function $\varphi(x)$

$$\int \varphi(x)\,\mu(dx)=\int\int \varphi(x)\,g(\theta',x)\,\nu(dx)\,\sigma(d\theta').$$

Therefore under the assumptions stipulated above we have

$$\varrho_{\mu}(a,x)=\frac{\int \varrho^{\theta}(a,x)\,g(\theta,x)\,\sigma(d\theta)}{\int g(\theta,x)\,\sigma(d\theta)}.$$

This result can be generalized as follows:

Let the family of measures $\mu^{\alpha,\theta}$ depend on two parameters α and θ varying in two separable complete metric spaces $\mathscr{A}$ and Θ correspondingly and let a family of measures μ^{θ} exist such that the conditions of Theorem 7 are satisfied and $\mu^{\alpha,\theta}\ll\mu^{\theta}$ for all $\alpha\in\mathscr{A}$. Assume that for each continuous bounded function $f(x)$ the integral $\int f(x)\,\mu^{\alpha,\theta}(dx)$ is a continuous function of α and θ. Denote by $\mathfrak{A}$ and $\mathfrak{S}$ the σ-algebras of Borel sets in $\mathscr{A}$ and Θ correspondingly. Let $\sigma(d\alpha,d\theta)$ be a probability measure on $\mathfrak{A}\times\mathfrak{S}$ and let the measure μ on $\mathfrak{B}$ be defined by the relation

$$\mu(E)=\int \mu^{\alpha,\theta}(E)\,\sigma(d\alpha,d\theta).$$

Then $M_{\mu}\supset\bigcap_{\alpha,\theta} M^{\alpha,\theta}$, where $M^{\alpha,\theta}$ is the set of admissible shifts of measure $\mu^{\alpha,\theta}$ and for $a\in\bigcap_{\alpha,\theta} M^{\alpha,\theta}$ there exist functions $\varrho^{\alpha,\theta}(a,x)$ and $g(\alpha,\theta,x)$ measurable in α, θ, x on $\mathfrak{A}\times\mathfrak{S}\times\mathfrak{B}$ such that

$$\varrho^{\alpha,\theta}(a,x)=\frac{d\mu_a^{\alpha,\theta}}{d\mu^{\alpha,\theta}}(x)\;(\operatorname{mod}\mu^{\alpha,\theta}),$$

$$g(\alpha,\theta,x)=\frac{d\mu^{\alpha,\theta}}{d\mu^{\theta}}(x)\;(\operatorname{mod}\mu^{\theta}),$$

and $\varrho_\mu(a, x)$ is expressed in terms of these functions by the formula

$$\varrho_\mu(a, x) = \frac{\int \varrho^{\alpha, \theta(x)}(a, x)\, g(\alpha, \theta(x), x)\, \sigma(d\alpha \mid \theta(x))}{\int g(\alpha, \theta(x), x)\, \sigma(d\alpha \mid \theta(x))}, \tag{10}$$

where $\sigma(d\alpha \mid \theta)$ is the conditional measure defined by the relation

$$\int \psi(\alpha, \theta)\, \sigma(d\alpha \mid \theta)\, \sigma(\mathscr{A}, d\theta) = \int \psi(\alpha, \theta)\, \sigma(d\alpha, d\theta),$$

which is valid for any function $\psi(\alpha, \theta)$ measurable in variables α and θ and $\theta(x)$ is a measurable function for which $\mu^\theta(\{x : \theta(x) = \theta\})$. Formula (10) follows from the following string of equations

$$\int f(x+a)\, \mu(dx) = \int \int f(x+a)\, \mu^{\alpha, \theta}(dx)\, \sigma(d\alpha, d\theta) =$$

$$= \int \int \int f(x)\, \varrho^{\alpha, \theta}(a, x)\, g(\alpha, \theta, x)\, \mu^\theta(dx)\, \sigma(d\alpha \mid \theta)\, \sigma(A, d\theta) =$$

$$= \int \int \int f(x)\, \varrho^{\alpha, \theta(x)}(a, x)\, g(\alpha, \theta(x), x) \times$$

$$\times \sigma(d\alpha \mid \theta(x))\, \mu^\theta(dx)\, \sigma(A, d\theta) =$$

$$= \int f(x) \frac{\int \varrho^{\alpha, \theta(x)}(a, x)\, g(\alpha, \theta(x), x)\, \sigma(d\alpha \mid \theta(x))}{\int g(\alpha, \theta(x), x)\, \sigma(d\alpha \mid \theta(x))} \mu(dx),$$

and the existence of the functions $\varrho^{\alpha, \theta}(a, x)$ and $g(\alpha, \theta, x)$ is proved in exactly the same manner as the existence of function $\varrho^\theta(a, x)$ in Theorem 7.

As it follows from formula (9) the density $\varrho_\mu(a, x)$ of the measure given by formula (8) does not depend on measure σ. If μ_1 and μ_2 are two measures defined by the relation

$$\mu_k(A) = \int \mu^\theta(A)\, \sigma_k(d\theta)$$

and $\sigma_1 \sim \sigma_2$ then $\mu_1 \sim \mu_2$ also and moreover

$$\frac{d\mu_2}{d\mu_1}(x) = \frac{d\sigma_2}{d\sigma_1}(\theta(x)),$$

where $\theta(x)$ is a function such that $\mu^\theta(\{x : \theta(x) = \theta\}) = 1$. Since $\varrho_{\mu_1}(a, x)$

and $\varrho_{\mu_2}(a, x)$ coincide, we have $\frac{d\mu_2}{d\mu_1}(x-a)=\frac{d\mu_2}{d\mu_1}(x)$ and this equation is valid if for all $a\in M_{\mu_1}$ and almost all x the equality $\theta(x-a)=\theta(x)$ holds. It turns out that this case is general to a certain extent. Assume that measures ν and μ are equivalent and that for some a, $\lambda a\in M_\mu$ for all real λ and $\varrho_\mu(a, x)=\varrho_\nu(a, x)$. Set $\varphi(x)=\frac{d\nu}{d\mu}(x)$. Let $E_t=\{x:\varphi(x)=t\}$ and σ be the measure on the real line such that

$$\sigma((-\infty, t))=\mu(\bigcup_{s<t} E_s),$$

and the family of measures μ^t is determined by relation

$$\mu(A\cap\{x:\varphi(x)\in\Lambda\})=\int_\Lambda \mu^t(A)\,\sigma(dt), \tag{11}$$

which is valid for all $A\in\mathfrak{B}$ and all Borel sets Λ on the real line (i.e. μ^t is the conditional distribution of x on the probability space $(\mathscr{X}, \mathfrak{B}, \mu)$ given $\varphi(x)=t$). We show that a is an admissible shift of measures μ^t σ-almost for all t. It follows from Theorem 4 that measure μ can be represented in the form

$$\mu(A)=\iint\limits_{\substack{x+sa\in A,\\ x\in\mathscr{X}_0}} f(x, s)\,\bar{\mu}(dx)\,dF(s),$$

where $f(x, s)=\frac{d\mu}{d\mu^*}(x+sa)$, and $\bar{\mu}$ is the measure on the subspace $\mathscr{X}_0=\{x:(a, x)=0\}$ determined by the relation $\bar{\mu}(A)=\mu(P^{-1}A)$ and P is the projector on $\mathscr{X}_0$, while $F(s)=\mu(\{x:(a, x)<s\})$. Analogously to representation (11) we have

$$\bar{\mu}(A)=\int \bar{\mu}^t(A)\,\sigma(dt)$$

(Note that the equality $\varphi(Px)=\varphi(x)$ μ-almost for all x yields the relation

$$\bar{\mu}(\bigcup_{s<t} E_s)=\bar{\mu}(\{x:\varphi(x)<t\})=\mu(\{x:\varphi(x)<t\})=\sigma((-\infty, t)).$$

Let $\bar{f}(t, x, s)$ be a measurable function such that $\bar{f}(t, x, s)=f(x, s)$ for $\varphi(x)=t$. Then

$$\mu(A)=\iint\limits_{x+sa\in A} \bar{f}(t, x, s)\,\bar{\mu}^t(dx)\,\sigma(dt)\,dF(s)=$$

$$= \int \left[\int\limits_{x+sa \in A} \bar{f}(t, x, s)\, \bar{\mu}^t(dx)\, dF(s) \right] \sigma(dt).$$

Setting

$$\int\limits_{x+sa \in A} \bar{f}(t, x, s)\, \bar{\mu}^t(dx)\, dF(s) = \mu^{*t}(A),$$

we obtain

$$\mu(A) = \int \mu^{*t}(A)\, \sigma(dt).$$

Hence for each measurable set on the line Λ and each continuous function $g(x)$ the equality

$$\iint\limits_{A} g(x)\, \mu^t(dx)\, \sigma(dt) = \iint\limits_{A} g(x)\, \mu^{*t}(dx)\, \sigma(dt),$$

is satisfied, i.e. $\int g(x)\, \mu^t(dx) = \int g(x)\, \mu^{*t}(dx)$ σ-almost for all t. It follows from the representation of μ^{*t} that a is an admissible shift for this measure. Finally note that

$$\nu(A) = \int\limits_{A} \varphi(x)\, \mu(dx) =$$

$$= \iint \chi_A(x)\, \varphi(x)\, \mu^t(dx)\, \sigma(dt) = \int \mu^t(A)\, \sigma_1(dt),$$

where $\frac{d\sigma_1}{d\sigma}(t) = t$. We have thus proved the following

Theorem 8. *If μ and ν are two equivalent measures for which the sets M_μ and M_ν satisfy $M_\mu = M_\nu$ (these sets are linear manifolds) and $\vartheta_\mu(a, x) = \vartheta_\nu(a, x)$, then there exists a one-parametric family of measures μ^t such that $a \in M_{\mu^t}$ and a measurable function $\varphi(x)$ such that $\mu^t(\{x : \varphi(x) = t\}) = 1$ and also equivalent measures (on the line) σ_1 and σ such that*

$$\mu(A) = \int \mu^t(A)\, \sigma(dt), \quad \nu(A) = \int \mu^t(A)\, \sigma_1(dt).$$

Moreover the function $\varphi(x)$ satisfies the equality $\varphi(x - a) = \varphi(x)$ for all $a \in M_\mu$ and μ-almost for all x.

A sufficient condition for admissibility of the shift. Consider measure μ on $(\mathscr{X}, \mathfrak{B})$. Let $\{e_n, n=1,\ldots\}$ be an orthonormal basis in $\mathscr{X}$, $\mathscr{X}_n$ be the subspace spanned by $e_1,\ldots, e_n$, P_n be the projector on this subspace and μ^n the projection of measure μ on $\mathscr{X}_n$. Assume that μ^n is absolutely continuous with respect to Lebesgue's measure on $\mathscr{X}_n$ and that its density $f_n(x)$ with respect to this measure is positive. Then

$$\frac{d\mu_a^n}{d\mu^n}(x)=\frac{f_n(x-a_n)}{f_n(x)}, \qquad a_n=P_n a, \quad x\in\mathscr{X}_n.$$

In view of Theorem 1 in Section 1 the limit

$$\lim_{n\to\infty}\frac{f_n(\mathsf{P}_n(x-a))}{f_n(\mathsf{P}_n x)}=g(x, a) \text{ exists } \mu\text{-almost everywhere.}$$

If $a\in M_\mu$, then $g(x, a)=\varrho_\mu(a, x)$ and in order that $a\in M_\mu$ it is necessary and sufficient that $\int g(x, a)\,\mu(dx)=1$ (cf. Corollary 2 to Theorem 1 in Section 1). It is difficult to verify this condition. Below conditions are presented which guarantee that $ta\in M_\mu$ for all real t. These conditions are based on the remark following Theorem 2 in Section 1.

Assume that the density $f_n(x)$ is continuously diffentiable with respect to x and denote by $\nabla f_n(x)$ the gradient of $f_n(x)$ i.e. a vector in $\mathscr{X}_n$ such that $\frac{d}{dt}f_n(x+ta)\big|_{t=0}=(\nabla f_n(x), a)$ for all $a\in\mathscr{X}_n$. Set for all $x\in\mathscr{X}$ and $a\in\mathscr{X}$

$$h_n(x, a)=\frac{1}{f_n(P_n x)}(\nabla f_n(P_n x), P_n a).$$

We show that the functions $h_n(x, a)$ for a fixed a and a running n form a martingale on the probability space $(\mathscr{X}, \mathfrak{B}, \mu)$. Denote by $\mathfrak{B}_n$ the σ-algebra of cylindrical sets with bases in $\mathscr{X}_n$. Let $\alpha_k=(a, e_k)$, $t_k=(x, e_k)$, $f_n(x)=F_n(t_1,\ldots, t_n)$. Then

$$h_n(x, a)=\frac{1}{F_n(t_1,\ldots, t_n)}\sum_{k=1}^{n}\frac{\partial F_n}{\partial t_k}(t_1,\ldots, t_n)\,\alpha_k.$$

If $\varphi(x)$ is a $\mathfrak{B}_n$-measurable function then $\varphi(x)=\Phi(t_1,\ldots, t_n)$. First assume that $\Phi(t_1,\ldots, t_n)$ is continuously differentiable which vanishes everywhere except for some bounded region. Then

$\mathsf{E}h_{n+1}(x, a)\,\varphi(x)=$

$$=\int\frac{1}{F_{n+1}(t_1,\ldots, t_{n+1})}\sum_{k=1}^{n+1}\frac{\partial}{\partial t_k}F_{n+1}(t_1,\ldots, t_{n+1})\,\alpha_k\times$$

$$\times \Phi(t_1,\dots,t_n)\, F_{n+1}(t_1,\dots,t_{n+1})\, dt_1\dots dt_{n+1}=$$
$$= \mathsf{E}h_n(x,a)\,\varphi(x)+\alpha_{n+1}\int \frac{1}{\partial t_{n+1}} F_{n+1}(t_1,\ \dots, t_{n+1})\times$$
$$\times \Phi(t_1,\dots,t_n)\, dt_1\dots dt_{n+1} = \mathsf{E}h_n(x,a)\,\varphi(x)-$$
$$-\alpha_{n+1}\int F_{n+1}\frac{\partial}{\partial t_{n+1}}\Phi(t_1,\dots,t_n)\, dt_1\dots dt_{n+1} = \mathsf{E}h_n(x,a)\,\varphi(x).$$

Since the functions $\varphi(x)$ of this type are everywhere dense in the space of all bounded $\mathfrak{B}_n$-measurable functions then the equality

$$\mathsf{E}h_{n+1}(x,a)\,\varphi(x) = \mathsf{E}h_n(x,a)\,\varphi(x)$$

is valid for all bounded $\mathfrak{B}_n$-measurable functions. From the last equality follows relation

$$\mathsf{E}(h_{n+1}(x,a)\,|\,\mathfrak{B}_n) = h_n(x,a),$$

i.e. $h_n(x,a)$ is a martingale.

Denote by N the set of a such that

$$\sup_n \int h_n^2(x,a)\,\mu(dx)<\infty.$$

For all $a\in N$ the limit $h(x,a)=\lim\limits_{n\to\infty} h_n(x,a)$ exists μ-almost everywhere and moreover, the sequence $\{h_1(x,a),\dots, h_n(x,a),\dots, h(x,a)\}$ is also a martingale on the probability space $(\mathscr{X}, \mathfrak{B}, \mu)$ and

$$\lim_{n\to\infty}\int [h(x,a)-h_n(x,a)]^2\,\mu(dx)=0.$$

It is easy to verify that N is a linear manifold and moreover for each real t and for a and b in N the relations

$$h(x,ta)=th(x,a), \qquad h(x,a+b)=h(x,a)+h(x,b).$$

are valid μ-almost for all x. The next theorem presents sufficient conditions for the inclusion $a\in M_\mu$ in terms of the function $h(x,a)$ defined above.

Theorem 9. *Let the densities $f_n(x)$ be positive and continuously differentiable and let $h(x,a)$ be defined for some $a\in\mathscr{X}$. If for some $\delta>0$*

$$\int e^{\delta|h(x,a)|}\,\mu(dx)<\infty,$$

then $ta \in M_\mu$ for all real t and formula

$$\varrho(ta, x) = \exp\left\{-\int_0^t h(x - sa, a)\, ds\right\} \tag{12}$$

is valid.

Proof. Set

$$I_n(t) = \int_{\mathfrak{X}_n} \ln\left(1 + \frac{f_n(x)}{f_n(x+ta)}\right) f_n(x)\, dx.$$

In view of the assumptions of the theorem the derivative $I_n'(t)$ exists and moreover

$$I_n'(t) = -\int_{\mathfrak{X}_n} h_n(x+ta, a) \frac{f_n(x)}{f_n(x) + f_n(x+ta)} f_n(x)\, dx.$$

Hence

$$|I_n'(t)| \leqslant \int_{\mathfrak{X}_n} |h_n(x+ta, a)|\, f_n(x)\, dx.$$

We shall utilize the following inequality due to Young*: if $g(t)$ is a function defined for all $t \geqslant 0$, continuous, strictly increasing with $g(0) = 0$ and $g^{-1}(t)$ is the inverse of g then for all $a > 0$, $b > 0$ we have

$$ab \leqslant \int_0^a g(t)\, dt + \int_0^b g^{-1}(t)\, dt.$$

Putting $g(t) = \frac{1}{\alpha} \ln(1+t)$, with $\alpha > 0$, we obtain

$$ab \leqslant \frac{a}{\alpha} \ln(1+a) + \frac{e^{\alpha b} - 1}{\alpha}.$$

Utilizing this inequality we have

$$I_n'(t) \leqslant \int_{\mathfrak{X}_n} |h_n(x+ta, a)| \frac{f_n(x)}{f_n(x+ta)} f_n(x+ta)\, dx \leqslant$$

---

* W. H. Young, Sur la generalization du theoreme de Parseval, *Comp. Rendus* **155** (1912) 30–3 (or *Proc. Royal Soc.* (A) **87** (1912) 225–9.) *Translator's Remark.*

$$\leqslant \frac{1}{\delta}\int_{\mathscr{X}_n} \frac{f_n(x)}{f_n(x+ta)} \ln\left(1+\frac{f_n(x)}{f_n(x+ta)}\right) f_n(x+ta)\,dx+$$

$$+\frac{1}{\delta}\int_{\mathscr{X}_n} e^{\delta|h_n(x+ta,a|} f_n(x+ta)\,dx=$$

$$=\frac{1}{\delta} I_n(t)+\frac{1}{\delta}\int e^{\delta|h_n(x,a)|}\mu(dx)$$

where $\delta>0$.

From this inequality involving the derivative $I_n'(t)$ we have

$$I_n(t)\leqslant\left(1+\int e^{\delta|h_n(x,a)|}\mu(dx)\right)e^{\frac{1}{\delta}t}$$

Since $\{h_1(x,a),\dots,h_n(x,a),\dots,h(x,a)\}$ is a martingale, it follows that

$$\int e^{\delta|h_n(x,a)|}\mu(dx)\leqslant\int e^{\delta|h(x,a)|}\mu(dx)$$

so that

$$\sup_n\int_{\mathscr{X}_n}\ln\frac{f_n(x-tP_na)}{f_n(x)}\cdot\frac{f_n(x-tP_na)}{f_n(x)}f_n(x)\,dx\leqslant$$

$$\leqslant\sup_n\int_{\mathscr{X}_n}\ln\left(1+\frac{f_n(x)}{f_n(x+tP_na)}\right)f_n(x)\,dx\leqslant\sup_n I_n(t)<\infty.$$

To verify that $a\in M_\mu$ one need merely apply the remark following Theorem 2 of Section 1.

We now proceed with the derivation of formula (12). First note that in view of relation

$$\varrho((t+a)\,a,x)=\varrho(ta,x)\,\varrho(sa,x-ta)$$

it is sufficient to establish this formula for t arbitrarily small. On the other hand utilizing the homogeneity of $h(x,a)$ in a, we may assume that the condition of the theorem is satisfied for δ sufficiently large, for example $\delta=4$. Utilizing the definition of $h_n(x,a)$ we have

$$\frac{f_n(P_n(x)-ta))}{f_n(P_nx)}=\exp\left\{-\int_0^t h_n(x-sa,a)\,ds\right\}.$$

Therefore,

$$\varrho(ta, x) = \lim_{n\to\infty} \exp\left\{-\int_0^t h_n(x-sa, a)\, ds\right\}.$$

To verify the existence of the integral $\int_0^t h(x-sa, a)\, ds$ μ-almost for all x and the fact that

$$\lim_{n\to\infty} \int_0^t h_n(x-sa, a)\, ds = \int_0^t h(x-sa, a)\, ds$$

in measure μ, it is sufficient to show that relation

$$\lim_{n, m\to\infty} \int \int_0^t |h_n(x-sa, a) - h_{n+m}(x-sa, a)|\, ds\, \mu(dx) = 0 \tag{13}$$

is satisfied. (Indeed it follows from the $\mathfrak{B}$-measurability of $h(x, a)$ that $h(x-sa, a)$ is a Borel function in s, $h_n(x-sa, a) \to h(x-sa, a)$ for each s and μ-almost for all x; hence in view of Fubini's theorem this holds μ-almost for all x and almost for all s with respect to the Lebesgue measure; it follows from (13) that the integral

$$\int_0^t |h(x-sa, a)|\, ds$$

is finite and the equality

$$\lim_{n\to\infty} \int \int_0^t |h_n(x-sa, a) - h(x-sa, a)|\, ds\, \mu(dx) = 0 \text{ is valid.)}$$

We have

$$\overline{\lim_{m, n\to\infty}} \int \int_0^t |h_n(x-sa, a) - h_{n+m}(x-sa, a)|\, ds\, \mu(dx) =$$

$$= \overline{\lim_{n, m\to\infty}} \int_{\mathscr{X}_{n+m}} \int_0^t |h_n(x, a) - h_{n+m}(x, a)| \quad f_{n+m}(x + sP_{n+m}a)\, ds\, dx.$$

We now again utilize the following form of Young's inequality

$$ab \leqslant \frac{e^a - 1}{\alpha} + b \ln(1 + \alpha b).$$

Then

$$\int_{\mathscr{X}_{n+m}} \int_0^t |h_n(x, a) - h_{n+m}(x, a)| \frac{f_{n+m}(x + sP_{n+m}a)}{f_{n+m}(x)} f_{n+m}(x)\, ds\, dx \leqslant$$

$$\leqslant \frac{t}{\alpha} \int [\exp\{|h_n(x, a) - h_{n+m}(x, a)|\} - 1]\, \mu(dx) +$$

$$+ \int_{\mathscr{X}_{n+m}} \int_0^t f_{n+m}(x) \ln\left(1 + \alpha \frac{f_{n+m}(x)}{f_{n+m}(x + sP_{n+m}a)}\right) ds\, dx.$$

Note that

$$\exp\{2|h_n(x, a) - h_{n+m}(x, a)|\} \leqslant \tfrac{1}{2} \exp\{4|h_n(x, a)|\} + \tfrac{1}{2} \exp\{4|h_{n+m}(x, a)|\}.$$

Therefore the integral

$$\int (\exp\{|h_n(x, a) - h_{n+m}(x, a)|\})^2\, \mu(dx)$$

is uniformly bounded and hence the function $\exp\{|h_n(x, a) - h_{m+n}(x, a)|\}$ is uniformly integrable and the limiting transition under the sign of the integral is justified. Consequently,

$$\lim_{n, m \to \infty} \int \exp\{|h_n(x, a) - h_{n+m}(x, a)|\}\, \mu(dx) = 1,$$

since $|h_n(x, a) - h_{n+m}(x, a)| \to 0$ as $n \to \infty$ and $m \to \infty$ in measure μ.

Therefore

$$\overline{\lim_{n, m \to \infty}} \int \int_0^t |h_n(x, a) - h_{n+m}(x, a)|\, ds\, \mu(dx) \leqslant$$

$$\leqslant \overline{\lim_{n \to \infty}} \int_{\mathscr{X}_n} \int_0^t f_n(x) \ln\left(1 + \frac{\alpha f_n(x)}{f_n(x - sP_n a)}\right) ds\, dx.$$

Since the sequence $\eta_n = \dfrac{f_n(x)}{f_n(x - sP_n a)}$ is a martingale on the probability space $(\mathscr{X}, \mathfrak{B}, \mu_{sa})$ and $\eta_\infty = \varrho(-sa, x-sa) = \lim \eta_n$ is such that $\mathsf{E}\eta_\infty = \lim_{n\to\infty} \mathsf{E}\eta_n = 1$, this sequence is uniformly integrable and hence, in view of Theorem 3 in Section 2 of Chapter II, the sequence $[\eta_n, n=1,2,\dots,\infty]$ is also a martingale. Therefore the sequence $[\eta_n \ln(1+\alpha\eta_n); n=1,2,\dots,\infty]$ is a semi-martingale since $\sup_n \mathsf{E}\eta_n \ln(1+\alpha\eta_n) < \infty$ (the boundedness of $\sup_n \mathsf{E}\eta_n \ln(1+\alpha\eta_n)$ is proved analogously to the boundedness of $I_n(t)$; actually for $\alpha \leqslant 1$, which is the case here, it follows from the boundedness of $I_n(t)$). Therefore

$$\mathsf{E}\eta_n \ln(1+\alpha\eta_n) \leqslant \mathsf{E}\eta_\infty \ln(1+\alpha\eta_\infty),$$
$$\lim_{n\to\infty} \mathsf{E}\eta_n \ln(1+\alpha\eta_n) = \mathsf{E}\eta_\infty \ln(1+\alpha\eta_\infty)$$

(the mathematical expectation is taken in the probability space $(\mathscr{X}, \mathfrak{B}, \mu_{sa})$). Since

$$\mathsf{E}\eta_n \ln(1+\alpha\eta_n) = \int f_n(x) \ln\left(1+\alpha\frac{f_n(x)}{f_n(x-sP_n a)}\right) dx,$$

it follows that

$$\lim_{n\to\infty} \int_0^t \int f_n(x) \ln\left(1+\alpha\frac{f_n(x)}{f_n(x-sP_n a)}\right) dx\, ds =$$
$$= \int_0^t \int \ln(1+\alpha\varrho(-as, x-as))\, \mu_{sa}(dx)\, ds.$$

Hence

$$\overline{\lim_{n,m\to\infty}} \int \int_0^t |h_n(x-sa, a) - h_{n+m}(x-sa, a)|\, ds\, \mu(dx) \leqslant$$
$$\leqslant \int_0^t \int \ln(1+\alpha\varrho(-as, x-sa))\, \mu_{sa}(dx)\, ds. \tag{14}$$

It is easy to verify that the integral

$$\int \ln(1+\alpha\varrho(-as, x-sa))\, \mu_{sa}(dx)$$

approaches zero monotonically as $\alpha \downarrow 0$. Approaching the limit in (14) as $\alpha \downarrow 0$ we obtain (13). The theorem is proved. □

Corollary. *Let μ be the Gaussian measure with mean value* 0 *and correlation operator B. Then $M_\mu = B^{1/2}\mathscr{X}$.*

Proof. Since the characteristic functional of measure μ is of the form $\varphi(z)=\exp(-\frac{1}{2}(Bz, z)\}$ and $\varphi(z)\to 1$ as $(Bz, z)\to 0$ we have, in view of Theorem 2, the inclusion $M_\mu \subset B^{1/2}\mathscr{X}$. Denote by $e_1, e_2, \dots$ and $\lambda_1, \lambda_2, \dots$ the eigenvectors and eigenvalues of the operator B, respectively. If $\mathscr{X}_n$ is the subspace spanned by $e_1, \dots, e_n$ then

$$f_n(x)=(2\pi)^{-n/2}\left(\prod_{k=1}^{n} \lambda_k\right)^{1/2} \exp\left\{-\frac{1}{2}\sum_{k=1}^{n} \frac{(x, e_k)^2}{\lambda_k}\right\}.$$

Therefore

$$h_n(x, a)=-\sum_{k=1}^{n} \frac{(x, e_k)(a, e_k)}{\lambda_k}.$$

It is easy to verify that

$$\int (h_n(x, a))^2 \mu(dx)=\sum_{k=1}^{n} \frac{(a, e_k)^2}{\lambda_k},$$

so that $h(x, a)$ is defined and

$$h(x, a)=-\sum_{k=1}^{\infty} \frac{(x, e_k)(a, e_k)}{\lambda_k},$$

provided only $\sum_{k=1}^{\infty} \frac{(a, e_k)^2}{\lambda_k}<\infty$. Since (x, e_k) are independent Gaussian variables on the probability space $(\mathscr{X}, \mathfrak{B}, \mu)$, $h(x, a)$ is also a Gaussian variable. Therefore

$$\int e^{|h(x,a)|} \mu(dx) \leqslant \int [e^{h(x,a)}+e^{-h(x,a)}] \mu(dx)=$$

$$=2\exp\left\{\frac{1}{2}\int h^2(x, a) \mu(dx)\right\}=2\exp\left\{\frac{1}{2}\sum_{k=1}^{\infty} \frac{(a, e_k)^2}{\lambda_k}\right\}.$$

Thus in view of Theorem 9, $a\in M_\mu$ provided only $\sum_{k=1}^{\infty} \frac{(a, e_k)^2}{\lambda_k}<\infty$ i.e. if $a\in B^{1/2}\mathscr{X}$. Our assertion is proved. □

It follows from (12) that

$$\varrho_\mu(a, x)=\exp\left\{\sum_{k=1}^{\infty} \frac{(x, e_k)(a, e_k)}{\lambda_k}-\frac{1}{2}\sum_{k=1}^{\infty} \frac{(a, e_k)^2}{\lambda_k}\right\}=$$

$$=\exp\{(B^{-1/2}a, B^{-1/2}x)-\tfrac{1}{2}|B^{-1/2}a|^2\}.$$

Finally we present an example of a measure for which M_μ is a linear manifold everywhere dense in $\mathscr{X}$ and a vector a exists which does not belong to M_μ and such that μ_a is not orthogonal to μ. Consider Gaussian measures μ^1 and μ^2 with mean values 0 and correlation operators A and B. We assume that the eigenvectors for both operators are the same, denote these eigenvectors by $e_1, e_2, \ldots$ and the corresponding eigenvalues will be denoted by α_i and β_i. Let $\frac{\alpha_n}{\beta_n} \to 0$ as $n \to \infty$. It is easy to verify that $M_{\mu^1} \subset M_{\mu^2}$ and $M_{\mu^2} - M_{\mu^1}$ is a non-empty set. If $a \in M_{\mu^2} - M_{\mu^1}$, then $\mu_a^2 \sim \mu^2$ and $\mu_a^1 \perp \mu^1$ (cf. corollary to Theorem 5). We show that $\mu_a^1 \perp \mu^2$. The variables $(x, e_k) = \xi_k$ are Gaussian on each of the probability spaces $\{\mathscr{X}, \mathfrak{B}, \mu^2\}$ and $\{\mathscr{X}, \mathfrak{B}, \mu_a^1\}$ and, moreover $\mathsf{E}\xi_k = 0$ and $\operatorname{Var} \xi_k = \beta_k$ on the first and $\mathsf{E}\xi_k = (a, e_k)$ and $\operatorname{Var} \xi_k = \alpha_k$ on the second of the spaces. Therefore

$$\frac{1}{n} \sum_{k=1}^{n} \frac{(x, e_k)^2}{\beta_k} \to 1$$

in measure μ^2 and

$$\frac{1}{n} \sum_{k=1}^{n} \frac{(x, e_k)^2}{\beta_k} \to 0$$

in measure μ_a^1. The orthogonality of μ^2 and μ_a^1 is proved. Note that it follows from the presented proof that $\mu^2 \perp \mu^1$ also. But then, in view of Theorem 6, the set of admissible shifts of measure $\mu = \frac{1}{2}(\mu^1 + \mu^2)$ coincides with $M_{\mu^1} \cap M_{\mu^2}$, i.e. with M_{μ^1}. If, however, $a \in M_{\mu^2} - M_{\mu^1}$ then μ_a possesses an absolutely continuous component $\frac{1}{2}\mu_a^2$ with respect to measure μ so that μ satisfies the required conditions.

§3. Absolute Continuity of Measures under Mappings of Spaces

The basic problem considered in this section is the study of conditions under which the mapping of a Hilbert space $\mathscr{X}$ into itself transforms measure μ into an absolutely continuous measure with respect to μ. If $T(x)$ is a measurable mapping of $\mathscr{X}$ into $\mathscr{X}$, i.e. a mapping such that $T^{-1}(A) \in \mathfrak{B}$ for each $A \in \mathfrak{B}$, then the measure μ under such a transformation is translated into measure ν defined by equality

$$\nu(A) = \mu(T^{-1}(A)). \tag{1}$$

Below sufficient conditions will be found which assure the absolute continuity of ν with respect to μ and an expression for $\frac{d\nu}{d\mu}$ is obtained by means of characteristics of measure μ and the mapping T.

Before studying measures in infinitely dimensional spaces, we shall obtain a solution of the posed problem in the case of a finite-dimensional Euclidean space. Let the measure μ possess a density with respect to the Lebegue measure

$$\mu(A)=\int_A f(x)\,dx,$$

and, moreover f does not vanish. We shall also assume that the transformation T is one-to-one and continuously differentiable. Then for a measurable bounded function g

$$\int g(x)\,\nu(dx)=\int g(T(x))\,\mu(dx)=\int g(T(x))\,f(x)\,dx=$$
$$=\int g(y)\,f(T^{-1}(y))\left|\frac{DT^{-1}(y)}{Dy}\right|dy,$$

where $\frac{DT^{-1}(y)}{Dy}$ is the Jacobian of the inverse of T which is also differentiable in view of the imposed assumptions. The last integral in this chain of equalities can be written as an integral in measure μ;

$$\int g(y)\,f(T^{-1}(y))\left|\frac{DT^{-1}(y)}{Dy}\right|dy=\int g(y)\frac{f(T^{-1}(y))}{f(y)}\left|\frac{DT^{-1}(y)}{Dy}\right|\mu(dy).$$

Note that

$$\frac{f(T^{-1}(y))}{f(y)}=\frac{f(y-(y-T^{-1}(y)))}{f(y)}=\varrho(y-T^{-1}(y),y),$$

where $\varrho(a,x)$ is the density of measure μ_a with respect to measure μ (we utilize the notation of Section 2). Therefore in a finite-dimensional space the following formula

$$\int g(x)\,\nu(dx)=\int g(x)\,\varrho(x-T^{-1}(x),x)\left|\frac{DT^{-1}(x)}{Dx}\right|\mu(dx)$$

is valid. In this form the formula makes sense also in the case of a Hilbert space provided we assign a suitable meaning to the Jacobian of the transformation. Let V be a linear operator such that $V-I$, where I is the identity transformation, is completely continuous. Then VV^* is a symmetric nonnegative operator and VV^*-I is also a completely continuous operator. Let λ_k be a sequence of eigenvalues of the operator VV^* (this operator possesses a complete system of eigenvectors), $\lambda_k \geqslant 0$, $\lambda_k \to 1$. We set

$$|\det V| = \sqrt{\prod_{k=1}^{\infty} \lambda_k},$$

provided this infinite product is either convergent or divergent to 0 or to $+\infty$.

Let $S(x)$ be a mapping of $\mathscr{X}$ into $\mathscr{X}$. This mapping is differentiable at point x_0 if a linear operator $dS(x_0)$ exists such that relation

$$|S(x_0+x) - S(x_0) - dS(x_0)\, x| = o(|x|), \quad x \in \mathscr{X},$$

is satisfied; operator $dS(x_0)$ is called the differential of $S(x)$ at point x_0. If $\mathscr{X}$ is a finite-dimensional space then, as it is easily seen, the Jacobian of the transformation $S(x)$ coincides with $|\det dS(x)|$. The latter makes sense in a Hilbert space as well. Thus we arrive at formula

$$\int g(x)\, \nu(dx) = \int g(x)\, \varrho(x - T^{-1}(x), x)\, |\det dT^{-1}(x)|\, \mu(dx). \tag{2}$$

The validity of this formula for a sufficiently wide class of functions g leads to the equality

$$\frac{d\nu}{d\mu}(x) = \varrho(x - T^{-1}(x), x)\, |\det dT^{-1}(x)|. \tag{3}$$

The remainder of this section is devoted to the investigation of conditions under which formula (3) is valid in the general as well as in the case of Gaussian measures μ. In order that formula (3) make sense, certain general conditions must be imposed on measure μ and on transformation T.

Condition 1. Measure μ possesses a linear manifold of admissible shifts M and for each orthonormal basis $\{e_k\}$ of the projection μ^n of measure μ on $\mathscr{X}_n$ possesses a continuous density $f_n(x)$ with respect to the Lebesgue measure on $\mathscr{X}_n$ (here $\mathscr{X}_n$ is the subspace spanned by the vectors $e_1, e_2, \dots, e_n$). Moreover, for each $c > 0$ and each finite-dimensional subspace $N \subset M$,

$$\sup\left[\left|\frac{f_n(\mathsf{P}_n(x-a))}{f_n(\mathsf{P}_n x)} - \varrho(a, x)\right|;\ |a| \leqslant c,\ a \in N\right] \to 0$$

in measure μ, where P_n is the projector on $\mathscr{X}_n$ and

$$\varrho(a, x) = \frac{d\mu_a}{d\mu}(x).$$

Condition 2. The transformation $T(x)$ possesses the inverse denoted by $S(x)$; operators $T(x)$ and $S(x)$ are locally bounded and continuously differentiable; also quantities $|\det dT(x)|$ and $|\det dS(x)|$ are finite, non-zero, continuous and locally bounded.

Theorem 1. *Let conditions* 1 *and* 2 *be satisfied and let a finite dimensional subspace N exist in M such that $x-T(x)\in N$, $x-S(x)\in N$ for all $x\in\mathscr{X}$ and, moreover, let for the projector P on N, $PT(x)=T(P(x))$, and $P(S(x))=$ $=S(P(x))$. Then $\nu\sim\mu$ and formula* (3) *is valid.*

Proof. We choose the basis $e_1, e_2, \ldots$ such that for some m the vectors $e_1, e_2, \ldots, e_m$ form a basis in N. Let μ^n and ν^n be the projections of measures μ and ν on the subspace $\mathscr{X}_n$. Then for $n>m$ and for any measurable bounded function g defined on $\mathscr{X}_n$ we have

$$\int_{\mathscr{X}_n} g(x)\,\nu^n(dx)=\int_{\mathscr{X}} g(P_n x)\,\nu(dx)=\int_{\mathscr{X}} g(P_n T(x))\,\mu(dx)=$$

$$=\int_{\mathscr{X}} g(T(P_n x))\,\mu(dx)=\int_{\mathscr{X}_n} g(T(x))\,\mu^n(dx)=$$

$$=\int_{\mathscr{X}_n} g(T(x))\,f_n(x)\,dx.$$

Under our assumptions the transformations T and S map $\mathscr{X}_n$ into $\mathscr{X}_n$ for $n>m$. Changing the variables of integration by means of $x=S(y)$, we obtain

$$\int_{\mathscr{X}_n} g(x)\,\nu^n(dx)=\int_{\mathscr{X}_n} g(x)\frac{f_n(S(x))}{f_n(x)}\,|\det dS(x)|\,\mu^n(dx).$$

Since $S(x)=x+P(S(x)-x)$, the Jacobi transformation matrix for $S(x)$ is of the form $I+V$, where V has non-zero elements only in the first m rows and these elements do not depend on m. Thus for $n>m$ the absolute value of the Jacobian of the transformation $S(x)$ does not depend on n and coincides with $|\det dS(x)|$.

Hence

$$\frac{d\nu^n}{d\mu^n}(x)=\frac{f_n(S(x))}{f_n(x)}\,|\det dS(x)|.$$

Clearly measures ν^n and μ^n are equivalent and

$$\frac{d\mu^n}{d\nu^n}(x)=\frac{f_n(x)}{f_n(S(x))}\,\frac{1}{|\det dS(x)|}.$$

Therefore,

$$\frac{d\mu^n}{d\nu^n}(T(x))=\frac{f_n(T(x))}{f_n(x)}\,\frac{1}{|\det dS(T(x))|}=\frac{f_n(T(x))}{f_n(x)}\,|\det dT(x)|,$$

since according to the rule of differentiation of a composite function

$$I = dx = d[S(T(x))] = dS(T(x))\, dT(x).$$

In view of condition 1 the following limits

$$\lim_{n\to\infty} \frac{f_n(P_n S(x))}{f_n(P_n x)} = \varrho(x - S(x), x),$$

$$\lim_{n\to\infty} \frac{f_n(P_n T(x))}{f_n(P_n x)} = \varrho(x - T(x), x)$$

exist, in the sense of convergence in measure μ; moreover these limits are different from zero (here we utilize the facts that $x - S(x) \in N$ and $x - T(x) \in N$, that these functions are locally bounded and condition 1). Applying Corollary 1 to Theorem 3 in Section 1 we complete the proof of the theorem.

Remark 1. We have proved that under the conditions of Theorem 1 $\nu \sim \mu$ and in addition to formula (3) the following formula

$$\frac{d\mu}{d\nu}(T(x)) = \varrho(x - T(x), x)\, |\det dT(x)| \tag{4}$$

is valid.

Remark 2. Formulas (3) and (4) remain valid if for any y and δ and finite dimensional subspace $N^y \subset M$ can be found such that the conditions of the theorem are satisfied for $|x - y| \leqslant \delta$ provided N is replaced by N^y.

Theorem 2. *Let conditions* 1 *and* 2 *be satisfied and let the basis* $\{e_k\}$ *of vectors in* M *exist such that*

1) *for* n *sufficiently large the mappings* $T_n(x) = x + P_n(T(x) - x)$ *and* $S_n(x) = x + P_n(S(x) - x)$ *are invertable and* $|\det dT_n(x)| \to |\det dT(x)|$, $|\det dS_n(x)| \to |\det dS(x)|$ *in the sense of convergence in measure* μ.

2) *expressions* $\varrho(P_n(x - S(x)), x)$ *and* $\varrho(P_n(x - T(x)), x)$ *possess limits as* $n \to \infty$ *in measure* μ *which will be denoted by* $\varrho(x - S(x), x)$ *and* $\varrho(x - - T(x), x)$ *correspondingly; we can substitute* x *by* $T(x)$ *in* $\varrho(x - S(x), x)$ *and*

$$\varrho(T(x) - x, T(x))\, \varrho(x - T(x), x) = 1.$$

Then the measures μ *and* ν *are equivalent and the formulas* (3) *and* (4) *are valid.*

Proof. Let

$$S_n^m(x) = x + P_n(S(P_m x) - x).$$

For $m>n$

$$S_n^m(\mathsf{P}_m x)=\mathsf{P}_m S_n^m(x).$$

The mapping S_n^m and its inverse satisfy the conditions of Theorem 1. If ν_n^m is the measure determined by the equality

$$\nu_n^m(A)=\mu(S_n^m(A)),$$

then

$$\frac{d\nu_n^m}{d\mu}(x)=\varrho(x-S_n^m(x),x)\,|\det dS_n^m(x)|.$$

Note that condition 1 and the fact that $f_n(x)$ are continuous and positive for all n imply that $\varrho(a,x)$ is uniformly continuous in a for $a\in\mathscr{X}_m$, $|a|\leqslant c$, in measure μ, i.e.

$$\sup\{|\varrho(a_1,x)-\varrho(a_2,x)|;\ |a_1|\leqslant c,\ |a_2|\leqslant c,\ a_1,a_2\in\mathscr{X}_m,\ \ |a_1-a_2|<\delta\}\to 0$$

in measure μ as $\delta\to 0$. Since $x-S_n^m(x)\in\mathscr{X}_n$ for all m, $x-S_n^m(x)\to x-S_n(x)$ and $x-S_n^m(x)$ are bounded in measure μ, it follows that

$$\varrho(x-S_n^m(x),x)\to\varrho(x-S_n(x),x)$$

in measure μ as $m\to\infty$. Next it is obvious that $dS_n^m(x)$ is of the form $dS_n^m(x)=I+V_n^m(x)$ for all m, where $V_n^m(x)$ maps the whole space into $\mathscr{X}_n$. It can be verified that in the case when V maps $\mathscr{X}$ into $\mathscr{X}_n$, we have

$$|\det(I+V)|=|\det\|((I+V)e_i,e_j)\|_{i,j=1,\dots,n}|,$$

where $\|((I+V)e_i,e_j)\|_{i,j=1,\dots,n}$ is the matrix of order n with the i,j-th element being $((I+V)e_i,e_j)$. Since for all i and j

$$\lim_{m\to\infty}(dS_n^m(x)e_i,e_j)=(dS_n(x)e_i,e_j),$$

it follows that

$$\lim_{m\to\infty}|\det dS_n^m(x)|=|\det dS_n(x)|.$$

Therefore

$$\lim_{m\to\infty}\frac{d\nu_n^m}{d\mu}(x)=\varrho(x-S_n(x),x)\,|\det dS_n(x)|$$

in the sense of convergence in measure μ. Utilizing the equality $x-S_n(x)=$ $=P_n(x-S(x))$ and the conditions of the theorem we may assert that

$$\lim_{n\to\infty}\lim_{m\to\infty}\frac{d\nu_n^m(x)}{d\mu}=\varrho(x-S(x),x)\,|\det dS(x)|$$

in the sense of convergence in measure μ. Hence one can choose sequences n_k and m_k such that $m_k > n_k$ and measures $\nu_k^* = \nu_{n_k}^{m_k}$ which satisfy the relation

$$\lim_{k\to\infty} \frac{d\nu_k^*}{d\mu}(x) = \varrho(x - S(x), x)\, |\det dS(x)|$$

in measure μ.

Now let $T_n^m(x) = x + P_n(T(P_m x) - x)$. Analogously we can show that one may choose sequences n_k and m_k in such a manner that the measures

$$\tilde{\nu}_k(A) = \mu(\tilde{T}_k^{-1}(A)), \quad \text{where} \quad \tilde{T}_k = T_{n_k}^{m_k}$$

satisfy the following

$$\lim_{k\to\infty} \frac{d\mu}{d\tilde{\nu}_k}(\tilde{T}_k(x)) = \varrho(x - T(x), x)\, |\det dT(x)|.$$

in the sense of convergence in measure μ.

Finally,

$$\varrho(T(x) - x, T(x))\, |\det dS(T(x))|\, \varrho(x - T(x), x)\, |\det dT(x)| = 1,$$

since according to the rule of differentiation of a composite function:

$$I = dx = d(S(T(x))) = dS(T(x))\, dT(x),$$

and, hence, $|\det dS(T(x))| \cdot |\det dT(x)| = 1$, while

$$\varrho(T(x) - x, T(x))\, \varrho(x - T(x), x) = 1$$

in view of condition 2) of the theorem. Hence one may utilize Corollary 2 of Theorem 3 in Section 1 which yields the proof of the theorem. □

Consider now the case when the measure μ is Gaussian with the mean value 0 and correlation operator B^2. We have shown in Section 2 that in this case $M = B\mathscr{X}$ and if $a = Bb$, e_k are the eigenvectors and β_k are the corresponding eigenvalues of the operator B, then

$$\varrho(a, x) = \exp\left\{\sum_{k=1}^{\infty} \frac{(a, e_k)(x, e_k)}{\beta_k^2} - \frac{1}{2}\sum_{k=1}^{\infty} \frac{(a, e_k)^2}{\beta_k^2}\right\} =$$

$$= \exp\left\{\sum_{k=1}^{\infty} \frac{(b, e_k)(x, e_k)}{\beta_k} - \frac{1}{2}|b|^2\right\}.$$

Let the transformation $T(x)$ be of the form $T(x) = x + B\lambda(x)$, where $\lambda(x)$ is a continuous and continuously differentiable mapping. If T is invertable, then $S(x) = x + B\lambda^*(x)$, where $\lambda^*(x) = -\lambda(S(x))$ is also continuous and continuously differentiable.

Since in the case of a Gaussian measure the function $\ln \dfrac{f_n(x-a)}{f_n(x)}$ is a sum of quadratic and linear functionals (in a) it follows from the convergence of $\frac{f_n(x-a)}{f_n(x)}$ to $\varrho(a, x)$, that this convergence is uniform in a for $a \in \mathscr{X}_m$, $|a| \leqslant c$, for any m and c. Therefore for Gaussian measures condition 1 of Theorem 2 is always satisfied. Consider now condition 2 of the same theorem.

Since

$$\varrho(P_n(x-T(x), x) = \exp\left\{-\sum_{k=1}^{n} \frac{(\lambda(x), e_k)(x, e_k)}{\beta_k} - \tfrac{1}{2}|P_n\lambda(x)|^2\right\},$$

the existence of the limit

$$\lim_{n\to\infty} \varrho(P_n(x-T(x)), x)$$

(different from zero) is equivalent to the convergence in measure μ of the series

$$\sum_{k=1}^{\infty} (\lambda(x), e_k)\frac{(x, e_k)}{\beta_k},$$

and this limit is equal to

$$\varrho(x-T(x), x) = \exp\left\{-\sum_{k=1}^{\infty} \frac{(\lambda(x), e_k)(x, e_k)}{\beta_k} - \tfrac{1}{2}|\lambda(x)|^2\right\}.$$

In the same manner the $\lim_{n\to\infty} \varrho(P_n(x-S(x), x)$ exists, iff the series

$$\sum_{k=1}^{\infty} (\lambda^*(x), e_k)\frac{(x, e_k)}{\beta_k}$$

is convergent in measure μ and moreover

$$\varrho(x-S(x), x) = \exp\left\{-\sum_{k=1}^{\infty} \frac{(\lambda^*(x), e_k)(x, e_k)}{\beta_k} - \tfrac{1}{2}|\lambda^*(x)|^2\right\}.$$

Note finally that

$$\varrho(T(x)-S(T(x)), T(x)) =$$

$$= \exp\left\{-\sum_{k=1}^{\infty} \frac{(\lambda^*(T(x)), e_k)(T(x), e_k)}{\beta_k} - \tfrac{1}{2}|\lambda^*(T(x))|^2\right\} =$$

$$= \exp\left\{\sum_{k=1}^{\infty} \frac{(\lambda(x), e_k)(x+B\lambda(x), e_k)}{\beta_k} - \tfrac{1}{2}|\lambda(x)|^2\right\} =$$

$$= \exp\left\{\sum_{k=1}^{\infty} \frac{(\lambda(x), e_k)(x, e_k)}{\beta_k} + \right.$$

$$+\sum_{k=1}^{\infty}\frac{(\lambda(x), e_k)(\lambda(x), Be_k)}{\beta_k}-\tfrac{1}{2}|\lambda(x)|^2\Big\}=(\varrho(x-T(x), x))^{-1}.$$

We have thus proved the following theorem:

Theorem 3. *Let μ be a Gaussian measure with mean 0 and correlation operator B^2; let β_k and e_k be the eigenvalues and eigenvectors respectively of the operator B. If*

a) *the transformation $T(x)$ satisfies condition 2 and is of the form $T(x)=x+B\lambda(x)$ and $S(x)=T^{-1}(x)=x+B\lambda^*(x)$; the transformations*

$$T_n(x)=x+P_nB\lambda(x), \qquad S_n(x)=x+P_nB\lambda^*(x),$$

are invertable for n sufficiently large and finally

$$|\det dT_n(x)|\to|\det dT(x)|, \ |\det dS_n(x)|\to|\det dS(x)|,$$

b) *the series*

$$\sum_{k=1}^{\infty}\frac{(\lambda(x), e_k)(x, e_k)}{\beta_k} \quad\text{and}\quad \sum_{k=1}^{\infty}\frac{(\lambda^*(x), e_k)(x, e_k)}{\beta_k}$$

are convergent μ-almost everywhere, then the measures ν and μ are equivalent and

$$\frac{d\nu}{d\mu}(x)=|\det dS(x)|\exp\Big\{-\sum_{k=1}^{\infty}\frac{(\lambda^*(x), e_k)(x, e_k)}{\beta_k}-\tfrac{1}{2}|\lambda^*(x)|^2\Big\}. \tag{5}$$

We now discuss certain sufficient conditions for convergence of the series appearing in b). These will be used to check whether condition b) is fulfilled.

Lemma. *If measure μ is as in Theorem 3 and $\lambda(x)$ is a continuous mapping of $\mathscr{X}$ into $\mathscr{X}$, then for the convergence of series*

$$\sum_{k=1}^{\infty}\frac{(\lambda(x), e_k)(x, e_k)}{\beta_k} \tag{6}$$

in measure μ it is sufficient that one of the following conditions be fulfilled:

1) *the numbers α_k exist such that $\sum_{k=1}^{\infty}\alpha_k^2<\infty$ and the series $\sum\alpha_k^{-2}\times$ $\times(\lambda(x), e_k)^2$ is convergent μ-almost everywhere.*

2) *the series*

$$\sum_{k=1}^{\infty}(\lambda(\qquad \lambda(P_{k-1}x), e_k)\frac{(x, e_k)}{\beta_k}, \qquad \sum_{k=1}^{\infty}(\lambda(P_{k-1}x), e_k)^2.$$

are convergent in measure μ.

3) *the series*

$$\sum_{i,j} \int [(\lambda_{ij}(x), e_i)(\lambda_{ij}(x), e_j) - (\lambda(x), e_i)(\lambda(x), e_j)]\, \mu(dx),$$

$$\sum_{k=1}^{\infty} \int (\lambda(x), e_k)^2 \frac{(x, e_k)^2}{\beta_k^2}\, \mu(dx),$$

where $\lambda_{ij}(x) = \lambda(x - (x, e_i)\, e_i - (x, e_j)\, e_j)$ are convergent.

Proof. 1) follows from the inequality

$$\sum_{k=1}^{\infty} \left| \frac{(x, e_k)}{\beta_k} (\lambda(x), e_k) \right| \leqslant \sqrt{\sum_{k=1}^{\infty} \alpha_k^2 \frac{(x, e_k)^2}{\beta_k^2}} \sqrt{\sum_{k=1}^{\infty} \frac{(\lambda(x), e_k)^2}{\alpha_k^2}}$$

and from the fact that

$$\sum_{k=1}^{\infty} \int \alpha_k^2 \frac{(x, e_k)}{\beta_k^2}\, \mu(dx) = \sum_{k=1}^{\infty} \alpha_k^2 < \infty .$$

2) Since the convergence of series (6) on the set $\{x : |\lambda(x)|^2 \leqslant c\}$ for any $c > 0$ implies its convergence in measure μ, we may assume without loss of generality that $|\lambda(x)|^2 \leqslant c$. Denote by H_m the set of x such that

$$\sum_{k=1}^{\infty} (\lambda(P_{k-1} x), e_k)^2 \leqslant m .$$

Let $\lambda_m(x) = \lambda(x)$ for $x \in H_m$, $\lambda_m(x) = 0$ for $x \notin H_m$. Since the convergence of the series $\sum_{k=1}^{\infty} (\lambda(P_{k-1} x), e_k)^2$ implies that $\mu(H_m) \to 1$ as $m \to \infty$, in order the series (6) be convergent it is sufficient that the series

$$\sum_{k=1}^{\infty} (\lambda_m(x), e_k) \frac{(x, e_k)}{\beta_k}$$

be convergent for each m in measure μ. However,

$$\sum_{k=1}^{\infty} (\lambda_m(x), e_k) \frac{(x, e_k)}{\beta_k} =$$

$$= \sum_{k=1}^{\infty} (\lambda_m(x) - (\lambda_m(P_{k-1} x), \frac{(x, e_k)}{\beta_k} + \sum_{k=1}^{\infty} \lambda_m(P_{k-1} x))\, e_k) \frac{(x, e_k)}{\beta_k} .$$

Note that in view of the inequality

$$\sum_{k=1}^{\infty} (\lambda(P_{k-1} P_l x), e_k)^2 \leqslant \sum_{k=1}^{\infty} (\lambda(P_{k-1} x), e_k)^2 +$$

$$+\sum_{k=1}^{\infty}(\lambda(P_l x), e_k)^2 \leqslant \sum_{k=1}^{\infty}(\lambda(P_{k-1}x), e_k)^2 + |\lambda(P_1(x)|^2$$

for all l and $x \in H_m$ we have $P_l x \in H_{m+c}$. Let $x \in H_{m-c}$. Then

$$\sum_{k=1}^{\infty}(\lambda_m(x)-\lambda_m(P_{k-1}x), e_k)\frac{(x, e_k)}{\beta_k} = \sum_{k=1}^{\infty}(\lambda(x)-\lambda(P_{k-1}x), e_k\frac{(x, e_k)}{\beta_k},$$

and by assumption the last series is convergent. Since m can be chosen arbitrarily large it is sufficient to show that the series $\sum_{k=1}^{\infty}(\lambda_m(P_{k-1}x), e_k) \times \times \frac{(x, e_k)}{\beta_k}$ is convergent in measure μ. However convergence of this series follows from the fact that its marginal sums form a martingale on the probability space $\{\mathscr{X}, \mathfrak{B}, \mu\}$. Indeed (x, e_k) are independent Gaussian variables and moreover

$$\mathsf{E}\left(\sum_{k=1}^{n}(\lambda_m(P_{k-1}x), e_k)\frac{(x, e_2)}{\beta_k}\right)^2 = \mathsf{E}\sum_{k=1}^{n}(\lambda_m(P_{k-1}x), e_k)^2\frac{(x, e_k)^2}{\beta_k^2} =$$

$$= \sum_{k=1}^{n}\mathsf{E}(\lambda_m(P_{k-1}x), e_k)^2\,\mathsf{E}\frac{(x, e_k)^2}{\beta_k} \leqslant m.$$

3) We show that series (6) is mean-square convergent. Clearly,

$$\int\left(\sum_{k=n}^{m}\frac{(\lambda(x), e_k)(x, e_k)}{\beta_k}\right)^2\mu(dx) =$$

$$= \sum_{k=n}^{m}\int\frac{(\lambda(x), e_k)^2(x, e_k)^2}{\beta_k^2}\mu(dx) +$$

$$+2\sum_{n\leqslant i<j\leqslant m}\int\frac{(\lambda_{ij}(x), e_i)(\lambda_{ij}(x), e_j)(x, e_i)(x, e_j)}{\beta_i\beta_j}\mu(dx) +$$

$$+2\sum_{n\leqslant i<j\leqslant m}\int[(\lambda(x), e_i)(\lambda(x), e_j) - (\lambda_{ij}(x), e_i)(\lambda_{ij}(x), e_j)] \times$$

$$\times\frac{(x, e_i)(x, e_j)}{\beta_i\beta_j}\mu(dx) = \int\sum_{k=n}^{m}\frac{(\lambda(x), e_k)^2(x, e_k)^2}{\beta_k^2}\mu(dx) +$$

$$+2\int\sum_{n\leqslant i<j\leqslant m}[(\lambda(x), e_i)(\lambda(x), e_j) -$$

$$-(\lambda_{ij}(x), e_i)(\lambda_{ij}(x), e_j)]\frac{(x, e_i)(x, e_j)}{\beta_i\beta_j}\mu(dx).$$

Since

$$\int (\lambda_{ij}(x), e_i)\,(\lambda_{ij}(x), e_j)\,(x, e_i)\,(x, e_j)\,\mu(dx)=0$$

in view of the independence of the variables $\lambda_{ij}(x)$, (x, e_i) and (x, e_j) on the probability space $\{\mathscr{X}, \mathfrak{B}, \mu\}$. Hence

$$\int \left(\sum_{k=n}^{m} \frac{(\lambda(x); e_k)\,(x, e_k)}{\beta_k}\right)^2 \mu(dx) \to 0$$

as $n \to \infty$ and $m \to \infty$. The lemma is thus proved. □

We apply Theorem 3 to the case when the transformation $T(x)$ is only slightly different from the identical transformation. Let a family of transformations $T_\varepsilon(x)=x+\varepsilon\lambda(x)$ be given, then $S_\varepsilon(x)$ is of the form $S_\varepsilon(x)=x-\varepsilon\lambda(x)+O(\varepsilon^2)$ (only terms of order not higher than ε are considered). If $d\lambda(x)$ has a finite trace, then

$$\ln|\det dT_\varepsilon(x)| = \sum_{k=1}^{\infty} \frac{(-1)^{k-1}}{k}\,\varepsilon^k\, \mathrm{Sp}\,[d\lambda(x)]^k$$

for $\varepsilon>0$ sufficiently small. Consequently, taking into account only the terms of order no higher than ε, we may write

$$\frac{d\nu_\varepsilon}{d\mu}(x)=1-\varepsilon\,\mathrm{Sp}\,[d\lambda(x)]-\varepsilon \sum_{k=1}^{\infty} \frac{(x, e_k)\,(\lambda(x), e_k)}{\beta_k^2}+O(\varepsilon^2).$$

As it is seen from this formula the basic difficulty encountered in applying Theorem 3, namely verification of the convergence of the series appearing in condition b) of this theorem remains intact for transformations which are arbitrarily close to the identity also.

Now consider the case when transformation T is linear.

Theorem 4. *Let μ be a Gaussian measure with mean value* 0 *and a positive correlation operator B^2. If the linear operator T is invertable and is of the form $T=I+BCB^{-1}$ where $\mathrm{Sp}\,CC^*<\infty$ and if the $I+C$ possesses a bounded inverse, then*

$$\frac{d\nu}{d\mu}(x)=K\exp\{W(x)\}, \tag{7}$$

where

$$K=\lim_{n\to\infty} |\det(I+D_n)|\, e^{-\mathrm{Sp}\,D_n}, \qquad D_n=P_nDP_n, \tag{8}$$

and

$$W(x)=\lim_{n\to\infty}\left[-(DB^{-1}P_nx, B^{-1}P_nx)-\tfrac{1}{2}|P_nDB^{-1}P_nx|^2+\mathrm{Sp}\,D_n\right]. \tag{9}$$

This limit is taken in the m.s. sense in measure μ, P_n is the projector on $\mathscr{X}_n$ and $D=B^{-1}T^{-1}B-I$.

Proof. Operator T maps M into M so that D is defined at least on M. We show that $\operatorname{Sp} DD^* < \infty$. Since

$$T^{-1}=I-BCB^{-1}T^{-1}, \quad D=CB^{-1}T^{-1}B=C(I+C)^{-1}=CV,$$

where V is a bounded operator, we have

$$\operatorname{Sp} DD^* = \sum_{k=1}^{\infty} (D^*e_k, D^*e_k) = \sum_{k=1}^{\infty} (V^*C^*e_k, V^*C^*e_k) \leqslant$$
$$\leqslant \|VV^*\| \sum_{k=1}^{\infty} |C^*e_k|^2 = \|VV^*\| \operatorname{Sp} CC^*.$$

From this relation it follows that D is bounded. Set $D_n = P_n D P_n$. Then

$$\operatorname{Sp}(D-D_n)(D-D_n)^* = \operatorname{Sp} DD^* - \operatorname{Sp} D_n D_n^* =$$
$$= \sum_{k=1}^{\infty} |D^*e_k|^2 - \sum_{k=1}^{\infty} |D_n^* e_k|^2 =$$
$$= \sum_{k=1}^{\infty} \sum_{j=1}^{\infty} (D^*e_k, e_j)^2 - \sum_{k=1}^{\infty} \sum_{j=1}^{n} (D^*e_k, e_j)^2 \to 0$$

as $n \to \infty$.

We show that the limit

$$\lim_{n\to\infty} |\det(I+D_n)| \exp\{-\operatorname{Sp} D_n\}$$

exists which is different from zero. Let $U_n = D_n + D_n^* + D_n D_n^*$. Then

$$|\det(I+D_n)| \exp\{-\operatorname{Sp} D_n\} = \sqrt{\det(I+U_n)\, e^{-\operatorname{Sp} U_n}} \exp\{\tfrac{1}{2} \operatorname{Sp} D_n D_n^*\}.$$

Since $\operatorname{Sp} D_n D_n^* \to \operatorname{Sp} DD^*$ it is sufficient to show that the limit

$$\lim_{n\to\infty} \det(I+U_n)\, e^{-\operatorname{Sp} U_n}. \tag{10}$$

exists. Set $U = D + D^* + DD^*$.

It follows from condition $\operatorname{Sp}(D-D_n)(D-D_n)^* \to 0$ that $\operatorname{Sp}(U-U_n) \times (U-U_n)^* \to 0$ also. Denote by $\lambda_1^{(n)}, \dots, \lambda_n^{(n)}$ the eigenvalues of the operator U_n in $\mathscr{X}_n$ (U_n maps $\mathscr{X}_n$ into $\mathscr{X}_n$ and its orthogonal complement into zero); next, denote by $f_1^n, \dots, f_k^{(n)}$ the eigenvectors corresponding to $\{\lambda_i^{(n)}\}$ (it is assumed that $\lambda_k^{(n)}$ are ordered according to their absolute value). It follows from relation

$$\sum_{i=1}^{n} |Uf_i^n - \lambda_i^{(n)} f_i^n|^2 \leqslant \operatorname{Sp}(U-U_n)^2 \to 0$$

that $\lambda_i^{(n)} \to \lambda_i$, $f_i^{(n)} \to f_i$, where f_i are the eigenvectors of the operator U and λ_i are the corresponding eigenvalues. Next we have

$$\det(I+U_n)\, e^{-\operatorname{Sp} U_n} = \prod_{k=1}^{n} (1+\lambda_k^{(n)})\, e^{-\lambda_k^{(n)}} =$$

$$= \left(\prod_{k=1}^{m} (1+\lambda_k^{(n)})\, e^{-\lambda_k^{(n)}}\right)\left(1+O\left(\sum_{k>n} (\lambda_n^{(n)})^2\right)\right).$$

Since

$$\lim_{n\to\infty} \prod_{k=1}^{m} (1+\lambda_k^{(n)})\, e^{-\lambda_k^{(n)}} = \prod_{k=1}^{m} (1+\lambda_k)\, e^{-\lambda_k},$$

to prove the existence of the limit (10) it is sufficient to show that

$$\lim_{m\to\infty} \overline{\lim_{n\to\infty}} \sum_{m+1}^{\infty} (\lambda_k^{(n)})^2 = 0.$$

However

$$\lim_{m\to\infty} \overline{\lim_{n\to\infty}} \sum_{k=m+1}^{\infty} (\lambda_k^{(n)})^2 = \lim_{m\to\infty} \overline{\lim_{n\to\infty}} \left[\operatorname{Sp} U_n^2 - \sum_{k=1}^{m} (\lambda_k^{(n)})^2\right] =$$

$$= \lim_{m\to\infty} \left[\operatorname{Sp} U^2 - \sum_{k=1}^{m} \lambda_k^2\right] = 0.$$

The existence of limit (10) is thus established. Moreover, this limit is different from zero since

$$K = \sqrt{\prod_{k=1}^{\infty} (1+\lambda_k)\, e^{-\lambda_k}\, e^{\frac{1}{2}\operatorname{Sp} DD^*}}$$

and since $1+\lambda_k \neq 0$ in view of the invertability of the operator $(1+D)\times \times(1+D^*)$. We now prove the existence of limit (9). Let

$$W_n(x) = \left[-(D_n B^{-1}x, B^{-1}P_n x) - \frac{1}{2}|D_n B^{-1} P_n x|^2 + \operatorname{Sp} D_n\right].$$

We then have, in view of formula (7) in Section 6 of Chapter 5:

$$\int [W_n(x) - W_m(x)]^2\, \mu(dx) =$$

$$= \int \left[\left(\left\{D_n + \frac{1}{2} D_n^* D_n - D_m - \frac{1}{2} D_m^* D_m\right\} B^{-1}x, B^{-1}x\right) + \right.$$

$$+\mathrm{Sp}(D_n-D_m)\Big]^2 \mu(dx)=\left[\frac{1}{2}\mathrm{Sp}(D_n^*D_n-D_m^*D_m)\right]^2+$$

$$+\mathrm{Sp}\left(\frac{D_n-D_m+D_n^*-D_m^*}{2}+\frac{D_n^*D_n-D_m^*D_m}{2}\right)^2\leqslant$$

$$\leqslant\left[\frac{1}{2}\mathrm{Sp}(D_n^*D_n-D_m^*D_m)\right]^2+\frac{1}{4}\mathrm{Sp}(U_n-U_m)^2;$$

the last expression approaches zero as $n\to\infty$ and $m\to\infty$. The existence of limit (9) is thus established.

We now proceed to the proof of formula (7). Let the measure v_n be defined by equation $v_n(A)=\mu(T_n^{-1}A)$, where $T_n^{-1}=B(I+D_n)\,B^{-1}$. It then follows from Theorem 3 that

$$\frac{dv_n}{d\mu}(x)=|\det B(I+D_n)\,B^{-1}|\times$$

$$\times\exp\left\{-\sum_{k=1}^{n}\left[\frac{(D_nB^{-1}x,e_k)\,(x,e_k)}{\beta_k}-\frac{\frac{1}{2}|D_nB^{-1}P_nx|^2}{\beta_k^2}\right\}=\right.$$

$$=|\det(I+D_n)|\exp\left\{-(D_nB^{-1}P_nx,B^{-1}P_nx)-\frac{1}{2}|D_nB^{-1}P_nx|^2\right\};$$

since $|\det B(1+D_n)\,B^{-1}|$ coincides with the absolute value of the determinant of the transformation matrix $B(1+D_n)\,B^{-1}$ considered in $\mathscr{X}_n$ and written in the orthonormal basis, it follows that

$$|\det B(I+D_n)\,B^{-1}|=|\det(I+D_n)|.$$

Hence

$$\frac{dv_n}{d\mu}(x)=K_n\exp\{W_n(x)\},$$

where $W_n(x)$ is defined above and

$$K_n=|\det(I+D_n)|\,e^{-\mathrm{Sp}\,D_n}.$$

As we have shown

$$\lim_{n\to\infty}\frac{dv_n}{d\mu}(x)=K\exp\{W(x)\}$$

in the sense of convergence in measure μ. Now let measure $\tilde{v}_n$ be defined by the equality $\tilde{v}_n=\mu(\tilde{T}_n^{-1}(A))$, where $\tilde{T}_n=1+BP_nCP_nB^{-1}$. It can be shown analogously to the above, that

$$\lim_{n\to\infty}\frac{d\mu}{d\tilde{v}_n}(\tilde{T}_n(x))=\tilde{K}\exp\{\tilde{W}(x)\},$$

in measure μ, where $\tilde{K}$ and $\tilde{W}$ are defined by formulas (8) and (9) with T replaced by T^{-1} and B^2 by TB^2T^*.

Utilizing now Corollary 2 of Theorem 3 in Section 1 we obtain the proof of the theorem. □

§4. Absolute Continuity of Gaussian Measures in a Hilbert Space

Let two Gaussian measures μ_1 and μ_2 with mean values a_1 and a_2 and correlation operators B_1 and B_2 correspondingly be defined in the Hilbert space $(\mathscr{X}, \mathfrak{B})$. Below we establish necessary and sufficient conditions on a_1, a_2, B_1 and B_2 for the measure μ_2 to be absolutely continuous with respect to μ_1. It will be shown that the density $d\mu_2/d\mu_1$ is everywhere positive so that the absolute continuity of μ_2 with respect to μ_1 implies the equivalence of measures μ_1 and μ_2. Moreover, it turns out that the violation of the condition of absolute continuity implies the orthogonality of the measures, so that two Gaussian measures are either equivalent or orthogonal.

The case when the measures have different location parameters but $B_1 = B_2$ was partially studied in Section 2.

Theorem 1. *If $B_1 = B_2 = B$, then in order that $\mu_2 \ll \mu_1$ it is necessary and sufficient that $a_2 - a_1 \in B^{1/2}\mathscr{X}$; moreover*

$$\frac{d\mu_2}{d\mu_1}(x) = \exp\{(B^{-1/2}(x-a_1), B^{-1/2}(a_2-a_1)) - \tfrac{1}{2}|B^{-1/2}(a_2-a_1)|^2\}. \quad (1)$$

If $a_2 - a_1 \notin B^{1/2}\mathscr{X}$, then $\mu_1 \perp \mu_2$.

Proof. The first assertion and formula (1) were verified in Section 2. Now let

$$\frac{1}{\sqrt{\lambda_k}}(a_2 - a_1, e_k) = \alpha_k,$$

where λ_k are the eigenvalues of the operator B and e_k are the corresponding eigenvectors. If $a_2 - a_1 \notin B^{1/2}\mathscr{X}$ then $\sum_{k=1}^{\infty} \alpha_k^2 = +\infty$. Consider the function

$$g_n(x) = \left(\sum_{k=1}^{n} \alpha_k^2\right)^{-1} \sum_{k=1}^{n} \frac{\alpha_k}{\sqrt{\lambda_k}}(x - a_1, e_k).$$

Since

$$\int g_n(x)\,\mu_1(dx) = 0, \quad \int g_n(x)\,\mu_2(dx) = 1,$$

$$\int g_n^2(x)\,\mu_1(dx)=\int (g_n(x)-1)^2\,\mu_2(dx)=\left(\sum_{k=1}^{n}\alpha_k^2\right)^{-1},$$

it follows that $g_n \to 0$ in measure μ_1 and that $g_n \to 1$ in measure μ_2. This implies the orthogonality of measures μ_1 and μ_2. The theorem is proved. □

Consider the case when $a_1=a_2=0$. Assume that $\mu_2 \ll \mu_1$. Then the ratio $(B_2 z, z)/(B_1 z, z)$ is necessarily bounded for $z \in \mathscr{X}$. Indeed if one can find a sequence z_n such that

$$\lim_{n\to\infty}\frac{(B_2 z_n, z_n)}{(B_1 z_n, z_n)}=+\infty,$$

$$\text{then } \frac{(z_n, x)}{\sqrt{(B_2 z_n, z_n)}} \to 0 \text{ in measure } \mu_1,$$

$$\mu_2\left(\left\{x: \frac{|(z_n, x)|}{\sqrt{(B_2 z_n, z_n)}} \leqslant \varepsilon\right\}\right)=\frac{1}{\sqrt{2\pi}}\int_{|t|\leqslant\varepsilon} e^{-\frac{1}{2}t^2}\,dt \leqslant \frac{2\varepsilon}{\sqrt{2\pi}},$$

hence $\dfrac{(z_n, x)}{\sqrt{(B_2 z_n, z_n)}}$ does not tend to zero in measure μ_2. This contradicts the absolute continuity of μ_2 with respect to μ_1. It is easy to see that the unboundedness of $\dfrac{(B_2 z, z)}{(B_1 z, z)}$ implies that measures μ_1 and μ_2 are (even) singular. Hence this ratio is bounded from below by a positive number. It follows from here that the ranges of values of operators $B_1^{\frac{1}{2}}$ and $B_2^{\frac{1}{2}}$ coincide and that the operators $C=B_1^{\frac{1}{2}}B_2^{-\frac{1}{2}}$ and $C^{-1}=B_2^{\frac{1}{2}}B_1^{-\frac{1}{2}}$ are bounded. Note that the boundedness of the operator C implies that $(z, B_2^{-\frac{1}{2}}x)$ is a measurable functional not only in measure μ_2 (cf. Chapter V, Section 6) but in measure μ_1 as well. This is because

$$(z, B_2^{-\frac{1}{2}}x)=(z, C^* B_1^{-\frac{1}{2}}x)=(Cz, B_1^{-\frac{1}{2}}x).$$

Consider the self-adjoint operator $C^*C=B_2^{-\frac{1}{2}}B_1 B_2^{-\frac{1}{2}}$. We show that $C^*C=I+D$, where D is a completely continuous operator. To do this, it is sufficient to verify that if E_λ is the resolution of the identity for the operator D, then the projectors E_λ for $\lambda<0$ and $I-E_\lambda$ for $\lambda>0$ map $\mathscr{X}$ into a finite-dimensional subspace. We first show that there is no eigenvalue $\lambda \neq 0$ for the operator D such that an infinitely dimensional proper subspace will correspond to this eigenvalue. Otherwise an infinite orthonormal sequence z_k in this subspace can be found such that $\frac{1}{n}\sum_{k=1}^{n}(z_k, B_2^{-\frac{1}{2}}x)^2 \to 1$ in measure μ_2 and $\frac{1}{n}\sum_{k=1}^{n}(z_k, B_2^{-\frac{1}{2}}x)^2 \to 1+\lambda$ in measure μ_1, since on each one of the probability spaces $(\mathscr{X}, \mathfrak{B}, \mu_1)$ and $(\mathscr{X}, \mathfrak{B}, \mu_2)$ the variables $(z_k, B_2^{-\frac{1}{2}}x)$ form a sequence of independent Gaussian variables with means 0 and variances $1+\lambda$ and 1 correspondingly.

Indeed

$$\int (z_k, B_2^{-\frac{1}{2}}x)\,(z_j, B_2^{-\frac{1}{2}}x)\,\mu_2(dx)=(z_k, z_j)=\delta_{kj},$$

$$\int (z_k, B_2^{-\frac{1}{2}}x)\,(z_j, B_2^{-\frac{1}{2}}x)\,\mu_1(dx)=(B_1B_2^{-\frac{1}{2}}z_k, B_2^{-\frac{1}{2}}z_j)=$$

$$=(z_k, z_j)+(Dz_k, z_j)=(1+\lambda)\,\delta_{kj}.$$

The fact that

$$\frac{1}{n}\sum_{k=1}^{n}(z_k, B_2^{-\frac{1}{2}}x)$$

converges in measures μ_1 and μ_2 to different constants, implies the orthogonality of these measures. Now let

$$z_k\in(E_{\lambda_{k-1}}-E_{\lambda_k})\,\mathscr{X},$$

where $0>\lambda=\lambda_0>\lambda_1>\ldots>\lambda_k, (E_{\lambda_{k-1}}-E_{\lambda_k})\,\mathscr{X}$ are non-empty subspaces. Then $(z_k, B_2^{-\frac{1}{2}}x)$ are, as before, independent Gaussian variables on the probability spaces $(\mathscr{X}, \mathfrak{B}, \mu_1)$ and $(\mathscr{X}, \mathfrak{B}, \mu_2)$.

Indeed,

$$\int (z_k, B_2^{-\frac{1}{2}}x)\,(z_j, B_2^{-\frac{1}{2}}x)\,\mu_2(dx)=\delta_{kj},$$

$$\int (z_k, B_2^{-\frac{1}{2}}x)\,(z_j, B_2^{-\frac{1}{2}}x)\,\mu_1(dx)=\delta_{kj}+(Dz_k, z_j)=$$

$$=\delta_{kj}\left(1+\int_{\lambda_k}^{\lambda_{k-1}}\lambda d(E_\lambda z_k, z_k)\right).$$

Using the strong law of large numbers we can assert that $\frac{1}{n}\sum_{k=1}^{n}(z_k, B_2^{-\frac{1}{2}}x)^2\to 1$ in measure μ_2 and that

$$\varlimsup_{n\to\infty}\frac{1}{n}\sum_{k=1}^{n}(z_k, B_2^{-\frac{1}{2}}x)^2\leqslant\varlimsup_{n\to\infty}\frac{1}{n}\sum_{k=1}^{n}\left(1+\int_{\lambda_k}^{\lambda_{k-1}}\lambda d(E_\lambda z_k, z_k)\right)\leqslant 1+\lambda<1$$

in measure μ_1. From these two relations it again follows that the measure μ_1 and μ_2 are singular. To complete the proof one need merely note that for $\lambda<0$ the subspace $E_\lambda\mathscr{X}$ is infinitely dimensional in the case when either an infinitely dimensional proper subspace corresponds to some $\lambda<0$ or a finite number of disjoint intervals on $(-\infty, \lambda)$ exists such that the increment E_λ is non-zero on each one of them. We have thus shown that $E_\lambda\mathscr{X}$ is finite dimensional for $\lambda<0$. In the same manner one can show that $(I-E_\lambda)\mathscr{X}$ is finite-dimensional for $\lambda>0$.

Thus the operator D is completely continuous. Let $e_1, e_2, \ldots$ be the eigenvectors of the operator D and let $\delta_1, \delta_2, \ldots$ be the corresponding eigenvalues. We now show that the absolute continuity of μ_2 with respect μ_1 implies that $\sum_{k=1}^{\infty} \delta_k^2<\infty$.

Indeed if $\sum_{k=1}^{\infty} \delta_k^2=+\infty$, then we take the sequence of functions

$$g_n(x)=\left(\sum_{k=1}^{n} \delta_k^2\right)^{-1} \sum_{k=1}^{n} \delta_k[(e_k, B_2^{-\frac{1}{2}}x)^2-1].$$

We have already noted that if vectors z_k belong to distinct orthogonal proper subspaces of the operator D, then the variables $(z_k, B_2^{-\frac{1}{2}}x)$ are independent and Gaussian on the probability spaces $(\mathscr{X}, \mathfrak{B}, \mu_1)$ and and $(\mathscr{X}, \mathfrak{B}, \mu_2)$. It follows from the relations

$$\int g_n(x)\, \mu_2(dx)=0,$$

$$\int g_n^2(x)\, \mu_2(dx)=\left(\sum_{k=1}^{n} \delta_k^2\right)^{-2} \sum_{k=1}^{n} \delta_k^2 \int [(e_k, B_2^{-\frac{1}{2}}x)^4 - 2(e_k, B_2^{-\frac{1}{2}}x)^2+1]\, \mu_2(dx)=2\left(\sum_{k=1}^{n} \delta_k^2\right)^{-1},$$

$$\int g_n(x)\, \mu_1(dx)=\left(\sum_{k=1}^{n} \delta_k^2\right)^{-1} \sum_{k=1}^{n} \delta_k(De_k, e_k)=1,$$

$$\int (g_n(x)-1)^2\, \mu_1(dx)=\left(\sum_{k=1}^{n} \delta_k^2\right)^{-2} \times$$

$$\times \sum_{k=1}^{n} \delta_k^2\left\{\int (e_k, B_2^{-\frac{1}{2}}x)^4\, \mu_1(dx)-\left[\int (e_k, B_2^{-\frac{1}{2}}x)^2\, \mu_1(dx)\right]^2\right\}=$$

$$=2\left(\sum_{k=1}^{n} \delta_k^2\right)^{-2} \sum_{k=1}^{n} \delta_k^2(1+\delta_k^2)=o\left(\left(\sum_{k=1}^{n} \delta_k^2\right)^{-1}\right)$$

that $g_n(x)\to 0$ in measure μ_2 and that $g_n(x)\to 1$ in measure μ_1. Hence the condition $\sum_{k=1}^{\infty} \delta_k^2=+\infty$ implies the orthogonality of the measures μ_1 and

μ_2. Another necessary condition, satisfied by δ_k, follows from relation

$$1+\delta_k=1+\frac{(De_k, e_k)}{(e_k, e_k)}=\frac{(B_2^{-\frac{1}{2}}B_1B_2^{-\frac{1}{2}}e_k, e_k)}{(B_2^{-\frac{1}{2}}B_2B_2^{-\frac{1}{2}}e_k, e_k)}=\lim_{n\to\infty}\frac{(B_1z_n, z_n)}{(B_2z_n, z_n)}>0,$$

where z_n is a sequence of vectors in $B_2^{\frac{1}{2}}\mathscr{X}$ such that $B_2^{\frac{1}{2}}z_n\to e_k$. Hence, $\delta_k>-1$.

Now let $\delta_k>-1$ and $\sum_{k=1}^{\infty}\delta_k^2<\infty$. We show that the measures μ_1 and μ_2 are equivalent. To do this consider the measure $\tilde{\mu}$ defined by the equality

$$\tilde{\mu}(A)=\int_A \varrho(x)\,\mu_1(dx),$$

where

$$\varrho(x)=\exp\left\{-\tfrac{1}{2}\sum_{k=1}^{\infty}\left[(B_2^{-\frac{1}{2}}x, e_k)^2\frac{\delta_k}{1+\delta_k}-\ln(1+\delta_k)\right]\right\}.$$

The convergence of the series

$$\sum_{k=1}^{\infty}\left[(B_2^{-\frac{1}{2}}x, e_k)^2\frac{\delta_k}{1+\delta_k}-\ln(1+\delta_k)\right]$$

in measure μ_1 follows from the fact that on the probability space $(\mathscr{X}, \mathfrak{B}, \mu_1)$ this is a series in independent random variables such that the corresponding series of mathematical expectations and variances given by

$$\mathsf{E}\left[(B_2^{-\frac{1}{2}}x, e_k)^2\frac{\delta_k}{1+\delta_k}-\ln(1+\delta_k)\right]=\delta_k-\ln(1+\delta_k)=O(\delta_k^2),$$

$$\mathsf{V}\left[(B_2^{-\frac{1}{2}}x, e_k)\frac{\delta_k}{1+\delta_k}\right]=\frac{\delta_k^2}{(1+\delta_k)^2}\,2(1+\delta_k)^2=2\delta_k^2.$$

are convergent.

We find the characteristic functional of measure $\tilde{\mu}$:

$$\tilde{\chi}(z)=\int e^{i(z,x)}\,\tilde{\mu}(dx)=\int e^{i(z,x)}\,\varrho(x)\,\mu_1(dx).$$

For any $z\in\mathscr{X}$ the relation

$$(z, x)=(B_2^{\frac{1}{2}}z, B_2^{-\frac{1}{2}}x)=\sum_{k=1}^{\infty}(B_2^{\frac{1}{2}}z, e_k)(B_2^{-\frac{1}{2}}x, e_k)$$

is satisfied where the series on the right is convergent μ_1-almost everywhere.

Utilizing the fact that the variables $(B_2^{-\frac{1}{2}}x, e_k)$ on the probability space $(\mathscr{X}, \mathfrak{B}, \mu_1)$ are independent and Gaussian with mean value 0 and variances $1+\delta_k$ we obtain

$$\tilde{\chi}(z) = \mathsf{E} \exp\Big\{i \sum_{k=1}^{\infty} (B_2^{\frac{1}{2}}z, e_k)(B_2^{-\frac{1}{2}}x, e_k) - -\tfrac{1}{2}\sum_{k=1}^{\infty}\Big[(B_2^{-\frac{1}{2}}x, e_k)^2 \frac{\delta_k}{1+\delta_k} - \ln(1+\delta_k)\Big]\Big\} =$$

$$= \prod_{k=1}^{\infty} \mathsf{E} \exp\Big\{i(B_2^{\frac{1}{2}}z, e_k)(B_2^{-\frac{1}{2}}x, e_k) - -\frac{\delta_k}{2(1+\delta_k)}(B_2^{-\frac{1}{2}}x, e_k)^2\Big\}\sqrt{1+\delta_k} =$$

$$= \prod_{k=1}^{\infty} \frac{1}{\sqrt{2\pi}} \int e^{i(B_2^{1/2}z, e_k)t - \frac{\delta_k}{2(1+\delta_k)}t^2 - \frac{1}{2(1+\delta_k)}t^2}\, dt =$$

$$= \prod_{k=1}^{\infty} \exp\{-\tfrac{1}{2}(B_2^{\frac{1}{2}}z, e_k)^2\} = \exp\{-\tfrac{1}{2}(B_2 z, z)\}.$$

Since the characteristic functional of measure $\tilde{\mu}$ coincides with the characteristic functional of measure μ_2 we have $\mu_2 = \tilde{\mu}$ and $\dfrac{d\mu_2}{d\mu_1}(x) = \varrho(x)$.

Thus the following theorem is established:

Theorem 2. *Let μ_1 and μ_2 be two Gaussian measures with mean values 0 and correlation operators B_k, $k=1, 2$. In order that measures μ_1 and μ_2 be equivalent, it is necessary and sufficient that the operator $D = B_2^{-1/2} \times \times B_1 B_2^{-1/2} - I$ be a Hilbert-Schmidt operator and its eigenvalues δ_k satisfy the inequality $\delta > -1$. If this condition is violated then the measures μ_1 and μ_2 are orthogonal. In the case when measures μ_1 and μ_2 are equivalent the following formula*

$$\frac{d\mu_2}{d\mu_1}(x) = \exp\Big\{-\tfrac{1}{2}\sum_{k=1}^{\infty}\Big[(B_2^{-\frac{1}{2}}x, e_k)^2 \frac{\delta_k}{1+\delta_k} - \ln(1+\delta_k)\Big]\Big\} \tag{2}$$

is valid where e_k are the eigenvectors of the operator D corresponding to the eigenvalues δ_k.

Remark. Let μ_1 and μ_2 be two measures as defined in Theorem 2. Denote by $\mathscr{L}^{(1)}$ and $\mathscr{L}^{(2)}$ the Hilbert spaces of linear measurable functionals with respect to measures μ_1 and μ_2 (cf. Section 6, Chapter V). If a sequence of functionals $\{l_k(x), k=1, 2, \ldots\}$ exists which belongs to both spaces and which is a complete orthogonal system in each one of these spaces and

$$\delta_k^{(i)}=\int [l_k(x)]^2\,\mu_i(dx), \qquad i=1,2, \quad k=1,2,\ldots,$$

then $\mu_1\sim\mu_2$ under the condition that $\delta_k^{(i)}>0$, $\sum_k\left(1-\frac{\delta_k^{(1)}}{\delta_k^{(2)}}\right)^2<\infty$ and moreover

$$\frac{d\mu_2}{d\mu_1}(x)=\exp\left\{\frac{1}{2}\sum_{k=1}^{\infty}\left[l_k^2(x)\left(\frac{1}{\delta_k^{(1)}}-\frac{1}{\delta_k^{(2)}}\right)-\ln\frac{\delta_k^{(1)}}{\delta_k^{(2)}}\right]\right\}.$$

The proof of this assertion is completely analogous to the proof of the sufficiency of the conditions of Theorem 2. □

Now consider the general case: We introduce in addition to measures μ_1 and μ_2 a measure μ_{12} with mean value a_1 and correlation operator B_2. We show that the condition $\mu_2\ll\mu_1$ implies relation $\mu_2\ll\mu_{12}\ll\mu_1$ so that

$$\frac{d\mu_2}{d\mu_1}=\frac{d\mu_2}{d\mu_{12}}\cdot\frac{d\mu_{12}}{d\mu_1};$$

and one can also utilize formulas (1) and (2) for computing $d\mu_2/d\mu_{12}$ and du_{12}/du_1 respectively. It is sufficient to show that $\mu_{12}\ll\mu_1$ since in this case $\mu_{12}\sim\mu_2$ and hence $\mu_2\ll\mu_{12}$. If $\bar{\mu}_2$ is a measure with mean a_2-a_1 and correlation operator B_2 and $\bar{\mu}_1$ is a measure with mean 0 and correlation operator B_1, then $\bar{\mu}_2\ll\bar{\mu}_1$. Let the measure $\bar{\mu}_i^*$ be defined by relation $\bar{\mu}_i^*(A)=\bar{\mu}_i(\{x:-x\in A\})$. Clearly $\bar{\mu}_1^*=\bar{\mu}_1$. Therefore $\bar{\mu}_2^*\ll\mu_1$ and hence $\bar{\mu}_2*\bar{\mu}_2\ll\bar{\mu}_1*\bar{\mu}_1$. It is easy to see that $\bar{\mu}_2*\bar{\mu}_2^*$ is a Gaussian measure with mean 0 and correlation operator $2B_2$, and that the measure $\bar{\mu}_1*\bar{\mu}_1$ has the same mean but a different correlation operator (which equals $2B_1$). Consequently, $\nu_2\ll\nu_1$ where ν_k $(k=1,2)$ is a Gaussian measure with mean 0 and correlation operator B_k. But then $\mu_{12}\ll\mu_1$ also since μ_{12} and μ_1 are obtained from measures ν_2 and ν_1 by the translation in the amount a_1. Thus the following theorem is valid in the general case:

Theorem 3. *If the measures μ_1 and μ_2 are two Gaussian measures with characteristic functionals*

$$\varphi_k(z)=\exp\{i(a_k,z)-\tfrac{1}{2}(B_kz,z)\}, \qquad k=1,2,$$

then in order that the measures μ_1 and μ_2 be equivalent it is necessary and sufficient that the following conditions be satisfied:

1) *$a_2-a_1=B_2^{1/2}b$, where $b\in\mathscr{X}$.*

2) *the operator $D=B_2^{-1/2}B_1B_2^{-1/2}-I$ is a Hilbert-Schmidt operator*

and its eigenvalues δ_k satisfy the inequality $\delta_k > -1$. If at least one of the conditions is not satisfied the measures μ_1 and μ_2 are orthogonal. The following formula

$$\frac{d\mu_2}{d\mu_1}(x) = \exp\left\{-\tfrac{1}{2}\left[\sum_{k=1}^{\infty} (B_2^{-\frac{1}{2}}(x-a_1), e_k)^2 \frac{\delta_k}{1+\delta_k} - \ln(1+\delta_k)\right] + \right.$$
$$\left. + (B_2^{-\frac{1}{2}}(x-a_1), b) - \tfrac{1}{2}|b|^2\right\} \tag{3}$$

is valid for equivalent measures, where e_k are the eigenvectors of operator D corresponding to the eigenvalues δ_k.

We shall consider some sufficient conditions for absolute continuity of Gaussian measures. The conditions presented below may turn out to be more convenient for applications, since they don't involve fractional powers of correlation operators. Assume that the operators $B_1 B_2^{-1}$ and $B_2 B_1^{-1}$ are bounded. Set $V = B_1 B_2^{-1} - I$. Since $V = B_2^{1/2} D B_2^{-1/2}$ then $V^2 = B_2^{1/2} D^2 B_2^{-1/2}$. Let $\{f_k\}$ be an orthonormal sequence of eigenvectors of operator B_2. Then

$$\operatorname{Sp} D^2 = \sum_{k=1}^{\infty} (D^2 f_k, f_k) = \sum_{k=1}^{\infty} \left(D^2 \sqrt{\lambda_k}\, B_2^{-\frac{1}{2}} f_k, \frac{1}{\sqrt{\lambda_k}} B_2^{\frac{1}{2}} f_k\right) =$$
$$= \sum_{k=1}^{\infty} (D^2 B_2^{-\frac{1}{2}} f_k, B_2^{-\frac{1}{2}} f_k) = \sum_{k=1}^{\infty} (V^2 f_k, f_k), \quad \lambda_k = (B_2 f_k, f_k).$$

Hence $\operatorname{Sp} D^2 < \infty$ provided only that $\operatorname{Sp} V^2$ is finite. Since V^2 is an asymmetric operator the verification of conditions for the existence of $\operatorname{Sp} V^2$ may be complicated. However, utilizing the equality

$$\sum_{k=1}^{\infty} |(V^2 e_k, e_k)| = \sum_{k=1}^{\infty} |(V e_k, V^* e_k)| \leqslant \sum_{k=1}^{\infty} |V e_k|\, |V^* e_k| \leqslant$$
$$\leqslant \sqrt{\sum_{k=1}^{\infty} |V e_k|^2 \cdot \sum_{k=1}^{\infty} |V^* e_k|^2} = \sqrt{\operatorname{Sp} V^* V \cdot \operatorname{Sp} V V^*} = \operatorname{Sp} V^* V$$

(recall that $\operatorname{Sp} V^* V = \operatorname{Sp} V V^*$) one can formulate the condition for absolute continuity in terms of the trace (spur) of a symmetric nonnegative operator $V^* V$.

Theorem 4. *Let μ_1 and μ_2 be Gaussian measures with mean 0 and correlation operators B_1 and B_2. If bounded operator V exists satisfying the relations*

$$V B_2 = B_1 - B_2, \qquad \operatorname{Sp} V^* V < \infty,$$

and if -1 *does not belong to the spectrum of this operator then* $\mu_2 \ll \mu_1$.

Proof. It is sufficient to show that in the case when $I+V$ is invertible, i.e. $B_2B_1^{-1}$ is bounded, then $\delta_k > -1$. Let $\delta_m = -1$ for some m. Then putting $z = B_2^{1/2} e_m$ we have

$$(I+V)\, z = z + B_2^{1/2} D B_1^{-1/2} B_2^{1/2} e_m = z - B_2^{1/2} e_m = 0,$$

i.e. -1 is an eigenvalue of operator V which is impossible by the assumption of the theorem. □

We note another simple formula for the density of one Gaussian measure with respect to another in the case when the means are zero. This formula is meaningful under certain additional restrictions, but it is more convenient since it does not involve eigenvectors and eigenvalues of the operator D:

Remark. If the conditions of Theorem 4 are satisfied and $\operatorname{Sp} V$ is defined (i.e. the series $\sum (Ve_k, e_k)$ is convergent in any orthornormal basis) then the following formula

$$\frac{d\mu_2}{d\mu_1}(x) = \sqrt{\det(J+V)}\, \exp\left\{-\tfrac{1}{2}(B_1^{-1} Vx, x)\right\} \tag{4}$$

is valid. In view of the results presented in Section 6 of Chapter V, the quadratic functional $(B_1^{-1} Vx, x)$ is measurable with respect to μ since $\operatorname{Sp} V$ exists and $\operatorname{Sp} V^* V < \infty$. To prove formula (4) we note that the existence of $\operatorname{Sp} V$ implies the existence of $\operatorname{Sp} D$ and hence the convergence of the series

$$\sum_{k=1}^{\infty} \delta_k, \quad \sum_{k=1}^{\infty} \log(1+\delta_k), \quad \sum_{k=1}^{\infty} (B_2^{-1} x, e_k)^2 \frac{\delta_k}{1+\delta_k}.$$

Let P_n be the projector on the subspace spanned by $e_1, \dots, e_n$. Then

$$\begin{aligned}
\sum_{k=1}^{\infty} (B_2^{-\frac{1}{2}} x, e_k)^2 \frac{\delta_k}{1+\delta_k} &= \sum_{k=1}^{\infty} \left[(B_2^{-\frac{1}{2}} x, e_k)^2 - (B_2^{-\frac{1}{2}} x, e_k)^2 \frac{1}{1+\delta_k}\right] = \\
&= \sum_{k=1}^{\infty} \left[(B_2^{-\frac{1}{2}} x, e_k)^2 - (B_2^{-\frac{1}{2}} x, e_k)(B_2^{-\frac{1}{2}} x, B_2^{\frac{1}{2}} B_1^{-1} B_2^{\frac{1}{2}} e_k)\right] = \\
&= \sum_{k=1}^{\infty} \left[(B_2^{-\frac{1}{2}} x, e_k)^2 - (B_2^{-\frac{1}{2}} x, e_k)(B_2^{\frac{1}{2}} B_1^{-1} x, e_k)\right] = \\
&= \lim_{n\to\infty} \sum_{k=1}^{\infty} \left[(P_n B_2^{-\frac{1}{2}} x, e_k)^2 - (P_n B_2^{-\frac{1}{2}} x, e_k)(B_2^{\frac{1}{2}} B_1^{-1} x, e_k)\right] = \\
&= \lim_{n\to\infty} \left[(P_n B_2^{-\frac{1}{2}} x, B_2^{-\frac{1}{2}} x) - (P_n B_2^{-\frac{1}{2}} x, B_2^{\frac{1}{2}} B_1^{-1} x)\right] = \\
&= ((B_2^{-1} - B_1^{-1})\, x, x) = (B_1^{-1} Vx, x).
\end{aligned}$$

Furthermore,

$$\sum_{k=1}^{\infty} \log(1+\delta_k) = \log|\det(I+D)| = \log|\det(I+B_2^{\frac{1}{2}}DB_2^{-\frac{1}{2}})| = \log|\det(I+V)|.$$

Substituting the derived expressions for $\sum_{k=1}^{\infty} (B_2^{-\frac{1}{2}}x, e_k)^2 \frac{\delta_k}{1+\delta_k}$ and $\sum_{k=1}^{\infty} \log(1+\delta_k)$ in (2) we obtain formula (4).

§5. Equivalence and Orthogonality of Measures Associated with Stationary Gaussian Processes

Consider two real stationary Gaussian processes $\xi_1(t)$ and $\xi_2(t)$ on the interval $[-T, T]$. Associated with these processes are Gaussian measures μ_1 and μ_2 on the space $\mathscr{L}_2[-T, T]$ of all functions $x(t)$ which are square integrable on $[-T, T]$. It is more convenient to consider the space of complex-valued functions with the scalar product

$$(x, y) = \int_{-T}^{T} x(t)\overline{y(t)}\,dt.$$

Let $\mathsf{E}\xi_j(t) = a_j(t)$, and $R_j(t)$ be the correlation function of the process $\xi_j(t)$. Then $a_j(\cdot)$ is the mean value of the measure μ_j and its correlation operator B_j is defined by the equation

$$(B_j x, y) = \int_{-T}^{T}\int_{-T}^{T} R_j(t-s)\,x(t)\,\overline{y(s)}\,dt\,ds.$$

The purpose of this section is to study the conditions for equivalence and orthogonality of measures μ_1 and μ_2 of this special type.

Denote by $F_j(\lambda)$ the spectral function of the process $\xi_j(t)$:

$$R_j(t) = \int e^{i\lambda t}\,dF_j(\lambda).$$

Let the process $\xi_j(t)$ possess the following spectral representation

$$\xi_j(t) = a_j(t) + \int e^{i\lambda t}\,dy_j(\lambda), \tag{1}$$

where $y_j(\lambda)$ is the complex-valued Gaussian process with non-correlated increments for which

$$\mathsf{E}|y_j(\lambda_2)-y_j(\lambda_1)|^2=|F(\lambda_2)-F(\lambda_1)|.$$

Henceforth we shall make use of the space $\mathscr{W}_T$ of functions $g(\lambda)$ admitting representation

$$g(\lambda)=\int_{-T}^{T} e^{it\lambda}\varphi(t)\,dt,$$

where $\varphi(\cdot)\in\mathscr{L}_2[-T,T]$. The space $\mathscr{W}_T$ coincides with the space of the entire analytic functions of the exponential type not higher than T and square integrable on the real line. In what follows we shall consider functions belonging to $\mathscr{W}_T$ only on the real line. Denote by $\mathscr{W}_T(F_1)$ the closure of $\mathscr{W}_T$ in the metric

$$\|g\|_{F_1}^2=\int|g(\lambda)|^2\,dF_1(\lambda).$$

The space $\mathscr{W}_T(F_1)$ is the Hilbert space with the scalar product

$$(g_1,g_2)_{F_1}=\int g_1(\lambda)\overline{g_2(\lambda)}\,dF_1(\lambda).$$

First we investigate the condition of equivalence and orthogonality of measures which correspond to processes with the same $R(t)$ and different mean values.

Let $R_1(t)=R_2(t)$, $a_1(t)=0$, $a_2(t)=a(t)$.

Theorem 1. *In order that measures μ_1 and μ_2 be equivalent, it is necessary and sufficient that the function $a(t)$ admit representation*

$$a(t)=\int e^{-i\lambda t}\,b(\lambda)\,dF_1(\lambda) \tag{2}$$

for $t\in[-T,T]$ where $b(\lambda)\in W_T(F_1)$. If this condition is satisfied then

$$\frac{d\mu_2}{d\mu_1}(\xi_1(\cdot))=\exp\left\{\int b(\lambda)\,dy_1(\lambda)-\tfrac{1}{2}\int|b(\lambda)|^2\,dF_1(\lambda)\right\}; \tag{3}$$

here $y_1(\lambda)$ is the function appearing in the spectral representation of $\xi_1(t)$ as given by formula (1).

Proof. First note that $\mu_1\sim\mu_2$. As it follows from Theorem 1 of Section 4

$$\frac{d\mu_2}{d\mu_1}(x)=\exp\{l(x)-c\},$$

where $l(x)$ is a measurable linear functional in measure μ_1 and c is a constant. It was established in Section 6 of Chapter V that every measurable linear functional $l(\xi_1(\cdot))$ of a stationary Gaussian process $\xi_1(t)$ on $[-T, T]$ can be represented in the form

$$l(\xi_1(\cdot)) = \int b(\lambda)\, dy_1(\lambda),$$

where $b(\lambda) \in \mathscr{W}_T(F_1)$. To find the relation between $a(t)$, $b(\lambda)$ and the constant c we write the characteristic function of the variable $\xi_2(t)$ (for fixed t):

$$\exp\{ia(t)\, z - \tfrac{1}{2} z^2 R_1(0)\} = \mathsf{E}\, e^{iz\xi_2(t)} = \mathsf{E}\, e^{iz\xi_1(t)} \frac{d\mu_2}{d\mu_1}(\xi_1(\cdot)) =$$

$$= \mathsf{E} \exp\left\{\int [b(\lambda) + ize^{it\lambda}]\, dy_1(\lambda) - c\right\}.$$

We now utilize formula $\mathsf{E}\, e^{\xi} = \exp\{\frac{1}{2}\mathsf{E}\xi^2\}$ for $\mathsf{E}\xi = 0$, which is valid for any Gaussian variable (including a complex-valued one). Note that since $\xi(t)$ is real, it follows that $dy(\lambda) = \overline{dy(-\lambda)}$ and $dF(-\lambda) = dF(\lambda)$; therefore

$$\mathsf{E}\left\{\int [b(\lambda) + iz\, e^{it\lambda}]\, dy_1(\lambda)\right\}^2 =$$

$$= \mathsf{E} \int [b(\lambda) + iz\, e^{it\lambda}]\, dy_1(\lambda) \int [b(\lambda) + iz\, e^{it\lambda}]\, \overline{dy_1(-\lambda)} =$$

$$= \mathsf{E} \int [b(\lambda) + iz\, e^{it\lambda}]\, dy_1(\lambda) \int [b(-\lambda) + iz\, e^{-it\lambda}]\, \overline{dy_1(\lambda)} =$$

$$= \int [b(\lambda) + iz\, e^{it\lambda}]\, [b(-\lambda) + iz\, e^{-it\lambda}]\, dF_1(\lambda) =$$

$$= \int b(\lambda)\, b(-\lambda)\, dF_1(\lambda) + 2iz \int b(\lambda)\, e^{-it\lambda}\, dF(\lambda) - z^2 R_1(0).$$

Finally since $l(\xi_1(\cdot))$ is real it follows that

$$\int b(\lambda)\, dy_1(\lambda) = \int \overline{b(\lambda)}\, \overline{dy_1(\lambda)} = \int \overline{b(-\lambda)}\, dy_1(\lambda).$$

Therefore $b(\lambda) = \overline{b(-\lambda)}$, $b(-\lambda) = \overline{b(\lambda)}$ and

$$\int b(\lambda)\, b(-\lambda)\, dF_1(\lambda) = \int |b(\lambda)|^2\, dF_1(\lambda).$$

Thus

$$\exp\{iza(t)-\tfrac{1}{2}R_1(0)z^2\}=\exp\Big\{-c+\tfrac{1}{2}\int|b(\lambda)|^2\,dF_1(\lambda)+ +iz\int e^{-i\lambda t}\,b(\lambda)\,dF_1(\lambda)-\tfrac{1}{2}z^2R_1(0)\Big\}$$

and hence

$$c=\tfrac{1}{2}\int|b(\lambda)|^2\,dF_1(\lambda),\quad a(t)=\int e^{-i\lambda t}\,b(\lambda)\,dF_1(\lambda).$$

We have thus established the necessity of the conditions of the theorem and have verified formula (3).

We now proceed to the proof of the sufficiency of the conditions of the theorem.

Let formula (2) be satisfied. Introduce measure $\tilde{\mu}$ which is absolutely continuous with respect to measure μ_1 and having density $d\tilde{\mu}/d\mu_1$ which coincides with the right-hand-side of equality (3).. We show that measures μ_2 and $\tilde{\mu}$ coincide. For this purpose we compare their characteristic functionals (the characteristic functionals of measures μ and μ_i will be denoted by $\tilde{\chi}$ and χ_i respectively). Clearly,

$$\chi_2(z)=\exp\Big\{i\int_{-T}^{T}a(t)\,z(t)\,dt-\tfrac{1}{2}\int_{-T}^{T}\int_{-T}^{T}R(t-s)\,z(t)\,z(s)\,dt\,ds\Big\}.$$

Next,

$$\tilde{\chi}(z)=\mathsf{E}\exp\Big\{i\int_{-T}^{T}z(t)\,\xi_1(t)\,dt\Big\}\frac{d\tilde{\mu}}{d\mu_1}(\xi_1(\cdot))=$$

$$=\mathsf{E}\exp\Big\{\int\Big[b(\lambda)+i\int_{-T}^{T}z(t)\,e^{it\lambda}\,dt\Big]dy_1(\lambda)-$$

$$-\tfrac{1}{2}\int|b(\lambda)|^2\,dF_1(\lambda)\Big\}=\exp\Big\{-\tfrac{1}{2}\int|b(\lambda)|^2\,dF_1(\lambda)+$$

$$+\tfrac{1}{2}\mathsf{E}\Big(\int\Big[b(\lambda)+i\int_{-T}^{T}z(t)\,e^{it\lambda}\,dt\Big]dy_1(\lambda)\Big)^2\Big\}=$$

$$=\exp\left\{\frac{i}{2}\int_{-T}^{T} z(t)\int [b(\lambda)\,e^{-it\lambda}+b(-\lambda)\,e^{it\lambda}]\,dF_1(\lambda)\,dt+\right.$$

$$\left.+\frac{i^2}{2}\int\int_{-T}^{T} z(t)\,e^{it\lambda}\,dt\int_{-T}^{T} z(s)\,e^{-is\lambda}\,ds\,dF_1(\lambda)\right\}=\chi_2(z).$$

Since $\chi_2=\tilde{\chi}$, it follows that $\mu_2=\tilde{\mu}$. The theorem is thus proved. □

Corollary. *It follows from Theorem 1 that if for some T the measures* μ_1^T *and* μ_2^T*, associated with the processes* $\xi_1(t)$ *and* $\xi_1(t)+a(t)$ *on the interval* $[-T, T]$*, are equivalent then an extension of the functional* $a(t)$ *to the whole line always exists such that measures* μ_1^∞ *and* μ_2^∞ *corresponding to these processes on* $(-\infty, \infty)$ *are also equivalent. The right-hand-side of equality* (2) *defined for all values of* t *may serve as such an extension.*

Assume that the spectral function $F_1(\lambda)$ possesses spectral density $f_1(\lambda)$. Let $a^\infty(t)$ be the above stated extension of $a(t)$ for which the measures μ_1^∞ and μ_2^∞ are equivalent. If $a(\lambda)$ is the Fourier transform of the function $a^\infty(t)$, then

$$\tilde{a}(\lambda)=2\pi b(\lambda)\,f_1(\lambda).$$

Therefore in order that the measures μ_1 *and* μ_2 *be equivalent (these are the initial measures discussed in Theorem* 1*), it is necessary and sufficient that a continuation of the function* $a(t)$ *exist on* $(-\infty, \infty)$ *such that the Fourier transform of this extension* $\tilde{a}(\lambda)$ *will satisfy relation*

$$\int\frac{|\tilde{a}(\lambda)|^2}{f_1(\lambda)}\,d\lambda<\infty.$$

The function $\dfrac{\tilde{a}(\lambda)}{2\pi f_1(\lambda)}$ *can be taken in place of* $b(\lambda)$

Consider now processes $\xi_j(t)$ with the zero mean value and unequal correlation functions $R_1(t)$ and $R_2(t)$. Denote by W_T^2 the space of functions $b(\alpha, \beta)$ admitting representation in the form

$$b(\alpha, \beta)=\int_{-T}^{T}\int_{-T}^{T} e^{it\alpha-is\beta}\,\varphi(t, s)\,dt\,ds,$$

where φ is the function which is square integrable on $[-T, T]\times[-T, T]$. Denote by $\mathscr{W}_T^2(F_1)$ the closure of $\mathscr{W}_T^2$ in the metric generated by the scalar product

$$(b_1, b_2)=\iint b_1(\alpha, \beta)\, \overline{b_2(\alpha, \beta)}\, dF_1(\alpha)\, dF_1(\beta).$$

Theorem 2. *If* $\mathsf{E}\, \xi_j(t)=0, j=1, 2$, *then in order that the measures* μ_1 *and* μ_2 *be equivalent it is necessary and sufficient that a function* $b(\alpha, \beta)\in W_T^2(F_1)$ *exist such that the representation*

$$R_2(t-s)-R_1(t-s)=\iint e^{-i\alpha t+i\beta s}\, b(\alpha, \beta)\, dF_1(\alpha)\, dF_2(\beta) \tag{4}$$

be valid. Moreover

$$\frac{d\mu_2}{d\mu_1}(\xi_1(\cdot))=\exp\left\{\iint \Phi(\alpha, \beta)\, dy_1(\alpha)\, \overline{dy_1(\beta)}+c\right\}, \tag{5}$$

where the function $\Phi(\alpha, \beta)$ *is connected with* $b(\alpha, \beta)$ *by relation*

$$\int \Phi(\alpha, \beta)\, \overline{b(\beta, \gamma)}\, dF_1(\beta)=b(\alpha, \gamma)-\Phi(\alpha, \gamma), \tag{6}$$

$$c=-\ln \mathsf{E} \exp\left\{\iint \Phi(\alpha, \beta)\, dy_1(\alpha)\, \overline{dy_1(\beta)}\right\}. \tag{7}$$

Proof. Necessity. Assume that $\mu_1 \sim \mu_2$. Then the spaces of linear measurable functionals in measures μ_1 and μ_2 coincide: i.e. $\mathscr{L}(\mu_1)=\mathscr{L}(\mu_2)$ (see Section 6, Chapter V, re linear measurable functionals). As has already been mentioned, for a given stationary Gaussian process $\xi_j(t)$ with $\mathsf{E}\xi_j(t)=0$ every measurable linear functional $l(\xi_j)$ can be represented in the form

$$l(\xi_j)=\int g(\alpha)\, dy_j(\alpha),$$

where $g\in \mathscr{W}_T(F_j)$. In the course of the proof of Theorem 2 in Section 4 a sequence of measurable functionals was constructed which forms a complete orthogonal system in $\mathscr{L}(\mu_1)$ and $\mathscr{L}(\mu_2)$ simultaneously (these are the functionals $(B_2^{-\frac{1}{2}}x, e_k)=l_k(x)$ where e_k are the eigenvectors of the operator D).

Let

$$l_k(\xi_j)=\int g_k(\alpha)\, dy_j(\alpha).$$

It follows from the orthogonality of l_k with respect to measures μ_1 and

μ_2 that

$$0=\mathsf{E}\int g_k(\alpha)\,dy_j(\alpha)\overline{\int g_m(\alpha)\,dy_j(\alpha)}=\int g_k(\alpha)\,g_m(\alpha)\,dF_j(\alpha),\qquad k\neq m.$$

We normalize g_k so that

$$\int |g_k(\alpha)|^2\,dF_1(\alpha)=1,$$

and moreover, let

$$\int |g_k(\alpha)|^2\,dF_2(\alpha)=1+c_k.$$

It follows from Theorem 2 of Section 4 that $\sum_{k=1}^{\infty} c_k^2<\infty$. We set

$$b(\alpha,\beta)=\sum_{k=1}^{\infty} c_k\,\overline{g_k(\alpha)}\,g_k(\beta)$$

and show that $b(\alpha,\beta)$ satisfies relation (4). Consider the function

$$\psi(t,s)=\iint e^{-i\alpha t+i\beta s}\,b(\alpha,\beta)\,dF_1(\alpha)\,dF_1(\beta)+R_1(t-s)-R_2(t-s).$$

If $z(\alpha)=\int_{-T}^{T} e^{-i\alpha t}\,\varphi(t)\,dt$, then

$$\int_{-T}^{T}\int_{-T}^{T}\psi(t,s)\,\varphi(t)\,\overline{\varphi(s)}\,dt\,ds=$$

$$=\iint z(\alpha)\,\overline{z(\beta)}\,b(\alpha,\beta)\,dF_1(\alpha)\,dF_1(\beta)+$$

$$+\int |z(\alpha)|^2\,(dF_1(\alpha)-dF_2(\alpha))=$$

$$=\sum_{k=1}^{\infty} c_k\left|\int z(\alpha)\,\overline{g_k(\alpha)}\,dF_1(\alpha)\right|^2+\sum_{k=1}^{\infty}\left|\int z(\alpha)\,\overline{g_k(\alpha)}\,dF_1(\alpha)\right|^2-$$

$$-\int\sum_{k=1}^{\infty}\left|g_k(\alpha)\int\overline{g_k(\beta)}\,z(\beta)\,dF_1(\beta)\right|^2 dF_2(\alpha)=$$

$$=\sum_{k=1}^{\infty}\left(1+c_k-\int |g_k(\alpha)|^2\,dF_2(\alpha)\right)\left|\int z(\alpha)\,\overline{g_k(\alpha)}\,dF_1(\alpha)\right|^2=0.$$

Utilizing the equality $\psi(t, s)=\overline{\psi(s, t)}$ we verify that $\psi(t, s)=0$. The necessity part of the theorem is thus proved.

Sufficiency. We now proceed to prove the sufficiency of the conditions of the theorem and the derivation of formula (5). Assume that a function $b(\cdot, \cdot) \in W_1^2(F_1)$ exists which satisfies relation (4). Consider the integral operator

$$Vg(\beta)=\int b(\alpha, \beta)\, g(\alpha)\, dF_1(\alpha).$$

If $b(\cdot, \cdot) \in \mathscr{W}_T(F_1)$, this operator maps $\mathscr{W}(F_1)$ into $\mathscr{W}(F_1)$. This is easy to verify by noting that

$$Vg \in \mathscr{W}_T, \quad \text{if} \quad b(\cdot, \cdot) \in \mathscr{W}_T^2$$

for any bounded function g and that

$$\|V\|=\iint |b(\alpha, \beta)|^2\, dF_1(\alpha)\, dF_1(\beta).$$

Thus the operator V is a bounded self-adjoined operator on $\mathscr{W}_T(F_1)$. Being an integral operator with a square integral kernel also, it is therefore a completely continuous operator and a Hilbert-Schmidt operator. Denote by $g_k(\alpha)$ the complete orthonormal sequence of eigenfunctions of the operator V and by λ_k the corresponding eigenvalues. Then

$$b(\alpha, \beta)=\sum_{k=1}^{\infty} \lambda_k\, \overline{g_k(\alpha)}\, g_k(\beta), \qquad \sum_{k=1}^{\infty} \lambda_k^2<\infty.$$

By construction the functions $g_k(\alpha)$ are orthogonal in $\mathscr{W}_T(F_1)$. We now show that they are also orthogonal in $\mathscr{W}_T(F_2)$. Let $\varphi_k^n(t)$ be a sequence of functions in $\mathscr{L}_2[-T, T]$ such that

$$\int_{-T}^{T} e^{-i\lambda t}\, \varphi_k^n(t)\, dt \to g_k(\lambda)$$

as $n\to\infty$ in the sense of convergence in $\mathscr{W}_T(F_1)$. Then

$$\int_{-T}^{T}\int_{-T}^{T} R_2(t-s)\, \varphi_k^n(t)\, \overline{\varphi_j^n(s)}\, dt\, ds-$$

$$-\int_{-T}^{T}\int_{-T}^{T} R_1(t-s)\, \varphi_k^n(t)\, \overline{\varphi_j^n(s)}\, dt\, ds=$$

$$=\iint b(\alpha,\beta)\int_{-T}^{T}\varphi_j^n(t)\,e^{-i\alpha t}\,dt\int_{-T}^{T}\overline{\varphi_k^n(s)\,e^{-i\beta s}}\,ds\,dF_1(\alpha)\,dF_1(\beta).$$

Approaching the limit in this equality as $n\to\infty$ we obtain:

$$\int g_k(\alpha)\,\overline{g_j(\alpha)}\,dF_2(\alpha)-\int g_k(\alpha)\,\overline{g_j(\alpha)}\,dF_1(\alpha)=$$
$$=\iint b(\alpha,\beta)\,g_j(\alpha)\,\overline{g_k(\beta)}\,dF_1(\alpha)\,dF_1(\beta).$$

Since for $k\neq j$

$$\iint b(\alpha,\beta)\,g_j(\alpha)\,\overline{g_k(\beta)}\,dF_1(\alpha)\,dF_1(\beta)=0,$$

it follows that the following equality

$$\int g_k(\alpha)\,\overline{g_j(\alpha)}\,dF_2(\alpha)=\int g_k(\alpha)\,\overline{g_j(\alpha)}\,dF_1(\alpha)=0$$

is valid for $k\neq j$.

Thus the sequence $g_k(\alpha)$ is orthogonal in the space $\mathscr{W}_T(F_1)$ as well as in the space $\mathscr{W}_T(F_2)$. It follows from formula (4) that $b(\alpha,\beta)$ can be chosen in such a manner that $b(\alpha,\beta)=b(-\alpha,-\beta)$; in this case we may take $g_k(\alpha)$ satisfying $g_k(\alpha)=\overline{g_k(-\alpha)}$. Under this condition, the functionals $\int g_k(\alpha)\,dy_j(\alpha)$ are real linear functionals on the processes $\xi_j(t)$. In view of the remark after Theorem 2 in Section 4 we have $\mu_1\sim\mu_2$ and moreover

$$\frac{d\mu_2}{d\mu_1}(\xi_1(\cdot))=\exp\left\{\frac{1}{2}\sum_{k=1}^{\infty}\left[\frac{\lambda_k}{1+\lambda_k}\left|\int g_k(\alpha)\,dy_1(\alpha)\right|^2-\ln(1+\lambda_k)\right]\right\}.$$

Next note that in view of formula (15) in Section 6 of Chapter V the following equality

$$\sum_{k=1}^{\infty}\frac{\lambda_k}{1+\lambda_k}\left[\left|\int g_k(\alpha)\,dy_1(\alpha)\right|^2-1\right]=\iint\Phi(\alpha,\beta)\,dy_1(\alpha)\,\overline{dy_1(\beta)}$$

is valid, where

$$\Phi(\alpha,\beta)=\sum_{k=1}^{\infty}\frac{\lambda_k}{1+\lambda_k}\,\overline{g_k(\alpha)}\,g_k(\beta).$$

To complete the proof, one need merely note that

$$\int \Phi(\alpha, \gamma)\, \overline{b(\gamma, \beta)}\, dF_1(\gamma) = \sum_{k=1}^{\infty} \frac{\lambda_k^2}{1+\lambda_k}\, \overline{g_k(\alpha)}\, g_k(\beta) = b(\alpha, \beta) - \Phi(\alpha, \beta).$$

The theorem is thus proved. □

Assuming the existence of spectral densities $f_j(\lambda) = \frac{d}{d\lambda} F_j(\lambda)$ $(j=1, 2)$, we now present certain sufficient conditions for the equivalence of measures. For this purpose an auxiliary result on orthogonal bases in $\mathscr{W}_T(F_1)$ – for a special choice of F_1 – is required.

Lemma. *Let* $f_1(\lambda) = |\varphi_0(\lambda)|^2$ *where* $\varphi_0 \in \mathscr{W}_s$ *and* $g_k(\lambda)$ *is an arbitrary orthonormal basis in* $\mathscr{W}_T(F_1)$. *Then*

$$\sum_{k=1}^{\infty} |g_k(\lambda)|^2 \leqslant \frac{T+s}{\pi f_1(\lambda)}. \tag{8}$$

Proof. Since $\mathscr{W}_T$ is everywhere dense in $\mathscr{W}_T(F_1)$ it is sufficient to verify inequality (8) for the case when $g_k(\lambda) \in \mathscr{W}_T$. Under this assumption $g_k(\lambda)\, \varphi_0(\lambda) \in \mathscr{W}_{T-s}$ and hence

$$g_k(\lambda)\, \varphi_0(\lambda) = \int_{-T-s}^{T+s} e^{-i\lambda t}\, \psi_k(t)\, dt,$$

where $\psi_k(t) \in \mathscr{L}_2[-T-s, T+s]$. Since

$$\int_{-\infty}^{\infty} g_k(\lambda)\, \varphi_0(\lambda)\, \overline{g_j(\lambda)\, \varphi_0(\lambda)}\, d\lambda = \int_{-\infty}^{\infty} g_k(\lambda)\, \overline{g_j(\lambda)}\, f_1(\lambda)\, d\lambda,$$

it follows from Parseval's theorem that

$$\int_{-T-s}^{T+s} \psi_k(t)\, \overline{\psi_j(t)}\, dt = \frac{1}{2\pi}\, \delta_{kj}.$$

Hence $\sqrt{2\pi}\, \psi_k(t)$ form an orthonormal system of functions in $\mathscr{L}_2[-T-s, T+s]$. Therefore it follows from Bessel's inequality that

$$2\pi \sum_{k=1}^{\infty} \left| \int_{-T-s}^{T+s} e^{-i\lambda t}\, \psi_k(t)\, dt \right|^2 \leqslant \int_{-T-s}^{T+s} |e^{-i\lambda t}|^2\, dt = 2T + 2s,$$

or

$$2\pi \sum_{k=1}^{\infty} |g_k(\lambda)\,\varphi_0(\lambda)|^2 \leqslant 2(T+s).$$

The lemma is thus proved. □

Theorem 3. *Let μ_1 and μ_2 be measures associated with stationary Gaussian processes $\xi_j(t)$, $\mathsf{E}\xi_j(t)$, possessing spectral densities $f_j(\lambda)$ $(j=1,2)$. If a function $\varphi_0(\lambda)\in\mathscr{W}_s$ exists and constants c_1 and c_2 exist such that the inequality*

$$c_1|\varphi_0(\lambda)|^2 \leqslant f_1(\lambda) \leqslant c_2|\varphi_0(\lambda)|^2 \tag{9}$$

is satisfied and if moreover

$$\int \left[\frac{f_2(\lambda)-f_1(\lambda)}{f_1(\lambda)}\right]^2 d\lambda < \infty,$$

then the measures μ_1 and μ_2 are equivalent for any T.

Proof. Set

$$\begin{aligned}
\tilde{f}_1(\lambda) &= c_1|\varphi_0(\lambda)|^2,\\
\tilde{f}_2(\lambda) &= \begin{cases} \tilde{f}_1(\lambda); & f_2(\lambda)>f_1(\lambda),\\ f_1(\lambda); & f_2(\lambda)\leqslant f_1(\lambda),\end{cases}\\
\tilde{f}_3(\lambda) &= \begin{cases} \tilde{f}_1(\lambda)+f_2(\lambda)-f_1(\lambda); & f_2(\lambda)>f_1(\lambda),\\ f_2(\lambda); & f_2(\lambda)\leqslant f_1(\lambda),\end{cases}\\
\tilde{f}_4(\lambda) &= \begin{cases} f_1(\lambda)-\tilde{f}_1(\lambda); & f_2(\lambda)>f_1(\lambda);\\ 0; & f_2(\lambda)\leqslant f_1(\lambda).\end{cases}
\end{aligned}$$

Denote by $\tilde{\mu}_j$, $j=1,2,3,4$, the measures which are associated with Gaussian stationary processes on $[-T,T]$ possessing the spectral densities $\tilde{f}_j(\lambda)$ respectively. Since $\mu_j=\tilde{\mu}_{j+1}*\tilde{\mu}_4$, $j=1,2$ (the spectral density of a sum of independent processes is equal to the sum of spectral densities of the summands) to prove the equivalence of measures μ_1 and μ_2 it is sufficient to show that $\tilde{\mu}_2\sim\tilde{\mu}_3$ or that $\tilde{\mu}_j\sim\tilde{\mu}_1$, $j=2,3$. The proof of the last assertion is the same for both cases $j=2$ and $j=3$. Denote by $\tilde{F}_j(\lambda)$ the spectral function having the spectral density $\tilde{f}_j(\lambda)$. Let $\{g_k(\lambda)\}$ be an arbitrary orthonormal basis in $\mathscr{W}_T(F_1)$. Setting $\dfrac{\tilde{f}_j(\lambda)-\tilde{f}_1(\lambda)}{\tilde{f}_1(\lambda)}=h(\lambda)$, we obtain, in view of the lemma that

$$\sum_{k=1}^{\infty}\left[\int |g_k(\lambda)|^2\, d\tilde{F}_j(\lambda) - \int |g_k(\lambda)|^2\, d\tilde{F}_1(\lambda)\right]^2 =$$

$$=\sum_{k=1}^{\infty}\left[\int |g_k(\lambda)|^2 h(\lambda)\tilde{f}_1(\lambda)\,d\lambda\right]^2 \leqslant$$

$$\leqslant \sum_{k=1}^{\infty}\int |g_k(\lambda)|^2 h^2(\lambda)\tilde{f}_1(\lambda)\,d\lambda\cdot\int |g_k(\lambda)|^2 \tilde{f}_1(\lambda)\,d\lambda=$$

$$=\int\sum_{k=1}^{\infty}|g_k(\lambda)|^2 h^2(\lambda)\tilde{f}_1(\lambda)\,d\lambda\leqslant\frac{T+s}{\pi}\int h^2(\lambda)\,d\lambda\leqslant$$

$$\leqslant\frac{T+s}{\pi}\left(\frac{c_2}{c_1}\right)^2\int\left[\frac{f_2(\lambda)-f_1(\lambda)}{f_1(\lambda)}\right]^2 d\lambda\leqslant c,$$

$\left(\text{since } |h(\lambda)|\leqslant\frac{c_2}{c_1}\left|\frac{f_2(\lambda)-f_1(\lambda)}{f_1(\lambda)}\right|\right)$. Let V be a symmetric operator in $\mathscr{W}_T(\tilde{F}_1)$ such that

$$(Vg, g)=\int |g(\lambda)|^2\,d\tilde{F}_j(\lambda).$$

We have shown that for any orthonormal basis $\{g_k(\lambda)\}$ in $\mathscr{W}_T(\tilde{F}_1)$ the relation

$$\sum_{k=1}^{\infty}([V-I]\,g_k, g_k)^2\leqslant c$$

is satisfied. This relation yields that $V-I$ is a Hilbert-Schmidt operator.

Let g_k be a sequence of eigenfunctions of operator $V-I$ and let α_k be the corresponding eigenvalues. Then the function

$$b(\alpha,\beta)=\sum_{k=1}^{\infty}a_k g_k(\alpha)\,g_k(\beta)$$

is defined and belongs to $\mathscr{W}_T^2(\tilde{F}_1)$ in view of the fact that $\sum_{k=1}^{\infty}\alpha_k^2<\infty$.

Denote by $\tilde{R}_k$ the correlation function of the process with the spectral density $\tilde{f}_k$ and denote by $\psi_t(\lambda)$ the function $e^{i\lambda t}$, $|t|\leqslant T$ (this function belongs to $\mathscr{W}_T(\tilde{F}_1)$). Then

$$\tilde{R}_j(t-s)-\tilde{R}_1(t-s)=\int e^{i\lambda(t-s)}\,(d\tilde{F}_j(\lambda)-d\tilde{F}_1(\lambda))=([V-I]\,\psi_t,\psi_s)=$$

$$=\sum_{k=1}^{\infty}([V-I]\,\psi_t, g_k)\,(g_k,\psi_s)=\sum_{k=1}^{\infty}\alpha_k(\psi_t, g_k)\,(g_k,\psi_s)=$$

$$=\sum_{k=1}^{\infty}\alpha_k\int e^{it\alpha}\,\overline{g_k(\alpha)}\,d\tilde{F}_1(\alpha)\int e^{-is\beta}\,g_k(\beta)\,d\tilde{F}_1(\beta)=$$

$$=\int\int e^{it\alpha-is\beta}\, b(\alpha,\beta)\, d\tilde{F}_1(\alpha)\, d\tilde{F}_1(\beta).$$

To complete the proof one need merely utilize Theorem 2. □

Remark. Inequality (9) may be violated on a set Δ of finite measure such that

$$\int_\Delta \left[\frac{f_k(\lambda)}{|\varphi_0(\lambda)|^2}\right]^2 d\lambda<\infty, \qquad k=1,2.$$

Indeed in this case a measure μ_1^* can be introduced which corresponds to a stationary Gaussian process with the spectral density $f_1^*(\lambda)$ which already satisfies inequality (9). This spectral density is defined by

$$f_1^*(\lambda)=\begin{cases} f_1(\lambda), & \lambda\notin\Delta,\\ c_1|\varphi_0(\lambda)|^2, & \lambda\in\Delta.\end{cases}$$

In this case

$$\int\left[\frac{f_1(\lambda)-f_1^*(\lambda)}{f_1^*(\lambda)}\right]^2 d\lambda<\infty, \qquad \int\left[\frac{f_2(\lambda)-f_1^*(\lambda)}{f_1^*(\lambda)}\right]^2 d\lambda<\infty,$$

so that $\mu_1\sim\mu_1^*$, $\mu_2\sim\mu_1^*$ and $\mu_1\sim\mu_2$.

We now derive sufficient conditions for orthogonality of measures μ_1 and μ_2 under the assumption that $f_2(\lambda)\geqslant f_1(\lambda)$. First consider the case where $f_1(\lambda)=\frac{1}{1+\lambda^2}$. Let the measures μ_1 and μ_2 be equivalent. In view of Theorem 2, a function $b(\alpha,\beta)$ exists such that

$$R_2(t-s)-R_1(t-s)=\int\int e^{-i\alpha t+i\beta s}\, b(\alpha,\beta)\frac{d\alpha}{1+\alpha^2}\frac{d\beta}{1+\beta^2}$$

and

$$\int\int |b(\alpha,\beta)|^2\frac{d\alpha}{1+\alpha}\frac{d\beta}{1+\beta^2}<\infty.$$

Since the function

$$b(\alpha,\beta)\frac{\alpha}{1+\alpha^2}\frac{\beta}{1+\beta^2}$$

is square integrable, the derivative

$$\frac{\partial^2}{\partial t\,\partial s}[R_2(t-s)-R_1(t-s)]$$

exists and moreover,

$$\frac{\partial^2}{\partial t\,\partial s}[R_2(t-s)-R_1(t-s)]=\int\int e^{-i\alpha t+i\beta s}\frac{\alpha}{1+\alpha^2}\frac{\beta}{1+\beta^2}b(\alpha,\beta)\,d\alpha\,d\beta.$$

Setting $R(t)=R_2(t)-R_1(t)$ we obtain that

$$\int_{-T}^{T}\int_{-T}^{T}[R''(t-s)]^2\,dt\,ds=\tfrac{1}{4}\int_{-2T}^{2T}[R''(t)]^2\,|2T-t|\,dt<\infty.$$

Utilizing relation

$$\int_{-2T}^{2T}[R''(t)]^2\,|2T-t|\,dt=$$

$$=\int_{-\infty}^{\infty}\int_{-\infty}^{\infty}\frac{\sin^2 T(\alpha-\beta)}{(\alpha-\beta)^2}\alpha^2[f_2(\alpha)-f_1(\alpha)]\,\beta^2[f_2(\beta)-f_1(\beta)]\,d\alpha\,d\beta,$$

and also equality $\frac{1}{f_1(\lambda)}=1+\lambda^2$ we find that in the case when $f_1(\lambda)=\frac{1}{1+\lambda^2}$ the equivalence of measures yields the inequality

$$\int_{-\infty}^{\infty}\int_{-\infty}^{\infty}\frac{\sin^2 T(\alpha-\beta)}{(\alpha-\beta)^2}\frac{f_2(\alpha)-f_1(\alpha)}{f_1(\alpha)}\frac{f_2(\beta)-f_1(\beta)}{f_1(\beta)}\,d\alpha\,d\beta<\infty.$$

Hence if for some $T>0$

$$\int_{-\infty}^{\infty}\int_{-\infty}^{\infty}\frac{\sin^2 T(\alpha-\beta)}{(\alpha-\beta)^2}\frac{f_2(\alpha)-f_1(\alpha)}{f_1(\alpha)}\frac{f_2(\beta)-f_1(\beta)}{f_1(\beta)}\,d\alpha\,d\beta=+\infty,$$

then the measures μ_1 and μ_2 associated with stationary processes on $[-T,T]$ will be orthogonal $\left(\text{provided } f_1(\lambda)=\frac{1}{1+\lambda^2}\right)$.

Note that all the previous arguments remain valid if instead of the requirement $f_1(\lambda)=\frac{1}{1+\lambda^2}$, the inequality

$$\frac{c_1}{1+\lambda^2}\leqslant f_1(\lambda)\leqslant\frac{c_2}{1+\lambda^2}$$

is satisfied for some c_1 and c_2.

Now let $\varphi_0(\lambda)$ be an entire analytic function of exponential type at most s, such that

$$\int_{-\infty}^{\infty} \frac{d\lambda}{(1+\lambda^2)\,|\varphi_0(\lambda)|^2} < \infty .$$

Let the spectral density $f_1(\lambda)$ satisfy the inequality

$$\frac{c_1}{(1+\lambda^2)\,|\varphi_0(\lambda)|^2} \leqslant f_1(\lambda) \leqslant \frac{c_2}{(1+\lambda^2)\,|\varphi_0(\lambda)|^2}$$

for some c_1 and c_2.

Consider the processes

$$\tilde{\xi}_j(t) = \int_{-\infty}^{\infty} e^{i\lambda t}\, \varphi_0(\lambda)\, dy_j(\lambda),$$

where $y_j(\lambda)$ is defined in representation (1) (p. 499). It is easy to verify that $\varphi_0 \in \mathscr{W}_s(F_1)$. Therefore taking a sequence $\psi_n(\lambda) = \int_{-s}^{s} h_n(u)\, e^{i\lambda u}\, du$, convergent to $\varphi_0(\lambda)$ in $\mathscr{W}_s(F_1)$ we obtain

$$\tilde{\xi}_j(t) = \lim \int e^{i\lambda t} \int_{-s}^{s} e^{i\lambda u}\, h_n(u)\, du\, dy_j(\lambda) = \lim \int_{-s}^{s} \xi_j(t+u)\, h_n(u)\, du .$$

Thus the values of the process $\xi_j(t)$ on $[-T-s, T+s]$ determine the process $\tilde{\xi}_j(t)$ on $[-T, T]$. The spectral density of the process $\tilde{\xi}_j(t)$ is equal to $\tilde{f}_j(\lambda) = f_j(\lambda)\,|\varphi_0(\lambda)|^2$, so that $\frac{c_1}{1+\lambda^2} \leqslant \tilde{f}_1(\lambda) \leqslant \frac{c_2}{1+\lambda^2}$.

Since $\frac{f_2 - f_1}{f_1} = \frac{\tilde{f}_2 - \tilde{f}_1}{\tilde{f}_1}$, it follows from the above that the measures $\tilde{\mu}_1$ and $\tilde{\mu}_2$ are orthogonal provided

$$\iint \frac{\sin^2 T(\alpha-\beta)}{(\alpha-\beta)^2}\, \frac{f_2(\alpha) - f_1(\alpha)}{f_1(\alpha)}\, \frac{f_2(\beta) - f_1(\beta)}{f_1(\beta)}\, d\alpha\, d\beta = +\infty . \tag{10}$$

However in this case the measures μ_1 and μ_2* will also be orthogonal.

* We assume that μ_j are associated with processes $\xi_j(t)$ on the interval $[-T-s, T+s]$.

Finally note that the function $(1+i\lambda)\,\varphi_0(\lambda)$ is also an entire function of exponential type at most s. We have thus proved the following

Theorem 4. *If μ_1 and μ_2 are measures associated with stationary Gaussian processes on $[-T, T]$ with spectral densities $f_j(\lambda)$, $j=1, 2$ and then mean values 0 and an entire analytic function, of exponential type at most $s<T$, $\varphi_0(\lambda)$ exists such that for some $c_1>0$ and $c_2>0$ the inequality $c_1\leqslant|\varphi_0(\lambda)|^2 f_1(\lambda)\leqslant c_2$ is satisfied, then the relation*

$$\int\int \frac{\sin^2(T-s)(\alpha-\beta)}{(\alpha-\beta)^2}\,\frac{f_2(\alpha)-f_1(\alpha)}{f_1(\alpha)}\,\frac{f_2(\beta)-f_1(\beta)}{f_1(\beta)}\,d\alpha\,d\beta=+\infty$$

implies orthogonality of measures μ_1 and μ_2.

Remark. The function

$$\varphi_0(\lambda)=\int(|\theta|^\alpha+1)\,\frac{\sin^m\varepsilon(\theta-\lambda)}{(\theta-\lambda)^m}\,d\theta,$$

where $\alpha>0$ is arbitrary and $m>\alpha+1$, satisfies

$$0<\inf_\lambda\left(|\varphi_0(\lambda)|^2\,\frac{1}{1+|\lambda|^{2\alpha}}\right)<\sup_\lambda\left(|\varphi_0(\lambda)|^2\,\frac{1}{1+|\lambda|^{2\alpha}}\right)<\infty.$$

The function $\varphi_0(\lambda)$ is entire and of the exponential type not exceeding $m\varepsilon$. Functions of this kind may be utilized for verifying the conditions of Theorem 3 and 4.

Corollary. *If $f_1(\lambda)$ and $f_2(\lambda)$ are bi-linear (rational) functions then the condition*

$$\lim_{\lambda\to\infty}\frac{f_2(\lambda)}{f_1(\lambda)}=1$$

is necessary and sufficient for the equivalence of measures μ_1 and μ_2.

Proof. If $f_1(\lambda)>0$ then the corresponding condition on $f_1(\lambda)$ stated in Theorem 3 as well as in Theorem 4, is satisfied with functions $\varphi_0(\lambda)$ of the form stated in the previous remark. If

$$\lim_{\lambda\to\infty}\frac{f_2(\lambda)}{f_1(\lambda)}=1, \quad \text{then} \quad \frac{f_2(\lambda)-f_1(\lambda)}{f_1(\lambda)}=O(\lambda^{-1})$$

and one can utilize Theorem 3. If this condition is violated, one can apply Theorem 4. If $f_1(\lambda)$ vanishes one may use $f_1^*(\lambda)$ (in place of $f_1(\lambda)$) such that $f_1^*(\lambda)>0$ and $\lim_{\lambda\to\infty} f_1(\lambda)/f_1^*(\lambda)=1$.

§6. General Properties of Densities of Measures Associated with Markov Processes

Let two random procecesses $\xi_1(t)$ and $\xi_2(t)$ with the values in a certain space $\mathscr{X}$ with a σ-algebra of measurable sets $\mathfrak{A}$ be defined on a certain numerical set T. Denote by μ_i^τ the measures associated with random processes $\xi_i(t)$ defined on the σ-algebra $\mathfrak{F}_\tau$ generated by the cylinders over $(-\infty, \tau)\cap T$. The measures μ_i^∞ will be denoted by μ_i. Assume that $\mu_2 \ll \mu_1$ and $\varrho(\xi_1(\cdot))=\frac{d\mu_2}{d\mu_1}(\xi_1(\cdot))$. Then we also have $\mu_2^\tau \ll \mu_1^\tau$ for all τ and moreover

$$\varrho_\tau(\xi_1(\cdot))=\frac{d\mu_2^\tau}{d\mu_1^\tau}(\xi_1(\cdot))=\mathsf{E}(\varrho(\xi_1(\cdot))\mid \mathfrak{F}_\tau),$$

where the conditional mathematical expectation is taken with respect to the probability space $\{\mathscr{F}_T(\mathscr{X}), \mathfrak{F}_\infty, \mu_1\}$ where $\mathscr{F}_T(\mathscr{X})$ is the space of all functions on T with values in $\mathscr{X}$. It is easily seen that the process $\{\varrho_\tau, \mathfrak{F}_\tau\}$ is a martingale satisfying $\mathsf{E}\varrho_\tau=1$. On the other hand, any nonnegative martingale $\{\varrho_\tau, \mathfrak{F}_\tau\}$ satisfying $\mathsf{E}\varrho_\tau=1$ can serve as the density μ_2^τ with respect to μ_1^τ for some pair of processes $\xi_1(\cdot)$ and $\xi_2(\cdot)$. This result which is valid for any process is of little interest. More interesting results are obtained if it is assumed that $\xi_1(t)$ and $\xi_2(t)$ are processes belonging to a certain more restricted class of processes. In this section the case when both processes are Markovian is discussed.

Let $\xi_1(t)$ and $\xi_2(t)$ be Markov processes defined on the interval $[a, b]$ and taking on values in a separable metric space $(\mathscr{X}, \mathfrak{A})$ ($\mathfrak{A}$ is a σ-algebra of Borel sets). Denote by $\mathscr{F}_{[\alpha,\beta]}$ the space of all functions with values in $\mathscr{X}$ defined on $[\alpha, \beta]$ and by $\mathfrak{F}_{[\alpha,\beta]}$ the σ-algebra of the subsets of $\mathscr{F}_{[\alpha,\beta]}$ generated by the cylindrical sets. Let $\mu^i_{x,[\alpha,\beta]}(A)$ be the measure on $\mathfrak{F}_{[\alpha,\beta]}$ constructed from the transition probabilities of the process $\xi_i(t)$ given $\xi_i(\alpha)=x$. Since the process is Markovian it follows that for cylindrical sets A of the form $A=A_1\cap A_2\cap A_3$ where A_1 and A_3 are cylinders in $\mathfrak{F}_{[a,c]}$ and $\mathfrak{F}_{[c,b]}$ respectively and A_2 is a cylinder in $\mathfrak{F}_{[c]}$ (i.e. a set of the form $\{x(\cdot):x(c)\in E\}$) the following formula

$$\mu^i_{[a,b]}(A)=\int_{A_2}\mu^i_{[a,c]}(A_1;dy)\,\mu^i_{y,[c,b]}(A_3), \tag{1}$$

is valid. Here $\mu^i_{[a,c]}$ is the measure corresponding to the Markov process $\xi_i(t)$ on $[a, c]$ (it is a measure on $\mathscr{F}_{[a,c]}$) and $\mu^i_{[a,c]}(A_1;A_2)$ is the measure defined by the equality

$$\mu^i_{[a,c]}(A_1;A_2)=\mu^i_{[a,c]}(A_1\cap A_2)$$

(it is a measure with respect to A_2 on $\mathscr{F}_{[c]}$). Note that $A_1 \cap A_2 \cap A_3$ represents a cylindrical set in $\mathscr{F}_{[a,b]}$ containing all the functions $x(\cdot)$ whose restrictions to $[a, c]$, $[c, c]$ and $[c, b]$ belong to A_1, A_2 and A_3 respectively. The set $A_1 \cap A_2$ is defined analogously. We now establish an auxiliary result. Recall that the σ-algebra $\mathfrak{C}$ is called separable if a sequence of sets $A_1, A_2, \ldots$ exists such that $\mathfrak{C}$ coincides with the minimal σ-algebra containing all the sets A_k.

Lemma. *Let $(\mathscr{X}, \mathfrak{B})$ and $(\mathscr{Y}, \mathfrak{C})$ be two measurable spaces, let μ_1 and μ_2 be two probability measures on $\mathscr{X}$ and let the probability measures $\nu_1(x, C)$ and $\nu_2(x, C)$ be defined on $\mathfrak{C}$ for each $x \in \mathscr{X}$ such that $\nu_k(x, C)$ is $\mathfrak{B}$-measurable for all $c \in \mathfrak{C}$. Define on $\mathfrak{B} \times \mathfrak{C}$ the measures π_k by the equality*

$$\pi_k(B \times C) = \int_B \mu_k(dx)\, \nu_k(x, C), \qquad B \in \mathfrak{B},\ C \in \mathfrak{C}.$$

If $\pi_2 \ll \pi_1$ and if a separable σ-algebra $\mathfrak{C}_0$ exists whose completion with respect to measure $\nu_1(x, C)$ contains $\mathfrak{C}$ μ_1-almost for all x, then $\mu_2 \ll \mu_1$ and $\nu_2(x, \cdot) \ll \nu_1(x, \cdot)$ μ_2-almost for all x.

Proof. Set

$$\varrho(x, y) = \frac{d\pi_2}{d\pi_1}(x, y).$$

Then for all $B \in \mathfrak{B}$ and $C \in \mathfrak{C}$ we have

$$\pi_2(B \times C) = \int_B \left[\int_C \varrho(x, y)\, \nu_1(x, dy)\right] \mu_1(dx) =$$

$$= \int_B \int_C \frac{\varrho(x, y)}{\int_{\mathscr{Y}} \varrho(x, y')\, \nu_1(x, dy')}\, \nu_1(x, dy) \left[\int_{\mathscr{Y}} \varrho(x, y')\, \nu_1(x, dy')\right] \mu_1(dx). \tag{2}$$

Taking $C = \mathscr{Y}$, we obtain that

$$\frac{d\mu_2}{d\mu_1}(x) = \int_{\mathscr{Y}} \varrho(x, y)\, \nu_1(x, dy). \tag{3}$$

Utilizing (2) and (3) we have

$$\pi_2(B \times C) = \int_B \nu_2(x, C)\, \mu_2(dx) = \int_B \int_C \tilde{\varrho}(x, y)\, \nu_1(x, dy)\, \mu_2(dx),$$

where

$$\tilde{\varrho}(x, y) = \varrho(x, y)\left[\int_{\mathscr{Y}} \varrho(x, y')\, \nu_1(x, dy')\right]^{-1} \tag{4}$$

Hence, μ_2-almost for all x the relation

$$\nu_2(x, C) = \int_C \tilde{\varrho}(x, y)\, \nu_1(x, dy) \tag{5}$$

is satisfied for each C.

Let C_k be a sequence of sets generating $\mathfrak{C}_0$. Then one can find a set $B^* \subset \mathfrak{B}$ such that $\mu_2(B^*) = 1$ and such that for all $x \in B^*$ and all C_k the relation

$$\nu_2(x, C_k) = \int_{C_k} \tilde{\varrho}(x, y)\, \nu_1(x, dy)$$

is valid. In this case, however, relation (5) is satisfied for all $C \in \mathfrak{C}_0$ also and hence this equality is also satisfied for all C belonging to the completion of $\mathfrak{C}_0$ with respect to measure $\nu_1(x, \cdot)$. The lemma is proved. □

Note that for stochastically continuous processes with values in a separable space one can always find a separable σ-algebra $\mathfrak{F}^0$ (this is the σ-algebra generated by cylindrical sets over a countable set of values of the argument of the process, which is everywhere dense in its domain of definition) such that the completion of $\mathfrak{F}^0$, with respect to the measure associated with the process, contains $\mathfrak{F}$. Applying the lemma just proved to measures $\mu^i_{[a,b]}$ represented by equality (1) we observe that almost for all x – with respect to the measure which is the distribution of $\xi_2(c)$ – $\mu^2_{x,[c,b]} \ll \mu^1_{x,[c,b]}$. Let $\varrho_{[a,c]}$ denote the density of measure $\mu^2_{[a,c]}$ with respect to $\mu^1_{[a,c]}$ if the argument of this density is substituted by $\xi_1(t)$ (here we assume that all the processes $\xi_i(t)$ are defined on a certain fixed probability space $\{\Omega, \mathfrak{B}, \mathsf{P}\}$). Analogously denote by $\varrho_{y,[c,b]}$ the density of the measure $\mu^2_{y,[c,b]}$ with respect to $\mu^1_{y,[c,b]}$ with the same argument. Then it follows from formula (1), the lemma and formula (4) that

$$\varrho_{[a,b]} = \varrho_{[a,c]}\varrho_{\xi_1(c),[c,b]}. \tag{6}$$

Denote by $\mathfrak{B}_{[\alpha,\beta]}$ the subalgebra in the probability space $\{\Omega, \mathfrak{B}, \mathsf{P}\}$ generated by the variables $\xi_1(t)$ for $t \in [\alpha, \beta]$. The functions $\varrho_{[a,c]}$ and $\varrho_{\xi_1(c),[c,b]}$ are measurable relative to $\mathfrak{B}_{[a,c]}$ and $\mathfrak{B}_{[c,b]}$ respectively. Consider the subdivision of the interval $[a, t]: a = t_0 < t_1 \ldots < t_k = t$. It follows from formula (6) that

$$\varrho_{[a,t]}=\varrho_{[t_0,t_1]}\prod_{j=1}^{k-1}\varrho_{\xi(t_j),[t_j,t_{j+1}]}. \tag{7}$$

Utilizing (6) we now prove the following theorem:

Theorem 1. *If $\xi_1(t)$ and $\xi_2(t)$ are stochastically continuous Markov processes, then the composite process $\{\xi_1(t);\varrho_{[a,t]}\}$ is also Markovian.*

Proof. It is sufficient to show that for any continuous bounded function $f(x,s)$ of two variables $x\in\mathscr{X}$ and $s\in\mathscr{R}^1$ the relation

$$\mathsf{E}(f(\xi_1(t),\varrho_{[a,t]})\mid\mathfrak{B}_{[a,t_1]})=\mathsf{E}(f(\xi_1(t),\varrho_{[a,t]})\mid\xi_1(t_1),\varrho_{[a,t_1]}) \tag{8}$$

is satisfied for $a<t_1<t$.

Since $\varrho_{[a,t]}=\varrho_{[a,t_1]}\varrho_{\xi_1(t_1),[t_1,t]}$, we have

$$f(\xi_1(t),\varrho_{[a,t]})=\varphi(\xi_1(t),\varrho_{[a,t_1]},\varrho_{\xi_1(t_1),[t_1,t]}),$$

where $\varphi(x,s_1,s_2)$ is a continuous bounded function on $\mathscr{X}\times\mathscr{R}^1\times\mathscr{R}^1$. Assume first that $\varphi(x,s_1,s_2)=\varphi(x,s_2)\psi(s_1)$. Then utilizing the measurability of $\varphi(\xi_1(t),\varrho_{\xi_1(t),[t_1,t]})$ relative to $\mathfrak{B}_{[t_1,t]}$ and the Markovian property of $\xi_1(t)$, we have

$$\begin{aligned}\mathsf{E}(\varphi(\xi_1(t),\varrho_{\xi_1(t_1),[t_1,t]})\,\psi(\varrho_{[a,t_1]})\mid\mathfrak{B}_{[a,t_1]})&=\\ &=\psi(\varrho_{[a,t_1]})\,\mathsf{E}(\varphi(\xi_1(t),\varrho_{\xi_1(t_1),[t_1,t]})\mid\mathfrak{B}_{[a,t_1]})=\\ &=\psi(\varrho_{[a,t_1]})\,\mathsf{E}(\varphi(\xi_1(t),\varrho_{\xi_1(t_1),[t_1,t]})\mid\xi_1(t_1))=\\ &=\mathsf{E}(\psi(\varrho_{[a,t_1]})\,\varphi(\xi_1(t),\varrho_{\xi_1(t_1),[t_1,t]})\mid\xi_1(t_1),\varrho_{[a,t_1]}).\end{aligned}$$

Noting that both sides of formula (8) are linear in f, and that linear combinations of the form

$$\sum c_k\varphi_k(x,s_2)\,\psi_k(s_1)$$

can approximate any continuous function $\varphi(x,s_1,s_2)$, we have this verified formula (8) and proved the theorem. □

Remark. Assume that for all $t\in[a,b]$ and all subdivisions of the interval $[a,t]$ equation (7) is valid where $\varrho_{[a,t]}$ and $\varrho_{\xi_1(t_i),[t_i,t_{i+1}]}$ are certain variables measurable relative to $\mathfrak{B}_{[a,t]}$ and $\mathfrak{B}_{[t_i,t_{i+1}]}$. Then $\xi_2(t)$ is a Markov process provided $\xi_1(t)$ is such and, moreover, the transition probabilities of process $\xi_2(t)$ are determined by the equality

$$P^{(2)}(t_1,x,t_2\;A)=\mathsf{E}(\chi_A(\xi_1(t_2))\,\varrho_{\xi(t_1),[t_1,t_2]}\mid\xi_1(t_1))_{\xi_1(t_1)=x}.$$

Indeed, for any collection of sets $A_1,\dots,A_k$ in $\mathfrak{A}$ we have

$$\mathsf{E}\chi_{A_1}(\xi_2(t_1))\dots\chi_{A_k}(\xi_2(t_k))=$$

$$= \mathsf{E}\chi_{A_1}(\xi_1(t_1))\,\varrho_{[a,t_1]}\prod_{j=2}^{k}\chi_{A_j}(\xi_1(t_1))\,\varrho_{\xi_1(t_{j-1}),[t_{j-1},t_j]} =$$

$$= \mathsf{E}\chi_{A_1}(\xi_1(t_1))\,\varrho_{[a,t_1]}\prod_{j=2}^{k-1}\chi_{A_j}(\xi_1(t_j))\,\varrho_{\xi_1(t_{j-1}),[t_{j-1},t_j]}\times$$

$$\times\,\mathsf{E}(\chi_{A_k}(\xi_1(t_k))\,\varrho_{\xi_1(t_{k-1}),[t_{k-1},t_k]}\mid\mathfrak{B}_{[a,t_{k-1}]}) =$$

$$= \mathsf{E}\chi_{A_1}(\xi_1(t_1))\,\varrho_{[a,t_1]}\prod_{j=2}^{k-1}\chi_{A_j}(\xi_1(t_j))\,\varrho_{\xi_1(t_{j-1}),[t_{j-1},t_j]}\times$$

$$\times\,\mathsf{P}^{(2)}(t_{k-1},\xi_1(t_{k-1}),t_k,A_k).$$

The required assertion follows from this relation. □

Consider now the construction of function $\varrho_{[a,t]}$ by means of the transition probabilities of the processes $\xi_1(t)$ and $\xi_2(t)$. At the same time we shall obtain certain sufficient conditions for absolute continuity of the measures associated with these processes. It follows from the lemma (p. 516) that the absolute continuity of μ_2 with respect to μ_1 implies the absolute continuity of the transition probability $P^{(2)}(t,x,s,A)$ of the process $\xi_2(t)$ (as a function of A) with respect to the transition probability $P^{(1)}(t,x,s,A)$ of the process $\xi_1(t)$ almost for all x (with respect to the measure which is the distribution of $\xi_2(t)$). Set

$$\varrho(t,x,s,y) = \frac{dP^{(2)}(t,x,s,\cdot)}{dP^{(1)}(t,x,s,\cdot)}(y) \tag{9}$$

(In the case when $P^{(2)}$ is not absolutely continuous with respect to $P^{(1)}$ for a given x, ϱ denotes the derivative of the absolutely continuous component of $P^{(2)}$ with respect to $P^{(1)}$. Next let $\varrho_a(y)$ denote the density of the distribution of the variable $\xi_2(a)$ with respect to the distribution of the variable $\xi_1(a)$. If $a=t_0<t_1\ldots<t_n=b$ is a subdivision of $[a,b]$ then

$$\varrho_a(\xi_1(a))\prod_{k=0}^{n-1}\varrho(t_k,\xi_1(t_k),t_{k+1},\xi_1(t_{k+1}))$$

coincides with the density of the measure $\mu_2^{(n)}$ with respect to $\mu_1^{(n)}$, where $\mu_i^{(n)}$ is the contraction of the measure μ_i on the σ-algebra $\mathfrak{F}_{\Lambda_n}$ and Λ_n is a finite set of values of the argument $t\colon\Lambda_n=\{t_0,t_1,\ldots,t_n\}$. If $\Lambda_n\subset\Lambda_{n+1}$ then the limit

$$\lim_{n\to\infty}\varrho^n = \lim_{n\to\infty}\frac{d\mu_2^{(n)}}{d\mu_1^{(n)}}(\xi_1(\cdot))$$

exists (with probability 1). If $\Lambda=\bigcup_n\Lambda_n$ is every where dense in $[a,b]$, and the process $\xi_1(t)$ is stochastically continuous then the completion of $\mathfrak{F}_\Lambda$

with respect to μ_1 contains $\mathfrak{F}_{[a,b]}$ and hence the $\lim\limits_{n\to\infty} \varrho^n$ coincides with $\varrho_{[a,b]}$.

Theorem 2. *Let $P^{(i)}(t, x, s, A)$ $i=1, 2$ be the transition probabilities of two Markov processes $\xi_1(t)$ and $\xi_2(t)$ defined on $[a, b]$. If the following conditions are satisfied:*

a) *the distribution of $\xi_2(a)$ is absolutely continuous with respect to the distribution of $\xi_1(a)$ with density $\varrho_a(x)$;*

b) *for all $x \in \mathscr{X}$, $a \leqslant t < s \leqslant b$, the measure $P^{(2)}(t, x, s, \cdot)$ is absolutely continuous with respect to measure $P^{(1)}(t, x, s, \cdot)$ with density $\varrho(t, x, s, y)$;*

c) *there exists a constant c such that*

$$\int \log \varrho(t, x, s, y)\, P^{(2)}(t, x, s, dy) \leqslant c(s-t),$$

then the measure μ_2 is absolutely continuous with respect to μ_1 and

$$\frac{d\mu_2}{d\mu_1}(\xi_1(\cdot)) = \lim_{n\to\infty} \varrho_a(\xi_1(a)) \prod_{k=0}^{n-1} \varrho(t_{nk}, \xi_1(t_{nk}), t_{nk+1}, \xi_1(t_{nk+1})), \tag{10}$$

where $a = t_{n0} < \ldots t_{nn} = b$ and the sets $\Lambda_n = \{t_{nk}, k=0, \ldots, n\}$ satisfy the conditions: $\Lambda_n \subset \Lambda_{n+1}$ and $\bigcup\limits_n \Lambda_n$ is everywhere dense in $[a, b]$.

Proof. Introduce the process $\xi_3(t)$ with transition probabilities equal to the transition probabilities of the process $\xi_2(t)$ and let the distribution of $\xi_3(a)$ coincide with the distribution of $\xi_1(a)$. Let μ_3 be the measure which is associated with the process $\xi_3(t)$ and $\mu_i^{(n)}$ be, as above, the restriction of measure μ_i to the σ-algebra $\mathfrak{F}_{\Lambda_n}$, where $\Lambda_n = \{t_{nk}, k=0, \ldots, n\}$. Then as it is easy to verify

$$\frac{d\mu_2^{(n)}}{d\mu_3^{(n)}}(\xi_3(\cdot)) = \varrho_a(\xi_3(a)),$$

$$\frac{d\mu_3^{(n)}}{d\mu_1^{(n)}}(\xi_1(\cdot)) = \prod_{k=0}^{n-1} \varrho(t_{nk}, \xi_1(t_{nk}), t_{nk+1}, \xi_1(t_{nk+1})).$$

Clearly,

$$\lim_{n\to\infty} \frac{d\mu_2^{(n)}}{d\mu_3^{(n)}}(\xi_3(a)) = \varrho_a(\xi_3(a)).$$

Also the following limit

$$\lim_{n\to\infty} \prod_{k=0}^{n-1} \varrho(t_{nk}, \xi_1(t_{nk}), t_{nk+1}, \xi_1(t_{nk+1}))$$

exists with probability 1. Denote this limit by ϱ'. In order that the relation $\mu_3 \ll \mu_1$ be valid, it is sufficient that $\mathsf{E}\varrho'=1$. In turn, in view of the remark following Theorem 2 in Section 1 it is sufficient for the validity of $\mathsf{E}\varrho'=1$ that the following expression

$$I_n = \mathsf{E} \prod_{k=0}^{n-1} \varrho(t_{nk}, \xi_1(t_{nk}), t_{nk+1}, \xi_1(t_{nk+1})) \times \\ \times \log \prod_{k=0}^{n-1} \varrho(t_{nk}, \xi_1(t_{nk}), t_{nk+1}, \xi_1(t_{nk+1}))$$

be bounded. Utilizing the equality

$$\mathsf{E}(\varrho(t_{nk}, \xi_1(t_{nk}), t_{nk+1}, \xi_1(t_{nk+1}) \mid \xi_1(t_{nk}))=1,$$

we obtain that

$$I_n = \mathsf{E} \prod_{k=0}^{n-1} \varrho(t_{nk}, \xi_1(t_{nk}), t_{nk+1}, \xi_1(t_{nk+1})) \times \\ \times \sum_{k=0}^{n-1} \log \varrho(t_{nk}, \xi_1(t_{nk}), t_{nk+1}, \xi_1(t_{nk+1})) =$$

$$= \mathsf{E} \sum_{l=0}^{n-1} \prod_{k=0}^{l-1} \varrho(t_{nk}, \xi_1(t_{nk}), t_{nk+1}, \xi_1(t_{nk+1})) \times \\ \times \int \log \varrho(t_{nl}, \xi_1(t_{nl}), t_{nl+1}, y) P^{(2)}(t_{nl}, \xi_1(t_{nl}), t_{nl+1}, dy) \leqslant \\ \leqslant c \sum_{l=0}^{n-1} (t_{nl+1} - t_{nl}) = c(b-a).$$

Consequently, $\mu_3 \ll \mu_1$, $\mu_2 \ll \mu_3$. Thus $\mu_2 \ll \mu_1$.

Formula (10) is a corollary of relation

$$\frac{d\mu_2}{d\mu_1} = \frac{d\mu_2}{d\mu_3} \frac{d\mu_3}{d\mu_1}.$$

The theorem is proved. □

Consider the problem of constructing Markov process $\xi_2(t)$ for which the associated measure μ_2 is absolutely continuous with respect to measure μ_1 which is associated with the given Markov process $\xi_1(t)$.

Theorem 3. *Let $a = t_{n0} < \dots < t_{nn} = b$ be a sequence of subdivisions of the interval $[a, b]$ such that the sets $\Lambda_n = \{t_{nk}, k=0, \dots, n\}$ form an increasing sequence and $\bigcup_n \Lambda_n$ is everywhere dense on $[a, b]$. Let a function $\alpha_n(t, x, s, y)$ be defined for each n, measurable in x and y, where $x, y \in \mathscr{X}$ $a \leqslant t < s \leqslant b$, satisfying the following conditions:*

1) *the limit*

$$\eta = \lim_{n\to\infty} \sum_{k=0}^{n-1} \alpha_n(t_{nk}, \xi_1(t_{nk}), t_{nk+1}, \xi_1(t_{nk+1}))$$

exists in the sense of convergence in probability;

2) $$\int e^{\alpha_n(t,x,s,y)} \alpha_n(t,x,s,y)\, P^{(1)}(t,x,s,dy) = O(s-t)$$

uniformly in x;

3) $$\int e^{\alpha_n(t_{nk},x,t_{nk+1},y)} P^{(1)}(t_{nk},x,t_{nk+1},dy) = 1$$

for all n, k *and* x. *Then the measure* μ_2 *defined on* $\mathfrak{F}_{[a,b]}$ *by the equality*

$$\mu_2(A) = \mathsf{E}\chi_A(\xi_1(\cdot))\, e^{\eta}$$

will be associated with a certain Markov process on $[a, b]$.

Proof. First we show that the measure μ_2 is a probability measure, i.e. $\mathsf{E}e^{\eta} = 1$. Let

$$\eta_n = \sum_{k=0}^{n-1} \alpha_n(t_{nk}, \xi_1(t_{nk}), t_{nk+1}, \xi_1(t_{nk+1})).$$

Then it follows from condition 3) of the theorem that

$$\mathsf{E}\, e^{\eta_n} = \int \mathsf{P}\{\xi_1(a)\in dx_0\} \prod_{k=0}^{n-1} \int e^{\alpha_n(t_{nk},x_k,t_{nk+1},x_{k+1})} \times \\ \times \mathsf{P}^{(1)}(t_{nk}, x_k, t_{nk+1}, dx_{k+1}) = 1.$$

On the other hand, in view of condition 2)

$$\mathsf{E}\, e^{\eta_n} \ln e^{\eta_n} = \mathsf{E}\, e^{\eta_n}\eta_n = \sum_{k=0}^{n-1} \int \mathsf{P}\{\xi_1(a)\in dx_0\} \prod_{j=0}^{k-1} \int e^{\alpha_n(t_{nj},x_j,t_{nj+1},x_{j+1})} \times \\ \times P^{(1)}(t_{nj}, x_j, t_{nj+1}, dx_{j+1}) \int e^{\alpha_n(t_{nk},x_k,t_{nk+1},x_{k+1})} \times \\ \times \alpha_n(t_{nk}, x_k, t_{nk+1}, x_{k+1})\, P^{(1)}(t_{nk}, x_k, t_{nk+1}, dx_{k+1}).$$

Therefore $\mathsf{E}e^{\eta_n} \ln e^{\eta_n}$ is bounded and hence e^{η_n} is integrable uniformly with respect to n and hence one may approach the limit under the sign of mathematical expectation in the relation $\mathsf{E}\, e^{\eta_n} = 1$. To verify that the measure μ_2 is associated with a Markov process consider the function

$$\eta(t) = \lim_{n\to\infty} \sum_{t_{nk}<t} \alpha_n(t_{nk}, \xi_1(t_{nk}), t_{nk+1}, \xi_1(t_{nk+1}))$$

(here the limit is taken in the sense of convergence in probability). We show that this limit exists for all $t\in[a, b]$ and coincides with the expression

$$\ln \mathsf{E}(e^{\eta} \mid \mathfrak{B}_{[0,t]}) = \lim_{n\to\infty} \ln \mathsf{E}(e^{\eta_n} \mid \mathfrak{B}_{[0,t]})$$

(here the limiting transition under the sign of the mathematical expectation is justified in view of the uniform integrability of the function e^{η_n}). For $t\in\bigcup \Lambda_n$ we have the equality

$$\eta_n(t) = \sum_{t_{nk}<t} \alpha_n(t_{nk}, \xi_1(t_{nk}), t_{nk+1}, \xi_1(t_{nk+1})) = \ln \mathsf{E}(e^{\eta_n} \mid \mathfrak{B}_{[0,t]}).$$

Therefore the $\lim \eta_n$ exists for $t\in\bigcup_n \Lambda_n$. If, however, $t_{nj}<t<t_{nj+1}$, then

$$\ln \mathsf{E}(e^{\eta_n} \mid \mathfrak{B}_{[0,t]}) - \sum_{t_{nk}<t} \alpha_n(t_{nk}, \xi_1(t_{nk}), t_{nk+1}, \xi_1(t_{nk+1})) =$$
$$= \ln \mathsf{E}\left(e^{\alpha_n(t_{nj}, \xi_1(t_{nj}), t_{nj+1}, \xi_1(t_{nj+1}))} \mid \mathfrak{B}_{[0,t]}\right).$$

It follows from condition 2) and 3) that

$$\mathsf{E}\, e^{\alpha_n(t_{nj}, \xi_1(t_{nj}), t_{nj+1}, \xi_1(t_{nj+1}))} \alpha_n(t_{nj}, \xi_1(t_{nj}), t_{nj+1}, \xi_1(t_{nj+1})) \to 0$$

as $n\to\infty$ uniformly in j. Utilizing the fact that the variable $\exp\{\alpha_n(t_{nj}, \xi_1(t_{nj}), t_{nj+1}, \xi_1(t_{nj+1}))\}$ is uniformly integrable with respect to j as well as the convergence of this variable in probability to 1, we obtain that

$$\ln \mathsf{E}\left(e^{\alpha_n(t_{nj}, \xi_1(t_{nj}), t_{nj+1}, \xi_1(t_{nj+1}))} \mid \mathfrak{B}_{[\alpha,t]}\right) \to 0$$

in the sense of convergence in probability.

The existence of $\eta(t)$ is thus proved. Let $a<c<b$ and $\varrho_{[a,c]} = \exp\{\eta(c)\}$, $\varrho_{[c,b]} = \exp\{\eta(b)-\eta(c)\}$. Clearly $\varrho_{[c,d]}$ is measurable with respect to $\mathfrak{B}_{[c,d]}$. The proof of the theorem now follows by utilizing the remark following Theorem 1. □

Consider the particular case when $\xi_1(t)$ and $\xi_2(t)$ are stochastically continuous processes with independent increments defined on $[a, b]$ and let $\xi_i(a)=0$. Denote by $\bar{\mu}^{(i)}_{[\alpha,\beta]}$ the measure associated with process $\xi_i(t)-\xi_i(\alpha)$ for $t\in[\alpha, \beta]$ where $[\alpha, \beta]\subset[a, b]$. Then relation $\mu_2 \ll \mu_1$ implies relation $\bar{\mu}^{(2)}_{[\alpha,\beta]} \ll \bar{\mu}^{(1)}_{[\alpha,\beta]}$. Let $\mathfrak{B}_{[\alpha,\beta]}$ denote the σ-algebra generated by the variables $\xi_1(t)-\xi_1(\alpha)$ for $t\in[\alpha, \beta]$. The variable

$$\frac{d\bar{\mu}^{(2)}_{[\alpha,\beta]}}{d\bar{\mu}^{(1)}_{[\alpha,\beta]}}(\xi_1(\cdot))$$

is $\mathfrak{B}_{[\alpha,\beta]}$-measurable. Let $a=t_0<t_1<\ldots<t_n=b$ be an arbitrary subdivision of the interval $[a, b]$. Consider the product of measurable

spaces $(\mathscr{F}_{[t_k, t_{k+1}]}, \mathfrak{F}_{[t_k, t_{k+1}]})$ and the product of measures

$$\prod_{k=0}^{n-1} \bar{\mu}^{(i)}_{[t_k, t_{k+1}]}$$

defined on it.

Since the processes $\xi_i(t)$ have independent increments, it follows that these products of measures are mapped into measures μ_i under the $1-1$ measurable mapping of the product of the spaces $(\mathscr{F}_{[t_k, t_{k+1}]}, \mathfrak{F}_{[t_k, t_{k+1}]})$ into $(\mathscr{F}_{[a, b]}, \mathfrak{F}_{[a, b]})$ by means of the formula

$$x(t) = \sum_{k=1}^{m} x_k(t_k) + x_{m+1}(t), \qquad t_m \leqslant t \leqslant t_{m+1},$$

$$x(\cdot) \in \mathscr{F}_{[a, b]}, \qquad x_k(\cdot) \in \mathscr{F}_{[t_k, t_{k+1}]}.$$

Therefore

$$\frac{d\mu^{(2)}}{d\mu^{(1)}}(\xi_1(\cdot)) = \prod_{k=0}^{n-1} \frac{d\bar{\mu}^{(2)}_{[t_k, t_{k+1}]}}{d\bar{\mu}^{(1)}_{[t_k, t_{k+1}]}}(\xi_1(\cdot)).$$

The factors in the r.h.s. are independent. Assume that measures μ_1 and μ_2 are equivalent. In this case the densities are positive and we may take the logarithm of the last product. We thus obtain the following

Theorem 4. *In order that*

$$\varrho_{[a, t]} = \frac{d\mu^{(2)}_{[a, t]}}{d\mu^{(1)}_{[a, t]}}(\xi_1(\cdot))$$

be densities of equivalent measures – associated with processes with independent increments defined on $[a, t]$ *– it is necessary and sufficient that the composite process* $\{\xi_1(t), \ln \varrho_{[a, t]}\}$ *be a process with independent increments,* $\varrho_{[a, t]}$ *be* $\mathfrak{B}_{[a, t]}$*-measurable and that the equality* $\mathsf{E}\varrho_{[a, t]} = 1$ *be satisfied.*

Chapter VIII

Measurable Functions on Hilbert Spaces

§1. Measurable Linear Functionals and Operators on Hilbert Spaces

Consider a measurable Hilbert space $(\mathfrak{X}, \mathfrak{B})$ on which a measure μ is defined. Every continuous linear functional $l(x)$ defined on $\mathfrak{X}$ is clearly $\mathfrak{B}$-measurable. It is known that if a sequence of continuous linear functionals $l_n(x)$ converges to a certain limit $l(x)$ for all x then this limit will also be a continuous linear functional on $\mathfrak{X}$. The situation, however becomes different if we require that $l_n(x)$ possess the limit not for all x but only on a set D such that $\mu(D)=1$. It is natural to refer to such limiting functions $l(x)$ as $\mathfrak{B}$-measurable functionals. These functions, being limits of sequences of measurable functions will also be $\mathfrak{B}$-measurable. It follows from the relations

$$\lim_{n\to\infty} l_n(\alpha x+\beta y)=\alpha \lim_{n\to\infty} l_n(x)+\beta \lim_{n\to\infty} l_n(y)$$

that the domain of definition D_l of functional $l(x)$ is a linear manifold and that $l(x)$ is a linear (additive and homogeneous) functional. (We assume that the functional $l(x)$ is defined wherever the corresponding limit exists.) Hereafter we shall consider non-degenerate measures μ such that $\mu(L)=0$ for any proper subspace L of the space $\mathfrak{X}$. Since $\mu(D_l)=1$, the set D_l is dense in $\mathfrak{X}$. Thus if $l(x)$ is a measurable functional – in the sense stipulated above – then 1) it is defined on a $\mathfrak{B}$-measurable linear manifold D_l such that $\mu(D_l)=1$; 2) $l(x)$ is a $\mathfrak{B}$-measurable function 3) $l(x)$ is linear on D_l. It turns out that these conditions are sufficient for μ-measurability of $l(x)$. This follows from

Theorem 1. *If a function $l(x)$ satisfies conditions* 1)–3) *then a sequence of continuous functionals $l_n(x)$ exists such that*

$$l(x)=\lim_{n\to\infty} l_n(x) \quad (\operatorname{mod}\mu).$$

Proof. We shall construct a sequence of continuous functionals $l_n(x)$ which converge to $l(x)$ in measure μ. Since a subsequence can be extracted from this sequence converging μ-almost everywhere, the theorem will be proved. Let $S_c=\{x:|l(x)|<c\}$. Since $\lim_{c\to\infty}\mu(D_l-S_c)=0$, for each $\varepsilon>0$ one can find c and a compactum $K\subset S_c$ such that $\mu(D_l-K)<\varepsilon$. Without loss of generality we may assume that K is convex and symmetric since S_c is such. Since $\mu(\{0\})=0$ one can find a $\delta>0$ such that also

$$\mu(K-S_\delta(0))>\mu(D)-\varepsilon.$$

(here $S_\delta(0)$ is a sphere of radius δ with the center at point 0). For $x\in K-S_\delta(0)$ the inequality

$$\frac{|l(x)|}{|x|}\leqslant\frac{c}{\delta}$$

is satisfied. This inequality is also satisfied for all $x\in\mathscr{L}$ where $\mathscr{L}$ is the linear hull of the set $K-S_\delta(0)$. Hence $l(x)$ is a bounded linear functional on $\mathscr{L}$. In view of Hahn-Banach's theorem there exists a linear extension of l onto the whole space $\mathscr{X}$ with the same modulus of continuity. We denote this extension by $l_\varepsilon(x)$. Let K_1 be the convex hull of the set $K-S_\delta(0)$. Then

$$\mu(\{x:|l_\varepsilon(x)-l(x)|>0\})\leqslant\mu(\mathscr{X}-K_1)=\mu(\mathscr{X})-\mu(K_1)<\varepsilon.$$

From it follows the existence of a sequence of bounded linear functionals converging in measure μ to $l(x)$. The theorem is thus proved. □

Corollary. *If $l(x)$ is a measurable functional and if $\mathscr{L}_n$ is a sequence of finite-dimensional subspaces such that $\mathscr{L}_n\in D_l$ and $\cup\mathscr{L}_n$ is dense in $\mathscr{X}$ and P_n is the projector on $\mathscr{L}_n$, then $l(\mathsf{P}_n x)$ converges in measure μ to $l(x)$.*

Indeed, let K_1 be the compactum constructed in the proof of the theorem. If n is chosen in such a manner that $\mathscr{L}_n$ forms a $\varepsilon\delta/c$-net in K_1 (here δ and c are as in the proof of the theorem), then

$$|l(P_n x)-l(x)|\leqslant\sup_{y\in\mathscr{L}}\frac{|l(y)|}{|y|}|P_n x-x|\leqslant\frac{c}{\delta}|P_n x-x|<\varepsilon$$

for all $x\in K_1$. Therefore

$$\mu(\{x:|l(P_n x)-l(x)|>\varepsilon\})<\varepsilon. \quad \square$$

In order to construct the space of all μ-measurable functionals it is convenient to use the characteristic functional $\varphi(z)$ of measure μ. Let a sequence of continuous functionals (z_n, x) converge in measure μ to a certain measurable functional $l(x)$. Then for each real t

$$\lim_{n,m\to\infty}\exp\{it(z_n-z_m, x)\}=1.$$

Hence

$$\lim_{n,m\to\infty} \varphi(t(z_n - z_m)) = \lim_{n,m\to\infty} \int \exp\{it(z_n - z_m, x)\}\, \mu(dx) = 1. \qquad (1)$$

Let

$$k(z) = \int (1 - \varphi(tz)) \frac{1}{1+t^2}\, dt.$$

Then the necessary and sufficient condition for the existence of the limit in measure μ of the sequence (z_n, x) is the condition that

$$\lim_{n,m\to\infty} k(z_n - z_m) = 0.$$

The necessity of this condition follows from (1) and the theorem on the limiting transition under the sign of the integral. To establish sufficiency note that

$$k(z) = \int \left[\int (1 - e^{it(z,x)}) \frac{1}{1+t^2}\, dt\right] \mu(dx) = \pi \int (1 - e^{-|(z,x)|})\, \mu(dx).$$

Therefore for any $\varepsilon > 0$

$$\mu(\{x : |(z_n - z_m, x)| > \varepsilon\}) \leq \frac{1}{\pi} k(z_n - z_m)(1 - e^{-\varepsilon})^{-1},$$

which implies the convergence of (z_n, x) in measure μ.

Since

$$k(z_1 + z_2) = \pi \int (1 - e^{-|(z_1, x) + (z_2, x)|}\, \mu(dx) \leq$$

$$\leq \pi \int (1 - e^{-|(z_1, x)| - |(z_2, x)|})\, \mu(dx) \leq \pi \int (1 - e^{-|(z_1, x)|})\, \mu(dx) +$$

$$+ \pi \int (1 - e^{|(z_2, x)|})\, \mu(dx) = k(z_1) + k(z_2),$$

$\mathscr{X}$ may be regarded as a metric space with the metric

$$r(x, y) = k(x - y).$$

Let $\tilde{\mathscr{X}}$ denote the completion of $\mathscr{X}$ in metric r. Each element of $\tilde{\mathscr{X}}$ can be associated with a certain μ-measurable functional $l(x)$: $\tilde{x} \overset{S}{\leftrightarrow} l$, provided a sequence z_n exists in $\mathscr{X}$ such that $r(z_n, \tilde{x}) \to 0$ and $(z_n, x) \to l(x)$ in measure μ. Denote by $\mathscr{L}(\mu)$ the space of all μ-measurable functionals. We shall identify those functionals which coincide μ-almost everywhere. Then the correspondence S between $\tilde{\mathscr{X}}$ and $\mathscr{L}(\mu)$ becomes one-to-one. By introducing the distance

$$r(l_1, l_2)=\pi \int (1-e^{-|l_1(x)-l_2(x)|})\, \mu(dx)$$

in $\mathscr{L}(\mu)$, this correspondence becomes isometric. It is therefore natural to identify the spaces $\tilde{\mathscr{X}}$ and $\mathscr{L}(\mu)$ and we shall follow this procedure hereafter.

We note another special feature of the space $\tilde{\mathscr{X}}$ with metric r. The characteristic functional of measure μ can be extended by continuity in metric r to the whole $\tilde{\mathscr{X}}$. This extension can be written in the form

$$\varphi(l)=\int e^{il(x)}\, \mu(dx)$$

(here l is a measurable functional regarded as an element of $\tilde{\mathscr{X}}=\mathscr{L}(\mu)$). We show that $\tilde{\mathscr{X}}$ is, in a certain sense, the widest possible space to which $\varphi(z)$ can be extended by continuity.

Let $\mathscr{Y}$ be a linear metric space with metric ϱ such that $\varrho(x, y)=\varrho(0, x-y)$ and let $\varphi(z)$ be continuous in this metric ϱ on $\mathscr{X}$ and extendable by continuity onto $\mathscr{Y}$. Since φ is continuous in metric ϱ for any $\varepsilon>0$ a $\delta>0$ can be found such that $\operatorname{Re}(1-\varphi(z))<\varepsilon$ provided $\varrho(0, z)<\delta$. Then utilizing the inequality

$$|\varphi(z_1)-\varphi(z_2)| \leqslant \int |1-e^{(z_1-z_2, x)}|\, \mu(dx) \leqslant$$

$$\leqslant \sqrt{\int 2(1-\cos(z_1-z_2, x)\, \mu(dx)} \leqslant \sqrt{2 \operatorname{Re}(1-\varphi(z_1-z_2))},$$

we find that for $\operatorname{Re}(1-\varphi(z))<\varepsilon$

$$\operatorname{Re}(1-\varphi(nz)) \leqslant \sum_{k=1}^{n} |\varphi((k-1)\, z)-\varphi(kz)| \leqslant n\sqrt{2\varepsilon}.$$

Therefore for $\varrho(0, z)<\delta$

$$k(z)=\int_{-\infty}^{\infty} \operatorname{Re}(1-\varphi(tz)) \frac{dt}{1+t^2}=$$

$$=\int_{|t|\leqslant n} \operatorname{Re}(1-\varphi(tz)) \frac{dt}{1+t^2}+\int_{|t|>n} \operatorname{Re}(1-\varphi(tz)) \frac{dt}{1+t^2} \leqslant$$

$$\leqslant \pi n \sqrt{2\varepsilon}+\frac{4}{n}.$$

Hence if $\varrho(z_n, z_m)\to 0$ then $k(z_n - z_m)\to 0$, so that $\mathscr{Y}$ can be isometrically imbedded into a certain subset of $\tilde{\mathscr{X}}$.

The space $\tilde{\mathscr{X}}$ is significantly wider than $\mathscr{X}$, since it contains, for example, the spaces $\mathscr{X}^B_-$ obtained by completion of $\mathscr{X}$ in the scalar product $(x, y)_- = (Bx, y)$, where B is a kernel operator such that $\varphi(x)$ is continuous in the scalar product (Bz, z).

In addition to the space of all measurable functionals $\mathscr{L}(\mu)$ one may consider the space $\mathscr{L}^{(2)}(\mu)$ of all square integrable linear functionals. However, this space may consist of only the null element. If the measure μ possesses a finite correlation operator C, then $\mathscr{L}^{(2)}(\mu)$ will contain the completion of $\mathscr{X}$ in the scalar product (Cx, y), but may not necessarily coincide with this completion. Moreover, it may occur that $\int (x, z)^2 \mu(dx) = +\infty$ for all $z \neq 0$ while $\mathscr{L}^{(2)}(\mu)$ may contain elements different from the zero element.

As an example, consider the measure μ which is the distribution of a random element ξ of the form

$$\xi = \sum_{k=1}^{\infty} \lambda_k \eta_k e_k,$$

where $\{e_k\}$ is an orthonormal basis, η_k a sequence of identically distributed independent random variables with stable distributions:

$$\mathsf{E}\, e^{is\eta_k} = \exp\{i\gamma s - |s|^\alpha\}.$$

First let $\gamma = 0$ and $\alpha > 1$. Then (ξ, z) possesses a stable distribution with the same exponent α. Therefore for any functional $l(x)$ in $\mathscr{L}(\mu)$

$$\int e^{isl(x)}\, \mu(dx) = e^{-|s|^\alpha}.$$

Hence, $\int l^2(x)\, \mu(dx) < \infty$ only if $l(x) = 0$.

If $\alpha > 1$ and $\gamma \neq 0$ then by choosing a sequence $z_n = \frac{1}{n} \sum_{k=1}^{n} \frac{1}{\lambda_k} e_k$, we have: $(\xi, z_n) = \frac{1}{n} \sum_{k=1}^{n} \eta_k$. Therefore with probability 1, i.e. μ-almost everywhere, the limit

$$\lim_{n\to\infty} (x, z_n) = \mathsf{E}\eta_k = \gamma$$

exists. Clearly $l(x) = \lim_{n\to\infty} (x, z_n)$ belongs to $\mathscr{L}^{(2)}(\mu)$.

On the other hand (z, x) will have a stable distribution with exponent α and hence

$$\int (z, x)^2 \, \mu(dx) = +\infty \quad \text{for} \quad z \neq 0.$$

These examples show that in the general case, it is (somewhat) unnatural to consider the space $\mathscr{L}^{(2)}(\mu)$.

Measurable linear operators. As in the case of measurable functionals, measurable linear operators are defined naturally as limits – in measure μ – of a sequence of continuous operators. Since there may be a strong or weak convergence of the sequence $A_n x$, we may define the measurability as either strong or weak. Thus an operator A is called strongly (weakly) measurable with respect to measure μ if a sequence of continuous linear operators A_n exists such that $A_n x$ converges strongly (weakly) to $Ax \pmod{\mu}$. Obviously a strongly measurable operator is also a weakly measurable one. Let A be weakly measurable. Denote by D_A the set of x such that the weak limit of the sequence $A_n x$ exists. Then denoting by N a certain countable set dense in $\mathscr{X}$ we will have $D_A = \{x : \sup |A_n x| < \infty;\ \lim_{n\to\infty} (z, A_n x), z \in N, \text{exists}\}$. From this relation it follows that D_A is measurable. It is also clear that D_A is a linear manifold. The weak limit $Ax = \lim_{n\to\infty} A_n x$ exists for all $x \in D_A$ and moreover $A(\alpha x + \beta y) = \alpha Ax + \beta Ay$ for all real α and β and $x, y \in D_A$. Finally we find that $\mu(D_A) = 1$. We show that the above conditions are sufficient even for a strong measurability of operator A. This will also show that the notions of weak and strong measurability are equivalent.

Theorem 2. *Let a measurable function Ax with values in $\mathscr{X}$ satisfying the relation $A(\alpha x + \beta y) = \alpha Ax + \beta Ay$ for all $x, y \in D_A$ and real α, β be defined on a certain measurable linear manifold D_A such that $\mu(D_A) = 1$. Then a sequence of continuous linear operators A_n exists such that $Ax = \lim_{n\to\infty} A_n x$ $\pmod{\mu}$.*

Proof. Note that $|Ax|$ is a measurable function. Therefore

$$\lim_{c\to\infty} \mu(\{x : |Ax| > c\}) = 0,$$

and hence for each $\varepsilon > 0$ one can find a compactum K such that $|Ax| \leq c$ for $x \in K$ and $\mu(\mathscr{X} - K) < \varepsilon$. This compactum may be considered a convex and centrally symmetrical set.

We choose δ as in the proof of Theorem 1. Let K_1 and $\mathscr{L}$ be as in

Theorem 1. Then

$$\frac{|Ax|}{|x|} \leqslant \frac{c}{\delta}.$$

Let N be a finite-dimensional subspace of $\mathscr{L}$ such that $N \cap K_1$ forms a $\frac{\delta}{c}$ ε-net in K_1. Construct the operator A_N in the following manner: for $x \in N$, $A_N x = Ax$, if y is orthogonal to N, then $A_N y = 0$. This extension of A from N onto the whole $\mathscr{X}$ does not increase the modulus of continuity of A. Therefore $|A_N x - Ax| \leqslant \sup\limits_{y \in \mathscr{L}} \frac{|Ay|}{|y|} |P_n x - x| < \varepsilon$ for $x \in K_1$. Hence,

$$\mu(\{|A_N x - Ax| > \varepsilon\}) < \varepsilon.$$

Choosing sequences $\varepsilon_n \to 0$ we construct a sequence of bounded linear operators which converges to A in measure and a subsequence can then be extracted from such a sequence which converges almost everywhere. The theorem is thus proved. □

Henceforth we shall use the term "a measurable linear operator" without specification as to strong (or weak).

We now consider the notion of an absolutely measurable linear operator. A measurable linear operator A is called absolutely measurable if for each measurable linear functional $l(x)$ the expression $l(Ax)$ is also a measurable linear functional. The last assertion may be interpreted in two ways. First, since a sequence z_n exists such that $k(l - z_n) \to 0$, one can understand $l(Ax)$ as the limit in measure μ of the sequence of measurable functionals $l_n(x) = (z_n, Ax)$. Secondly, one may interpret $l(Ax)$ as the standard superposition of two measurable functions. This superposition is also measurable and the condition of additivity and homogeneity is fulfilled in the domain of the definition of this function. The set $\{x : Ax \in D_l\}$ where D_l is the domain of the definition of $l(x)$, – is the domain of the definition of this function. If Δ_A denotes the range of values of the operator A, then in order that $l(Ax)$ be a measurable functional it is necessary and sufficient that the equality

$$\mu(A^{-1}(\Delta_A \cap D_l)) = 1$$

be satisfied. Moreover since any measurable linear manifold L such that $\mu(L) = 1$ may serve as the set D_l the condition $\mu(A^{-1}(\Delta_A \cap L)) = 1$ must be satisfied provided $\mu(L) = 1$. Utilizing Theorems 1 and 2 one can verify the equivalence of both interpretations of the measurability of $l(Ax)$.

We now describe the structure of an absolutely measurable operator. Note that for any absolutely measurable operator A the convergence in measure μ of a sequence of measurable linear functionals $l_n(x)$ to $l(x)$

implies the convergence in measure μ of the sequence $l_n(Ax)$ to $l(Ax)$. To show this it is sufficient to consider the convergence almost everywhere in place of the convergence in measure. In this case, one can find a linear manifold L with $\mu(L)=1$ such that $l_n(x)\to l(x)$ for $x\in L$. Hence $l_n(Ax)\to$ $\to l(Ax)$ for all x such that $Ax\in L$ and this set is of measure 1 in view of the absolute measurability of the operator A. Thus an operator A^* can be associated with A which maps $\tilde{\mathfrak{X}}$ into $\tilde{\mathfrak{X}}$, acting according to the formula $[A^*l](x)=l(Ax)$. It follows from the above that this operator is continuous in metric r since the convergence of functionals in this metric is equivalent to their convergence in measure μ. We show that the converse is also true: if A is a measurable linear operator such that operator A^* defined by the relation $A^*z(x)=(z, Ax)$ for all $z\in\mathfrak{X}$ is extended by continuity in metric r over the whole $\tilde{\mathfrak{X}}$, then A is an absolutely measurable operator. If A^* is extendable by continuity on $\tilde{\mathfrak{X}}$ then $\varphi(A^*z)$ will be an r-continuous positively definite functional. Therefore the series

$$Ax=\sum_{k=1}^{\infty}[A^*e_k](x)\,e_k$$

is convergent μ-almost everywhere in any orthogonal basis $\{e_k\}$ and moreover $\varphi(A^*z)$ will be the characteristic functional of the variable Ax defined in this manner. Let l be a measurable functional, $\{f_k\}$ an orthonormal basis in D_l. Then

$$Ax=\sum_{k=1}^{\infty}(Ax,f_k)f_k=\sum_{k=1}^{\infty}(A^*f_j,x)f_j.$$

Let P_n be the projector on the subspace spanned by $f_1,\dots,f_n$. We show that $l(P_nAx)$ is convergent in measure to a certain limit. Indeed,

$$l(P_nAx)=\sum_{j=1}^{n}(A^*f_j,x)\,l(f_j)=\left[A^*\sum_{j=1}^{n}l(f_j)f_j\right](x),$$

and since $\left(\sum_{j=1}^{n}l(f_j)f_j,\ x\right)=l(P_nx)$ converges in measure μ to $l(x)$ in view of the corollary of Theorem 1, we have

$$\left[A^*\sum_{j=1}^{n}l(f_j)f_j\right](x)\to(A^*l)(x)$$

in measure μ in view of the continuity of A^* in $\tilde{\mathfrak{X}}$. Hence

$$l(Ax)=[A^*l](x)$$

is a measurable linear functional for any linear functional l. The absolute measurability of A is thus proved. □

We now consider measurable linear mappings of one Hilbert space $\mathscr{X}$ into another Hilbert space $\mathscr{Y}$.

We shall discuss only strongly measurable mappings. It turns out that the study of such mappings easily reduces to the study of measurable mappings of $\mathscr{X}$ into $\mathscr{X}$, i.e. of measurable linear operators. Indeed, let R be an isometric one-to-one mapping of $\mathscr{Y}$ onto $\mathscr{X}$ (we assume that both spaces are separable). Let V be a measurable mapping of $\mathscr{X}$ into $\mathscr{Y}$, then a sequence of continuous linear mappings V_n can be found such that $V_n x \to Vx$ (in $\mathscr{Y}$) in measure μ. But RV_n is a sequence of continuous linear mappings of $\mathscr{X}$ into $\mathscr{X}$ convergent in measure μ to RV. Hence RV is a measurable linear operator. Conversely, if U is a measurable linear operator mapping $\mathscr{X}$ into $\mathscr{X}$, then UR^{-1} is a measurable linear mapping of $\mathscr{X}$ into $\mathscr{Y}$. We thus have completely described all the measurable linear mappings of $\mathscr{X}$ into $\mathscr{Y}$.

Denote by ν the measure defined on $\mathscr{Y}$ by the relation $\nu(E)=\mu(V^{-1}(E))$, where V is a measurable linear mapping of $\mathscr{X}$ into $\mathscr{Y}$. We find the characteristic functional of measure ν. For this purpose the notion of a conjugate mapping of V will be required. Let D be the domain of definition of mapping V with $\mu(D)=1$. The expression (Vx, y) is defined for all $y\in\mathscr{Y}$ and $x\in D$ and is a measurable functional on D. Therefore an element $l_y\in\tilde{\mathscr{X}}$ exists such that $(Vx, y)=l_y(x)$. Set $l_y=V^*y$; V^*y defines a homogeneous and additive mapping of $\mathscr{Y}$ into $\tilde{\mathscr{X}}$ which is continuous in the following sense: $r(V^*y_1, V^*y_2)\to 0$ as $|y_1-y_2|\to 0$. This mapping V^* is called the mapping *conjugate* to V. The case when V^* can be regarded as a measurable mapping (with respect to measure ν) of $\mathscr{Y}$ to $\mathscr{X}$ is of special interest. Let $\{e_k\}$ be an orthonormal basis in $\mathscr{X}$. In order for V^*y to belong to $\mathscr{X}$ it is necessary and sufficient that

$$V^*y=\sum_{k=1}^{\infty}(V^*y, e_k)\,e_k=\sum_{k=1}^{\infty}(y, Ve_k)\,e_k$$

and that the series $\sum_k (y, Ve_k^2)^2$ be convergent ν-almost everywhere. The last assertion is equivalent to the μ-almost everywhere convergence of the series $\sum_k (Vx, Ve_k)^2$. Finally we find $\varphi_\nu(y)$ the characteristic functional of measure ν. Denote by $\varphi_{\underline{\mu}}(l)$ the extension by continuity of the characteristic functional $\varphi_\mu(z)$ on $\tilde{\mathscr{X}}$. Then

$$\varphi_\nu(y)=\int e^{i(y,Vx)}\,\mu(dx)=\int e^{i(V^*y,x)}\,\mu(dx)=\varphi_\mu(V^*y).$$

§ 2. Measurable Polynomial Functions. Orthogonal Polynomials

Despite the fact that, in the previous section, we have established – even for the case of linear measurable functions – the basic difference between the spaces of measurable functions and the spaces of square integrable functions, we shall, nevertheless, when studying polynomial functions of higher degrees, confine ourselves to the case of square integrable functions which arise as mean square limits of continuous polynomial functions. The reasons for this restriction are, on one hand due to the complexity of the structure of square measurable functions (let alone one higher polynomial function), and on the other hand, due to the ease with which the mean square integrable functions can be used in various analytic applications, in particular for the construction of orthogonal expansions. However, to assure the existence of nontrivial square integrable continuous polynomials, certain restrictions should be imposed on the measure μ.

Denote by M_n the class of measures μ such that

$$\int |x|^n \mu(dx) < \infty \quad \text{and let} \quad M_\infty = \bigcap_n M_n$$

Gaussian measures are examples of measures in M_∞. In the case of measures μ belonging to M_∞ every polynomial of degree n will belong to $\mathscr{L}_2[\mu]$ which is the space of measurable functions, square integrable with respect to μ. Moreover, if $\mu \in M_\infty$, then $\mathscr{L}_2[\mu]$ contains all the continuous polynomials. We recall the definition of a polynomial function (or simply a polynomial.) A function $\Phi(x)$ which can be represented in the form

$$\Phi(x) = H(x, \ldots, x),$$

where $H(x_1, \ldots, x_n)$ is an n-linear form on $\mathscr{X}$ is called a *homogeneous polynomial* of degree n and the function of the form

$$T_n(x) = \sum_{k=0}^{n} \Phi_k(x),$$

where Φ is a homogeneous polynomial of degree k is called a *polynomial* of degree n. For each homogeneous polynomial Φ of degree k there exists a k-linear continuous symmetric function $\tilde{H}(x_1, \ldots, x_k)$ generating the polynomial (or associated with the polynomial). Such a function is uniquely determined. Let $\{e_k\}$ be an orthonormal basis in $\mathscr{X}$. The numbers

$$\alpha_{i_1, \ldots, i_k} = \tilde{H}(e_{i_1}, \ldots, e_{i_k})$$

are called the *coefficients* of the function $\tilde{H}$ and the form Φ in this basis.

The form Φ is expressed in terms of its coefficients in the following manner: $\Phi(x)=\sum_{i_1,\dots,i_k}\alpha_{i_1,\dots,i_k}(x,e_{i_1})\dots(x,e_{i_k})$. Consider the expression for the integral $\int T_n(x)\,T'_{n'}(x)\,\mu(dx)$, where T_n and $T'_{n'}$ are polynomials in terms of the characteristics of measure μ. Since $T_n(x)\cdot T'_{n'}(x)$ is a polynomial it is sufficient to be able to determine integrals of a single polynomial and to do this, it is sufficient to determine integrals of homogeneous forms. Let $\mu\in M_n$ and $\Phi(x)$ be a homogeneous form of degree n. Denote by H the corresponding n-linear continuous function. If P is the projector on a certain space, then

$$|\Phi(x)-\Phi(Px)|=|H(x,\dots,x)-H(Px,\dots,Px)|\leqslant$$
$$\leqslant\sum_{k=1}^{n}|H(\underbrace{x,\dots,x}_{k},Px,\dots,Px)-$$
$$-H(\underbrace{x,\dots,x}_{k-1},Px,\dots,Px)|=$$
$$=\sum_{k=1}^{n}|H(\underbrace{x,\dots,x}_{k-1},x-Px,Px,\dots,Px)|\leqslant$$
$$\leqslant nC|x-Px|\,|x|^{n-1},$$

where $C=\sup[H(x_1,\dots,x_n);\,|x_i|\leqslant 1]$. Since $\Phi(x)-\Phi(Px)\to 0$ as $P\to I$ and is bounded by the quantity $2nC|x|^n$ integrable with respect to measure μ, it follows that

$$\int\Phi(x)\,\mu(dx)=\lim_{P\uparrow I}\int\Phi(Px)\,\mu(dx).$$

We choose an arbitrary orthonormal basis $\{e_m\}$ and denote by P_m the projector on the subspace spanned by $e_1,\dots,e_m$. Then

$$\int\Phi(x)\,\mu(dx)=\lim_{m\to\infty}\int\Phi(P_mx)\,\mu(dx).$$

If $\alpha_{i_1,\dots,i_n}$ are the coefficients of the form Φ in the basis then

$$\Phi(P_mx)=\sum_{\substack{i_k\leqslant m\\ k=1,\dots,n}}\alpha_{i_1,\dots,i_n}(x,e_{i_1})\dots(x,e_{i_n})$$

and

$$\int\Phi(P_mx)\,\mu(dx)=\sum_{\substack{i_k\leqslant m\\ k=1,\dots,n}}\alpha_{i_1,\dots,i_n}\int(x,e_{i_1})\dots(x,e_{i_n})\,\mu(dx).$$

Let

$$S_\mu^{(n)}(z_1, \ldots, z_n) = \int (x, z_1) \ldots (x, z_n)\, \mu(dx)$$

be an n-moment form for measure μ. Then

$$S_\mu^{(n)}(e_{i_1}, \ldots, e_{i_n}) = \int (x, e_{i_1}) \ldots (x, e_{i_n})\, \mu(dx)$$

are the coefficients of this form in basis $\{e_k\}$. Consequently,

$$\int \Phi(x)\, \mu(dx) = \lim_{m \to \infty} \sum_{\substack{i_k \leqslant m \\ k = 1, \ldots, n}} \alpha_{i_1, \ldots, i_n} S_\mu^{(n)}(e_{i_1}, \ldots, e_{i_n}).$$

To prove that the expression on the right of the equation is the sum of a convergent series, note that relation

$$\int \Phi(x)\, \mu(dx) = \lim_{m_1 \to \infty, \ldots, m_n \to \infty} \int H(P_{m_1} x, \ldots, P_{m_n} x)\, \mu(dx) =$$

$$= \lim_{m_1 \to \infty, \ldots, m_n \to \infty} \sum_{i_1 = 1}^{m_1} \ldots \sum_{i_n = 1}^{m_n} H(e_{i_1}, \ldots, e_{i_n})\, S_\mu^{(n)}(e_{i_1}, \ldots, e_{i_n})$$

is satisfied. This implies the convergence of the series

$$\sum_{i_1, \ldots, i_n = 1}^{\infty} H(e_{i_1}, \ldots, e_{i_n})\, S_\mu^{(n)}(e_{i_1}, \ldots, e_{i_n}). \tag{1}$$

If for two n-linear symmetric continuous functions H and $S_\mu^{(n)}$ the series (1) is convergent in any orthonormal basis, then the sum of this series (which is independent of the choice of the basis) will be denoted by $\operatorname{Sp} H * S_\mu^{(n)}$ and will be called the trace of the product of these forms. Thus formula

$$\int \Phi(x)\, \mu(dx) = \operatorname{Sp} H * S_\mu^{(n)}$$

is verified where H is an n-linear form corresponding to the homogeneous form Φ and $S_\mu^{(n)}$ is the n-moment form of measure μ.

Construction of an orthogonal system of polynomial functions. Henceforth we shall assume that $\mu \in M_\infty$. The mean square limit of continuous polynomials of degree n will be referred to as a measurable polynomial of degree at most n. For construction of all the measurable polynomials it is convenient to utilize an orthogonal system of polynomials. Let $\mathscr{P}_n$ be the set of all measurable polynomials of degree at most n; $\mathscr{P}_n$ is a subspace in the Hilbert space $\mathscr{L}_2[\mu]$. Clearly $\tilde{\mathscr{P}}_0 \subset \tilde{\mathscr{P}}_1 \subset \ldots \subset \tilde{\mathscr{P}}_n$. The subspace of $\tilde{\mathscr{P}}_n$ which is the orthogonal complement to $\tilde{\mathscr{P}}_{n-1}$ is denoted by $\mathscr{P}_n$. The subspaces $\mathscr{P}_0, \mathscr{P}_1, \ldots, \mathscr{P}_n$ are mutually orthogonal and are called an orthogonal system of polynomials. Every measurable polynomial can be uniquely represented in the form $\sum c_k g_k(x)$, where $g_k \in \mathscr{P}_k$. To construct all the measurable polynomials it is sufficient to construct the whole subspace $\mathscr{P}_k$. Such a construction is normally carried out by induction.

Let $T(x)$ be a homogeneous form of degree n generated by a symmetric n-linear function H. Then

$$T(x) = P_n(H, x) - \sum_{k=0}^{n-1} Q_k(H, x), \tag{3}$$

where $P_n(H, x) \in \mathscr{P}_n$ and $Q_k(H, x) \in \mathscr{P}_k$. Clearly $P_n(H, x)$ and $Q_k(H, x)$ are linearly dependent on H. Denote by $\tilde{\Phi}^n$ the space of all n-linear continuous functions. Introduce in $\tilde{\Phi}^n$ the scalar product

$$\langle H, H' \rangle_n = \int P_n(H, x)\, P_n(H', x)\, \mu(dx) \tag{4}$$

and complete the space $\tilde{\Phi}^n$ with respect to this scalar product. The Hilbert space obtained will be denoted by Φ^n and its elements will be called generalized forms. Note that the correspondence $H \leftrightarrow P_n(H, x)$ is isometric and therefore can be extended over the whole Φ^n. The function in $\mathscr{P}_n$ which corresponds to $H \in \Phi^n$ will also be denoted by $P_n(H, x)$. Functions $Q_k(H, x)$ in formula (3) can be represented in the form $Q_k(H, x) = P_k(H_k, x)$ where $H_k \in \Phi^k$. The linear operator which maps $H \in \tilde{\Phi}^n$ into $H_k \in \Phi^k$ will be denoted by V_{nk}. We thus obtain from (3)

$$P_n(H, x) = T(x) + \sum_{k=0}^{n-1} P_k(V_{nk} H, x). \tag{5}$$

The last formula shows that in order to determine $P_n(H, x)$ for $H \in \tilde{\Phi}^n$ it is sufficient to know operators V_{nk}, while to extend $P_n(H, x)$ onto Φ^n it is necessary to know the scalar product $\langle \cdot, \cdot \rangle_n$. If these two characteristics are known one can reduce the construction of $P_n(H, x)$ to the construction of $P_k(H, x)$ for $k < n$.

Denote for an arbitrary $H_n \in \tilde{\Phi}^n$ and $H_k \in \tilde{\Phi}^k$ the $(n+k)$-linear form $H_n(x_1, \ldots, x_n)\, H_k(x_{n+1}, \ldots, x_{n+k})$ (this form is asymmetric) by $H_n \times H_k$. The following recursion relation follows from formulas (2), (4) and (5):

$$\langle H, H' \rangle_n = \mathrm{Sp}(H \times H' * S_\mu^{(2n)}) - \sum_{k=0}^{n-1} \langle V_{nk} H, V_{nk} H' \rangle_k. \tag{6}$$

To determine V_{nk} we introduce bilinear forms $A_{nk}(H_n, H_k)$ defined for

$H_n \in \tilde{\Phi}^n$ and $H_k \in \tilde{\Phi}^k$:

$$A_{nk}(H, H_k) = \int T(x)\, P_k(H_k, x)\, \mu(dx), \tag{7}$$

where $T(x)$ is a homogeneous form of degree n and $H \in \tilde{\Phi}^n$ is the corresponding n-linear function. To determine A_{nk} we have the recursion relation

$$A_{nk}(H_n, H_k) = \mathrm{Sp}(H_n \times H_k * S_\mu^{(n+k)}) + \sum_{j=0}^{k-1} A_{nj}(H_n, V_{kj} H_k), \tag{8}$$

which follows from (5) and (7). Finally to determine V_{nk} we have the following equality which is valid for all $H \in \tilde{\Phi}^n$ and $H_k \in \tilde{\Phi}^k$:

$$A_{nk}(H, H_k) = -\langle V_{nk} H, H_k \rangle_k, \tag{9}$$

(To obtain this relation, multiply (5) by $\mathsf{P}_k(H^k, x)$ and then integrate.) If $\{H_m^k\ m = 1, 2, \ldots\}$ is an orthonormal basis in Φ^k, then

$$V_{nk} H = -\sum_{m=1}^{\infty} A_{nk}(H, H_m^k)\, H_m^k. \tag{10}$$

Relations (6), (8) and (9) enable us to determine successively $\langle \cdot, \cdot \rangle_0$, A_{10}, V_{10}, $\langle \cdot, \cdot \rangle_1$, A_{20}, V_{20}, A_{21}, $V_{21}, \ldots$ and so on.

As an example consider the construction of subspaces P_n for the Gaussian measure with mean 0 and the correlation operator B. All the formulas will be significantly simplified if the scalar product

$$(x, y)_+ = (Bx, y)$$

is utilized in which the traces are calculated. The moment forms of measure μ are especially simple in this scalar product – they are $S_n^+(z_1, \ldots, z_n) = 0$ for n odd and

$$S_n^+(z_1, \ldots, z_n) = \sum \prod_{k=1}^{n/2} (z_{i_k}, z_{j_k})_+$$

for n even. Here the summation is carried over all possible partitions of numbers $1, 2, \ldots, n$ into $n/2$ pairs (i_k, j_k). The last formula follows from the equality

$$\begin{aligned} S_n^+(z_1, \ldots, z_n) &= \\ &= i^{-n} \frac{\partial^n}{\partial \alpha_1 \ldots \partial \alpha_n} \int \exp\left\{ i\left(x, \sum_1^n \alpha_k, z_k\right)\right\} \mu(dx) \Bigg|_{\substack{\alpha_1 = 0 \\ \vdots \\ \alpha_n = 0}} = \\ &= i^{-n} \frac{\partial^n}{\partial \alpha_1 \ldots \partial \alpha_n} \exp\left\{ -\tfrac{1}{2}\left(\sum_1^n \alpha_k z_k, \sum_1^n \alpha_k z_k\right)_+ \right\} \Bigg|_{\substack{\alpha_1 = 0 \\ \vdots \\ \alpha_n = 0}} \end{aligned}$$

The traces in this scalar product $(\cdot, \cdot)_+$ will be denoted by Sp_+. We introduce the mapping of $\tilde{\Phi}^n$ into $\tilde{\Phi}^{n-2}$ defined by the formula

$$\mathrm{Sp}_+^2 H(z_1, \dots, z_{n-2}) = \sum_k H(e_k, e_k, z_1, \dots, z_{n-2}),$$

where $\{e_k\}$ is a certain orthonormal basis (with respect to the scalar product $(\cdot, \cdot)_+$). We shall construct successively the operators V_{nk} and the scalar products $\langle \cdot, \cdot \rangle_n$. First we compute

$$\mathrm{Sp}_+ (H_n \times H_k * S_{n+k}^+).$$

The quantity is zero for $n+k$ odd. Let $n+k=2m$. Note that for the evaluation of $\mathrm{Sp}_+ (H_{n+k} * S_{n+k}^+)$, where H_{n+k} is an $(n+k)$-linear form, one should subdivide the arguments of H_{n+k} into all possible pairs, then convolve in each pair (i.e. substitute in place of this pair of arguments the identical vectors from the orthogonal basis and to sum up over the basis) and finally add the results. Let the arguments of $H_n \times H_k$ be subdivided into pairs in such a manner that there are S pairs containing the arguments of H_n and H_k. Convolving each form separately over the remaining pairs we obtain

$$(\mathrm{Sp}_+^2)^{\frac{n-s}{2}} H_n \quad \text{and} \quad (\mathrm{Sp}_+^2)^{\frac{k-s}{2}} H_k.$$

The number of such partitions is

$$C_n^s k(k-1)\dots(k-s+1)(n-s-1)!!(k-s-1)!! = \frac{n!k!}{s!(n-s)!!(k-s)!!}.$$

Hence putting $\frac{k-s}{2} = j, \frac{n-k}{2} = r$ we obtain

$$\mathrm{Sp}_+ (H_n \times H_k * S_{n+k}^+) =$$
$$= \sum_{j \leqslant k/2} \frac{n!k!}{s!(n-s)!!(k-s)!!} \mathrm{Sp}_+ \{(\mathrm{Sp}_+^2)^{j+r} H_n * (\mathrm{Sp}_+^2)^j H_k\}.$$

Utilizing this formula we determine

$$V_{2n,0} H_{2n} = -(2n-1)!! (\mathrm{Sp}_+^2)^n H_{2n},$$
$$V_{2n+1,1} H_{2n+1} = -(2n+1)!! (\mathrm{Sp}_+^2)^n H_{2n+1}.$$

Next

$$A_{2n,2}(H_{2n}, H_2) = \frac{(2n)!}{(2n-2)!!} \mathrm{Sp}_+ ((\mathrm{Sp}_+^2)^{n-1} H_{2n} \times H_2) +$$
$$+ (2n-1)!! (\mathrm{Sp}_+^2)^n H_{2n} (\mathrm{Sp}_+^2 H_2) - A_{2n,0}(H_{2n}, V_{20} H_2) =$$
$$= \frac{(2n)!}{(2n-2)!!} \mathrm{Sp}_+ ((\mathrm{Sp}_+^2)^{n-1} H_{2n} * H_2).$$

Consequently

$$V_{2n,2} = -\frac{(2n)!}{(2n-2)!!\,2}(\mathrm{Sp}_+^2)^{n-1} H_{2n}.$$

It can be verified by induction that

$$A_{2n,2k}(H_{2n}, H_{2k}) = \frac{(2n)!}{(2n-2k)!!}\mathrm{Sp}_+((\mathrm{Sp}_+^2)^{n-k} H_{2n} \times H_{2k}),$$

$$\langle H_{2k}, \tilde{H}_{2k}\rangle_{2k} = (2k)!\,\mathrm{Sp}_+ H_{2k} * \tilde{H}_{2k}$$

and hence

$$V_{2n,2k} H_{2n} = -\frac{(2n)!}{(2n-2k)!!\,(2k)!}(\mathrm{Sp}_+^2)^{n-k} H_{2n}.$$

In the same manner

$$V_{2n+1,2k+1} H_{2n+1} = -\frac{(2n+1)!}{(2n-k)!!\,(2k+1)!}(\mathrm{Sp}_+^2)^{n-k} H_{2n+1}.$$

Thus we finally obtain

$$\langle H_n, \tilde{H}_n\rangle_n = n!\,\mathrm{Sp}_+ H_n * \tilde{H}_n,$$

$$V_{nk} H_n = \begin{cases} 0; & n+k \quad \text{odd} \\ -\dfrac{n!}{(n-k)!!\,k!}(\mathrm{Sp}_+^2)^{\frac{n-k}{2}} H_n; & n+k \quad \text{even}. \end{cases}$$

We now investigate the question of which measurable polynomials are dense in $\mathscr{L}_2[\mu]$. A sufficient condition is given by the following lemma.

Lemma. *If the characteristic functional $\varphi(z)$ of a measure μ is such that for each z the function $\varphi(tz)$ is an analytic function of t in a certain neighborhood zero, then the set of measurable polynomials is dense in $\mathscr{L}_2[\mu]$.*

Proof. Denote the closure of the set of all measurable polynomials by $\mathscr{P}$. We show that $e^{i(x,z)}$ regarded as a function of x belongs to $\mathscr{P}$. For this purpose it is sufficient to prove that for some sequence of real-valued polynomials $q_n(t)$

$$\lim_{n\to\infty} \int |e^{i(z,x)} - q_n((z,x))|^2 \,\mu(dx) = 0. \tag{11}$$

Denote by $F(t)$ the distribution function

$$F(\lambda) = \mu(\{x : (z,x) < \lambda\}).$$

Relation (11) is equivalent to the following:

$$\lim_{n\to\infty} \int |e^{i\lambda}-q_n(\lambda)|^2\, dF(\lambda)=0. \tag{12}$$

Since

$$\varphi(zt)=\int e^{it\lambda}\, dF(\lambda),$$

it follows from the analyticity of this function in the neighborhood of zero that for some $\delta>0$

$$\int e^{\delta|\lambda|}\, dF(\lambda)<\infty.$$

Let $\mathscr{L}$ be the space of complex-valued functions $g(\lambda)$ such that $\int |g(\lambda)|^2\, dF(\lambda)<\infty$ with the scalar product

$$\int g_1(\lambda)\, \overline{g_2(\lambda)}\, dF(\lambda).$$

Let $\mathscr{L}'$ be the closure in $\mathscr{L}$ of the set of all polynomials and let $g(\lambda)$ be the projection of function $e^{i\lambda}$ on $\mathscr{L}'$. Then

$$\int (e^{i\lambda}-g(\lambda))\, \lambda^n\, dF(\lambda)=0$$

for all $n\geqslant 0$.

Utilizing the inequalities

$$\left|\sum_1^m \frac{(i\lambda t)^n}{n!}\right|\leqslant e^{\lambda|t|},\quad 2\left|(e^{i\lambda}-g(\lambda))\sum_1^m \frac{(i\lambda t)^n}{n!}\right|\leqslant |e^{i\lambda}-g(\lambda)|^2+e^{|\lambda|\delta}$$

for $|t|<\frac{\delta}{2}$ we obtain that

$$0=\lim_{m\to\infty}\int (e^{i\lambda}-g(\lambda))\sum_{n=1}^m \frac{(i\lambda t)^n}{n!}\, dF(\lambda)=\int (e^{i\lambda}-g(\lambda))\, e^{i\lambda t}\, dF(\lambda).$$

Differentiating this relation with respect to t we find that for $|t|<\frac{\delta}{2}$

$$\int (e^{i\lambda}-g(\lambda))\, e^{i\lambda t}\, \lambda^n\, dF(\lambda)=0.$$

Hence we have for $|t|<\frac{\delta}{2}$ and $|u|<\frac{\delta}{2}$

$$0=\int (e^{i\lambda}-g(\lambda))\, e^{i\lambda t} \sum_{n=0}^{\infty} \frac{(i\lambda u)^n}{n!}\, dF(\lambda)=\int (e^{i\lambda}-g(\lambda))\, e^{i\lambda t+i\lambda u}\, dF(\lambda),$$

or

$$0=\int (e^{i\lambda}-g(\lambda))\, e^{i\lambda t}\, dF(\lambda) \quad \text{for} \quad |t|<\delta.$$

Continuing the previous arguments we observe that for all t

$$\int (e^{i\lambda}-g(\lambda))\, e^{i\lambda t}\, dF(\lambda)=0$$

Setting $t=-1$ we obtain

$$\int (e^{i\lambda}-g(\lambda))\, e^{-i\lambda}\, dF(\lambda)=0,$$

which together with the equality

$$\int (e^{i\lambda}-g(\lambda))\, \overline{g(\lambda)}\, dF(\lambda))=0$$

yields the relationship

$$\int |e^{i\lambda}-g(\lambda)|^2\, dF(\lambda)=0,$$

which implies the limit (12). We now show that any continuous bounded function belongs to the set $\mathcal{P}$. Let $f(x)$ be such a function. We choose a compactum K such that $\mu(\mathcal{X}-K)<\varepsilon$. The function $f(x)$ is uniformly continuous in K hence for each $\varepsilon>0$ a $\delta>0$ can be found such that $|f(x)-f(y)|<\varepsilon$ for $|x-y|<\delta$.

Let N be a finite-dimensional subspace which is a δ-net in K and denote by P the projector on N and let K' be the projection of K on N. There exists a trigonometric polynomial T on N such that $|f(Px)-T(x)|<\varepsilon$ for $x\in K'$ and which does not exceed $\sup_x |f(x)|$ in its absolute value. It is easy to see that

$$\int |f(x)-T(Px)|^2\, \mu(dx)=O(\varepsilon).$$

Hence in addition to trigonometric polynomials all the bounded continuous functions belong to $\mathscr{P}$ and hence also $\mathscr{L}_2[\mu]\subset\mathscr{P}$, since the bounded continuous functions form a dense set in $\mathscr{L}_2[\mu]$. The lemma is thus proved. □

We now present an example which shows that even on the real line there exists a measure for which all the polynomials are square integrable but are not dense in $\mathscr{L}_2[\mu]$. Let the measure μ on $\mathscr{R}^1$ be defined by the density

$$f(x)=\begin{cases}\dfrac{1}{\sqrt{2\pi}}\dfrac{1}{x}\,e^{-\frac{1}{2}\log^2 x}, & x>0,\\ 0, & x\leqslant 0.\end{cases}$$

Consider the following function $g(x)$ belonging to $\mathscr{L}_2[\mu]$:

$$g(x)=\begin{cases}\exp\{\varepsilon\log^2 x\}\sin\pi(1-2\varepsilon)\log x, & x>0,\\ 0, & x\leqslant 0, \quad \varepsilon<\frac{1}{4}.\end{cases}$$

For all integer valued k we have the equality:

$$\int x^k g(x)\,f(x)\,dx=\frac{1}{\sqrt{2\pi}}\int_{-\infty}^{\infty} e^{kt}\,e^{-\frac{1}{2}(1-2\varepsilon)t^2}\sin\pi(1-2\varepsilon)\,t\,dt=$$

$$=\frac{1}{2i\sqrt{2\pi}}\int_{-\infty}^{\infty}\exp\{(k+i\pi(1-2\varepsilon))\,t-\tfrac{1}{2}(1-2\varepsilon)\,t^2\}\,dt-$$

$$-\frac{1}{2i\sqrt{2\pi}}\int_{-\infty}^{\infty}\exp\{(k-i\pi(1-2\varepsilon))\,t-\tfrac{1}{2}(1-2\varepsilon)\,t^2\}\,dt=$$

$$=\frac{1}{2i\sqrt{1-2\varepsilon}}\left(\exp\left\{\frac{(k+i\pi(1-2\varepsilon))^2}{2(1-2\varepsilon)}\right\}-\right.$$

$$\left.-\exp\left\{\frac{(k-i\pi(1-2\varepsilon))^2}{2(1-2\varepsilon)}\right\}\right)=0.$$

Thus $g(x)$ is orthogonal to all the polynomials.

§3. Measurable Mappings

Let $\mathscr{X}$ and $\mathscr{Y}$ be two Hilbert spaces with the σ-algebras of Borel sets $\mathfrak{A}$ and $\mathfrak{B}$ respectively. The function $R(x)$ defined on a $\mathfrak{A}$-measurable set

D_R and taking on values in $\mathscr{Y}$ is called a *measurable mapping* of $\mathscr{X}$ into $\mathscr{Y}$ provided for all $B \in \mathfrak{B}$, $R^{-1}(B) \in \mathfrak{A}$. Such a mapping is called *μ-measurable* or *measurable with respect to measure μ* provided $\mu(D_R)=1$. Henceforth the notion of measurable mapping is used in this section in the sense of a mapping measurable with respect to the corresponding measure. If a mapping R is μ-measurable then it maps measure μ into a measure ν on $(\mathscr{Y}, \mathfrak{B})$ defined by the formula

$$\nu(B)=\mu(R^{-1}(B)).$$

Evaluation of integrals with respect to measure ν are reduced to their evaluation with respect to measure μ: for any $\mathfrak{B}$-measurable function $f(y)$

$$\int f(y)\, \nu(dy)=\int f(R(x))\, \mu(dx),$$

provided only one of these integrals is well defined. An important question in many applications is the problem of determining the characteristics of the measure ν (i.e. the characteristic functional or moment-functions) from the known characteristics of the measure μ.

The simplest example of a measurable mapping is a continuous mapping of $\mathscr{X}$ into $\mathscr{Y}$. The theorem below shows the connection between continuous and measurable mappings.

Theorem. *For any μ-measurable mapping $R(x)$ one can find a sequence of continuous mappings $R_n(x)$ such that*

$$R(x)=\lim_{n\to\infty} R_n(x) \quad (\operatorname{mod} \mu).$$

Proof. It is sufficient to show that for any $\varepsilon>0$ one can find a continuous mapping $\bar{R}(x)$ of $\mathscr{X}$ into $\mathscr{Y}$ such that

$$\mu(\{x:|\bar{R}(x)-R(x)|>\varepsilon\})<\varepsilon.$$

Denote by ν the measure on $(\mathscr{Y}, \mathfrak{B})$ which is the image of μ under R. Let K be a compactum in $\mathscr{Y}$ such that $\nu(\mathscr{Y}-K)<\frac{\varepsilon}{2}$. Denote by K' the preimage of K under R. Then $\mu(\mathscr{X}-K')<\frac{\varepsilon}{2}$. Let N be the finite-dimensional linear subspace in $\mathscr{Y}$ which is a $\frac{\varepsilon}{2}$-net in K and $y_1, \ldots, y_m$ be a basis in N. Then for all $x \in K$

$$|R(x)-\sum_1^m (R(x), y_k)\, y_k|<\frac{\varepsilon}{2}.$$

Since $(R(x), y_k)$ is a measurable function (with respect to μ) a continuous function $\varphi_k(x)$ exists such that

$$\mu\left(\left\{x: |\varphi_k(x)-(R(x), y_k)| > \frac{\varepsilon}{2m}\right\}\right) < \frac{\varepsilon}{2m}.$$

Then

$$\mu\left(\left\{x: \left|R(x)-\sum_1^m \varphi_k(x)\, y_k\right| > \varepsilon\right\}\right) <$$

$$< \mu\left(\left\{x: \left|R(x)-\sum_1^m (R(x), y_k)\, y_k\right| > \frac{\varepsilon}{2}\right\}\right) +$$

$$+\sum_1^m \mu\left(\left\{x: |(R(x), y_k)-\varphi_k(x)| > \frac{\varepsilon}{2m}\right\}\right).$$

To complete the proof, we note that the mapping $\bar{R}(x) = \sum_{k=1}^m \varphi_k(x)\, y_k$ is continuous. □

When studying measurable mappings of $\mathscr{X}$ into $\mathscr{Y}$, it is sufficient to consider measurable mappings of $\mathscr{X}$ into $\mathscr{X}$ since every measurable mapping of $\mathscr{X}$ into $\mathscr{Y}$ can be represented as a composition of a measurable mapping of $\mathscr{X}$ into $\mathscr{X}$ and a continuous mapping of $\mathscr{X}$ into $\mathscr{Y}$.

Polynomial mappings. Consider a mapping of $\mathscr{X}$ into $\mathscr{X}$. The mapping R is called *polynomial* if $(R(x), z)$ is a polynomial in x for any z. If for some z the expression $(R(x), z)$ is a homogeneous polynomial form of degree n, then we say that $R(x)$ is a homogeneous polynomial mapping of degree n. When studying homogeneous polynomial mappings the standard mappings – to be described below – play an important role.

Denote by $\mathscr{X}^{0k}$ the space of k-linear symmetric continuous forms S satisfying the condition

$$\operatorname{Sp} S * S < \infty.$$

The space $\mathscr{X}^{0k}$ with scalar product

$$(S, T) = \operatorname{Sp} S * T$$

is a separable Hilbert space. Denote by $\mathfrak{B}^k$ the σ-algebra of Borel sets in $\mathscr{X}^{0k}$. Consider the mapping of $\mathscr{X}$ into $\mathscr{X}^{0k}$ defined by the relation

$$x \overset{T}{\leftrightarrow} T_x(z_1, \ldots, z_k) = \prod_{j=1}^k (z_j, x).$$

This mapping is continuous since

$$\operatorname{Sp}(T_{x_1}-T_{x_2})*(T_{x_1}-T_{x_2})=\sum_{i_1,\ldots,i_k}\left[\left(\prod_{j=1}^{k}(x_1,e_{i_j})-\prod_{j=1}^{k}(x_2,e_{i_j}\right]^2=$$

$$=\sum_{i_1,\ldots,i_k}\left\{\sum_{l=1}^{k}\prod_{j=1}^{l-1}(x_1,e_{i_j})(x_1-x_2,e_{i_j})\prod_{j=l+1}^{k}(x_2,e_{i_j})\right\}^2\leqslant$$

$$\leqslant k^2|x_1-x_2|^2\max_{l\leqslant k}\{|x_1|^{2l-2}+|x_2|^{2l-2}\}.$$

Consequently the mapping is measurable. Introduce on $\mathscr{X}^{\text{o}k}$ the measure $\mu^{\text{o}k}$ by means of the formula

$$\mu^{\text{o}k}(A)=\mu(T^{-1}(A)).$$

The measure $\mu^{\text{o}k}$ is called the k-th power of the measure μ. Note that $\mu^{\text{o}k}$ is the distribution of the random variable T_x with values in $\mathscr{X}^{\text{o}k}$ in the probability space $(\mathscr{X}, \mathfrak{B}, \mu)$.

Denote by $\varphi_k(T)$ the characteristic functional of measure $\mu^{\text{o}k}$. Since $\operatorname{Sp} T_x*S=S(x,\ldots,x)$ we have that

$$\varphi_k(S)=\int e^{i\operatorname{Sp}T*S}\mu^{\text{o}k}(dT)=\int e^{iS(x,\ldots,x)}\mu(dx).$$

Thus $\varphi_k(S)$ is determined by the measure μ and hence by the characteristic functional $\varphi(z)$ of measure μ. We now describe some methods for constructing $\varphi_k(S)$ by means of $\varphi(z)$. Let $V_{u_1,\ldots,u_k}$ be a form in $\mathscr{X}^{\text{o}k}$ of the type $V_{u_1,\ldots,u_k}(z_1,\ldots,z_k)=\prod_1^k(u_j,z_j)$. Then

$$\varphi_k(V_{u_1,\ldots,u_k})=\int\exp\left\{i\prod_1^k(x,u_j)\right\}\mu(dx).$$

To evaluate the integral appearing in the r.h.s. we note that $\varphi(t_1u_i+\ldots+t_ku_k)$ is the joint characteristic function of the variables $(x,u_1),\ldots,(x,u_k)$ and

$$\int\exp\left\{is\prod_1^k(x,u_j)\right\}\mu(dx)$$

is the characteristic functional of the variable $\prod_{j=1}^k(x,u_j)$ on the probability space $(\mathscr{X},\mathfrak{B},\mu)$. If $\varphi(t_1u_1+\ldots+t_ku_k)$ is absolutely integrable with respect to $t_1,\ldots,t_k$, then

$$\varphi_k(V_{u_1,\ldots,u_k})=\left(\frac{1}{2\pi}\right)^k\int\ldots\int\exp\{is_1\ldots s_k-is_1t-\ldots-is_kt_k\}\times$$

$$\times\varphi(t_1u_1+\ldots+t_ku_k)\,dt_1\ldots dt_k\,ds_1\ldots ds_k. \qquad (1)$$

Computing the integral in (1) by means of a certain regularization procedure (for example, by introducing the factor $\exp\{-\varepsilon \sum t_j^2\}$ under the sign of the integral and then approaching the limit as $\varepsilon \to 0$) we thus show that formula (1) is valid for arbitrary characteristic functionals φ.

Consider the joint characteristic functional

$$\varphi_{k,1}(T, z) = \int \exp\{iT(x, \dots, x) + i(z, x)\}\, \mu(dx).$$

Assume that $\int |x|^k \mu(dx) < \infty$. Then for any form S in $\mathscr{X}^{0k}$ the relation

$$\operatorname{Sp} d_z^k \varphi_{k,1}(T, z)\, ^r S = i^k \int S(x, \dots, x) \exp\{iT(x, \dots, x) + i(z, x)\}\, \mu(dx)$$

is valid, where $d_z^k \varphi_{k,1}(T\ z)$ is the k-th differential of the function $\varphi_{k,1}(T\ z)$ with respect to z (such a differential is a k-linear form). On the other hand

$$i \int S(x_1, \dots, x) \exp\{iT(x, \dots, x) + i(z, x)\}\, \mu(dx) = \operatorname{Sp} d_T \varphi_{k,1}(T, z) * S,$$

where $d_T \varphi_{k,1}(T, z)$ is the first differential of the function $\varphi_{k,1}(T, z)$ with respect to T.

Thus the function $\varphi_{k,1}$ satisfies the differential equation

$$i^{k-1} d_T \varphi_{k,1}(T, z) = d_z^k \varphi_{k,1}(T, z). \tag{2}$$

Note also that $\varphi_{k,1}(V_{u_1, \dots, u_k}, z)$ can be evaluated by means of formula (1) if we substitute $\varphi(t_1 u_1 + \dots t_k u_k)$ by $\varphi(z + t_1 u_1 + \dots + t_k u_k)$ in the r.h.s. of (1).

Let V be an arbitrary measurable linear mapping of $\mathscr{X}^{0k}$ into $\mathscr{X}$. The composition of the mappings

$$\mathscr{X} \xrightarrow{T} \mathscr{X}^{0k} \xrightarrow{V} \mathscr{X}$$

will be called a *measurable polynomial mapping of the k-th degree*. Let $R(x)$ be such a mapping. Then $(R(x), z)$ is a homogeneous polynomial of degree k for each z or is the limit of such polynomials in measure μ.

Indeed if $R(x) = VT_x$ where V is a measurable linear mapping of $\mathscr{X}^{0k}$ into $\mathscr{X}$, then $(R(x), z) = V_z^*(T_x)$, where V^* is the conjugate of V (cf. Section 1) (this mapping maps measurable linear functionals on $\mathscr{X}$ into measurable linear functionals on $\mathscr{X}^{0k}$), V_z^* is a functional on $\mathscr{X}^{0k}$ which is the image of $(z, \cdot)$ on $\mathscr{X}$, and $V_z^*(S)$ is the value obtained when V_z^* is applied to $S \in \mathscr{X}^{0k}$.

If V_z^* is a continuous functional, then $V_z^*(T_x)$ is a homogeneous polynomial of degree k. If, however, V_z^* is the limit in measure μ^{0k} of continuous linear functionals g_n defined on $\mathscr{X}^{0k}$, then $g_n(T_x) \to V_z^*(T_x)$ in measure μ, and $g_n(T_x)$ is a homogeneous polynomial of degree k. We now prove the converse. Assume that $R(x)$ is a homogeneous polynomial mapping of $\mathscr{X}$ into $\mathscr{X}$. Denote by S_z the k-linear form such that

$$(R(x), z) = S_z(x, \ldots, x).$$

If T_x is an element in $\mathscr{X}^{0k}$ defined by the equality

$$T_x(z_1, \ldots, z_k) = (x, z_1) \ldots (x, z_k),$$

then $(R(x), z) = \operatorname{Sp} S_z * T$. Clearly S_z is a linear mapping of $\mathscr{X}$ into $\mathscr{X}^{0k}$, we denote it by U: $S_z = Uz$. Then

$$(R(x), z) = \operatorname{Sp} Uz * T_x = (z, U^* T_x),$$

where U^* is the conjugate of U. Hence $R(x) = U^* T_x$, where U^* is a continuous linear mapping of $\mathscr{X}^{0k}$ into $\mathscr{X}$.

Now let $R(x)$ be the limit in measure μ of continuous mappings $R_n(x)$ of degree k. Then $R_n(x) = V_n T_x$ and $R_n(x)$ converges μ-almost everywhere. Therefore a measurable linear mapping V of the space $\mathscr{X}^{0k}$ into $\mathscr{X}$ exists to which $V_n T$ are convergent in measure μ^{0k} and such that $R(x) = VT_x$.

Using function $\varphi_k(T)$ one can easily find the characteristic function of the measure ν to which μ is mapped under the measurable polynomial mapping $R(x)$ of degree k:

$$\nu(A) = \mu(R^{-1}(A)).$$

Indeed,

$$\varphi_\nu(z) = \int e^{i(z, R(x))} \mu(dx) = \int e^{i(z, VT_x)} \mu(dx) =$$

$$= \int e^{i(V^* z, T)} \mu^{0k}(dT) = \varphi_k(V_z^*). \tag{3}$$

Expansion of measurable mappings in terms of orthogonal systems of polynomials. Let the measure μ be such that the set of all polynomials be dense in $\mathscr{L}_2[\mu]$ and let $R(x)$ be a measurable mapping of $\mathscr{X}$ into $\mathscr{X}$ satisfying condition

$$\int |R(x)|^2 \mu(dx) < \infty.$$

Then for each $z \in \mathscr{X}$ the expression $(R(x), z)$ can be expanded into orthogonal subspaces $\mathscr{P}_k$ constructed in Section 2: $(R(x), z) = \sum_{k=0}^{\infty} P_k(z, x)$, where $P_k(z, x) \in \mathscr{P}_k$. Clearly, $P_k(z, x)$ depends linearly on z. We choose an orthonormal basis $\{e_k\}$ in $\mathscr{X}$. Then

$$\int \sum_{i=1}^{\infty} |P_k(e_i, x)|^2 \mu(dx) \leq \sum_{i=1}^{\infty} \int (R(x), e_i)^2 \mu(dx) = \int |R(x)|^2 \mu(dx).$$

Hence the series

$$\sum_{i=1}^{\infty} \mathsf{P}_k(e_i, x)\, e_i = R_k(x)$$

is convergent in measure μ and

$$\mathsf{P}_k(z, x) = \sum_{i=1}^{\infty} \mathsf{P}_k(e_i, x)\,(z, e_i) = (R_k(x), z).$$

Thus under the assumptions above a measurable mapping $R(x)$ can be represented in the form

$$R(x) = \sum_{k=1}^{\infty} R_k(x),$$

where each one of the mappings R_k is a measurable polynomial mapping of degree k and the series is mean-square convergent. For $k \neq i$ the mappings R_k and R_i are orthogonal in the following sense: for each bounded operator B the equality

$$\int (BR_k(x), R_i(x))\, \mu(dx) = 0$$

is satisfied. Indeed,

$$\int (BR_k(x), R_i(x))\, \mu(dx) = \int \sum_{j=1}^{\infty} (BR_k(x), e_j)\,(e_j, R_i(x))\, \mu(dx) =$$

$$= \int \sum_{j=1}^{\infty} (R_k(x), Be_j)\,(R_i(x), e_j)\, \mu(dx) = 0,$$

since $(R_k(x), z)$ and $(R_i(x), u)$ are orthogonal for $k \neq i$. Therefore

$$\int |R(x)|^2 \mu(dx) = \sum_{k=1}^{\infty} \int |R_k(x)|^2 \mu(dx)$$

and

$$\int (R(x), z)^2 \mu(dx) = \sum_{k=1}^{\infty} \int (R_k(x), z)^2 \mu(dx). \tag{4}$$

Formula (4) enables us to calculate the correlation operator of measure ν which is the image of μ under the mapping $R(x)$.

§4. Calculation of Certain Characteristics of Transformed Measures

In this section, results are presented which enable us to determine characteristic functionals and some other characteristics of measures obtained from the given measure by means of measurable mappings. Some results of this nature were presented previously in this volume. For example, in Section 3 of Chapter VII, a formula for the density of the transformed measure in terms of the original one was obtained and in Section 1 of the present chapter, a formula for the characteristic functional of a measure obtained from the given one by means of a measurable linear transformation was presented, while in Section 3, a similar formula for a characteristic functional of a measure obtained from the given one by means of a measurable polynomial homogeneous transformation. Obviously the results presented below are not sufficient for solving all the problems occurring in the theory of measurable transformations of probability measures, however, for a great number of important problems in applications, computational algorithms for solutions may be constructed on the basis of the results presented below.

Groups of transformations. Let a group of transformations $R_t(x)$ depending on a parameter t by the differential equation

$$\frac{d}{dt} R_t(x) = \mathscr{G}(R_t(x)) \tag{1}$$

be defined in a Hilbert space $\mathscr{X}$, where $\mathscr{G}(x)$ is a continuous mapping of $\mathscr{X}$ into $\mathscr{X}$ which assures the existence and uniqueness of the solution of equation (1) satisfying the initial condition $R_0(x) = x$ for all $x \in \mathscr{X}$. Then $R_t(x)$ will also be a continuous mapping of $\mathscr{X}$ into $\mathscr{X}$ for all t. Let μ be a probability measure on $(\mathscr{X}, \mathfrak{B})$ and let ν_t be the measure obtained from μ under the mapping $R_t(x)$. Denote by $\varphi_t(z)$ and $\varphi(z)$ the characteristic functionals of the measures ν_t and μ. Then

$$\varphi_t(z) = \int e^{i(z, R_t(x))} \mu(dx). \tag{2}$$

Assume that $\int |\mathscr{G}(R_t(x))| \mu(dx) < \infty$. Then one can differentiate the integral appearing in the r.h.s. of (2) with respect to t and obtain

$$\frac{\partial}{\partial t} \varphi_t(z) = i \int (z, \mathscr{G}(R_t(x))) e^{i(z, R_t(x))} \mu(dx) = i \int (z, \mathscr{G}(x)) e^{i(z,x)} \nu_t(dx).$$

Under certain additional assumptions on $\mathscr{G}(x)$ the r.h.s. of the last equality can be expressed in terms of $\varphi_t(z)$. Let

$$\mathscr{G}(x)=\mathscr{G}_1(x)+\mathscr{G}_2(x),$$

where $\mathscr{G}_1(x)$ is a polynomial mapping and $\mathscr{G}_2(x)$ admits the following representation

$$\mathscr{G}_2(x)=\int e^{i(u,x)}\,\varrho(du), \tag{3}$$

where $\varrho(du)$ is a countably-additive set function of bounded variation defined on $(\mathscr{X}, \mathfrak{B})$ and taking on values in $\mathscr{X}$. In this case if

$$(\mathscr{G}_1(x), z)=\sum_{k=0}^{n} H_k^z(x,\dots,x),$$

where $H_k^z(x_1,\dots,x_k)$ are continuous k-linear forms, then

$$\int (\mathscr{G}_1(x), z)\, e^{i(z,x)}\, \nu_t(dx)=\sum_{k=0}^{n} i^{-k}\, \operatorname{Sp} d^k\varphi_t(z)*H_k^z.$$

On the other hand,

$$\int (z, \mathscr{G}_2(x))\, e^{i(z,x)}\, \nu_t(dx)=\int\int e^{i(u,x)}\, e^{i(z,x)}\,(z, \varrho(du))\,\nu_t(dx)=$$

$$=\int\left[\int e^{i(u+z,x)}\,\nu_t(dx)\right](z,\varrho(du))=\int \varphi_t(u+z)\,(z,\varrho(du)).$$

Thus, under the assumptions imposed, $\varphi_t(z)$ satisfies the following integro-differential equation

$$\frac{\partial}{\partial t}\varphi_t(z)=i\sum_{k=0}^{n} i^{-k}\,\operatorname{Sp} d^k\varphi_t(z)*H_k^z+i\int \varphi_t(u+z)\,(z,\varrho(du)). \tag{4}$$

Transformations closely related to the linear. The representation of mappings by means of Fourier transforms of couñtably-additive functions of bounded variation, with values in $\mathscr{X}$ may be utilized for the determination of the characteristic functional of the transformed measure provided the mapping is close to a linear one. Let $R(x)=Vx+\varepsilon\mathscr{G}_2(x)$, where V is a continuous linear operator, ε a sufficiently small number and let $\mathscr{G}_2$ admit representation (3). Then $\mathscr{G}_2$ is bounded and hence

$$\varphi_\nu(z)=\int e^{i(z,x)}\,\nu(dx)=\int e^{i(z,Vx+\varepsilon\mathscr{G}_2(x))}\,\mu(dx)=$$

$$=\int\sum_{k=0}^{\infty}\frac{i^k\varepsilon^k(z,\mathscr{G}_2(x))^k}{k!}\,e^{i(z,Vx)}\,\mu(dx)=$$

$$= \sum_{k=0}^{\infty} \frac{i^k \varepsilon^k}{k!} \int (z, \mathscr{G}_2(x))^k e^{i(V^*z, x)} \mu(dx).$$

Utilizing representation (3) we find

$$(z, \mathscr{G}_2(x))^k = \left[\int e^{i(u, x)} (z, \varrho(du)) \right]^k =$$

$$= \int \dots \int e^{i(u_1 + \dots + u_k, z)} (z, \varrho(du_1)) \dots (z, \varrho(du_k)).$$

Thus,

$$\int (z, \mathscr{G}_2(x))^k e^{i(V^*z, x)} \mu(dx) =$$

$$= \int \dots \int \varphi(V^*z + u_1 + \dots + u_k)(z, \varrho(du_1)) \dots (z, \varrho(du_k)).$$

Finally we obtain

$$\varphi_v(z) = \sum_{k=0}^{\infty} \frac{i^k \varepsilon^k}{k!} \underbrace{\int \dots \int}_{k \text{ times}} \varphi(V^*z + u_1 + \dots + u_k)(z, \varrho(du_1)) \dots (z, \varrho(du_k)). \tag{5}$$

Note that under our assumptions this series is convergent for all ε and is an entire analytic function in ε. Formula (5) can be substantially simplified if we assume that $(z, \varrho(du))$ is a non-negative measure and put $i\varepsilon = \lambda$ (provided $\lambda > 0$). Let $\pi_z(du)$ be a measure in $\mathscr{X}$ which is infinitely-divisible with the characteristic functional

$$\int e^{i(z^*, u)} \pi_z(du) = \exp\left\{ \lambda \int (e^{i(z^*, u)} - 1)(z, \varrho(du)) \right\}.$$

Then

$$\sum_{k=0}^{\infty} \frac{\lambda^k}{k!} \underbrace{\int \dots \int}_{k \text{ times}} \varphi(V^*z + u_1 + \dots + u_k)(z, \varrho(du_1)) \dots (z, \varrho(du_k)) =$$

$$= \int \varphi(V^*z + x) \pi_z(dx).$$

Clearly measures ϱ could have been defined not on $\mathscr{X}$ but on a certain extension of it to which the function $\varphi(z)$ is extendable by continuity.

The duality formula and other expansions in powers of a small parameter. When evaluating various integrals with respect to measures in Hilbert spaces (in particular, measures associated with square integrable processes), it is sometimes convenient to use a very simple formula which replaces integration with respect to one measure by integration with respect to another. Let ξ and η be two independent random variables with values in $\mathscr{X}$ and let μ and ν be the distributions of these variables and φ_μ and φ_ν their characteristic functionals. Evaluating the integral $\mathsf{E}\, e^{i(\xi,\eta)} = \int\int e^{i(x,y)}\,\mu(dx)\,\nu(dy)$ in two possible ways (first with respect to μ and then with respect to ν and conversely) we obtain

$$\int \varphi_\nu(x)\,\mu(dx) = \int \varphi_\mu(y)\,\nu(dy). \tag{6}$$

This formula will be called the duality formula. A particular case of this formula is:

$$\int e^{-\frac{1}{2}(Bx,\,x)}\,\mu(dx) = \int \varphi_\mu(y)\,\nu(dy),$$

where ν is a Gaussian measure with the correlation operator B. Formula (6) is especially convenient to use when the measure ν is an infinitely-divisible distribution. Let $h(x)$ be a functional of the form $\frac{1}{i}\log\varphi_\nu(z)$, where ν is an infinitely divisible distribution. We introduce the family of random variables η_t possessing the characteristic functional $e^{ith(z)} = \mathsf{E}\, e^{i(z,\eta_t)}$. Then $\mathsf{E}\, e^{ith(\xi)}$ – where ξ is a random variable with values in $\mathscr{X}$ distributed according to μ – can be evaluated by means of formula (6):

$$\mathsf{E}\, e^{ith(\xi)} = \mathsf{E}\,\varphi_\mu(\eta_t). \tag{7}$$

This formula is valid only for $t>0$, in the case of negative t (or in the case of complex-valued t appearing in the Laplace transform) one can utilize the analytic continuation in t of the expression obtained.

The duality formula can be used to obtain an expansion of the characteristic functions of functionals of random variable ξ_ε in powers of a small parameter ε if the expansion in powers of ε of the characteristic *functional* of the variable ξ_ε is known. Let

$$\mathsf{E}\, e^{i(z,\,\xi_\varepsilon)} = \varphi_\varepsilon(z) = \sum \varepsilon^k \chi_k(z) \tag{8}$$

be given and it is required to evaluate the characteristic function of the variable $h(\xi_\varepsilon)$ where h satisfies formula (7). Then $\mathsf{E}\, e^{ith(\xi_\varepsilon)} = \sum \varepsilon^k \mathsf{E}\chi_k(\eta_t)$.

As an example for application of the formula consider the case when $h(x)=(Bx, x)$ where B is a positive-definite operator. We have the equality $t^2(Bx, x)=\log \mathsf{E}\, e^{t(x,\eta)}$, where η is a Gaussian variable with the correlation operator B (note that this variable can be defined in a certain extension of the space $\mathscr{X}$). Therefore

$$\mathsf{E}\, e^{is(B\xi_\varepsilon,\xi_\varepsilon)}=\mathsf{E}\, e^{\sqrt{is}\,(\xi_\varepsilon,\eta)}\quad \mathsf{E}\,\varphi_\varepsilon\left(\sqrt{\frac{s}{i}}\,\eta\right).$$

The last formula is valid provided $\varphi_\varepsilon(tz)$ is an entire analytic function in t. Substituting $s=it$, in this formula, we obtain the Laplace transform of the variable $(B\xi_\varepsilon, \xi_\varepsilon)$:

$$\mathsf{E}\, e^{-t(B\xi_\varepsilon,\xi_\varepsilon)}=\mathsf{E}\varphi_\varepsilon(\sqrt{t}\,\eta)=\sum \varepsilon^k \mathsf{E}\chi_k(\sqrt{t}\,\eta).$$

Let the family of random variables $\bar{\xi}_\varepsilon$ depending on a positive parameter ε be such that $\bar{\xi}_\varepsilon\to 0$ in probability as $\varepsilon\to 0$. Next assume that for the characteristic functional $\varphi_\varepsilon(z)$ of the variable $\xi_\varepsilon=\frac{1}{\varepsilon}\,\bar{\xi}_\varepsilon$, expansion (8) is valid. Furthermore, let the functional $h(x)$ admit representation of the form $h(x)=\sum_{k=1}^{\infty} P_k(x)$ where $P_k(x)$ is a homogeneous polynomial of degree k and H_k is its generating k-linear form. We find the expansion in powers of ε of the characteristic functional of the variable $\frac{1}{\varepsilon}\,h(\bar{\xi}_\varepsilon)$ under the assumption that expansion (8) is termwise infinitely differentiable. We have

$$\begin{aligned}\mathsf{E}\exp\left\{\frac{is}{\varepsilon}\,h(\bar{\xi}_\varepsilon)\right\}&=\mathsf{E}\exp\left\{\frac{is}{\varepsilon}\sum_{k=1}^{\infty}P_k(\bar{\xi}_\varepsilon)\right\}=\mathsf{E}\exp\left\{is\sum_{k=1}^{\infty}\varepsilon^{k-1}P_k(\xi_\varepsilon)\right\}=\\&=\mathsf{E}\exp\{is\,P_1(\xi_\varepsilon)\}\sum_{n=0}^{\infty}\frac{1}{n!}\left[is\sum_{k=2}^{\infty}\varepsilon^{k-1}P_k(\xi_\varepsilon)\right]^n=\\&=\mathsf{E}\exp\{is\ P_1(x)\ \sum_{n=0}^{\infty}\varepsilon^n\sum_{k=0}^{2n}Q_{nk}(\xi_\varepsilon)\,r_{nk}(is).\end{aligned}$$

Here $Q_{nk}(x)$ are homogeneous polynomials of degree k, and $r_{nk}(t)$ are numerical (real-valued) polynomials of degree at most $k/2$. These polynomials are uniquely determined by means of the relations

$$\sum_{k=0}^{2n} Q_{nk}(x)\, r_{nk}(t)=\frac{\partial^n}{\partial\varepsilon^n}\exp\left\{t\sum_{k=2}^{\infty}\varepsilon^{k-1}P_k(x)\right\}\bigg|_{\varepsilon=0}.$$

Assume that $P_1(x)=(a, x)$ where $a\neq 0$ and $Q_{nk}(x)=T_{nk}(x,\dots,x)$, where

T_{nk} is a k-linear form. Then

$$\mathsf{E}\exp\{is\,P_1(\xi_\varepsilon)\}\,Q_{nk}(\xi_\varepsilon)=\mathsf{E}\,e^{is(\xi_\varepsilon,a)}\,T_{nk}(\xi_\varepsilon,\dots,\xi_\varepsilon)=$$
$$=i^{-k}\operatorname{Sp}d^k\varphi_\varepsilon(sa)*T_{nk}=i^{-k}\sum_m\varepsilon^m\operatorname{Sp}d^k\chi_m(sa)*T_{nk}.$$

Consequently,

$$\mathsf{E}\exp\left\{\frac{is}{\varepsilon}h(\xi_\varepsilon)\right\}=\sum_{n=0}^{\infty}\varepsilon^n\sum_{k=0}^{2n}r_{nk}(is)\,i^{-k}\sum_{m=0}^{\infty}\varepsilon^m\operatorname{Sp}d^k\chi_m(sa)*T_{nk}. \qquad (9)$$

Expansion (9) should be rewritten collecting the coefficients with the common powers of ε. In the same manner, one can also obtain a finite expansion with a bound on the remainder if in place of a series in (8) we take a finite expansion with a remainder and represent $h(x)$ in the form

$$h(x)=\sum_{k=1}^{N}P_k(x)+o(T_{N+1}(x)),$$

where T_{N+1} is a homogeneous polynomial of the $N+1$-th degree.

A formula analogous to (9) can be used to determine the characteristic functional of the transformed measure provided the transformation $R(x)$ admits representation

$$R(x)=\sum_{k=1}^{\infty}R_k(x), \qquad (10)$$

where $R_k(x)$ are homogeneous polynomial transformations of the k-th degree. Let V_k be a linear mapping of $\mathscr{X}$ into $\mathscr{X}^{0k}$ which maps z into the form $V_k(z)$ which generates the polynomial $(R_k(x), z)$. Then

$$\int e^{i(z,R(x))}\mu(dx)=\int e^{i(V_1z,x)}\sum_{n=0}^{\infty}\frac{1}{n!}\left(i\sum_{k=2}^{\infty}(V_k(z,x,\dots,x)\right)^n\mu(dx).$$

If we denote by φ_ν the characteristic function of the measure ν, then

$$\varphi_\nu(z)=\sum_{k=0}^{\infty}\operatorname{Sp}d^k\varphi_\mu(V_1z)*T_k^z, \qquad (11)$$

where T_k^z is a k-linear form given by

$$T_k^z=\sum_j\sum_{n_1+2n_2+\dots+jn_j=k}\frac{i^{n_1+\dots+n_j}}{n_1!\dots n_j!}V_1^{0n_1}(z)\dots V_j^{0n_j}(z)^*. \qquad (12)$$

---

* V^{0k} stands for $\underbrace{V\dots V}_{k\text{ times}}$.

Expansion (11) is meaningful if measure μ satisfies the following condition: for all m and t

$$\int e^{t\left|\sum_{m}^{\infty} R_k(x)\right|} \mu(dx) < \infty .$$

An application of orthogonal polynomials. Let μ be a measure for which the orthogonal polynomials $P_k(H_k, x)$ are constructed (cf. Section 2). We shall investigate how this fact assists us in finding the characteristic functional φ_ν of measure ν obtained from μ by transformation $R(x)$. Assume that we have an expansion of function $e^{i(z, R(x))}$ into a series of orthogonal polynomials: $e^{i(z, R(x))} = \sum_{k=0}^{\infty} P_k(H_k^z, x)$. Then $\varphi_\nu(z) = \int e^{i(z, R(x))} \mu(dx) = P_0(H_0^z)$. Hence the problem of expanding $e^{i(z, R(x))}$ is not simpler than determining $\varphi_\nu(z)$. It is more natural to use orthogonal polynomials in the case when ν is absolutely continuous with respect to μ and when the density $\frac{d\nu}{d\mu}(x) = \varrho(x)$ belongs to $\mathscr{L}_2[\mu]$. Assume that the expansion of $\varrho(x)$ in terms of orthogonal polynomials is known: $\varrho(x) = \sum_{k=0}^{\infty} P_k(H_k, x)$. Then

$$\varphi_\nu(z) = \int e^{i(z, x)} \varrho(x) \mu(dx) = \int e^{i(z, x)} \sum_{k=0}^{\infty} P_k(H_k, x) \mu(dx) =$$
$$= \sum_{k=0}^{\infty} \int e^{i(z, x)} P_k(H_k, x) \mu(dx).$$

Clearly, $\int e^{i(z, x)} Q(x) \mu(dx)$, where Q is a polynomial which can be easily expressed in terms of $\varphi_\mu(z)$ as a differential operator of the form

$$\sum \frac{1}{i^k} \operatorname{Sp} d^k \varphi_\mu(z) * H^k .$$

Therefore it can be assumed that the functions

$$\chi_k(H_k, z) = \int e^{i(z, x)} P_k(H_k, x) \mu(dx),$$

are determined. Then $\varphi_\nu(z)$ is expressed in terms of these functions by formula

$$\varphi_\nu(z) = \sum_{k=0}^{\infty} \chi_k(H_k, z) . \tag{13}$$

Expressions for the density of the transformed measure in terms of the original one are presented in Section 3 of Chapter VII. This density is expressed in terms of the function $\varrho(a, x)$ which is the density of the shifted measure μ_a relative to the initial measure μ.

We now present a method for obtaining an expansion of $\varrho(a, x)$ into a series of orthogonal polynomials. Assume that

$$\varrho(a, x)=\sum_{k=0}^{\infty} P_k(H_k^a, x), \tag{14}$$

where H_k^a is a k-linear form dependent on a. Then for every polynomial $P_l(H_l, x)$ in $\mathscr{P}_l$ (cf. Section 2) the following relationship

$$\int P_k(H_k, x)\, P_k(H_k^a, x)\, \mu(dx)=\int P_k(H_k, x)\, \varrho(a, x)\, \mu(dx)=$$

$$=\int P_k(H_k, x+a)\, \mu(dx)=S_a(H_k)$$

is satisfied. The linear functional $S_a(H_k)$ on $\tilde{\Phi}^k$ can be easily evaluated if we expand the polynomial $P_k(H_k, x+a)$ by means of the Taylor theorem and use the expansion of each one of the polynomials obtained into a series of orthogonal polynomials. Obviously, the linear function $S_a(H_k)$ on $\tilde{\Phi}^k$ can be represented by means of the scalar product

$$S_a(H_k)=\langle H_k, H_k^a\rangle_k$$

(cf. Section 2). This relation uniquely determines H_k^a and thereafter the function $\varrho(a, x)$ is determined by formula (14). In order that $\varrho(a, x)$ exist and belong to $\mathscr{L}_2[\mu]$ it is necessary and sufficient that series (14) be convergent, i.e. that the inequality

$$\sum_{k=1}^{\infty} \langle H_k^a, H_k^a\rangle_k<\infty$$

be satisfied.

Historical and Bibliographical Remarks

The remarks presented below contain a number of references to the literature dealing with the problems discussed in the book. They are not intended to present a complete bibliography or to sketch the history of the basic ideas in the theory of random processes. In many cases we do not refer to the original obscure publications but rather to more recent textbooks and monographs which contain a bibliography on the topics under consideration.

Chapter I

§1. The exposition is based on the by now commonly accepted – set – theoretical axiomatization of probability theory as suggested by A. N. Kolmogorov in 1929 and presented in his monograph [60] and [58]. In connection with the measure – theoretic results and the theory of integrations (used in this volume) see the texts by A. N. Kolmogorov and S. V. Fomin [70], P. Halmos [42], I. I. Gihman and A. V. Skorohod [33], J. Neveu [80] and P. A. Meyer [77].

§2. The general 0–1 law was established by A. N. Kolmogorov {60}.

§3. The theory of conditional probabilities and conditional mathematical expectations was introduced by A. N. Kolmogorov [60]. It was further developed by J. L. Doob [20]. See also M. Loéve [74] and J. Neveu [80].

§4. The basic theorem is due to A. N. Kolmogorov.

Chapter II

§2. Martingales were discussed by various authors but the systematic theory of this notion is due to J. L. Doob [20]. He first derived the basic inequalities for martingales, proved the theorem on the existence of the limit, introduced the notion of a semimartingale and also obtained other results. More information on martingales can be found in the books by J. L. Doob, M. Loéve and P. A. Meyer quoted above.

§3. The basic ideas and results presented in this section are due to A. N. Kolmogorov and A. Ya. Khinchin [54] and to A. N. Kolmogorov [57]. Series of independent random variables are discussed in more detail in the books by J. L. Doob [20], M. Loéve [74] and A. V. Skorohod [101].

§4. Markov chains with a finite number of states were introduced in (1906) and studied by A. A. Markov [76]. The general definition of Markov chains and processes is due to A. N. Kolmogorov [64]. More general approaches are developed in E. B. Dynkin's monographs [23] and [24].

§5. Markov chains with a countable number of states were first studied in the works of A. N. Kolmogorov [62, 63], W. Doeblin [17] and later were investigated by numerous authors. See W. Feller [28], K. L. Chung [12], E. B. Dynkin and A. A. Yushkevich [25], J. S. Kemeny, J. L. Snell and A. W. Knapp [52],

§6. Random walks were studied by various authors and many results are known in this field. See W. Feller, E. B. Dynkin and A. A. Yushkevich [28], A. V. Skorohod and N. P. Slobodenyuk [103] and F. Spitzer.

§7. B. V. Gnedenko [36] was the first to study the local limit theorems for lattice one-dimensional distributions. See B. V. Gnedenko and A. N. Kolmogorov, I. A. Ibragimov and Yu. V. Linnik [38], A. V. Skorohod and N. P. Slobodenyuk [103].

§8. Ergodic theorems originated in relation to problems in statistical mechanics. See A. Ya. Khinchin's book [55] in this connection. The first ergodic theorems due to J. von Neumann and G. Birkhoff served as the beginning of an intensive development of the theory. A survey of the first period of the development of ergodic theory is contained in E. Hopf's monograph [44]. A simple proof of the Birkhoff-Khinchin theorem was given by A. N. Kolmogorov [65]. Further developments in ergodic theory are discussed in books by P. Halmos [43], K. Jacobs [48] and P. Billingsley [6].

Chapter III

§1. A multi-dimensional generalization of the central limit theorem was first considered by S. N. Bernstein [5]. B. de Finneti [29] originated the systematic study of processes with independent increments. The characteristic function of a process with independent increments in the case of finiteness of the second-order moment was obtained by A. N. Kolmogorov [59] and in the general case by P. Levy [73] (both are univariate). See also remarks to sec. 4 in Chapter II in connection with the general definitions of Markov processes.

§§2 and 3. The feasibility of constructing a random process stochastically equivalent to the given one with sample functions satisfying certain regularity conditions was first investigated by E. E. Slutzky and A. N. Kolmogorov (cf. E. E. Slutzky's paper [105]). Many substantial results are due to J. L. Doob in connection with further developments and various versions of the axiomatic definition of random functions. References to earlier papers are found in Doob's monograph [20]. The basic theorems in section 2 and 3 are due to J. L. Doob. See also E. E. Slutzky [105].

§4. Theorem 1 in a somewhat weaker form was proved by N. N. Chenčov [11]; theorem 2 by J. H. Kinney [56] (in the case of Markov processes). P. Lévy [73] established the absence of discontinuities of the second kind in the case of stochastically continuous processes with independent increments. J. L. Doob [20] studied the properties of sample functions of martingales.

§5. Theorem 2 was proved by E. B. Dynkin [22] and independently by J. H. Kinney [56] (for Markov processes). A somewhat weaker version of theorem 6 is due to A. N. Kolmogorov and was first published in E. E. Slutzky's work [105]. Yu K. Belyaev [3, 4] studied local properties of Gaussian processes. See also the book by H. Cramér and M. R. Leadbetter [15].

Chapter IV

§§1 and 2. A. Ya. Khinchin [53] introduced the notion of a wide-sense stationary process. In the same paper the spectral representation of the correlation function of a wide-sense stationary process was presented. F. Riesz and G. Herglotz obtained in 1911 the spectral representation of a positive definite sequence and S. Bochner [8] obtained in 1932 the representation for positive definite functions. J. L. Schönberg's work contains the spectral representation of a homogeneous and isotropic random field in Euclidean and Hilbert spaces.

§3. E. E. Slutzky [104] and M. Loéve [74].

§4. The theory of stochastic integrals was introduced by H. Cramér [13]. A. N. Kolmogorov was the first to clarify the connection between stochastic integrals, spectral representations and methods in the theory of Hilbert spaces [68]. See also J. L. Doob [20].

§5. Theorem 1 is due to K. Karhunen [50] and Theorem 2 is due to H. Cramér [13].

§6. Using the theory of filters the spectral representation of a stationary process can be easily obtained (A. Blanc-Lapierre and R. Fortet [7]). A more general theory of linear transformations of random processes may be constructed by means of the theory of generalized random process introduced by I. M. Gelfand and K. Ito (I. M. Gelfand and N. Ya. Vilenkin [30], K. Ito [47]).

§7. Basic results for the case of stationary sequences are due to A. N. Kolmogorov [68] and for continuous parameter processes are due to K. Karhunen [51] (see J. L. Doob [20] and Yu A. Rozanov [91]).

§8. The general formulation of the problems of linear forecasting (for stationary sequences) its connection with the geometry of Hilbert spaces and its reduction to a problem in the theory of functions is due to A. N. Kolmogorov. N. Wiener developed efficient methods for solving problems of linear forecasting and filtering for continuous-parameter processes. A. M. Yaglom's method with a large number of examples is presented in his monograph [116].

§9. The theorem on decomposition of a stationary process and the notions of determinate and undeterminate processes are due to H. Wold. The general solution of the problem of forecasting a stationary sequence from its past was obtained by A. N. Kolmogorov and for continuous parameter processes by M. G. Krein [71, 72]. The problem of forecasting a vector-valued stationary sequence was discussed by Yu A. Rozanov [89], N. Wiener and P. Masani [115]. Details on forecasting continuous parameter processes are given in the books by J. L. Doob [20] and Yu A. Rozanov [91].

Chapter V

The construction of a measure in a functional space was first carried out by N. Wiener [112]. The general method of constructing such measures is due to A. N. Kolmogorov [29]. Measures in Banach and complete metric spaces were studied in works of A. N. Kolmogorov, E. Mourier [79], Yu V. Prohorov [85] and K. R. Parthasarathy [81].

§3. The construction of an extension of the initial spaces on which there exists a measure with a given positive definite function for its characteristic functional is due to L. Gross [40]; the theorem in Section 3 is due to E. Mourier [79].

The theorem in Section 5 is due to V. V. Sazonov [94] and R. A. Minlos [78]; generalized measures on a Hilbert space were introduced by Yu A. Daletzkii [16].

The theorem in Section 5 is due to V. V. Sazonov [94] and R. A. Minlos [78]; general-Vershik [110]; he also studied linear and quadratic functionals measurable with respect to these measures. Multiple stochastic integrals were constructed by K. Ito [46]. Yu A. Rozanov [93] obtained the general form of linear and quadratic functions on stationary Gaussian processes.

Chapter VI

§1. The proof of the sufficiency of conditions of Theorem 1 is due to Yu V. Prohorov [85].

§2. The condition of weak compactness of measures on a Hilbert space was established by K. R. Parthasarathy [81].

§3. The general form of an infinitely divisible distribution and the condition for convergence of a distribution of a sum of independent random variables with value in a Hilbert space to such a distribution were obtained by S. R. S. Varadhan [108]. Conditions for convergence to a Gaussian distribution were studied by N. A. Kandelaki and V. V. Sazonov [49]. A sufficiently detailed exposition of these results is given in K. R. Parthasarathy's book [81].

§4. M. Donsker [18] initiated the study of general limit theorems for random processes; his result is stated in Theorem 4. Theorems 1–3 are due to Yu V. Prohorov.

§5. The first limit theorem for processes without discontinuity of the second kind is due to I. I. Gihman [31]. The space $D_{[0,1]}$ and the limiting theorems for processes on this space were studied by A. V. Skorohod [98]. The convergence in $D_{[0,1]}$ was investigated by A. N. Kolmogorov [68] and Yu V. Prohorov [85]. An interesting limit theorem was obtained by N. N. Chenčov [11]; theorem 3 is a minor modification of this theorem. The convergence of processes with independent increments and Markov processes was studied by A. V. Skorohod [98, 99, 100]. An application of limit theorems to statistical problems was considered by M. Donsker and I. I. Gihman [19].

Chapter VII

Various problems of absolute continuity of measures in functional spaces are discussed in I. I. Gihman and A. V. Skorohod's [34] paper.

§2. Measures with an everywhere dense set of admissible shifts were considered by V. N. Sudakov [107]. The structure of the set of admissible shifts was studied by T. S. Pitcher. Theorem 4 is stated in A. M. Vershik's paper.

§3. Certain results on absolute continuity of Gaussian measures in Hilbert spaces under non-linear transformations are given in the paper by V. V. Baklau and A. D. Shatashvili [2].

§4. The condition for absolute continuity and the formula for the density of a Gaussian measure under a shift were obtained by U. Grenander [39]. General conditions for absolute continuity and singularity for Gaussian measures are found in Ja. Hajek's, J. Feldman's and Yu A. Rozanov's papers [41, 26, 92].

§5. Theorems 1 and 2 are due to Yu A. Rozanov. Basic results in this field are presented in Yu A. Rozanov's book [93].

§6. Absolutely continuous mappings of certain classes of Markov processes were considered by I. V. Girsanov [35] and A. V. Skorohod [100]. General theorems on absolute continuity measures associated with processes with independent increments and with Markov processes are presented in I. I. Gihman and A. V. Skorohod's paper [34].

Chapter VIII

§1. Measurable linear operators and linear functionals are studied in the book by G. E. Shilov and Fan Dyk Tan [96].

§2. The orthogonal system of polynomials for the Wiener measure was constructed in the paper by K. Ito [46]; various applications of these polynomials are given in N. Wiener's book. Orthogonal polynomials for the Gaussian measure were constructed by A. M. Vershik [110].

Bibliography

1. Akhiezer, N. I., Glazman, I. M.: Theory of Linear Operations in Hilbert Spaces, N.Y.: Frederick Ungar Publishing Co., 1966 [English translation].
2. Baklan, V. V., Shatashvili, A. D.: Transforms of Gaussian Measures under non-linear transformations in Hilbert spaces. Dopovidi, A. N. URSR **9**, 1115–1117 (1965) [in Ukrainian].
3. Belyaev, Yu. K.: Local properties of sample function of stationary Gaussian processes. Theor. Probability Appl. **5**, 128–131 (1960).
4. Belyaev, Yu. K.: Continuity and Holder's conditions for sample functions of stationary Gaussian processes. Proc. Fourth Berk. Symp. on Math. Stat. and Probability **2**, 23–33 (1961).
5. Bernstein, S.: Sur l'extension du théorème limite du calcul des probabilités aux sommes de quantités dépendentes. Math. Ann. **97**, 1–59 (1926). [Russian translation: Uspehi Mat. Nauk **10**, 65–114 (1944)].
6. Billingsley, P.: Ergodic Theory and Information. N.Y.: J. Wiley 1965.
7. Blanc-Lapierre, A., Fortet, R.: Théorie des fonctions aléatoires. Paris: Masson et Cie. 1953.
8. Bochner, S.: Lectures on Fourier Integrals, Princeton, 1959.
9. Cameron, R. H., Martin, W. T.: Transformations of Wiener integrals under a general class transformation. Trans. Amer. Math. Soc. **58**, 184–219 (1945).
10. Cameron, R. H., Martin, W. T.: Transformation of Wiener integrals by nonlinear transformation. Trans. Amer. Math. Soc. **66**, 253–283 (1949).
11. Chenčov, N. N.: Weak convergence of stochastic processes whose trajectories have no discontinuities of the second kind and the heuristic approach to the Kolmogorov-Smirnov tests. Theor. Probability Appl. **1**, 140–149 (1956).
12. Chung, K. L.: Markov chains with stationary transition probabilities. Berlin – Göttingen – Heidelberg: Springer 1960.
13. Cramér, H.: On the theory of stationary random processes. Ann. Math. **41**, 215–230 (1940).
14. Cramér, H.: On stochastic processes whose trajectories have no discontinuities of the second kind. Ann. di Matematica (iv) **71**, 85–92 (1966).
15. Cramér, H., Leadbetter, M. R.: Stationary and related stochastic processes, N.Y.: J. Wiley 1967.
16. Daletskii, Yu. L.: Infinite-dimensional elliptic operators and parabolic equations connected with them. Russian Math. Surveys **22**, No. 4, 3–53 (1967).
17. Doeblin, W.: Sur les propriétés asymptotiques de mouvement régis par certain types de chaînes simples. Bull. Math. Soc. Sci. Math. R. S. Roumanie **39**, No. 1, 57–115; No. 2, 3–61 (1937).
18. Donsker, M.: An invariance principle for certain probability limit theorems. Mem. Amer. Math. Soc. **6**, 1–12 (1951).

19. Donsker, M.: Kolmogorov-Smirnov theorems. Ann. Math. Statist. **23**, 277–281 (1952).
20. Doob, J. L.: Stochastic processes. N.Y.: J. Wiley 1953.
21. Dunford, N., Schwartz, J. T.: Linear operators I, II. New York: Interscience Publishers 1958, 1963.
22. Dynkin, E. B.: Criterion for continuity and absence of discontinuities of the second kind for trajectories of a Markov random process. Izv. Akad. Nauk Armjan. SSR Ser. Mat. **16** (1952).
23. Dynkin, E. B.: Foundations of the theory of Markov processes. Fismatgiz. (1959).
24. Dynkin, E. B.: Markov processes. Vols. I and II. N.Y.: Academic Press 1965.
25. Dynkin, E. B., Yushkevich, A. A.: Theorems and Problems in Markov Processes. Moscow: Nauka 1967.
26. Feldman, J.: Equivalence and perpendicularity of Gaussian processes, Pacific J. Math. **8**, 699–708 (1958).
27. Feldman, J.: Some classes of equivalent Gaussian processes on interval. Pacific J. Math. **10**, 1211–1220 (1960).
28. Feller, W.: An introduction to probability theory and its application. N.Y.: J. Wiley, Vol. I (1957), Vol. II (1966).
29. Finneti, B.: Sulle funzioni a incremento aleatorio. Rend. Acad. Naz. Lincei, Cl. Sci. Fis. Mat. Nat. (6), **10**, 163–168 (1929).
30. Gelfand, I. M., Vilenkin, N. Ya.: Applications of harmonic analysis; Saţurated Hilbert spaces. Fizmatgiz. (1961).
31. Gihman, I. I.: On a theorem of Kolmogorov. Nauch. Zap., Kiev, Un-ta, Mat. Sborn. **7**, 76–94 (1953).
32. Gihman, I. I.: Markov processes in problems of mathematical statistics. Ukrainian Math. J. **6**, 28–36 (1954).
33. Gihman, I. I., Skorohod, A. V.: Introduction to the theory of random processes. Fizmatgiz. (1965).
34. Gihman, I. I., Skorohod, A. V.: On densities of probability measures in functions spaces, Russian Math. Surveys (Uspehi Mat. Nauk), XXI, **6**, 83–156 (1966).
35. Girsanov, I. V.: On transforming a class of random processes by absolutely continuous substitutions of measures. Theory prob. and its Applic. **5**, 314–334 (1960).
36. Gnedenko, B. V.: On a local theorem for the limit stable distributions. Ukrainian Math. J. **1**, 3–15 (1949).
37. Gnedenko, B. V.: The Theory of Probability. English translation of the fourth edition, New York: Chelsea 1967.
38. Gnedenko, B. V., Kolmogorov, A. N.: Limit Distributions for sums of Independent Random Variables, Reading, Mass.: Addison Wesley 1954.
39. Grenander, U.: Stochastic processes and statistical inference. Ark. Mat. **1**, 195–277 (1950).
40. Gross, L.: Harmonic analysis on Hilbert space. Mem. Amer. Math. Soc. **46**, 1–62 (1963).
41. Hajek, J.: On a property of normal distribution of a stochastic process. Czechoslovak. Math. J. **8**, 610–618 (1958) [Russian-English summary].
42. Halmos, P. R.: Measure Theory. Princeton, N.J.: D. van Nostrand 1950.
43. Halmos, P. R.: Lectures on Ergodic Theory. J. Math. Soc. Japan, No. 3 (1956).
44. Hopf, E.: Ergoden Theory. Ergebnisse der Math., Vol. 2. Berlin: J. Springer 1937 (Reprinted by Chelsea Publishing Co., N.Y. 1948).
45. Ibragimov, I. A., Linnik, Yu. V.: Independent and stationary associated variables. Moscow: Nauka 1965 [English translation].
46. Ito, K.: Multiple Wiener Integral. J. Math. Soc. Japan, **3**, 157–169 (1951).
47. Ito, K.: Stationary random distributions. Mem. Coll. Sci. Univ. Kyoto **28**, 206–223 (1954).

48. Jacobs, K.: Neuere Methoden und Ergebnisse der Ergodentheorie. Berlin – Göttingen – Heidelberg: Springer 1960.
49. Kandelaki, N. P., Sazonov, V. V.: On the central limit theorem for random elements with values in Hilbert space. Theory of Prob. and Applic. **9**, No. 1, 38–46 (1964).
50. Karhunen, K.: Über lineare Methoden in der Wahrscheinlicherechnung, Ann. Acad. Sci. Fennicae, Ser. A. Math. Phys. **37**, 3–79 (1947).
51. Karhunen, K.: Über die Struktur stationaeren zufaelliger Funktionen, Ark. Math. **1**, 141–160 (1950).
52. Kemeny, J. G., Snell, J. L., Knapp, A. W.: Denumerable Markov chains. N.Y.-L.: Van Nostrand 1966.
53. Khinchin, A.: Correlation theory of stationary random processes. Usp. Mat. Nauk, **5**, 42–51 (1938).
54. Khinchin, A., Kolmogorov, A. N.: Über Konvergenz von Reihen deren Glieder durch den Zufall bestimmt werden. Matem. Sb. **32**, 668–677 (1925).
55. Khinchin, A.: Mathematical Foundations of Statistical Mechanics. N.Y.: Dover Publications 1949.
56. Kinney, J. H.: Continuity properties of sample functions of Markov processes. Trans. Amer. Math. Soc. **74**, 280–302 (1953).
57. Kolmogorov, A. N.: Über die Summen durch den Zufall bestimter unabhängiger Größen. Math. Ann. **99**, 309–319 (1928); **100**, 484–488 (1929).
58. Kolmogorov, A. N.: General Measure theory and calculus of probabilities. Trudy Komm. Akad., Math. Division **1**, 8–21 (1929).
59. Kolmogorov, A. N.: Sulla forma generale di un processo stocastico omogeneo, Atti. Accad. Lincei **15**, 805–808, 866–869 (1932).
60. Kolmogorov, A. N.: Foundations of the theory of probability. N.Y.: Chelsea Press. [The German original appeared in 1933; Russian version 1936].
61. Kolmogorov, A. N.: La transformation de Laplace dans les linéaires. Compt. Rend. Acad. Sci. (Paris) **200**, 1717 (1935).
62. Kolmogorov, A. N.: Anfangsgründe der Theorie der Markoffschen Ketten mit unendlichen vielen möglichen Zuständen, **1**, 607–610 (1936).
63. Kolmogorov, A. N.: Markov chains with a countable number of possible states. Bull. Math. Univ. Moscow **1**, 1–16 (1937) [in Russian].
64. Kolmogorov, A. N.: Über die analytischen Methoden in Wahrscheinlichkeitsrechnung. Math. Ann. (104), 1931 [Russian transl.: Usp. Mat. Nauk **5**, 5–41 (1938)].
65. Kolmogorov, A. N.: Simplified proof of the Birkhoff Khinchin ergodic theorem. Uspekhi Math. Nauk **5**, 52–56 (1938) [in Russian].
66. Kolmogorov, A. N.: Curves in Hilbert spaces invariant relative to one-parametric group of motions. Dokl. Akad. Nauk **26**, 6–9 (1940).
67. Kolmogorov, A. N.: Wiener's spiral and some other interesting curves in Hilbert spaces. Dokl. Akad. Nauk **26**, 115–118 (1940).
68. Kolmogorov, A. N.: Stationary sequences in Hilbert space. Bull. Math. Univ. Moscow **2**, No. 6, 1–40 (1941) [in Russian].
69. Kolmogorov, A. N.: On Skorohod's convergence*. Theor. Probability Appl. **1**, 239–247 (1956).
70. Kolmogorov, A. N., Fomin, S. V.: Elements of the theory of functions and functional analysis. Sec. ed. Moscow: Nauka 1968.
71. Krein, M. G.: On an Extrapolation Problem of A. N. Kolmogorov. Dokl. Akad. Nauk SSSR **46**, 306–309 (1944).
72. Krein, M. G.: On the Basic Approximation Problem in the Theory of Extrapolation and Filtering of Stationary Random Processes. Dokl. Akad. Nauk SSSR **94**, 13–16.
73. Lévy, P.: Sur les intégrales dont les éléments sont des variables aléatoires indépendentes, Ann. Scuola Norm. Sup. Pisa **2**, No. 3, 337–366 (1934).
74. Loéve, M.: Probability Theory, 2nd Ed. Princeton, N.J.: D. van Nostrand, 1960.

75. Lyusternik, L. A., Sobolev, V. I.: Elements of Functional Analysis. N.Y.: Ungar 1961.
76. Markov, A. A.: Extension of the law of large numbers to dependent events. Bull. Soc. Phys. Math. Kazan (2) **15**, 155–156 (1906) [in Russian].
77. Meyer, P. A.: Probability and Potentials. 1966.
78. Minlos, R. A.: Generalized stochastic processes and their extension with respect to measure. Trudy Moscow Math. Soc. **8**, 497–518 (1959).
79. Mourier, E.: Eléments aléatoires dans un espace de Banach. Ann. Inst. H. Poincaré **13** (1953).
80. Nevue, J.: Mathematical Foundations of the Calculus of Probability. San Francisco: Holden-Day 1965.
81. Parthasarathy, K. R.: Probability Measures on Metric Space. N.Y.-L.: Academic Press 1967.
82. Pinsker, M. S.: Information and information stability of random variables and processes. San Francisco: Holden Day 1964 [English transl.].
83. Pitcher, T. S.: The admissible mean values of stochastic process. Trans. Amer. Math. Soc. **108**, 538–546 (1963).
84. Privalov, I. I.: Boundary properties of Analytic Functions. Moscow-Leningrad 1950 [German translation: VEB Deutscher Verlag der Wissenschaften, Berlin, 1956].
85. Prohorov, Yu. V.: Convergence of random processes and limit theorems in probability theory. Theory of Prob. and its applic. **1**, 187–214 (1956).
86. Prohorov, Yu. V.: The method of characteristic functionals. Proc. 4th Berkley Symp. **2**, 403–419 (1961).
87. Prohorov, Yu. V., Sazonov, V. V.: Some results associated with Bochner's theorem. Theor. Probability Appl. **6**, 82–87 (1961).
88. Prohorov, Yu. V., Fish, M.: A characterization of normal distributions in Hilbert space. Theor. Probability Appl. **2**, 468–470 (1957).
89. Rozanov, Yu. A.: Spectral theory of multi-dimensional stationary random processes with discrete time. Uspehi Mat. Nauk **13**, 2 (80), 93–142 (1958).
90. Rozanov, Yu. A.: On the density of one Gaussian measure in relation to another. Theor. Probability Appl. **7**, 84–89 (1962).
91. Rozanov, Yu. A.: Stationary Random Processes. San Francisco; Holden Day 1967.
92. Rozanov, Yu. A.: On the density of Gaussian distributions and Wiener-Hopf integral equations. Dokl. Akad. Nauk SSSR **165**, 1000–1002 (1965), [Soviet Math. Dokl. **6**, 1551–1553 (1965)].
93. Rozanov, Yu. A.: Gaussian infinitely dimensional distributions, Steklov Math. Institute Public, **108**, 1–136 (1968).
94. Sazonov, V. V.: A remark on characteristic functionals. Theor. Probability Appl. **3**, 188–192 (1958).
95. Schoenberg, J. L.: Metric spaces and completely monotone functions. Ann. Math. **39**, 811–841 (1938).
96. Shilov, G. E., Fan Dyk Tan: Integral, measure and derivatives on linear spaces. Moscow: Nauka 1967.
97. Skorohod, A. V.: Limit theorems for stochastic processes. Theor. Probability Appl. **1**, 261–290 (1956).
98. Skorohod, A. V.: Limit theorems for stochastic processes with independent increments. Theor. Probability Appl. **2**, 138–171 (1957).
99. Skorohod, A. V.: Limit theorems for Markov processes. Theor. Probability Appl. **3**, 202–246 (1958).
100. Skorohod, A. V.: Studies in the theory of random processes. Reading, Mass.: Addison Wesley: 1965.
101. Skorohod, A. V.: Random processes with independent increments. Fizmatgiz. 1964.
102. Skorohod, A. V.: On the densities of probability measures in functional space. Proc. 5th Berkley symp. **2**, 163–182 (1965).

103. Skorohod, A. V., Slobodenyuk, N. P.: Limit theorems for random walks, UkrSSSR; Naukova Dumka 1970.
104. Slutzky, E. E.: Sur les fonctions éventuelles continues, intégrables et dérivables dans le sens stochastique, Comptes Rendus Acad. sci. **187**, 370–372 (1928).
105. Slutsky, E. E.: Some statements concerning the theory of random processes. Publ. of Middle-Asian University. Mat. Series (5), **31**, 3–15 (1949).
106. Spitzer, F.: Principles of random walk. Princeton, N. J.: D. van Nostrand 1964.
107. Sudakov, V. N.: Linear spaces with quasi-invariant measure. Dokl. Akad. Nauk **127**, 524–525 (1959).
108. Varadhan, S. R. S.: Limit theorems for sums of independent random variables with values in a Hilbert space. Sankhya **24**, 213–238 (1962).
109. Vershik, A. M.: On the theory of normal dynamic systems, Dokl. Akad. Nauk **144**, 9–12 (1962).
110. Vershik, A. M.: General theory of Gaussian measures in linear spaces. Uspekhi Math. Nauk **19**, 210–212 (1964).
111. Vershik, A. M.: Duality in the theory of measure in linear spaces. Dokl. Akad. Nauk **170**, 497–500 (1966).
112. Wiener, N. Differential space, J. Math. Phys. Mass. Inst. Tech. **2**, 131–174 (1923).
113. Wiener, N.: Extrapolation, interpolation and smoothing of stationary time series. N.Y. 1949.
114. Wiener, N.: Nonlinear problems in random theory. M.I.T. and J. Wiley 1958.
115. Wiener, N., Masany, P.: Prediction theory of multivariate stochastic processes. Acta Math. **98**, 111–150 (1957); **99**, 93–137 (1958).
116. Yaglom, A. M.: An Introduction to the theory of Stationary Random Functions, Englewood Cliffs, N.J.; Prentice-Hall 1962.

Corrections

Pp. 340–341. The proof of the boundedness of $m(z)$ presented on pp. 340–341 is erroneous. Below we present a corrected proof.

Since $m_n(z)\uparrow m(z)$ and $m_n(z)$ is continuous, $m(z)$ is continuous from below. Moreover, by virtue of Minkowski's inequality,

$$\begin{aligned}[m(z_1+z_2)]^{1/k} &= [\int |(x, z_1+z_2)|^k \mu(dx)]^{1/k} \\ &\leqslant [\int |(x, z_1)|^k \mu(dx)]^{1/k} + [\int |(x, z_2)|^k \mu(dx)]^{1/k} \\ &\leqslant [m(z_1)]^{1/k} + [m(z_2)]^{1/k}.\end{aligned}$$

Thus $[m(z)]^{1/k}$ is a semiadditive function continuous from below and hence in view of the theorem by I. M. Gelfand (see, e.g., L. V. Kantorovich and G. P. Akilov, *Functional Analysis in Normed Spaces*, Moscow, Fizmatgiz, 1959, p. 233) there exists a constant M such that $[m(z)]^{1/k} \leqslant M|z|$.

P. 408. Omit the last line on this page.

Pp. 525–531. In Section 1 of Chapter VIII two theorems on measurable linear functions and their operators are presented. As stated therein the theorems are not correct. We now present the correct statements and proofs.

A function $l(x)$ is called a measurable linear functional with respect to measure μ on a Hilbert space $(X, \mathfrak{B})$ if $l(x)$ is the limit in measure μ of a sequence of continuous linear functionals $l_n(x)$.

Theorem 1. *In order that a $\mathfrak{B}$-measurable function $l(x)$ be a measurable linear functional with respect to measure μ it is necessary and sufficient that a symmetric convex compact set K exist such that the following conditions be satisfied:*

1) *if $\mathfrak{D}$ is a linear hull of K, then $\mu(\mathfrak{D})=1$;*
2) *$l(x)$ is linear on $\mathfrak{D}$;*
3) *$l(x)$ is continuous on K.*

Proof. Necessity. If $l(x)$ is a measurable linear functional then a sequence of continuous functionals $l_n(x)$ exists such that

$$l(x)=\mu\text{-}\lim_{n\to\infty} l_n(x)$$

and

$$\mu\left(\left\{x: |l_n(x)-l_{n+1}(x)|>\frac{1}{n^2}\right\}\right)\leqslant\frac{1}{n^2}.$$

Set $\mathcal{G}_k=\bigcap_{n=k}^{\infty}\{x: |l_n(x)-l_{n+1}(x)|\leqslant 1/n^2\}$. Clearly, $\mathcal{G}_k$ is a symmetric convex closed set and

$$\mu(\mathcal{G}_k)=1-\sum_{n\geqslant k}\mu\left(\mathscr{X}-\left\{x: |l_n(x)-l_{n+1}(x)|\leqslant\frac{1}{n^2}\right)\geqslant 1-\sum_{n\geqslant k}^{\infty}\frac{1}{n^2}.$$

Since $l_n(x)$ is uniformly convergent to $l(x)$ on each one of the sets $\mathcal{G}_k$, it is convergent to $l(x)$ on the linear hull $\tilde{\mathcal{G}}_k$ of the set $\mathcal{G}_k$ and hence $l(x)$ is linear on $\tilde{\mathcal{G}}_k$. Let F_k be a symmetric convex compact such that $F_k\subset\mathcal{G}_k$, $\mu(\mathcal{G}_k-F_k)<1/k$, and $\mathcal{D}_k$ be the linear hull of F_k. Choose a sequence $\rho_k\downarrow 0$ such that

$$\sum\rho_k(\sup_{x\in F_k}|x|+\sup_n\sup_{x\in F_k}|l_n(x)|)<\infty$$

and let

$$K=\{x: x=\sum\rho_k x_k,\ x_k\in F_k\}.$$

It is easy to see that: a) K is a symmetric convex compact set; b) the linear hull $\mathcal{D}$ of the set K contains all the sets $\mathcal{D}_k$ and hence $\mu(\mathcal{D})=1$; c) $l_n(x)$ converges to $l(x)$ uniformly on K and therefore $l(x)$ is linear on $\mathcal{D}$. The necessity of the theorem's conditions is thus verified.

Sufficiency. Denote $K_n=\{x: (1/n)x\in K\}$. Then $\mathcal{D}=\bigcup_n K_n$ and hence for any $\varepsilon>0$ there exists n such that $\mu(K_n)>1-\varepsilon$. Clearly K_n is a symmetric convex compact set. We show that for any $\delta>0$ a continuous linear functional $\varphi(x)$ exists such that $|\varphi(x)-l(x)|<\delta$ for $x\in K_n$. Set

$$S_1=\left\{x: l(x)\geqslant\frac{\delta}{2}\right\}\cap K_n,\qquad S_2=\left\{x: l(x)\leqslant-\frac{\delta}{2}\right\}.$$

In view of conditions 2) and 3), S_1 and S_2 are convex compacts symmetrical with respect to the origin 0. Therefore there exists a hyperplane which passes through the origin and separates these sets. Let this hyperplane be represented by $\{x: (a,x)=0\}=L$. Denote by $\varphi(x)$ the functional

$$\varphi(x)=l(x_0)\frac{(a,x)}{(a,x_0)},$$

where $x_0 \in S_1$ is a point such that (a, x_0) is maximal. $\varphi(x)$ is a continuous functional. Furthermore,

$$|l(x)-\varphi(x)| = \left|l(x)-l\left(x_0\frac{(a, x)}{(a, x_0)}\right)\right| = \left|l\left(x-x_0\frac{(a, x)}{(a, x_0)}\right)\right|.$$

However,

$$\frac{1}{2}\left(x-x_0\frac{(a, x)}{(a, x_0)}\right) \in K_n \cap L \subset K_n \setminus S_1 \setminus S_2,$$

since $|(a, x)/(a, x_0)| \leq 1$. Hence

$$\left|l\left(\frac{1}{2}\left(x-x_0\frac{(a, x)}{(a, x_0)}\right)\right)\right| \leq \frac{\delta}{2}, \qquad |l(x)-\varphi(x)| < \delta. \quad \square$$

Remark. If $l(x)$ is a μ-measurable linear functional, then there exists an orthonormal basis $\{e_k\}$ in $\mathscr{D}$ such that $l(x)$ is the limit in measure μ of the sequence $l(P_n x)$. Indeed, let $l(x) = \lim_{n\to\infty} (a_n, x)$ in measure μ. There exists an everywhere positive function $\rho(x)$ such that

$$\lim_{n\to\infty} \int [(a_n, x)-l(x)]^2 \rho(x)\mu(dx) = 0, \qquad \int |x|^2 \rho(x)\mu(dx) < \infty.$$

Let A be a symmetric operator satisfying

$$(Az, z) = \int (z, x)^2 \rho(x)\mu(dx).$$

As follows from the lemma in Volume I, Chapter V, Section 5, A is a symmetric kernel operator. Denote its eigenvectors by $\{e_k\}$ and the corresponding eigenvalues by λ_k. Then

$$\begin{aligned}\int [(a_m, x)-(a_n, x)]^2 \rho(x)\mu(dx) &= (A(a_n - a_m), a_n - a_m)\\ &= \sum_{k=1}^{\infty} \lambda_k [(a_n, e_k)-(a_m, e_k)]^2.\end{aligned}$$

Therefore the limits $\lim_{n\to\infty} (a_n, e_k) = \alpha_k$ exist such that $\sum_{k=1}^{\infty} \alpha_k^2 \lambda_k < \infty$, $l(x) = \sum_{k=1}^{\infty} \alpha_k (x, e_k)$, where the series is convergent in measure μ (cf. Volume I, Chapter V, Section 6).

A measurable function $A(x)$ defined on $\mathscr{X}$ with values in $\mathscr{X}$ is called a *measurable linear operator* if a sequence of linear operators A_n exists such that $A_n x$ is weakly convergent to $A(x)$ in measure μ, i.e., for all $y \in \mathscr{X}$ the numerical sequence $(A_n x, y)$ converges in measure μ to $(A(x), y)$.

Theorem 2. *In order that $A(x)$ be a measurable linear operator it is necessary and sufficient that there exist a symmetric compact set K such that the following conditions are satisfied:*

1) *if $\mathscr{D}$ is a linear hull of K, then $\mu(\mathscr{D})=1$;*
2) *$A(x)$ is linear on $\mathscr{D}$;*
3) *$A(x)$ is continuous on K.*

Proof. The necessity of the theorem's conditions is established in exactly the same manner as in Theorem 1. We now prove their sufficiency. Let $\mathscr{L}_n$ be a monotone sequence of finite-dimensional subspaces such that $\bigcup \mathscr{L}_n$ is dense in $\mathscr{X}$ and let P_n be the projection operator on $\mathscr{L}_n$. Since for all x for which $A(x)$ is defined $P_nA(x) \to A(x)$ as $n \to \infty$ it is sufficient to show that the operator $P_nA(x)$ for each n is a limit of a sequence of operators $A_m^{(n)}(x)$ convergent in measure μ. However, in that case $P_nA_m^{(n)}x$ also converges in measure μ to $P_nA(x)$. In order that the latter be fulfilled it is sufficient that $(A_m^{(n)}x, e_k)$ converge in measure μ to $(P_nA(x), e_k)$, where $\{e_k\}$ is a basis such that its intercepts are bases in $\mathscr{L}_n$. Since $(P_nA(x), e_k)=(A(x), e_k)$ is a μ-measurable linear functional, in view of Theorem 1 one can find a sequence of vectors $a_k^{(m)}$ such that

$$(x, a_k^{(m)}) \to (A(x), e_k)$$

as $m \to \infty$ in measure μ. However, then

$$\sum_{k=1}^{\infty} (x, a_k^{(m)}) P_n e_k \to P_nA(x)$$

also in measure μ (only a finite number of summands are nonzero in the sum on the left of the last relationship). Thus the required sequence of operator $A_m^{(n)}$ is defined by the equation

$$(A_m^{(n)}x, e_k) = \begin{cases} (x, a_k^{(m)}), & e_k \in \mathscr{L}_n, \\ 0, & e_k \notin \mathscr{L}_n, \end{cases}$$

and the sufficiency of the conditions of Theorem 2 is verified. □

Subject Index

CIM

M. Aigner Combinatorial Theory ISBN 978-3-540-61787-7
A. L. Besse Einstein Manifolds ISBN 978-3-540-74120-6
N. P. Bhatia, G. P. Szegő Stability Theory of Dynamical Systems ISBN 978-3-540-42748-3
J. W. S. Cassels An Introduction to the Geometry of Numbers ISBN 978-3-540-61788-4
R. Courant, F. John Introduction to Calculus and Analysis I ISBN 978-3-540-65058-4
R. Courant, F. John Introduction to Calculus and Analysis II/1 ISBN 978-3-540-66569-4
R. Courant, F. John Introduction to Calculus and Analysis II/2 ISBN 978-3-540-66570-0
P. Dembowski Finite Geometries ISBN 978-3-540-61786-0
A. Dold Lectures on Algebraic Topology ISBN 978-3-540-58660-9
J. L. Doob Classical Potential Theory and Its Probabilistic Counterpart ISBN 978-3-540-41206-9
R. S. Ellis Entropy, Large Deviations, and Statistical Mechanics ISBN 978-3-540-29059-9
H. Federer Geometric Measure Theory ISBN 978-3-540-60656-7
S. Flügge Practical Quantum Mechanics ISBN 978-3-540-65035-5
L. D. Faddeev, L. A. Takhtajan Hamiltonian Methods in the Theory of Solitons ISBN 978-3-540-69843-2
I. I. Gikhman, A. V. Skorokhod The Theory of Stochastic Processes I ISBN 978-3-540-20284-4
I. I. Gikhman, A. V. Skorokhod The Theory of Stochastic Processes II ISBN 978-3-540-20285-1
I. I. Gikhman, A. V. Skorokhod The Theory of Stochastic Processes III ISBN 978-3-540-49940-4
D. Gilbarg, N. S. Trudinger Elliptic Partial Differential Equations of Second Order ISBN 978-3-540-41160-4
H. Grauert, R. Remmert Theory of Stein Spaces ISBN 978-3-540-00373-1
H. Hasse Number Theory ISBN 978-3-540-42749-0
F. Hirzebruch Topological Methods in Algebraic Geometry ISBN 978-3-540-58663-0
L. Hörmander The Analysis of Linear Partial Differential Operators I – Distribution Theory and Fourier Analysis ISBN 978-3-540-00662-6
L. Hörmander The Analysis of Linear Partial Differential Operators II – Differential Operators with Constant Coefficients ISBN 978-3-540-22516-4
L. Hörmander The Analysis of Linear Partial Differential Operators III – Pseudo-Differential Operators ISBN 978-3-540-49937-4
L. Hörmander The Analysis of Linear Partial Differential Operators IV – Fourier Integral Operators ISBN 978-3-642-00117-8
K. Itô, H. P. McKean, Jr. Diffusion Processes and Their Sample Paths ISBN 978-3-540-60629-1
T. Kato Perturbation Theory for Linear Operators ISBN 978-3-540-58661-6
S. Kobayashi Transformation Groups in Differential Geometry ISBN 978-3-540-58659-3
K. Kodaira Complex Manifolds and Deformation of Complex Structures ISBN 978-3-540-22614-7
Th. M. Liggett Interacting Particle Systems ISBN 978-3-540-22617-8
J. Lindenstrauss, L. Tzafriri Classical Banach Spaces I and II ISBN 978-3-540-60628-4
R. C. Lyndon, P. E Schupp Combinatorial Group Theory ISBN 978-3-540-41158-1
S. Mac Lane Homology ISBN 978-3-540-58662-3
C. B. Morrey Jr. Multiple Integrals in the Calculus of Variations ISBN 978-3-540-69915-6
D. Mumford Algebraic Geometry I – Complex Projective Varieties ISBN 978-3-540-58657-9
O. T. O'Meara Introduction to Quadratic Forms ISBN 978-3-540-66564-9
G. Pólya, G. Szegő Problems and Theorems in Analysis I – Series. Integral Calculus. Theory of Functions ISBN 978-3-540-63640-3
G. Pólya, G. Szegő Problems and Theorems in Analysis II – Theory of Functions. Zeros. Polynomials. Determinants. Number Theory. Geometry ISBN 978-3-540-63686-1
W. Rudin Function Theory in the Unit Ball of $\mathbb{C}^n$ ISBN 978-3-540-68272-1
S. Sakai C*-Algebras and W*-Algebras ISBN 978-3-540-63633-5
C. L. Siegel, J. K. Moser Lectures on Celestial Mechanics ISBN 978-3-540-58656-2
T. A. Springer Jordan Algebras and Algebraic Groups ISBN 978-3-540-63632-8
D. W. Stroock, S. R. S. Varadhan Multidimensional Diffusion Processes ISBN 978-3-540-28998-2
R. R. Switzer Algebraic Topology: Homology and Homotopy ISBN 978-3-540-42750-6
A. Weil Basic Number Theory ISBN 978-3-540-58655-5
A. Weil Elliptic Functions According to Eisenstein and Kronecker ISBN 978-3-540-65036-2
K. Yosida Functional Analysis ISBN 978-3-540-58654-8
O. Zariski Algebraic Surfaces ISBN 978-3-540-58658-6